ALLE ZEIT WACH 1842

R. P. Pharis S. B. Rood (Eds.)

Plant Growth Substances 1988

With 229 Figures and 115 Tables

Springer-Verlag
Berlin Heidelberg New York London
Paris Tokyo Hong Kong Barcelona

Professor Richard P. Pharis
Plant Physiology Research Group
Biological Sciences Department
University of Calgary
Calgary, Alberta T2N 1N4
Canada

Professor Stewart B. Rood
Department of Biological Sciences
University of Lethbridge
Lethbridge, Alberta T1K 3M4
Canada

ISBN-13: 978-3-642-74547-8 e-ISBN-13: 978-3-642-74545-4
DOI: 10.1007/978-3-642-74545-4

Softcover reprint of the hardcover 1st edition 1990

Typesetting: International Typesetters Inc., Makati, Philippines
2131/3145(3011)-543210 – Printed on acid-free paper

Preface

The 13th International Conference on Plant Growth Substances was held from the 17th to the 26th July, 1988 in Calgary, Alberta Canada under the auspices of the IPGSA (International Plant Growth Substances Association) and the University of Calgary.

Over 550 participants from all over the world attended, along with 70 Associates and 25 University of Calgary graduate students who assisted in audiovisual presentations when not attending the scientific sessions.

Fine weather prevailed, as was usual for summer on the eastern slopes of the Canadian Rockies, and participants arriving early visited the famous Calgary Stampede. A hosted buffet opened the Conference on Sunday evening. On Wednesday evening, following an afternoon field trip into the mountains of the Kananaskis Valley, the IPGSA traditional banquet became a western barbecue on Richards' Ghost River Ranch in the foothills of the Rockies, with folk and country and western music provided by the Great Western Orchestra. The fine Alberta weather continued through the weekend, and the Conference ended with a field trip to Sunshine Meadows, a World Heritage Site in Banff National Park.

A major emphasis of the 13th IPGSA Conference was the use of appropriate mutants as research tools and the use of molecular biology and molecular techniques to more fully explore mechanisms by which plant growth substances evoke the many and varied responses we see throughout the plant kingdom. This emphasis is seen not only in the Symposia papers presented herein, but was also reflected by many of the oral and poster contributions.

The program began with plenary lectures by M. Bopp and N. Takahashi. Following these lectures, the Symposia series began with a Symposium chaired by J. B. Reid on Uses of Mutants to Explore Hormonal Control Systems. This was followed by Symposia on the Role of Hormones in Embryo, Seed and Fruit Development (chaired by R. Quatrano) and Auxin Physiology (chaired by M. Jacobs). On Tuesday morning and early afternoon a Symposium, emphasizing Hormones and Gene Expression (six speakers), was chaired by J. Jacobsen. Later in the afternoon, Symposia on Calcium, Its Role in Hormone-Regulated Processes (chaired by A. C. Leopold) and Abscisic Acid (chaired by B. Milborrow) ended the day. Wednesday morning presented us with a Symposium on

Plant Hormone Binding (chaired by L. Rappaport), with the field trip to the Kananaskis Valley in the afternoon. On Thursday morning Symposia on Flowering (chaired by G. Bernier) and Practical Applications of PGRs (chaired by W. Rademacher) shared the stage. Symposia on Cytokinin Metabolism and Mode of Action (chaired by R. Horgan), Gibberellin Biochemistry (chaired by R. Pharis), and Ethylene Metabolism and Physiology (chaired by D.M. Reid) were held Thursday afternoon. On Friday morning Symposia on IAA Biochemistry, Metabolism and Physiology (chaired by R. Bandurski) and Fusion of Molecular and Analytical Techniques for Studies of IAA Metabolism (chaired by J. Cohen) were presented. Contributed paper sessions were held concurrently with most Symposia, and Poster sessions were held in the evenings, and on Friday afternoon, with posters being available for viewing throughout the Conference. The meeting ended formally on Friday afternoon.

All told, 432 individuals contributed oral papers (247) or posters (185) to the 13th triennial IPGSA Conference, and 72 of the Symposia presentations make up this volume.

Special thanks are due to the University of Calgary's Conference Office, especially Mrs. Madeleine Aldridge and Ms. Lois Kokoski. Support, moral and real, was provided by the Biological Sciences Department, and especially Mrs. Norma Herrington and her colleagues.

The present volume of Proceedings in the Life Sciences series follows three volumes from earlier meetings (1976, 1979 and 1985), and should be useful for critically assessing recent progress in individual topics, or in the broader subject areas. Additionally, when taken in the context of the earlier volumes, *Plant Growth Substances 1988* will provide a guide to the recent development of research in plant growth substances.

Financial Support was provided by the Natural Sciences and Engineering Research Council of Canada and the University of Calgary through special Conference Grants, Government of Alberta – Tourism Promotion, BASF Aktiengesellschaft, Abbott Laboratories, DuPont de Nemours, Imperial Chemical Industries Ltd., Westbridge Agricultural Products, Shell Research Ltd., Blackwell Scientific, Elsevier Science, Kluwer Academic and Springer-Verlag.

Finally, we have enjoyed working with members of Springer-Verlag in the preparation of this volume, and we appreciate their support and help in this endeavour.

Calgary, Alberta, Canada 1990

Richard P. Pharis
Stewart B. Rood

Contents

Plant Hormones in Lower Plants
M. Bopp (With 3 Figures) 1

Endogenous Plant Hormones in Rice in Relation
to the Regulation of Its Life Cycle
N. Takahashi (With 6 Figures) 11

I Genetic Approaches

The Benefit of Biosynthesis and Response Mutants
to the Study of the Role of Abscisic Acid in Plants
C. M. Karssen, H. W. M. Hilhorst, and M. Koornneef
(With 1 Figure) 23

A Genetic Approach to Auxins and Cytokinins
P. J. King, A. D. Blonstein, Y. Fracheboud, J. Oetiker, and
M. Suter (With 6 Figures) 32

Development of Genetic and Analytical Systems
for Studies of Auxin Metabolism
J. D. Cohen, K. Bialek, J. P. Slovin, B. G. Baldi,
and K.-H. Chen (With 4 Figures) 45

Developmental Studies of *Physcomitrella patens* Using
Auxin and Cytokinin Sensitivity Mutants
N. W. Ashton, D. J. Cove, T. L. Wang, and M. J. Saunders
(With 1 Figure) 57

Dwarf Mutants of Maize – Research Tools
for the Analysis of Growth
B. O. Phinney and C. R. Spray (With 8 Figures) 65

Gibberellin Synthesis and Sensitivity Mutants in *Pisum*
J. B. Reid (With 5 Figures) 74

Gibberellin-Insensitive and Overgrowth Mutations
in Temperate Cereals
J. L. Stoddart (With 1 Figure) 84

II Molecular Aspects

Auxin and Gene Expression
G. Hagen, B. McClure, C. Brown, M. Gee, and
T. Guilfoyle (With 1 Figure) 95

IAA Perception and Auxin-Regulated Gene-Expression
A.M. Mennes, C.J.M. Boot, K.R. Libbenga,
E.J. van der Zaal, and A.C. Maan 100

Membrane-Bound Auxin Receptors
M.A. Venis and R.M. Napier 106

Wound-Induced ACC Synthase, an Immunochemical
Comparison of the Wound-Induced and Auxin-Induced
Enzymes
H. Imaseki, N. Nakagawa, and N. Nakajima
(With 6 Figures) 113

Polyphosphoinositide Turnover and Signal Transduction
of Auxin on Isolated Membranes of *Daucus carota* L.
B. Zbell, C. Walter-Back, H. Hohenadel, and
I. Schwendemann (With 6 Figures) 122

Complexity of the Barley α-Amylase Gene Families:
Determination of Gene Numbers for Each Family and
mRNA Levels for Individual Members
J.C. Rogers and B. Khursheed 130

Regulation of Gene Expression by Abscisic Acid
in Barley Aleurone Layers
T.H.D. Ho, B. Hong, R.C. Nolan, S.J. Uknes, and
L.-S. Lin (With 4 Figures) 138

Novel Affinity Probes for Gibberellin Receptors
in Aleurone Protoplasts of *Avena Fatua*
R. Hooley, M.H. Beale, S.J. Smith, and J. MacMillan
(With 5 Figures) 145

Fusicoccin-Binding Sites in Higher Plants
E.W. Weiler, C. Meyer, A. Mithöfer, and M. Feyerabend
(With 2 Figures) 154

Cytokinin Mode of Action – Problems and Perspectives
D.E. Hanke, H. Davies, M. Biffen, R.J.A. Connett, and
T.C. Freathy (With 3 Figures) 161

A Functional Cytokinin-Binding Protein in Photochemical Reactions of Chloroplast
H. Huang, S.Y. Yang, and Y.W. Tang (With 4 Figures) .. 173

Structure-Activity Relationship Studies and Development of *s*-Triazine and Carbamate Anticytokinins
H. Iwamura (With 5 Figures) 179

Construction of a System for Regulated Alterations of Endogenous Cytokinins
J.I. Medford, J.A. Winter, and H.J. Klee
(With 5 Figures) 186

Ethylene Binding Receptors – Is There More Than One?
E.C. Sisler
(With 4 Figures) 192

III Hormones and Calcium

Calcium and Second Messengers in Hormonal Regulation
A.C. Leopold 203

The Role of Calcium in the Response of Roots to Auxin and Gravity
M.L. Evans, C.L. Stinemetz, L.M. Young, and W.M. Fondren (With 3 Figures) 209

The Role of Calcium in Stimulus-Response Coupling
B.W. Poovaiah and A.S.N. Reddy (With 8 Figures) 216

IV Hormone Synthesis and Metabolism

Recent Studies of the Metabolism of Abscisic Acid
J.A.D. Zeevaart, D.A. Gage, and R.A. Creelman
(With 2 Figures) 233

Recent Investigations of the Biochemistry of Abscisic Acid
B.V. Milborrow (With 7 Figures) 241

Tissue-Specific Metabolism of Both Abscisic Acid and Gibberellins in Legume Seeds
M.L. Brenner, S.K. Stombough, and P.R. Birnberg 254

An Overview of Cytokinin Biosynthesis
L.M.S. Palni, S.K. Nandi, S. Singh, and D.S. Letham
(With 2 Figures) 258

Genetic Differences in the Enzymatic Regulation
of Zeatine Metabolism in *Phaseolus* Embryos
D.W.S. Mok, M.C. Mok, G. Shaw, S.C. Dixon, and
R.C. Martin (With 3 Figures) 267

Xylem-Translocated Cytokinin: Metabolism and Function
D.S. Letham, C.W. Parker, R. Zhang, S. Singh,
M.N. Upadhyaya, P.J. Dart, and L.M.S. Palni 275

Cytokinin Oxidase and the Degradative Metabolism
of Cytokinins
R. Horgan, L.R. Burch, and L.M.S. Palni
(With 5 Figures) 282

Metabolic Aspects of Ethylene Biosynthesis
S.F. Yang, W.-K. Yip, S. Satoh, J.H. Miyazaki, X. Jiao,
Y. Liu, L.-Y. Su, and G.D. Peiser (With 2 Figures) 291

Conjugation of Gibberellins in *Zea mays*
G. Schneider and J. Schmidt (With 3 Figures) 300

Metabolism of Gibberellins A_{20} and A_9 in Plants:
Pathways and Enzymology
J. MacMillan (With 2 Figures) 307

The Dioxygenases in Gibberellin Biosynthesis
after Gibberellin A_{12}-Aldehyde
J.E. Graebe and T. Lange (With 4 Figures) 314

The Action of Plant Growth Retardants at the
Biochemical Level
P. Hedden (With 5 Figures) 322

Measurement of the in Vivo Rate of Indole-3-Acetic
Acid Turnover
H.M. Nonhebel and T.P. Cooney 333

Relationship Between Stimuli, IAA and Growth
R.S. Bandurski, A. Schulze, M. Desrosiers, P. Jensen,
B. Epel, and D. Reinecke (With 4 Figures) 341

Oxidation of Indole-3-Acetylaspartic Acid in *Vicia*
S. Tsurumi and S. Wada (With 5 Figures) 353

Indole-3-Ethanol Metabolism and Its Possible Role in the Regulation of Indole-3-Acetic Acid Biosynthesis
V. Magnus and G. Laćan (With 3 Figures) 360

The Oxindole-3-Acetic Acid Pathway in *Zea mays*
D. Reinecke 367

Metabolism of Jasmonic Acid
G. Sembdner, A. Meyer, O. Miersch, and C. Brückner
(With 3 Figures) 374

V Hormones: Physiology and Effects

How Abscisic Acid Causes Depressions of the Photosynthetic Capacity of Leaves
K. Raschke (With 5 Figures) 383

The Role of Abscisic Acid in Chilling Resistance
Rui-chi Pan (With 6 Figures) 391

Dormancy in Cereals – Levels of and Response to Abscisic Acid
M. Walker-Simmons (With 3 Figures) 400

Auxin Regulation of Cell Differentiation in Moss Protonema
M.M. Johri and J.S. D'Souza (With 6 Figures) 407

The Second Messenger in Apical Dominance Controlled by Auxin
W. Russell and K.V. Thimann (With 8 Figures) 419

Auxin Transport and Its Regulation by Flavonoids
P.H. Rubery and M. Jacobs (With 4 Figures) 428

GC-MS Quantifications of Free and Ester Indol-3-Acetic Acid in Relation to Root Growth and Gravitropism
M. Saugy and L. Rivier (With 5 Figures) 441

Hemmstoff und Wachstum: Growth Inhibitors, Not Auxin, Regulate Phototropism
J. Bruinsma, M. Sakoda, and K. Hasegawa
(With 2 Figures) 450

Cytokinins as Metabolic Stimulants
Which Induce Pod Set
D. Dyer, J.C. Cotterman, C.D. Cotterman,
P.S. Kerr, and D.R. Carlson (With 7 Figures) 457

Gibberellins in Embryo Development
A. Alpi 468

Gibberellins and Flower Initiation in Herbaceous
Angiosperms
J.D. Metzger 476

Cytokinins in Flower Initiation
G. Bernier, P. Lejeune, A. Jacqmard and J.-M. Kinet
(With 3 Figures) 486

Hormones and *Cuscuta* Development: Influence of
Hormones on Secondary Xylem Differentiation,
Phenylalanine Ammonia Lyase Activity and Lignification
I. Rajagopal, S. Ramachandiran, and S. Mahadevan
(With 2 Figures) 492

The Regulation of Tumor Morphology in Crown Gall
W.L. Pengelly, L.-Y. Su, and B.R. Campell
(With 2 Figures) 503

Cytokinin: Evidence for Spatial Control of Signal
Transduction
M.J. Saunders (With 4 Figures) 511

Plant Wound Signals and Translation
E. Davies (With 3 Figures) 519

The Effect of Growth Regulators on the Winter Survival
of Winter Wheat
L.V. Gusta, B.J. O'Connor and M.J.T. Reaney 531

Hormonal Control of Senescence
L.D. Noodén, J.J. Guiamét, S. Singh, D.S. Letham,
J. Tsuji, and M.J. Schneider (With 1 Figure) 537

The Possible Role of Protein Kinases in the Plant Cell
Response to Phytohormones
O.N. Kulaeva (With 2 Figures) 547

VI Practical Applications and Economic Implications

New Types of Plant Growth Regulators of Microbial Origin: The Likelihood of Practical Use
H.G. Cutler (With 11 Figures) 555

Can PGRs Alleviate the Recently Evolved Cross Resistances to Herbicides?
J. Gressel (With 2 Figures) 566

Economic Aspects of Plant Growth Regulators
A.D. Baylis (With 2 Figures) 576

Practical Considerations in Using Growth Regulators on Turfgrass
J.E. Kaufmann (With 5 Figures) 585

New Aspects for the Practical Use of Ethylene-Releasing Compounds
M.S. Reid 595

Commercial Uses of Gibberellins and Cytokinins and New Areas of Applied Research
R.D. Carlson and A.J. Crovetti (With 1 Figure) 604

New Types of Plant Growth Retardants: Additional Perspectives for Practical Application in Agriculture and Horticulture
W. Rademacher (With 1 Figure) 611

Subject Index 619

Abbreviations

(1,2) DAG	1,2-diacylglycerol
(1,4) IP_2	inositol 1,4-bisphosphate
(1,4,5) IP_3	inositol 1,4,5-trisphosphate
2D	two-dimensional
2,4-D	2,4-dichlorophenoxyacetic acid
2,4-DP	2,4-dichlorophenoxy-1-propyne
2-OH-OXIAA-Asp	2-indolone-3-(2-hydroxy)acetylaspartic acid
[7G]Z	zeatin-7-glucoside
7-OH-OxIAA	7-hydroxyoxindole-3-acetic acid
7-OH,OxIAA	7-hydroxyoxindole-3-acetic acid
7-OH-OxIAA-glc	7-hydroxy oxindole-3-acetic acid-7^{i}-O-β-D-glucopyranoside
7-OH-DiOxIAA-glc	7-hydroxy dioxindole-3-acetic acid-7^{i}-O-β-D-glucopyranoside
[9R]Z	zeatin-9-riboside
[9R]iP	N^6-isopentenyladenosine
ABA	abscisic acid
ACC	1-aminocyclopropane-1-carboxylate
AEC	1-amino-2-ethylcyclopropane-1-carboxylic acid
Alar	daminozide or B-995 (*N*-dimethylamino-succinamic acid)
AMO-1618	4-hydroxyl-5-isopropyl-2-methylphenyl trimethyl-ammonium chloride, 1-piperidine carboxylate
AMP	adenosine monophosphate
amu	atomic mass unit(s)
ATG	start codon for mRNA translation
ATP	adenosine triphosphate
AVG	2-aminoethoxyvinyl-glycine
BA	^{6}N-benzyladenine
Br	brassinosteroid(s)
BSA	bovine serum albumin
BuOH	butanol
CCC	2-chloroethyltrimethyl ammonium chloride
CCKBP	chloroplast cytokinin binding protein

CD	circular dichroism
cDNA	complementary deoxyribonucleic acid
Chlormequat	2-chloroethyltrimethyl ammonium chloride
CI	chemical ionization
CKs	cytokinins
Con A-Sepharose	Concanavalin A Sepharose 4B (Sigma source)
CRP	cold-requiring plant
DAG	diacylglycerol
DAPI	diamidino-2-phenylindole
DIA	dioxindole-3-acetic acid
DIA-Asp	3-hydroxy-2-indolone-3-acetylaspartic acid
(diH)Z	dihydrozeatin
(diH)[9R]Z	dihydrozeatin riboside
diOxIAA	dioxindole-3-acetic acid (3 hydroxy-2-indolone-3-acetic acid)
DiOxIAA	dioxindole-3-acetic acid (3 hydroxy-2-indolone-3-acetic acid)
DJA	9,10-dihydrojasmonic acid
DLE	delayed light emission
DMO	5,5-dimethyloxazolidine-2,4-dione
DMSO	dimethylsulfoxide
DPA	dihydrophaseic acid
DPG	diphosphatidylglycerol
DPU	1,3-diphenylurea
DTE	dithioerythritol
DTT	dithiothreitol
dw	dry weight
DZ	DZR, DZMP, dihydroderivatives of Z, ZR, ZMP
ECD	electron capture detection
EDTA	ethylenediaminetetraacetic acid
EI	electron impact
Ethephon	2-chloroethylphosphonic acid
EtOH	ethanol
FAD	flavin adenine dinucleotide
FC	fusicoccin
FPLC	fast protein liquid chromatography
FR	far-red light
FT-IR	Fourier transform-infrared
fw	fresh weight
-G	glucoside, or glucopyranoside
G_1	pre-synthetic phase of the cell cycle
G_2	post-synthetic phase of the cell cycle
GA(s)	gibberellin(s)

GA_n	gibberellin A_n
GA_{12}ald	gibberellin A_{12} aldehyde
GC	gas chromatography
GC-ECD	gas chromatography-electron capture detection
GC-MS	GC-mass spectrometry
GC-SIM	GC-MS-selected ion monitoring
-GE	glucosyl ester
GI	guanylyl imidophosphate
glc	glucoside
GTP	guanosine triphosphate
HOAc	acetic acid
HPLC	high performance liquid chromatography
hs-ipt	chimeric heat shock isopentenyl transferase gene
IAA	indole-3-acetic acid
IAA-L-ala	indole-3-acetyl-L-alanine
IAA-asp	indole-3-acetylaspartic acid
IBA	indole-3-butyric acid
ICA	indole-3-carboxylic acid
IEF	isoelectric focussing
IgG	immunoglobulin G
ILA	indole-3-lactic acid
$Ins(1,4,5)P_3$	inositol 1,4,5-triphosphate
iP	N^6-isopentenyladenine
IP_3	inositol 1,4,5-tris-phosphate
iPA	N^6-isopentenyladenosine
IPA	indole-3-pyruvic acid
IPP	isopentenylpyrophosphate
ipt	isopentenyl transferase gene
IR	infrared radiation, or infrared spectroscopy
JA	(−)jasmonic acid
K	kilo
kb	kilobase
kD	kilodalton
KRI	Kovat's Retention Index
LD	long day(s)
LDP	long-day plant(s)
LSDP	Long→short-day plant
MACC	N-malonyl-ACC
MCPA	4-chloro-2-methylphenoxyacetic acid
Me	methyl
MeOH	methanol
M_r	molecular weight
mRNA	messènger ribonucleic acid
MS	mass spectrometry

MTA	methylthioadenosine
MTR	methylthioribose
MVA	mevalonic acid
NAA	naphthaleneacetic acid
NAD	nicotine adenine dinucleotide
NADPH	reduced nicotine adenine dinucleotide phosphate
NMR	nuclear magnetic resonance
NPA	1-naphthylphthalamic acid
(OG diH)Z	O-glucosyldihydrozeatin
[OG]Z	zeatin-O-glucoside
OGZ, OGZR	O-glucosides of the corresponding cytokinins
ORD	optical rotation dispersion analysis
OxIAA	oxindole-3-acetic acid (2-indolinone-3-acetic acid)
2-OH-OXIAA-Asp	2-indolone-3-(2-hydroxy)acetylaspartic acid
(OX)Z	O-xylosylzeatin
PA	phaseic acid
PA	phosphatidic acid (Hanke et al. chapter)
PAA	phenylacetic acid
Paclobutrazol	or PP333 (1-(4-chlorophenyl)4,4-dimethyl-2-(1,2,4-triazoll-yl)pentan-3-ol
PAGE	polyacrylamide gel electrophoresis
PAL	phenylalanine ammonia lyase
PBS	phosphate buffered saline
PC	phosphatidylcholine
PE	phosphatidylethanolamine
Phosphon D	tributyl(2,3-dichlorobenzyl)phosphonium chloride
PG	phosphatidylglycerol
PGA	3-phosphoglycerate
PGR(s)	plant growth regulator(s)
Pi	inorganic phosphate
pI	the pH at which a protein has no net charge
PI	phosphatidylinositol
PI response	phosphoinositide response
PI(4)P	phosphatidylinositol 4-phosphate
(4)PIP	phosphatidylinositol 4-monophosphate
$PI(4,5)P_2$	phosphatidylinositol 4,5 biphosphate
PIP_2	phosphatidylinositol 4,5 biphosphate
PS	photosystem
PS	phosphatidylserine (Hanke et al. chapter)
$PtdIns4,5P_2$	phosphatidyl-inositol 4,5-biphosphate
PtdOH	phosphatidic acid

QCA	2-quinolone-4-carboxylic acid
R	red light
-RC	radiochromatography, radiochromatogram scanning
RH	relative humidity
RIA	radioimmunoassay
RNA	ribonucleic acid
rRNA	ribosomal ribonucleic acid
Rt	retention time
RuBP	ribulose biphosphate
SAM	S-adenosylmethionine
SD	short day(s)
SDP	short day plant(s)
SDS	sodium dodecyl sulfate
SDS-PAGE	sodium dodecylsulfate polyacrylamine gel electrophoresis
SLDP	short→long-day plant
ss DNA	single-stranded DNA
T-DNA	transferred DNA
TIBA	2,3,5-triiodobenzoic acid
TLC	thin-layer chromatography
tRNA	transfer ribonucleic acid
tryptophol	indole-3-ethanol
UDP	uridine diphosphate
UTP	uridine triphosphate
UV	ultraviolet radiation
Z	*trans*-zeatin
ZMP	zeatin monophosphate
ZR	*trans*-zeatin riboside

List of Contributors

You will find the addresses at the beginning of the respective contribution

Alpi, A. 468
Ashton, N.W. 57
Baldi, B.G. 45
Bandurski, R.S. 341
Baylis, A.D. 576
Beale, M.H. 145
Bernier, G. 486
Bialek, K. 45
Biffen, M. 161
Birnberg, P.R. 254
Blonstein, A.D. 32
Boot, C.J.M. 100
Bopp, M. 1
Brenner, M.L. 254
Brown, C. 95
Brückner, C. 374
Bruinsma, J. 450
Burch, L.R. 282
Campell, B.R. 503
Carlson, D.R. 457
Carlson, R.D. 604
Chen, K.-H. 45
Cohen, J.D. 45
Connett, R.J.A. 161
Cooney, T.P. 333
Cotterman, C.D. 457
Cotterman, J.C. 457
Cove, D.J. 57
Creelman, R.A. 233
Crovetti, A.J. 604
Cutler, H.G. 555
Dart, P.J. 275
Davies, H. 161, 519
Desrosiers, M. 341
Dixon, S.C. 267
D'Souza, J.S. 407
Dyer, D. 457
Epel, B. 341
Evans, M.L. 209
Feyerabend, M. 154
Fondren, W.M. 209
Fracheboud, Y. 32
Freathy, T.C. 161
Gage, D.A. 233
Gee, M. 95
Graebe, J.E. 314
Gressel, J. 566
Guiamét, J.J. 537
Guilfoyle, T. 95
Gusta, L.V. 531
Hagen, G. 95
Hanke, D.E. 161
Hasegawa, K. 450
Heeden, P. 322
Hilhorst, H.W.M. 23
Ho, T.H.D. 138
Hohenadel, H. 122
Hong, B. 138
Hooley, R. 145
Horgan, R. 282
Huang, H. 173
Imaseki, H. 113
Iwamura, H. 179
Jacobs, M. 428
Jacqmard, A. 486
Jensen, P. 341
Jiao, X. 291
Johri, M.M. 407
Karssen, C.M. 23
Kaufmann, J.E. 585
Kerr, P.S. 457
Khursheed, B. 130
Kinet, J.-M. 486
King, P.J. 32
Klee, H.J. 186
Koornneef, M 23

Kulaeva, O.N. 547
Laćan, G. 360
Lange, T. 314
Lejeune, P. 486
Leopold, A.C. 203
Letham, D.S. 258, 275, 537
Libbenga, K.R. 100
Lin, L.-S. 138
Liu, Y. 291
Maan, A.C. 100
MacMillan, J. 145, 307
Magnus, V. 360
Mahadevan, S. 492
Martin, R.C. 267
McClure, B. 95
Medford, J.I. 186
Mennes, A.M. 100
Metzger, J.D. 476
Meyer, A. 374
Meyer, C. 154
Miersch, O. 374
Milborrow, B.V. 241
Mithöfer, A. 154
Miyazaki, J.H. 291
Mok, D.W.S. 267
Mok, M.C. 267
Nakagawa, N. 113
Nakajima, N. 113
Nandi, S.K. 258
Napier, R.M. 106
Nolan, R.C. 138
Nonhebel, H.M. 333
Noodén, L.D. 537
O'Connor, B.J. 531
Oetiker, J. 32
Palni, L.M.S. 258, 275, 282
Pan, Rui-chi 391
Parker, C.W. 275
Peiser, G.D. 291
Pengelly, W.L. 503
Phinney, B.O. 65
Poovaiah, B.W. 216
Rademacher, W. 611
Rajagopal, I. 492
Ramachandiran, S. 492
Raschke, K. 383
Reaney, M.J.T. 531
Reddy, A.S.N. 216
Reid, J.B. 74
Reid, M.S. 595
Reinecke, D. 341, 367
Rivier, L. 441
Rogers, J.C. 130
Rubery, P.H. 428
Russell, W. 419
Sakoda, M. 450
Satoh, S. 291
Saugy, M. 441
Saunders, M.J. 57, 511
Schmidt, J. 300
Schneider, G. 300
Schneider, M.J. 537
Schulze, A. 341
Schwendemann, I. 122
Sembdner, G. 374
Shaw, G. 267
Singh, S. 258, 275, 537
Sisler, E.C. 192
Slovin, J.P. 45
Smith, S.J. 145
Spray, C.R. 65
Stinemetz, C.L. 209
Stoddart, J.L. 84
Stombough, S.K. 254
Su, L.-Y. 291, 503
Suter, M 32
Takahashi, N. 11
Tang, Y.W. 173
Thimann, K.V. 419
Tsuji, J. 537
Tsurumi, S. 353
Uknes, S.J. 138
Upadhyaya, M.N. 275
Venis, M.A. 106
Wada, S. 353
Walker-Simmons, M. 400
Walter-Back, C. 122
Wang, T.L. 57
Weiler, E.W. 154
Winter, J.A. 186
Yang, S.Y. 173
Yang, S.F. 291

Yip, W.-K. 291
Young, L.M. 209
Zaal, E.J. van der 100
Zbell, B. 122
Zeevaart, J.A.D. 233
Zhang, R. 275

Plant Hormones in Lower Plants

M. Bopp[1]

1 Introduction

Why should we discuss phytohormones in lower plants? In flowering plants hormonal systems are well established, so that one begins to understand the primary steps of reaction on a molecular level using standard objects which offer the opportunity to study all aspects of this story in detail. Therefore one may ask: will including lower plants in the research programme bring any advantages and can we expect new aspects regarding lower plants as model systems for plant hormones? I hope that I may show that this is really the case, at least for some groups of the lower plants, although in this meeting as well only about 20 contributions of over 400 concern them.

Two reasons may justify more clearly the use of lower plants:

1. The detailed analysis of the phytohormones and its actions will give some hints toward understanding the evolutionary background of hormone systems.
2. The simplicity of the morphology makes systems easier to understand in some respects, e.g. sensitivity, competence, etc. [73]. Thus, lower plants can serve as models for the mechanism of hormones. This is not so far-fetched, because we start to understand the molecular basis of hormonal activated signal chains by comparison with the signal chains in animal cells [80]. And if animals and plants have similar or even identical signal chains, one must assume that such signals, if acting in lower plants, may also act along the same chains as in higher plants.

The limits of what we mean when we use the term hormone or phytohormone are not exactly fixed in flowering plants and even less so in the huge and heterogenous sphere of lower plants. Therefore, we will restrict our attention to the five well-defined groups: auxins, gibberellins, cytokinins, abscisic acid and ethylene, generally accepted as signal substances. Many more endogenous substances may have comparable functions, but these are perhaps restricted to only few families or orders in lower plants and microorganisms. These other substances can have stimulating or inhibiting effects on growth, and many are formed [58] either as secondary metabolites (with unknown functions) or as signal substances in the lower plants themselves, such as pheromones [35], sexual hormones, etc.

[1] Botanisches Institut der Universität, Im Neuenheimer Feld 360, 6900 Heidelberg, FRG

To analyze the hormone system with all its aspects three things have to be considered:

1. Do lower plants contain phytohormones or similar substances?
2. Do exogenously applied hormones induce physiologically significant responses in lower plants?
3. Can we find evidence for a causal connection between the endogenous substance and the response?

If these questions can be answered positively, experimental approaches are necessary to study the anabolic and catabolic pathway of the substances, their transport and, as an important question, the reaction chain between signal and response.

This, however, can be done only in a few cases because of the large diversity of the lower plants. Not only do the lower plants include autotrophic and heterotrophic plants far from the green evolutionary tree, but the orders of this tree may be quite diverse as well.

2 Phytohormones in Heterotrophic Plants

Many of the heterotrophic plants are pathogens, and are thus highly dependent on higher plants. This occurs in three different ways: direct, indirect, and transgenic. Very often the pathogen changes growth or morphogenesis of the host, producing a phytohormone as a signal which is "understood" by the host. If this molecule of the pathogen is slightly modified in its synthesis or chemical structure the effect on the host may be even stronger, because the host itself cannot regulate the amount of such a "strange" substance. Table 1 shows a fragmentary summary of examples where pathogenic organisms produce phytohormones which can interfere with the host's normal function and induce alterations characteristic for the particular diseases.

The second possibility is that the signal substances like cytokinins from microorganisms stimulate the host plant to synthesize a wide spectrum of similar substances [27, 59]. The pattern of these cytokinins can be completely different in the inducing microorganism and the attached host. Where this is the case only the products of the host may act as signals [17].

Transgenic transformation of host cells is carried out by plasmid genes from the pathogen organism, and, as in *Agrobacterium* species, is a third mechanism. It was this type of hormonal regulation that provided a very important tool in attaining our present level of knowledge of cytokinin and auxin synthesis [46].

3 Non-Green Algae

A separate group represents the large non-green algae, e.g. red and brown algae. Different reports exist concerning their cytokinins which can be found in extracts or concentrations of commercially available preparations of seaweed [44, 65]. These preparations are used as plant nutrient and/or soil conditioner. The algal cyto-

Table 1. Examples of pathogenic organisms which produce phytohormones

Species	Hormones	Literature
Gibberella fujikuroi	Mainly GA3 in those strains infecting the Graminae	5, 26
Sphaceloma manihoticola	Mainly GA4	26, 54
Fusarium and other species	ABA	18
Fusarium culmorum	Cytokinin-like substances	42
Ceratocystis coerulacens	ABA	18
Agrobacterium tumefaciens	Zeatin 2iP	14, 45
Rhizobium leguminosarum	IAA Cytokinin	71, 72
Rhizobium phaseoli	IAA, GAs	3
Azospirillum lipoferum	IAA GAs	12, 28
Pseudomonas savastanoi	IAA Cytokinins	31, 40, 51
Plasmodiophora brassicae	IAA	39, 47
Corynebacterium fascians	Cytokinins	48, 51
Alternaria brassicae	Cytokinins	60
Pisolithus tinctorius (Mycorrhiza)	IAA	19
Streptomyces mutabilis	IAA	20
Ramelina duriaei (Lichen)	IAA Ethylene	21

kinins do not seem to be breakdown products of t-RNA, but rather they reflect the actual, relatively low, endogenous cytokinin content of the seaweed [65]. Much other, mostly older, literature demonstrates the presence of auxin-, gibberellin- and cytokinin-like substances in many marine algae [64]. Even ethylene and its precursor, ACC, have been found in those organisms [49]. However, there is little evidence that any or all of these substances have a "signal" purpose in seaweed itself.

In red algae a hormone-like substance, rhodomorphin from *Griffithsia pacifica*, was analyzed by Waaland and Coworker [69, 75] as a glycoprotein of 14000–17500 Kd with internal sulphydrylbounds and α-mannosyl groups. It induces cell division, regulates cell elongation, and controls cell differentiation in the very low concentration range of 10^{-13} to 10^{-14}M. This morphogenetic substance is involved in the process of cell repairing and fusion.

4 Lower Green Plants

If we now focus our interest on the plants included in the evolutionary tree of "green plants" [32] we realize that since 1985 almost all substances mentioned above have been characterized not only by bioassays but also by more rigorous analytical methods. Thus, IAA in algae, liverworts and mosses, ethylene in algae, liverworts and mosses, ABA in hornworts, mosses and green algae, cytokinin in mosses, ferns and *Equisetum* and GAs in ferns and *Psilopsida* have been demonstrated (Fig. 1). However again, presence per se does not tell us whether they truly have a signal function for the particular species.

Just four examples, shown in detail, should give a preliminary but clear answer to this question of function.

4.1 Caulerpa

Among the green algae, those with a large and differentiated thallus are better candidates for hormonal regulation than unicellular or uniform filamentous types. *Caulerpa* [32, 34] is such a coenocytic macroscopic plant with different polar organized organs. Using several different procedures, IAA was detected at about 1 μg/g fw. In contrast, GA-like substances detected in the dwarf rice bioassay could

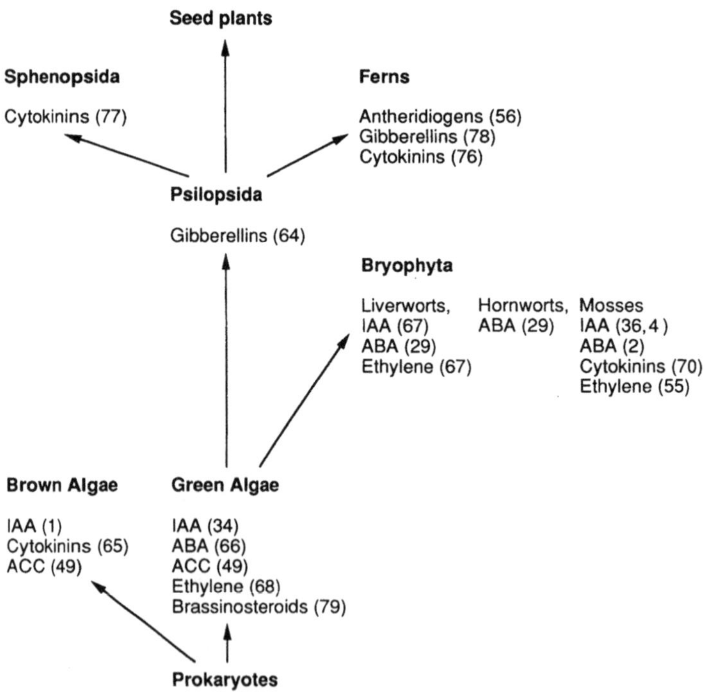

Fig. 1. Evolutionary tree with the main groups of lower plants with phytohormones determined in the different groups. Literature indicated by the numbers. To compare with Fig. 1 in [32]

not be identified by GC-MS [34], hence the unknown substances may be molecules other than GAs.

The response of *Caulerpa* to auxins was described earlier as a stimulation of the cauloid part of the plant and an inhibition of the "rhizome" elongation [43]. However, the reaction is not very specific, since exogenous GAs can stimulate rhizome elongation and rhizoid initiation nearly 2-fold [33].

There seems to be no polar or local distribution of IAA after ^{14}C IAA application [32]. The localized effect of the response may thus not come from actual polar transport in the large coenocytic cell. Altogether this system is not yet confirmed definitely to be hormonally regulated.

4.2 Anthocerotae

For a long time ABA seemed to be absent in liverworts, as in mosses [9, 53]. However, use of the ELISA after HPLC separation has demonstrated ABA-like immune reactive material in Marchantiales and in hornwort-species [29], and shown that ABA concentration depends on external conditions. In wet-grown plants the amount of ABA was significantly lower than in plants collected from drier habitats. Furthermore, as in higher plants, ABA concentration increases after drying stress within 20 h from 30 to 300 nmol kg^{-1} fw (Fig. 2). Finally the stomata-bearing sporophytes and small tubers contain more ABA then the stomata-free-thallus. If this has any significance, one should expect that the stomata of *Anthoceros* respond to ABA by closing, and this was found [29] as had been noted several years earlier for the stomata of the moss *Funaria* [24]. These facts support the hypothesis that ABA in the *Anthocerotae* has the function of a water-stress hormone. Therefore, if ABA is found in mosses and ferns [30] and induces a response antogonistic to other hormones [30, 41, 57] we can postulate that ABA in lower plants will probably have a purpose similar to that in higher plants. The function is already fully developed, as the morphological structure exists soon in the evolution, at least for stomata regulation.

4.3 Fern Prothalli

Many ferns produce "antheridiogens", substances which induce precocious formation of antheridia in very young gametophytes. In several families at least four different antheridiogens are produced [50, 57], and these are not always strongly species selective [25]. The chemical structure is closely related to GAs; GA9 is an antheridiogen of Lygodium, and GA7 can be chemically converted into antheridic acid, the antheridiogen of *Anemia phyllitidis* [23]. In fern sporophytes at least 10 GAs occur [76]. Exogenously applied GAs and antheridiogens cause two independent responses: spore germination in the dark and antheridium formation. Exactly the same thing happens with the endogenous substances delivered from older gametophytes in the neighbourhood, producing a concentration gradient of the antheridiogen in the substrate, which triggers the two responses there [57]. These responses, however, are different, depending on the type of antheridiogen. Whereas

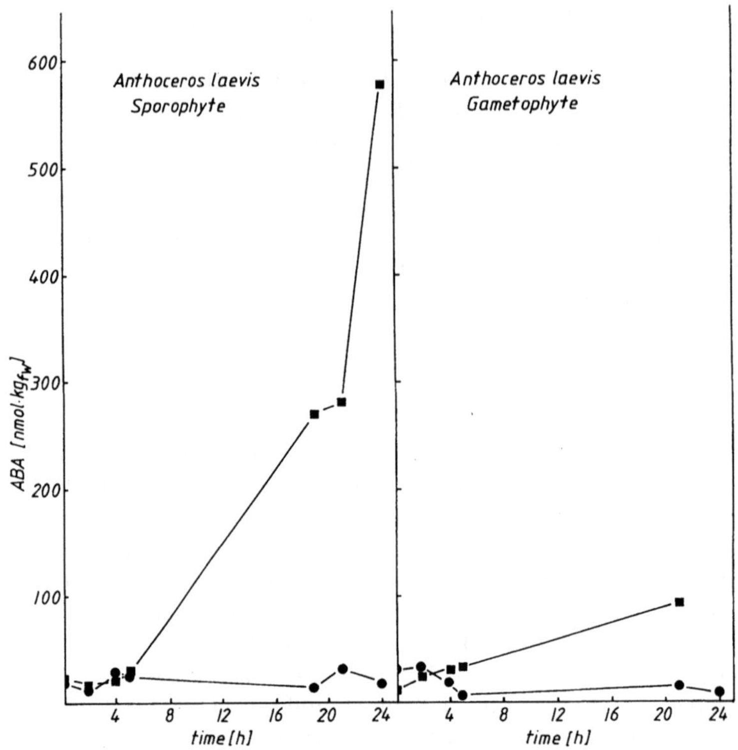

Fig. 2. *Anthoceros laevis*: Endogenous ABA in moist —●— and dry —■— cultivated sporophytes and gametophytes (After [29] with permission)

the antheridiogen of *Pteris vittata* can induce antheridium formation in other species, it can not stimulate the dark germination [25].

Different signals, therefore, have different "values" for the two responses, perhaps because different receptors are involved in both processes. Furthermore, the quite unspecific compensation of antheridiogens by GAs may show that we are possibly studying the evolutionary starting point of a hormone system, since a hormonal function of gibberellins for plant groups below the ferns has not been documented unequivocally.

4.4 Mosses

The last example concerns mosses: *Funaria* and *Physcomitrella*. We confine our view to the protonema, a morphogenetic system with a clear structure consisting of three cell types: chloronema, caulonema and the side branches of caulonema. The probable role of the auxin and cytokinin class of phytohormones in the regulation of morphogenesis has only recently been elucidated. ABA, ethylene and cAMP have been identified from mosses [2, 13, 55] but data about their participation in any hormone system are missing or incomplete.

Auxins and cytokinins, however, seem to be actual "signals" in mosses, based on the results of several experimental approaches [8]:

1. The presence of IAA has been demonstrated using different methods in *Funaria* and *Physcomitrella* [4, 36], as is also the case for cytokinins. A 2iP-like substance was isolated from the substrate on which a moss callus culture was grown [6], and this cytokinin could also be identified in the substrate of certain mutants [70]. Repeated analyses with ELISA has confirmed that cytokinins are constituents of normally growing protonemata, where the concentration of 2iP is always ($\pm$) 10 times greater than zeatin, in contrast to higher plants. The wild type protonema always contains fewer cytokinins than the just-mentioned Ove-mutants which produce more gametophores as the wild type.
2. An unequivocal response to exogenous auxin was shown for the first time by Johri [37], who found that protonema cultivated in conditions which prevent spontaneous caulonema formation can be stimulated from caulonema when treated with auxin. This was repeatedly confirmed for a variety of conditions which do not allow the spontaneous chloronema/caulonema transition [15, 41]. Thus, all experimental results [38] agree with the hypothesis that auxin is the signal for the transition from the chloronema to the caulonema stage.

   Exogenously applied cytokinins enhance the number of buds and accelerate bud formation in all moss species tested so far. It has also been shown that cytokinins can stimulate cell division, especially side branch formation at low concentrations [10, 61], an important finding for the detailed analysis of the reaction chain of cytokinins in mosses [62,63].
3. Evidence for the causal connection between concentrations of endogenous phytohormones and the physiological response comes from studies of the metabolism of the hormones in wild types and mutants [7, 9, 15]. Such mutants exist for *Funaria* as well as for *Physcomitrella*.

One group of mutants always remains in the chloronema stage, thereby resembling an auxin-deficiency; other mutants appear to have a changed cytokinin system.

For the auxin "resistent" mutants a deficiency of auxin is found, as well as other changes in sensitivity [22] not yet identified. If, however, the endogenous auxin content is less than the wild type, two possibilities exist: either the production of auxin is reduced or its degradation is more rapid. The latter was demonstrated for the mutant NAR 2 of *Funaria* [7]. All auxin precursors have the same effect as IAA itself, indicating that the synthetic pathway from tryptophan to IAA is not interrupted. But the activity of IAA oxidases is enhanced, whereas, quite surprisingly, the peroxidase activity is appreciably diminished. We have evidence that both activities belong to the same isoenzyme, so that the mutation may concern one single polypeptide, which is changed in a way to stimulate IAA oxidase activity while reducing peroxidase activity. This may explain why this mutant has less auxin and thus remains in the chloronema stage.

Because caulonema cells or, if already formed, their side branches, are the target cells for cytokinin, the response to auxin is a prerequisite for cytokinin action. Therefore, the auxin-deficient mutants form buds only after an auxin treatment. Also important are mutants which look like wild type protonemata that have been treated with high levels of exogenous cytokinin. These Ove-mutants [15] contain a surplus of endogenous cytokinins, mainly 2iP. But it is not known whether the synthesis of cytokinins is enhanced or the degradation reduced. Both are possible. In this, Ove-mutants t-RNA metabolism is not changed [16], so it is unlikely to be a major source of the 2iP.

To understand the cytokinin degradation process we studied its course in *Funaria* in vitro and in vivo. We were able to isolate an "oxidase" which removes the side chain of kinetin with high efficiency, producing adenine and derivatives. This enzyme did not, however, degrade BA or 2iP. This result explains the response of *Funaria* to exogenously applied cytokinins. The *Funaria* response is stable for BA, but only temporary for kinetin. In comparison, the pattern of metabolism in higher plants is clearly different from that in mosses [52], yet another deviation between the hormones of mosses and higher plants. Nevertheless, the moss system is a good tool to analyze the action chain for cytokinins.

A last and quite important aspect of cytokinin action in mosses concerns the fact that we can follow the response of single cells. As a whole the number of buds induced by cytokinin is proportional to the logarithm of the concentration between 0.1 and 10 μM under standard conditions [11]. But, if we analyze the distribution of the buds, we can see that with the lowest concentration only one cell in a filament of *Funaria*, cell number six, responds (Fig. 3). However, with increasing concentration younger cells in the same filament also respond. This means that the younger cells are less sensitive, and this can also be demonstrated with isolated single cells. Thus, the proportionality to the concentration of Cytokinin is only an expression of the change of sensitivity which happens for each cell during its development. Therefore, the moss system may provide a good tool to study the problem of hormone sensitivity [22].

5 Conclusion

Now, when we finish the quick journey through the hormonal systems found in lower plants, we may draw a few conclusions: Even when we assume that phytohormones are originally products of the general metabolism long before they are

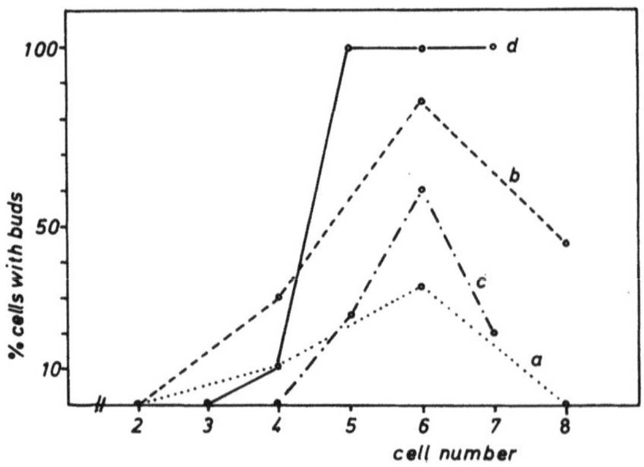

Fig. 3. *Funaria hygrometrica*: Bud induction by kinetin in isolated single cells (*a*, *b*) and intact filaments (*c*, *d*). The cells are numbered from the tip (cell nr 1) to the base (cell nr 8). The cells were isolated either with (*b*) or without (*a*) side branches. Kinetin treatment was 50 μM. The intact filaments are grown on kinetin for 24 h (*c* = 0.2 μM, *d* = 10 μ M)

used, we find experimental data that they gained, early in the evolution, the character of "signal substances". This may depend on the formation of adequate receptors [56, 57]. It seems important that the response to signals in lower plants is comparable to the same in higher plants. This leads to the hypothesis that hormone systems which had already formed in lower plants somewhere in the evolutionary tree in such a way that a special signal chain connects the signal with the desired response are conserved in the higher plants in the same way. However, it can be expected that in higher plants additional new connections between signals and responses exist, such as in complex processes like seed germination, flower formation, abscision, etc.

References

1. Abe H, Uchiyama M, Salo R (1972) Agric Biol Chem 36:2259
2. Atzorn R, Busch S, Maltry B, Bopp M (1990) Planta (Berl) (submitted)
3. Atzorn R, Crozier A, Wheeler CT, Sandberg G (1988) Planta 175:532
4. Ashton NW, Schulze A, Hall P, Bandurski RS (1985) Planta (Berl) 164:142
5. Bearder JR (1980) In: MacMillan J (ed) Encyclopedia of plant physiology. New Series Vol. 9. Springer, Berlin Heidelberg New York, p 681
6. Beutelmann P, Bauer L (1978) Planta (Berl) 133:215
7. Bhatla SC, Bopp M (1985) J Plant Physiol 120:233
8. Bopp M (1990) In: Chopra RN, Bhatla SC (eds) Bryophytes: physiology and biochemistry. CRC, Boca Raton, Fl (in press)
9. Bopp M, Gerhäuser D, Kessler U (1986) In: Bopp M (ed) Plant growth substances 1985. Springer, Berlin Heidelberg New York Tokyo, p 263
10. Bopp M, Jacob HJ (1986) Planta (Berl) 169:462
11. Bopp M, Markmann-Mulisch U, Kessler U (1988) Beitr Biol Pflanz 63:41
12. Bottini R, Fulchieri M, Pearce D, Pharis RP (1989) Plant Physiol 89:1
13. Chopra RN, Bhatla SC (1983) Physiol Plant 53:383
14. Claeys M, Messens E, van Montagu M, Schell J (1978) Fresenius' Z Anal Chem 290:125
15. Cove DJ, Ashton NW (1984) In: Dyer AF, Duckett JG (eds) Experimental biology of bryophytes. Academic Press, London, p 177
16. Cove DJ, Perry KC (1986) Physiol Plant 67:680
17. Dekhuizen HM (1981) Plant Cell Rep 1:18
18. Dörffling K, Petersen W, Sprecher E, Urbasch J, Hansen HP (1984) Z Naturforsch 39c:683
19. Ek M, Ljungquist PO, Stenström E (1983) New Phytol 94:401
20. El-Sayed MA, Valadon LRG, El-Shanshoury A (1987) Microbios 36:85
21. Epstein E, Sagee O, Cohen JD, Garty J (1986) Plant Physiol 82:1122
22. Firn RD (1986) Physiol Plant 62:267
23. Furber M, Mader LN (1987) J Am Chem Soc 109:6389
24. Garner LB, Paolillo DJ (1973) Bryologist 73:423
25. Gemmerich AR (1986) J Plant Physiol 125:157
26. Graebe JE (1987) Annu Rev Plant Physiol 38:419
27. Greene EM (1980) Bot Rev 46:25
28. Hartman A, Singh M, Klingmuller W (1983) Can J Microbiol 29:916
29. Hartung W, Weiler EW, Volk OH (1987) Bryologist 90:393
30. Hichok LG (1985) Can J Bot 63:1582
31. Hutcheson S, Kosuge T (1985) J Biol Chem 260:6281
32. Jacobs WP (1986) In: Bopp M (ed) Plant growth substances 1985. Springer, Berlin Heidelberg New York Tokyo, p 249
33. Jacobs WP, Davis W (1983) Ann Bot 52:39
34. Jacobs WP, Falkenstein K, Hamilton RH (1985) Plant Physiol (Bethesda) 78:844
35. Jaenicke L (1988) Bot Acta 101:149

36. Jayaswal RK, Johri MM (1985) Phytochemistry (Oxf) 24:1211
37. Johri MM, Desai S (1973) Nature 245:223
38. Johri MM, this volume
39. Kavanagh JA, Williams PH (1981) Trans Br Mycol Soc 77:125
40. Kosuge T, Heskett MG, Wilson EE (1966) J Biol Chem 241:3738
41. Lehnert B, Bopp M (1983) Z Pflanzenphysiol 110:379
42. Michniewicz M, Michalski L, Rosej B, Kuszha G (1986) Acta Physiol Plant 8:85
43. Mishra AK, Kefford NP (1969) J Phycol 5:103
44. Mooney PA, van Staden J (1987) Bot Mar 30:323
45. Morris RO (1986) Annu Rev Plant Physiol 37:509
46. Morris RO, Powell GK, Beaty JS, Durley RC, Hommes NG, Lica L, MacDonald EMS (1986) In: Bopp M (ed) Plant growth substances 1985. Springer, Berlin Heidelberg New York Tokyo, p 185
47. Müller P, Hilgenberg W (1986) Physiol Plant 66:245
48. Murai N, Skoog F, Doyle ME, Hanson RS (1980) Proc Natl Acad Sci USA 77:619
49. Nelson WR, van Staden J (1985) Bot Mar 28:415
50. Nester JE, Veysey S, Coolbaugh RC (1987) Planta (Berl) 170:26
51. Nester EW, Kosuge T (1981) Annu Rev Microbiol 33:531
52. Palmer MV, Palni LMS (1987) Plant Physiol 126:365
53. Price RJ (1972) Phytochemistry 11:1759
54. Rademacher W, Graebe JE (1979) Biochem Biophys Res Commun 91:35
55. Rohwer F, Bopp M (1985) J Plant Physiol 117:331
56. Schraudolf H (1985) In: Dyer AF, Page CN (eds) Biology of Pteridophytes. Publ Royal Society of Edinburgh, Edinburgh, p 75
57. Schraudolf H (1986) In: Bopp M (ed) Plant growth substances 1985. Springer, Berlin Heidelberg New York Tokyo, p 270
58. Sembdner G, Gross D (1986) In: Bopp M (ed) Plant growth substances 1985. Springer, Berlin Heidelberg New York Tokyo, p 139
59. Sequeira L (1973) Annu Rev Plant Physiol 24:353
60. Suri RL, Mandahar CL (1985) Plant Sci 41:105
61. Saunders MJ, Hepler PK (1981) Planta (Berl) 152:272
62. Saunders MJ, Hepler PK (1982) Science 217:943
63. Saunders MJ, Hepler PK (1983) Dev Biol 99:41
64. Takahashi M, Yamane H, Satoh Y, Takahashi N, Iwatsuki K (1984) Phytochemistry (Oxf) 23:681
65. Tay SAB, Palni LMS, Macleod JK (1987) J Plant Growth Regul 5:133
66. Tietz A, Kasprik W (1986) Biol Biochem Pflanz 181:269
67. Thomas RJ, Harrison MA, Taylor J, Kaufman PB (1983) Plant Physiol 73:391
68. Vanden Driessche Th, Kevers C, Collet M, Gaspar Th (1989) J Plant Physiol 133:635
69. Waaland SD (1986) In: Bopp M (ed) Plant growth substances 1985. Springer, Berlin Heidelberg New York Tokyo, p 257
70. Wang TL, Cove DJ, Beutelmann P, Hartmann E (1980) Phytochemistry (Oxf) 10:1103
71. Wang TL, Wood EA, Brewin NJ (1982) Planta (Berl) 155:345
72. Wang TL, Wood EA, Brewin NJ (1982) Planta (Berl) 155:350
73. Wareing PF (1986) In: Bopp M (ed) Plant growth substances 1985. Springer, Berlin Heidelberg New York Tokyo, p 1
74. Watanaba M, Kando (1976) Plant Cell Physiol 17:1159
75. Watson BA, Waaland SP (1986) Plant Cell Physiol 27:1043
76. Yamane H, Takahashi N, Takeno K, Furuya M (1979) Planta (Berl) 147:251
77. Yamane H, Watanabe M, Sato Y, Takahashi N, Iwatsuki K (1980) Plant Cell Physiol 24:1027
78. Yamane H, Yamaguchi J, Kobayashi M, Takahashi M, Sato Y, Takahashi N, Iwatsuki K, Phinney BO, Spray CR, Gaskin P, MacMillan J. (1985) Plant Physiol 78:899
79. Yokota T, Kim SK, Fukui Y, Takahashi N, Takeuchi Y, Takematzu T (1987) Phytochemistry 26:503
80. Zbell B, Walter C (1987) In: Klämbt D (ed) Nato ASI Genes Vol H 10. Springer, Berlin Heidelberg New York Tokyo, p 142

Endogenous Plant Hormones in Rice in Relation to the Regulation of Its Life Cycle

N. TAKAHASHI[1]

1 Introduction

It is very important to clarify the roles of endogenous plant hormones in the regulation of the life cycle of higher plants from the standpoint of not only plant physiology but agricultural application as well.

Rice is known to be one of the most important and productive crops. Its life cycle, which consists of multiple stages, i.e. germination, growth and development of leaf, shoot and tiller, panicle initiation, internode elongation, heading, anthesis, pollination, fertilization and seed development, has been extensively investigated from the agronomical standpoint. However, in the early 1970s very little was known about the endogenous plant hormones in rice. We started the investigation of endogenous plant hormones of rice from 1975 in terms of the following research subjects:

1. Roles of plant hormones in the regulation of the life cycle of rice.
2. Relation between endogenous gibberellin level and the expression of dwarfism.

2 Roles of Endogenous Plant Hormones in the Regulation of the Life Cycle of Rice

2.1 Plant Materials and Identification Methods of Plant Hormones

Oryza sativa L. cv Nihonbare (tall, *japonica* type), Tan-ginbozu and Waito-C (dwarf, *japonica* type), and Tong-il (dwarf, a hybrid between *japonica* and *indica* types) were grown in a paddy field or pots, and tissues such as shoots and leaves, roots and ears were harvested at appropriate stages. The extracts of tissues were purified by the procedure established by our group [10] and the final identification was made by GC-MS and/or GC-SIM. The quantification was made by GC-SIM with internal standards.

2.2 Auxins

IAA and ICA were identified in both shoot and ear of cv Nihonbare. Fluctuation patterns of IAA in the shoot and ear are shown in Fig. 1. The content of IAA in the

[1] Department of Agricultural Chemistry, The University of Tokyo, Bunkyo-ku, Tokyo 113, Japan

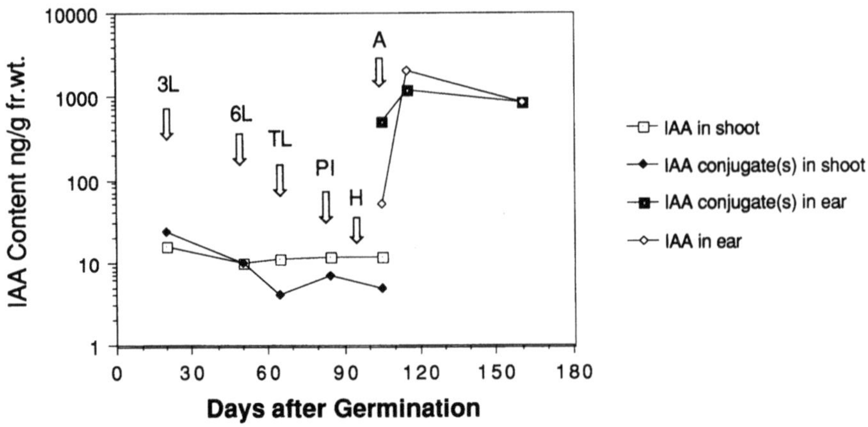

Fig. 1. Seasonal change of IAA and its conjutage content in rice. *3L* 3rd-leaf stage; *6L* 6th-leaf stage; *TL* tillering stage; *PI* panicle initiation stage; *H* heading; *A* anthesis. These abbreviations are used hereafter in other figures

shoots was about 10–20 ng/g fw. and no significant changes were observed throughout their growth. The content in the ear, however, was much higher than that in the shoot, the highest (1200 ng/g fw.) being observed about one week after anthesis. Figure 1 also shows the fluctuation pattern of IAA-conjugate(s) which was quantified by measuring IAA after hydrolysis with 1N NaOH. IAA-conjugate(s) was present in the immature and mature ear at the same level as that of free IAA. Rapid accumulation of IAA and its conjugate(s) in the ear just after anthesis suggests that reproductive growth of rice, namely anthesis or embryo development, may possibly be regulated by IAA and/or seed germination.

2.3 Gibberellins

As shown in Table 1, GAs ($GA_{1,4,8,9,12,17,19,20,24,29,34,44,51,53}$, see Fig. 3) were identified in various tissues at various growth stages of cv Nihonbare [4, 5, 6, 9].

Figure 2 shows the fluctuation pattern of $GA_{1,4,19,20,29,34}$ in the leaf and shoot from the 6th-leaf stage to the flowering stage. GA_{19} was present at the highest level, 2 to 7 ng/g fw. The concentrations of other GAs were rather low, less than 1 ng/g fw. The peak of GA_{19} was observed at the vigorous tillering stage, but its level was rather low at the internode elongation stage.

It has been shown that GA_{19} has very weak activity on the growth of the dwarf rice cv Waito-C, in which the conversion of GA_{19} to GA_1 appears to be genetically blocked [8]. In contrast, GA_{19} promotes the growth of the normal cv Nihonbare and the dwarf cv Tan-ginbozu, in which the conversion of GA_{19} to GA_1 takes place. Finally, GA_1 promotes the growth of all three cultivars. This phenomenon indicates that GA_{19} itself is inactive per se, even though it is the major GA present, while GA_1, a minor GA in quantity, probably functions as an active GA for shoot elongation. The striking change in the level of GA_{19} in rice during its life cycle indicates that

Table 1. Gibberellins identified in various tissues of rice

Shoot and Leaf	
C-13-OH-GA	$GA_{1,8^a,19,20,29,53}$
C-13-H-GA	$GA_{4,34}$
Ear	
C-13-OH-GA	$GA_{1,17,19,20,29,53}$
C-13-H-GA	$GA_{4,9,12,24,34,51}$
Seed	
C-13-OH-GA	$GA_{1^b,19,20,29,44,53}$
C-13-H-GA	$GA_{34,51}$

[a]Tentative identification in seedlings.
[b]Tentative identification.

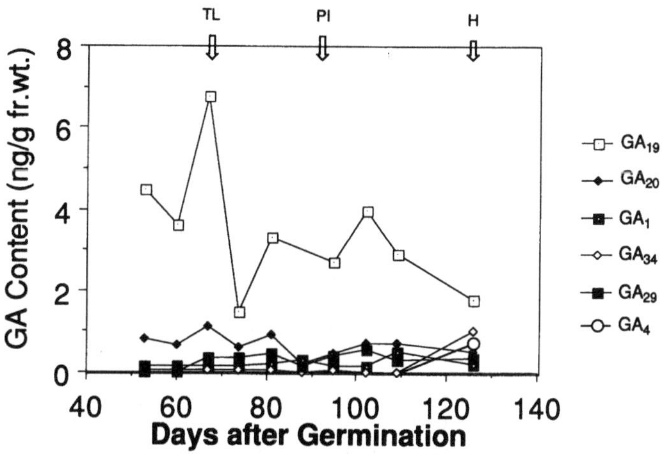

Fig. 2. Seasonal change of GA content in the shoot of rice

GA_{19} may function as a pool GA, whereas GA_1, which is a product of GA_{19}, most probably regulates the growth and development of plant at the vegetative growth stage.

It should be noted that the most GAs identified in vegetative tissues (i.e. $GA_{1,8,19,20,29,53}$) have a C-13 hydroxyl group, while in the shoot and leaf at the flowering stage, GAs lacking the C-13 hydroxyl group, such as $GA_{4,34}$, were detected only at low levels. In reproductive organs such as the whole ear, spikelet and rachis, and further, the lemma and anther at the flowering stage, both of C-13 hydroxy-GAs (i.e. $GA_{1,17,19,20,29,44,53}$) and C-13 non-hydroxy-GAs (i.e. $GA_{4,9,12,24,34,51}$) were identified. Their contents are shown in Table 2. In the ear, GA_{19} was also detected at a high level, but GA_1 content was rather low. GA_4 was one of the most abundant GAs in reproductive organs, along with GA_{19}, and GA_4 was present in the anther at an extremely high concentration. The occurrence of many C-13 non-hydroxy-GAs at fairly high levels in reproductive organs is in contrast with the fact that they either are not present, or are present only at very low levels in vegetative tissues.

Table 2. Content of endogenous GAs in reproductive tissue (ng/g fw.)

(a) In rachis, spikelet and ear

Organ	13-OH GAs GA_1	GA_{19}	GA_{20}	GA_{29}	GA_{53}	
Rachises	0.05	8	0.1	–	–	
Spikelets	0.23	5	0.3	–	–	
Whole Ear	0.27	13	0.6	0.2	–	
Organ	**13-H GAs GA_4**	**GA_9**	**GA_{12}**	**GA_{24}**	**GA_{34}**	**GA_{51}**
Rachises	13	–	–	–	–	–
Spikelets	33	–	–	–	0.7	1.0
Whole Ear	32	0.6	–	4.3	0.4	2.0

(b) In lemmas, and anther

Organ	13-OH GAs GA_1	GA_{19}	GA_{20}	GA_{29}	GA_{53}	
Lemmas	0.13	15	1.1	–	–	
Anthers	10	42	0.3	0.4	–	
Organ	**13- GAs GA_4**	**GA_9**	**GA_{12}**	**GA_{24}**	**GA_{34}**	**GA_{51}**
Lemmas	24	1	–	–	0.6	1
Anthers	3700	32	+	150	9	29

+, Identified but not quantified; –, Not detected.

These findings strongly suggest that GA_4 in the ear is mainly contained in the anther, and that the biosynthetic site of C-13 non-hydroxy-GAs may also be localized in the anther. It is possible that GA_4 plays a significant role in the regulation of the physiology of anther, and especially the pollen grain.

It can thus be deduced that there are two independent pathways in GA biosynthesis in rice (Fig. 3). The pathway in which C-13 hydroxy-GAs are involved may operate throughout the whole life cycle of rice, and C-13 hydroxylation may occur at a rather early stage of the pathway, probably at the C_{20} GA formation stage. The other pathway is for formation of C-13 non-hydroxy-GAs. This pathway may only operate at the reproductive growth stage, and the biosynthetic site may be localized in reproductive organs such as anther and pollen.

2.4 Abscisic acid

ABA was identified in every tissue. Its content was not particularly high in the shoot and leaf throughout the vegetative growth stages. Rather, it maintained a uniform pattern at a few ng/g fw. [9]. This suggests that the internode elongation is not affected very much by ABA content. ABA content in the ear was much higher than that in shoot and leaf, and this high level, 200–250 ng/g fw., was maintained for a long period after

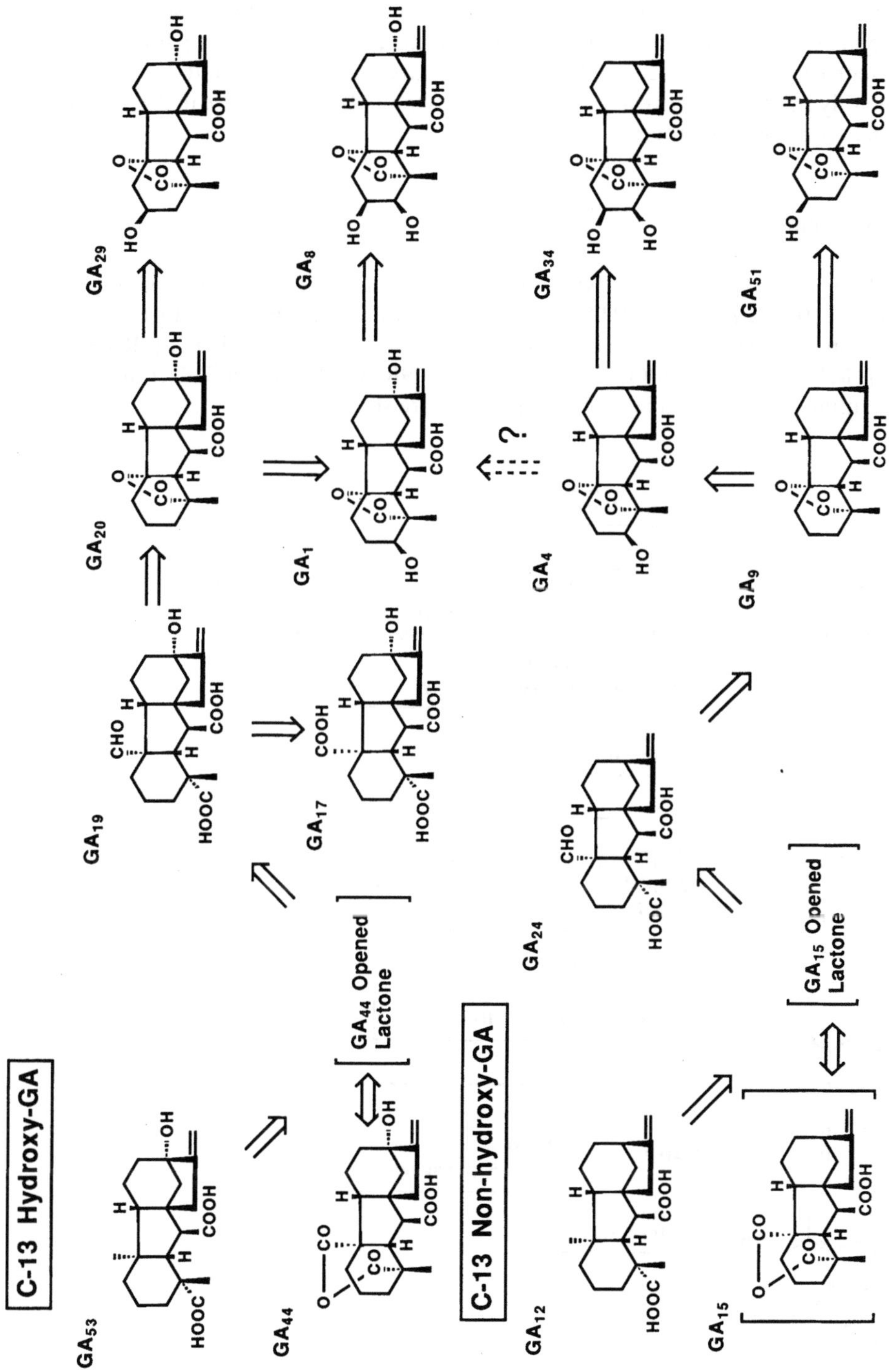

Fig. 3. Biosynthetic pathway of GA in rice. GAs without brackets were identified and those in brackets not identified

anthesis ABA concentration reached a maximum at the full ripening stage, at which time the seed reaches its maximum weight. Since a similar changing pattern of ABA content was also observed in the dwarf cultivars, Tan-ginbozu- and Tong-il, this is probably a general feature in rice, and suggests that ABA plays a role in the regulation of the seed ripening process.

2.5 Cytokinins

In the shoot and root of cv Nihonbare, *trans(t)*- and *cis(c)*-Z and their respective ribosides (*t*-[9R]Z and *c*-[9R]Z) as well as iPA were identified [10]. In the seed, in addition to these cytokinins, iP, iPA monophosphate (iPAMP), *t*-[9R]ZMP *c*-[9R]ZMP, *t*-(OG)Z, *c*-(OG)Z, *t*-[9R](OG)Z, *c*-[9R](OG)Z and (9G)Z, (*trans/cis* not determined) were identified. Their structures are shown in Fig. 4.

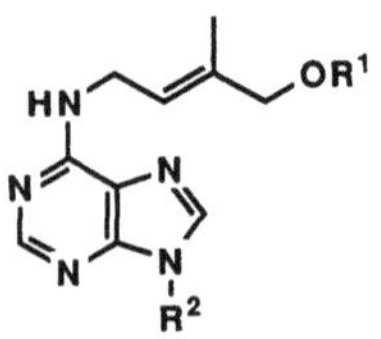

t-Zeatin Homologs

	R^1	R^2
t-Z	H	H
t-[9R]Z	H	ribosyl
t-[9R]ZMP	H	ribosyl-5'-phosphate
t-(OG)Z	glucosyl	H
t-[9R](OG)Z	glucosyl	ribosyl

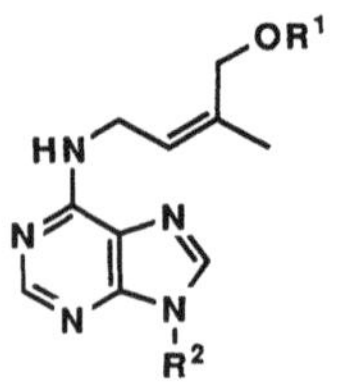

c-Zeatin Homologs

	R^1	R^2
c-Z	H	H
c-[9R]Z	H	ribosyl
c-[9R]ZMP	H	ribosyl-5'-phosphate
c-(OG)Z	glucosyl	H
c-[9R](OG)Z	glucosyl	ribosyl

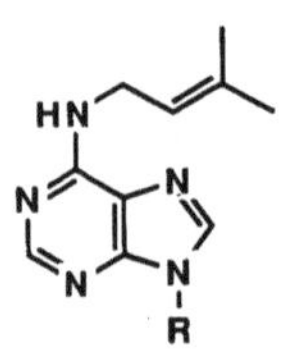

Isopentenyl Adenine Homologs

	R
*i*P	H
*i*PA	ribosyl
*i*PAMP	ribosyl-5'-phosphate

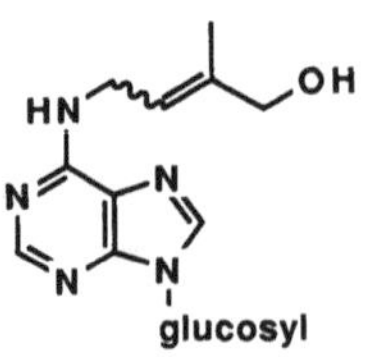

Zeatin-9-glucoside

(9G)Z

Fig. 4. Structures of endogenous cytokinins in rice

Fluctuation of the cytokinin level in the shoot and root is shown in Fig. 5. It was found that the contents of *t*-[9R]Z and *t*-Z in the shoot and root reached a maximum at the maximum tillering stage and then declined steeply. With reference to the data of Harrison and Kaufman [2, 3], this marked change seems to indicate that *t*-[9R]Z and *t*-Z may correlate with a breaking of the apical dominance of rice and, hence, their tiller bud release. The level of *c*-[9R]Z in the root was one order of magnitude higher than in the shoot, but did not change significantly. The levels of *c*-[9R]Z and iP in the shoot and root were rather low. Since it is known that these cytokinins have a relatively low activity in several bioassays, their per se contribution to the regulation of growth and differentiation of rice, therefore, is considered to be small.

In the ear, all the cytokinins reached a maximum concentration at the early growth stages, i.e. either heading, flowering or milky-ripening stage, suggesting that cytokinins may also play important roles in the development and growth of the grain. The earlier experiment in which the whole ear was analyzed gave different results (in part) from the present experiment, in that the highest level of *t*-Z was observed at a later stage, namely, the dough ripening stage. This discrepancy, however, remains to be examined. In cases of Z-type cytokinins in rice, *cis*-isomers are always coexistent and more abundant than *trans*-isomers. The occurrence of glucosides of *c*-Z and *c*-[9R]Z indicates that *cis* isomers may not be artefacts derived from *t*-RNAs during extraction, as was earlier concluded by Tay et al. [12].

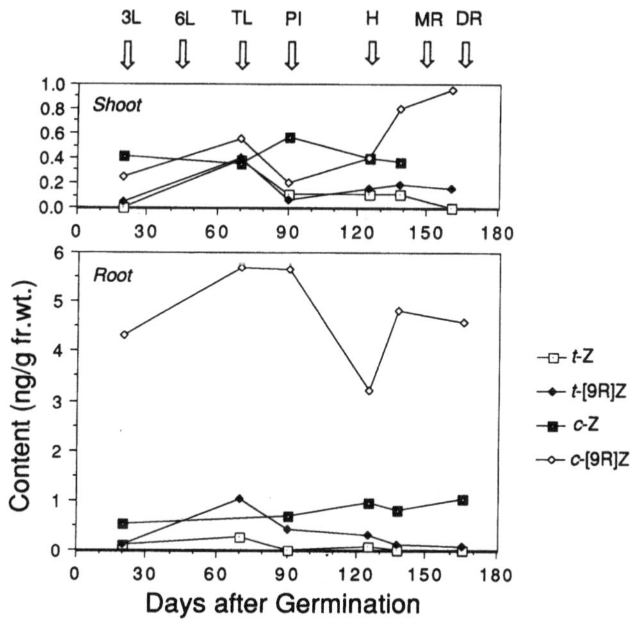

Fig. 5. Seasonal change of cytokinin content in the shoot and root of rice. MR milky ripening stage, DR dough ripening stage

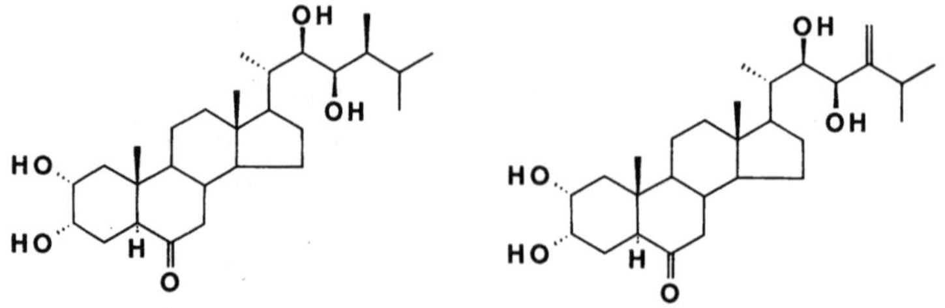

Castasterone Dolichosterone

Fig. 6. Structures of endogenous brassinosteroids in rice (cv Nihonbare and cv Arborio J1)

2.6 Brassinosteroids

Typical Br-type biological activity based on the rice leaf inclination assay was found in extracts of shoot, root and ear of cv Nihonbare. Br in the root and ear were further purified [11], and the chromatographic behavior was in good agreement with those of typical known Br. The ear extract was further analyzed by HPLC and GC-SIM, resulting in the identification of castasterone (Fig. 6) as the major Br. Abe et al. [1] have reported the occurrence of castasterone and dolichosterone (Fig. 6) in shoots of Arborio Jl (*indica* type). Since Br have been known to specifically elicit pronounced bending [7] and unrolling [13] of the rice leaf, it is quite likely that endogenous castasterone and its congeners function as PGRs of rice.

3 Relation Between Endogenous GA Level and Expression of Dwarfism

Clarification of the mechanism for the expression of dwarfism is one of most important research targets in terms of both molecular biology and agricultural application. In our previous work [9], GA and ABA levels were determined in the normal cultivar, Nihonbare and two dwarf cultivars, Tan-ginbozu and Tong-il. Dwarfism in these dwarf mutants was correlated with a low GA level in vegetative tissues, but was not correlated with ABA levels. Our recent results on investigations of GA level in two *japonica* type dwarf cultivars, Tan-ginbozu and Waito-C, are presented herein. Tan-ginbozu (*dx* gene) and Waito-C (*dy* gene), reach about 80 cm and 50 cm in height respectively, at heading stage, and Nihonbare reaches about 100 cm in height. Table 3 shows endogenous levels of GA_1, GA_4, GA_{19} and GA_{20} in whole plants of these two dwarf cultivars, and in the normal one, Nihonbare, throughout their life cycles. The levels of GA_1, GA_{19} and GA_{20} in the shoot of Tan-ginbozu were much lower than those in Nihonbare. In contrast to Tan-ginbozu, the level of GA_{20} in the shoot of Waito-C was much higher than that in Nihonbare, while the level of GA_1 was lower than in Nihonbare. These findings indicate that GA biosynthesis is blocked at an early step in Tan-ginbozu, and at the 3β-hydroxylation step in Waito-C. The lower levels of GA_1 in these dwarf cultivars relative to Nihonbare suggests that dwarfism of Tan-ginbozu and Waito-C is caused by the blockage of biosynthesis of GA_1.

Table 3. Seasonal change of endogenous GA content in the shoot of dwarf cultivars (ng/g fw.)

Stage	Cultivar	GA_1[a]	GA_{19}
6th leaf stage (6L)			
	Nihonbare	0.16	2.5
	Tan-ginbozu	0.09	0
	Waito-C	0.06	6.0
Tillering stage (TL)			
	Nihonbare	0.17	11.9
	Tan-ginbozu	0.04	< 0.01
	Waito-C	0.10	11.5
Panicle initialization stage (PI)			
	Nihonbare	0.12	6.9
	Tan-ginbozu	0	< 0.01
	Waito-C	0.07	4.7
Heading stage (H)			
	Nihonbare	0.27	4.9
	Tan-ginbozu	–	0
	Waito-C	–	4.3

[a] – Not detected due to interference of GA_{34} peak.

However, the level of GA_4, which was identified only in the reproductive organs of these dwarf cultivars, was somewhat higher in the dwarfs than that in Nihonbare, suggesting that GA biosynthesis is not blocked in reproductive organs of the dwarf rices.

For the analysis of GA biosynthesis in reproductive organs of rice, we prepared cell-free extracts from anthers of Nihonbare and Waito-C, and [^{3}H]- or [^{14}C]-labeled GAs were added to the extracts. The conversion of [^{14}C]GA_{12} to a GA_4-like substance was observed in cell-free extracts of both cultivars. This supports the early finding that the biosynthetic pathway of C-13 non-hydroxy-GAs exists in anther at the heading stage. Additionally, [^{3}H]GA_9 was converted to a GA_4-like substance and [^{3}H]GA_{20} to a GA_1-like substance in the cell-free extract of Waito-C. This shows the presence of 3β-hydroxylation activity in reproductive tissues of Waito-C, even though this step is blocked in vegetative tissues. Although further investigation is needed for confirmation, GA metabolism in cell-free extracts of reproductive tissues of normal and dwarf cultivars is likely to be similar.

4 Conclusions

1. The following plant hormones were definitively identified in various tissues of rice:
   Auxin: IAA, ICA
   Gibberellins: $GA_{1,4,8,9,12,17,19,20,24,29,34,44,51,53}$
   Cytokinins: *t*-Z, *c*-Z, *t*-[9R]Z, *c*-[9R]Z, iP, iPA, *t*-[9R]ZMP, *c*-[9R]ZMP, iPMP, *t*-(OG)Z, *c*-(OG)Z, *t*-[9R](OG)Z, *c*-[9R](OG)Z, (9G)Z.
   Abscisic acid
   Brassinosteroids: castasterone.

2. Fluctuation patterns of endogenous levels of plant hormones in various tissues of rice at different growth stages were investigated. In the case of GAs, two independent biosynthetic pathways, i.e. the C-13 hydroxy-GA and C-13 non-hydroxy-GA pathways, were postulated to be important in regulating the life cycle of rice, the former operating through the all growth stages and the latter only at the reproductive growth stage.
3. The correlation between endogenous GA level and dwarfism of rice was investigated. In the vegetative growth stages of dwarf rice, Tan-ginbozu, the GA biosynthetic pathway is blocked at the early stage, while in Waito-C, the 3β-hydroxylation stage is blocked. In reproductive tissues, the C-13 non-hydroxy-GA pathway was present, as shown by the use of cell-free systems obtained from dwarf cv Waito-C.

We have demonstrated that many kinds of PGRs play very important roles in regulating the life cycle of rice. This type of research should be useful for clarification of regulatory mechanisms in other crops and in important weeds. The information obtained from these investigations should also give us the opportunities develop new technologies to control growth and differentiation of various crops and weeds, resulting in the enhancement of productivity of higher plants.

Acknowledgements. This research has been conducted through a collaboration between our group at the University of Tokyo, Dr. M. Kobayashi, Dr. S. Kurogochi, Dr. N. Murofushi, Dr. Y. Suzuki, Ms. M. Takagi, Dr. I. Yamaguchi and Dr. T. Yokota, and the group at the National Institute of Agrobiological Resources, Dr. Y. Ota and Dr. H. Saka, to whom the author wishes to express his thanks.

References

1. Abe H, Nakamura K, Morishita T, Uchiyama M, Takatsuto S, Ikekawa N (1984) Agric Biol Chem 48:1103
2. Harrison MA, Kaufman PB (1980) Plant Physiol 66:1123
3. Harrison MA, Kaufman PB (1983) J Plant Growth Regul 2:215
4. Kobayashi M, Yamaguchi I, Murofushi N, Ota Y, Takahashi N (1984) Agric Biol Chem 48:2725
5. Kobayashi M, Yamaguchi I, Murofushi N, Ota Y, Takahashi N (1988) Agric Biol Chem 52:1189
6. Kurogochi S, Murofushi N, Ota Y, Takahashi N (1979) Planta 146:185
7. Maeda E (1964) Physiol Plant 18:813
8. Murakami Y (1972) In: Carr DJ (ed) Plant growth regulation. Springer, Berlin Heidelberg New York, p 166
9. Suzuki Y, Kurogochi S, Murofushi N, Ota Y, Takahashi N (1981) Plant Cell Physiol 22:1085
10. Takahashi N, Yamaguchi I (1986) Acta Hortic 179:45
11. Takagi M, Yokota T, Murofushi N, Ota Y, Takahashi N (1985) Agric Biol Chem 49:3271
12. Tay SAB, MacLeod JK, Palni LMS (1986) Plant Sci 43:131
13. Wada K, Kondo N, Kondo H, Marumo S (1985) In: Proceedings of internal meeting of the Society for Chemical Regulation of Plants held in Utunomiya, Japan 1985, p 58
14. Yokota T, Baba J, Koba S, Takahashi N (1984) Agric Biol Chem 48:2529

I Genetic Approaches

The Benefit of Biosynthesis and Response Mutants to the Study of the Role of Abscisic Acid in Plants

C.M. KARSSEN[1], H.W.M. HILHORST[1], and M. KOORNNEEF[2]

1 Introduction

In the 21 years since the birth of ABA in Ottawa, at the 6th International Conference of Plant Growth Substances, many effects have been attributed to it. In line with its origins most of the effects of applied ABA are inhibitory to growth and development, but a few stimulative effects have also been reported. Until the isolation of ABA mutants it was only possible to identify roles for ABA by studying the effects of its application or through physiologically correlated changes in its endogenous levels.

In 1980 Walton [23] was forced to conclude in his review of the biochemistry and physiology of ABA that detailed knowledge about the functional role of ABA was then still lacking. Today, Zeevaart and Creelman [26] speak of a renaissance in ABA studies. This transformation in our understanding of both the function of ABA and its biosynthetic pathway is, for the most, due to the use of ABA mutants.

Herein we will focus on the new insights that have been gained and those developments that may possibly be expected in the near future regarding the use of ABA mutants.

2 Mutants

Two types of mutations have been isolated in a range of species. Those having low levels of ABA in their tissues, termed ABA-deficient, and those with reduced sensitivity to ABA, the ABA-response mutants. The isolation procedures and the characterization of these mutants have been reviewed recently [12, 18] and only a brief summary will be given here.

2.1 ABA-Deficient Mutants

These mutants were isolated by their increased tendency to wilt and/or a reduction in seed dormancy. Application of ABA caused a reversion to wild-type characteristics.

[1] Department of Plant Physiology, Agricultural University, Arboretumlaan 4, 6703 BD Wageningen, The Netherlands

[2] Department of Genetics, Agricultural University, Dreijenlaan 2, 6703 HA Wageningen, The Netherlands

ABA-deficient mutants are described in tomato: *notabilis (not), flacca (flc)* and *sitiens (sit)*; potato: *droopy (dr)*; pea: *wilty (wil)* and *Arabidopsis thaliana (aba)* [cited in 12, 18]. All these mutants are monogenic recessive. ABA levels in non-stressed leaves of the tomato mutants range from 49% of wild-type for *not*, 26% for *flc* and 15% for *sit* [16]. Thus, most likely all three mutants are leaky. Young seeds of *sit* contained, halfway through development, maximally 1 ng ABA $seed^{-1}$ in contrast to 36 ng $seed^{-1}$ in wild-type tomato seeds cv Moneymaker [3]. In mature seeds the ABA contents were 0.2 and 1.8 ng $seed^{-1}$ respectively.

Immature seeds of the most viable ABA mutant of *Arabidopsis* (*aba^3*) contained maximally 8 ng g^{-1} fw ABA in contrast to 600 ng g^{-1} in wild-type seed [10]. In the alleles (*aba^1* and *aba^4*) that showed more extreme wilting, the ABA levels were below the detection level. A common characteristic of *dr* potato, *sit* and *flc* tomato, *wil* pea and *aba Arabidopsis* is their inability to accumulate ABA in response to water stress [see 26].

In maize, various mutants have been described with primary lesions at different steps in the carotenoid pathway, and they all have reduced ABA levels ranging from 7 to 70% of the wild type [12].

2.2 ABA-Response Mutants

In *Arabidopsis*, mutants with reduced sensitivity to exogenous ABA were selected on the basis of their good growth in a solution containing 10 μM ABA [14]. The mutations were at three different loci and were termed *abi1*, *abi2* and *abi3*. Phenotypically these mutants resembled the *aba* mutants in several responses, yet they contained ABA levels similar or even somewhat higher than in the wild type.

Borkird and Sung [1] isolated seven ABA-insensitive cell lines in a somatic embryo culture of carrot in which ABA failed to arrest the development of torpedo-stage embryos into plantlets, as occurs in wild type when ABA is added. All lines also showed reduced auxin sensitivity. Characterization of three lines showed lower levels of ABA uptake as a possible cause of ABA insensitivity. However, the uptake of 2,4-D was higher than in wild-type embryos. In tobacco, cell-lines resistant to inhibition of growth by ABA were also isolated [25], and cell proliferation continued in the presence of what would be growth-inhibiting ABA concentrations for normal cell lines.

3 Benefit of Mutants to Biosynthesis Studies

Recent studies with the three ABA-deficient tomato mutants have contributed extensively to the clarification of the biosynthesis of ABA. These studies support the so-called indirect pathway of ABA biosynthesis in which the C40 xanthophyll violaxanthin is the likely precursor of ABA, with xanthoxin as an intermediate [22, 26]. It was concluded that the *flc* and *sit* mutations are impaired in the conversion of xanthoxin to ABA, and that the lesion in *not* is at a step between xanthoxin and violaxanthin.

The biochemical characterization of the ABA-deficient mutants in other species is not yet available, but presumably the lesions are in the last part of the biosynthetic pathway, as carotenoid levels are normal.

4 Benefit of Mutants to Physiological Studies

4.1 Reliability of the Method

ABA-response and -biosynthesis mutants provide a good mechanism to establish or reject the many tentative functions of ABA in growth and development of plants. Essentially, the method consists simply in observing which processes are influenced by a genetically induced decrease of the endogenous ABA content, or a decrease in the ABA-sensitivity, and which processes are not affected. However, caution has to be exercised.

Quarrie [18] emphasized that in contrast to the lesions in *flc* and *sit* other mutations may affect ABA synthesis only indirectly. As mentioned above, the *vp* mutants in maize are devoid of carotenoids and consequently ABA deficient. Mutations may also occur simultaneously in two different genes, one directly affecting ABA biosynthesis and the other having an unrelated effect. In the case of close linkage, separation of such mutations by back crossing is not always successful.

However, the reliability of the conclusions from mutant studies increases substantially when similarities occur between phenotypes of different ABA mutants, particularly those with independently isolated alleles of one locus, such as the allelic series of the *aba* locus of *Arabidopsis* [10] and the *sit* and *flc* loci of tomato (Koornneef, unpublished). Indeed, many effects of both synthesis- and response-mutants show an encouraging consistency for several responses across species and across mutants that have arisen either naturally (potato), or by X-ray or chemical mutagenesis. But even then interpretation of the results of mutant studies is complex, because a single mutation may cause a multitude of pleiotropic effects, i.e. it influences a wide range of seemingly unrelated developmental processes. Some pleiotropic effects are very indirect. For instance, the ABA mutants in tomato show leaf epinasty, swelling of the upper stem and formation of adventitious roots along the stem, effects which were thought to be characteristic for auxin or ethylene overproduction [21]. However, Neill et al. [17] found no differences in auxin and ethylene levels between *flc* and wild-type leaves.

4.2 Established Roles of Endogenous ABA

4.2.1 Stomatal Control

The isolation of tomato mutants that contain low levels of ABA and show an ABA-reversible tendency to wilt [20] strongly favours a role of ABA in the regulation of stomatal aperture in water-stressed plants. The hypothesis originated from the observation of ABA accumulation in stressed leaves. All other ABA-deficient mutants isolated so far show a wilty character. From the ABA-response mutants in *Arabidopsis* only the *abi1* and *abi2* mutations show a disturbance of normal water relations. The *abi3* mutant did not deviate from wild type in this respect [14].

The *flc* tomato mutant is frequently used to study the mechanism of ABA action in stomata, and work with it has shown that ABA overrides the action of several environmental factors, such as light and CO_2 [2].

4.2.2 Seed Dormancy

Studies with developing seeds of *Arabidopsis* provided the first proof that ABA had been properly named 'dormin' in the first years after its discovery. Developing seeds of different genotypes were tested for their ability to germinate precociously. Half-way through development a variable percentage of wild-type seeds were able to germinate precociously (Fig. 1). However, during maturation the ABA-producing wild-type seeds developed dormancy. Dormancy was broken during dry storage (weeks), or incubation at low temperatures (days) [9]. In contrast to wild type, seeds of the *aba* and *abi* mutations reached the capacity for full germination some time before maturation (Fig. 1) [10, 11]. The *abi1* seeds did develop some dormancy in the last phase of seed development, particularly in summer cultivation, and *abi3* seeds developed dormancy soon after harvest (E. Lacka, unpublished results). These forms of dormancy were also relieved by chilling or dry storage.

Reciprocal crosses between wild type and *aba* mutants showed that ABA in *Arabidopsis* had both a maternal and an embryonic origin [10]. However, it was

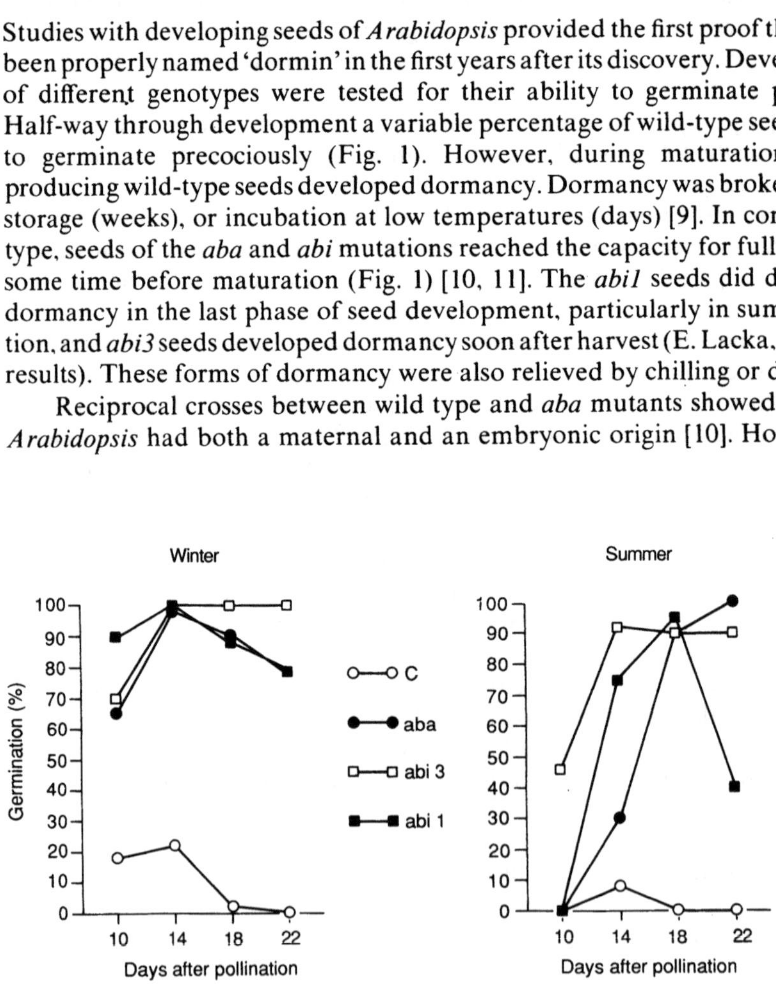

Fig. 1. Precocious germination of seeds of four different genotypes of *Arabidopsis thaliana* in water under continuous light at 24°C. The seeds were isolated from the fruits at the indicated time, and subsequently transferred to the germination conditions. Germination was counted after 14 days. Seeds developed either in winter or summer

exclusively the embryonic ABA that induced seed dormancy. The presence of a high content of maternal ABA, or ABA originating from exogenous spray application to the plants, or applications to the roots did not influence dormancy induction. Nevertheless, ABA from both sources apparently reached the testa of the developing seeds, because normal development of mucilage-producing cells was stimulated, whereas development of these cells is reduced in the *aba* mutants.

Tomato seeds are less suitable for dormancy studies than *Arabidopsis*. Dormancy develops in wild-type seeds only under certain cultivation conditions (higher temperatures, no standard defoliation) and disappears again after short periods of dry storage [3]. Seeds of the *sit* mutant in the background of cv Moneymaker do not even possess that feeble form of dormancy, and therefore support conclusions obtained from the *Arabidopsis* studies. Also, in tomato the induction of seed dormancy depended on ABA synthesized in the embryo and/or endosperm. The presence of ABA in those tissues was shown by pollinating *sit* plants with *Sit* pollen, leading to *Sit/sit* seeds on a mutant plant [8]. At maturity only 15% of wild-type seeds germinated, whereas *Sit/sit* and *sit/sit* seeds from mutant plants had germination percentages of 57 and 97% respectively. These results indicate that in wild-type tomato, in contrast to *Arabidopsis*, maternal ABA is also involved in dormancy control.

4.2.3 Vivipary and Seed Development

Since the precocious germination of isolated seeds or embryos can be inhibited in vitro by applied ABA, it has often been suggested that endogenous ABA has a similar role in developing seeds. It was also suggested that ABA could have promotive effects on seed development, in particular on the accumulation of certain reserve proteins [see 19 for recent review]. Originally, however, our experiments with *Arabidopsis* and tomato mutants did not support these suggestions. Seeds of *aba, abi1* and *abi3 Arabidopsis* and *sit* tomato developed normal fresh and dry weights, and also formed the normal set of reserve proteins [3]. Germination did not occur during seed development of the mutants. We thus concluded that other factors prevented early germination and that ABA was not involved in the regulation of seed development, apart from dormancy induction [11]. In *Arabidopsis* a lack of water in maturing seed and fruits prevented germination during development. Vivipary only occurred when the plants were kept at very high relative humidity or when fruits were detached and incubated on wet filter paper. In *sit* tomato mutants vivipary only occurred in overripe fruits. Analysis of the germination characteristics showed that *sit* seeds, in contrast to wild-type seeds, were able to germinate at the high osmotic potential maintained by the fruit tissues.

However, recent experiments with seeds of the recombinants of the *aba1* and *abi3* mutations of *Arabidopsis* changed our conclusions drastically [12a]. Recombinant seeds that resulted from selfing of the double recessive genotypes showed a deviation in seed development: seeds did not lose water during maturation, fruits and seeds were still partly green after the normal developmental period, they did not survive dehydration and they did not accumulate the late abundant 12S and 2S proteins (Table 1). This pattern never occurred in the recombinant of *aba* and *abi1*,

Table 1. The presence (+) or absence (–) of phenotypic aberrations from the wild type in the recombinants of the *abi1* or *abi3* mutations with *aba* in *Arabidopsis thaliana* [summary of data 12a]

Phenotype	*abi3, aba*[1]	*abi1, aba*[1]
Reduced seed dormacy	+	+
Green/brown seeds	+	–
Reduced loss fw	+	–
Desiccation intolerance	+	–
Lack of 12S and 2S protein	+	–

nor in the single mutants *aba, abi3* and *abi1*. It was also not observed when the *aba abi3* seeds developed on plants heterozygous for either *aba* or *abi3*. Application to the roots of either ABA or the experimental compound LAB 173711 (BASF, Ludwigshafen, FRG), an ABA analogue [7], antagonized the abnormal pattern of seed development.

We thus concluded that seeds only fail to develop normally when a deficiency of both maternal and embryonic ABA is combined with reduced sensitivity to ABA due to the *abi3* mutation. Evidently, the ABA sensitivity of these seed development processes must be very high.

4.3 Questionable Roles of Endogenous ABA

In a number of cases the ABA-deficient mutants did not show the phenotypic deviations that were expected from earlier ABA application experiments. Therefore, some putative roles of ABA are questionable. This applies for instance to root gravitropism where a role for ABA must be questioned because the ABA-deficient *w3, vp5* and *vp7* mutants of maize, and the *flc* mutant of tomato showed normal gravicurvature [15, 24]. However, the experiments with the *aba abi3* recombinant (described above) show that the aberrant phenotype is probably absent due to a combination of the leakiness of the mutated gene and a very high sensitivity of the response systems. More definitive answers may be obtained when experiments are repeated, either with recombinants of ABA-deficient and -insensitive mutants, or with double tomato mutants like *flc not* or *sit not* [22].

The reversibility by applied ABA of the reduction in growth of leaves and stems of ABA-deficient mutants suggests, at first view, a growth-stimulatory effect of ABA [26]. However, Quarrie [18] rightly remarks that all those observations [25] were made in greenhouses without a proper humidity control. At a 95% RH in a controlled environment cabinet, plants of *Arabidopsis* wild type and the mutants with the least serious ABA deficiency (*aba*[3]) were of the same height and those of the severely deficient *aba1* mutant were 10–12% shorter. Hence, more detailed studies are required to unravel direct and indirect effects of ABA deficiency on the growth response of stems and leaves.

5 Concept of Regulation

5.1 Synthesis or Sensitivity

The strong phenotypic resemblance between the ABA-deficient and -response mutants of *Arabidopsis* underlines once more that both the presence of the hormone and the responsiveness of the test system may modulate the eventual hormone effect. However, it is a matter of extended debate whether endogenously or environmentally induced changes in the rate or phase of developmental processes are generated by changes in hormone synthesis and/or hormone sensitivity.

The availability of mutations with a different degree of ABA deficiency is thus an important tool to test the relevance of ABA concentration on development. The three mutations of the *aba* locus in *Arabidopsis* indicate a positive interaction between the ABA concentration and the degree of deviation of all but one of the phenotype characters (Table 2). Seed development is the exception. The experiments with the *aba abi3* recombinant (described above) showed that the responsiveness of seed development to ABA is so high that endogenous ABA concentration during seed development was never limiting. However, in other responses a reduction of ABA concentration below wild-type levels does limit normal development.

The mutant studies do not yet present a definite answer to the question of whether or not natural regulation of development is causally related to changes in either ABA synthesis or the responsiveness to ABA, and as such is in contrast to studies with GA mutants. For seeds of *Arabidopsis* it has clearly shown that light stimulates both GA-synthesis and GA-responsiveness [5].

Our knowledge regarding the environmental control of ABA levels in plants is restricted to the water-stress induced rise in leaves. At present it is thought that loss of turgor is the critical parameter of cell water relations which initiates ABA biosynthesis. At zero turgor a relaxation of the plasmalemma and associated conformational changes may be the signal for increased ABA synthesis [26].

Table 2. Phenotypic observations on the allelic series of mutations of the *aba* locus of *Arabidopsis thaliana* [from 10 and 13]

	Wild type	aba^3	aba^1	aba^4
Withered main stem (% of plants)	0	11	56	69
Total plant height (cm)	20.3	14.3	8.2	6.4
Germination (%) light	0	95	95	100
dark	0	22	58	–
Thickness mucilage layer (μm)	67	52	27	25
Seed weight (μg)	19	18	17	21
ABA content of developing seeds (pg/seed)	0.54	0.14	<0.02	<0.02

5.2 Responsiveness to ABA

It is yet not known at what level in the ABA response mechanism the mutations are located that reduce the responsiveness to ABA. A likely candidate is a reduction in the number of receptors. However, reduced affinity of the receptor might also be involved. Moreover, the lesions may also be located in the chain of events subsequent to the hormone-receptor interaction, or be related to uptake mechanisms of the hormone. The latter is the case in the ABA-insensitive cell lines isolated in carrot [1].

The characterization of receptors for plant hormones has just started, in spite of the description of different proteins or membrane fractions able to bind hormones [4]. For ABA only Hornberg and Weiler [6] have succeeded so far in such a receptor characterization. They used photoaffinity labeling to cross-link radiolabeled ABA to putative binding sites at the plasmalemma of *Vicia faba* guard cell protoplasts and obtained high-affinity (K_D = 3–4 nM) binding sites which were specific for those protoplasts.

Progress in this area is expected from the use of mutants with reduced ABA-responsiveness. *Arabidopsis*, with its excellent properties for experimentation in molecular genetics, is a particularly likely candidate for further progress.

The experiments with the *abi1* and *abi3* mutants of *Arabidopsis* mentioned above indicate the complexities that may be expected. The summary of the phenotypic characteristics of the *aba, abi1* and *abi3* mutants shows (Table 3) that at least three different response mechanisms may exist in *Arabidopsis* (*abi2* resembles *abi1*): (a) A first mechanism is related to the ABA inhibition of growth responses in germination, including seed dormancy, and seedling growth. It is reduced by in both mutations *abi1* and *abi3*. (b) Wilting is increased either when mutations *aba* reduces ABA levels, or *abi1* reduces responsiveness; *abi3* is not involved. (c) With respect to seed development it is the opposite: only the *abi3* mutation, in combination with *aba*, causes a blockage of seed development. Thus, *Arabidopsis* shows, with respect to ABA responsiveness, a clear tissue specificity that may indicate different receptor systems for different responses.

Table 3. The presence (+) or absence (−) of phenotypic aberrations from the wild type in the *aba, abi1* and *abi3* mutations of *Arabidopsis thaliana* seeds during development [from 14, 12a]

Phenotype	*aba*	*abi1*	*abi3*
ABA deficiency	+	−	−
Reduced ABA sensitivity			
of germination	−	+	+
of seedling growth	−	+	+
Reduced seed dormancy	+	+	+
Increased wilting	+	+	−
Blockage seed development	−	− (+ *aba*)	+ (+ *aba*)

5.3 Molecular Action of ABA

Studies with ABA mutants have not yet contributed substantially to the expansion of our knowledge concerning the molecular action of ABA. Studies with tomato mutants indicate an important role for ABA in regulating membrane and cell wall structure [see 18 for review]. During seed development applied ABA stimulates protein synthesis both by maintaining levels of mRNA stability and promoting RNA synthesis [19]. The *aba abi3* recombinant may allow these experiments to be shifted to an in vivo situation.

References

1. Borkird C, Sung ZR (1987) Plant Physiol 84:1001
2. Bradford KJ, Sharkey TD, Farquahar GD (1983) Plant Physiol 72:245
3. Groot SPC (1987) PhD thesis Agric University of Wageningen, Wageningen, p 107
4. Guern J (1987) Ann Bot 60 Suppl 4:75
5. Hilhorst HWM, Karssen CM (1988) Plant Physiol 86:591
6. Hornberg C, Weiler EW (1984) Nature 310:321
7. Jung J, Grossman K (1985) J Plant Physiol 121:361
8. Karssen CM, Groot SPC (1987) In: Pinfield NJ, Black M (eds) Growth regulators and seeds, British Plant Growth Regulator Group, Long Ashton, p 17
9. Karssen CM, Laçka E (1986) In: Bopp M (ed) Plant growth substances 1985. Springer, Berlin Heidelberg New York Tokyo, p 315
10. Karssen CM, Brinkhorst-van der Swan DLC, Breekland AE, Koornneef M (1983) Planta 157:158
11. Karssen CM, Groot SPC, Koornneef M (1987) In: Thomas H, Grierson D (eds) Developmental mutants in higher plants. Cambridge University Press, Cambridge, p 119 (Soc Exp Biol Semin Ser 32)
12. Koornneef M (1986) In: Blonstein AD, King PJ (eds) A genetic approach to plant biochemistry, Springer, Vienna New York, p 35

12a. Koornneef M, Hanhart CJ, Hilhorst HWM, Karssen CM (1989) Plant Physiol 90:463

13. Koornneef M, Jorna ML, Brinkhorst-van der Swan DLC, Karssen CM (1982) Theor Appl Genet 61:385
14. Koornneef M, Reuling G, Karssen CM (1984) Physiol Plant 61:377
15. Moore R, Smith JD (1984) Planta 162:342
16. Neill SJ, Horgan R (1985) J Exp Bot 36:1222
17. Neill SJ, Horgan R, Parry AD (1986) Planta 169:87
18. Quarrie SA (1987) In: Hoad GV, Lenton JR, Jackson MB, Atkin R (eds) Hormone action in plant development, A critical appraisal. Butterworth, London, p 201
19. Quatrano RS (1987) In: Davies PJ (ed) Plant hormones and their role in plant growth and development. Nijhoff, Dordrecht, p 494
20. Tal M, Imber D (1970) Plant Physiol 46:373
21. Tal M, Imber D, Erez A, Epstein E (1979) Plant Physiol 63:1044
22. Taylor IB (1987) In: Thomas H, Grierson D (eds) Developmental mutants in higher plants, Cambridge University Press, Cambridge, p 197 (Soc Exp Biol Semin Ser 32)
23. Walton DC (1980) Annu Rev Plant Physiol 31:453
24. Weyers JDB (1985) J Plant Physiol 121:475
25. Wong JR, Sussex IM (1980) Planta 148:103
26. Zeevaart JAD, Creelman RA (1988) Annu Rev Plant Physiol Plant Mol Biol 39:439

A Genetic Approach to Auxins and Cytokinins

P.J. King, A.D. Blonstein, Y. Fracheboud, J. Oetiker, and M. Suter[1]

1 Introduction

The biochemical analysis of monogenic mutant plants differing from the wild type in either content or response to plant hormones would clearly help answer some of the outstanding questions about hormone biosynthesis and its regulation [1, 6]. Such mutants would allow more definitive experiments to be carried out on the role of hormones both at the cellular level and in plant development, and afford a direct route to cloning genes related to hormones and their action.

Useful progress has been made in the genetics of GAs, ABA and, more recently, ethylene [11]. However, relatively few mutations have been reported involving auxins or cytokinins. Both recessive and dominant mutants have been isolated with increased resistance to synthetic auxins [8, 14, 16], and two mutants have been described with significantly altered levels of IAA or IAA conjugates [20, 22]. Mutations have been reported that affect the cytokinin requirement of tobacco leaf cells in culture [15], and cytokinin resistance and production in the moss *Physcomitrella patens* [23].

In our laboratory we have been using M2 seed families and mutagenized haploid protoplasts of *Nicotiana plumbaginifolia* in a combined approach to search both for mutants resistant to toxic concentrations of auxins and cytokinins, and for temperature-sensitive auxin auxotrophs. The hope has been that variation isolated using these rather unspecific selection procedures would include mutations in hormone uptake, biosynthesis, catabolism and reception. The isolation methods, particularly for the cloning of auxotrophs, are laborious, and during the last five years we have slowly assembled a small collection of mutants which we feel will allow us to test several current theories about auxin and cytokinin synthesis and mode of action. This paper briefly reviews our work on the mutants and their characterization.

2 Auxin- and Cytokinin-Resistant Mutants

2.1 Isolation

Seed of *N. plumbaginifolia* treated with GA_3 to break dormancy and with EMS for mutagenesis was sown at high density in soil in the greenhouse. After the M1 plants

[1] Friedrich Miescher-Institut, Postfach 2543, 4002 Basel, Switzerland

had selfed, a single seed capsule was harvested per plant, and the seed preserved as individual M2 families. Mutagen doses were applied that resulted in 49–86% germination of M2 seed and the appearance of chlorophyll mutants in 4–20% of the M2 families (Blonstein, Van Hai, King, unpublished).

Testing M2 families for the segregation of auxin- or cytokinin-resistant mutants was done either individually, by placing seeds of single M2 families onto agar media containing inhibitory auxin or cytokinin concentrations, or in groups, by combining seed from ten M2 families and sowing them onto filter papers soaked in auxin-containing medium. Concentrations of hormones were chosen that produced characteristic disturbances in the development of young seedlings (see Fig. 1).

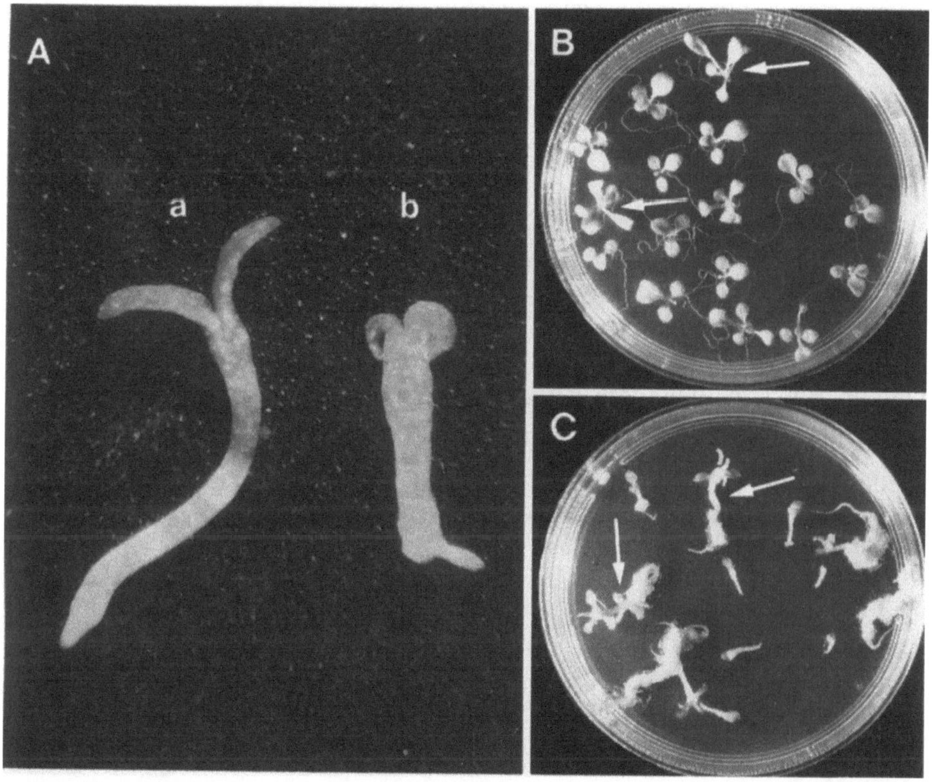

Fig. 1A-C. Auxin- and cytokinin-resistant mutants of *Nicotiana plumbaginifolia*. **A** Seventeen-day-old *aur1* (**a**) and wild-type (**b**) seedlings of *N. plumbaginifolia* germinated in the presence of 20 μM NAA; **B** segregation of the cytokinin-resistant (*ckr1*) mutant of *N. plumbaginifolia* (*arrowed*) on control medium without hormones; **C** segregation of *ckr1* (*arrowed*) on selection medium containing 20 μM BAP. Note the blocked development of the wild type compared to the leaf and root formation by the mutant

2.2 Mutants Resistant to Auxin

Group screening of about 5000 M2 families for resistance to 20 μM NAA or 80 μM IAA ethyl ester has yielded eleven auxin-resistant mutants (Blonstein, Stirnberg, King, unpublished). Four mutants have been confirmed as recessive and two are dominant. *aur1*, a single-gene recessive mutation (Fig.1A) is 5–10 times more resistant to NAA than the wild type, and shows cross resistance to other auxins. Homozygous mutant plants are fully fertile but exhibit several slight disturbances to normal growth and development, including impaired rooting, abnormal gravitropism, reduced apical dominance and late flowering.

Our auxin-resistant *N. plumbaginifolia* mutants add to the small collection of mutations that have been described during the last few years in tobacco [16] and in *Arabidopsis* [9], which define at least two and perhaps up to four genes involved in responses to applied auxin. The tobacco, *Arabidopsis* and *N. plumbaginifolia* mutants resemble each other in having disturbed root growth characteristics. No complete biochemical explanation of the auxin resistance of any of these mutants has been established.

2.3 A Cytokinin-Resistant Mutant

During the individual screening of 2500 M2 families we found one single-gene recessive mutation (designated *ckr1*) that allows germination and seedling growth at concentrations of BA (and other cytokinins) inhibitory to wild-type development (Fig. 1B and C) [3]. The mutant shows more rapid germination and seedling growth than the wild type under non-selective conditions. Despite a marked tendency to wilt, *ckr1* plants survive in the greenhouse and are fully fertile. Cytokinin-resistant lines of the moss *P. patens* have been reported, but unfortunately they were sexually sterile and, therefore, not analyzed genetically [22]. The *ckr1* mutant of *N. plumbaginifolia* appears to be the only higher plant mutant so far described with increased resistance to cytokinins.

3 Auxin Auxotrophs

Mutations leading to loss of a function and thus an auxotrophic phenotype are of great importance to biochemical genetics. Although we now know the plant phenotype of GA- and ABA- auxotrophic mutants, no auxin auxotrophs are known and indeed the phenotype of such a mutant is difficult to predict. We make the assumption that auxin auxotrophy prevents cell growth and division and that the trait is lethal at all stages of plant growth. For this reason, we have been searching for temperature-sensitive auxin auxotrophic mutants in vitro among clones derived from mutagenized, haploid leaf protoplasts of *N. plumbaginifolia.*

Treatment with auxin and cytokinin promotes initial division of isolated mesophyll protoplasts in vitro, and the continued presence of the hormones is often necessary to maintain the products of the initial divisions in the proliferative, tissue culture state. However, the cells have no permanent, genetically-determined requirement for exogenous hormones, since viable "hormone-independent" plants can be regenerated from the cells following removal of the hormones. Thus, the wild-type phenotype of leaf protoplast-derived cell clones following withdrawal of auxin from the medium is survival, proliferation and morphogenesis. This is the basic protocol for plant regeneration from leaf protoplasts from many species [7], and a further example of the principles established by Skoog and Miller [19]. Large populations of protoplast-derived clones thus offer a system for screening out rare auxin-auxotrophic mutants whose growth and differentiation is inhibited by auxin deprivation.

3.1 Isolation

The isolation of growth-inhibited and/or rapidly lethal plant auxotrophs is still fraught with great difficulty [2]. We have applied both the tedious process of individual colony testing (*total isolation*) [4], and the use of nucleoside analogues for enrichment [21], to find temperature-sensitive (ts) auxin auxotrophs. Briefly, using the total isolation technique, colonies derived from haploid protoplasts after UV or N-methyl-N'-nitro-N-nitrosoguanidine mutagenesis and culture on auxin- and cytokinin-containing medium are placed individually on medium without auxin and incubated at 33 °C (*restrictive conditions*). Clones showing abnormal growth or development are rescued at the earliest opportunity by transfer to auxin-containing medium at 26 °C. After further growth each clone is then retested and transferred to plant regeneration media. Enrichment for ts auxin-auxotrophic clones is achieved by transferring all cells en masse to restrictive conditions, allowing time for the cessation of DNA replication of auxotrophic cells, and then treating the population with 5-bromodeoxyuridine and 5-fluorodeoxyuridine, which preferentially destroy replicating wild-type cells. Colonies developing after washing and replating the cells are retested as putative auxotrophs [21].

Table 1 summarizes the results of our auxin-auxotroph isolation experiments to date. Auxin auxotrophs were found by total isolation at a frequency of about 10^{-4}, but no ts auxin auxotrophs were recovered using this method. However, seven ts auxin auxotrophs were isolated after enrichment. In addition to these auxotrophs of *N. plumbaginifolia*, one ts auxin auxotroph was isolated earlier from *Hyoscyamus muticus* [10].

Table 1. The isolation of auxin auxotrophs of *N. plumbaginifolia* by *total isolation* or *enrichment*

Method	Colonies tested	Auxin auxotrophs	ts auxin auxotrophs
Total isolation	56 000	8	0
Enrichment	13 700	13	7

3.2 Auxin Auxotrophs

The cellular phenotype of one of the auxin auxotrophs is shown in Fig. 3. Whereas the wild type survives (Fig. 3A) and initiates shoot morphogenesis when transferred to auxin-minus medium, growth of the auxin auxotrophic clone ceases (Fig. 3B). Such auxotrophic clones exhibit characteristic dose/response relationships with auxin (Fig. 2A-C), but unlike the wild type their growth is not promoted by cytokinin in the absence of auxin (data not shown).

There are already a number of cell lines in daily use that appear to be auxin dependent, e.g. the *Acer pseudoplatanus* culture originally isolated by Lamport in 1964 [12]. These cultures have been selected over many years for rapid, friable growth in suspension in the presence of high auxin concentrations and would seem to be phenotypically auxin auxotrophic due to epigenetic alterations. Auxin independent subclones can be isolated at a high frequency by plating at high density without auxin or by transfer to decreasing auxin concentrations.

An important distinction between such long-established cell lines and the auxin auxotrophs of *N. plumbaginifolia* we have isolated recently is that the latter can be induced to regenerate shoots which themselves appear to be auxin auxotrophic [5]. Repeated subculture of *N. plumbaginifolia* auxin auxotrophic clones on a medium with relatively high cytokinin (4 μM) and low auxin (0.5 μM) led to the regeneration of slowly growing shoots with abnormal morphology such as bunched, strap-like leaves and aborted apices (Fig. 2E and F). These shoots continue to proliferate under these conditions but cease growth altogether when transferred to auxin-minus medium. The appearance of the shoots was a characteristic of the clone from which they arose. Plants regenerated from 150 non-selected control clones after mutagen treatment were normal and viable when transplanted to soil.

With the limited number of shoots available it was not possible to devise a method for rescuing non-growing shoots by feeding them auxin in vivo. As an alternative, developing shoots of three independently isolated auxotrophs were grafted to *N. tabacum* plants to test whether auxin from the wild-type stock would support the growth of the variant scions. Whereas grafted shoots derived from wild-type control cultures of *N. plumbaginifolia* grew up and flowered (Fig. 3C), shoots from auxin auxotrophic clones did not survive [5]. However, the details of their degeneration are interesting. In each case there was proliferation of shoot buds to a degree far greater than that achieved in vitro, but as expanding leaves reached a certain size, discrete necrotic spots appeared that enlarged to cover the whole surface. Thus, variant grafts persisted for quite some time (often until the stock aged and died) as a green centre of proliferating small buds surrounded by an increasing mass of brown, dried leaves (Fig. 3D). The response of these shoots may be a first indication of the phenotype of auxin auxotrophic plants, and suggests that auxin is required by all cells, either for some essential aspect of primary metabolism or for the regulation of otherwise lethal secondary metabolism.

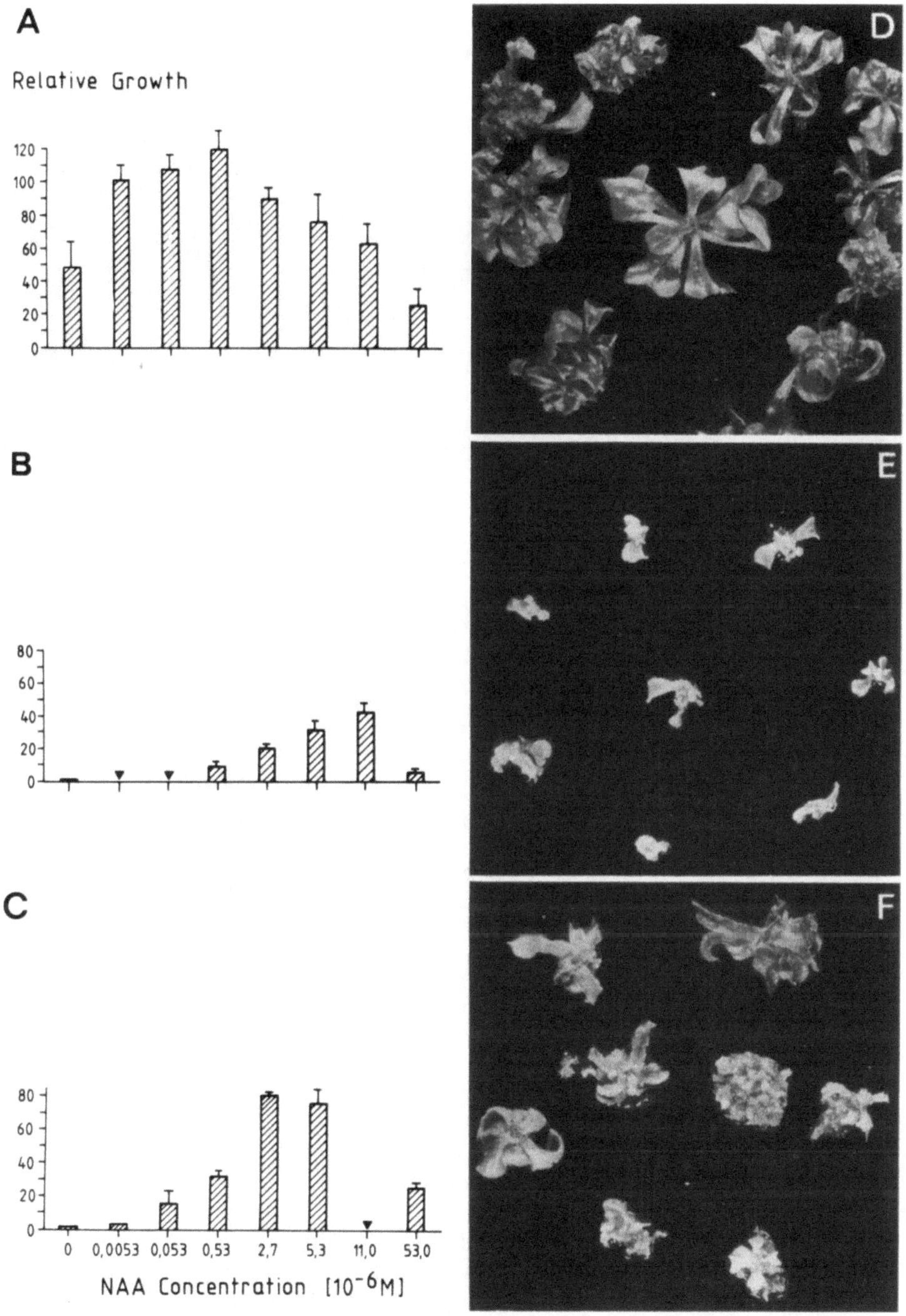

Fig. 2A-F. *Left* The effect of NAA on the growth of wild type (**A**) and two auxin-requiring variants of *N. plumbaginifolia*, CN446 (**B**) and AB58 (**C**). The data are the means (and SEM) of five replicate callus pieces per treatment on a medium containing 0.44 μM BAP. Relative growth = $(W_t - W_0)/W_0$. The *inverted arrows* indicate concentrations not tested. *Right* Regenerated shoots from wild-type (**D**), CN446 (**E**) and AB58 (**F**) cultures in vitro. The terminal in vitro development of CN446 and AB58 on medium without auxin is shown, together with wild-type shoots of equivalent age

Fig. 3A-D. *Top* Callus cultures of wild type *N. plumbaginifolia* (**A**) and an auxin-requiring variant, AB58 (**B**) on medium (**a**) with and (**b**) without NAA. *Bottom* Regenerated shoots of wild type (**C**) and an auxin-requiring variant, BN14 (**D**) after grafting to the top of *N. tabacum* stocks. The wild type is growing normally and will flower and set seed. The BN14 scion at this stage consists mostly of older, brown, desiccated leaves interspersed at the base with younger, green leaves (*arrowed*)

3.3 Temperature-Sensitive Auxin Auxotrophs

We are presently characterizing three ts auxin-auxotrophic variants which have a tissue-culture phenotype like that shown in Fig. 4A. Plants regenerated from all three clones may be grown up at low temperature but they die at high temperature (Fig. 4B). Here also it appears that the older tissues are most temperature sensitive. We have shown for two of the variants that the temperature sensitivity is repaired by auxin application (Fracheboud, Suter and King, unpublished) (Fig. 4C). These observations again suggest that there is some aspect of the metabolism of all cells of a plant for which auxin is an essential co-factor.

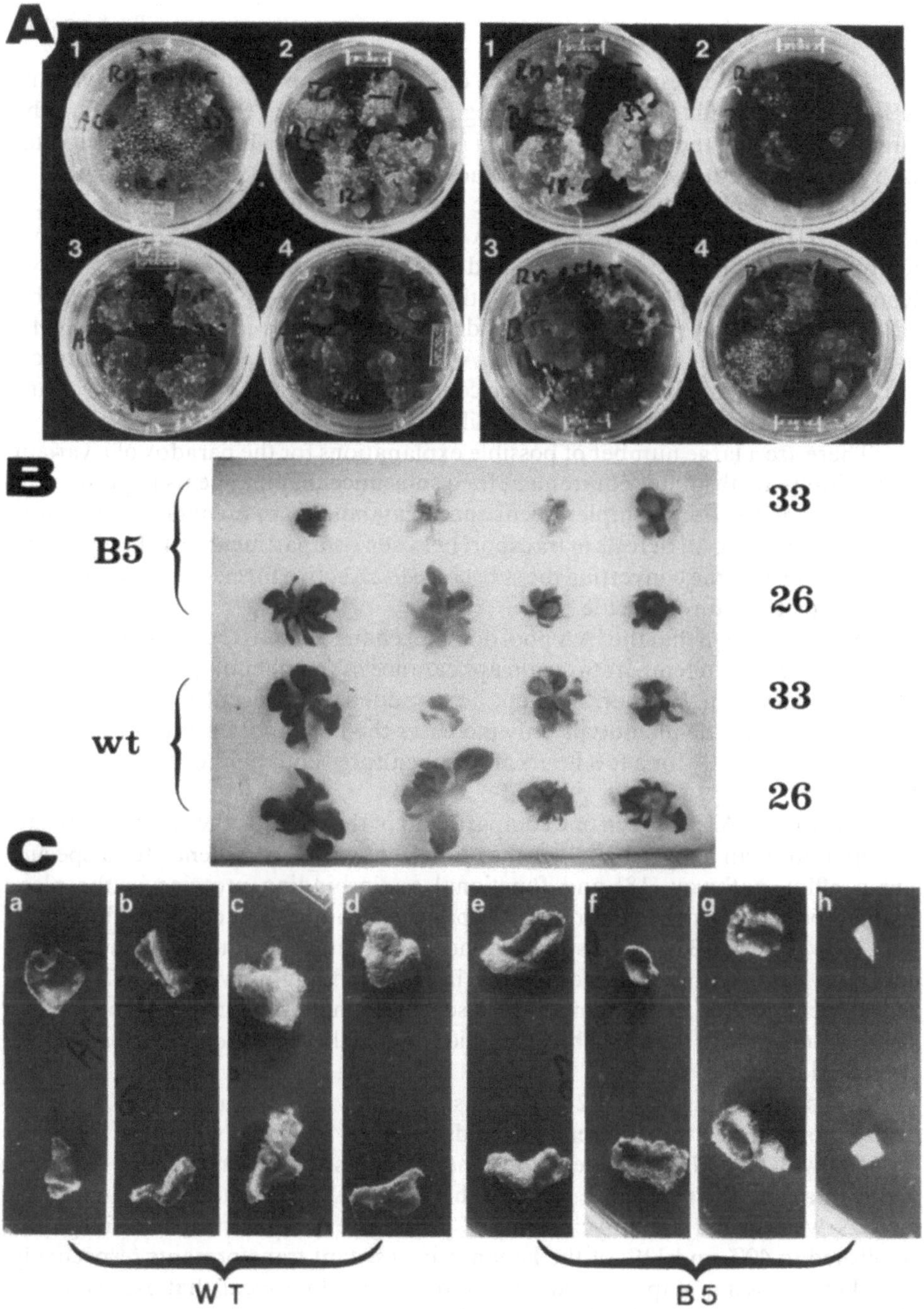

Fig. 4. **A** Callus cultures of wild type (*left*) and a ts auxin-auxotrophic variant, B5 (*right*) cultured under the following conditions: (1) 33°C, + auxin; (2) 33°C, – auxin; (3) 26°C, + auxin; (4) 26°C, – auxin. **B** The temperature sensitivity of shoots regenerated from a ts auxin-auxotrophic culture line, B5 compared to the wild type. The shoots of B5 shown here have grown very little at the high temperature. Not clearly visible on this black and white photograph is the fact that almost all the leaves of B5 at 33°C are either brown or bleached; only one or two of the very youngest leaves remain green for a longer time. **C** A demonstration of the "correction" of the temperature sensitivity of B5 leaves by auxin. Leaf sections from wild type and B5 were incubated under the following conditions: (*a*+*e*) 26°C +auxin (2 μM NAA); (*b*+*f*), 26°C -auxin; (*c*+*g*) 33°C +auxin; (*d*+*h*) 33°C -auxin. Whereas the leaf pieces shown in *a*–*g* are green and mostly callused, those of B5 in picture (*h*) show no callusing reaction and have lost all chlorophyll

An auxin dose/response curve for one of the ts auxin auxotrophs (variant XIIB2) and the wild type is shown in Fig. 5A and B. As for the auxin auxotrophs already described, the growth inhibition is relieved by any active auxin (including IAA itself) but not by inactive analogues or by cytokinin. A common cause of auxotrophy is a block in the biosynthetic pathway of an essential substrate. To test whether XIIB2 suffers from a block in IAA biosynthesis expressed at the restrictive temperature, extracts of wild-type and XIIB2 cells growing at the restrictive temperature in the presence of a synthetic auxin (Fig. 5C) were assayed by HPLC/RIA for IAA (Fig. 5D). The validity of the RIA was confirmed by GC-MS using [$^2H_4$-]IAA as an internal standard. As both the biomass and the concentration of IAA increased in XIIB2 under these conditions, we concluded that this variant has a capacity for IAA biosynthesis similar to that of the wild type [17].

There are a large number of possible explanations for the paradox of a variant that exhibits an absolute requirement for a substance that it appears to produce in normal amounts. For example: a) enhanced catabolism, b) breakdown of a positive-feedback loop, c) defects in transport between compartments, d) change in the affinity of an enzyme converting the substance to an active form, or e) change in the affinity of a receptor molecule.

We have shown that the IAA pool does not change in XIIB2 cells in the interval between increase in temperature and appearance of the phenotype in the absence of exogenous auxin. Further, synthetic auxin added at a concentration that blocks the expression of the phenotype only increases the IAA pool by a factor of 2, and then only transiently for a few hours after subculture. Thus neither a) nor b) seems to apply.

Whether IAA is made in one compartment of the cell (e.g. the chloroplast) and transported to an active site in another is not known. The evidence for a specific auxin efflux pathway [18] and functional auxin-binding proteins in the plasmamembrane [13] suggest that the compartment in question could be the whole cell. However, XIIB2 appears to have no defect either in inter- or intracellular transport because there is no difference in NPA binding between XIIB2 and the wild type, and the XIIB2 temperature-sensitive phenotype can be functionally complemented with the auxin biosynthetic genes of the *Agrobacterium tumefaciens* Ti plasmid.

Cell suspensions of XIIB2 were transformed by co-cultivation with *Agrobacterium* containing two different plasmids: 1) *pIGVTi23neo,* wild-type nopaline T-DNA with the APH(3′)II gene for kanamycin resistance, or 2) *pEND4K:LHR tms*, containing the two auxin genes of the T_B-DNA of a limited host range plasmid as well as the kanamycin marker gene. The temperature sensitivity was abolished in 40% and 11% of the kanamycin-resistant transformants respectively (Oetiker and King, unpublished). Therefore, it would appear that expression of the auxin genes *inside* XIIB2 cells is sufficient to satisfy their auxin requirement under restrictive conditions. Preliminary data suggest that rescue of XIIB2 by expression of Ti-plasmid auxin genes is not correlated with changes in the pool of IAA.

Although the concentration of IAA in the cells appears to be tightly regulated, such that XIIB2 and wild type have similar IAA pool sizes whatever the conditions, it still seems to be the case that the amount of auxin supplied is important for

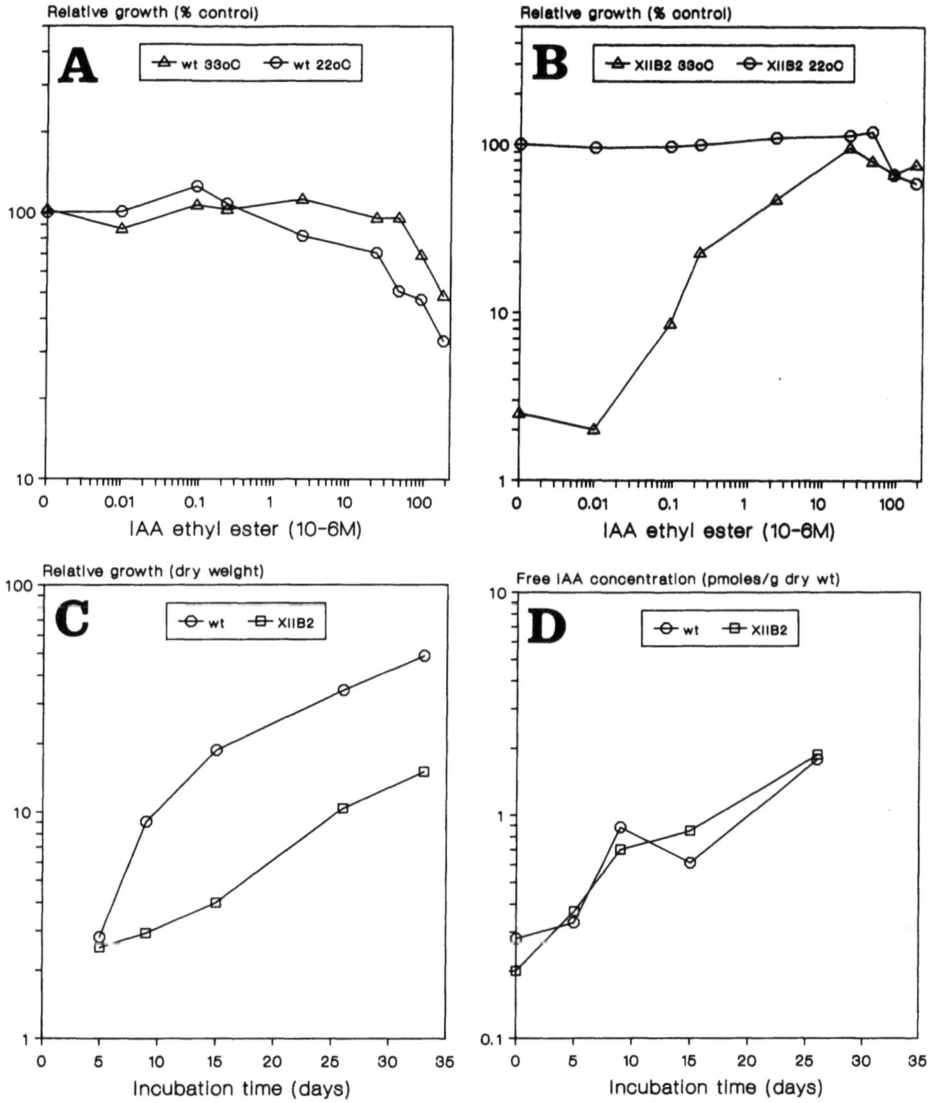

Fig. 5A-D. The effect of IAA ethyl ester concentration on the callus growth of *H. muticus* wild-type (**A**) and a ts auxin-auxotrophic variant, XIIB2 (**B**) at 22 and 33 °C. Relative growth is plotted as a percentage of the growth minus auxin at 22 °C. **C** The relative growth during one passage of wild-type and XIIB2 suspension cultures at 33 °C in a medium containing NAA at 2×10^{-5} M. **D** Changes in the IAA content with time of the cultures shown in **C**

rescuing XIIB2. It is possible, therefore, that the affinity of a receptor or a derivatizing enzyme for IAA is reduced. We are currently examining auxin-binding proteins in the wild type and the variant, and studying the metabolism of IAA by HPLC analysis of [2-^{14}C]-IAA ethyl ester feeding experiments and [2-^{14}C]-IAA pulse/chase experiments under various conditions (Fig. 6).

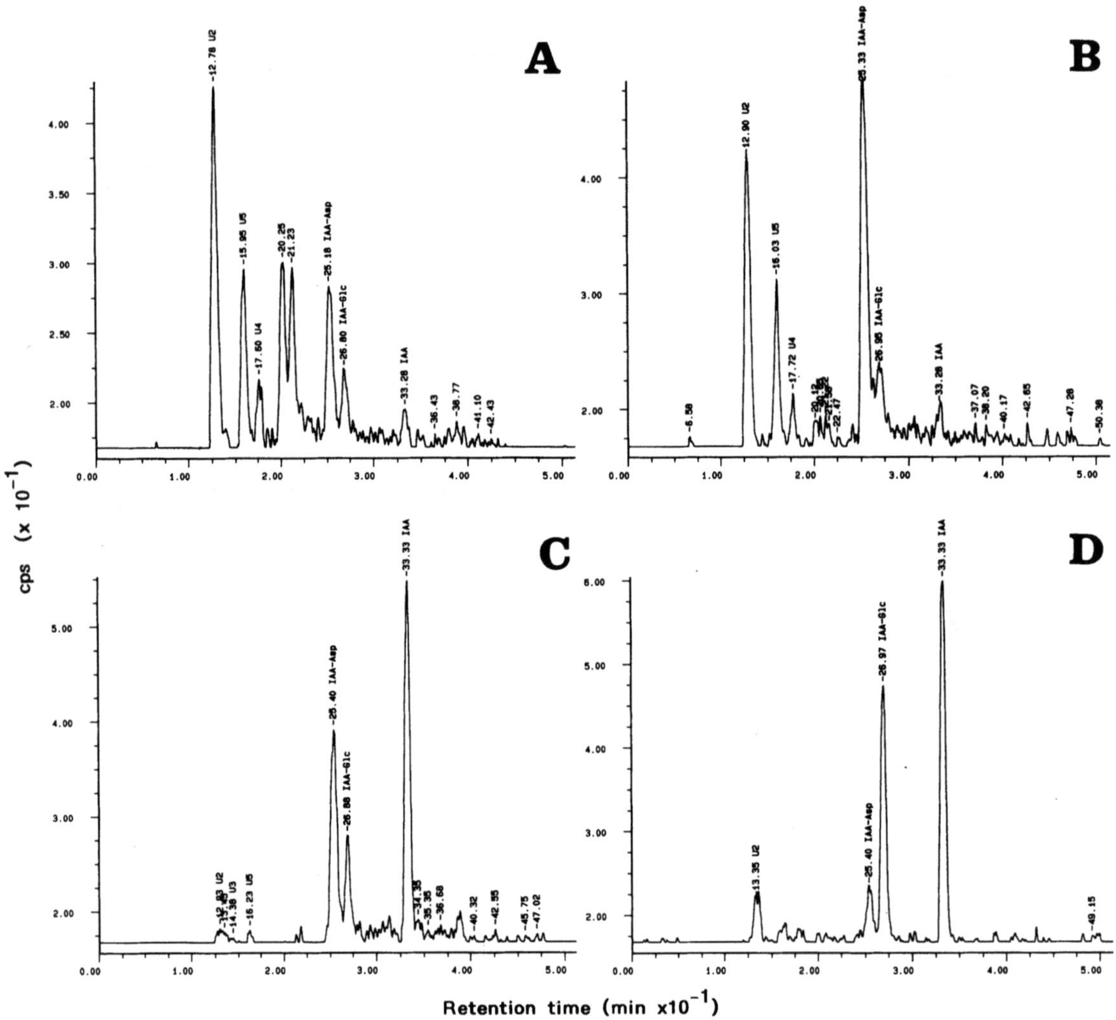

Fig. 6A-D. The metabolism of a short pulse of IAA by wild-type (**A** + **C**) and a ts auxin-auxotrophic variant (**B** + **D**) of *H. muticus*. The suspension cultures were precultured at 26 °C without auxin and subcultured just prior to the experiment into fresh medium with 3.75 μM 2,4-D (**A** + **B**) or without 2,4-D (**C** + **D**). After 2 days incubation at 33 °C (sufficient time for the expression of the XIIB2 phenotype), all cultures were pulsed for 10 min with 0.45 Ci/g [ring-2]-^{14}C-IAA (CEA; 49 mCi/mmol) and chased for 1 min with a large volume of W5 medium containing IAA at 10^{-4} M. Tissues were extracted in 80% methanol, the extracts were reduced in vacuo to the aqueous phase and an aliquot containing 5.0×10^4 dpm was separated on a 25 × 0.46 cm HPLC column packed with 55 Nucleosil C18. The buffers were (*a*) 10% methanol or (*b*) 60% methanol both in 20 mM acetic acid/triethylamine pH 3.5. The gradient was 30 min linear 0–100% (*b*) at a flow rate of 0.8 ml/min. The metabolites were monitored using an on-line radioactivity detector in the heterogenous mode. Full scale is 100 cps and the plots are fitted to the highest peak. The peaks labelled IAA-Asp and IAA-Glc were further purified and identified by GC-MS or treatment with endo-b-1,4-glucanohydrolase and further HPLC respectively

The results to date of the IAA metabolism experiments and the conclusions to be derived may be summarized as follows:

1. IAA ethyl ester is rapidly taken up and hydrolysed by wild type and variant cells. The half-life of IAA ethyl ester in the medium at a variant-rescuing concentration (i.e. an amount that will protect the variant for at least three weeks) is 5 h.
2. The half-life of IAA in the cells is about 20 min in both wild type and variant precultured without auxin (Figs. 6C and D). In cells subcultured at an IAA ethyl ester concentration sufficient to maintain the auxin auxotroph XIIB2 for several weeks, the endogenous IAA pool is only increased a maximum of twofold for a maximum of 8 h after subculture. Thus, neither the IAA in the medium nor intracellular IAA is the ultimate source of auxin for the cells; one or more of the further metabolites must be active or available for reconversion to IAA.
3. The major early derivatives of IAA metabolism are IAA-aspartate and an IAA-glucosyl ester (Figs. 6A and B).
4. In the absence of auxin in the medium, there is a distinct difference in the further metabolism of IAA between wild type and XIIB2 (Figs. 6C and D). Whereas the half-life of the glucoside conjugate is somewhat increased in the variant, IAA-aspartate is much more rapidly metabolized ($t1/2 = 27$ min) than in the wild type ($t1/2 = 9.5$ h). As a consequence, metabolic products of IAA-aspartate appear much earlier in XIIB2 cells than in the wild type.
5. We have clearly established that the half-life of IAA in cells in auxin-containing medium is < 2min and in auxin-free medium 20 min. Further, preculture in auxin-containing medium alters the IAA metabolism of XIIB2 cells such that the metabolites present after a short pulse/chase tend to be the same as those in the wild type (Figs. 6A and B). Preliminary experiments indicate that IAA-aspartate is more stable in XIIB2 cells grown in the presence of auxin (i.e. in permissive conditions) than in its absence.

We have not yet shown whether the alteration in IAA metabolism found in XIIB2 is causal for the auxin requirement. Because the deviant metabolism is constitutive, i.e. it is not induced by elevated temperature, it has yet to be explained why such as alteration would make the cells more temperature sensitive. Although large changes in IAA metabolism occur in the presence of auxin in the medium, it may be difficult to decide whether the changes per se or simply the added auxin rescues the variant. In this respect, it will be interesting to examine the IAA metabolism of phenotypically normal, *Agrobacterium* transformants of XIIB2.

Whatever the outcome, the availability of a variant like XIIB2 with an altered metabolism is proving useful for investigating questions about the significance of IAA and its metabolites in cells. We hope that further characterization of the broad range of resistant and auxotrophic mutants described in this paper will lead us onto firmer ground, particularly in the areas of hormone biosynthesis and reception.

Acknowledgements. We are grateful to Georg Aeschbacher, Nguyen Van Hai and Vreni Schnebli for excellent technical assistance, and to our colleagues at the FMI for helpful discussion.

References

1. Bandurski RS, Schulze A, Reinecke DM (1986) In: Bopp M (ed) Plant growth substances 1985. Springer, Berlin Heidelberg New York Tokyo, p 83
2. Blonstein A (1986) In: Blonstein A, King PJ (eds) A genetic approach to plant biochemistry. Springer, Vienna New York, p 259
3. Blonstein AD, Parry AD, Horgan R, King PJ (1990) Planta, in press
4. Blonstein AD, Vahala T, Fracheboud Y, King PJ (1988) Mol Gen Genet 211:252
5. Blonstein AD, Vahala T, Koornneef M, King PJ (1988) Mol Gen Genet 215:58
6. Cohen JD, Bialek K (1984) In: Crozier A, Hillman JR (eds) The biosynthesis and metabolism of plant hormones. Cambridge University Press, Cambridge, UK, p 165
7. Eriksson TR (1985) In: Fowke LC, Constabel F (eds) Plant protoplasts. CRC, Boca Raton, Fl, p 1
8. Estelle MA, Somerville CR (1987) Mol Gen Genet 206:200–206
9. Finkelstein R, Estelle M, Martinez-Zapater J, Somerville C (1988) In: Verma DPS, Goldberg RB (eds) Temporal and spatial regulation of plant genes. Springer, Vienna New York, p 1
10. Gebhardt Ch, Schnebli V, King PJ (1981) Planta 153:81
11. King PJ (1988) Trends Genet 4:157
12. Lamport DTA (1964) Exp Cell Res 33:195
13. Löbler M, Klämbt D (1985) J Biol Chem 260:9854
14. Maher EP, Martindale SJB (1980) Biochem Genet 18:1041
15. Meins F, Foster R (1986) Dev Genet 7:159
16. Muller JF, Goujaud J, Caboche M (1985) Mol Gen Genet 199:194
17. Oetiker J, Gebhardt Ch, King PJ (1990) Planta 180:220
18. Rubery PH (1985) In: Bopp M (ed) Plant growth substances 1985. Springer, Berlin Heidelberg New York Tokyo, p 197
19. Skoog F, Miller CO (1957) Soc Exp Biol Symp 11:118
20. Slovin JP, Cohen JD (1987) Plant Physiol 86:522
21. Suter M, Schnebli V, King PJ (1988) Theor Appl Genet 75:869
22. Torti G, Lombardi L, Manzocchi LA, Salamini F (1984) Maydica XXIX:335
23. Wang TL, Futers TS, McGeary F, Cove DJ (1984) In: Crozier A, Hillman JR (eds) The biosynthesis and metabolism of plant hormones. Cambridge University Press, Cambridge, UK, p 135

Development of Genetic and Analytical Systems for Studies of Auxin Metabolism

J.D. COHEN[1], K. BIALEK[1], J.P. SLOVIN[1,2], B.G. BALDI[1,3], and K.-H. CHEN[2,4]

1 Introduction

Early investigations of auxin conjugates concerned the general "release" of auxin in vivo or in situ [e.g., 8]. More recent studies have examined either IAA released by hydrolysis of extracts of plant tissue, or have studied specific conjugates formed by the covalent attachment of IAA to other molecules. Conjugated forms of IAA can be classified by size, type of covalent linkage, or the molecule to which the IAA is attached. Low molecular weight conjugates include esters such as IAA-glucose and IAA-*myo*-inositol and amides such as IAA-aspartate and IAA-glutamate. The higher molecular weight conjugates include esters where the IAA is linked to the carbohydrate portion of a glycoprotein [17], or is linked to a glucan [18]. Higher molecular weight amide conjugates are also known, where IAA is linked directly to a peptide or protein [3]. These higher molecular weight conjugates have been difficult to study due to the lack of suitable methods for macromolecular separations and structure determination. Improvements in available methods for studies of macromolecules now make it practical to examine these types of compounds in more detail and begin to ask questions as to their role in the hormonal relationships within the plant.

In addition to the techniques for the isolation and subsequent study of IAA conjugates, progress in the development of methods for the study, generation, and selection of metabolic mutants has been made. One such mutant, the jsR_1 line of *Lemna gibba*, was selected by its large size relative to the parent line. Its unusually high level of IAA and low amounts of conjugates during later periods of culture make it uniquely suitable for isotopic studies of IAA biosynthesis. Other mutant systems in higher plants also need to be examined for their utility for studies of auxin metabolism such that a pool of useful genetic material can be obtained which enhances our biochemical approaches to these problem areas.

[1,2] USDA-ARS Plant Hormone Laboratory, Beltsville Agricultural Research Center, Beltsville, MD 20705 and Department of Botany, University of Maryland, College Park, MD 20742 USA
[3] Current address: Friedrich Miescher Institut, Postfach 2543, 4002 Basel, Switzerland
[4] Current address: Department of Horticulture, National Taiwan University, Taipei 10764, Taiwan ROC

2 Analytical Studies

2.1 IAA Peptide Conjugates from *Phaseolus*

An amide-linked conjugate between IAA and a small peptide is one of the major forms of IAA in *Phaseolus* seed. We have characterized this molecule by GC-MS analysis of the IAA liberated after basic hydrolysis, by analysis of the amino acids released following acid hydrolysis, by the FT-IR spectra of the intact peptide, and by SDS-PAGE [3]. Further studies have required the development of techniques for the isolation of the peptide and for the study of related peptides and proteins.

Chromatographic methods for the isolation of the intact peptide have focused on taking advantage of the extreme hydrophobic character of this peptide and its higher molecular weight relative to other IAA conjugates. A preparative Hamilton PRP-1 reverse phase column was used for the purification of the initial extract and this has allowed us to work on much larger extracts. We have also used a preparative PAGE system using a BRL Prep Gel method. These two techniques, coupled with the isolation methods used previously [3], have allowed us to prepare larger amounts of highly purified peptide.

The purified peptide has been used for additional studies of its physical properties. The peptide has certain unusual characteristics (such as its solvent solubility) that render it difficult to sequence. We found that the peptide has a free N-terminus. However, we have only been able to sequence through the first four amino acids (Phe-Leu-Cys-Ala. . .). After these residues the analysis stopped, probably due either to peptide substitutions or to the hydrophobic nature of this peptide. Current studies in our laboratory, therefore, are centered on using plasma desorption MS and microbore HPLC/dynamic FAB-MS to obtain physical data on this peptide. Such studies have shown that this peptide has a molecular mass of 3627.4 based on the singly charged molecular ion.

Immunological studies of the peptide have utilized two approaches. First, we found that the highly purified peptide competed with methyl-IAA in the ELISA assay previously described [26]. Thus, we have experimented with using this technique for analysis of IAA-containing peptides using Western blotting techniques. The 3.6 kD peptide is somewhat difficult to handle for such studies, since it is soluble in methanol and easily diffuses out of the gel. However, by rapid blotting using semi-dry methods we were able to significantly increase the efficiency of blotting the peptide. The antibody originally described for use with ELISA [26] was found to be of too low titer for use in these expriments. Therefore, we have obtained several higher titer antibodies to the antigen formed by the linkage of IAA through the carboxyl to proteins. We have a polyclonal antibody from Dr. Michael Bausher, USDA/ARS Orlando Florida, and we also have two monoclonals. One was from Dr. John Caruso, University of Cincinnati. The second monoclonal was from ascites fluid obtained using the same cell lines used for ELISA detection of IAA [26], and was obtained from Idetek.

We have developed a rapid method for the preparation of the 3.6 kD peptide from bean seed using selective solvent extractions and preparative slab gel electrophoresis followed by electroelution. This method produces a product free of any detectable protein or peptide contamination. Using peptide produced in this way we collaborated with Dr. Bausher to produce polyclonal rabbit antibodies. The antibody produced in this way had sufficiently high titer for Western blot analysis, and shows specificity for the peptide in the preparation injected. We have used the antibody to determine if there are immunologically related proteins being present in bean seeds. A total protein extract from bean seed was separated on SDS-PAGE and blotted onto nitrocellulose. Staining for total protein on the blot showed numerous protein bands and is dominated by the storage proteins of the seed. Immunological detection using the antibody to 3.6 kD peptide showed only a pair of closely running proteins at 25 kD, and two other faster running bands. The proteins at 25 kD form sharper bands and stain more intensely than these other proteins. This was the first indication of the existence of such larger proteins. It is possible that the 3.6 kD peptide is produced by processing these larger proteins and we are doing in vivo labeling experiments to determine if this is so. Additional studies using Western techniques showed that these same bands could be detected using monoclonal antibodies produced against IAA linked through the carboxyl to proteins, as discussed previously.

2.2 Hydrolysis of IAA Peptides

To study these proteins further we developed methods for the hydrolysis of IAA-peptides without their extraction from the plant material. This was necessitated also in order to evaluate extraction efficiency. Other macromolecules containing IAA might not be extracted under the particular conditions employed, and without the ability to determine a true value for "total IAA" the presence of such species was difficult to predict. In addition, we were interested in extending these investigations to other plant tissues and to other species where the solubility of macromolecular conjugates, if present, would be unknown (Fig. 1). Instead of extraction of the tissue followed by hydrolysis of the extracted conjugate, we

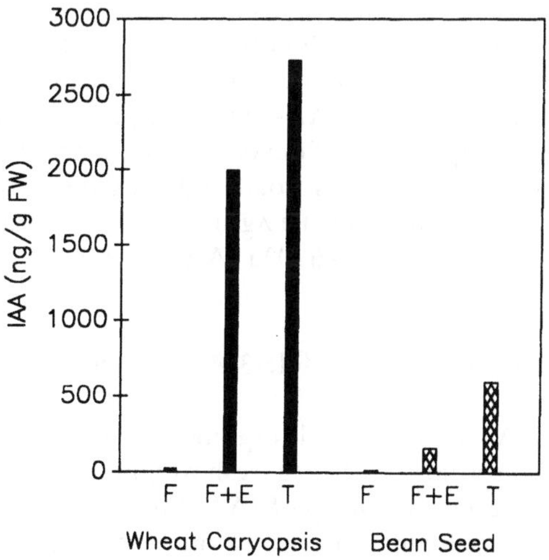

Fig. 1. Indole-3-acetic acid levels in two different plant tissues, wheat caryopsis and bean seed. The data show the difference between levels of free IAA (*F*), free plus "extractable conjugate" (*F+E*) and total IAA (*T*). Values for free plus "extractable conjugate" were obtained by quantitative analysis of the aqueous acetone extract by GC-SIM-MS [7]. The values for total were obtained by direct tissue hydrolysis using 1 N NaOH at room temperature (wheat caryopsis [6]) or 7 N NaOH at 100°C (bean seed[4]). Hydrolysis of the wheat tissue with 7 N NaOH did not release additional IAA

hydrolyzed the entire crude seed powder using 7 N NaOH at 100°C. This, of course, produces a very crude and dark-looking hydrolyzate to be purified. Use of $^{13}C_6$-IAA as an internal standard with final analysis by GC-SIM-MS [11], however, provides for convenient quantitation even in such extreme situations. As a control to check for tryptophan conversion to IAA, as was reported many years ago [21], we included ^{15}N-[indole]-tryptophan as a marker in some experiments. Our earlier hydrolysis of the 70% acetone extract from bean seed had indicated an IAA content of 160 μg/g fw. However, direct hydrolysis of the crude seed powder at 100°C showed an increasing amount of total IAA with time of hydrolysis (over 600 μg/g fw after 3 h) and no significant conversion of added ^{15}N-tryptophan [4].

In these initial studies on the direct hydrolysis of *Phaseolus* seed, the material was ground in a Wiley mill to a fine mesh and treated directly with 7 N NaOH in a sealed Teflon container flushed with 2–5 ml/min water saturated N_2 (99.998%). Isotope dilution analysis showed that the "release" of IAA continued for much longer than 3 h even though this was sufficient time for the quantitative hydrolysis of standards of IAA-amino acid conjugates. This additional IAA obtained past three h was not due to IAA released from conjugates, however, since it was found that additional care to exclude O_2, by adding an O_2 scrubbing cartridge to the gas line, prevented the formation of this additional IAA [4]. Thus, for such direct hydrolysis it is very important to take extra precautions to exclude oxygen from the reaction vessel in order to avoid conversion of other compounds in the plant residues to IAA. This is especially true if hydrolysis times in excess of 3 h are to be utilized, since the secondary production of IAA is slow compared to amide conjugate hydrolysis.

These results show that a major fraction of the amide bound IAA in bean seed is not extractable from finely ground seed powder by aqueous acetone. Lower molecular weight conjugates or precursors cannot account for the non-extractable portion since these compounds are readily extracted by aqueous acetone from plant material. If care is used concerning conditions for hydrolysis, strong basic hydrolysis of crude plant tissue can provide new information on the IAA content of these materials and can be used to further guide more detailed investigations into the role of macromolecular conjugates in the hormone economy of plants. For example, we have followed the production of IAA-peptide conjugates throughout bean seed development (Fig. 2). These studies not only are useful to help document the dynamics of IAA-conjugate production, but have also guided our selection of the proper stage of development for the production of a cDNA expression library from poly-A RNA isolated during the period of peak peptide accumulation. We have prepared such a λgtll expression library and will screen the library using the antibody to the 3.6 kDa IAA-peptide.

2.3 Analysis of Indole-3-Acetic Acid

IAA is a highly labile organic compound present in plant materials at typical levels of around 10–20 μg/g fresh weight. Even with specialized handling and separation materials, recovery of IAA during purification and analysis is never quantitative and is highly variable. Isotope dilution techniques [1, 2, 7, 15, 16, 19, 20] are

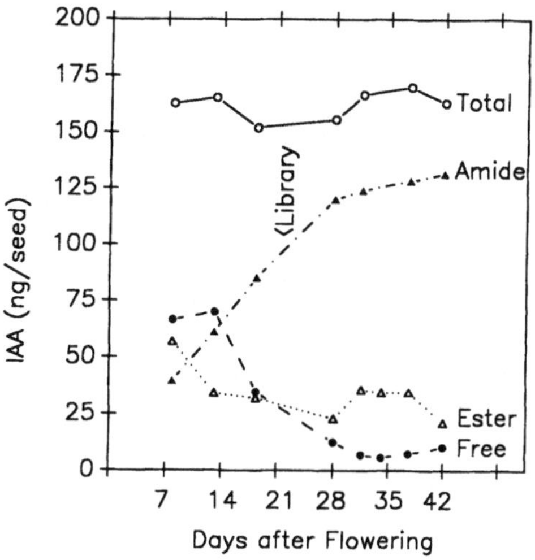

Fig. 2. Levels of free, ester and amide IAA in bean seeds during seed development (Bialek and Cohen, unpublished). Values for amide IAA were obtained by direct tissue hydrolysis [4]. The only amide conjugates detectable in this tissue are IAA-peptides. The stage of seeds used for production of a cDNA library is indicated

particularly useful for IAA analysis because quantitative determination is then not based on amount detected, but rather on the recovered ratio of an isotopic species to that of the natural unlabeled compound. The assumption implicit in isotope dilution analysis is that the behavior of the isotopic species and that of the natural unlabeled compound are the same under the conditions employed for separation and analysis. Thus, the selection of the proper isotope labeled standard is an important aspect of such procedures. We previously described both the use and the many advantages of $^{13}C_6$-IAA[5,6] as an internal standard for GC-MS analysis of IAA [11].

The problem of selectivity of detection has been a major aspect of plant hormone analysis since the early days of bioassay, as shown by the plethora of studies on naturally occurring "auxin synergists" and "auxin antagonists". Two complementary approaches have been taken to improve the selectivity of detection. The first approach has been the application of high resolution techniques of separation science in order to purify the phytohormone away from interfering compounds. The second approach has been the use of selective detectors following chromatographic purification. The quality of an analytical method is dependent on the interplay of these two factors, resolution and detector selectivity.

Working in the realm of biological systems adds two additional criteria to development of analytical systems: sample size and sample number. The "bottom line" is, of course, to be able to process the minimum amount of plant material and the maximum number of samples in the least amount of time! These demands of

[5] The internal standard, $^{13}C_6$-IAA, has recently become commercially available (Cambridge Isotope Laboratories, #CLM1896) or can be prepared by synthesis from $^{13}C_6$-aniline as previously described [11].

[6] Mention of a trademark, proprietary product, or vendor does not constitute a guarantee or warranty of the product by the United States Department of Agriculture, and does not imply its approval to the exclusion of other products or vendors that may be suitable.

the developmental biologist or biochemist were formerly at odds with the requirements of the analytical chemist, but as methods have improved, the number of biological problems that can be studied has dramatically increased.

Several different approaches to plant hormone analysis have been proposed and the appropriate method to use for each problem is not always obvious. One of the major problems in the field of development of analytical methods for IAA has been the failure to establish reference methods for the evaluation of new techniques. It was consideration of this situation that lead us to produce a multiple labeled, highly enriched, ^{13}C-IAA and to develop methods for its use in routine analysis [11]. We feel that our current methods, using GC-MS, meet six important criteria: 1) they are applicable to a wide variety of plant tissues with only minor modification, 2) they are highly selective in that ions specific to the target molecule are selected, 3) by use of multiple ion pairs they are self validating, 4) both in terms of tissue sample size and levels of detection these GC-MS techniques offer high sensitivity, 5) the techniques are rapid, simple, and easy to teach to those not trained in the problems of phytohormone analysis, and 6) they are *absolute methods*, thus requiring little complex standardization or calibration.

Our method for the quantitative analysis of IAA consists of four major steps: 1) extraction, 2) prepurification, 3) HPLC, and 4) GC-MS. The major time-consuming steps are the purification steps prior to GC-MS. The traditional prepurification [1, 11] has involved solvent partitioning steps and, in some protocols, open column liquid chromatography. Use of high resolution bonded phase capillary GC columns has allowed increased sensitivity at the mass spectral step, and the use of 3 and 5 μm HPLC packings has made it reasonable to use shorter columns, thus reducing the time required for HPLC. The improvement in recovery afforded by the shorter HPLC columns and the improved sensitivity of capillary GC-MS suggested to us that it was possible to scale down the sample size to a level that made practical the use of *Sep-Pak*-like disposable mini-columns for sample preparation [2, 6, 7].

Plant tissue samples in the range of 30 mg to 3 g are ground in a homogenizer with 4 ml per gram of sample of 60% 2-propanol:0.2 M imidazole buffer, pH 7.0, containing $^{13}C_6$-IAA (0.1–1 μg/g sample)[5] and about 50 000 dpm 5-^{3}H-IAA (20–30 Ci/mM). Isotope equilibration occurs during a 1 h incubation period at 4 C following the homogenation and then the extract is clarified by centrifugation at 10 000 × g for 5 min.

The sample is next diluted to a final 2-propanol content of about 10% with imidazole buffer. The diluted sample is then applied at a rate of 5 ml/min to a preconditioned amino anion exchange minicolumn (Baker-10 SPE[4] 3 ml, Fisher Prep-Sep 2 ml, or, for larger samples, a Lida Extra-Sep 6 ml). After sample application, the column is aspirated for 30 s and then washed sequentially with 2 ml each of hexane, ethyl acetate, acetonitrile, and methanol. The IAA is eluted by a further wash with 3 ml of methanol containing 2% acetic acid. The sample is evaporated in vacuo, then HPLC is used for further purification by injecting the sample onto a 5 μm Whatman ODS-3 RAC column (12.5 × 4.6 mm) with elution by 20% acetonitrile containing 1% acetic acid. The fractions from the HPLC column containing radioactivity from the ^{3}H-IAA tracer are then pooled, reduced to dryness, methylated with diazomethane [10] and analyzed by selected ion current monitoring GC-MS, as was previously reported [2, 7, 11].

2.4 Analysis of IAA Released by Hydrolysis of Conjugates

Quantitative studies of IAA conjugates can be carried out by two approaches, the direct quantitation of specific conjugates or the quantitation of IAA released following chemical hydrolysis of the conjugates [9, 14]. Each approach has advantages. However, hydrolysis is the only technique applicable to those plant materials for which the conjugates have not been identified. Ester conjugates of IAA are hydrolyzed by treatment in 1 N NaOH for 1 h at room temperature, while amide conjugates require stronger treatment. The amide conjugates are stable to the 1 N NaOH treatment, however they are completely hydrolyzed by treatment with 7 N NaOH for 3 h at 100°C. While the hydrolysis of ester conjugates can generally be accomplished without special apparatus or conditions, the hydrolysis of amide conjugates requires special precautions to exclude oxygen if reasonable yields are to be obtained. Use of internal standards of IAA with isotopic atoms in non-exchangeable positions is, of course, mandatory for such hydrolysis experiments [11, 16]. Details of the exchange that can occur when isotopes of hydrogen are used for labeling IAA were reported [11, 16].

Hydrolysis at 100°C for 3 h can be done in either a sealed container in vacuo or in a closed vessel purged with water-saturated N_2. The yield from such techniques is highly dependent on the amount of IAA present, the conditions of hydrolysis, and the nature of the extract being hydrolyzed. Most work in our laboratory is done using Teflon containers (Tuf-Tainers, Pierce) equipped with silicon/Teflon seals. Teflon tubing is inserted for N_2 purging and the containers are heated to 100°C in a dry block device. These conditions have proven to be excellent for work with extracts from a variety of plant materials, however additional precautions are necessary for hydrolysis of crude tissue samples (as discussed above).

Samples from hydrolysis, once neutralized, have a salt content that renders them inappropriate for the ion exchange techniques used for quantitative GC-MS of free IAA, as described above. Such samples are desalted by absorption to C-18 minicolumns, eluted with methanol, and then prepared as with the samples for free IAA analysis.

3 Use of Mutants to Study Auxin Metabolism

3.1 Genetic Analysis of Auxin Metabolism by Mutant Selection Using *Lemna gibba*

One of the characteristics of phytohormone metabolism that makes studies of the reactions difficult is the low level of these compounds normally present in plant tissues. Biochemical investigations are difficult, due both to the low amount of substrate and to the problem of showing that low activities measured in vitro are related to the processes being studied in vivo. Axenic cultures of *Lemna gibba* provide an excellent experimental system for studies of hormone metabolism. Problems of bacterial metabolism are avoided and, due to rapid uptake by the non-cuticularized lower frond surface, pulse-chase labeling with hormones, amino acids, and metabolites are possible. For the past few years we have been developing a system for doing simple genetic analysis using *Lemna* and have established techniques for selection of mutant lines of interest for our studies of auxin metabolism [13]. Another major advantage of the use of *Lemna* for such studies is the ability to obtain large numbers of clonal plants by vegetative propagation. By manipulation of the photoperiod it is also possible to induce flowering for genetic analysis of selected lines and for the establishment of highly inbred lines of *Lemna*.

Regeneration of plants from tissue culture resulted in several variant plant lines [22]. We have been successful in obtaining callus from *Lemna* using the 2,4-D containing medium described by Chang and Chiu [5]. This callus was capable of regenerating *Lemna* plants and we isolated seven different plant lines from such callus (lines jsR_1–jsR_7). All seven lines show significant morphological differences from the parent line. Plants of jsR_1 have very large, dark green, dense appearing fronds [24]. They are capable of flowering, setting seed, and the large phenotype occurs in some of the F_1s following selfing. The jsR_1 line is about 1.8 times the size of the parent line and this difference is accounted for by an increase in cell size rather than cell number. This difference in size is noted when the fronds are grown on enriched media as well as on minimal salts medium, which suggests that differences in nutritional requirements are probably not involved in the phenotype. Microspectrodensitometry using DAPI staining showed that both the jsR_1 and the parent line have approximately the same DNA content per nucleus and, therefore, the jsR_1 line is not a polyploid of the parent line [24].

Using quantitative GC-MS the levels of IAA during the log-linear phase of the growth cycle was measured in both the jsR_1 line and the parent line [24]. The level of IAA varied considerably for both lines over the culture period [24]. While the level of IAA was always greater in the jsR_1 line, the differences were most notable at the very beginning and toward the end of the 45-day growth period. At the latest stages, the level of IAA in jsR_1 was about 10^2 greater than that of the parent line. In addition, at the latest stage there were essentially no detectable conjugates of IAA in the jsR_1 line, while the parent line had 93% of the IAA conjugated (ester 75%, amide 18%). The jsR_1 line does appear to be able to form at least some types of conjugates, since both lines metabolize exogenous IAA in a way that is qualitatively and quantitatively the same (unpublished data). Since the level of IAA is so high in mature cultures, it should be an excellent plant for studies of IAA metabolism, where the ability to detect isotopes incorporated into the product is often limiting.

More recent work in our laboratory has greatly extended our ability to generate *Lemna* mutants, since we are now able to obtain very high mutation frequencies using low level chemical mutagenesis of intact plants, primarily utilizing derivative lines of *L. gibba* G-3 obtained from selfing experiments through 6–9 generations [23, 25]. The techniques for mutant generation allow a number of different selection approaches to be tried for obtaining useful lines. Current work in our laboratory in this area involves three approaches: 1) Selection of lines for ability to hydrolyze IAA conjugates using the toxic auxin analog, 5-Br-IAA, conjugated to L-alanine.

Our studies have shown that 5-Br-IAA is quite toxic to *Lemna* and we expect that mutant plants unable to hydrolyze such a conjugate will survive or grow using such a screen; 2) Screening lines for variations in their metabolism of radioactive IAA supplied in the culture medium using either TLC and autoradiography or HPLC-RC; and 3) Selections using toxic analogs of tryptophan or tryptophan precursors.

3.2 Isotope "Tracking" Using the jsR$_1$ Line of *Lemna gibba* G-3

The use of typical isotope methods for the study of IAA biosynthesis have been limited due to the poor ability to detect significant isotope incorporation as well as the lack of good methods for positional analysis of the incorporated label. We have begun a program to use stable isotopes to study IAA biosynthesis. Central to this objective is the use of the high IAA mutant of *Lemna* which has increased levels of IAA, and the availability of $^{13}C_6$-IAA for standardization. In addition, the use of *Lemna* allows the work to be done under axenic conditions and uptake problems are lessened due to the non-cuticularized lower frond surface. High resolution MS offers unique advantages for such studies since it allows minor enrichments of individual isotopes to be studied in selected compounds. Thus, labeled precursors, e.g. tryptophan, can be fed to plant materials at levels approximating endogenous pool sizes, and the incorporation of label into IAA detected even at extremely low rates. It is possible, for example, to determine minor enrichments of a single ^{15}N in the presence of a like compound enriched in ^{13}C. This is possible since, although the presence of a single ^{13}C or ^{15}N will both increase the mass of a particular IAA molecule by ~ 1 a.m.u. from its nominal mass of 175.063329, high resolution analysis shows that the ^{15}N-labeled species will have a mass of 176.060364, and the ^{13}C-labeled molecule will be 176.066684; a difference of 0.00632 a.m.u. (Fig. 3). This ability of high resolution MS to differentiate and quantify individual isotopes of the same nominal mass can be a powerful tool for studies of biosynthetic

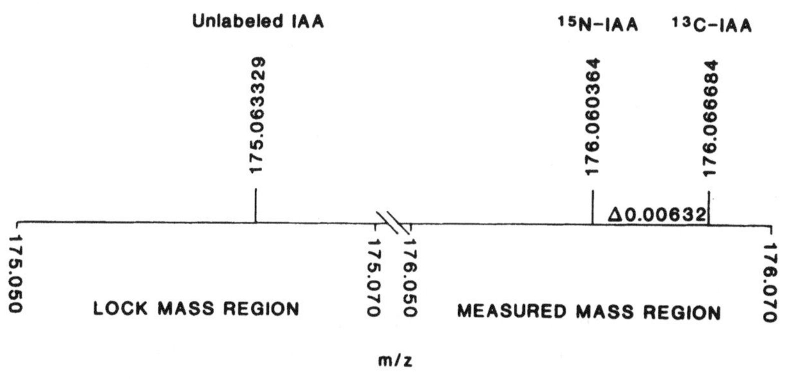

Fig. 3. The utility of high resolution mass spectrometry to determine the difference in labeling between two isotopic forms of IAA with the same nominal mass. The mass spectrometer used the ion of the unlabeled IAA (m/z 175.063329) as a reference to maintain calibration ("Lock Mass") and then is able to resolve the 6 millimass difference in the two labeled forms

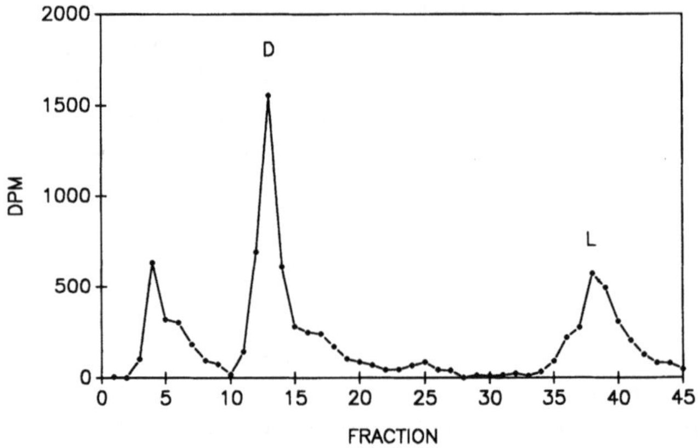

Fig. 4. Resolution of D and L forms of ^{14}C-tryptophan using a Chiralpak-WH (Diacel Chemical Industries, Ltd.) HPLC column. The extreme difference in Rt of the two forms allows full resolution even with the large (100:1) difference in amounts found in plant material

pathways. Two studies using this approach are in progress. First, we are interested in the problem of the possible involvement of D-tryptophan in IAA biosynthesis. Prior studies are difficult to interpret, due to the failure to correct for the tremendous difference in the size of the internal pools of the D and L forms. Thus we have applied isotope dilution techniques to determine pool sizes of both isomers. We then feed a mixture of ^{13}C-[methylene]-D-tryptophan and ^{15}N-[indole ring]-L-tryptophan set to the ratio found by analysis (Fig. 4). The enrichment of both isotopes into each pool is followed with time, as is the enrichment into IAA. In this way it is possible to follow the rate of uptake, the rate of racemization, and the rate of IAA biosynthesis in a single experiment. In a similar way, we plan to look at biosynthesis from much earlier possible precursors such as anthranilic acid, shikimic acid, chorismic acid, serine, and indole. We currently have three of the five compounds with stable isotope labels and will prepare the others as the work progresses. We feel that it is now time to do a careful reevaluation of the direct pathways so far proposed from tryptophan using methods that allow kinetic analysis of isotope incorporation and pool size correction. Mutant plants obtained from our selections of *Lemna* have made these studies possible and additional variants should further enhance such studies.

4 Conclusions

Methods are now available for the routine and rapid GC-MS quantitation of IAA levels in plant samples. These methods have unsurpassed precision and excellent chemical certainty as to what is being measured. In addition to their direct use for biological studies, they are an absolute method which should serve as a reference for the evaluation of the suitability of other, less certain, methods [12, 26].

Advanced techniques for measurement of dynamic metabolic events and the availability of suitable genetic material should now allow experiments to be done to determine in exact detail the pathway of auxin biosynthesis in plants. It should be emphasized, however, that both the identification of suitable genetic material as well as the development of analytical techniques to take advantage of such materials are necessary. A general trend, we feel, in this research area has been to seek metabolic mutants, but to give only limited thought and attention to how such germplasm should be used in metabolic experiments. Prior methods used for studies of IAA metabolism often were ill suited for this complex task. The use of such ill-suited methods with new genetic material will limit the potential of such material for advancing our knowledge of IAA metabolism. In addition, many workers seem to expect (we think naively) that simple analysis of hormone levels will be sufficient to define mutations in biosynthesis. It is far more likely that techniques for measurements of *rates* of biosynthesis or turnover and analysis of pool sizes of precursors, conjugates and products will be necessary to fully understand indolic metabolism in selected lines. Techniques for studies of the higher molecular weight conjugates now allow a more complete study of this aspect of IAA metabolism, and other new techniques are rapidly becoming available. With these tools the prospect for obtaining a better understanding of the complexities of IAA metabolism is excellent. An understanding of hormone metabolism and its regulation will probably not lead directly to an appreciation of how hormonal mechanisms control development and growth. However, without such knowledge it is impossible for us to fully understand how plants control developmental aspects of their own growth through hormonal messengers.

Acknowledgments. Work reported was supported, in part, by research grants from the United States National Science Foundation (DMB-86-17171), US-Israel BARD Fund (US-842-84; US-1362-87), and the United States Department of Agriculture Competitive Research Grants Office (Molecular and Cellular Mechanisms of Growth and Development 85-CRCR-1-1718). We thank Ms. Jo Etta Hubbard for her technical assistance with much of the work reported.

References

1. Bandurski RS, Schulze (1977) Plant Physiol 60:211
2. Baraldi R, Chen K-H, Cohen JD (1988) J Chromatogr 442:301
3. Bialek K, Cohen JD (1986) Plant Physiol 80:99
4. Bialek K, Cohen JD (1989) Plant Physiol 90:398
5. Chang W-C, Chiu P-L (1978) Z Pflanzenphysiol 89:91
6. Chen K-H (1987) Analysis of indole-3-acetic acid in tobacco genetic tumors and in wheat GA_3 insensitive mutant "Tom Thumb", Ph D Thesis. University of Maryland, College Park, pp 130
7. Chen K-H, Miller AN, Patterson GW, Cohen JD (1988) Plant Physiol 86:822
8. Cholodny NG (1935) Planta 23:289
9. Cohen JD (1982) Plant Physiol 70:749
10. Cohen JD (1984) J Chromagtogr 303:193
11. Cohen JD, Baldi BG, Slovin JP (1986) Plant Physiol 80:14
12. Cohen JD, Bausher MG, Bialek K, Buta JG, Gocal GFW, Janzen LM, Pharis RP, Reed AN, Slovin JP (1987) Plant Physiol 84:982
13. Cohen JD, Slovin JP, Bialek K (1985) In: Randall DD, Blevins DG, Larson RL (eds) Current topics in plant biochemistry and physiology 1985, vol 4. University of Missouri-Columbia, p 75
14. Epstein E, Baldi BG, Cohen JD (1986) Plant Physiol 80:256
15. Hamilton RH, Bandurski RS, Grigsby BH (1961) Plant Physiol 36:354

16. Magnus V, Bandurski RS, Schulze A (1980) Plant Physiol 66:775
17. Percival FW, Bandurski RS (1976) Plant Physiol 58:60
18. Piskornik Z, Bandurski RS (1972) Plant Physiol 50:176
19. Rivier L (1986) In: Linskens HF, Jackson JF (eds) Modern methods of plant analysis, new series. Vol 3 Gas Chromatography/Mass Spectrometry, Springer, Berlin Heidelberg New York Tokyo, p 146
20. Sandberg G, Crozier A, Ernstsen A (1987) In: Rivier L, Crozier A (eds) Principles and practice of plant hormone analysis. Academic Press, London, p 169
21. Schoken V (1949) Arch Biochem 23:198
22. Slovin JP, Cohen JD (1985) (Abstract) Plant Physiol 77:S-11
23. Slovin JP, Cohen JD (1986) (Abstract) Plant Physiol 80:S-33
24. Slovin JP, Cohen JD (1988) Plant Physiol 86:522
25. Slovin JP, Cohen JD (1989) (Abstract) Plant Physiol 89:S-111
26. Weiler EW, Jourdan PS, Conrad W (1981) Planta 153:561

Developmental Studies of *Physcomitrella patens* Using Auxin and Cytokinin Sensitivity Mutants

N.W. Ashton[1], D.J. Cove[2], T.L. Wang[3], and M.J. Saunders[4]

1 Introduction

Substantial progress towards understanding at least some developmental processes in animals has been made by working with species (model systems) that are often of no direct economic import. To date, the use of model systems for plants has been less popular. However, for the study of plant morphogenesis and/or plant cell differentiation, the moss, *Physcomitrella patens*, offers many advantageous traits. These include:

1. Ease of axenic culture on simple, completely defined, solid or liquid media [1, 23].
2. Small size and a short life cycle, i.e. spore to spore in approximately 10 weeks.
3. A small number of cell types and morphogenesis of a kind which enables the vast majority of cells present at any stage of development to be observed with a minimum of disturbance.
4. Spores and gametophytic tissues which are haploid. This has made straightforward the isolation of a wide range of biochemical [1, 14, 23], tropically-abnormal [11, 12, 19, 21, 22], morphological [3, 8, 23] and hormone-insensitive [4, 23] mutants.
5. Established methodologies for genetic analysis by means of conventional sexual crossing [1, 6, 9] and somatic hybridization following protoplast fusion induced either chemically [6, 17, 18, 19] or electrically [6, 28].
6. A relatively small genome size. The best estimate yet available, from microdensitometry (J.-P. Zrd, unpublished data), gives a haploid genome size of 0.7 pg (= approx. 6×10^5 kbp). This latter feature will facilitate application of modern techniques of molecular genetics to this plant.

Studies of wild-type and cytokinin-overproducing mutants have revealed that iPA is the major endogenous cytokinin of *P. patens* and that Z is also present [16, 25, 26, 27]. Definitive proof of the presence of IAA in gametophytic tissue of *P. patens* has been obtained [5], although we do not know whether IAA is the only or even major auxin in *P. patens*.

[1]Department of Biology, University of Regina, Regina, Saskatchewan, S4S 0A2, Canada
[2]Department of Genetics, University of Leeds, Leeds, LS2 9JT, UK
[3]John Innes Institute and IPSR, Norwich, NR4 7UH, UK
[4]Biology Department, University of South Florida, Tampa, Florida, FL 33620, USA

Herein we describe progress to date towards elucidating the roles of these two classes of hormones and their interaction with light in regulating the development of *P. patens*. Special emphasis is given to the use of auxin and/or cytokinin sensitivity mutants, and future prospects for understanding the molecular basis of cell differentiation, morphogenesis and hormone action in this model system are discussed.

2 Cell Differentiation and Morphogenesis in Wild-Type *Physcomitrella patens* Gametophytes

In the presence of Ca^{++} (Ashton, unpubl. data; McClelland and Cove, unpubl. data) and R [12], spores germinate to produce primary chloronemata, i.e. filaments of primary chloronemal cells, which elongate by apical cell division. Sub-apical cells divide, usually only once or twice, to produce additional apical cells from which side-branches, composed of further branching chloronemata, are derived. Primary chloronemal apical cells divide about every 20 h (McClelland and Cove, unpubl. data) and the cross walls formed are perpendicular to the long axis of the chloronema, the cells of which contain many plump chloroplasts.

About 6 days after germination under standard conditions (25 °C, continuous white light (WL) at > 60 μmol $m^{-2}s^{-1}$), some chloronemal apical cells divide to produce a second cell type, caulonemal apical cells, which divide about every 6 h (McClelland and Cove, unpubl. data) forming cross walls at an oblique angle to the long axis of the filaments. The caulonemal filaments generated in this way are comprised of cells containing few and spindle-shaped chloroplasts. Older caulonemal cells have walls impregnated with a red pigment. The sub-apical cells of caulonemata usually divide once to form single-cell side-branch initials, which have the following possible fates:

1. They may show no further development.
2. They may give rise to compound filaments of secondary chloronemal cells, which are morphologically similar to primary chloronemata.
3. They may give rise to further caulonemal filaments.
4. They may develop into gametophore buds which subsequently become the leafy shoots which bear the gametangia.

The probability of commitment to these various developmental fates is affected by many factors but especially by nutritional status, plant hormones, and light quality and quantity. Nutritional status will not be dealt with in this article. Light quality and quantity will be discussed briefly at this point and subsequently the roles of auxin and cytokinin will be examined in some detail.

Most of our studies on the regulation of side-branch fate by light have been of two types. Both have yielded similar results:

1. Cultures were grown from the outset under various levels of continuous monochromatic R (MRL). Data acquired in this manner have been reported elsewhere [2].

2. Dark-grown caulonemal filaments were exposed to MRL. When WL-grown cultures are transferred to darkness, caulonemal apical cells continue to divide, leading to the formation of negatively gravitropic caulonemata. The sub-apical cells formed soon after transfer to darkness may divide once, giving rise to caulonemal side branches. However, this process soon ceases and most cells in dark-grown caulonemata have no side-branches or initials. The effects of continuous MRL upon such caulonemata depends upon the photon flux employed [13].

At 3 nmol $m^{-2}s^{-1}$, about 50% of the cells in a caulonemal filament produce side-branch initials which grow into further caulonemal filaments. Almost all the remaining cells in the axial caulonemal filaments form one-cell side-branch initials. As the photon flux is increased, the proportion of initials which develop into caulonemata decreases. At about 200 nmol $m^{-2}s^{-1}$ only approximately 3% show this fate, the rest of the side-branches remain as initials. As the light level is further increased the proportion of initials becoming caulonemata remains unchanged. As the photon flux is raised between 200 and 500 nmol $m^{-2}s^{-1}$ the proportion of initials which grow into unbranched secondary chloronemal filaments increases. At 500 nmol $m^{-2}s^{-1}$ most initials form chloronemata which, at higher levels of MRL, themselves branch to form the compound filaments typical of cultures grown in standard WL conditions. At about 500 nmol $m^{-2}s^{-1}$ a few initials develop into gametophore buds; at higher levels of MRL, gametophore production is greater and a maximum of 1–2% of the initials become buds. We can conclude from these observations that light is necessary for the formation of side-branch initials on caulonemata. It also seems reasonable to conclude that, under standard WL, light promotes the production (from initials) of chloronemata, and also their subsequent branching, while inhibiting the formation of secondary caulonemata. Light also allows the formation of gametophores from a minority of initials.

3 Establishment of Roles for Auxin and Cytokinin and Their Interaction with Light in Normal Development Using Hormone Sensitivity Mutants

Cytological and physiological analysis of auxin and/or cytokinin sensitivity mutants of *P. patens* induced by the mutagens N-methyl-N′-nitro-N-nitrosoguanidine (NTG) and ethyl methane sulphonate (EMS), and isolated selectively in the presence of high concentrations of NAA or BA, have enabled us to classify the mutants into several categories, and also to construct a model of the roles of auxin and cytokinin in the early development of this moss [4, 10]. In brief, the morphology of BA-resistant (BAR) category 4 mutants, which consist of primary chloronemata, is repaired to near normality by low levels of exogenous auxin. This suggests that the formation of caulonemata requires auxin. Category 5 BAR strains, which may be a leaky version of category 4 mutants, produce a normal number of caulonemata and overproduce secondary chloronemata, but form few or no gametophores. Like category 4 mutants, category 5 strains are repaired by low levels of exogenous auxin. This suggests that the formation of gametophores and inhibition of the growth of secondary chloronemata also require auxin, albeit probably at a somewhat higher level than is needed to produce caulonemata. Morphologically, category 2, NAA-resistant (NAR) mutants resemble category 5 BAR strains. However, category 2 strains are resistant to exogenous auxin and are repaired instead by low levels of exogenous cytokinin. This suggests that, in addition to auxin, cytokinin is required for bud formation and also to inhibit secondary chloronemal growth. These findings also indicate that sensitivity of *P. patens* tissues to each type of hormone requires the presence of the other type; this was demonstrated directly in the case of cytokinin sensitivity using a category 4 mutant [4]. When grown under standard culture conditions for 3 weeks, category 1 mutants (consisting only of primary chloronemata) are resistant to both NAA and BA. Since we had no evidence that cytokinin is required for caulonema formation, we originally proposed that category 1 strains might be affected either in their auxin reception or in their process of signal transduction following hormone reception and leading to auxin-requiring developmental transitions.

3.1 Further Analysis of Category 1 and 2 Strains

In order to ascertain whether strains, isolated selectively by their resistance to NAA, are also insensitive to IAA, these mutants were cultured with IAA under continuous polychromatic R (PRL) with a photon flux of approximately 6 μmol $m^{-2}s^{-1}$ at the surface of the culture medium. PRL was obtained by using the same light source as for standard WL [1, 17] and covering the Petri plates with one layer of red filter (Roscolux, No. 27). Under these conditions, IAA in the growth medium is not photodegraded (Ashton, unpubl. data). Sensitivity to exogenous NAA and to BA has also been tested in PRL. All category 1 and 2 mutants are insensitive or strongly resistant in PRL to both auxins at concentrations ranging from 200 nM to 5 μM. A surprising discovery is that some category 1 strains, which we had previously found to be resistant to BA in WL, are clearly as sensitive in PRL to BA as the wild-type strain and their mutant morphology is essentially repaired by a low level (100 nM) of BA in PRL. If cultured for an extended period of time (5 weeks) under WL, these same category 1 mutants exhibit some sensitiviy, though less than in PRL, to and are partially repaired by 100 nM BA and are as sensitive as wild-type to 1 μM BA. We now propose that cytokinin-repairable category 1 (re-designated category 2A) strains are non-leaky versions of category 2 (re-designated 2B) mutants. Other category 1 strains have been shown to be insensitive to 100 nM and 1 μM BA in PRL as well as in WL. All category 2A mutants consist exclusively or almost exclusively of primary chloronemata. Indeed, they produce much more chloronemal tissue under both PRL and WL than wild-type. Since they are repaired (at least under PRL) by BA, it now seems likely that inhibition of the overproduction of primary chloronemata, as well as promotion of the formation of primary caulonemal filaments, require cytokinin, although possibly at a lower level than is needed for the induction of gametophores and suppression of overproduction of secondary chloronemal tissue. Cytokinin-repairable category 2A and category 2B strains may thus be deficient in endogenous cytokinin. An alternative explanation is that they overproduce a cytokinin antagonist. Hence, category 1 mutants, which are insensitive to both hormone types, may be affected in auxin or cytokinin reception or signal transduction processes.

Light is known to be required for gametophore formation and for sensitivity to cytokinin [3]. Also, the light-induced transition of side-branch initials, derived from subapical caulonemal cells, into gametophore buds appears to be phytochrome-mediated [15]. Data [2] derived from the culture in continuous MRL [20] of wild-type and a selection of cytokinin-repairable category 2A and 2B mutants, in the presence and absence of exogenous cytokinin (100 nM BA), reinforce our conclusions drawn from similar experiments using PRL. Furthermore, they confirm that MRL is sufficient for gametophore formation in the wild-type and for sensitivity to exogenous cytokinin in all the strains tested. For wild-type, grown in the absence of exogenous cytokinin, the numbers of gametophores observed are in good agreement with more detailed unpublished data (Jenkins and Cove), the latter indicating an approximately linear relationship between gametophore production and photon flux in the range, 0.5 to 16 μmol $m^{-2}s^{-1}$. The data also suggest that the level of light required for gametophore formation is reduced by exogenous cytokinin.

There is an enhanced production of primary chloronemal tissue by auxin-insensitive, cytokinin-repairable category 2A mutants and of secondary chloronemata by category 2B strains, even in low levels of MRL at which the wild-type forms no or few primary or secondary chloronemal cells [2]. Similarly, there is a greater production of chloronemal cells by cytokinin insensitive, auxin-repairable mutants grown in a low level of continuous PRL (Ashton, unpubl. data). These results suggest that auxin and cytokinin are required during normal development to antagonize a light-promoted tendency of (1) primary chloronemal apical cells to remain chloronemal, i.e. not to differentiate into caulonemal apical cells, and (2) of the side-branch initials formed on caulonemal filaments to become secondary chloronemal apical cells. Such an antagonism between light and these hormones might account for the following observations [2] with NAR 87, which has a phenotype consistent with its being partially deficient in endogenous cytokinin, namely that at the highest level of MRL employed NAR 87 consists exclusively of primary chloronemata (Fig. 1a), while at substantially lower levels of MRL it produces caulonemata. In the presence of exogenous cytokinin, NAR 87 forms caulonemata and gametophores but no or little chloronemal tissue over almost the entire range of MRL levels employed [2] (Fig. 1b). If wild-type cultures, grown for about one week in WL on medium supplemented with 0.5% sucrose, are transferred to darkness, long unbranched caulonemata are produced. The observation that some, though not all, auxin and/or cytokinin-insensitive strains also give rise to at least some caulonema-like filaments when grown in the dark, supports the above model and might also indicate that, in the absence of light, auxin and cytokinin are no longer required for the formation and/or growth of caulonemata from existing apical cells.

Clearly, in the case of gametophore formation for which auxin, cytokinin and light at a relatively high level are all required (> 16 μmol $m^{-2}s^{-1}$ of MRL for maximal production in the absence of exogenous cytokinin [Jenkins and Cove,

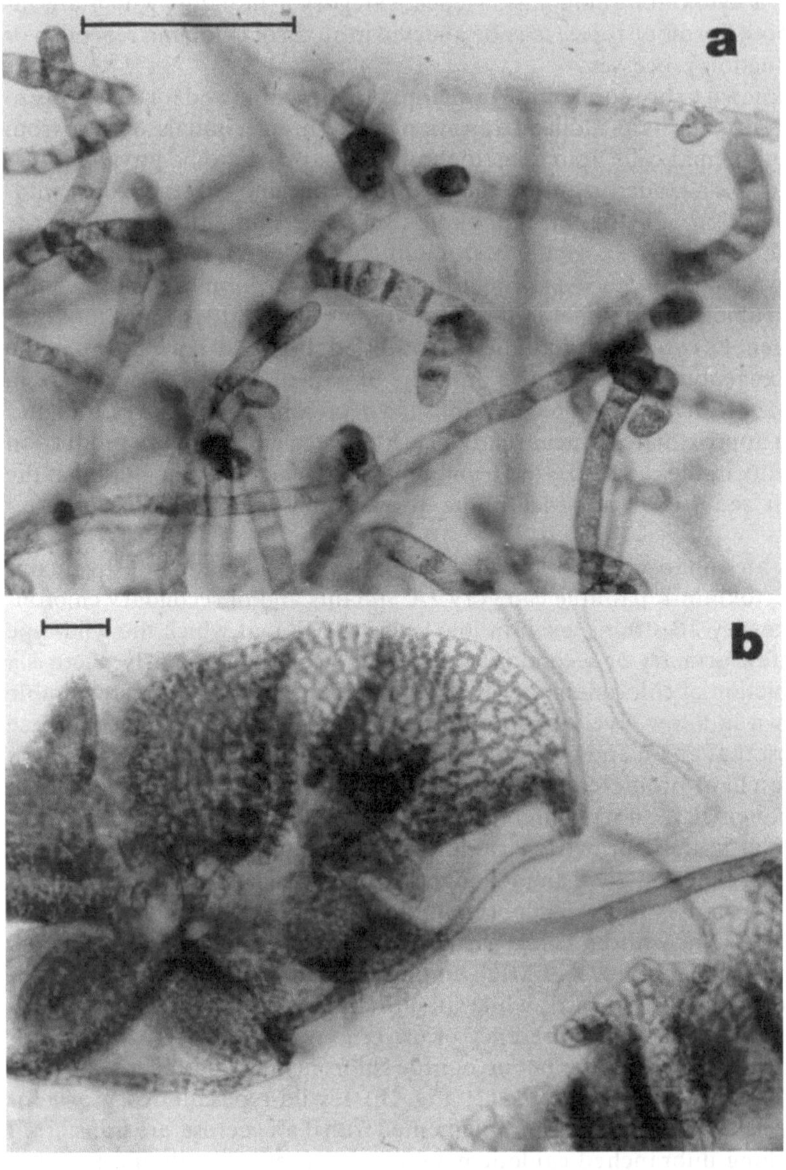

Fig. 1. **a** NAR 87 primary chloronemata formed in monochromatic R (MRL) (8.2 μmol $m^{-2}s^{-1}$) in the absence of exogenous cytokinin. **b** NAR 87 caulonemal filaments and leafy gametophores formed under MRL (3.4 μmol $m^{-2}s^{-1}$) in the presence of 100 nM BA. Scale *bars*: 100 μm

unpubl. data] and approximately 5 μmol $m^{-2}s^{-1}$ for maximal production in the presence of exogenous cytokinin [Futers, McClelland and Cove, unpubl. data], a different kind of interaction must be occurring between these hormones and light.

4 Prospects for Understanding the Molecular Basis of Development in *Physcomitrella patens*

There are many possibilities for the use of hormone sensitivity mutants in the analysis of plant development at the molecular level. Herein we discuss one approach toward understanding hormone-regulated developmental transitions in *P. patens.*

Preliminary data (unpubl. results, Knight, Long, Mohammed, Kumar, Radford, Martin, McClelland, Ashton and Cove) suggest that we have genetically transformed *P. patens* protoplasts with plasmids carrying a selectable marker gene. The expression of this marker confers a resistance to concentrations of kanamycin, which would otherwise inhibit growth. We subject protoplasts, mixed with plasmid DNA in the presence of polyethylene glycol, to a brief heat shock followed by the selection of kanamycin-resistant moss regenerants. This yielded stable kanamycin-resistant strains which are now being analyzed by Southern blotting to determine how much of the plasmid DNA has been inserted into the moss genome.

We have now obtained kanamycin-resistant regenerants using several different plasmids:

1. pKC7, a 5.8 kbp plasmid with a single kanamycin resistance gene prefixed by a bacterial promoter [24].
2. pBin19, an approximately 10 kbp plasmid derived from the Ti plasmid of *Agrobacterium tumefaciens* [7]. It contains a wide host range replicon, a kanamycin resistance marker with a bacterial promoter for direct selection in bacteria as well as a second kanamycin resistance gene flanked by the *nos* promoter and poly A addition site. It also possesses a multiple cloning region within a β-galactosidase gene which allows the detectable insertion of additional DNA sequences into the plasmid.
3. pSS1 (Schofield, unpubl. data), in which the *tmr* gene from the Ti plasmid of *A. tumefaciens* has been inserted into the multiple cloning site of pBin19. The *tmr* gene encodes isopentenyl transferase which catalyses the formation of N^6-(Δ^2-isopentenyl) AMP within suitable plant host cells. This substance can be dephosphorylated to give iPA which can be detected readily in *P. patens*, even at very low concentrations, by its effect on gametophytic development. The kanamycin-resistant moss which we have obtained following our transformation procedure using pSS1 exhibits abnormal morphogenesis of a kind which is consistent with the overproduction of cytokinin within its cells.

It is likely that genetic transformation will soon prove to be a very powerful tool with which to investigate moss development. In our approach it entails the production of new developmental mutants, including hormone-insensitive types, by insertion of identifiable plasmid DNA into the *P. patens* genome. The region of the moss genome affected will be recognizable since it will flank the transforming DNA on both sides. Isolation, cloning, sequencing and further analysis of the developmental gene that is affected should then be feasible.

References

1. Ashton NW, Cove DJ (1977) Mol Gen Genet 154:87
2. Ashton NW, Cove DJ (1990) In: Chopra RN, Bhatla SC (eds) Physiology and biochemistry of development in bryophytes. CRC, Boca Raton, Fl (in press)
3. Ashton NW, Cove DJ, Featherstone DR (1979) Planta 144:437
4. Ashton NW, Grimsley NH, Cove DJ (1979) Planta 144:427
5. Ashton NW, Schulze A, Hall P, Bandurski RS (1985) Planta 164:142
6. Ashton NW, Boyd PJ, Cove DJ, Knight CD (1988) In: Glime JM (ed) Methods in bryology. Proc Bryol Methods Workshop Mainz, Hattori Bot Lab Nichinan, p 59
7. Bevan M (1984) Nucleic Acids Res 12:8711
8. Courtice GRM, Cove DJ (1983) J Bryol 12:595
9. Courtice GRM, Ashton NW, Cove DJ (1978) J Bryol 10:191
10. Cove DJ, Ashton NW (1984) In: Dyer AF, Duckett JG (eds) The experimental biology of bryophytes. Academic Press, New York, p 177
11. Cove DJ, Knight CD (1987) In: Thomas H, Grierson D (eds) Developmental mutants of higher plants. Cambridge University Press, Cambridge, England, p 181
12. Cove DJ, Schild A, Ashton NW, Hartmann E (1978) Photochem Photobiol 27:249
13. Doonan JH (1983) PhD Thesis, University of Leeds
14. Engel PP (1968) Am J Bot 55(4):438
15. Futers TS (1984) PhD Thesis, University of Leeds
16. Futers TS, Wang TL, Cove DJ (1988) Mol Gen Genet 203:529
17. Grimsley NH, Ashton NW, Cove DJ (1977) Mol Gen Genet 154:97
18. Grimsley NH, Ashton NW, Cove DJ (1977) Mol Gen Genet 154:103
19. Grimsley NH, Featherstone DR, Courtice GRM, Ashton NW, Cove DJ (1979) In: Advances in protoplast research. Proc 5th Int Protoplast Symp. Akadémiai Kiadó, Budapest, Hungary, p 363
20. Jenkins GI, Cove DJ (1983) Planta 158:357
21. Jenkins GI, Cove DJ (1983) Planta 159:432
22. Jenkins GI, Courtice GMR, Cove DJ (1986) Plant Cell Environ 9:637
23. Knight CD, Cove DJ, Boyd PJ, Ashton NW (1988) In: Glime JM (ed) Methods in bryology. Proc Bryol Meth Workshop Mainz, Hattori Bot Lab Nichinan, p 47
24. Rao RN, Rogers SG (1979) Gene 7:79
25. Wang TL, Cove DJ, Beutelmann P, Hartmann E (1980) Phytochemistry 19:1103
26. Wang TL, Horgan R, Cove DJ (1981) Plant Physiol 68:735
27. Wang TL, Beutelmann P, Cove DJ (1981) Plant Physiol 68:739
28. Watts JW, Doonan JH, Cove DJ, King JM (1985) Mol Gen Genet 199:349

Dwarf Mutants of Maize – Research Tools for the Analysis of Growth

B.O. PHINNEY and C.R. SPRAY[1]

1 Introduction

The purpose of this paper is to present and analyze selected information concerning genetic dwarfism in maize. This includes the current status of the GA mutants and the description and evaluation of selected dwarf mutants that do not respond to GAs. Additional information on GAs and genetic dwarfism is available in a number of recent reviews [e.g. 5, 15, 17, 23].

The usefulness of the genetic approach to the analysis of growth and development is based on the following rationale. The control of a specific phenotype by a single gene implies unitary control at the biochemical level. A single gene mutant that is recessive is usually associated with the loss of a primary function which is expressed as an altered phenotype; the analysis of the biochemical basis for the altered phenotype can reveal critical information on a specific (chemical) factor controlling the phenotype (in this case, short growth). The rationale also forms the basis for the molecular biologists' studies on gene expression. The approach is now becoming widely accepted as a research tool for plant physiologists.

In maize, more than 50 single gene mutants have been described, the phenotypes of which involve the stature, or height, of the plant [5, 16]. These mutants are non-allelic to each other, and the majority are simple recessives. The dwarf phenotype may be expressed from the seedling stage to maturity, or, depending on the mutation, only during the seedling stages, or only in later stages of growth. The phenotype may be expressed in the dark as well as in the light. Seed for a number of dwarf mutants are available from The Maize Genetics Cooperation Stock Center, c/o E.B. Patterson, S-116 Turner Hall, Agronomy Department, University of Illinois, Il 61801, USA (Table 1).

The actual number of dwarf mutants in maize is probably in the hundreds, since current studies with mutagens (e.g. EMS) are continually producing new dwarf mutants [e.g. 4, 17]. The majority of these new mutants have yet to be tested for allelism to the known dwarfs.

Most of the dwarf mutants in maize are non-GA responders[2] (Figs. 1–5); the physiological/biochemical basis for their reduced growth has yet to be defined. For instance, they could be receptor mutants to either known or unknown classes of plant hormones; they could be mutants blocking steps in the biosynthesis of, as yet,

---

[1] UCLA Department of Biology, 405 Hilgard Avenue, Los Angeles, CA 90024–1606, USA

[2] The maize non-GA responding mutants could be called "insensitive mutants". The term is a current buzzword, with exciting, interesting, but undefinable implications. If the word "insensitive" is to be used, it should *always* be associated with *insensitive to what*?

Fig. 1

Fig. 2

Fig. 3

Fig. 4

Fig. 5. *Dwarf-8 (D8)*, a non-GA-responder. This mutant is probably a GA mutant (see text)

Table 1. A selected list of dwarf mutants in maize for which stocks are available that can be used to increase seed for use in the laboratory[a]

Mutant	Linkage	Mutant	Linkage
(1) *anther ear-1 (an1)*	1-L	(12) *dwarf-3 (d3)*	9-S
(2) brachytic-1 (br1)	1-L	(13) *dwarf-5 (d5)*	2-S
(3) brachytic-2 (br2)	1-L	(14) *dwarf-8 (D8)*	1-L
(4) brachytic-3 (br3)	5	(15) midget-1 (mi1)	1
(5) brevis-1 (bv1)	5-L	(16) nana-1 (na1)	3-L
(6) brevis-2 (bv2)	—	(17) nana-2 (na2)	5-L
(7) compact-1 (ct1)	8	(18) pigmy-1 (py1)	1-L
(8) compact-2 (ct2)	1-S	(19) reduced-1 (rd1)	6-L
(9) crinkly-1 (cr1)	3-S	(20) reduced-2 (rd2)	6-L
(10) *dwarf-1 (d1)*	3-S	(21) thick tassel dwarf (td1)	5
(11) *dwarf-2 (d2)*	3		

[a]Italicized mutants are GA-mutants. [Also see Maize Genetics Cooperation News Letters, ref. 16]

Figs. 1-4. Examples of non-GA-responding dwarf mutants of maize unknown classes of plant hormones; they could involve inhibitors of the action of plant hormones. The important point to be made here is that there is a wealth of genetic material in maize that is available for study by plant physiologists and plant biochemists interested in the control of growth and development in flowering plants.

Gibberellins (GAs) are part of an isoprenoid pathway (see Fig. 6) that originates from MVA. Condensation of dimethylallylpyrophosphate (DMAPP) with isopentenylpyrophosphate (IPP) forms geranylpyrophoshate (GPP) which condenses with IPP to give first farnesylpyrophosphate (FPP) and then geranylgeranylpyrophosphate (GGPP). Cyclization of GGPP gives copalylpyrophosphate (CPP) which undergoes further cyclization to *ent*-kaurene, the first tetracyclic diterpene in the pathway; stepwise oxidation followed by ring contraction gives GA_{12}-aldehyde, the common precursor to all known gibberellins. Several pathways diverge from GA_{12}-aldehyde. One of them is the non-3,13-hydroxylation pathway; a second is the early-3-hydroxylation pathway, and a third is the early-13-hydroxylation pathway (see Fig. 6). This latter pathway is unique to higher plants; it is the major pathway in the shoots of maize. [For reviews on the chemistry and biosynthesis of GAs, see 2, 3, 9]. The major GAs in the early-13-hydroxylation pathway (Fig. 6) are GA_{53}, GA_{44}, GA_{19}, GA_{20}, and GA_1. Gibberellin A_{17}, GA_{29}, and GA_8 are inactive branch metabolites from the main pathway. All the evidence supports the position that GA_1 is the only active gibberellin (i.e. active per se) in this series [5, 22, 25].

Recently, trace amounts of GA_3 and GA_5 have been identified by GC-MS from vegetative shoots of maize [8]. Feeding studies with $[^{13}C, ^{3}H]GA_{20}$ now demonstrate conclusively that this GA_3 originates from GA_{20} via GA_5, not via GA_1 [7]. While the levels of GA_1 are appreciably higher than those of GA_3 in maize shoots, and the conversion of GA_{20} to GA_1 is higher than the conversion of GA_{20} to GA_5 [1, 7], such data do not really resolve the relative roles of GA_1 and GA_3 in the control of shoot elongation in maize. In the same study trace amounts of other GAs (GA_9, GA_{15}, GA_{24}, GA_4, and GA_7) were also identified [8]. Their relative roles in the control of shoot growth are also unknown.

2 Gibberellin Mutants

Each of the mutants, *d1*, *d2*, *d3*, *d5*, and *an1* (Figs. 7 and 8), control a specific and different step in the GA biosynthetic pathway leading to GA_1 (Fig. 6), the GA that presumably controls shoot elongation in maize. A sixth mutant, Dominant Dwarf (*D8*) (Fig. 5), albeit a "non-responder" to GA_1, has been classified as a GA mutant controlling a step in the pathway (see section on the *Dwarf*-8 lesion). *D8* could be a mutant involving either a receptor for the active GA, or it could control a step downstream from the receptor.

The dwarf-5 (d5) *lesion* (Figs. 7 and 8)

The *dwarf-5* mutant blocks the cyclization of CPP to *ent*-kaurene [12] (Fig. 6). This early step in the pathway is catalyzed by the B activity of *ent*-kaurene synthetase

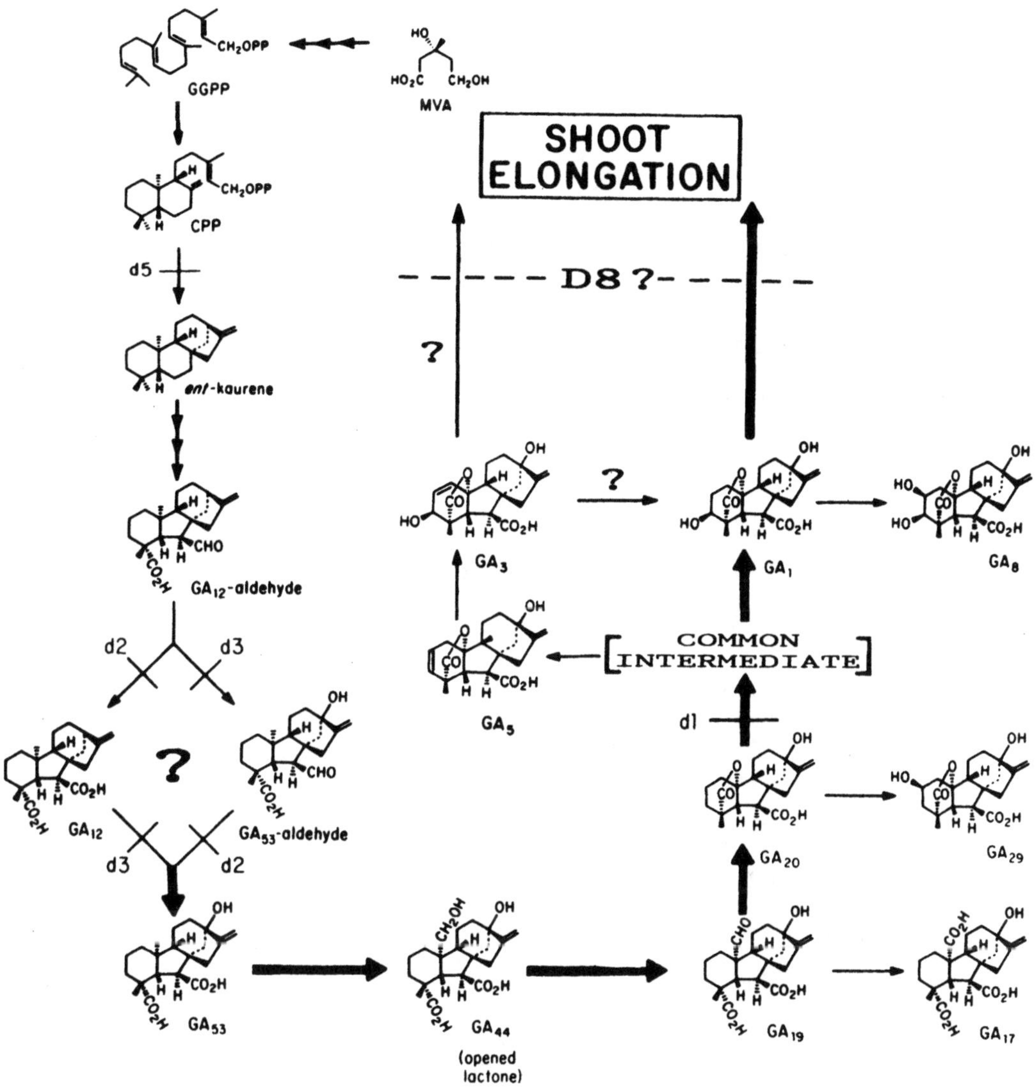

Fig. 6. The gibberellin biosynthetic pathway in *Zea mays* shoots. The position in the pathway controlled by each of the dwarf mutants, *d1*, *d2*, *d3*, and *d5*, is shown by a *cross line* on the appropriate *arrow*. This early-13-hydroxylation pathway leads to GA_1, the bioactive GA controlling shoot elongation in maize. The absence of GA_1, GA_5, and GA_3 in the *d1* mutant, together with the accumulation of GA_{20} [6], suggest that the *d1* gene controls a common step leading to GA_1 and GA_5

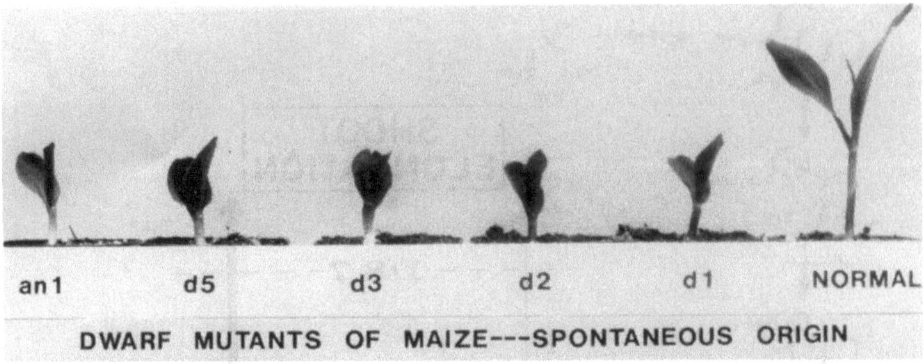

Fig. 7. Ten-day-old seedlings of normal and the five GA mutants (*d1*, *d2*, *d3*, *d5*, and *an1*)

Fig. 8. Mature plants of normal and the five GA mutants (*d1*, *d2*, *d3*, *d5*, and *an1*)

[24]. Location of the position of the genetic block in the pathway is based on the following information:

1. All maize GAs in the main pathway and their precursors, including *ent*-kaurene, are active when assayed on *d5* seedlings (and inactive, except for GA_1, when assayed on *d1* seedlings – a mutant blocked late in the pathway) [13, 22].
2. In the *d5* mutant, GGPP and CPP are metabolized to *ent*-kaurene at a rate one-fifth that of the normal [13].
3. Endogenous GAs are absent (or present in trace amounts) in *d5* seedlings and present in normal seedlings [6].

The anther ear-1 (an1) *lesion* (Figs. 7 and 8)

The *anther ear-1* mutant probably controls a step prior to the cyclization of CPP to *ent*-kaurene, although only preliminary data are available for this conclusion. It has been shown that *ent*-kaurenol, *ent*-kaurenoic acid, and GA_3 are bioactive when added to *an1* seedlings [14]. This mutant is currently under investigation. While the mutant is indistinguishable from other GA-mutants in the early seedling stages, the degree of dwarfism may become less apparent in later stages of growth. As a result, mature mutant plants range from half to full height of normals. All *an1* mutants have anthers in the ears, regardless of their final height at maturity; all progeny (seed) from these mutants give seedlings that are uniformly short and indistinguishable from each other and from the other GA dwarf mutants.

The dwarf-1 (d1) *lesion* (Figs. 7 and 8)

The *d1* gene controls a step late in the pathway, the 3β-hydroxylation of GA_{20} to GA_1 (see Fig. 6). The evidence for location of the genetic block is as follows:

1. Gibberellin A_{20} has less than 1% the activity of GA_1 when assayed on *d1* seedlings, whereas GA_{20} is as active as GA_1 on *d2, d3,* and *d5* seedlings [22], mutants blocked early in the pathway.
2. The level of endogenous GA_1 (and its inactive metabolite, GA_8) in *d1* seedlings is less than 2% that present in normal seedlings [6].
3. Endogenous GA_{20} (and its inactive branch metabolite, GA_{29}) accumulates in *d1* seedlings to a level ten times that found in normals [6].
4. Gibberellin A_{20} is not metabolized to GA_1 by *d1* seedlings whereas normals do metabolize GA_{20} to GA_1 [25].

While the GA data from the dwarf mutants have been interpreted to mean that the mutated gene codes for an altered enzyme, direct evidence is still lacking that would distinguish between lowered enzyme activity due to an altered enzyme, as opposed to lowered enzyme activity resulting from the presence of an inhibitor of the enzyme.

On reflection, it is interesting to review the earlier (pre-gibberellin) studies on the physiology of genetic dwarfism in maize. For example, in 1938 van Overbeek [19] reported lower levels of "diffusible auxin" from *d1, d2, d3,* and *d5* coleoptiles. In 1951, Harris [11] confirmed the lower levels from *d1* coleoptiles based on "ether extractable auxin"; he also found that addition of IAA to decapitated coleoptiles resulted in renewed elongation for normal coleoptiles but not for *d1* coleoptiles. Also coleoptile elongation was observed following replacement of normal tips to normal decapitated coleoptiles, but not to dwarf decapitated coleoptiles. Harris [11] concluded that auxin was not the factor limiting growth in the *d1* mutant of maize (GAs were not available to the western world for research studies until 1955 ±).

The Dwarf-8 (D8) *lesion* (Fig. 5).

While the *D8* mutant is a GA non-responder, it has been classified as a GA-mutant based on its phenotypic resemblance to the GA-responders [8, 10, 21]. Thus, this dwarf phenotype includes a series of morphological characters, including short, relatively broad and often wrinkled leaves, shortened internodes, small tassels and ears, and the presence of well-developed anthers in the ears. The presence of these anthers is unique to the GA mutants. The *D8* mutant could have all the properties of GA mutants and not respond to exogenous GA if *D8* controls a step subsequent to the bioactive GA_1 (and GA_3?), e.g. the binding of the active GA to a protein (i.e. a receptor mutant), or the control of some subsequent step downstream from the receptor. The presence in *D8* seedlings of all members of the early-13-hydroxylation pathway found in normal seedlings, and the accumulation of relatively high levels of GA_{20}, GA_1, and GA_8, in *D8*, supports the conclusion that the *D8* mutant blocks a step subsequent to GA_1 [8].

3 Non-GA-Responding Maize Mutants (Figs. 1–4; Table 1)

Most of the dwarf mutants of maize are GA non-responders. Unfortunately much of this non-response information is not in print. Brief mention has been made by Coe and Neuffer [5], Pelton [20], and Phinney [21], Likewise, there is little information in the literature on the response or lack of response of dwarf mutants to other classes of plant hormones, e.g. auxin. There have been few serious screening studies looking for auxin dependent dwarf mutants of maize, e.g. Cohen, personal communication.

It has been reported that the cause of the reduced growth in the *nana-1* (*na1*) mutant (Fig. 1) is correlated with a lower production of auxin due to a higher rate of destruction [18, 19]. The data have little significance when considered in terms of present day auxinology.

Four non-allelic, simple recessive, non-GA-responding dwarf mutants of maize are shown in Figs. 1–4. Nothing is known about the physiology of these mutants. Dominant Dwarf (*D8*), although a non-GA-responding dwarf, is considered to be a GA mutant (see above); it is shown in Fig. 5.

In summary, it is hoped that the information in this paper will encourage plant physiologists and biochemists to investigate the basis for the many non-GA-responding dwarf mutants of maize, many of which are available to the investigator. The same can be said for the GA-responding dwarf mutants of maize.

Acknowledgements. Financial support from the National Science Foundation, grant DMB85-06998, is gratefully acknowledged.

References

1. Albone KS, Gaskin P, MacMillan J, Smith VA, Weir J (1989) Planta 177:108
2. Bearder JR (1980) In: MacMillan J (ed) Encyclopedia of plant physiology. New Series, Vol 9. Springer, Berlin Heidelberg New York, p 9
3. Bearder JR (1983) In: Crozier A (ed) The biochemistry and physiology of gibberellins Vol 1. Praeger, New York, p 251
4. Bird RMcK, Neuffer MG (1987) In: Janick J (ed) Plant breeding reviews Vol 5. Van Nostrand Reinhold, New York, p 139
5. Coe EH Jr, Neuffer MG (1977) In: Sprague GF (ed) Corn and corn improvement. American Society of Agronomy, Inc, Madison, Wisconsin, p 111
6. Fujioka S, Yamane H, Spray CR, Gaskin P, MacMillan J, Phinney BO, Takahashi N (1988) Plant Physiol 88:1367
7. Fujioka S, Yamane H, Spray CR, Phinney BO, Gaskin P, MacMillan J, Takahashi N (1990) Plant Physiol (in press)
8. Fujioka S, Yamane H, Spray CR, Katsumi M, Phinney BO, Gaskin P, MacMillan J, Takahashi N (1988) Proc Natl Acad Sci USA 85:9031
9. Graebe JE (1987) Annu Rev Plant Physiol 38:419
10. Harbard N, Freeling M (1989) Genetics 121:827
11. Harris RM (1953) Ph D Thesis, University of California, Los Angeles
12. Hedden P, Phinney BO (1979) Phytochemistry 18:1475
13. Katsumi M (1964) Ph D Thesis, University of California, Los Angeles
14. Katsumi M, Jefferies PR, Henrick CA, Phinney BO (1964) Science 144:849
15. MacMillan J, Phinney BO (1987) In: Cosgrove DJ, Knievel DP (eds) Physiology of cell expansion during plant growth. American Society of Plant Physiologists, Rockville, MD, p 156
16. Maize Genetics Cooperation News Letters (1927–1988) Curtis Hall – Agronomy Department, University of Missouri, Columbia, MO 65211, USA
17. Neuffer MG (1978) In: Walden DB (ed) Maize breeding and genetics, Wiley, New York, p 579
18. van Overbeek J (1935) Proc Natl Acad Sci USA 21:292
19. van Overbeek J (1938) Plant Physiol 13:587
20. Pelton J (1964) Bot Rev 30:479
21. Phinney BO (1956) Proc Natl Acad Sci USA 42:185
22. Phinney BO, Spray CR (1982) In: Wareing PF (ed) Plant growth substances 1982. Academic Press, London, p 101
23. Reid JB (1987) In: Davies RJ (ed) Plant hormones and their role in plant growth and development. Nijhoff, Dordrecht, p 318
24. Shechter I, West CA (1969) J Biol Chem 244:3200
25. Spray CR, Phinney BO, Gaskin P, Gilmour SJ, MacMillan J (1984) Planta 160:464

Gibberellin Synthesis and Sensitivity Mutants in *Pisum*

J. B. Reid[1]

1 Introduction

In the last few years the number of physiological papers using mutants as research tools has increased dramatically. There is no better demonstration of this trend than the study of plant mutants influencing the synthesis of, or sensitivity to, plant growth substances. The GA-deficient genotypes have proved of immense value in understanding which responses are influenced by the levels of endogenous GAs and the development by Phinney [13] of the proposition that GA_1 may be the only endogenous GA controlling stem elongation per se in plants possessing the early 13-hydroxylation GA biosynthetic pathway.

The major advantage of mutants still stems from the classical one gene, one primary action hypothesis. Consequently, they allow the control of complex developmental processes to be analyzed as a set of individual partial processes. To illustrate this technique the control of stem elongation in peas is described.

At the present time, the genes at 12 of the more than 30 loci reported to influence internode length in pea (*Pisum sativum* L.) have been examined in some detail. Nine of the 12 mutations result in reduced internode length, while three lead to increased elongation (Table 1). However, a more useful and probably natural separation is into synthesis mutants and sensitivity mutants. A brief review of the synthesis genes and their action will be given and then recent results with sensitivity mutants will be examined to show how GA-sensitivity genes may operate and how factors such as light relate to natural changes in GA-sensitivity.

2 Synthesis Mutants

Four of the dwarfing mutants *le, na, lh,* and *ls*, block steps leading to the production of GA_1, the GA thought to control internode length in peas [6, 9]. They can be distinguished most easily from other mutants by the fact that they become true phenocopies of wild-type tall plants after treatment with appropriate quantities of GA_1 (Fig. 1). However, proof can only be obtained at the biochemical level where they are distinguished by their significantly reduced levels of GA_1 [6, 9]. Depending on the position of the genetic block in the GA_1 biosynthetic pathway, they may possess normal or increased levels of precursors including "biologically active"

[1]Department of Botany, University of Tasmania, Box 252C, GPO, Hobart, Tasmania 7001, Australia

Table 1. The major internode length genes in peas divided into groups that either reduce the biosynthesis of GA_1 (synthesis mutants) or alter the response to applied GA_1 (sensitivity mutants)

Gene(s)	Phenotype	Action
GA-synthesis mutants		
le	Dwarf	Reduced conversion of GA_{20} to GA_1
le[d]	Nana	More severe blockage of GA_{20} to GA_1
na	Nana	Blocks *ent*-7 α hydroxy kaurenoic acid to GA_{12}-aldehyde
lh	Dwarf	Blocks prior to *ent*-kaurene
ls	Dwarf/nana	Blocks prior to *ent*-kaurene
GA-sensitivity mutants		
la cry[s]	Slender[a]	Plants behave as if saturated with GA_1 regardless of endogenous GA_1 level
la cry[c]	Crypto[a]	Plants behave as if possessing substantial GA_1
lk	Erectoides	Very short, ridged and brittle stems, petioles and peduncles. Possibly more ethylene
lka	Short	Similar to *lk* but less severe, not a true phenocopy of dwarf plants
lkb	Short	Similar to *lk* but less severe, not a true phenocopy of dwarf plants
lm	Micro	Reduced size of all parts, including roots
lw	Short	Delayed flowering and increased water congestion
lv	Long	GA-hypersensitive, behaves as if partially etiolated

[a] *La* and *Cry are duplicate genes.*

Fig. 1. Plants possessing GA-synthesis mutants, such as *ls*, may be turned into phenocopies of wild-type (*Ls*) plants by treatment with GA_1 (10μg applied in ethanol to leaf 3 as indicated by tape)

GAs such as GA_{20} [9]. To date, none of the mutants with increased elongation in peas has been proven to result from the overproduction of biologically-active GAs [7, 19].

2.1 GA Biosynthesis

The pea plant possesses a wide range of GAs. The detailed work of Sponsel, MacMillan and Graebe [see 3, 23] has shown that the dominant pathway in peas from GA_{12}-aldehyde, the precursor for all GAs in higher plants, is probably the early 13-hydroxylation pathway (Fig. 2) leading from GA_{53} to GA_{29}-catabolite [3,

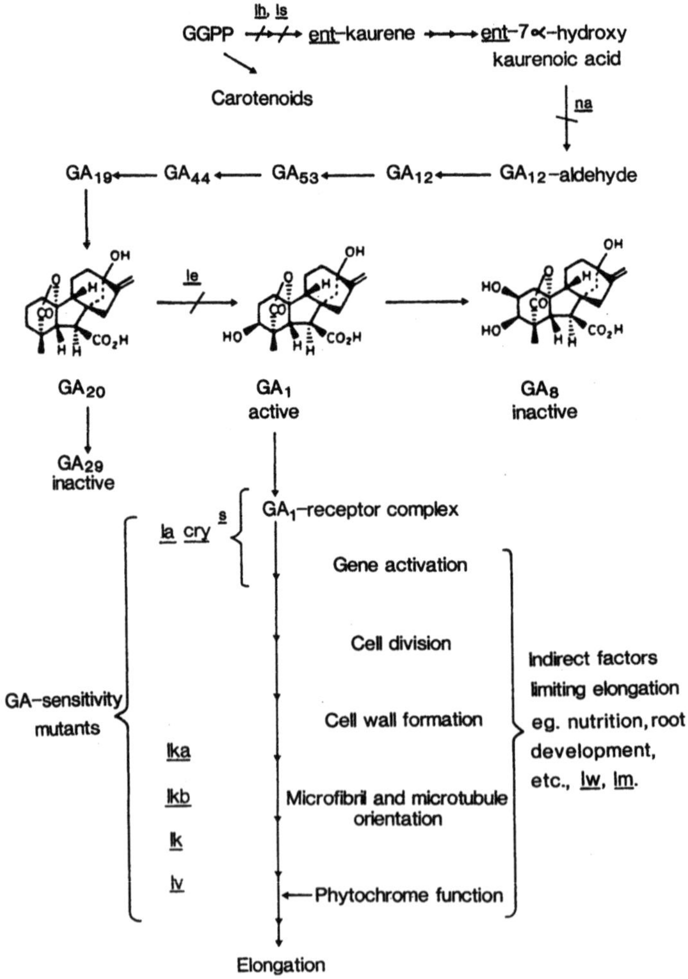

Fig. 2. The proposed sites of action of the GA-synthesis and sensitivity genes in the GA-biosynthetic pathway, and the possible sequence of events leading from GA_1 to elongation

23]. In addition, the non-13-hydroxylated pathway from GA_{12} to GA_{51}-catabolite occurs [3].

These pathways were initially determined from developing seeds where GAs occur at high levels but may be of little biological significance. None of these GAs in the seed are hydroxylated in the 3β position, a feature often associated with high biological activity. However, in the young expanding shoot (the tissue capable of elongation) GA_{20} has been shown to undergo 3β-hydroxylation to the highly active GA_1 [9]. The alternative path from GA_{20} is 2β-hydroxylation to GA_{29}, a step which leads to biological inactivity prior to catabolism (Fig. 2) [see 23]. GA_1 is thought to be the only native GA possessing biological activity per se in peas [8, 13].

2.2 Site of Action of the Synthesis Genes *le, na, lh* and *ls*

Gene *le* partially blocks the 3β-hydroxylation of GA_{20} to GA_1 (Fig. 2). However, it is a "leaky" mutant since a very small amount of [^{3}H, ^{13}C]GA_1 and [^{3}H, ^{13}C]GA_8 have been identified after large scale feeds of [^{3}H, ^{13}C]GA_{20} to *le* plants [8]. Further, a more severe allele at the *le* locus, le^d, has been identified which shows a reduced response to applied GA_{20}, compared to the response of *le* plants [22]. Plants possessing le^d also contain lower levels of endogenous GA_1 (Fig. 3).

The gene *na* probably blocks GA-biosynthesis prior to GA_{12}-aldehyde, since precursors earlier in the pathway (e.g. *ent*-7α hydroxy kaurenoic acid) show no biological activity in *na* plants but show some activity on comparable *Na* plants [6]. Further, *na* plants fed *ent* [^{3}H] kaurenoic acid produce no metabolites co-eluting with the C_{19}-GAs, GA_8, GA_{29}, GA_{20} or GA_1, while *Na* plants do. Plants possessing *na* can metabolize [^{2}H]GA_{12}-aldehyde to these C_{19}-GAs and the products contained no detectable dilution by endogenous [^{1}H] GAs [6].

Plants possessing genes *lh* and *ls* contain very low levels of GA-like activity [17] and respond as well as wild-type plants (which were dwarfed with the GA-synthesis inhibitor, AMO1618) to all GA-precursors examined, including *ent*-kaurene [6], suggesting that these genes block GA-biosynthesis prior to *ent*-kaurene. Neither *ls* nor *lh* plants appear to possess impaired carotenoid production, suggesting that the

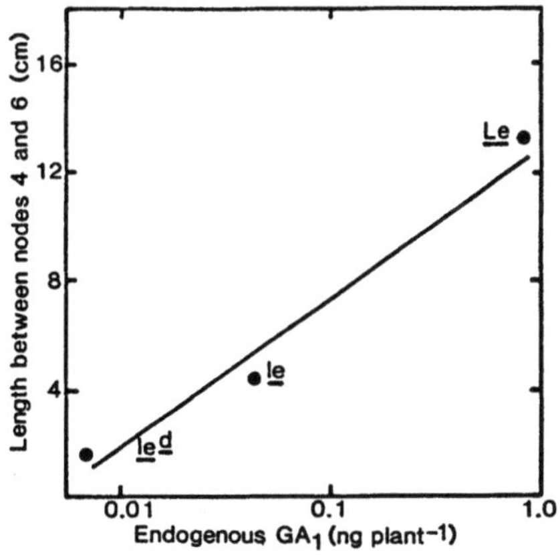

Fig. 3. The length between nodes 4 and 6 plotted against $\log_{10}$ of the endogenous level of GA_1 for plants of genotypes *Le*, *le* and le^d. GA_1 levels were determined by GC-MS after reverse isotope dilution (Ross, Reid, Gaskin and MacMillan, unpubl.)

blocks are after geranylgeranylpyrophosphate. However, both genes *lh* and *ls* appear to be "leaky" mutants, since the double recessive genotype *lh ls* is substantially shorter than either single recessive type [15].

2.3 Relationship Between Stem Elongation and the Level of GA_1

Three lines of evidence suggest a direct quantitative relationship between the level of GA_1 and elongation. Firstly, all the dwarf synthesis mutants are deficient in GA_1 but not necessarily its precursors such as GA_{44}, GA_{19} and GA_{20} [e.g. *le* plants, 2, 9, 17]. However, it has been argued that in tall (wild-type) plants (as opposed to dwarf mutants) GA_1 may never be limiting or acting to regulate elongation. This is unlikely since tall (*Le*) peas can double their length in response to 10 μg of the native gibberellin, GA_1 (Fig. 4). The double mutants, such as *na ls* or *lh ls*, which have even shorter internodes than the corresponding single dwarf mutants [15], suggest that all the mutations involved are "leaky" to some extent, and that the double block causes on even greater deficiency in GA_1 and therefore, stem elongation. If GA_1 is added at the appropriate level we get true phenocopies of normal types.

Secondly, if $[^3H]GA_{20}$ is applied to *na le* and *na Le* plants we find a direct log-linear relationship for both genotypes between the level of radioactivity in the expanding apical tissue and elongation. However, the *le* plants are less sensitive because the conversion of GA_{20} to GA_1 is partially blocked in these plants (i.e. the line is parallel but shifted to the right along the X-axis compared with the line for

Fig. 4. The response of wild-type tall plants (cv. Torsdag) to 10μg of GA_1 placed on the leaf indicated. The plants were grown under a 10-h photoperiod

Le plants, 8). Saturation of the growth response occurs over two orders of magnitude, comparable to the dose-response curves for animal hormones. By comparison the dose response curve is not shifted dramatically when the level of the 3β-hydroxylated GA, GA_8 (a metabolite of GA_1) is monitored [8]. In this experiment the level of GA_1 could not be directly measured due to our inability to separate GA_1 from its relatively inactive epimer, 3-epi-GA_1.

Thirdly, the allelic series *Le, le, le^d* results in reduced internode elongation [22]. It also results in reduced 3β-hydroxylation of [^{3}H]GA_{20}. When the logarithm of the amount of GA_1 is plotted against shoot elongation a linear relationship is apparent (Fig. 3). This evidence argues against GA_{20} possessing activity per se. The 2β-hydroxylated products of GA_{20} and GA_1, GA_{29} and GA_8 respectively, possess no or little activity, respectively [see 4]. These results taken as a whole leave little doubt that the level of GA_1 controls stem elongation in pea plants grown in white light.

3 Sensitivity Mutants

The term "sensitivity" has historically been used in a variety of ways [5]. Herein sensitivity mutants refer to those mutants which show an altered response to applied GA_1 relative to comparable plants (usually near-isogenic lines) possessing the wild-type allele. This group of mutants possesses widely differing phenotypes and includes plants with extremely short internodes and a reduced response to applied GA_1, as well as plants with longer internodes than normal, and plants which are hyper-responsive to GA_1. This range of phenotypes is not surprising given the multitude of limiting factors (steps) that may occur between reception of the GA signal and the phenotypic response of elongation. These steps are not presently known for any plant but must include effects directly related to the GAs, such as the level of the hypothesized receptor and the receptor affinity for GA_1, and the subsequent chain of events leading to elongation [5] such as cell division and cell elongation (Fig. 2). The investigation of such mutants should eventually allow the processes from GA-perception to elongation to be determined.

3.1 The Slender Gene Combination, *la cry^s*

The duplicate genes *la* and *cry^s* result in the extremely long slender (*la cry^s*) type. This genotype is phenotypically similar to normal plants treated with saturating levels of GA_1 or GA_3 for all characters (internode length, rate of internode production and leaf expansion, leaf size and colour, flowering behaviour, flower and pod development, etc.). However, internode elongation is entirely independent of GA levels in slender plants, since they show little or no response to applied GA, or to levels of GA synthesis inhibitors (e.g. AMO1618 and paclobutrazol) which have a marked influence on elongation in normal plants [7, 14]. Further, at the biochemical level the synthesis genes *le* and *na* still block GA_1 biosynthesis in slender plants but do not result in any large effect at the phenotypic level [7]. On the basis of these results the slender gene combination (*la cry^s*) has been suggested to prevent the production of a functional repressor. In normal (*La* and/or *Cry*) plants, GA_1 acts as the effector of this repressor [14] allowing elongation and other GA-mediated processes to proceed.

3.2 The Genes *lk*, *lka* and *lkb*

All three genes result in reduced elongation, little or no alteration in the level of GA-like substances, and a reduced response to applied GA_1 [17, 20]. Although gene *lk* is more severe in effect than genes *lka* and *lkb*, the range of pleiotropic effects is similar in each genotype, suggesting a similar mode of action. For example, the stems and petioles become ridged, swollen and brittle in all cases with reduced leaflet size (especially width). The peduncles are also dramatically reduced in length [20, 21]. Gene *lk* is epistatic to the slender gene combination (*la* cry^s) [15], the reverse of the situation for the GA-deficiency gene *na*, suggesting that *lk* may operate later than *la* or cry^s in the sequence of events leading to elongation.

Plants possessing *lk* show significant elongation and increased sensitivity to applied GA_1 after treatment with the ethylene synthesis inhibitor AVG. The *lk* plants also produced more ethylene than comparable *Lk* plants under the conditions used [21]. However, whether the involvement of ethylene is a primary action of this gene, or merely a secondary consequence, is not clear. However, it does highlight the need to look at more than one group of growth substances.

3.3 Genes *lm* and *lw*

The gene *lm* results in the "micro" phenotype in which all aspects of the plants' growth are reduced. Plants possessing *lm* also have a reduced response to applied GA_1 and this genotype can therefore be considered to be a sensitivity mutant [18]. However, all available evidence suggests that *lm* may reduce the capacity of the plant to respond by altering some basic aspect of cell development. The first visible effects of gene *lm* occur in the roots where root elongation and root fresh weight are reduced within one week of germination. If such an indirect effect is the basis for reduced GA sensitivity in this genotype, then further investigation of this system is unlikely to determine the mechanism of GA-perception and primary action.

A similar situation may hold for gene *lw*, which reduces stem elongation, delays flowering and increases the susceptibility of plants to the disease known as water congestion [10]. It does not appear to reduce GA-like activity, but does cause a reduced response to GA_1. This complex range of pleiotropic effects suggests *lw* may act via a mechanism that partially overrides the response to GA_1 [10].

3.4 Gene *lv* and the Action of Light

Gene *lv* results in increased elongation and an increased response to all levels of applied GA_1 in plants grown in white light [19]. It does not alter the level of endogenous GA-like substances or the metabolism of $[^3H]GA_{20}$. However, *lv* plants are dramatically shortened by the GA-synthesis inhibitor paclobutrazol [19]. Plants possessing gene *lv* therefore behave as if GA-hypersensitive. This contrasts markedly with the slender (*la cry*^{s}) type, which is elongated regardless of the level of GA_1.

The gene *lv* has little or no effect when plants are grown in complete darkness or under continuous FR light, but does exert a substantial effect when plants are grown under continuous R or white light (Fig. 5). Mature plants possessing *lv* also possess reduced levels of chlorophyll compared with *Lv* plants (Steane and Reid, unpubl.). This is similar to the response shown by long-hypocotyl mutants of tomato (*au*^{w} and *yg-2*) which possess substantially reduced levels of phytochrome [11, 12]. Consequently, gene *lv* requires further examination to see if it reduces phytochrome levels or function. These results suggest that *lv* plants grown in white or red light behave as if still partially etiolated. Measurements of cell number and cell length are consistent with this view (Steane and Reid, unpubl.).

This raises the question of whether dark etiolation results from increased GA-sensitivity in the same manner as the gene *lv* appears to act in white light. When the sensitivity of very short, GA-deficient *na* plants to GA_1 was compared in the light and dark, the dark-grown plants possessed markedly increased sensitivity [16]. This suggests that light, probably via the phytochrome system, influences elongation primarily by altering GA-sensitivity, rather than by influencing the level of GA_1, as was previously suggested [1].

4 Discussion

The range of so-called sensitivity mutants clearly shows the multitude of different factors that can influence the response to GA_1. Some mutants have indirect effects, simply overriding the normal response to GA_1 (e.g. *lm* and *lw*). Others result in a

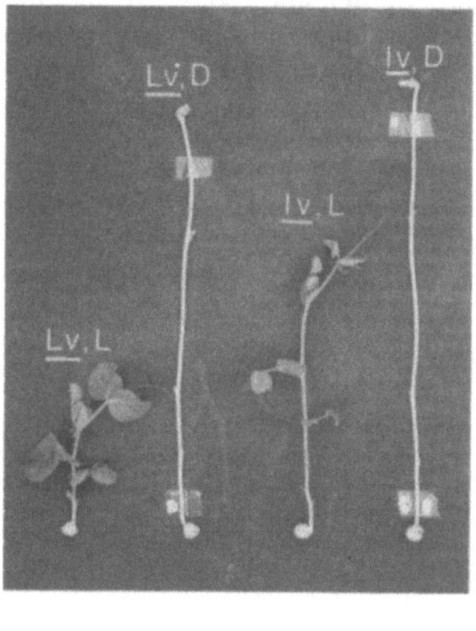

Fig. 5. In white light (*L*) *Lv* and *lv* plants possess markedly different internode lengths and total shoot lengths, whereas in darkness (*D*) little or no difference is present

similar phenotype and may block steps in a pathway between reception and elongation (*lk, lka, lkb*), just as the synthesis mutants result in similar phenotypes by blocking steps leading to the production of GA_1 (Fig. 2). It is possible to envisage such mutants acting by partially blocking events such as the correct orientation of microtubules or microfibrils during elongation. However, the phenotypes of plants possessing genes *lk, lka* and *lkb* are not the same as those caused by the GA-synthesis-blocking genes. Consequently, these genes are unlikely to be acting by simply reducing the perceived level of GA_1 (i.e. by either reducing the level of the GA-receptor or reducing the affinity of the receptor for GA_1). Other mutants appear to influence the whole range of GA-mediated responses, suggesting that they may directly influence reception of the GA-signal (e.g. *la* and cry^s). Finally, the mutant *lv* appears to increase the ability to respond to GA_1 in a manner, at least superficially, similar to darkness. This suggests a possible interaction with phytochrome function.

In the past, sensitivity mutants have frequently been assumed to arise simply from altered reception of the hormone by a modification to the affinity of the receptor, or the amount of the receptor present. However, such mutants are probably a distinct minority of sensitivity mutants and of the eight detailed in peas, probably only *la* and cry^s possess characteristics suggesting such a primary and central mode of action. Consequently, if research on these major facets of hormone physiology are to be probed using mutants, care must be taken to select and characterize the mutants far more carefully than in the past.

Some sensitivity mutants in peas suggest that certain mutants may, either directly or indirectly, influence other plant growth substances (e.g. *lk*), or influence the action of photomorphogenic pigments (e.g. *lv*). This highlights the interactions which can occur, and also the need for future work on the control of development to be concerned with more than just one group of plant growth substances, or one pigment system.

Given these limitations the sensitivity mutants may still be the best tools we have to understand the sequence of steps between hormone reception and the developmental response, an area in which we have a distinct lack of knowledge at present. They allow the partial processes to be examined one at a time. In fact, they have the potential to allow us to gain as detailed a view of "GA sensitivity" as we currently have of the role that GA_1 plays in the control of stem elongation in peas.

Acknowledgments. I wish to thank my colleagues, Drs. Ingram, Murfet and Ross for helpful discussion, and the Australian Research Grants Scheme for financial support.

References

1. Campell BR, Bonner BA (1986) Plant Physiol (Bethesda) 82:909
2. Davies PJ, Emshwiller E, Gianfagna TJ, Proebsting WM, Noma M, Pharis RP (1982) Planta 154:266
3. Graebe JE (1987) Annu Rev Plant Physiol 38:419
4. Graebe JE, Ropers HJ (1978) In: Letham DS, Goodwin PB, Higgins TJV (eds) Phytohormones and related compounds – a comprehensive treatise, vol 1. Elsevier/North Holland, Amsterdam, p 107
5. Firn RD (1986) Physiol Plant 67:267

6. Ingram TJ, Reid JB (1987a) Plant Physiol (Bethesda) 83:1048
7. Ingram TJ, Reid JB (1987b) J Plant Growth Regul 5:235
8. Ingram TJ, Reid JB, MacMillan J (1986) Planta 168:414
9. Ingram TJ, Reid JB, Murfet IC, Gaskin P, Willis CL, MacMillan J (1984) Planta 160:455
10. Jolly CJ, Reid JB, Ross JJ (1987) Physiol Plant 69:489
11. Koornneef M, Cone JW, Dekens RG, O'Herne-Robers EGO, Spruit CJP, Kendrick RE (1985) J Plant Physiol 120:153
12. Parks BM, Jones AM, Adamse P, Koornneef M, Kendrick RE, Quail PH (1987) Plant Mol Biol 9:97
13. Phinney BO (1984) In: Crozier A, Hillman JR (eds) The biosynthesis and metabolism of plant hormones. Cambridge University Press, Cambridge, p 17
14. Potts WC, Reid JB, Murfet IC (1985) Physiol Plant 63:357
15. Reid JB (1986) Ann Bot 57:577
16. Reid JB (1988) Physiol Plant 74:83
17. Reid JB, Potts WC (1986) Physiol Plant 66:417
18. Reid JB, Ross JJ (1988) Physiol Plant 72:547
19. Reid JB, Ross JJ (1988) Physiol Plant 72:595
20. Reid JB, Ross JJ (1989) Physiol Plant 75:81
21. Ross JJ, Reid JB (1986) Physiol Plant 67:673
22. Ross JJ, Reid JB (1987) Ann Bot 59:107
23. Sponsel VM (1983) In: Crozier A (ed) The biochemistry and physiology of gibberellins, vol 1. Praeger, New York, p 151

Gibberellin-Insensitive and Overgrowth Mutations in Temperate Cereals

J.L. STODDART[1]

1 Introduction

In theory, single point mutations should be invaluable tools in the elucidation of GA action. In practice, they have yet to fulfill their potential. It is easy to draw parallels with the elegant work on GA biosynthesis, but this ignores the essential difference between the problems. Namely, that a biosynthetic sequence is known to be enzymatic and to proceed by discrete steps. It must, therefore, bear a direct relationship to primary gene products. On the other hand, we are still awaiting information on the general nature of response mechanisms (let alone their fine detail) and consequently cannot make firm statements about the unitization of the process.

Molecular biology now provides a suite of precise tools which seem poised to provide guidance regarding GA action and the results of these approaches are evident elsewhere in this volume. Nevertheless, it seems certain that insensitive and overgrowth mutants will be vital assets for confirming putative mechanistic schemes and for adding fine detail once the framework has been established. This chapter briefly reviews the state of knowledge on GA response mutations in cereals.

2 GA-Insensitive Wheat (*Triticum aestivum* L.) Mutants

2.1 Genetic Basis

The earliest descriptions of the insensitivity trait were related to the dwarf genotypes Norin 10 and Tom Thumb where it was noted that their response to applied GA_3 differed from that characterizing most other tall and dwarf genotypes then in commerce [4]. A total of 10 *Rht* loci have been described to date, exerting varying effects on stature and with actions modified by genetic background. All *Rht* types are regulated by single genes located on chromosomes 4A and 4D. In most instances they are considered to be non-allelic and dominance relationships vary. For example, *Rht 1* and *Rht 2* are incompletely recessive whereas *Rht 3* is partially dominant [5]. Pleiotropic effects are also evident, exemplified by *Rht 3* effects on both culm elongation and aleurone function, as well as complex influences on ear number, grain weight and composition [5]. It should be noted that the designation *Rht* (= reduced height) is not synonymous with GA-insensitivity.

[1]Biochemistry Department, Welsh Plant Breeding Station, Plas Gogerddan, Aberystwyth, Dyfed Sy 23 3Eb, Wales, UK

Rht alleles are thought to have allelic or homeoallelic relationships with "grass-clump" or hybrid dwarfs. This latter form of extreme dwarfism is based upon the presence of at least one allele of the *D1* to *D4* gene group. Presence of two alleles (e.g. *D1* + *D2*) produces semi-dwarfs, three alleles not including *D3* yield extreme dwarfs with reduced heading and three alleles (including *D3*) produce grass clumps and suppressed heading [5]. Genes *D1, D2* and *D4* are located on the group 2 chromosomes, *D3* is on the long arm of chromosome 4B.

Uniculm ('stunting') genes (*US1* and *US2*) resulting in rosette formation and premature death have been described [5]. Expression is environmentally sensitive and, in some cases, is associated with enhanced ABA lèvels.

The foregoing data indicate the presence of a significant number of stature-determining loci in the wheat genome. Only a minority exhibit GA-insensitivity characteristics and may, therefore, be presumed to relate to hormonal response.

Multiple loci for insensitivity provide evidence for the inadvisability of conceptualizing hormone response as a single-step process.

2.2 Hormonal Relationships

Vegetative tissues of *Rht 1, Rht 2* and *Rht 3* dwarfs contain significantly more extractable endogenous, GA than is present in wild-type material. RIA of *Rht 3* seedling extracts indicated GA_1 pool sizes which were 13 × those of co-segregating wild types [16].

Recent quantitative analyses based upon GC-MS have shown a 10 × increase in GA_1 in *Rht 1* vegetative tissues and a 30 × enhancement in GA_1 for *Rht 3*, suggesting an inverse relationship between stature and pool size [10]. In contrast, no variations in the level of the predominant GA_{54} were found for developing grains [9] of the two dwarfs.

The interpretation of this phenomenon is unclear. It is suggestive of the necessity for biological action to drive GA turnover: association with the active site being part of metabolic de-activation and, therefore, pool depletion. In the context of GA_1 this could be envisaged as the insertion of a 2ß hydroxyl group to form biologically inactive GA_8. Thus, it might be inferred that *Rht 3* should have a reduced flux from GA_1 to GA_8. Metabolic studies with radio-labelled GA_1 [16] have shown that the rates of GA_8 and polar conjugate formation are higher in wild-type than in *Rht 3* dwarf seedlings. However, when the dilution effect of the larger GA_1 pool size in the dwarfs is compensated for, there is no significant difference in the flux of radioactivity between GA_1 and GA_8 in dwarfs and controls. An alternative is to postulate a growth-dependent sequestration, possibly exemplified by irreversible protein binding or immobilization in the cell wall matrix or endomembrane system. Growth-dependent cell wall binding of GAs has been previously reported [15]. Compartmentation in the vacuole would not affect extractability and is not, therefore, a candidate explanation. The high active GA levels present in the mutants are indicative that feed-back controls operating on the GA biosynthetic pathway (if they exist) are relatively insensitive. Interestingly, the high GA_3 levels detected in dwarf leaf tissues do not appear to be present in actively growing stem internode tissues [5].

The genetics of the situation do, however, suggest that the dwarfing genes are active rather than negative. Chromosome deletion lines indicate that the recessive, non-dwarfing, alleles are equivalent to the null state [5]. The production of unusual metabolites, GA-antagonists, modifications of binding proteins and structural change in components of the response chain are all candidates for positive effects.

Plant hormones other than GAs have been implicated in the action of *Rht* genes. Some evidence has been advanced to show changes in GA/IAA interactions and increases in ABA contents in *Rht 3* seedlings have been reported and discounted [5].

Rht 1, *Rht 2* and *Rht 3* also have effects on the GA responsiveness of aleurone tissue in mature seeds and of the amylase content of whole seeds during grain maturation [3]. The largest effects on aleurones are produced by *Rht 3* where responses to saturating doses of GA_1 are only 25% of those obtained with wild-type seeds and flatter dose/response curves are also characteristics. Whilst *Rht 1* and *Rht 2* are similarly less sensitive than controls, the effects are less marked than in *Rht 3*. This property leads to greatly reduced pre-maturity sprouting in seeds of varieties containing GA-insensitive dwarfing genes [6].

A period of grain pretreatment at 5°C is reported to overcome GA-insensitivity of *Rht 1*, *Rht 2* and *Rht 3* aleurone layers [14]. Changes in phospholipid composition were detected during this period and correlated with the development of sensitivity, suggesting that membrane fluidity may have a regulatory role in the processes α-amylase secretion. As yet, there are no reports of low-temperature treatment overcoming the action of *Rht* genes in vegetative tissues.

During the later stages of maturation (ca. 30–50 days post anthesis) grain α-amylase values fall to relatively low levels (20–25 mU/grain). This reduction has been shown to be greater in the case of *Rht 1* and *Rht 3* where final enzyme levels can be around ¼th of those in wild-type seeds [6].

2.3 Other Physiological Effects

Coleoptile length is reduced in the presence of all three GA-insensitivity alleles, but with a marked effect of background genotype [5]. No consistent effects on root weight, rooting depth, apical growth rates, differentiation or primordium number have been found. Both cell number and cell size have been implicated in height reductions [5]. Generally, there is a reduction in biomass but a compensating adjustment of dry matter partitioning towards the grain.

The growth/temperature relationships in GA-sensitive dwarfs of *Oryza* and *Zea* are modified towards higher temperatures for growth cessation [17]. Similar studies on *Rht 3* have revealed no differences in growth/temperature relationships, indicating that the dwarfism has a different physiological basis from that pertaining in maize and rice.

Basic metabolism is, apparently, unaffected in *Rht 3* aleurone layers, as instanced by comparable rates of respiration, amino-acid uptake, GA uptake, protein synthesis and ATP content [5].

3 Overgrowth Mutations

Overgrowth or 'slender' mutants of *Pisum*, characterized by excessive elongation of stem internodes, were first described in 1927 [7] and are considered in detail elsewhere in this volume. Slender mutants of barley have also been characterized [2]. The *Cb 3014* barley mutation is attributed to a single recessive allele *sln* at a locus on chromosome 4, probably close to the centromere. Due to the extreme extension of the rachis and floral structures, these mutants are sterile and must be multiplied by the intercrossing of heterozygotes. The mutants are pale, mechanically weak and exhibit extreme overgrowth of leaf sheaths and stem internodes. Frequently, sub-apical internodes will elongate, carrying the shoot apex well clear of the soil surface. As with the pea mutants, these plants behave as though saturated with GA. Emerging leaves of wild-type seedlings grow at 1.5 mm h^{-1} whilst slender seedlings achieve rates in excess of 3.0 mm h^{-1}. The mutation has a marked effect on the growth/temperature relationships [17]. High resolution transducer traces for mutant and wild-type seedlings show radically different responses.

On cooling, third-leaf normal seedlings gave a biphasic curve with a slope discontinuity at 17°C and growth cessation at 5.5°C. On rewarming, there was strong positive hysteresis with higher growth rates at all temperatures. Slender segregates showed an essentially linear decline in growth rate to growth cessation at −5°C and, on rewarming, showed negative hysteresis with reduced growth rates throughout. This behaviour is indicative of lethal frost damage. The single gene recessive slender mutation clearly has a fundamental effect on the ability to growth at low temperatures and removes the normal adaptive constraint providing protection against frost damage. In contrast to the situation described for *Rht* genes in wheat, this behaviour also suggests a positive action of the dominant *Sln* allele in suppressing growth rate in barley.

3.1 Hormonal Relationships

Bioassays of *la cry*^{s} and wild-type pea seedlings have shown that the slender phenotype is associated with lower endogenous pools of GA_{20}-like activity, particularly in immature tissues [13]. Determinations of GA pool sizes in two leaf barley seedlings exhibiting slender or normal phenotypes are detailed in Table 1. Levels of the C20, non-lactonic GA_{19} are elevated in *sln/sln* seedlings but, for all other GAs, markedly lower contents are characteristic of the slender phenotype. Maximum depressions were found in the actively extending lower leaf segments. The largest effects were for GA_1, considered to be the biologically active form for shoot elongation in many higher plants [11]. Here, the levels in the immature slender tissue were one-seventh of the wild-type values and a similar order of reduction applied to GA_{20}, the immediate precursor of GA_1. These results suggest that the slender mutation is associated with a reduced flux through the $GA_{19} \rightarrow GA_{20}$ interconversion but specific turnover data will be required to confirm this indication. The lower absolute amounts of GA_8 in slender tissues could be regarded as being suggestive of reduced de-activation by 2β hydroxylation. However, when

Table 1. Endogenous gibberellin (GA) contents of mature and expanding portions of slender and wild-type barley seedling leaves (ng/g fresh weight[a])

	Wild-type (*Sln/sln* + *Sln/Sln*) Mature	Extending	Slender (*sln/sln*)[b] Mature	Extending
GA_{19}	4.99	8.00	6.25 (125)	11.50 (144)
GA_{20}	0.21	0.58	0.07 (33)	0.10 (17)
GA_1	0.35	0.21	0.09 (26)	0.03 (14)
GA_8	4.43	7.27	1.88 (42)	2.67 (37)
Sample weights (g)	116	90	161	134

[a] Measurements by GC-MS, quantitated by reference to deuterated and tritiated internal standards.
[b] ( ) = Values as percentages of corresponding wild-type contents.

GA_8 contents are related to the levels of precursor GA_1, in each case it is evident that the GA_8:GA_1 ratio is substantially increased for both mature and extending zones (Table 2). This may be taken as circumstantial evidence that GA_1 turnover is higher in extending zones and that the increased growth rates in slender tissues are accompanied by an enhanced turnover of GA_1. It must be borne in mind, however, that the absolute levels of GA_8 are lower in the mutant.

Does the lower pool size of active GA species suggest that growth rate is related to a higher consumption of newly biosynthesized GA_1, possibly by turnover to metabolites other than GA_8 or by irreversible conjugation with macromolecules? If this is the case, it should be possible to demonstrate effects of inhibition of GA biosynthesis on the expression of the slender mutation.

When seedlings are grown in the presence of varying concentrations of paclobutrazol or CCC (GA biosynthesis inhibitors), growth is progressively inhibited in the wild type whilst, over an extended range, there is no measurable effect on growth in slender seedlings. For example, with 1×10^{-5} M paclobutrazol, plant stature in normal seedlings was reduced by 50% with respect to controls, and slender was unaffected. Measurements of GA_1 pool sizes showed an 80% reduction in wild-type plants and a 20% drop in slender. The relative insensitivity of GA biosynthesis in slender is surprising but is probably a reflection of already low rates of GA_1 production (about one-tenth of controls). An hypothesis that *sln/sln*

Table 2. Ratios of GA_8 to GA_1 in extracts of leaves from slender and wild-type barley seedlings (values for GA_8 with GA_1 expressed as unity)[a]

	Mature	Extending
Wild-type (*Sln/Sln* + *Sln/sln*)	12.7	34.6 (172)
Slender (*sln/sln*)	20.9	89.0 (326)
	(64)	(157)

[a] ( ) = Percentage increases.

conferred insensitivity to paclobutrazol would be inconsistent with the other facts. Confirmation that the inhibitor is indeed acting in the mutant is provided by reductions in GA_1 contents to around 16% of control values, suggesting that there is a relatively small flux through the $GA_{19} \rightarrow GA_{20}$ biosynthetic step.

It is, therefore, supportable to suggest that the low GA_1 contents of *sln/sln* seedlings are not indicative of enhanced consumption due to rapid growth but are, rather, evidence that the growth process in the mutant is independent of GA supply. Exogenous applications of GA_1, which have no effect on barley slender seedlings but increase growth in wild types, confirm this contention.

3.2 Growth/Turgor Relationships

A mathematical description of steady-state growth is provided by recent restatements of the Lockhart equation [1]. This relates internal and external osmotic potentials, cell wall properties and hydraulic conductance to growth rate. Co-variations of growth rate and turgor pressure at various growth temperatures are detailed for wild-type and slender seedlings in Fig. 1.

Between 2 and 20°C there is a linear relationship between temperature and growth rate with a greater slope in the case of slender seedling. Pressure probe measurements of turgor pressure in individual meristematic cells do not, however, show either a marked temperature response or any correlation with growth rate. The physical basis of the differences in cell elongation must, therefore, reside in other components of the Lockhart equation. Initial measurements of hydraulic conductivity provide similar values for all genotypes, thus placing emphasis on cell wall properties.

3.3 Cell Wall Properties

Determinations of wall rheological properties have shown marked differences between mutant and wild type. Extensibility (mm/g^{-1} applied load) was measured as 0.11 for normals and 0.33 for slender seedlings whilst load/relaxation hysteresis values (under 10 g applied load) were respectively 3.3×10^{-6} and 8.4×10^{-6} J. Thus, there are marked differences in the apparent plasticity of the mutant cell wall, but these were abolished when measurements were made with killed tissue, suggesting that the differences were mediated by dynamic wall-based processes.

Cell wall linkage turnover has been studied using [^{14}C]-labelling and self-autolysis techniques [8]. Normal genotypes showed higher overall levels of autolysis over a 5-h period following wall isolation, mainly releasing arabinose and arabinose-rich oligosaccharides. Slender genotypes released more oligogalactose [12]. Interpretation of such data is complex but they do support the suggestion of functional cell wall differences induced by the *sln* mutation.

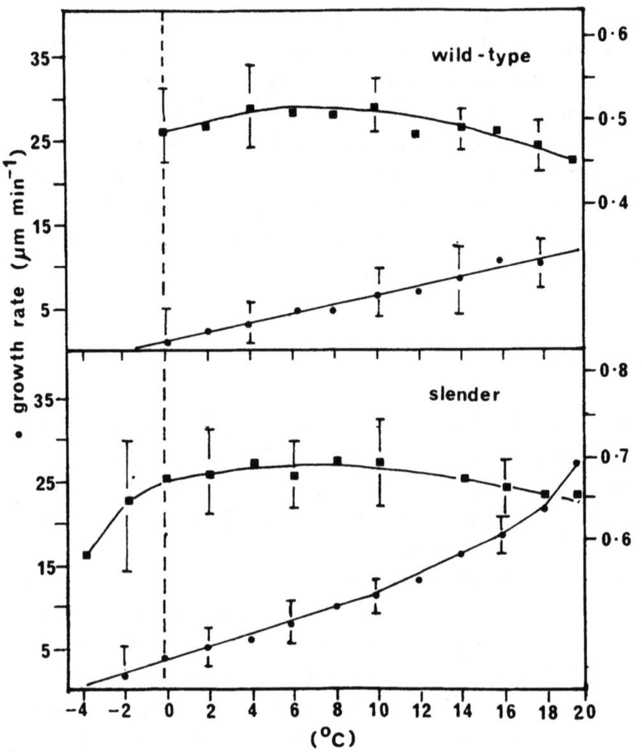

Fig. 1. Growth rate (●) and turgor pressure (■) changes with respect to meristem temperature for 3-leaf slender (*sln/sln*) and wild-type seedlings. Growth rates determined using high-resolution displacement transducers connected to emerging leaves and turgor pressures measured by pressure probe for individual meristem cells. *Bars* indicate SEM

3.4 Aleurone Behaviour

Although the obvious manifestation of the "slender" phenotype is as a markedly enhanced extension growth, there are accompanying changes in enzyme secretion by embryoless half-seeds and, by inference, aleurone cells. Seeds with the *sln/sln* genotype are essentially independent of GA supply for synthesis and secretion of α-amylases. Detailed enzymological studies (Chandler, per. comm.) suggest that the rate of α-amylase production by slender seeds equates to that in normals treated with 10^{-5} M GA_3 and that both high and low pI groups are secreted.

4 Discussion

It is evident from the growing body of evidence concerning GA metabolism in both the *Rht 3* wheats and the various mutants exhibiting slender phenotypes (barley, pea, tomato) that the idea of a direct relationship between the endogenous pool size of active GA structures and the overall growth rate of the tissue is no longer tenable. If anything, GA insensitive and overgrowth plants provide an inverse relationship, which could be argued as being supportive of the concept that growth is related to the rate of consumption of GA. The high GA_1 level in *Rht 3* and the diminished pool sizes in slender material would be consistent with this view. However, we lack data on the fluxes through the later stages of GA biosynthetic pathway, particularly the $GA_{19} \rightarrow GA_{20}$, $GA_{20} \rightarrow GA_1$ and $GA_1 \rightarrow GA_8$ steps, which would allow an objective assessment of the hypothesis.

In the case of slender barley there are strong indications that even this interpretation may not be a proper reflection of the GA/growth relationship. The extremely small GA_1 pool size, coupled with the insensitivity of growth to retardant treatments, suggests that extension growth is independent of both pool size and rate of GA synthesis. This is not to suggest that a causal relationship does not exist between hormone and cell enlargement but, rather, to indicate that interference with post-receptor events can disable the response mechanisms in both negative and *positive* senses. If a cascade of steps is envisaged between receptor/ligand association and ultimate biological response, and if each step is regarded as having two reversible states, then any mutation which locks a "gate" in either the open or closed state will either permanently disable or enable response to the hormone, regardless of receptor affinity or hormone pool size. Insensitive and overgrowth mutations can be rationalized on these grounds, and support is provided by the existence (at the Welsh Plant Breeding Station) of at least two further slender barley phenotypes resulting from mutations at loci differing from Cb 3014, and by the observation of multiple polypeptide changes during in vitro translations. Clearly, molecular biology will enable a rapid elucidation of the preceding questions and is presently being applied in several laboratories. Finally, the designation "GA-insensitive", normally applied to *Rht* wheats, seems appropriate in some cases for certain slender genotypes (e.g. Cb 3014 barley), but may be a less definitive classification than previously envisaged.

Acknowledgements. This work is supported by the UK Agricultural and Food Research Council. Permission from Dr A.D. Tomos, UCNW, Bangor, to reproduce data contained in Fig. 1 is gratefully acknowledged.

References

1. Cosgrove DJ (1981) Plant Physiol 68:1439
2. Foster CA (1977) Barley Genet Newsl 7:24
3. Gale MD, Lenton JR (1987) Aspects Appl Biol 15
4. Gale MD, Marshall GA (1973) Ann Bot NS 37:729
5. Gale MD, Youssefian S (1983) In: Russell GE (ed) Progress in plant breeding. Butterworth, London, p 1
6. Gale MD, Salter AM, Lenton JR (1987) In: Mares DJ (ed) 4th Internat symp on pre-harvest sprouting in cereals. Westview Press, p 273
7. de Haan H (1927) Genetica 12:481
8. Heyn ANJ (1986) Plant Sci 45:77

9. Lenton JR, Gale MD (1987) In: Mares DJ (ed) 4th Internat symp on pre-harvest sprouting in cereals. Westview Press, p 253
10. Lenton JR, Hedden P, Gale MD (1987) In: Hoad GV, Lenton JR, Jackson MB, Atkin RK (eds) Hormone action in plant development – a critical appraisal. Butterworth, London, p 145
11. Phinney BO, Freeling M, Robertson DS, Spray CR, Silverthorne J (1985) In: Bopp M (ed) Plant growth substances. Springer, Berlin Heidelberg New York Tokyo, p 55
12. Pollock CJ, Pomos AD, Thomas A, Smith CJ, Lloyd EJ, Stoddart JL (1989) in press
13. Potts WC, Reid JB, Murfet IC (1985) Physiol Plant 63:357
14. Singh SP, Paleg LG (1985) Aust J Plant Physiol 12:269
15. Stoddart JL (1979) Planta 146:353
16. Stoddart JL (1984) Planta 161:432
17. Stoddart JL, Lloyd EJ (1986) Planta 167:364

II Molecular Aspects

Auxin and Gene Expression

G. HAGEN, B. MCCLURE, C. BROWN, M. GEE, and T. GUILFOYLE[1]

1 Introduction

Auxins are thought to play a role in diverse growth processes such as cell extension, cell division and differentiation. The initial observation that the exposure of plant cells to auxin alters nucleic acid metabolism was made over 30 years ago [11]. Current research efforts have focused on examining primary auxin responses, with a major emphasis being placed on studying the very early and most rapid alterations in gene expression following auxin treatment. These studies have been facilitated by the development of molecular techniques that allow a high degree of resolution.

2 Auxin-Regulated mRNAs

Evidence for the rapid modulation of specific mRNAs came from 2D gel analyses of in vitro synthesized translation products from control and auxin-treated organs of soybean [18, 19, 20], pea [12] and maize [17]. The results indicated that although the overall patterns of translation products from mRNA populations isolated from control and auxin-treated plant organs were similar, several (up to 10 in soybean) products increased in amounts following a 1- h auxin treatment. Some of the changes were observed within 10–15 min of auxin treatment [12, 18, 19].

Auxin-regulated mRNAs have been pursued further using recombinant DNA technology. cDNA copies of mRNAs from soybean [6, 8, 15], pea [13] and tobacco [14] have been produced and auxin-induced sequences within each "library" of cDNA clones have been identified (Table 1). When radiolabeled, these cDNAs become sensitive, molecular probes for the characterization of the induction of the mRNA levels by auxin.

In most cases, a detectable induction of mRNA levels is observed within 15–30 min following auxin exposure. The levels of three soybean mRNAs (detected by the cDNA clones 6, 10A and 15) are induced by 2.5 min of treatment of excised elongating soybean hypocotyl sections with the synthetic auxin 2,4-D [8]. These mRNAs and others identified in soybean [5, 16], pea [13] and tobacco [14] are specifically induced by auxins such as IAA and NAA, and not induced by non-auxin analogs. With some exceptions, these auxin-induced sequences do not accumulate in response to other plant growth regulators such as GA, ABA, cytokinins and ethylene, or to environmental stresses such as heat shock or cold

[1] Department of Biochemistry, University of Missouri, Columbia, MO 65211, USA

Table 1. Auxin-regulated cDNA clones

Plant	Organ source[a] (RNA/cDNA)	Reference
Soybean		
pGH1	IH	6
pGH2/4	"	"
pGH3	"	"
6	EHS	8
10A	"	"
15	"	"
pJCW1	EZ-IH	15
pJCW2	"	15
Pea		
pIAA 4/5	ETIS	13
pIAA 6	"	13
Tobacco		
pCNT 103	CSC	14
pCNT 107	"	14
pCNT 114	"	14
pCNT 115	"	14
pCNT 116	"	14
pCNT 117	"	14
pCNT 123	"	14

[a] IH, intact hypocotyl; EHS, elongating hypocotyl section; EZ-IH, elongating zone-intact hypocotyl; ETIS, epicotyl third internode section; CSC, cell suspension culture.

shock [5, 8, 13, 16]. In vitro transcription assays using nuclei isolated from auxin-treated soybean organs [5, 9] and tobacco cells [14] indicate that at least part of the accumulation of these mRNAs results from increased transcription rates on the corresponding genes.

3 Auxins-Regulated cDNA and Gene Structure

Several auxin-regulated cDNAs have been sequenced [1, 9]. Two of the soybean sequences (pJCW1 and pJCW2) have several co-linear regions of high homology at the nucleotide level (77–80%) and deduced amino acid level (80–100%), suggesting that these sequences are part of a related family of genes [1]. The three soybean cDNAs (designated 6, 10A and 15) corresponding to the small (0.5 kb*) mRNAs (SAURs, small auxin-up RNAs), likewise, contain regions of high homology [9]. There are no apparent homologies between the cDNAs pJCW1/pJCW2 and the cDNAs 6, 10A and 15. The genes corresponding to these five soybean cDNAs have been isolated and sequenced. The genes for pJCW1 and

*kb = kilobase

pJCW2 (designated Aux 28 and Aux 22 respectively) are present in one to two copies per haploid genome and contain four (Aux 28) and two (Aux 22) intervening sequences (introns) within the protein coding region [1]. The genes for cDNAs 6, 10A and 15 are clustered within 5 kb of soybean genomic DNA [Fig. 1; 9]. The DNA sequence analysis of over 7 kb, which includes this cluster, has revealed the presence of two additional, homologous sequences (designated X10A and X15, Fig. 1). It is unknown if these two genes are expressed or regulated by auxin. None of the five genes within the cluster contain introns within the protein coding region. The genes are transcribed in alternate directions and are spaced at intervals of approximately 1.25 kb within the cluster. Hybridization results indicate that this gene cluster is present at one copy per haploid genome.

DNA sequence analysis of the regions flanking the protein coding portion of the Aux 28 and Aux 22 soybean genes has identified two sequences (9 and 11 bp) that occur at similar distances and spacing upstream from the transcription start site in each gene [1]. Alignment of the five SAUR genes has revealed two regions of homology upstream of the open reading frame (ORF) and one region downstream of the ORF in each gene [9]. There is no striking homology in the upstream elements when Aux 28, Aux 22 and the SAUR gene sequences are compared [9]. The significance of these elements is unknown, but the fact that they are located in the same position in related genes suggests that they may play some role in the regulation of gene expression.

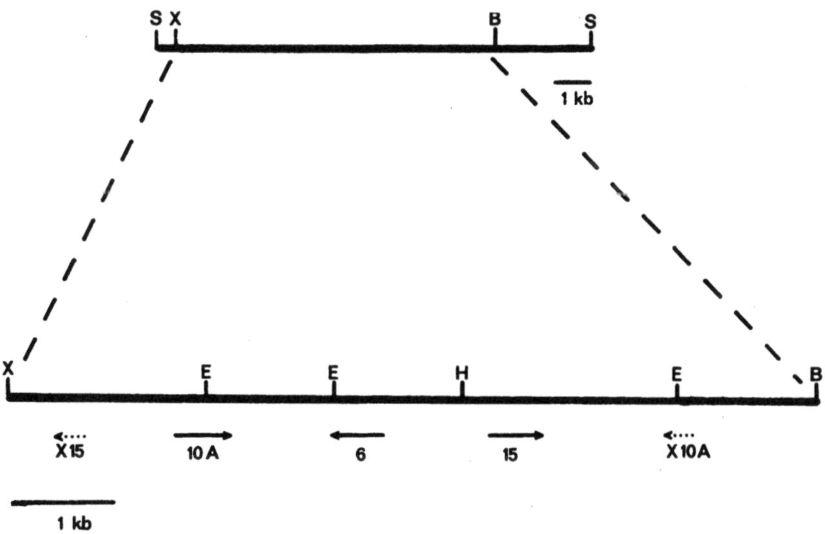

Fig. 1. Genomic organization of the small, auxin-up regulated (SAUR) locus in soybean. The *top line* indicates the 12 kb Sau 3A (S) genomic clone from which the 7.7 kb Xba I (X)- BamH I (B) fragment was subcloned. *Solid arrows* show the location of the genes for the cDNA clones 10A, 6 and 15. The direction of transcription is indicated by the arrows. *Dashed arrows* show regions of the genomic sequence which are highly homologous to genes 10A and 15, and are designated X10A and X15

4 Current and Future Directions

With the gene sequence known, it is possible to define the nucleotide elements that are critical for auxin-regulated gene expression. It has been shown that when the SAUR gene cluster is transferred into petunia via *Agrobacterium*-mediated transformation, expression of the genes can be induced by auxin treatment of the transgenic plants (Wright et al., manuscript submitted). This result indicates that nucleotide elements (cis elements) which are necessary for auxin-induced gene expression are contained in the cluster. Deletion of various upstream and downstream sequences surrounding the genes and assay of auxin inducibility of these deletions in transgenic petunia and in transient expression systems should provide insights into the sequence elements involved in the regulation of these genes by auxin. In addition, it will be of interest to characterize nuclear protein factors that bind to auxin-regulated genes and are important in the regulation of gene expression.

An additional goal of the studies on auxin-induced gene expression will be to determine the role of the polypeptides encoded by the auxin-induced mRNAs. One of the auxin-induced soybean mRNAs has been shown to encode a heat shock protein [4]. This gene is also expressed in response to heat [3] and heavy metals [4, 7] and may represent a general stress protein. The function of the polypeptides encoded by the other auxin-regulated mRNAs is unknown. Computer comparison of the deduced amino acid sequences of the soybean auxin-induced genes with other known protein sequences has failed to reveal any homologies.

One approach being used to determine protein function and localization is to produce antibodies to the proteins encoded by the auxin-induced mRNAs. Using the soybean cDNA pGH3 in an *E. coli* expression vector, a fusion protein has been isolated and antibodies have been produced. These antibodies have been used to study auxin-induced GH3 protein production in soybean [17].

It has been suggested that several of the auxin-induced sequences may function in the process of cell elongation [8, 13, 15], based on the rapid induction of these sequences predominantly in the elongating region of the hypocotyl or epicotyl. This suggestion has received further support for the SAUR sequences in recent gravitropism experiments [9]. For these studies, soybean seedlings were placed in a horizontal position for various lengths of time. The beans were sectioned and blotted directly onto nylon membrane filters, which imprint an image of the tissues within an organ [2]. (^{35}S)-labeled probes with homology to the SAUR genes were hybridized to these tissue blots to detect the presence of SAUR RNAs. The results indicate that there is a rapid redistribution of the SAUR RNAs in response to gravistimulation towards the lower or rapidly elongating side of the hypocotyl. This redistribution can be detected within 20 min of gravistimulation and occurs before visible bending of the hypocotyl is detected (45 min). This redistribution is not observed when tissue prints are hybridized to nonauxin-inducible (control) probes. Tissue prints of cross sections indicate that the SAUR RNAs are expressed in the cortical cells of the elongating region of the hypocotyl.

Clearly, major questions remain to be answered concerning the regulation of gene expression by auxin. Research in this field has been greatly advanced with the isolation of molecular probes. A concerted, multifaceted approach, however, will be required to elucidate the events leading to the physiological responses that have been attributed to auxins.

References

1. Ainley WM, Walker JC, Nagao RT, Key JL (1988) J Biol Chem 263:10658
2. Cassab GI, Varner JE (1987) J Cell Biol 105:2581
3. Czarnecka E, Edelman L, Schoeffl F, Key JL (1984) Plant Mol Biol 3:45
4. Czarnecka E, Nagao RT, Key JL, Gurley WB (1988) Mol Cell Biol 8:1113
5. Hagen G, Guilfoyle T (1985) Mol Cell Biol 5:1197
6. Hagen G, Kleinschmidt A, Guilfoyle T (1984) Planta 162:147
7. Hagen G, Uhrhammer N, Guilfoyle TJ (1988) J Biol Chem 263:6442
8. McClure BA, Guilfoyle TJ (1987) Plant Mol Biol 9:611
9. McClure BA, Hagen G, Brown CS, Gee MA, Guilfoyle TJ (1989) The Plant Cell 1:229
10. McClure BA, Guilfoyle TJ (1989) Science 243:91
11. Silberger J, Skoog F (1953) Science 118:443
12. Theologis A, Ray PM (1982) Proc Natl Acad Sci USA 79:418
13. Theologis A, Huynh TV, Davis RW (1985) J Mol Biol 183:53
14. Van der Zaal EJ, Memelink J, Mennes AM, Quint A, Libbenga KR (1987) Plant Mol Biol 10:145
15. Walker JC, Key JL (1982) Proc Natl Acad Sci USA 79:7185
16. Walker JC, Legocka J, Edelman L, Key JL (1985) Plant Physiol 77:847
17. Wright RM, Hagen G, Guilfoyle T (1987) Plant Mol Biol 9:625
18. Zurfluh LL, Guilfoyle TJ (1982) Planta 156:525
19. Zurfluh LL, Guilfoyle TJ (1982) Plant Physiol 69:332
20. Zurfluh LL, Guilfoyle TJ (1982) Plant Physiol 69:338

IAA Perception and Auxin-Regulated Gene-Expression

A.M. MENNES, C.J.M. BOOT, K.R. LIBBENGA, E.J. VAN DER ZAAL, and A.C. MAAN[1]

1 Introduction

In the study of auxin perception and transduction we have initially focused on auxin perception. In this respect we have tried to identify soluble IAA-binding proteins which, in analogy with the model of steroid hormone action, may interact with gene expression at the transcriptional level. The main characteristics of such binding sites from tobacco have been described [10] and it was found that addition of partially purified binding proteins to isolated nuclei from tobacco callus gave an IAA-dependent stimulation of transcription [15]. Such reconstitution experiments are very important in the study of transduction of the auxin signal. However, this system [15] was too crude to draw definite conclusions. One of the first steps in improving the system should be to better specify auxin-dependent stimulation of nuclear transcription in vivo. For this purpose we used an auxin-dependent batch-cultured cell line from tobacco as auxin-target system [16]. We were able to demonstrate that readdition of auxin to auxin-starved cells rapidly induced at least 7 classes of mRNAs.

Using cDNA clones corresponding to the auxin-induced mRNAs we were able to study their induction in some detail [17]. Genes, corresponding to a number of these auxin-responsive mRNAs, have been cloned and the promoter regions are now being studied, e.g. to look for cis-acting auxin-responsive elements as possible binding sites for trans-acting (auxin-binding?) proteins [18].

The results obtained thus far have enabled us to study more precisely the transcriptional activity in nuclei isolated from the auxin-target cells. Only isolated nuclei from auxin-activated cells transcribed the 7 auxin-responsive genes. Interestingly, the nuclei not only showed run-off transcription, but we could demonstrate that also new RNA chains were initiated [9]. We will now use this in vitro transcription assay for testing of transacting (auxin-binding?) proteins from nuclear lysates on target nuclei. In previous experiments we showed that such nuclear lysates contain the soluble auxin-binding protein as described earlier [4].

Herein we provide a short review of our progress to date, and discuss the state of the art with respect to identification and characterization of soluble auxin-binding proteins, including results obtained in other laboratories.

---

[1] Department of Plant Molecular Biology, State University of Leiden, Botanical Laboratory, Nonnensteeg 3, 2311 VJ Leiden, The Netherlands

2 Results

2.1 Experimental System

All experiments dealing with auxin-induced gene-expression used a batch-cultured cell line from tobacco (*Nicotiana tabacum L.* var. White Burley), which requires only 2,4-D as an exogenous hormone for cell proliferation. When early stationary phase cells are transferred to a fresh medium containing 2.2×10^{-7} M 2,4-D they enter a lag-phase of ca. 1 day and reach the stationary phase after 7 days. However, when cells are transferred to an auxin-free medium they reach a new stationary phase after 5 days when any remaining auxin has been metabolized (Fig. 1). At this point cell division can be restored by injecting a small volume of 2,4-D into the culture medium, giving a final concentration of 2.2×10^{-6} M. When the same volume of water is injected as a control, cell division is not restored.

This batch-cultured cell line apparently lacks the membrane-bound auxin-binding protein [6]. It thus provides us with a good auxin-target system to study the role of soluble auxin-binding sites in the auxin perception-transduction mechanism. Auxin-starved early-stationary phase cells respond relatively rapidly to added auxin with cell division.

2.2 Effect of Auxin on Gene Expression

To know whether auxin affects nuclear gene transcription prior to the onset of cell division, we extracted RNA at 1, 2 and 4 h after the addition of 2,4-D or water to early stationary phase cells in auxin-lacking medium. These RNAs were used in an in vitro reticulocyte translation system. Analysis of the [^{35}S]-methionine-labelled polypeptide spots on 2-dimensional gels showed a rapid increase in at least 3 mRNA species within 1 h after 2,4-D application as compared to the water control [16].

A cDNA library from mRNAs isolated 4 h after 2,4-D or water addition was constructed and differentially screened with single-stranded cDNA. This resulted in the selection of 7 cDNA clones from 2,4-D-induced mRNAs showing no

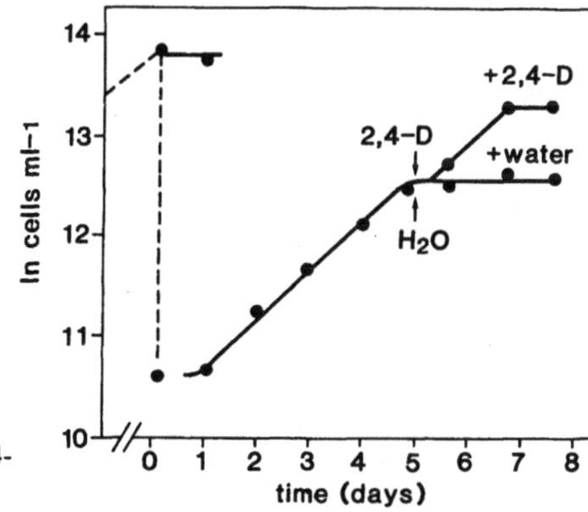

Fig. 1. Growth of tobacco cells after transfer to 2,4-D-free medium [Adapted from 8]

Table 1. Plasmids containing cDNA inserts to 2,4-D-induced mRNAs

Plasmid	Insert length (bp)[a]	Length of the corresponding mRNA (bp)	MW of the proteins (D)[b]
pCNT103	850	1000	27 500
pCNT107	950	1100	27 000
pCNT114	550	950	27 000
pCNT115	1400	1450	31 000/38 000
pCNT116	1750	2200	67 000
pCNT117	850	1050	?
pCNT123	650	1000	?

[a] Insert length was determined using DNA digested with EcoRI/HindIII as MW markers.
[b] After hybrid-selected translation [adapted from 17].

cross-hybridization [Table 1 and 17]. Analysis of the induction kinetics of the 2,4-D-induced mRNAs on Northern blots hybridized with the 7 [^{32}P]-labelled cDNA clones showed that all mRNAs were induced by auxin within 30 min. pCNT103 and pCNT107 showed that most rapid induction, less than 15 min. After 1 day, when cell division was restored (see Fig. 1) the levels of most mRNAs were lower than after 4 h of auxin treatment.

We have also studied the dose-dependent accumulation of the 2,4-D-induced mRNAs by treating the cells for 4 h with 2,4-D concentrations ranging from 2.2×10^{-9} to 2.2×10^{-5} M. An interesting clone is pCNT103. Its dose-response curve apparently exhibits simple saturation kinetics, and saturation is reached within two orders of magnitude (Fig. 2). If we express cell division in the same figure, then the

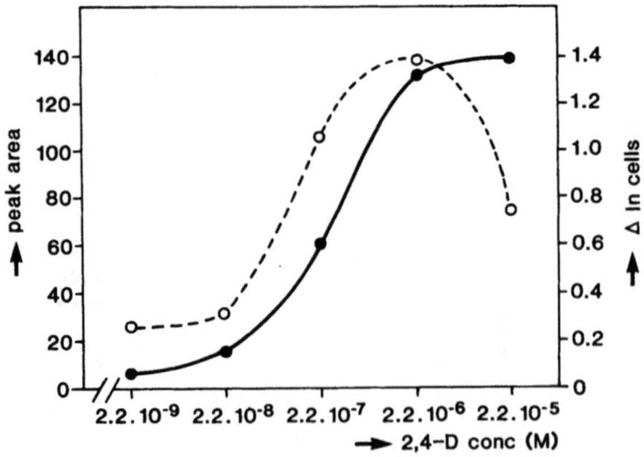

Fig. 2. Correlation between pCNT103 mRNA accumulation (—) and cell division (. . .) after treatment of hormone-starved cells with various 2,4-D concentrations. Cell density was determined at the start of the experiment (t = 0) and 72 h after hormone addition. The increase in cell density is given as ln cells/ml (t = 72) – ln cells/ml (t = 0). The level of pCNT103 mRNA was determined 4 h after hormone treatment by Northern blot hybridization and densitometric scanning of the autoradiographs [Adapted from 17]

dose-response curves for pCNT103 induction and cell division correlate well (Fig. 2), except at the very high (and probably unphysiological) concentration of 2.2×10^{-5} M 2,4-D.

In fact the specificity of the induction of pCNT103 mRNA and cell division in cells treated with various hormones at 2.2×10^{-6} M for 4 h also shows a good correlation (Fig. 3).

It thus seems that pCNT103 mRNA may be involved in the transduction of the auxin-signal to initiate cell division. The induction of mRNA accumulation in 4 of the 7 cDNA clones (pCNT103, pCNT107, pCNT114 and pCNT115) is not influenced by cycloheximide and is thus presumably independent of protein synthesis. Hence genes coding for these mRNAs are good candidates for target genes that are directly regulated by auxin. These four cycloheximide-independent cDNA clones have been sequenced. They show, except for pCNT115, some homology. Using the coding sequence of pCNT103, pCNT107 and pCNT114 respectively, genomic clones have been obtained and the promoter regions will be analyzed for auxin-responsive elements.

2.3 In Vitro Reinitiation of Auxin-Regulated Gene Transcription

Since the auxin-signal is apparently transduced to the nucleus to regulate the transcription of specific genes, it is therefore of interest to study in vitro transcription in isolated nuclei. Nuclei were isolated according to [7] 15 min after addition of 2,4-D or water to hormone-starved stationary-phase cells. These nuclei were used in an in vitro transcription system in the presence of [^{32}P]-UTP. After extraction, the labelled RNA was hybridized to membranes containing the plasmid DNA of the seven cDNA clones mentioned before. In all experiments nuclei isolated from 2,4-D-treated cells transcribed the mRNAs complementary to the seven cDNA clones; only nuclei from water treated-cells gave no hybridization signal [17]. Therefore, these nuclei provide an excellent system with which transacting factors that may play a role in auxin-induced gene expression can be analyzed (i.e. the soluble auxin receptor [10, 15]).

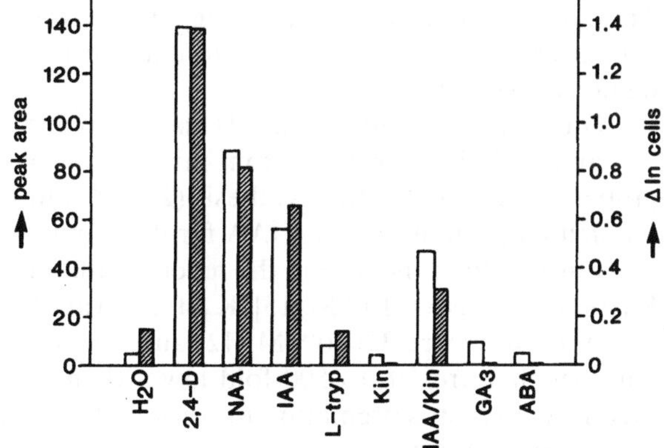

Fig. 3. Correlation between pCNT103 mRNA accumulation (*open bars*) and cell division (*hatched bars*) after treatment of hormone-starved cells with different (non-) hormonal compounds as indicated. The increase in cell density and mRNA level is represented as in Fig. 2 [Adapted from 17]

However, it is necessary to show that the nuclei are capable of correctly reinitiating RNA transcription after isolation in order to identify regulation of gene expression.

To detect initiation we employed [guanosine-5′-O-(3-thiotriphosphate) (GTP-y-S) and adenosine-5′-O-(3-thiotriphosphate)(ATP-y-S) instead of GTP and ATP to specifically label any newly initiated RNA in vitro. [y-S]-labelled RNA was subsequently isolated using a Hg-agarose affinity column. By using [α-^{32}P]UTP in the transcription assay, the RNA eluted from the Hg-agarose column was both newly initiated and [^{32}P]-labelled. When this RNA was hybridized to the blotted cDNA clones all seven mRNAs were reinitiated.

3 Discussion

We have shown that auxin, in an appropriate experimental system, induces specific genes that may be linked to the initiation and long-term effects of cell division. Most especially the auxin-specificity (Fig. 3) and the dose-response relationship (Fig. 2) argue in favor of the involvement of pCNT103 in cell division. Other groups have also shown rapid effects of auxin on transcription and translation (e.g. [20] – soybean hypocotyls; [14] – pea epicotyls), but these experiments were carried out with tissues that respond to auxin treatment with cell elongation rather than cell division. For a better understanding of auxin perception and transduction in our experimental system we are investigating the role of auxin-binding proteins. We were able to demonstrate the presence of high-affinity auxin-binding proteins in both cytoplasmic/nuclear [10, 15] and membraneous preparations [5, 19]. It seems unlikely, however, that the membrane-bound auxin-binding protein is directly involved in cell division, since this binding protein is absent from 2,4-D-grown tobacco cell suspension or callus cultures [6]. Indeed, the cytoplasmic/nuclear auxin-binding protein is a much better candidate, since it has been shown that a partially purified receptor gives a hormone-dependent increase in mRNA-transcription (42% increase in RNA-polymerase-II activity in isolated nuclei) [15].

Similar results were obtained using two partially purified (by chromatofocusing) auxin-binding proteins from pea-epicotyls [1, 2]. Both proteins stimulated transcription in isolated pea nuclei in the presence of 10^{-8} M IAA [3]. They also isolated two auxin-binding proteins from soybean cell suspensions with similar characteristics, although to date no influence on transcription has been demonstrated [3].

On the other hand, Sakai [11] and Sakai and Hanagate [12] partially purified two auxin-binding proteins from mung bean by affinity chromatography. Both proteins stimulated RNA-synthesis (24–39%) in isolated mung bean hypocotyls, but neither protein required IAA for this stimulatory effect [13].

Further differences lie in the reported affinities for the auxin-binding proteins: K_d for tobacco is 6×10^{-9} M [15]; K_ds for pea are 10^{-8} M and 6×10^{-8} M [3] while K_ds for mung bean are 3×10^{-6} M [12] and 10^{-5} M [11]. The latter values represent affinities that are 50- to 500-fold lower than for pea and tobacco material. We therefore think it is premature to consider the auxin-binding proteins from mung bean as true receptors.

The next step in the perception-transduction mechanism of auxin should be more specific. We plan to use the reinitiating nuclei from tobacco in reconstitution experiments with partially purified auxin-binding protein. We have shown in vivo that several genes are specifically induced by auxin. We hope to be able to show the same induction in vitro by addition of auxin-receptor complex to nuclei isolated from auxin-starved cells, where these genes are normally not expressed.

This approach, i.e. attacking the problem of auxin perception and transduction from both mRNA and auxin receptor sides, will give us a better understanding of the events that take place between the entry of auxin into a cell, and cell division, some 30 h later.

References

1. Jacobsen H-J (1984) Plant Cell Physiol 25:867
2. Jacobsen H-J, Hajek K (1985) Biol Plant (Praha) 27:110
3. Jacobsen H-J, Hajek K, Mayerbacher R, Herber B (1987) In: Klämbt D (ed) Plant hormone receptors. Springer, Berlin Heidelberg New York Tokyo, p 63
4. Libbenga KR, van Telgen HJ, Mennes AM, van der Linde PCG, van der Zaal EJ (1987) In: Fox JE, Jacobs M (eds) Molecular biology of plant growth control. Liss, New York, p 229
5. Maan AC, Vreugdenhil D, Bogers RJ, Libbenga KR (1983) Planta 158:10
6. Maan AC, van der Linde PCG, Harkes PAA, Libbenga KR (1985) Planta 164:376
7. Mennes AM, Bouman H, van der Burg MPM, Libbenga KR (1978) Plant Sci Lett 13:329
8. Mennes AM, Nakamura C, van der Linde PCG, van der Zaal EJ, van Telgen H-J, Quint A, Libbenga KR (1987) In: Klämbt D (ed) Plant hormone receptors. Springer, Berlin Heidelberg New York Tokyo, p 51
9. Mennes AM, et al. (in prep.)
10. Oostrom H, Kulescha Z, van Vliet ThB, Libbenga KR (1980) Planta 149:44
11. Sakai S (1985) Plant Cell Physiol 26:185
12. Sakai S, Hanagate T (1983) Plant Cell Physiol 24:685
13. Sakai S, Seki J, Imaseki H (1986) Plant Cell Physiol 27:635
14. Theologis A, Huynh TV, Davis RW (1985) J Mol Biol 183:53
15. Van der Linde PCG, Bouman H, Mennes AM, Libbenga KR (1984) Planta 160:102
16. Van der Zaal EJ, Mennes AM, Libbenga KR (1987) Planta 172:514
17. Van der Zaal EJ, Memelink J, Mennes AM, Quint A, Libbenga KR (1987) Plant Mol Biol 10:145
18. Van der Zaal EJ, et al. (in prep.)
19. Vreugdenhil D, Burgers A, Libbenga KR (1979) Plant Sci Lett 16:115
20. Zurfluh LL, Guilfoyle TJ (1982) Plant Physiol 69:332

Membrane-Bound Auxin Receptors

M.A. Venis and R.M. Napier[1]

1 Introduction

At present we know very much more about auxin binding to membrane proteins than to soluble proteins, and most of this information is from work with one species, maize (*Zea mays* L.). It was with maize membranes that the first convincing binding of a PGṘ, the auxin transport inhibitor NPA, was detected [10], and similar membrane preparations were soon thereafter shown to exhibit saturable binding of the auxin, NAA [5].

A different type of 'binding', first characterized in zucchini [6] and more recently in lupin [16] and in maize [7], consists of pH-driven auxin accumulation by sealed membrane vesicles and mimics the characteristics of in vivo auxin transport. Sites of auxin transport and of auxin action may have some relationship [4], but are operationally distinguishable in that binding of the kind thought to represent receptor binding does not require either a pH gradient or intact vesicles.

Auxin binding has been detected in several other species, [reviewed in 21], but apart from the time- and temperature-dependent particulate binding in tobacco cultures and leaves [see 21 and 14] these have mainly been single reports, not pursued further. We have been unable to repeat the observations of Zažimalova and Kutacek [23] that auxin binding to wheat membranes also shows strong temperature-dependence (unpublished data).

Without doubt, auxin binding sites in maize membranes have been studied the most extensively in a number of laboratories. Several lines of correlative evidence suggest that the sites have a receptor function [21] as do the more direct observations of inhibition of auxin-induced growth by an affinity-purified antibody to the auxin binding protein [12] and functional reconstitution of binding protein and ATPase in a bilayer lipid membrane [18]. The binding sites can be readily solubilized from the membranes using an acetone procedure and a native M_r of 40–45 kDa is obtained on gel filtration [19].

Reliable and reproducible procedures are needed to purify auxin receptor protein for antibody production and for further biochemical characterization. Two auxin affinity methods have been reported [11, 17], but such methods have yielded indifferent results in our hands [22]. This communication describes auxin receptor purification using commercially available chromatographic materials and the initial characterization of monoclonal and polyclonal antibodies produced against the resulting preparations.

[1]Institute of Horticultural Research, East Malling, Maidstone, Kent, ME19 6BJ, UK

2 Materials and Methods

2.1 Receptor Purification from Maize Membranes

The initial steps of solubilization, DEAE chromatography and gel filtration were slight modifications of reported procedures [19]. The pooled fractions from a Sephacryl S-200 (Pharmacia) column were applied to a Mono Q HR5/5 FPLC column (Pharmacia) and eluted in a 20 ml gradient from 0–350 mM NaCl in 0.25 M sucrose-20 mM Tris-HCl pH 7.3. Fractions of 1 ml were collected and assayed for binding of NAA-1-[^{14}C] (61 mCi/mmol, Amersham) by one of three methods [20]. The most active fractions were pooled, desalted and lyophilized. This preparation (approx. 50% receptor) was used either for monoclonal antibody production or was fully purified by native PAGE in a neutral pH discontinuous system [3]. The gel was briefly electroblotted (5 min, 10 mA) to nitrocellulose and the small fraction of transferred proteins visualized by rapid staining [8]. This blot was then used to locate precisely the bulk protein bands remaining in the gel.

2.2 Antibody Production

Polyclonal antiserum was produced by immunization of a single rabbit with native PAGE gel slices containing pure auxin receptor.

For monoclonal antibodies, inbred rats were immunized with post-Mono Q receptor. Fusions and hybrid myeloma selection procedures were based on those of Galfre and Milstein [2]. Initial screening of culture supernatants was by ELISA using pure receptor. Strong positives were further screened by immunoblotting against post-DEAE receptor after SDS-PAGE.

2.3 SDS-PAGE and Immunoblotting

Electroblots from SDS-PAGE [9] in 12% gels were stained either for total protein [9], glycoprotein [1] or immunochemically using Tween blocking, primary antiserum diluted 1:1000 (rabbit) or 1:10 (culture supernatant) and peroxidase-labelled second antibody.

2.4 Immunoprecipitation

Post-DEAE fraction (105 μg total protein, containing approx. 5 μg receptor) was incubated for 15 h at 4 °C in phosphate-buffered saline at pH 7.0 with 0–40 μg of IgG purified from immune or pre-immune rabbit serum (by ammonium sulfate fractionation and anion exchange chromatography). Binding of [^{14}C]-NAA (saturable by 0.1 mM unlabelled NAA) was then measured either directly or in the supernatant obtained after removal of immunoprecipitate by centrifugation (100 000 g, 5 min), using the ammonium sulfate precipitation method [20]. The immunoprecipitates were analyzed by SDS-PAGE.

3 Results

3.1 Purification

As expected from earlier gel filtration data [19] the auxin binding peak from Sephacryl S-200 elutes close to the position of ovalbumin (45 kDa). Monitoring of the Mono Q elution profile by SDS-PAGE (Fig. 1) reveals a predominant 22 kDa polypeptide at the binding peak (fraction 19, approx. 0.2 M NaCl). Although coleoptile tissue is routinely used for receptor preparation, processing of mesocotyl tissue through the same purification procedure gives identical 22 kDa polypeptide enrichment in the auxin binding fractions (Fig. 1). In many preparations (from either tissue) a minor 21 kDa band is also seen.

When the active post-Mono Q eluate is resolved by native PAGE, auxin binding assays reveal that activity is associated only with the protein region of highest mobility. On SDS-PAGE this band migrates as a single polypeptide at 22 kDa.

3.2 Polyclonal Antibodies

Gel slices from native PAGE, showing only the single 22 kDa band on SDS-PAGE, were used to generate a polyclonal antiserum. When this antiserum is used to probe nitrocellulose blots of a post-DEAE receptor fraction after SDS-PAGE, a band at

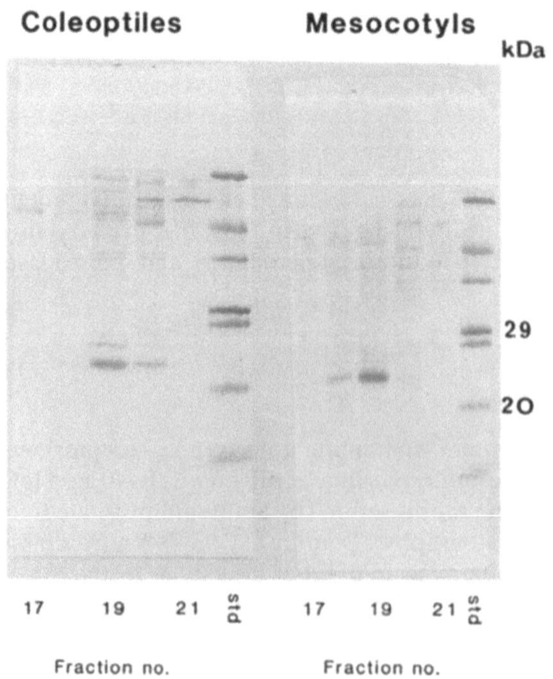

Fig. 1. SDS-PAGE analysis of fractions obtained from Mono Q separation of coleoptile and mesocotyl extracts. Peak auxin binding activity was in fraction 19 in both runs

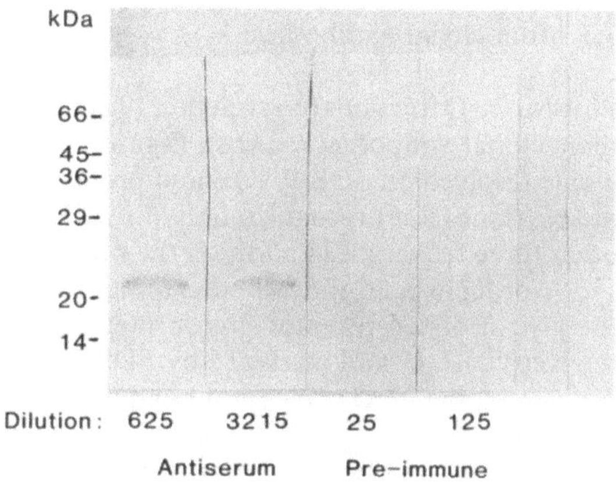

Fig. 2. Immunoblots of post-DEAE maize membrane extracts using different dilutions of rabbit pre-immune or anti-auxin receptor serum

22 kDa is selected, whereas pre-immune serum is completely unreactive (Fig. 2). Receptor in as little as 3 mg of coleoptile tissue can be detected. Immunoblots of membrane preparations from other species show the presence of immunologically similar 22 kDa polypeptides in wheat and barley as well as in the broad leaf species tomato, pea and zucchini. The same band is also found in maize supernatant preparations.

Immune, but not pre-immune, serum is able to inhibit auxin binding (Fig. 3b) and remove binding activity from solution (Fig. 3a). This removal coincides with the appearance of a 22 kDa polypeptide in the immunoprecipitate when analyzed by SDS-PAGE (not shown).

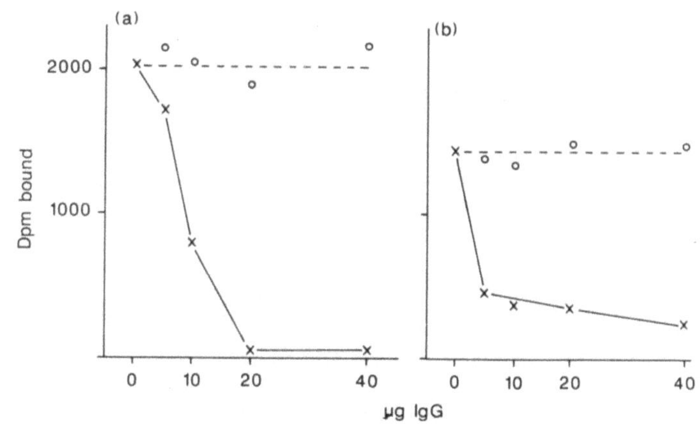

Fig. 3. Effect of rabbit pre-immune (o- - - -o) or anti-auxin receptor serum (x- - - -x) on saturable [^{14}C] NAA binding by post-DEAE maize membrane extracts either after (*a*) or without (*b*) removal of immunoprecipitate by centrifugation

3.3 Monoclonal Antibodies

Post-Mono Q fractions (e.g. fraction 19 in Fig. 1) were used to immunize rats to raise monoclonal antibodies (mAbs). Figure 4 shows an immunoblot of a post-DEAE eluate resolved on SDS-PAGE and probed with the five mAbs raised. Two of the mAbs, (lanes 2, 5) recognize only the 22 kDa receptor polypeptide, whereas the other three recognize in addition the minor 21 kDa polypeptide.

In order to determine whether the antibodies recognize the polypeptide and not the glycan of the receptor glycoprotein [13] the receptor was digested with endoglycosidase H and analyzed by SDS-PAGE and electroblotting. As digestion progresses, the 22 kDa receptor band disappears to yield a single band at 20 kDa, consistent with the removal of one glycan unit of M_r 2 kDa approx. This is confirmed in complementary tracks probed with concanavalin A which binds to the band at 22 kDa but not to the band at 20 kDa. Immunoblot analysis reveals that all the mAbs as well as the polyclonal antiserum recognise deglycosylated receptor at least as efficiently as they do the glycosylated form (data not shown).

One of the mAbs has been used to quantify the auxin receptor present in membranes from the coleoptile, enclosed leaf, mesocotyl and roots of etiolated maize seedlings by immunoblot analysis. The receptor is most abundant in coleoptile membranes, with leaf membranes containing 65%, mesocotyls 25% and roots 2.5% of the receptor concentration of coleoptiles, on a FW basis. Similar relative concentrations are obtained if the polyclonal serum is used.

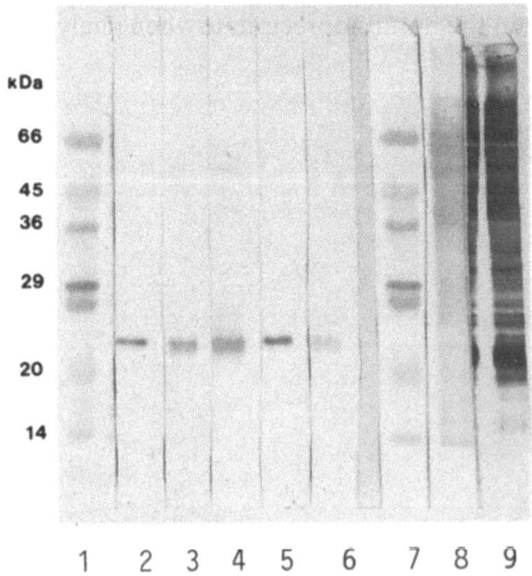

Fig. 4. Western blots of post-DEAE maize membrane extracts probed with five mAbs to the auxin receptor (lanes *2–6*). Similar tracks were stained for total protein [8] or glycoprotein [9]. MW markers in lanes *1, 7*

4 Discussion

The purification procedure described provides a reproducible means of obtaining highly enriched auxin receptor preparations from maize membranes, receptor purity normally ranging from 40–80% after the Mono Q step. Fractions of about 50% purity were used to immunize rats in order to produce mAbs. This high degree of enrichment was dictated by the failure to generate a suitable antigenic response in earlier attempts using preparations of lower (approx. 10%) purity, when non-receptor antigens proved to be strongly immunodominant. Native PAGE provided pure receptor showing only a single 22 kDa band on SDS-PAGE and an excellent rabbit polyclonal antiserum was produced.

Löbler and Klämbt [11] reported a single 20 kDa polypeptide by SDS-PAGE of their receptor preparations. Our findings are more in accord with those of Shimomura et al. [17], who found major 21 kDa and minor 20 kDa subunits. In our hands, migration of major and minor bands corresponds to M_r values of approx. 22 and 21 kDa respectively. Since the peak of auxin binding activity corresponds to a native molecular mass close to 45 kDa, we conclude that the major receptor species is a dimer of 22 kDa subunits. It is likely that the 21 kDa species is a breakdown product of the 22 kDa receptor subunit. We assume that the two mAbs which recognize only the 22 kDa polypeptide can be mapped to a terminal epitope within (or substantially within) the 1 kDa peptide that is lost from the 22 kDa species.

Immunoblot analysis of membrane digests after SDS-PAGE shows that the polyclonal antiserum to the maize receptor recognizes a 22 kDa polypeptide in several other species, including dicotyledenous ones. This suggests that even in species where auxin binding has not been convincingly demonstrated, homologous auxin receptors are present, albeit in lower abundance. Murphy [15] showed that maize supernatant fraction also contains auxin-binding activity and our immunoblot analysis suggests that this is probably attributable to the same protein that is present in the membranes. The fact that the membrane receptor can be solubilized through the acetone method and also that it can be partially removed by high salt treatment (unpublished data) suggests that it is a peripheral, rather than an intrinsic, membrane protein. Therefore, at present it is not clear whether the "soluble" auxin binding protein represents cytoplasmic/nuclear receptor, or whether it simply derives from the membranes by shear during pestle and mortar homogenisation.

Dr. C. Lazarus (Bristol) has used our polyclonal antiserum successfully to screen a maize cDNA expression library in λ gt11. Hybrid-release translation followed by immunoprecipitation and SDS-PAGE has shown at least one cDNA that specifically selects a mRNA encoding an approx. 22 kDa translation product (non-glycosylated plus presumed signal peptide). Recently, a full length cDNA encoding the maize auxin receptor gene has been isolated by oligonucleotide screening from a λ gt10 library in Prof. J. Schell's Köln laboratory, after earlier attempts to screen a λ gt11 library were unsuccessful (reported by Dr. K. Palme at an EEC meeting in Bonn, April 11–15, 1988). The sequence encodes 201 amino acid residues, including a long 38 amino acid leader sequence. The latter is the only hydrophobic region, consistent with the view that the receptor is an extrinsic membrane protein.

Thus, molecular tools (antibodies and cDNA probes) to investigate auxin action are now available. We are now in the process of utilizing the polyclonal and monoclonal antibodies for immunopurification and also to probe receptor structure and function in maize and in other species. For example, preliminary results from Prof. J. Guern's laboratory (by H. Barbier-Brygoo, Gif) show that the antiserum blocks auxin-induced hyperpolarization of tobacco protoplasts. We are also seeking to use the antibodies to examine receptor features in auxin response mutants of several species and to explore signal transduction pathways.

Note added in proof: Data on antibody blockade of the protoplast auxin response have now been published [24]. Furthermore, Shimomura reports on receptor gene cloning [25].

Acknowledgement. This work was partly supported under the Biotechnology Action Programme of the European Economic Communities.

References

1. Faye L, Chrispeels MJ (1985) Anal Biochem 149:218
2. Galfrè G, Milstein C (1981) Methods Enzymol 73:3
3. Hames BD (1981) In: Hames BD, Rickwood D (eds) Gel electrophoresis of proteins: a practical approach. IRL Press, London, p 1
4. Hertel R (1983) Z Pflanzenphysiol 112:53
5. Hertel R, Thomson K-St, Russo VEA (1972) Planta 107:325
6. Hertel R, Lomax TL, Briggs WR (1983) Planta 157:193
7. Heyn A, Hoffmann S, Hertel R (1987) Planta 172:285
8. Kumar BV, Lakshmi MV, Atkinson JP (1985) Biochem Biophys Res Commun 131:883
9. Laemmli UK (1970) Nature (Lond) 227:680
10. Lembi CA, Morré DJ, Thomson K-St, Hertel R (1971) Planta 99:37
11. Löbler M, Klämbt D (1985) J Biol Chem 260:9848
12. Löbler M, Klämbt D (1985) J Biol Chem 260:9854
13. Löbler M, Simon K, Hesse T, Klämbt D (1987) In: Fox JE, Jacobs M (eds) Molecular biology of plant growth control. Liss, New York, p 279
14. Maan AC, Van der Linde PCG, Hawkes PAA, Libbenga KR (1985) Planta 164:376
15. Murphy GJP (1980) Plant Sci Lett 19:157
16. Sabater M, Sabater F (1986) Planta 167:76
17. Shimomura S, Sotobayashi T, Futai M, Fukui T (1986) J Biochem (Tokyo) 99:1513
18. Thompson M, Krull UL, Venis MA (1983) Biochem Biophys Res Commun 110:300
19. Venis MA (1977) Nature (Lond) 66:268
20. Venis MA (1984) Planta 162:502
21. Venis MA (1985) Hormone binding sites in plants. Longman, London
22. Venis MA (1987) In: Klämbt D (ed) Plant hormone receptors (Proc. NATO Advanced Workshop, Bonn, FRG 1986) Springer, Berlin Heidelberg New York Tokyo, p 27
23. Zazimalova E, Kutacek M (1985) Plant Growth Regul 3:15
24. Barbier-Bryoo H, Ephritikhine G, Klämbt D, Ghislain M, Guern J (1989) Proc Nat Acad Sci 86:891
25. Inohara N, Shimomura S, Fukui T, Futai M (1989) Proc Nat Acad Sci USA 86:3564

Wound-Induced ACC Synthase, an Immunochemical Comparison of the Wound-Induced and Auxin-Induced Enzymes

H. Imaseki, N. Nakagawa, and N. Nakajima[1]

1 Introduction

Ethylene biosynthesis in higher plants is developmentally regulated and its rate changes transiently in various tissues at certain stages of plant development. Hook and plumule opening of dicot seedlings, leaf or fruit abscission, leaf or flower petal senescence and fruit ripening are associated with changes in the ethylene production rate [1]. At least in vegetative tissues, elevated cellular concentration of auxin increases the ethylene biosynthetic rate, and the other PGRs such as cytokinins and ABA regulate auxin action [9]. The biosynthetic rate of ethylene is also greatly changed by various environmental stimuli that are irregularly imposed upon plants, and tissue wounding, for example, causes a remarkable increase in ethylene production in many tissues [9, 13]. In both cases, ethylene is formed by enzymatic cleavage of ACC which is produced from S-adenosyl-methionine by ACC synthase. The rate of ethylene biosynthesis is well correlated with the endogenous activity of ACC synthase [5, 8, 14], and this enzyme plays a role as the rate-limiting enzyme in ethylene biosynthesis [13]. Thus, cellular activity of ACC synthase can be increased by two very different stimuli; auxin, a chemical stimulus and tissue wounding, a physical stimulus. Studies with inhibitors of RNA and protein synthesis [15] and with density labeling [2] have suggested that the increased enzyme activity was due to de novo synthesis of the enzyme protein. More recently, Imaseki et al. [7] showed, using antibody to a highly purified ACC synthase from wounded mesocarp of winter squash (*Cucurbita maxima* Duch. cv. Ebisu), that translatable mRNA for the enzyme increased in proportion to enzyme activity after wounding. Bleecker et al. [4] also showed, using monoclonal antibody, that the enzyme protein increased after wounding of tomato (*Lycopersicon esculentum* Mill.) pericarp. However, rigorous proof that if ACC synthase is synthesized through activation of the gene is presently lacking.

Another question is whether the wound- and auxin-induced enzymes in one plant species are the product of a single gene, or of two different genes. The enzymological properties of the wound- and auxin-induced enzymes are not very different and relative molecular size of the wound-induced enzymes from winter squash mesocarp [12] and tomato [3, 4] is reported to be 50–55 kDa by SDS-PAGE. Similarly, that of the auxin-induced enzyme in mung bean hypocotyls is estimated as about 55 kDa by gel filtration (unpublished observation). These results imply

[1]Research Institute for Biochemical Regulation, School of Agriculture, Nagoya University, Chikusa, Nagoya 464, Japan

that the wound- and auxin-induced enzymes are the same protein which could thus originate from a single gene. However, one cannot rule out the possibility of two separate genes for the enzymes induced by the two stimuli.

We have approached these problems immunochemically using a specific antibody against wound-induced ACC synthase purified from fruits of winter squash (*Cucurbita maxima* Duch. cv. Ebisu).

2 Purification of Wound-Induced ACC Synthase

Slices of the mesocarp of winter squash fruits increased their ACC synthase activity during incubation at 29 °C to a peak at 8 to 10 h, at which time the activity reached $2.5\text{–}3.5 \times 10^{-4}$ units/g fresh weight ($1\text{–}1.5 \times 10^{-4}$ units/mg protein; 1 enzyme unit is 1 μmol ACC formed per min). As freshly sliced tissue contains no detectable level of the enzyme activity, a simple slicing (wounding) thus caused the increase in ACC synthase. The enzyme was purified about 1300-fold to 0.22 units/mg protein with conventional methods as well as HPLC with various column media [11]. The preparation was still not completely pure. Thus, we raised antibody against the highly, but partially, purified preparation [7] and prepared an immunoaffinity column with the antibody to concentrate the enzyme from crude preparations at an about 300-fold purification. Protein eluted from the immunoaffinity column with 50 mM $K_2HPO_4//KH_2PO_4$ buffer, pH 11, showed significant enzyme activity (Fig. 1). The enzyme was further purified by HPLC with a MonoQ column followed by PAGE, giving a single polypeptide band on SDS-PAGE by silver staining [12]. The final preparation having 2.4 units/mg protein was used as an antigen to raise a specific antibody. The molecular size of the purified enzyme was estimated to be 45–50 kDa by SDS-PAGE.

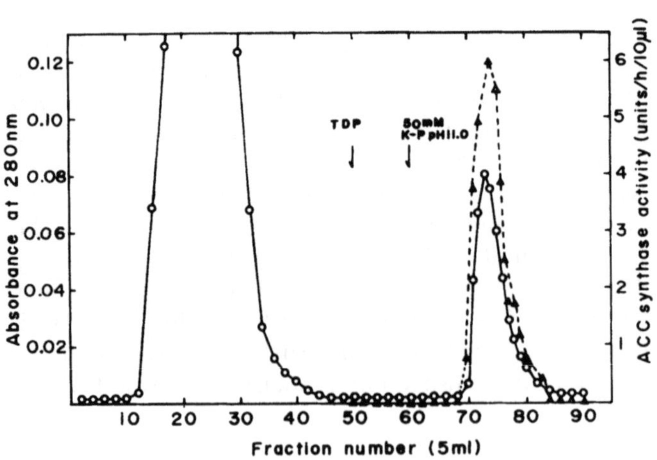

Fig. 1. Separation of ACC synthase by immunoaffinity column chromatography. *Solid line*, absorbance at 280 nm; *dashed line*, ACC synthase activity [From 12]

3 Changes in mRNA Coded for ACC Synthase After Wounding

The antiserum raised against the purified enzyme effectively bound ACC synthase of wounded mesocarp, whereas the non-immune serum did not (Fig. 2). The in vitro translation products in a wheat germ extract directed by poly(A) $^+$RNA isolated from wounded mesocarp tissue, but not from fresh mesocarp tissue, contained one major 55 kDa polypeptide which was bound by the antiserum (Fig. 3, inset). The translatable mRNA for ACC synthase increased after wounding in proportion to the increase of endogenous enzyme activity (Fig. 3).

Previously we showed that the increase in auxin-induced ACC synthase activity could be partially suppressed by ethylene. If ethylene synthesis in auxin-treated tissue is inhibited by AVG, the endogenous activity of ACC synthase greatly increases [15]. A similar effect of ethylene was also found in wound-induced ACC

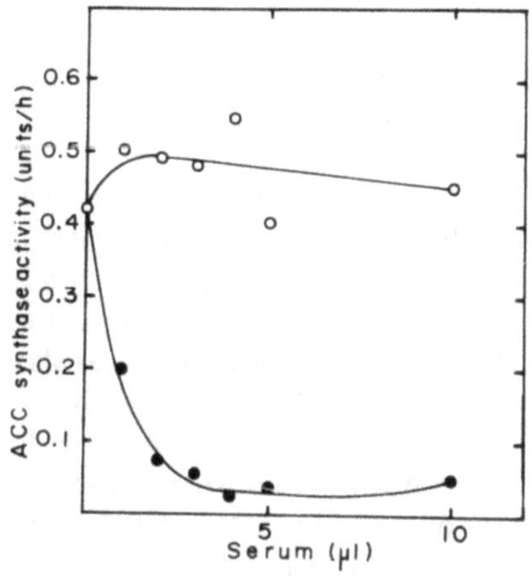

Fig. 2. Immunotitration of a partially purified ACC synthase with an antibody against purified ACC synthase from wounded mesocarp of *Cucurbita maxima* Duch. After mixing an enzyme fraction with various amounts of the antibody, the antigen-antibody complex was precipitated with protein A-Sepharose. The enzyme activity remaining in the supernatant was then assayed [From 12]

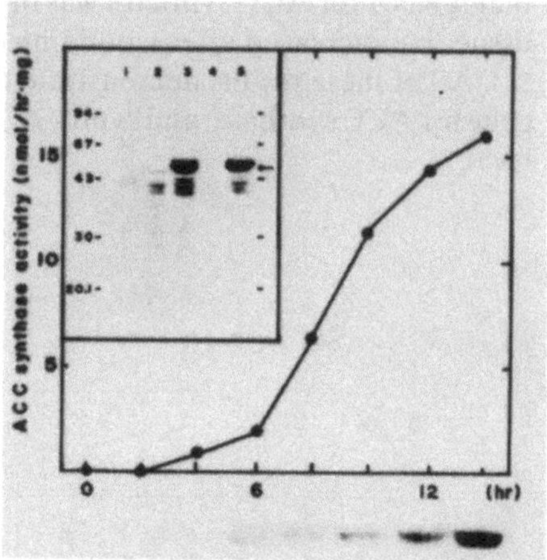

Fig. 3. Changes in translatable mRNA coded for ACC synthase and enzyme activity after wounding of the mesocarp of fruits of *Cucurbita maxima* Duch. Inset, fluorogram of in vitro translation product bound to an antibody against purified ACC synthase from wounded mesocarp. Poly(A) $^+$RNA from fresh tissue (lane *1*), 8 h (lane *2*) and 12 h (lanes *3* and *5*) after wounding was used as a template. Lane *4*, RNA from 12 h after wounding was used, but it precipitated with non-immune serum [Adapted from 7]

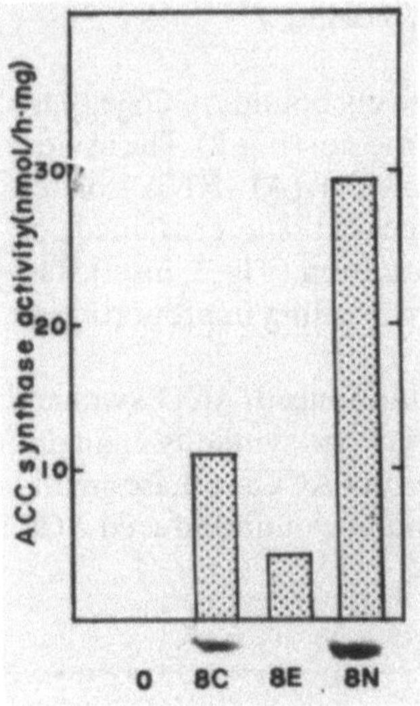

Fig. 4. Effect of ethylene on changes in ACC synthase activity and translatable mRNA for the enzyme from the wounded mesocarp of winter squash. Tissue slices were incubated for 4 h in the presence of ethylene or 2,5-norbornadiene. *C* control; *E* treated with 10 μl ethylene/1; *N* treated with 2,5-norbornadiene [Adapted from 7]

synthase (Fig. 4). When slices of winter squash mesocarp were incubated in air with ethylene at 10 ppm, the increase in ACC synthase activity was reduced. However, incubation of the slices in air containing 2,5-norbornadiene, a specific inhibitor of ethylene action, resulted in a remarkable increase in the enzyme activity (Fig. 4). Under these conditions, translatable mRNA for ACC synthase changed in proportion to the enzyme activity.

We have obtained a cDNA clone of the wound-induced enzyme by screening a cDNA library constructed in an expression vector, pTTQ18, with the antibody. Size of the insert DNA in the clone (pCMW33) was 1.8 kbp, and the transformant bacteria harboring this plasmid produced under the inductive condition a protein of about 58 kDa exhibiting ACC synthase activity. Northern blot analysis of RNA obtained from incubated mesocarp slices with the insert DNA as a probe showed that mRNA for ACC synthase was of 1.9 kb. This mRNA was not detected in fresh tissue, but increased after wounding (unpublished results).

All of these results demonstrate that tissue wounding eventually activates the gene for ACC synthase, and synthesis of the enzyme is regulated at a transcriptional level.

4 Molecular Size of ACC Synthase

The molecular size of the ACC synthase purified from wounded winter squash mesocarp was about 50 kDa, whereas that of the in vitro translation product was apparently larger, about 55 kDa. A question thus arises if ACC synthase is processed after translation. Western blot analysis of crude extracts from wounded tissues revealed that two bands (about 55 and 50 kDa), which were stained by the antibody but not by the non-immune serum, increased after wounding. The amount of 55 kDa polypeptide was always greater than the amount of 50 kDa polypeptide (the latter was the same size as the purified enzyme). When the wounded tissue slices were labeled with [^{35}S]methionine, two labeled proteins (45–50 and 55–58 kDa) recognized by the antibody were detected by SDS-PAGE/flurorography, and on incubation of the labeled crude extract in vitro at 30 °C, label in the 55 kDa polypeptide gradually decreased, while that in the 50 kDa polypeptide increased (Fig. 5), thereby indicating that the 55 kDa polypeptide was transformed to the 50 kDa polypeptide. Whether the 55 kDa protein has enzyme activity per se is not known, and the possibility exists that the 55 kDa protein is a precursor of ACC synthase, which is post-translationally processed. However, this is not likely because the 55 kDa protein was always more abundant in wounded tissue than was the 50 kDa protein, and the change occurred relatively slowly in the extract. For the time being, we assume that the actual size of the ACC synthase subunit is about 55 kDa, and the 50 kDa polypeptide is a limited hydrolysis product, both in vivo and in vitro, although it does retain enzyme activity.

N-terminal amino acid sequencing of the purified 50 kDa enzyme did not give any amino acid, but a peptide fragment obtained from it after cyanogen bromide degradation gave a definite amino acid sequence (data not shown). Therefore, the N-terminal amino acid of ACC synthase is blocked and the limited hydrolysis of the native enzyme probably occurred at the C-terminal side to produce the 50 kDa

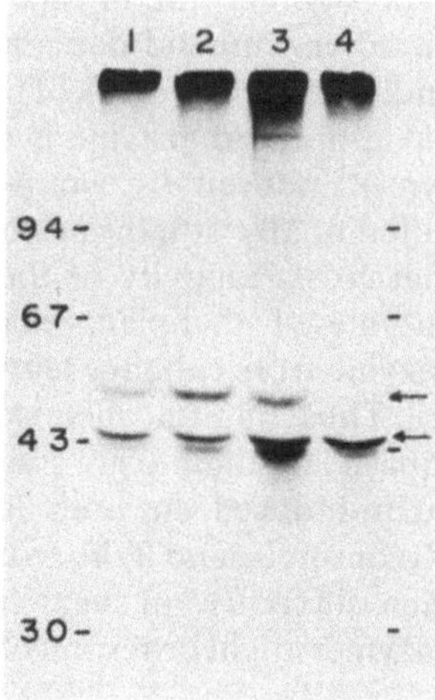

Fig. 5. Change in the molecular size of ACC synthase in vitro. A crude extract of wounded and [^{35}S[methionine-labeled mesocarp slices was incubated at 30 °C for O (lane *1*), 1 (lane *2*), 8 (lane *3*) and 14 h (lane *4*). ACC synthase was collected by the antibody and subjected to SDS-PAGE/fluorography

enzyme. This interpretation also supports the view that the 55 kDa protein is not a precursor of the enzyme, since processing of precursor proteins in many proteins is known to occur at the N-terminal side.

5 Immunochemical Comparison of Wound- and Auxin-Induced ACC Synthase

The enzymological properties of auxin- and wound-induced enzymes reported thus far in the literature are not very different. Although tissues used for auxin treatment and wounding are generally different, plant tissues respond to both auxin and wounding with increases in ACC synthase activity. Mesocarp of winter squash thus increases enzyme activity after wounding and hypocotyls of the same plant increase enzyme activity in response to auxin. The same is true for tomato. An intriguing question is whether the auxin-induced enzyme and the wound-induced enzyme in one species are the same protein or different one. Figure 6 illustrates cross-reactivities of the antibody against the wound-induced enzyme of winter squash to the wound- and auxin-induced enzymes of winter squash and tomato [101]. The antibody reacted with both of wound-induced ACC synthase from tomato fruit pericarp (panel 2) and from winter squash hypocotyls (panel 3), although the immunochemical reaction with the tomato pericarp enzyme was weaker than with the winter squash mesocarp or hypocotyl enzyme. In contrast, the auxin-induced enzymes of hypocotyls of winter squash (panel 4), tomato (panel 5) and mung bean (panel 6) did not bind with the antibody. The possible presence of materials which interfere with the antigen-antibody reaction in an auxin-induced enzyme preparation was examined by mixing the wounded mesocarp enzyme and the auxin-treated hypocotyl enzyme in different activity ratios (panel 1). In this experiment, the wounded mesocarp enzyme and the IAA-treated hypocotyl enzyme of winter squash were mixed in 1:0 (line A, a, only wound-induced enzyme), 2:1 (line B, b), 1:1 (line C, c) and 0:1 (line D, d, only auxin-induced enzyme) ratios. The mixtures were then treated with the antibody (line A-D), or with the non-immune serum (line a-d). The residual enzyme activities could be accounted for by the IAA-treated hypocotyl enzyme in each case. This indicates that the preparation of auxin-induced enzyme did not contain substances that interfered with immunoreaction, and, therefore, the lack of immunochemical reaction between the antibody and the auxin-induced enzyme is genuine. Particularly important is the finding that enzymes induced by wounding or auxin in winter squash hypocotyls reacted differentially with the antibody (panels 3 and 4). It should be mentioned, however, that cross-reactivity of the antibody was not universal among wound-induced enzymes of other plants examined. The antibody did not recognize wound-induced enzyme from cabbage leaves or grapefruit albedo.

Thus, immunochemical properties of the wound-induced enzymes of winter squash and tomato are similar, but can apparently be distinguished from those of auxin-induced enzymes in hypocotyls of these species. If ACC synthase is a glycoprotein, and if the antibody was generated only for sugar residues as epitopes, then differences in sugar residues alone between the wound- and auxin-induced enzymes might be responsible for these results. However, this possibility is ruled out

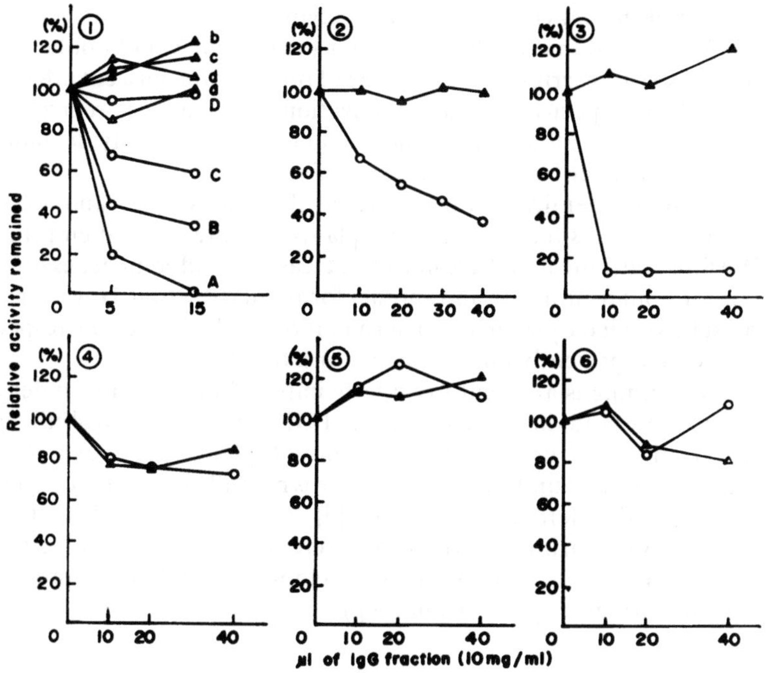

Fig. 6. Immunoprecipitation of wound- and auxin-induced ACC synthases from winter squash and tomato. Enzyme preparations were mixed with an antibody against winter squash wound-induced ACC synthase (*open symbol*) or non-immune serum (*closed symbol*), and the antigen-antibody complex was precipitated by protein A-Sepharose. ACC synthase activity remaining in the supernatant was assayed. *Panel 1* the mesocarp wound-induced enzyme (*A, a*), and the hypocotyl IAA-induced enzyme (*D, d*) of winter squash. The wound- and IAA-induced enzymes were mixed in 2:1 (*B, b*) and 1:2 (*C, c*) activity ratios. *A-D*, treated with the antibody; *a-d*, with non-immune serum. *Panel 2* The wound-induced enzyme of tomato pericarp; *panel 3* the wound-induced enzyme of winter squash hypocotyls; *panel 4* the IAA-induced enzyme of winter squash hypocotyls; *panel 5* the IAA-induced enzyme of tomato hypocotyls and *panel 6* the IAA-induced enzyme of mung bean hypocotyls [From 10]

as the antibody recognized an in vitro translation product, which should not have been glycosylated even were the enzyme to be a glycoprotein. Therefore, we conclude that the wound- and auxin-induced ACC synthase enzymes, at least in winter squash and tomato plants are different in their primary structures, and thus originate from different genes.

6 Conclusion

It has been unequivocally demonstrated that wound induction of ACC synthase is regulated at a transcriptional level. When tissue is wounded, endogenous mRNA coded for ACC synthase (as determined both by its translatability and RNA-DNA hybridization with a cDNA clone as a probe) increases concomitantly with both content of the enzyme protein and enzyme activity. As Northern blot analysis of RNA indicated that mRNA coded for the enzyme could not be detected in fresh tissue, post-transcriptional activation of pre-existant, nascent mRNA may be ruled out. Likewise, post-translational activation of an inactive proenzyme by wounding is not probable, although the endogenous enzyme is easily degraded in vivo or in vitro to a smaller but enzymatically active form.

Immunochemical studies revealed that the wound- and auxin-induced enzymes in winter squash and tomato plants are different in their primary structures. We thus conclude that there must be at least two different genes for ACC synthase, one for wound induction and another for auxin induction. However, this conclusion must be verified by sequence determination of cDNAs for the respective enzymes; this work is presently under way in our laboratory.

Wounding is an event irregularly imposed upon plants under natural conditions, whereas auxin is an endogenous PGR which can control ethylene production rate during the normal development of plants. Although the physiological significance of wound-induced ethylene is not fully understood, it has been thought to be related to defence mechanisms [6]. It is thus reasonable that plants may have evolved by preparing genes that are differently regulated for an enzyme which plays important roles in both normal development, and other processes which are activated irregularly (e.g. wounding).

Abbreviations

ABA	Abscisic acid
ACC	1-Aminocyclopropane-1-carboxylic acid
AVG	Aminoethoxyvinylglycine
cDNA	Complementary deoxyribonucleic acid
HPLC	High performance liquid chromatography
kDa	Kilodalton
mRNA	Messenger ribonucleic acid
PAGE	Polyacrylamide gel electrophoresis
PGR	Plant growth regulator
SDS	Sodium dodecylsulfate

Acknowledgements. Work presented in this article was supported in part by grants from the Ministry of Education, Science and Culture, Japan, and from Toray Science Foundation.

References

1. Abeles FB (1973) Ethylene in plant biology. Academic Press, New York
2. Acaster MA, Kende H (1983) Plant Physiol 72:139
3. Bleecker AB, Kenyon WH, Sommerville SC, Kende H (1986) Proc Natl Acad Sci USA 83:7755
4. Bleecker AB, Robinson G, Kende H (1988) Planta 173:385
5. Hyodo H, Tanaka K, Watanabe K (1983) Plant Cell Physiol 27:969
6. Imaseki H (1985) Hormonal control of wound-induced responses. In: Pharis RP, Reid DM (eds) Hormonal regulation of development III. Springer, Berlin Heidelberg New York Tokyo, p 489
7. Imaseki H, Nakajima N, Todaka I (1988) Biosynthesis of ethylene and its regulation in plants. In: Steffens GL, Rumsey TR (eds) Biomechanisms regulating growth and development. Kluer Academic, Dordrecht, p 205
8. Kende H, Boller T (1981) Planta 151:476
9. Lieberman M (1979) Annu Rev Plant Physiol 30:533
10. Nakagawa N, Nakajima N, Imaseki H (1988) Plant Cell Physiol 29:1255
11. Nakajima N, Imaseki H (1986) Plant Cell Physiol 27:969
12. Nakajima N, Nakagawa N, Imaseki H (1988) Plant Cell Physiol 29:(989)
13. Yang SF, Hoffman NE (1984) Annu Rev Plant Physiol 35:155
14. Yoshii H, Imaseki H (1981) Plant Cell Physiol 22:369
15. Yoshii H, Imaseki H (1982) Plant Cell Physiol 23:639

Polyphosphoinositide Turnover and Signal Transduction of Auxin on Isolated Membranes of *Daucus carota* L.

B. ZBELL, C. WALTER-BACK, H. HOHENADEL, and I. SCHWENDEMANN[1]

1 Introduction

Plant growth and morphogenesis are controlled by external stimuli such as light and gravity, but also by internal chemical signals like phytohormones. Despite considerable efforts in research into the signal transduction mechanisms in plants, however, our knowledge is as yet limited and fulfils a more phenomenologic description than an adequate biochemical or molecular understanding of the processes involved. This statement is also valid for the action of phytohormones, as is obviously expressed in the title of a recent critical review on this topic: *Regulation from Within: the Hormone Dilemma* [10].

In contrast to the situation concerning plants, the understanding of the signal transduction processes in animal cells is much more detailed and somewhat sophisticated. Besides the signal transduction mechanism involving the adenylate cyclase [14], another important mechanism concerns the signal dependent phosphoinositide metabolism, which is initiated by distinct types of receptors for light, various hormones and neurotransmitters [3, 5]. The extracellular signal is percepted by its specific receptor at the outside face of the plasma membrane, and it is obviously transduced across the membrane to the cytoplasm by an amplification process involving high-affinity GTP-binding proteins (G-proteins) [8]. On the cytoplasmic face of the plasma membrane the G-protein activates a (4,5)PIP_2-specific phosphoinositidase C which leads by cleavage of the phospholipid to a release of membrane-bound (1,2)DAG and free (1,4,5)IP_3 into the cytoplasm [3, 5]. These compounds act as second messengers for the stimulation of the protein kinase C and the release of Ca^{2+} from the endoplasmic reticulum or the calciosome, respectively [3, 5, 25]. While the increase of the cytosolic Ca^{2+} level functions as a transient the protein phosphorylation act as persistent signal for the control of cellular processes [2]. Some years ago, an auxin-promoted PI metabolism was suggested to be a function in the auxin action on plant membranes [27]. Since then the experimental evidences increased with time for the occurrence of the functional elements to be involved in the signal dependent PI metabolism. The polyphosphoinositides (4)PIP and (4,5)PIP_2 as lipids [26], their specific lipid kinases [23] as well as the phosphoinositidase C [15] were found to be localized at the plasma membrane of plant cells. Moreover, there are recent reports on the detection of high-affinity GTP binding proteins, whose functions are as yet not identified [6, 12, 13]. Consequently, the PI response is discussed now also for plants [21], though as

[1]Botanical Institute, Ruprecht-Karls-University, Im Neuenheimer Feld 360, 6900 Heidelberg, FRG

yet reports are rare which can prove the cooperation of the single elements in a defined process of signal transduction in plants. Putative signals for the initiation of the PI response in plants include the actions of gravity [19, 20], light [17], and phytohormones [7, 18, 28, 29]. An auxin-mediated loss of PI was explained previously by the hormone's action on the inositol exchange reaction [16], but recent findings point rather to an auxin action on the PI metabolism via a phosphoinositidase C-like reaction [7, 28, 29]. This paper summarizes some of our recent results regarding the auxin-mediated PI response and its signal transduction on microsomal membranes derived from carrot suspension cells.

2 Material and Methods

2.1 Plant Material and Membrane Preparation

Microsomal membranes were prepared as described [28, 29] from carrot cells being grown in suspension cultures. Briefly, after homogenization of the cells the extract was centrifuged at 50 000 × g for 60 min, and after resuspension the membranes were fractionated in discontinuous Renografin density gradients by centrifugation in a swinging-bucket rotor for 2 h. The membrane fractions were washed three times with buffer, and a concentrated membrane suspension was stored in liquid nitrogen until use.

2.2 Phosphoinositidase C Assay

The standard assay contained 93 kBq [γ^{32}P]ATP, 100 μM Na_2·ATP, 10 μM GTP, 10 mM $MgSO_4$, 25 mM LiCl, IAA as indicated, or only its solvent DMSO (control assay), and 125 μg membrane proteins in a final volume of 500 μl buffer at pH 7.5. In experiments to look for a G-nucleotide effect on the PI turnover, GTP was replaced by GTPγS, and the hormone was omitted. The reaction was started by the addition of the membranes, incubated at room temperature, and terminated by the addition of 1 ml ice-cold stop solution containing 2-propanol/conc. HCl (100/1; v/v). The lipids were extracted from the acidified propanolic solutions with n-hexane as organic solvent [11]. After lipid extraction the inositol phosphates of the aqueous phases were separated by anion exchange on Dowex AG 1-X2 (200–400 mesh) resin [28, 29]. [^{32}P]-label of the extracts was measured via the Cerenkov-radiation in a liquid scintillation counter.

2.3 GTP Binding Assay

High-affinity GTP binding was assayed in a final volume of 40 μl according to the procedure as described [24]. Briefly, membranes were treated for 60 min at +25°C in 25 mM TRIS-Cl buffer at pH 7.5 with 250mM sucrose, 20mM $MgCl_2$, 100 mM NaCl, 1mM EDTA, and 1mM DTT in the presence of 1.5–2.0 kBq [^{35}S]GTPγS and various concentrations of unlabelled nucleotides as indicated. The binding reaction was started by the addition of the membranes and stopped by dilution with excess amount of ice-cold buffer and subsequent rapid vacuum filtration of the assay through nitrocellulose filters (0.45μm). After solubilization of the filters the radioactivity was determined by liquid scintillation counting.

2.4 Auxin Binding Assay

The capacity for auxin binding of the membrane fractions was proved by a sedimentation assay with the use of an oil phase [9]. Since the assays were performed in small scintillation vials (Beckman BIO-Vial) it was possible to carry out in the same vessel all sequential steps of the auxin binding experiment including incubation, the sedimentation of membranes, the extraction of the membrane-bound auxin, and also the determination of its radioactivity by liquid scintillation counting. The binding assay contained in a final volume of 2 ml the membrane fraction, 3.7kBq [^{3}H]-IAA (0.55 or 1.07 TBq mmol^{-1}) and, if stated, unlabelled auxin in excess concentration with DMSO as a solvent. The binding reaction was started by the addition of the membranes to the assay, and after an incubation period of 30 min at +25°C an aliquot (0.5 ml) of a precooled solution containing dibutyl- and dinonylphthalate in a volume ratio of 9:4 and with a density of ρ = 1.028 g cm^{-3} was pipetted to the assay just before the centrifugation was performed with a fixed-angle rotor at 50 000 g at +2°C for 20 min. The resulting supernatant including the oil phase was discarded, and, after careful wiping of the inside of each vial with tissue paper, 0.5 ml ethanol was placed in each vial in order to extract overnight all the radioactivity from the membrane pellets. The ethanolic extract was mixed with a scintillant and the radioactivity was determined by liquid scintillation counting. The binding assay was performed with triplicate samples, and results are expressed as the mean and standard deviations.

3 Results

3.1 Auxin-Mediated Phosphoinositide Response

As described in previous reports [28, 29] microsomal membranes prepared from carrot suspension cells have the enzymatic activities for the phosphorylation of endogenous lipids utilizing [γ^{32}P]ATP as energy source. These reactions were saturated in the range of physiological ATP concentrations and depend on the presence of Mg^{2+} indicating Mg•ATP as the true substrate. The analysis of the in vitro -phosphorylated membrane lipids by thin-layer chromatography and subsequent autoradiography reveals the presence of strong label in PA, but lower incorporations in (4)PIP, LPIP and (4,5)PIP_2. During incubation periods up to 60s the presence of 1μM IAA led to a significant reduction of the [^{32}P]-labelled phospholipids, which was not caused by a hormone-mediated inhibition of the phosphorylating reactions, but was generated by a fast hydrolysis of phosphoinositides and a simultaneous release of [^{32}P]IP_3 in high and of [^{32}P]IP_2 in low amounts (Fig. 1). A much higher absolute [^{32}P]-label was detected for the

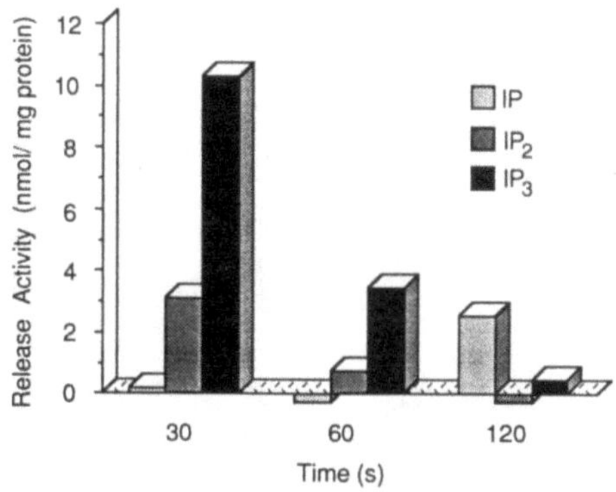

Fig. 1. Kinetics of the auxin-stimulated release of [^{32}P]-labelled compounds derived from microsomes. The membranes were incubated under the condition for the in vitro lipid phosphorylation in the absence and presence of 1 μM IAA. After lipid extraction [^{32}P]-labelled inositol phosphates were separated by anion exchange chromatography. The radioactivity of the control assays was subtracted systematically

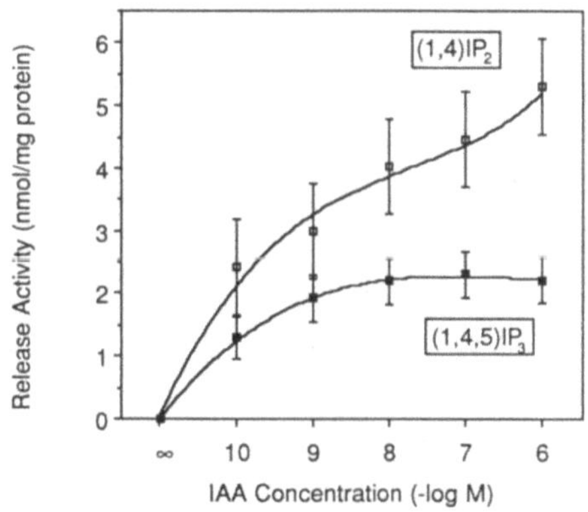

Fig. 2. Dose response effect of IAA on the release of [^{32}P]-labelled inositol phosphates during a 30-s incubation of microsomal membranes under the condition for the in vitro lipid phosphorylation in the absence and presence of the hormone. The radioactivity of the control assays was subtracted systematically

released inositol phosphates than for the hydrolyzed phospholipids indicating a replenishment of the pool of PIP/PIP$_2$ by successive phosphorylation of PI as it was described to go off on membranes from animal cells [1]. The IAA-stimulated release of [^{32}P]IP$_2$ and [^{32}P]IP$_3$ exhibits also a dose response relationship (Fig. 2).

3.2 G-Nucleotide Action and High-Affinity Binding

Since it is suggested that the signal dependent stimulation of the phosphoinositidase C in animal cells is processed by the involvement of an as yet unknown G-protein [4], the effect of G-nucleotides on the in vitro lipid phosphorylation of carrot microsomal membranes was tested. Whereas only a weak release of inositol phosphates and a drastic inhibition of lipid phosphorylation could be observed in the presence of high GTP concentrations, a prominent production of [^{32}P]IP$_2$ and [^{32}P]IP$_3$ was caused by micromolar concentration of GTP$_\gamma$S (Fig. 3). That the auxin-mediated reaction was not found to be absolutely dependent on GTP can be a result either of the presence of sufficient amounts of endogenous G-nucleotides bound to a membrane protein or of the introduction of GTP as a contamination of commercially available ATP used as energy substrate in the assays. Although the

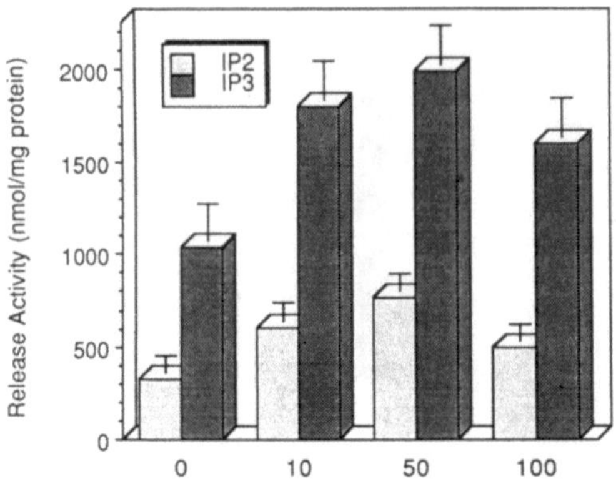

Fig. 3. Dose response effect of GTPγS on the release of [^{32}P]-labelled inositol phosphates during a 30-s incubation of microsomal membranes under the condition for the in vitro lipid phosphorylation

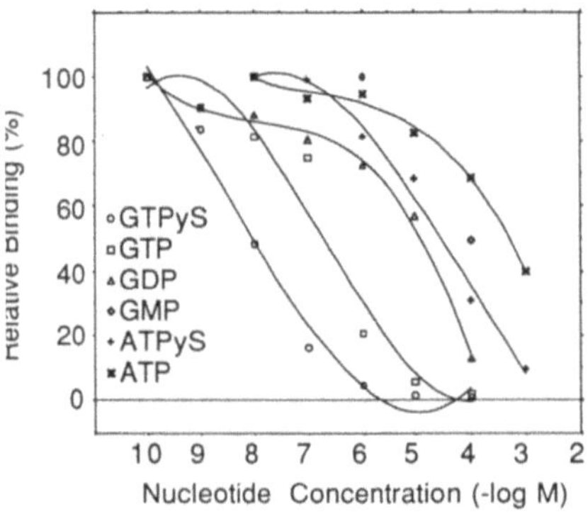

Fig. 4. Displacement of bound [^{35}S]GTPγS on carrot microsomal membranes by increasing concentrations of various nucleotides

results presented are not a direct proof of a G-nucleotide dependence of the IAA promoted PI response, another indication for the presence of a G-protein on the membranes is the detection of a G-nucleotide specific binding with a $K_d \approx 1$ nM and a capacity ≤ 5 pmol mg^{-1} protein. The order of the effective displacement of [^{35}S]GTPγS binding was determined to be GTPγS > GTP ≫ GDP > ATPγS > GMP > ATP (Fig. 4).

3.3 High-Affinity Auxin-Binding Sites

In experiments to prove the presence of high-affinity auxin binding it was necessary to carry out the hormone binding assay with [^{3}H]-IAA, a labelled compound of high specific radioactivity and with DMSO as a solvent for the addition of high concentrations of unlabelled auxin [27]. Considering these prerequisites, a very low concentration of labelled auxin (< 5 nmol l^{-1}) was displaced continuously with increasing concentrations of unlabelled hormone (Fig. 5). Analyzing the data by a Scatchard plot [22] reveals a biphasic curve indicating high and low affinity binding (Fig. 5, insert). In the case presented the dissociation constant and capacity for the

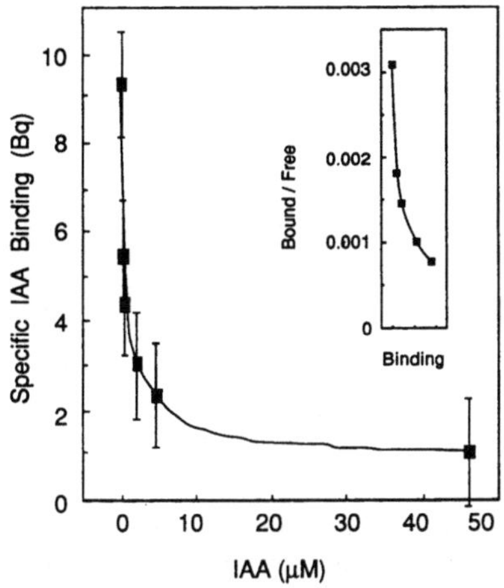

Fig. 5. Displacement of bound [5-^{3}H]IAA on carrot microsomal membranes by increasing amounts of unlabelled IAA. Results are given as saturable auxin binding, i.e. the residual unspecific binding at 100 μM IAA was subtracted systematically. The insert exhibits the Scatchard analysis of the same data, but only those of the high-affinity binding

high-affinity binding were calculated as Kd $\approx$ 280 nM and 2 pmol g^{-1} fresh weight of carrot cells, respectively. The magnitude of the dissociation constant differs from that of the hormone concentration needed for the half-maximal stimulation of the PI response. This phenomenon might be explained by the different experimental conditions which are used for auxin binding and auxin response with respect to the time scale and the phosphorylating conditions in the appropriate assays. Consequently, it is as yet not proven that the high-affinity auxin binding is functionally identical with the auxin receptor to be involved in the hormone-mediated PI response.

6 Conclusions

All the functional elements which are known to be involved in a signal dependent PI response in animal cells [3,4,5] were detected also on the microsomal membranes prepared from carrot suspension cells, i.e. the enzymes for the phosphoinositide phosphorylation, a signal dependent phosphoinositidase C, high-affinity binding sites for a hormone as well as for GTP. The detection of the functional elements can be evaluated only as a further indication for an occurrence of an auxin-mediated PI response in plants [7, 27, 28, 29]. However, it must be clarified in future work, if in plants it is a similar (Fig. 6) or distinct signal transduction mechanism in comparison to the well-known animal process [3]. For this purpose the in vitro system presented by us is certainly a promising experimental tool for the further analysis of the signal transduction processes of auxin on membranes. There are, however, a few important questions to be answered in the future. First, an auxin-mediated stimulation of the high-affinity GTP-binding as well as of the GTPase activity is a prerequisite so that the GTP binding protein can be classified as a G-protein. Such a finding together with the vice versa reaction, i.e. a G-nucleotide dependent modulation of the affinity state of the binding site for auxin, will prove the coupling between the auxin receptor and the G-protein. In

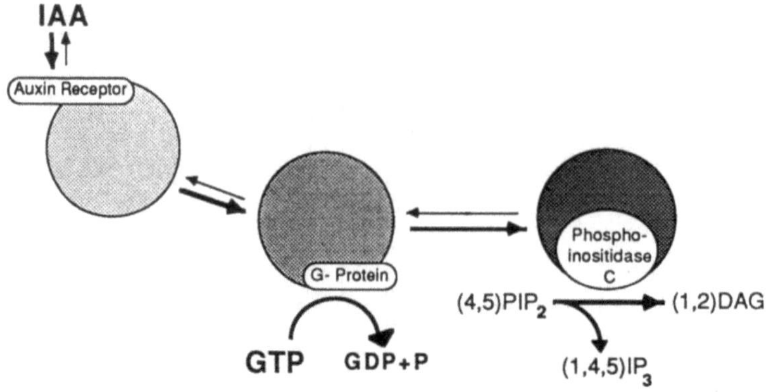

Fig. 6. Scheme summarizing recent suggestions on the signal transduction mechanism of auxin action on membranes. After binding of auxin to its specific receptor the hormone-receptor complex is activated to couple on the G-protein. This process leads to a GTP/GDP exchange at the G-protein and a subsequent activation of the phosphoinositidase C. The auxin-mediated response is terminated by the dissociation of the ligand from its receptor and by subsequent steps like GTP hydrolysis by the endogenous GTPase of the G-protein and its uncoupling from the phosphoinositidase C

order to understand whether auxin acts as an extra- or intracellular signal, the subcellular localization as well as the vectorial orientation of its receptor in the membrane must be determined.

Abbreviations

(1,2)DAG	1,2-diacylglycerol
$(1,4)IP_2$	inositol 1,4-bisphosphate
$(1,4,5)IP_3$	inositol 1,4,5-trisphosphate
PI	phosphatidylinositol
PI response	phosphoinositide response
(4)PIP	phosphatidylinositol 4-monophosphate
$(4,5)PIP_2$	phosphatidylinositol 4,5-bisphosphate

Acknowledgements. This work was supported by the Deutsche Forschungsgemeinschaft (DFG). We thank Prof. Dr. K.H. Jakobs and Dr. P. Gierschik for their stimulating discussions as well as for their unique cooperation in GTP binding research.

References

1. Berridge MJ (1983) Biochem J 212:849
2. Berridge MJ (1987) Biochim Biophys Acta 907:33
3. Berridge MJ (1987) Annu Rev Biochem 56:159
4. Cockroft S (1987) Trends Biochem Sci 12:75
5. Downes CP, Michell RH (1985) In: Cohen P, Houslay MD (eds) Molecular mechanisms of transmembrane signalling. Elsevier, Amsterdam, pp 3–56
6. Drøbak BK, Allan EF, Comerford JG, Roberts K, Dawson AP (1988) Biochem Biophys Res Commun 150:899
7. Ettlinger C, Lehle L (1988) Nature 331:176
8. Gilman AG (1987) Annu Rev Biochem 56:615
9. Gliemann J, Østerlind K, Vinten J, Gammeltoft S (1972) Biochim Biophys Acta 286:1
10. Guern J (1987) Annu Bot 60 (Suppl. 4):75
11. Hara A, Radin NS (1978) Anal Biochem 90:420
12. Hasunuma K, Funadera K (1987) Biochem Biophys Res Commun 143:908
13. Hasunuma K, Furukawa K, Tomita K, Mukai C, Nakamura T (1987) Biochem Biophys Res Commun 148:133
14. Levitzki A (1987) FEBS Lett 211:113
15. Melin P-M, Sommarin M, Sandelius AS, Jergil B (1987) FEBS Lett 223:87
16. Morré DJ, Gripshover B, Monroe A, Morré JT (1984) J Biol Chem 259:15364
17. Morse MJ, Crain RC, Satter RL (1987) Proc Natl Acad Sci USA 84:7075
18. Owen JH (1988) Physiol Plant 72:637
19. Perdue DO, LaFavre AK, Leopold AC (1988) Plant Physiol 86:1276
20. Poovaiah BW, McFadden JJ, Reddy ASN (1987) Physiol Plant 71:401
21. Poovaiah BW, Reddy ASN, McFadden JJ (1987) Physiol Plant 69:569
22. Scatchard G (1949) Annu NY Acad Sci 51:660
23. Sommarin M, Sandelius AS (1988) Biochim Biophys Acta 958:268
24. Sternweis PC, Robishaw JD (1984) J Biol Chem 259:13806
25. Volpe P, Krause K-H, Hashimoto S, Zorzato F, Pozzan T, Meldolesi J, Lew DP (1988) Proc Natl Acad Sci USA 85:1091
26. Wheeler JJ, Boss WF (1987) Plant Physiol 85:389
27. Zbell B (1983) Über die molekulare Wirkung von Auxin. Biochemische Untersuchungen an isolierten Membranen aus in vitro kultivierten Zellen von *Daucus carota* L. Thesis, The Free University, Berlin
28. Zbell B, Walter C (1987) In: Klämbt D (ed) Plant hormone receptors. Springer, Berlin Heidelberg New York Tokyo, pp 141–153
29. Zbell B, Walter-Back C (1988) J Plant Physiol 133:353

Complexity of the Barley α-Amylase Gene Families: Determination of Gene Numbers for each Family and mRNA Levels for Individual Members

J. C. ROGERS and B. KHURSHEED[1]

1 Introduction

The barley α-amylase genes have served as a model system for investigating the roles of the phytohormone GA_3 in regulating gene expression. These genes are expressed in the aleurone layer, a tissue that surrounds the starchy endosperm of the grain. During germination, in response to GA synthesized by the embryo, the aleurone cells switch protein synthesis towards production and secretion of large quantities of hydrolytic enzymes that degrade the storage components in the starchy endosperm into products used for growth by the embryo. Although the magnitude of response to GA may vary when aleurone layers from different cultivars of barley are compared, the mechanism appears to be a general one and applies to other cereals, such as wheat [1] and oats [30]. Our experiments, as well as those of many other investigators in the field, use aleurone layers from *H. vulgare* L. cv Himalaya because this cultivar has a dramatic response to the exogenous application of GA_3.

Alpha-amylase is the hydrolytic enzyme produced in largest quantities by GA-treated aleurone tissue. These quantities are remarkable; aleurone layers from 10 Himalaya barley grains (representing about 10^6 cells) synthesize and secrete on the order of 300 μg α-amylase in 24 h [29], and some malting varieties make three times more. Research into the molecular biology of this system thus involves efforts to define how GA affects expression of the α-amylase genes, as well as efforts to identify mechanisms by which the protein products are expressed at such high levels.

2 Organization and Structure of Barley α-Amylase Genes

The analysis of barley α-amylase gene regulation is complicated by the fact that there are multiple genes located in two different chromosomal loci; the genes encoding the low pl (or type A) isozymes are on chromosome 1, while the genes encoding the high pl (or type B) isozymes are on chromosome 6 [2]. The protein products from these genes can easily be separated into groups by virtue of their different isoelectric points [2]. Similar isozyme groups can be identified for α-amylases from wheat, triticale, and rye, but oats and maize have only isozymes of

[1] Division of Hermatology-Oncology, Washington University School of Medicine, 660 S. Euclid Avenue – Box 8125, St. Louis, MO 63110, USA

the low-pI type [17]. For wheat [16, 17] and oats [9], molecular data demonstrate that the genes for the different protein isozymes are correspondingly similar to those from barley.

We cloned full-length cDNAs for mRNAs representing low-pI [25] and high-pI [23] mRNAs in order to study mechanisms regulating the change in mRNA levels. The sequence of our high-pI cDNA is identical to pHV 19 described by Chandler et al. [3] except for a single nucleotide difference [13], and to cDNA 1-28 cloned by Deikman and Jones [4] except for a different site for poly(A) addition (Deikman and Jones, personal communication).

We used these cDNAs as specific probes to study expression of the two different α-amylase gene types. In order to avoid any problem of cross-hybridization between the two different types, we compared the sequences of the two cDNAs in order to construct probes that would indeed be specific. The sequences of the cDNAs corresponding to regions encoding the mature proteins are about 77% identical, while sequences for regions encoding the signal peptides are only 54% identical; the 5′ untranslated regions have little or no homology [23].

This general pattern is what would be expected for members of a multigene family. Such families arise from a gene duplication event in the evolutionary past. The parts of the duplicated genes that are important for specific functions (e.g. the mature protein coding sequences) tolerate fewer mutations and maintain a greater similarity, while other parts that have less stringent requirements for their functions (e.g. the untranslated sequences and signal peptide coding sequences) tolerate more mutations and diverge. Thus, probes derived from the 5′ untranslated-signal peptide coding regions would be most useful in detecting specific types; we used them to define responses of the two different gene types to GA at the level of mRNA abundance (see below) and to estimate the potential number of genes for each type by hybridization to digests of barley DNA. Results from these studies provided a standard to which other approaches could be compared; such comparisons validated the use of probes derived from the mature protein coding sequences where hybridization is performed at high stringency to ensure specificity for a given gene type.

Probes derived from the 5′ untranslated/signal peptide regions of the two cDNAs hybridized to multiple restriction fragments on Southern blots of barley DNA, and the patterns were different for the two cDNA types [25, 23]. Since the restriction enzymes used to digest the barley genomic DNA in these studies did not cut within the cDNAs for either type of α-amylase, we inferred that the multiple hybridizing bands represented multiple different genes or pseudogenes. An alternative explanation would suggest that the restriction sites generating multiple fragments were found within intervening sequences in the genes, so that one gene could generate several DNA fragments all of which would hybridize to one cDNA probe. The fact that other restriction enzymes also generated multiple bands [20] made the latter explanation less likely. Different observations indicated, however, that each of the two α-amylase isozyme types represented products from multigene families. For the low-pI isozymes, our cloning data demonstrated that there were two genes from this family expressed at high levels, the gene from which the mRNA corresponding to the clone E cDNA [25], and the gene corresponding to our genomic clone, Amy32b [26]. The complexity of the high-pI gene family was signaled by several apparently different small cDNA fragments cloned and sequenced by Huang et al. [10].

Direct evidence that multiple, apparently intact, genes are present in the barley genome for the high-pl isozymes has come from analysis of different high-pl α-amylase genomic clones [15, 13], and from expriments where the hybridization probes were derived from specific parts of these clones [13]. We characterized two different high-pl genomic clones [13, 14]. The first, Amy6–4, corresponds exactly to the gene from which our high-pl cDNA [23] was derived; the second, Amy46, is very similar to a clone from Sundance barley (identified as p141.117) characterized by Knox et al. [15]. The regions within Amy6–4 and Amy46 encoding the protein sequences are highly conserved, and the sequences encoding the 5′ untranslated regions of the two mRNAs differ by only 16/57 nucleotides. The latter circumstance contrasts to the 5′ untranslated regions of the two low-pl genes expressed at high levels [26]; in those genes, the 27 nucleotides immediately 5′ to the ATG initiation codons are conserved, but the remainder of the sequences (97 and 82 nucleotides respectively) have no apparent homology.

When the nucleotide sequences of the promoter regions for the two genes were compared, it was observed that they are very similar for about 240 nucleotides upstream from the sites for transcription initiation. The sequences then diverge; Amy6–4 contains about 250 nucleotides that are missing from the Amy46 sequence. Further upstream, at the end of this divergent region, the sequences again are very similar [13]. In short, the two genes are very similar for about 200 nucleotides upstream from the promoter TATA boxes; where they diverge, the primary reason appears to be that Amy46 has suffered a deletion of 250 nucleotides that are present in Amy6–4. This observation is pertinent to interpretation of hybridization results and, possibly, to explaining substantially different steady-state mRNA levels derived from the two genes.

When a probe from the coding sequence of our high-pl cDNA was used to hybridize to barley DNA digested with the restriction enzyme *Hin*d III under stringency conditions where it would only identify high-pl sequences, at least 7 different fragments were identified [13]. These all probably represent intact α-amylase genes because they also carry closely related promoter sequences; this was shown by their hybridization to a 210 base pair probe representing the region of the Amy46 gene promoter that is also conserved in Amy6–4. Six of the 7 fragments, but not the fragment known to carry the Amy46 gene, also hybridize to a promoter/upstream probe from Amy6–4 that includes the region deleted in Amy46 [13]. These results are striking because they show that the high- pl gene family has retained close sequence similarity, not only in the protein coding regions but also in a large region of flanking DNA. To emphasize this fact, the low-pl promoter/upstream regions showed no homology with the same type of hybridization assay [13]. This information is relevant when interpreting experiments designed to identify mRNAs from different members of each family (see below). It also suggests that the high-pl gene family resulted from duplications that occurred relatively recently during evolution; in contrast, the differences among the three low-pl gene family members probably result from divergence of functionally unimportant sequences during a much longer time period following earlier gene duplications.

All of the genes, both high-pl and low-pl types, would be expected to share short nucleotide sequences within their promoter/upstream regions that might specify, for example, responses to phytohormones and/or tissue specificity of expression. The hybridization experiments described would not be likely to detect such short regions of similarity. This is illustrated by the fact that the two low-pl genes, that for clone E and Amy32b, respond coordinately to GA_3 [26] even though their promoter/upstream sequences do not cross-hybridize [13]. Computer-assisted comparisons of similar regions from low-pl and high-pl genes show only a few short homologous sequences [13]; the functional importance of such similarities is unknown at present.

3 Restriction Fragment Length Polymorphism Studies (RFLP Analysis) of α-Amylase Gene Variation in Different Wild and Domesticated Barley Isolates

Holwerda [8, manuscript in preparation] used our two α-amylase cDNAs to analyze the DNA sequences flanking the high-pl and low-pl genes in different strains of wild and domesticated barleys (*H. vulgare* and *H. spontaneum*). These represent accruals to a germplasm bank from a number of different locations in the Mediterranean and Middle East, as well as a malting cultivar, Bonanza, grown extensively in Canada. His strategy was to digest genomic DNA from each of the strains with different restriction enzymes, and then analyze the variability in number and size of fragments that hybridized specifically to each of the two probes. He found that all of the different DNAs gave almost identical hybridization patterns when probed with the low-pl cDNA sequence, patterns that were indistinguishable to our results obtained with DNA from Himalaya barley [13]. In striking contrast, the patterns obtained with the high-pl probe were unique for each DNA sample; the differences involved both alterations in size and in number of hybridizing fragments [8, manuscript in preparation). These results indicate that, although the sequences of regions flanking the low-pl genes have diverged so that they will not cross-hybridize, the organization of restriction enzyme sites at some distance from the genes is highly conserved. The situation with regard to the high-pl genes seems to be exactly opposite. This would lead to the speculation that the high-pl gene family is still undergoing a rapid process of recombination and mutation, perhaps because of a peculiar physical organization of the genes. The low-pl genes appear, in contrast, to have reached a position of stability. Perhaps that is due to a physical organization that is highly unfavorable to recombination among the members. A related, but speculative, possibility is that the function of the low-pl genes is so critical to the organism that it will not tolerate major rearrangements of the different members, while the function of any one high-pl member may be less important. It should be emphasized that no information exists regarding the physical organization of the gene families, other than data localizing them to two different chromosomal loci [2].

4 RNA Complementary to α-Amylase mRNA in Barley

One of the many complexities involving the study of α-amylase genes is the presence of RNA that can hybridize across essentially the full length of α-amylase mRNA in developing endosperm and in aleurone tissue that is treated with ABA [24]. Because the probe for such antisense RNA is the equivalent of α-amylase mRNA, it was so diluted out in RNA preparation from GA-treated aleurone tissue that it was not possible to determine if antisense RNA was present under those conditions. This RNA was shown to be imperfectly complementary to mRNA representing either the low-pl or high-pl cDNAs; thus, it was not a perfect copy of either [24]. These results are difficult to interpret because we do not know how the antisense RNA is generated. If, somehow, all α-amylase mRNAs are copied into antisense RNAs, since the population of templates is diverse, then the antisense products would be diverse, and only a minor fraction would be perfectly complementary to any single mRNA. Another explanation would be that possibly RNA is transcribed from the opposite strand of an α-amylase pseudogene. It may have physiologic relevance, however, because antisense RNA that can hybridize to a specific mRNA in plant cells has been shown to interfere with expression of that particular gene [27]; it is not clear whether this is due to instability of the RNA duplex or to interference with translation of the hybridized mRNA. It is theoretically possible that RNA that can hybridize to α-amylase mRNA might be an example of "fine-tuning" on the part of the barley plant, such that mRNA transcribed under conditions where expression of the enzyme would be deleterious could be neutralized before its protein product could be produced.

5 Expression of Barley α-Amylase Genes

The increased synthesis of α-amylase protein by aleurone cells in response to GA is due, at least in part, to an increase in the steady-state abundance level of α-amylase mRNA. Whether translational control of protein synthesis is also involved is controversial [18, 7] and remains to be clarified. On a level that addresses only total α-amylase mRNA, it is clear that GA-mediated increased rate of transcription of α-amylase genes is one mechanism responsible for increasing mRNA abundance [30, 12]. These technically difficult experiments using nuclei from barley aleurone protoplasts [12] could not have distinguished between responses of low-pl versus high-pl genes, while the similar experiments using oat aleurone protoplast nuclei [30] involved only a low-pl gene family. No information is available regarding the mechanism(s) by which GA may cause changes in gene transcription, and, indeed, the fact that a time period of several hours after application of GA is required before any increase in α-amylase mRNA can be measured argues that GA probably acts through several intermediates. Once the GA-mediated increase in abundance is established, barley α-amylase mRNAs for either isozyme type appear to be very stable [21] so it is likely that mechanisms controlling this increase act early in the process. This discussion will assume that GA-mediated changes in transcription rates for the various genes are primarily responsible for the increases in mRNA abundance, although we have speculated that other mechanisms, such as changes in the efficiency of processing RNA transcripts, might also play a role [14].

Our hybridization expriments using probes specific for the two different types of α-amylase mRNA [23] demonstrated that the two gene families had different responses to GA. The low-pl mRNAs were present at a detectable level in aleurone layers that had not been treated with exogenous GA and increased about 20-fold after GA_3 treatment [26, 23]. In contrast, the high-pl mRNAs were not detected in untreated tissue, but increased over 200-fold to levels approximately the same as for low-pl mRNAs after GA treatment [23]. These findings agree with results from other workers [5, 22] who used full-length cDNA probes at high stringency. Others have suggested that the high baseline for low-pl mRNAs in untreated tissue is due to a remarkable sensitivity to very low levels of GA present [10, 22]. These observations indicate that the high-pl genes are much more rigorously controlled by GA. The two gene families also differ in the time after GA addition at which mRNA levels reach a maximum. The high-pl mRNAs peak at about 16–18 hours and decline thereafter, while the low-pl mRNAs peak at around 20–24 h, after which their levels remain stable [22, 23].

We wanted to know if all genes in one isozyme family responded similarly to GA. In the case of the low-pl genes this answer was relatively easy to approach. Primer extension experiments were performed. In these experiments [26] a DNA fragment that could hybridize to the signal peptide coding sequence of the low-pl cDNA was labeled with ^{32}P at one end, allowed to hybridize to RNA isolated from GA-treated aleurone tissue, and then used as a primer by reverse transcriptase to synthesize DNA out to the end of RNAs to which it was hybridized. The products were characterized according to size by gel electrophoresis; individual fragments were purified from the gel and sequenced by the chemical degradation method for definitive identification. The assumptions underlying the experiments were that a primer from our low-pl cDNA would be capable of hybridizing to any low-pl mRNA, that the sequences of the 5′ untranslated regions for low-pl mRNAs would be sufficiently different to allow them to be identified, and that the relative amount of a given primer extension product would accurately reflect the abundance of the mRNA from which it was derived. The two major sets of primer extension products identified [26] represented mRNAs for the clone E cDNA itself, and for a mRNA corresponding to our low-pl genomic clone, Amy32b. These were present in approximately equal amounts relative to each other in both untreated and GA_3-treated tissue, indicating that they responded in a coordinate manner to that hormone. No other extension products were present in amounts sufficient to allow characterization [26]. In retrospect, mRNA corresponding to the low-pl genomic clone identified as p155.3 characterized by Knox et al. [15] should also have hybridized to the same primer; the fact that no products from this gene with a different 5′ untranslated region sequence were identified suggests that its mRNA must be expressed at substantially lower levels than the other two genes.

The results from similar primer extension experiments using a labeled primer specific for the high-pl cDNA signal peptide coding sequence were more difficult to explain. In these experiments, all primer extension products identified had the same sequence for a distance 50 nucleotides in the 5′ direction from the ATG translation initiation codon (it was not possible to obtain sequence data further upstream because of the limited amounts of material available) [23]. These results indicated that perhaps only one gene was expressed at high levels, or that the signal peptide coding sequences of the different genes were so divergent that the primer could hybridize to only its own mRNA, or perhaps that the different genes were so closely related that their 5′ untranslated sequences could not be distinguished. Our data comparing the sequences of Amy6–4 and Amy46, and the genomic hybridization experiments described above [13], both suggest that the latter explanation is the correct one.

6 Quantitative Comparisons of mRNA Levels for Individual Genes

The availability of different genomic clones for both high and low-pI α-amylase genes made it possible to design experiments that would compare mRNA levels for the individual genes in a quantitative manner using S1 nuclease protection assays. In these experiments, the DNA strand complementary to mRNA from each of the genes was labeled at one end, corresponding to a restriction enzyme site within the first part of the mature protein coding sequence. Each fragment was prepared so that the unlabeled end represented DNA extending upstream beyond the RNA transcription initiation site. When such a fragment was hybridized to mRNA and then digested with S1 nuclease, an enzyme that selectively degrades single stranded DNA, only the portion of the labeled DNA that was essentially perfectly hybridized to its own mRNA would remain intact. Fragments that hybridized to related mRNAs would be digested at internal positions where the nucleotides did not match perfectly. The size of the remaining large fragments would accurately measure the distance from the labeled end to the 5′ end of the specific mRNA. In addition, if probes for each of the genes were prepared in an identical fashion so that their specific activities could be determined to be the same, they could be used in parallel hybridizations for quantitation: the amount of radioactive DNA for a given probe protected from S1 nuclease would be proportional to the amount of mRNA available for hybridization. We knew that the sequences of Amy6–4 and Amy46 were sufficiently different in their 5′ untranslated regions to allow them to be distinguished in this assay. The unresolved problem is that other mRNAs from genes closely related to Amy6–4 probably could not be distinguished. Thus the results, if anything, may over-estimate the amount of Amy6–4 mRNA. We found [13] that each of the low-pI mRNAs was present at levels approximately five times higher than those for mRNA from Amy6–4; mRNA from Amy46 was present at levels at least 10-fold lower than Amy6–4. The experiments were done with RNA preparations from both 19 and 24 h after GA_3-treatment to ensure that the two types of mRNA would be measured at approximately their maximal levels.

These results indicate that probably mRNA from all of the (at least) seven high-pI genes together would be required to reach a level comparable to that from the two low-pI genes. The large differences in steady-state mRNA levels between the high-pI genes Amy6–4 and Amy46 are remarkable because Amy46 is the most divergent of members in that family. It is tempting to speculate that the apparent deletion of a large section of DNA in the upstream/promoter region of that gene may be responsible; this is reasonable if transcription rates are primarily responsible for the differences in mRNA abundance observed. So far, that supposition remains to be proven. It is also worth noting that both genes are tightly regulated by GA. If DNA sequences in the upstream/promoter region are responsible for GA effects, they should be sought in the sequences that are conserved between the two. As expression systems become available, where marker genes carrying different parts of α-amylase gene promoters can be introduced into hormonally responsive aleurone cells, it should be possible to identify functionally important sequences from those genes.

References

1. Baulcombe DC, Buffard D (1983) Planta 157:493
2. Brown AHD, Jacobsen JV (1982) Genet Res Camb 40:315
3. Chandler PM, Zwar JA, Jacobsen JV, Higgins TJV, Inglis AS (1984) Plant Mol Biol 3:407
4. Deikman J, Jones RL (1985) Plant Physiol 78:192
5. Deikman J, Jones RL (1986) Plant Physiol 80:672
6. Higgins TJV, Zwar JA, Jacobsen JV (1976) Nature 260:166
7. Higgins TJV, Jacobsen JV, Zwar JA (1982) Plant Mol Biol 1:191
8. Holwerda BC (1987) Genetic variation in *Hordeum vulgare* and *Hordeum spontaneum*. Ph D Thesis. University of Saskatchewan, Saskatoon Saskatchewan, Canada
9. Hooley R, Chandler PM (1988) cDNA cloning of Avena fatua α-amylase genes. Abstract 340. 13th International Conference on Plant Growth Substances. Calgary, Alberta Canada.
10. Huang J-K, Swegle M, Dandekar AM, Muthukrishnan S (1984) J Mol Appl Genet 2:579
11. Huttly AK, Martienssen RA, Baulcombe DC (1988) Mol Gen Genet 214:232
12. Jacobsen JV, Beach LR (1985) Nature 316:275
13. Khursheed B, Rogers JC (1988) J Biol Chem 263:8953
14. Khursheed B, Rogers JC (1989) Proc Natl Acad Sci USA 86:3987
15. Knox CAP, Sonthayanon B, Ram Chandra G, Muthukrishnan S (1987) Plant Mol Biol 9:3
16. Lazarus CM, Baulcombe DC, Martienssen RA (1985) Plant Mol Biol 5:13
17. MacGregor AW, Marchylo BA, Kruger JE (1988) Cereal Chem 65:326
18. Mozer TJ (1980) Cell 20:479
19. Mundy J, Rogers JC (1986) Planta 169:51
20. Muthukrishnan S, Chandra GR, Maxwell ES (1983) J Biol Chem 258:2370
21. Nolan RC, Lin L-S, Ho T-hD (1987) Plant Mol Biol 8:13
22. Nolan RC, Ho T-hD (1988) Planta 174:551
23. Rogers JC (1985) J Biol Chem 260:3731
24. Rogers JC (1988) Plant Mol Biol 11:125
25. Rogers JC, Milliman C (1983) J Biol Chem 258:8169
26. Rogers JC, Milliman C (1984) J Biol Chem 259:12234
27. Smith CJS, Watson CF, Ray J, Bird CR, Morris PC, Schuch W, Grierson D (1988) Nature 334:724
28. Whittier RF, Dean D, Rogers JC (1987) Nucleic Acids Res 15:2515
29. Yomo H, Varner JE (1971)In: Moscona AA, Monroy A (eds) Current topics in developmental biology. Academic Press, New York, p 111
30. Zwar JA, Hooley R (1986) Plant Physiol 80:459

Regulation of Gene Expression by Abscisic Acid in Barley Aleurone Layers

T.H.D. Ho, B. Hong, R.C. Nolan, S.J. Uknes, and L.-S. Lin[1]

1 Introduction

The aleurone layers of certain small cereal grains, such as barley, wheat and oat, have been used as a convenient system to study the mode of action of two phytohormones, GA_3 and ABA. It has been well documented that GA induces the synthesis and secretion of several hydrolytic enzymes, including α-amylases, proteases, β-1,3;1,4 glucanase and nuclease in isolated barley aleurone layers [for review see 7, 12]. At least for α-amylases, the GA_3 induction of their synthesis is mainly regulated at the transcription level [11]. Essentially all the GA_3 effects in barley aleurone layers can be blocked by ABA [7]. However, kinetic studies by Chrispeels and Varner [2] indicate that these two hormones do not directly compete with each other for a common site. This observation seems to suggest that ABA may have a unique mode of action instead of being the mere reversal of GA_3 action. As part of our attempt to investigate the mode of action of ABA in barley aleurone layers, we have studied the regulation of ABA metabolism and the effect of ABA metabolites on the synthesis of GA_3-induced α-amylase. In addition, we have shown that certain effects of ABA in aleurone layers are dependent on RNA and protein synthesis. We also report here the characterization of several ABA-induced proteins and their potential physiological roles.

2 Regulation of ABA Metabolism

The metabolism of ABA in isolated barley aluerone layers follows the ABA → PA → DPA pathway that has been shown to exist in many plant tissues [3, 23]. The biological activities of isolated PA and DPA have been tested in this system. Although DPA has little or no biological activity, PA is as active as ABA, i.e. the GA_3-induced α-amylase is effectively inhibited by either PA and ABA, but not by DPA [3]. It has been reported that in animal tissues vitamin D has to be metabolized by hydroxylation in order to become biologically active [4]. Thus, the observation that PA is biologically active has raised the question of whether PA is the active component in ABA action. The metabolism from ABA to PA is catalyzed by a cytochrome P_{450} type monooxygenase with a short-lived intermediate, 6′-hydroxymethyl ABA [23]. A pretreatment of barley aleurone layers with 10^{-5} M ABA for 24 h enhances the tissue's ability to convert [^{3}H]ABA to [^{3}H]PA by 3- to 5-fold

[1]Department of Biology, Washington University, Campus Box 1137, St. Louis, Missouri 63130, USA

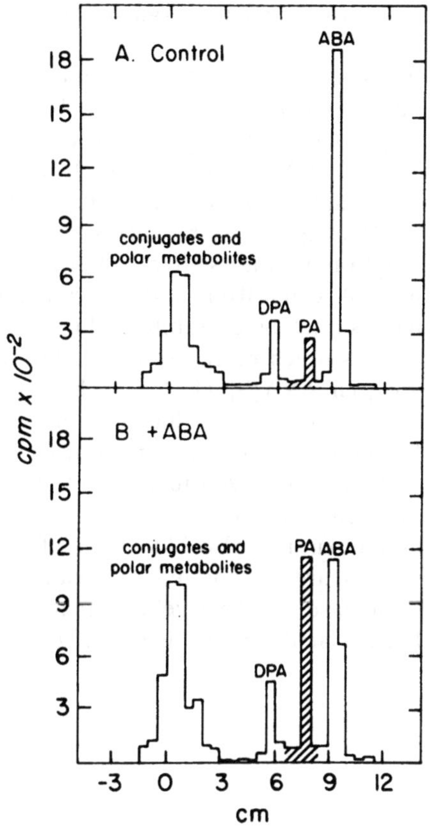

Fig. 1. Effect of pretreatment of barley aleurone layers with ABA on the metabolism of [^{3}H]ABA. Barley aleurone layers were incubated with or without 10^{-5} M ABA for 24 h. After incubation, fresh medium containing 1 μCi [^{3}H]ABA/ml was added and the tissue was incubated for an additional 4 h before ABA and its metabolites were extracted and analyzed by TLC. The cpm values were plotted against the distance travelled in cm (Quenching was constant among the TLC fractions). The negative values on the distance travelled are the preadsorbent area on the TLC plate. "0" is the junction between the preadsorbent area and the silica gel. The *shaded area* indicates the amount of [^{3}H]PA. *A* Control, no ABA pretreatment; *B* +ABA, pretreated with 10^{-5} M ABA. From Uknes and Ho [21]

[Fig. 1; 21]. An ABA concentration as low as 10^{-8} M is sufficient to enhance its own metabolism, and this effect can be observed within 2 h of ABA treatment [21]. The formation of the next stable metabolite, DPA, is not enhanced by pretreatments with either ABA or PA [21]. Thus, the enhanced PA formation is unlikely a scavenging mechanism to remove excessive ABA because the tissue would have to enhance the formation of DPA in order to eliminate the biological activities. The self-induction of ABA metabolism can be prevented by transcription and translation inhibitors, suggesting that ABA induces the monooxygenase (or a cofactor for this enzyme) responsible for PA formation. The regulation of ABA metabolism in barley aleurone layers is similar to the induction of nitrate reductase by its substrate, nitrate. In this latter case, treatment of a plant tissue with nitrate enhances its ability to metabolize nitrate.

We are currently searching for specific inhibitors that block the conversion of ABA to PA. This type of inhibitor should be useful to firmly establish the physiological role of PA in aleurone layers.

3 Transcription and Translation Dependency of ABA Action

It has been reported that the action of ABA can be inhibited by the RNA synthesis inhibitor, cordycepin [9]. This observation was recently confirmed by Nolan et al. [19] who have studied the effect of ABA on the stability of α-amylase mRNA. The level of mRNA for both the high- and low-pI groups of α-amylase isozymes reaches a maximum around 20 h of gibbrellic acid (GA_3) treatment [17]. These α-amylase mRNAs appear to be very stable because their levels are not affected by transcription inhibitors after 20 h of GA_3 treatment [8]. However, ABA added after 20 h of GA_3 can still effectively decrease the level of α-amylase mRNA, suggesting that ABA somehow destabilizes the normally stable α-amylase mRNA. The effect of ABA is much more profound with the high-pI α-amylase mRNA than with the low-pI mRNA [9]. The level of high-pI mRNA starts to decrease within 0.5 h of ABA application and less than 10% of the original mRNA is left after 4 h (Fig. 2) [19]. Intriguingly, although the transcription inhibitor, cordycepin, does not affect α-amylase mRNA after 20 h of GA_3 treatment, it could block the action of ABA on the reduction of α-amylase mRNA [Fig. 3; 19]. Similarly, the translation inhibitor, cycloheximide, also blocks the effect of ABA [18]. Taken together, these observations are interpreted to mean that the action of ABA on the reduction of α-amylase mRNA level is dependent on the synthesis of new RNA and proteins. Phaseic acid can mimic the effect of ABA on the reduction of α-amylase mRNA

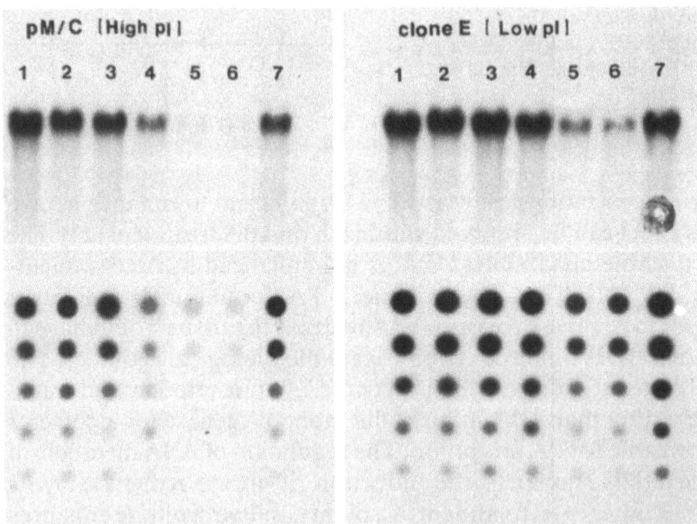

Fig. 2. The time course for decreases in α-amylase mRNA after midcourse ABA addition. Aleurone layers were incubated in the presence of 10^{-6} M GA_3 for 20 h (lane 1) or 26 h (lane 7); or 2×10^{-5} M ABA was added 20 h after GA_3 and then the layers were ground 0.5, 1, 2, 4 of 6 h later (lanes 2, 3, 4, 5, 6, respectively). For the Northern gels (*upper part* of the figure), 10 μg of total RNA from each treatment was loaded per lane. For the dot blots (*lower part* of the figure), 8, 4, 2, and μg of total RNA were used. The filters were hybridized to the entire clone E (low-pI μ-amylase cDNA) or pM/C (high-pI μ-amylase cDNA) plasmids. From Nolan et al. [19]

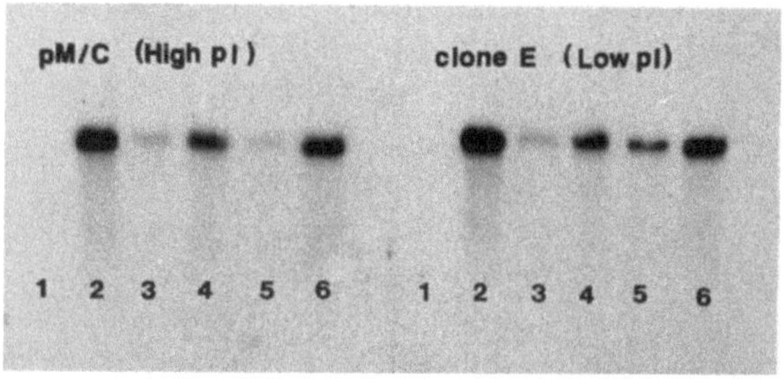

Fig. 3. Effect of the transcription inhibitor, cordycepin, on the action of ABA in barley aleurone layers. Northern blot analysis is the same as described under Fig. 2. Aleurone layers were incubated for 25 h in the absence of GA_3 (lane *1*), 25 h in the presence of 10^{-6} M GA_3 (lane *2*), 24 h in the presence of GA_3 and 2×10^{-5} M ABA with a second addition of ABA at 12 h (lane *3*), 25 h in the presence of GA_3 with ABA added 10 h after GA_3 (lane *4*), 25 h in the presence of GA_3 with ABA added 20 h after GA_3 (lane *5*), or 25 h in the presence of GA_3 with ABA and 10^{-4} M cordycepin added 20 h after GA_3 (lane *6*). Note that cordycepin addition prevented the reduction of α-amylase mRNA induced by ABA (cf lanes *5* and *6*). From Nolan et al. [19]

level, and the effect of PA can also be blocked by cordycepin or cycloheximide [18]. Therefore, the new proteins necessary for ABA action are unlikely to be involved in the conversion of ABA to PA.

4 Characterization of ABA-Induced Proteins

In light of the transcription and translation dependency of ABA action and the self-induction of ABA metabolism, we [13] and other groups [6, 14] have compared the profiles of newly synthesized proteins in barley aleurone layers treated with or without ABA. Using typical 2-D PAGE analysis, we can observe that at least 16 polypeptides are induced by ABA treatment [13]. A 35 kD protein reacts with antibodies against a barley lectin specific for aminosugars [13, 20], and a 21 kD Protein appears to be an α-amylase inhibitor [15, 22] (Fig. 4). The most abundant among the ABA-induced proteins is an acid (0.1M HCl) soluble 29 kD protein located in organelles that sedimented at 1000 and 13 000 g. However, most of the other ABA-induced proteins are present in the 80 000 g supernatant [13]. The induction of these proteins is suppressed by GA_3, and none of them are secreted. More recently, we have isolated 30 different ABA-induced cDNA clones, and one of them has been sequenced and characterized [Fig. 5]. The level of mRNA encoded by clone pHV A1 is tripled within 40 min, and reaches a maximum 8–12 h of ABA treatment [10]. A hybrid-select translation experiment indicates that pHV A1 encodes a protein that migrates as a 27 kD protein on SDS PAGE, yet the sequence analysis indicate that pHV A1 has an open reading frame for a 22 kD protein. The deduced amino acid sequence reveals several interesting features. First, it has high alanine (22%), threonine (16%) and lysine (14%) content [10]. The unusually high content of a basic amino acid, lysine, probably accounts for the slower than expected mobility on gels during electrophoresis. Second, it has at least nine 11-amino-acid imperfect repeats [10]. The consensus sequence of these repeats is

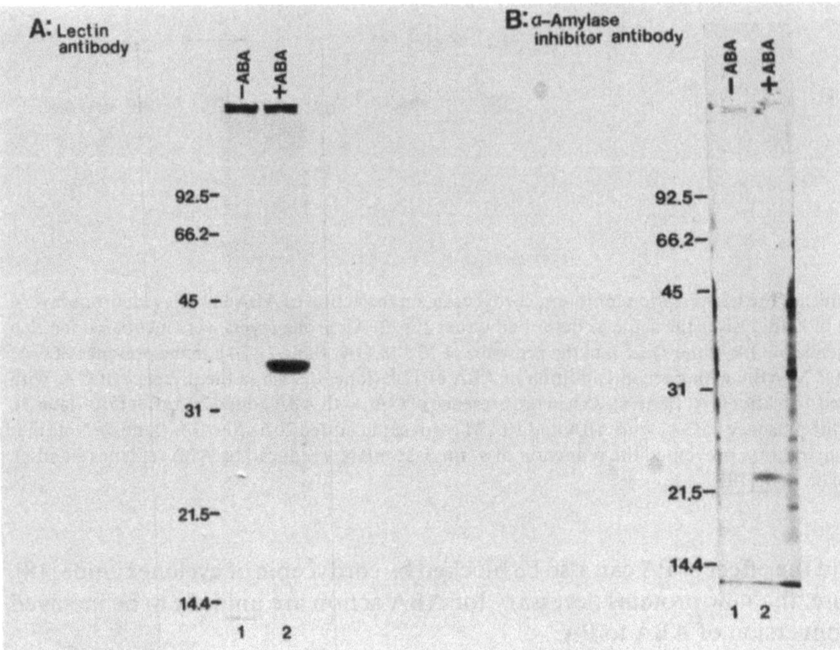

Fig. 4. Immunoprecipitation of ABA-induced proteins in barley aleurone layers with antibodies against barley lectin and α-amylase inhibitor. Aleurone layers were incubated with or without 2×10^{-5} M ABA for 24 h. Labeling with 50 μCi/ml [^{35}S] methionine took place in the final h of incubation. In *A* proteins were extracted with 0.1 M Na acetate (pH 5.5); in *B* proteins were extracted with 2% Triton X-100 at 40°C for 30 min. Aliquots of the extracted proteins were reacted with the antibodies and then precipitated with *Staphylococcus aureus* Cowan strain 1. The pellets were washed and analyzed by SDS-PAGE. Fluorograms of the gels are shown. Mol wt marker are indicated by *bars at left*. L.-S. Lin and T.H.D. Ho, unpublished

Thr-Glu-Ala-Ala-Lys-Gln-Lys-Ala-Ala-Glu-Thr. A similar sequence has been observed in one of the ABA-induced proteins in cotton embryogenesis (Lea 7) [1]. ABA-induced proteins with a repeating sequence and unusually high content of specific amino acids have been observed in several plant systems [1, 5, 16]. In barley aleurone layers, pHV A1 as well as several other ABA-induced proteins can be induced by osmotic or salt stress (LS Lin and THD Ho, unpublished observation). At least in osmotic stress, the level of ABA is increased, and the osmotic stress induction of ABA proteins is prevented by an ABA biosynthesis inhibitor, fluridone. Thus, the stress-induction of these proteins is most likely the consequence of stress-induced synthesis of ABA.

The function of these ABA-induced proteins remains unclear. Since the α-amylase inhibitor is also synthesized during seed development, it is likely to play a mop-up role in getting rid of unwanted α-amylase activity [15, 22]. Some of these ABA proteins may be related to a plant's tolerance to water stress, yet a definite proof is still lacking. In barley aleurone layers, some of these proteins could be

```
-119  GTGCCGGTAGTAAATCATGAGCATCTCTTGC
-88   GACTCGAAACGTAGTACAGCAACAGCCTAAAGCGAGTCCGAGTGGTGATTCCAGTTCGTGTTTGTTTGAGCTAGATCGTGAGACGAAG

1     ATG GCC TCC AAC CAG AAC CAG GGG AGC TAC CAC GCC GGC GAG ACC AAG GCC CGC ACC GAG
      Met Ala Ser Asn Gln Asn Gln Gly Ser Tyr His Ala Gly Glu Thr Lys Ala Arg Thr Glu
61    GAG AAG ACC GGG CAG ATG ATG GGC GCC ACC AAG CAG AAG GCG GGG CAG ACC ACC GAG GCC
      Glu Lys Thr Gly Gln Met Met Gly Ala Thr Lys Gln Lys Ala Gly Gln Thr Thr Glu Ala
121   ACC AAG CAG AAG GCC GGC GAG ACG GCC GAG GCC ACC AAG CAG AAG ACC GGC GAG ACG GCC
      Thr Lys Gln Lys Ala Gly Glu Thr Ala Glu Ala Thr Lys Gln Lys Thr Gly Glu Thr Ala
181   GAG GCC GCC AAG CAG AAG GCC GCC GAG GCC AAG GAC AAG ACG GCG CAG ACG GCG CAG GCG
      Glu Ala Ala Lys Gln Lys Ala Ala Glu Ala Lys Asp Lys Thr Ala Gln Thr Ala Gln Ala
241   GCC AAG GAC AAG ACG TAC GAG ACG GCG CAG GCG GCC AAG GAG CGC GCC GCC CAG GGC AAG
      Ala Lys Asp Lys Thr Tyr Glu Thr Ala Gln Ala Ala Lys Glu Arg Ala Ala Gln Gly Lys
301   GAC CAG ACC GGC AGC GCC CTC GGC GAG AAG ACG GAG GCG GCC AAG CAG AAG GCC GCC GAG
      Asp Gln Thr Gly Ser Ala Leu Gly Glu Lys Thr Glu Ala Ala Lys Gln Lys Ala Ala Glu
361   ACG ACG GAG GCG GCC AAG CAG AAG GCC GCC GAG GCA ACC GAG GCG GCC AAG CAG AAG GCG
      Thr Thr Glu Ala Ala Lys Gln Lys Ala Ala Glu Ala Thr Glu Ala Ala Lys Gln Lys Ala
421   TCC GAC ACG GCG CAG TAC ACC AAG GAG TCC GCG GTG GCC GGC AAG GAC AAG ACC GGC AGC
      Ser Asp Thr Ala Gln Tyr Thr Lys Glu Ser Ala Val Ala Gly Lys Asp Lys Thr Gly Ser
481   GTC CTC CAG CAG GCC GGC GAG ACG GTG GTG AAC GCC GTG GTG GGC GCC AAG GAC GCC GTG
      Val Leu Gln Gln Ala Gly Glu Thr Val Val Asn Ala Val Val Gly Ala Lys Asp Ala Val
541   GCA AAC ACG CTG GGC ATG GGA GGG GAC AAC ACC AGC GCC ACC AAG GAC GCC ACC ACC GGC
      Ala Asn Thr Leu Gly Met Gly Gly Asp Asn Thr Ser Ala Thr Lys Asp Ala Thr Thr Gly
601   GCC ACC GTC AAG GAC ACC ACC ACC ACC ACC AGG AAT CAC TAG ACGCATGCGTTCGCGCTTAATTTCCG
      Ala Thr Val Lys Asp Thr Thr Thr Thr Thr Arg Asn His End
669   TTCCTTTAGTCGTGTTTGGTCGTTCGAGGGCCTTCTACATATTTCATATTTGTATGTTTCCACTCTTTCATGATTTCCGCTCATTTAGTGTAA
762   GTTTGCCTCCGATTTGATGTACTCGTCTCTGGTTCTGTAATGAGTTATAATCCATGGGCTTTGGTGTAAATGGATAACGAGGACACTCGA
852   AGGCGGCAATAAAGTTGTATGTGATCGAAAAAAAAAAAAAA
```

Fig. 5. DNA sequence and deduced amino acid sequence of a cDNA clone, pHV Al, encoding an ABA-induced mRNA. Nucleotide + 1 is assigned to the predicted translation initiation codon, and the negative numbers refer to the 5′-noncoding region. The nine imperfect 11-amino-acid repeats are *underlined* and *numbered*. The polyadenylation signal, AATAAA, is shown by *dots*. From Hong et al. [10]

involved in ABA metabolism and in the action of ABA regulation of α-amylase induction. To further elucidate the potential physiological roles of these ABA-induced proteins, experiments are underway to localize these ABA-induced proteins in aleurone cells with specific antibodies.

References

1. Baker J, Steele C, Dure L III (1988) Plant Mol Biol 11:277
2. Chrispeels MJ, Varner JE (1966) Nature 212:1066
3. Dashek WV, Singh BN, Walton DC (1979) Plant Physiol 64:43
4. DeLuca HF, Schnoes HK (1983) Annu Rev Biochem 52:411
5. Gomez J, Sanchez-Martinez D, Stiefel V, Rigau J, Puigdomenech P, Pages M (1988) Nature 334–262
6. Higgings TJV, Jacobsen JV, Zwar JA (1982) Plant Mol Biol 1:191
7. Ho THD (1988) In: Kung SD, Arntzen CJ (eds) Plant biotechnology. Butterworth, New York, p 207
8. Ho THD, Varner JE (1974) Proc Natl Acad Sci USA 71:4783
9. Ho THD, Varner JE (1976) Plant Physiol 57:175
10. Hong B, Uknes SJ, Ho THD (1988) Plant Mol Biol 11:495
11. Jacobsen JV, Beach LR (1986) Nature 316:275
12. Jacobsen JV, Chandler PM (1986) In: Davies PJ (ed) Plant hormones and their roles in plant growth and development. Nijhoff, Dordrecht p 164
13. Lin LS, Ho THD (1986) Plant Physiol 82:289
14. Mozer TJ)1980) Cell 20:479
15. Mundy J (1984) Carlsberg Res Commun 49:439
16. Mundy J, Chua NH (1988) EMBO J 7:2279
17. Nolan RC, Ho THD (1988) Planta 174:551
18. Nolan RC, Ho THD (1988) Plant Physiol 88:588
19. Nolan RC, Lin LS, Ho THD (1987) Plant Mol Biol 8:13
20. Patridge J, Shannon L, Grumpf D (1976) Biochim Biophys Acta 451:470
21. Uknes SJ, Ho THD (1984) Plant Physiol 75:1126
22. Westlake RJ, MacGregor AW, Hill RD, Duckwirth HW (1983) Plant Physiol 73:1008
23. Zeevart JAD, Creelman RA (1988) Annu Rev Plant Physiol 39:439

Novel Affinity Probes for Gibberellin Receptors in Aleurone Protoplasts of *Avena Fatua*

R. HOOLEY[1], M.H. BEALE[2], S.J. SMITH[1], and J. MACMILLAN[2]

1 Introduction

Our understanding of the molecular mechanism of action of GA in aleurone has advanced considerably over recent years. In order to establish the precise link between GA perception and the observed increase in rates of α-amylase gene transcription [7, 19], it is important to identify and understand the action of GA receptors in aleurone.

Conventional hormone-binding studies have led to the discovery of GA-binding proteins in the cytosol of cucumber hypocotyls and pea epicotyls. These proteins have been partially purified and display some of the characteristics expected of a receptor [11, 12, 18, 17, 14]. In vivo and in vitro GA-binding studies have been performed with aleurones of barley and wheat [10, 8] but, at present, no GA-binding proteins have been characterised, nor has a candidate GA-receptor emerged.

GA-responsive isolated aleurone protoplasts [6] offer advantages over the intact tissue in receptor studies and we are exploiting these using novel GA-derivatives [3] as probes for receptors. We report here our progress in photoaffinity labelling of GA-binding proteins in aleurone protoplasts, and consider the feasibility of this technique for identifying GA-receptors. We also describe a strategy for GA-receptor identification employing functional assays and report the probable subcellular location of GA-receptors in aleurone protoplasts.

2 Photoaffinity Probes for GA-Binding Proteins and Receptors

Photoaffinity labelling is a powerful technique that has been used successfully in identifying and characterizing receptors in mammalian cells [2, 15, 4]. The technique has also been applied to the identification of binding sites for ABA on guard cell protoplasts [5]. On the basis of these, apparently, most promising results with ABA we have investigated the design and use of GA-photoaffinity probes.

[1]Molecular Biology Group, Department of Agricultural Sciences, University of Bristol, Institute of Arable Crops Research, Long Ashton Research Station, Long Ashton, Bristol, BS18 9AF, UK
[2]School of Chemistry, The University, Bristol, BS8 1TS, UK

2.1 Design and Characteristics

Preliminary experiments with two enones derived from GA_3 [3] were unsuccessful because the conditions required to photolyze the enones kill aleurone protoplasts. In order to achieve successful photoaffinity labelling there is a need to address certain criteria in the design and use of photoaffinity probes (Table 1). Photoaffinity GA-derivatives based on 17-thiols [3] have been evaluated using these criteria.

GA_4-17-S-propyl-S-phenacyl azide (structure I, Fig. 3) induces α-amylase in aleurone protoplasts (Fig. 1), although it is less active than the parent compound (GA_4). In aqueous solution it has a λ_{max} of 294 nm (Fig. 2). Solutions are stable in the dark but are photolyzed within seconds of exposure to intense light of wavelengths above 300 nm (Fig. 2). Aryl azides photolyse to highly reactive aryl nitrenes that can covalently attach to amino acids at the binding site. See [15] for a detailed account of the reactions of aryl nitrenes.

When designing photoaffinity probes to a known and characterized receptor, candidates can be screened for their ability to inhibit binding of the natural ligand

Table 1. Criteria for an effective photoaffinity probe

1.	Should be biologically active/bind to receptor
2.	Should be stable under experimental conditions
3.	Photolysis should be efficient and rapid
4.	Highly reactive short-lived intermediates should be formed on photolysis
5.	Photolysis conditions should not significantly damage protoplasts or denature proteins

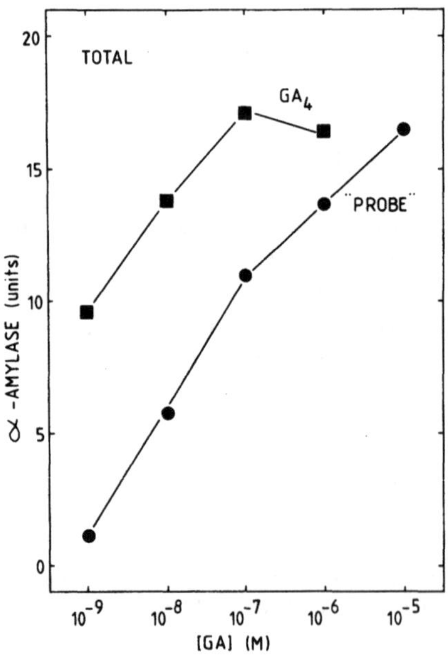

Fig. 1. Dose response curves of GA_4 and GA_4-17-S-propyl-S-phenacyl azide (structure I, Fig. 3) in aleurone protoplasts

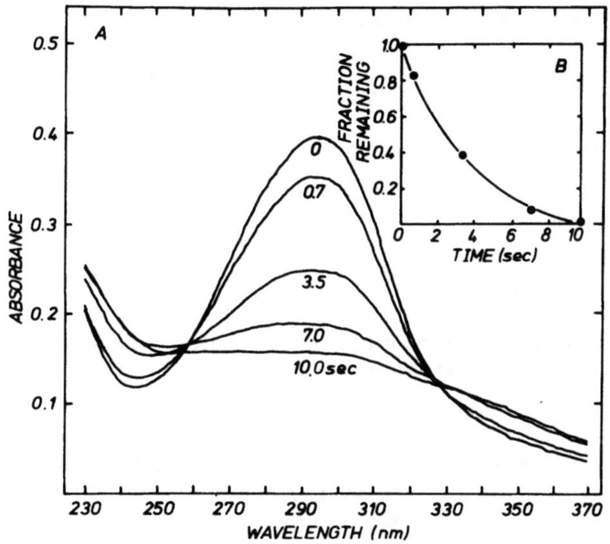

Fig. 2A,B. Photolysis of GA_4-photoaffinity probe. Absorbance spectra (**A**) before and during exposure to light of wavelengths above 300 nm. Rate of photolysis (**B**)

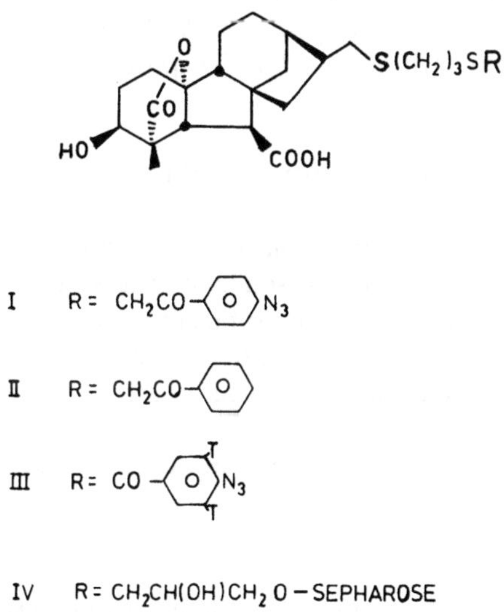

Fig. 3. Structures of C-17-derivatives GA_4

to the receptor. Thus, photoinactivation assays can be performed with unlabelled compound as an integral part of the design strategy prior to radiolabelling [9]. We have performed a variation of the photoinactivation assay based on inhibition by the candidate probe of α-amylase induction in aleurone protoplasts (Table 2). Protoplasts incubated with GA_4-17-S-propyl-S-phenacyl azide produce approximately 50% less α-amylase after photolysis compared with the non-irradiated control. This photoinhibition of α-amylase induction is not reversible by washing and then transferring the protoplasts to fresh media. α-Amylase induction by a control compound without the azide group (structure II, Fig. 3) was unaffected by

Table 2. Photoinhibition of α-amylase induction

Treatment	Minus GA	Total α-Amylase (units) GA_4-17-S-propyl-S-phenacyl azide[a]	GA_4-17-S-propyl-S-phenacyl[b]
No irradiation	0.2	8.8	11.8
Irradiated	0.2	4.2	11.5
Irradiated[c]	0.2	4.7	11.7

[a] Structure I.
[b] Structure II, Fig. 3.
[c] Protoplasts were irradiated, washed, transferred to fresh medium containing the indicated GA and incubated in the dark.

the treatments (Table 2). These data suggest that the photoaffinity probe may bind to and impair GA-receptors, and confirm that criterion 5 (Table 1) has been satisfied. On the basis of these observations, tritium labelling was deemed justified.

For convenience of synthesis the preparation of a tritiated photoactive GA-derivative involved slight modification of the C-17 side chain. The GA_4-photoaffinity probe, GA_4-17-S-propyl-S-benzoyl azide (structure III, Fig. 3) was thus tritiated to 1.79 TBq/mmol.

2.2 Photoaffinity Labelling

Aleurone protoplasts were irradiated with the [^{3}H] GA_4-photoaffinity probe in the presence and absence of excess unlabelled GAs. Binding of the probe to aleurone proteins appears to be covalent and is photolysis dependent (Table 3). Binding is reduced by 15% in the presence of excess unlabelled GA_4, by 19% in the presence of excess unlabelled photoaffinity probe, but not reduced by excess unlabelled GA_8, which is known to be inactive in aleurone protoplasts [3]. At best these data *may* indicate a low level of specific binding against a high level of nonspecific, non-saturable binding.

Table 3. Incorporation of the tritiated GA_4 photoaffinity probe into aleurone proteins

Treatment[a] [^{3}H]-Photoaffinity Probe[b]	GA	TCA precipitable counts (dpm per 5×10^5 protoplasts)
2 nM	–	14 694
2 nM	1 μM GA_4	12 424
2 nM	1 μM GA_8	15 092
2 nM	1 μM cold probe[c]	11 948

[a] Non-irradiated control = 1782 d pm.
[b] Structure III, Fig. 3.
[c] Structure I, Fig. 3.

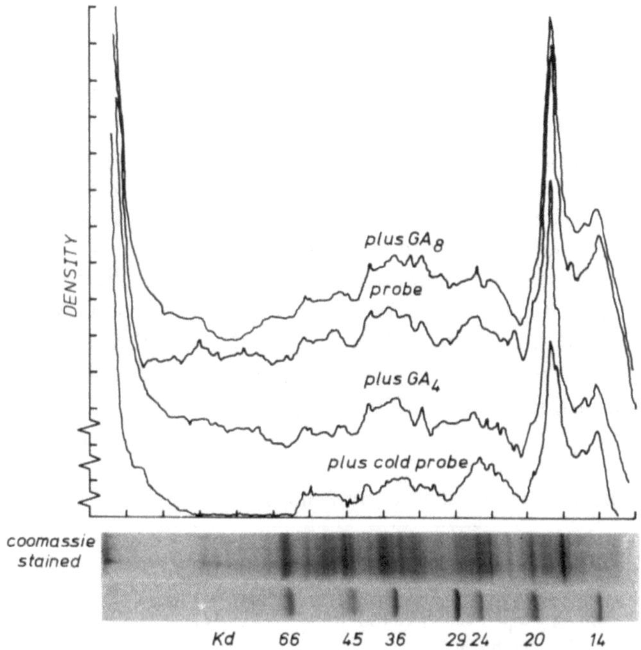

Fig. 4. Coomassie-stained SDS-PAGE and densitometric scans of fluorograph of [^{3}H]-GA-photoaffinity labelled aleurone polypeptides

A proportion of the incorporated radioactivity could be solubilized in SDS and were analyzed by SDS-PAGE and fluorography (Fig. 4). Three polypeptides (19.0, 18.5 and 14.1 kD) that did not correspond to any particularly abundant aleurone proteins were photoaffinity labelled. Other polypeptides were labelled to a lesser extent. Photoaffinity labelling of the 19.0 and 18.5 kD polypeptides did not appear to be specific. Labelling of the 14.1 kD polypeptide may indicate some specificity towards GA_4 but not GA_8. Incorporation into other polypeptides could not be quantified.

It would appear, therefore, that while the photoaffinity probe may indeed bind to, and impair, GA-receptors it also binds in a non-specific manner to other aleurone proteins.

3 Functional Assays for GA-Receptors

Determining the subcellular location of putative GA-receptors might allow photoaffinity labelling to be performed using subcellular fractions that possess more acceptable backgrounds of non-specific binding. More importantly, it would, in fact, mark a substantial step forward in our understanding of the mechanism of GA-action in aleurone.

3.1. α-Amylase Induction by Immobilized GAs

GA-receptors might be located on the external face of the aleurone plasma membrane. If so, a GA immobilized on a support matrix may interact with them when presented to aleurone protoplasts, but would be prevented from doing so in intact aleurones by the cell wall. A 17-thiol derivative of GA_4 that was linked to Sepharose 6B (structure IV, Fig. 3) [3] induces α-amylase in isolated aleurone protoplasts, but not in intact aleurones (Table 4). A similar response is observed using a GA_{4+7}-derivative linked to oxirane acrylic beads (data not presented). This functional assay therefore suggests that putative GA-receptors are located on the external face of the aleurone plasma membrane. A similar approach using a fusicoccin derivative has indicated its interaction with the protoplast plasma membrane from tobacco mesophyll [1].

Table 4. α-amylase induction by immobilized GA_4

Treatment	Response[a] (%) Protoplasts	Cells
10^{-3} M GA_4-Sepharose[b]	100	11
10^{-4} M GA_4-Sepharose	85	10
10^{-4} M GA_4-Sepharose plus 10^{-6} M GA_4	100	91
Sepharose 6B	2	9
No additions	1	2

[a] Relative to the maximum α-amylase induced by GA_4.
[b] Structure IV, Fig. 3.

3.2 Inhibition of GA-Induced α-Amylase by Anti-Idiotypic Antibodies

Anti-receptor antibodies raised through anti-idiotypes can be used to probe cell surface receptors [16]. The monoclonal antibody MAC182 displays a highly specific recognition of the A/B ring area of GA_4 [13]. MAC182 has been used as an immunogen in rat to generate anti-idiotypic sera. Potential anti-idiotypes were screened for their ability to compete with [^{3}H] GA_4 for binding to MAC182. Positives were then tested for their ability to influence GA_4-induction of α-amylase in aleurone protoplasts.

The anti-idiotypic serum 455 2 inhibits GA_4-induction of α-amylase in aleurone protoplasts (Table 5). This inhibition is of high titre and can be reversed by MAC182, but not by pre-immune serum. Another anti-idiotypic serum, 455 1, behaves similarly but is of lower titre (data not presented). Both these sera will agglutinate aleurone protoplasts (Fig. 5), indicating that they probably recognize determinants at the protoplast surface which are involved in GA-perception.

Thus, two functional assays, based on non-permeant inducing and competing ligands, suggest that GA-receptors are present, and are located on the external face of the aleurone plasma membrane.

Table 5. Inhibition of GA_4-induced α-amylase by an anti-idiotypic antibody

Antiserum	Dilution	Inhibition[a] (%)
MAC182	1/250	0
Pre-immune	1/250	0
455 2	1/1000	100
	1/5000	46
	1/10 000	28
455 2	1/1000	99
Plus	1/5000	49
Pre-immune	1/10 000	36
455 2	1/1000	100
Plus	1/5000	32
MAC182	1/10 000	0

[a] Relative to protoplasts incubated without antisera.

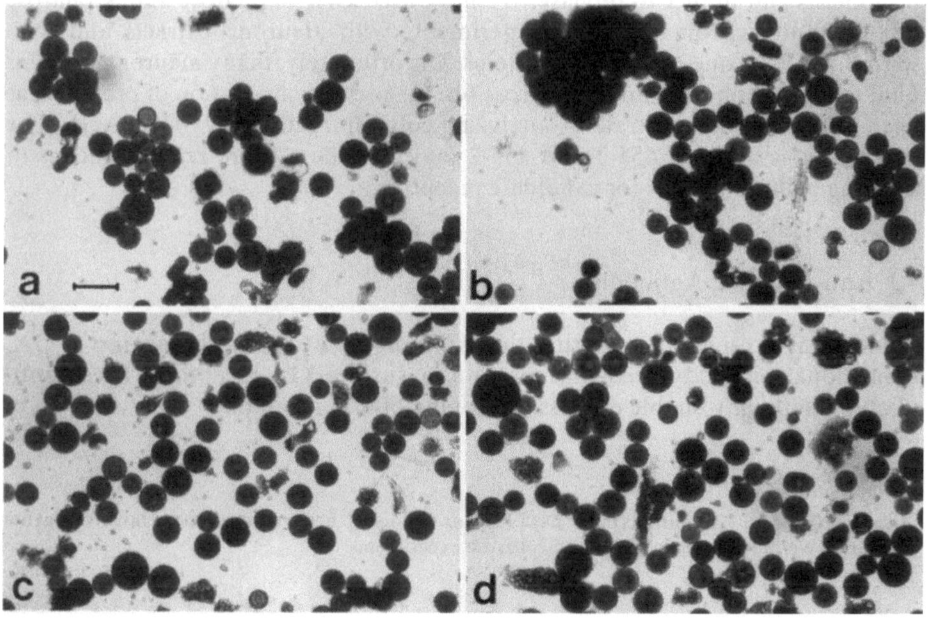

Fig. 5a-d. Agglutination of aleurone protoplasts by anti-idiotypic antibodies. Protoplasts incubated with 455 1(**a**). 455 1 and MAC 182 (**b**). MAC 182 (**c**) and pre-immune serum (**d**). Scale *bar* 100 μm

4 Conclusions and Prospects

4.1 Photoaffinity Probes

By systematic design, synthesis and evaluation, a tritiated GA-photoaffinity derivative has been produced with many of the characteristics of an effective photoaffinity probe. The compound may indeed photoaffinity label GA-receptors, but experiments with isolated aleurone protoplasts suffer from problems of a high degree of non-specific binding. Recently the probe has been successfully iodinated and photoaffinity labelling of aleurone membrane preparations with [^{125}I]-labelled photoaffinity GA-derivatives is in progress. The higher specific activity and emission energy of ^{125}I may allow a greater flexibility in the design and execution of photoaffinity labelling experiments.

4.2 Immobilized GAs

Immobilized GAs appear to interact with GA-receptors at the surface of aleurone protoplasts and induce α-amylase. We have used GA_4-Sepharose and Sepharose 6B in affinity chromatography experiments with aleurone extracts and with solubilized aleurone membrane proteins. Unfortunately, many aleurone proteins bind to GA_4-Sepharose and Sepharose 6B. Ligand specific elution of proteins has been difficult to demonstrate by analyzing column fractions by SDS-PAGE. The anti-idiotypic antisera 455 1 and 455 2 may provide a more realistic means of assaying column eluates for candidate receptors.

4.3 Anti-Idiotypic Antibodies

Evidence is presented suggesting that anti-idiotypic sera raised against a GA_4 monoclonal antibody recognize protoplast surface GA-receptors. These anti-idiotypic sera may enable expression cloning of GA-receptors.

Acknowledgements. The authors thank Dr. J.P. Knox for raising the antibodies used in this investigation and Dr. J. Weir for suggesting the GA-Sepharose experiments.

References

1. Aducci P, Federico R, Ballio A (1980) Phytopathol Mediterr 19:187
2. Bayley H (1983) In: Work TS, Burdon RH (eds) Laboratory techniques in biochemistry and molecular biology, vol 12. Elsevier, Amsterdam, p 187
3. Beale MH, Hooley R, MacMillan J (1986) In: Bopp M (ed) Plant growth substances 1985. Springer, Berlin Heidelberg New York Tokyo, p 64
4. Gronemeyer H (1985) Trends Biochem Sci 10:264
5. Hornberg C, Weiler EW (1984) Nature 310:321
6. Hooley R (1982) Planta 154:29
7. Jacobsen JV, Beach LR (1985) Nature 316:275
8. Jelsema CL, Ruddat M, Morré JD, Williamson FA (1977) Plant Cell Physiol 18:1009
9. Katzenellenbogen JA, Kilbourn MR, Carlson KE (1980) In: Tometsko MA, Richards FM (eds) Applications of photochemistry in probing biological targets. Ann NY Acad Sci 346:18
10. Keith B, Boal R, Srivastava LM (1980) Plant Physiol 66:956
11. Keith B, Foster NA, Bonettemaker M, Srivastava LM (1981) Plant Physiol 68:344
12. Keith B, Brown S, Srivastava LM (1982) PNAS 79:1515
13. Knox JP, Beale MH, Butcher GW, MacMillan J (1987) Planta 170:86
14. Lashbrook CC, Keith B, Rappaport L (1987) In: Fox JE, Jacobs M (eds) Molecular biology of plant growth control. UCLA Symposia on Molecular and Cellular Biology NS. Liss, New York, 44:299
15. Ruoho AE, Rashidbaigi A, Roeder PE (1984) In: Venter JC, Harrison LC (eds) Receptor biochemistry and methodology, vol 1: membranes, detergents and receptor solubilization. Liss, New York, p 119
16. Sege K, Peterson PA (1978) PNAS 75:2443
17. Srivastava LM (1987) In: Klambt D (ed) Plant hormone receptors. NATO ASI Series, vol H10. Springer Berlin Heidelberg New York Tokyo, p 199
18. Yalpani N, Srivastava LM (1985) Plant Physiol 79:963
19. Zwar JA, Hooley R (1986) Plant Physiol 80:459

Fusicoccin-Binding Sites in Higher Plants

E.W. WEILER, C. MEYER, A. MITHÖFER, and M. FEYERABEND[1]

1 Introduction

The physiological effects of FC are manifold and well documented [16]. Most, if not all of them can be seen in conjunction with the marked acidification of the extracellular space and the hyperpolarization of the membrane potential observed almost immediately after addition of the toxin [16]. Drastic changes in solute transport across the plasma membrane occur (e.g. the potassium uptake by guard cells is stimulated [11]) concomitant with (as a consequence of ?) the FC-induced increase in proton motive force. High-affinity FC-binding sites were characterized several years ago [for review see: 2] in membranes from a number of plants, but purification of the sites proved impossible. As a consequence perhaps, interest in the binding sites faded after 1982, but has now resumed in several laboratories and this has resulted in the recent identification of the presumptive binding protein as well as its partial purification.

2 Fusicoccin-Derived Radioligands

Binding studies with FC became possible after the introduction of [^{3}H]-dihydrofusicoccin ([^{3}H]DHFC) of high specific radioactivity [7]. This compound is chemically unreactive however, and functionalization of the toxin required a different chemical approach. It turned out that oxidative removal of C-9′ from the de-t-pentenyl moiety attached to C-6′ of the glucose [13] provided a reactive aldehyde intermediate from which the corresponding 8′-alcohol ([^{3}H]FCol) was accessible in high specific radioactivity through reduction with NaB^3H_4. The compound had only a slightly weaker biological activity than FC in stimulating stomatal opening in *Commelina communis* L. and *Vicia faba* L. in the dark [14]. Moreover, the aldehyde intermediate allows facile access to further FC-derivatives such as protein conjugates [13] or photoaffinity labels with retention of biological activity (see Sect. 5).

3 Properties of Fusicoccin-Binding Sites

The general properties of FC-binding sites determined in earlier studies using [^{3}H]DHFC and in our experiments using [^{3}H]FCol are closely similar. Fusicoc-

[1] Lehrstuhl für Pflanzenphysiologie, Ruhr-Universität Bochum, Fakultät für Biologie, Postfach 102148, 4630 Bochum, FRG

cin-Binding can be detected in the microsomal fraction from a variety of plant tissues such as maize coleoptiles [8], oat roots [21] and spinach leaves [7]. Little, if any, binding activity seems to be in the soluble supernatant. Most of the binding activity occurs in the plasma membrane fraction [6], an observation which was extended recently using plasma membranes of *V. faba* [14] and *Arabidopsis thaliana* (L.) Heynh. [17] purified by aqueous two-phase partitioning. Proteolysis experiments and the use of membrane-impermeant FC-protein conjugates in the *Vicia* system which consists of sealed, predominantly (75%) right-side out oriented vesicles suggest that the FC-binding domain is accessible from the apoplastic face of the membrane [14]. This orientation needs to be established for other plants. However, the data are in accord with those of Aducci et al. [3, 4] who reported preliminary information on the aggregation of tobacco leaf protoplasts in the presence of FC-protein conjugates. However, the general observation that the pH optimum for FC-binding is around 6, i.e. close to the normal range of cell wall pH, seems to support an apoplastic orientation of the FC-binding domain. A fraction of the binding sites (oriented to the medium) is readily degraded when plasmalemma vesicles are treated with proteases such as trypsin, papain, chymotrypsin and thermolysin [17], indicating the presence of both hydrophobic and hydrophilic amino acids in the exposed domains. Several lines of evidence indicate a glycoprotein nature of the binding sites, namely their diffuse appearance on SDS-polyacrylamide gels (see below), the strong reduction in FC-binding when the plasmalemma vesicles are pretreated with mannosidase [1, 2], $NaIO_4$ and with several commercial cellulases [17]. It also follows from these findings that the glycosidic moiety is apparently required for proper binding of the toxin. Table 1 gives a comparison of the properties of the fusicoccin-binding proteins from leaf plasma membranes of *V. faba* and *A. thaliana*.

4 Solubilization and Purification of the Fusicoccin-Binding Protein

As a prerequisite for a study of the molecular structure and also the function of the FCBP, a suitable clean-up procedure is required. The stability of FCBP-FC complexes at low temperature opens the possibility to purify the FCBP-radioligand complex and, thus, to trace the binding site during the process of purification. Solubilization of the FCBP can be achieved with acetone [2], Triton X-100 [18, 21], but in low yields. Non-charged detergents of the octylglucoside-type [12] and, in particular, Mega-9 extract most (60–90%) of the FCBP-radioligand complexes intact [15]. The high detergent: protein ratio ($\geq$25:1) required to solubilize the FCBP suggests that the protein is deeply inserted into the plasma membrane. Starting from here, we have devised a clean-up procedure which involves a combination of different anion exchange and gel permeation chromatographic steps [15]. These can be used sequentially, and, if performed using FPLC equipment at 4–10°C, yield an enriched fraction of the FCBP in less than 6–8 hrs. SDS-PAGE analysis of the fractions allowed to correlate the radioactivity profile with the protein profile leading to the identification of a protein band with an apparent relative molecular mass of 33–36 kDa (*V. faba*) which coeluted with the protein-bound radiolabel. A second, minor, band with a slightly lower molecular weight (approx. 30 to 31 kDa) was also detected. On gel filtration columns, the

Table 1. Characteristics of fusicoccin-binding proteins from *Vicia faba* and *Arabidopsis thaliana* leaf tissue

Parameter	*V. faba*	*A. thaliana*
Abundance in plasma membrane rich vesicles[a] [pmol/mg protein]	20–30	10–15
pH-optimum for the binding of [^{3}H]FCol	5.5–6.0	5.5–6.0
Temperature optimum for the binding of [^{3}H]FCol	25–30	20–25
t/2 (min) for the association of [^{3}H]FCol at 25°C	≈6	≈12
t/2 (h) for the dissociation of FCBP-radioligand complexes at 25°C	≈2	≈3
K_{a1}, approx. (M^{-1})[b]	0.2×10^9	0.7×10^9
K_{a2}, approx. (M^{-1})[b]	3.6×10^7	2.4×10^7
Apparent molecular mass (kDa)		
– gel filtraton of solubilized FCBP-[^{3}H]FCol complexes	80 ± 20	n. d.[c]
– SDS-PAGE of photoaffinity labeled FCBP	35 ± 1	34 ± 1

[a] $U_3 + U_3'$ phases, data are given as ranges observed in a series of typical experiments.
[b] Apparent association constants for [^{3}H]FCol, obtained from Scatchard analysis.
[c] n. d. = Not determined.

solubilized FCBP-radioligand complexes exhibit an apparent molecular mass of 80 ± 20 kDA [15, 21]. These data suggest the occurrence of the FCBP as a di- or trimeric complex. However, a direct evidence for this hypothesis is still lacking.

5 Photoaffinity Labeling of the Fusicoccin-Binding Protein

Further proof of the identity of the 33–36 kDa band with the FCBP came from photoaffinity labeling studies. The label used was a derivative of the FC-aldehyde and 4-azido-(3,5-^{3}H)benzoic acid linked through a short spacer bridge (^{3}H-ABE-FC, Fig. 1). This compound is biologically highly active (e.g. 89% activity relative to FC in inducing stomatal opening in *V. faba* in the dark [15]), activated at sufficiently long wavelenghts (λ_{max} = 270 nm) to avoid serious protein photodegration. Due to the fact that the azido group is attached at some distance from the FC-moiety, the photoactive group has a high mobility in the FCBP-ligand complex, thus resulting in high yields of photoaddition products (7–10% of the initially bound radiotracer is immobilized to the FCBP as a result of irradiation). Figure 2 shows the activity profile obtained after SDS-PAGE of plasma membranes from *A. thaliana* preincubated with ^{3}H-ABE-FC in the dark followed by photoactivation. The major labeled band has an apparent molecular mass of 34 ± 1 kDa [17] (*V.*

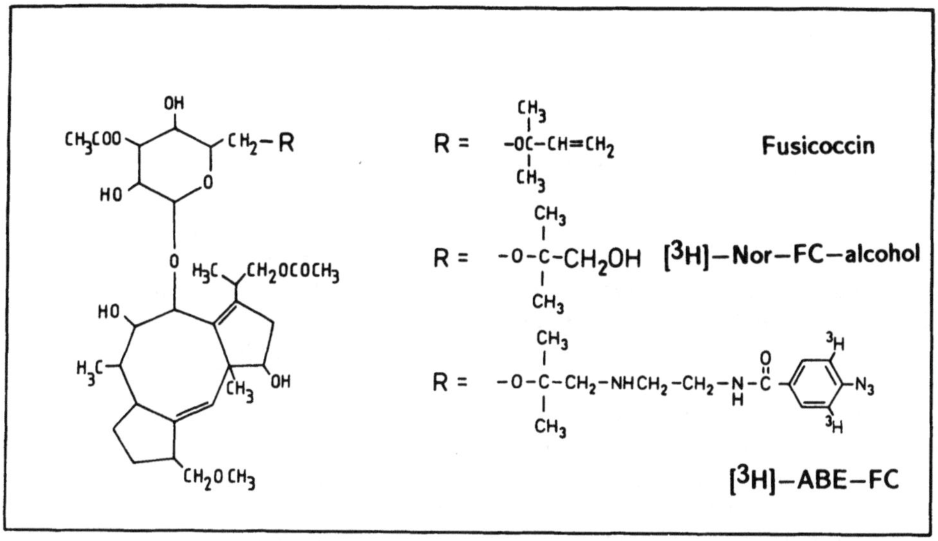

Fig. 1. Structure of fusicoccin and fusicoccin-derivatives

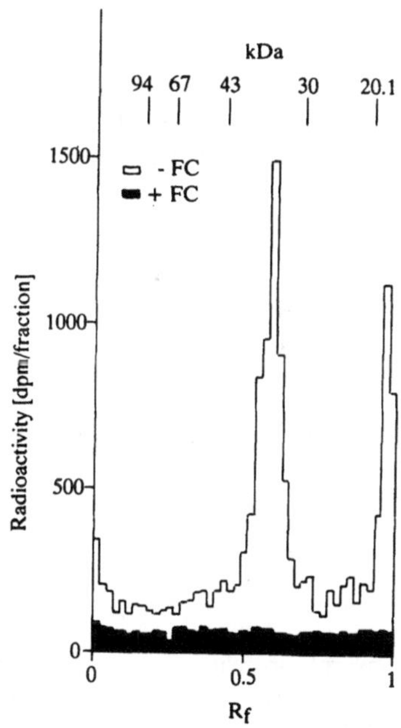

Fig. 2. Distribution of radioactivity in plasma membrane proteins of *Arabidopsis thaliana* incubated for 1 h at 25°C in 1 nM [^{3}H]ABE-FC, followed by centrifugation, resuspension of the membranes and 10 s irradiation at $\lambda \geq 305$ nm, then separation on 11% SDS-polyacrylamide gels. Gels were cut into 2 mm pieces, these were eluted followed by scintillation counting of the eluates. The *closed bars* represent the control for unspecific reactions (co-incubation of 1 nM [^{3}H]-ABE-FC and 40 μM FC)

faba: 35 ± 1 kDa [15]). Some label was obtained in a low molecular mass area, most likely reflecting reaction of the photoactivated nitrene with lipid components (see Fig. 2).

6 Physiological Significance of the FCBP

The picture of the FCBP now emerging is that of plasma membrane glycoprotein, deeply inserted into the membrane. The FC-binding domain faces the apoplastic space and apparently requires an intact glycosidic part for recognition of FC. The photoaffinity labeled FC-binding polypeptide has an apparent molecular mass of 34–35 kDa, and there is at least some indirect evidence that it might be part of a larger complex *in situ*.

What about its physiological significance and potential function? In the absence of any direct proof of function and/or significance, we have to resort to indirect evidence. Table 2 gives a comparison of biological activities, and the ability to compete with [^{3}H]FCol for the binding sites, of a number of FC-related structures. It can be seen that there is generally a good correlation between the two sets of data. Similar data were obtained for other systems [5]. This argues in favour of a physiological significance of the FCBP but does not, of course, constitute proof. This should ultimately come through the analysis of mutants with an altered or non-functional FCBP. Such mutants are, however, not yet available.

Despite considerable effort spent, we still have no clear answer to the question of how FC brings about its effects on the plasmamembrane. It has been suggested that FC activates the vanadate-sensitive proton-pumping ATPase [16]. However, the effect of FC on the activity of this enzyme, when measured in cell-free systems, are slight and only detectable at sub-optimum pH [10, 19]. A direct interaction of FC with the ATPase can now be ruled out for several reasons. This enzyme appears

Table 2. Relative activities of the FC-related structures in inducing stomatal opening in isolated epidermal strips of *Vicia faba* in the dark (bioassay) and in competing with the radioligand, [^{3}H]FCol, for the microsomal binding sites.

	Relative activities (%)[a]		
Compound	Bioassay *V. faba*	Microsomal binding assay *V. faba*	Microsomal binding assay *A. thaliana*
Fusicoccin (FC)	100	100	100
Dihydrofusicoccin	112	107	100
Monodeacetyl-FC	79	98	96
Dideacetyl-FC	9	51	14
Norfusicoccin alcohol	79	92	96
de-t-Pentenyl-di-cleacetyl-FC	0.1	4.2	3
FC aglycone	1.6	0.2	0.04
Cotylenin A	112	37	26
Cotylenin C	71	12	7

[a]All data were calculated from dose-response curves at 50% effect and are given on a relative basis (FC = 100%) [14, 17].

as a 100–105 kDa polypeptide on SDS-PAGE and is thus clearly different from the FCBP. Furthermore, the photoaffinity labeling experiments have not resulted in any detectable labeling of the 105 kDa region, while the enzyme activity was clearly present in the membrane vesicles. Finally, the solubilized ATPase still hydrolyzes ATP, but this activity is unaffected by FC [10]. There is also the possibility that the FCBP represents a regulatory subunit of the ATPase which is not tightly associated with the catalytic subunit and thus lost during the purification of the enzyme. This possibility cannot be ruled out at present, but there is also no direct support in favour of it. The availability of highly enriched FCBP [15] as well as plasma membrane ATPase [20] should in the future allow a co-reconstitution of the two purified components in phospholipid vesicles in order to study the potential interaction of the two systems in detail. Obviously, other alternative hypotheses are feasible and will have to be tested. These include the possibility of an indirect action of the FCBP on effector systems, through an enzymatic activity (kinase?) of the FCBP itself or mediated by a signal transduction pathway. We should also not restrict ourselves to regarding the H^+-ATPase as the only possible target system to explain the action of FC. In recent voltage-clamp whole cell current analyses on guard cell protoplasts, Blatt [9] has shown only a small influence of FC on the pump current (presumably, this current mainly originates from the H^+-pumping, vanadate-sensitive plasmamembrane ATPase) while FC reduced leak currents by a factor of 1.5–2.6. Symport or antiport systems which return H^+ to the cell could thus, among others, be targets of FC action.

7 Conclusions

The FCBP has now been identified in plasma-membrane-rich preparations of several higher plants. It can be obtained in functional form (at least with respect to FC-binding) and a high degree of purity. This opens the possibility of reconstituting the pure FCBP into phospholipid vesicles and of studying the properties of these reconstituted vesicles. Co-reconstitution with the purified H^+-ATPase should allow a critical test of the nature of the presumed interaction, if any, between the two proteins. It should furthermore be possible now to obtain antibodies against the FCBP and to use these to study the localization of the FCBP as well as to identify its gene from suitable expression libraries. There is an urgent need now for mutants with altered properties of the FCBP. These would allow to clearly establish the physiological role of the FCBP. Its identification in *A. thaliana* has opened the door to a molecular characterization of this interesting membrane protein.

Acknowledgements. Work done in the author's laboratory was supported by the Deutsche Forschungsgemeinschaft (Sonderforschungsbereich 171-C10/C9) and the Fonds der Chemischen Industrie (literature provision).

References

1. Aducci P, Ballio A, Fiorucci L, Simonetti E (1984) Planta 160:422
2. Aducci P, Ballio A, Federico R (1982) In: Marme D, Marre E, Hertel R (eds) Plasmalemma and tonoplast. Their function in the plant cell. Elsevier Biomedical, Amsterdam, p 279
3. Aducci P, Ballio A, Federico R, Montesano L (1982) In: Wareing PF (ed) Plant growth substances 1982. Academic Press, London, p 395
4. Aducci P, Federico R, Ballio A (1980) Phytopathol Mediterr 19:187
5. Ballio A (1977) In: Marre E, Ciferri O (eds) Regulation of cell membrane activities in plants. Elsevier Biomedical, Amsterdam, p 217
6. Ballio A (1982) Adv Exp Med Biol 148:223
7. Ballio A, Federico R, Pessi A, Scalorbi D (1980) Plant Sci Lett 18:39
8. Ballio A, Federico R, Scalorbi D (1981) Physiol Plant 52:476
9. Blatt M (1988) Planta 174:187
10. Blum W, Key G, Weiler EW (1988) Physiol Plant 72:279
11. Clint GM, MacRobbie EAC (1984) J Exp Bot 35:180
12. De Boer AH, Lomax TL, Sandstrom RP, Cleland RE (1987) In: Wirtz KWA (ed) Membrane receptors, dynamics, and energetics. Plenum New York, p 181
13. Feyerabend M, Weiler EW (1987) Plant Physiol 85:835
14. Feyerabend M, Weiler EW (1988) Planta 174:115
15. Feyerabend M, Weiler EW (1989) Planta 178:282
16. Marre E (1979) Annu Rev Plant Physiol 30:273
17. Meyer C, Feyerabend M, Weiler EW (1989) Plant Physiol 89:692
18. Pesci P, Tognoli L, Beffagna N, Marre E (1979) Plant Sci Lett 15:313
19. Rasi-Caldogno F, Pugliarello MC (1985) Biochem Biophys Res Commun 133:280
20. Serrano R (1984) Biochem Biophys Res Commun 121:735
21. Stout R, Cleland RE (1980) Plant Physiol 66:353

Cytokinin Mode of Action – Problems and Perspectives

D.E. Hanke[1], H. Davies[1], M. Biffen[1], R.J.A. Connett[2], and T.C. Freathy[1]

1 Introduction

Recent results have modified the expectation that responses to cytokinin, along with the other growth substances, would all turn out to be mediated via cytosolic Ca^{2+} by mechanisms closely resembling those proposed (but not yet proved) for the responses of animal cells to hormones. The big friendly giant is a fertile source of ideas (such as the alarming fact that some animal hormones may have as many as five different receptors [1]) but given the ease with which it is possible to discover only what we expect to find, rigorous and critical testing is called for. As the extreme complexity of the cell's signal transduction machinery slowly comes to light, we discover how easily the interactive network of biochemical components can generate artifacts and how difficult it is to reconstruct its in vivo operation.

Some of the pioneering work implicating Ca^{2+} in responses to cytokinin needs to be re-examined in the light of what we now know.

As Kauss [2] has pointed out, effects obtained using external Ca^{2+} at concentrations higher than 1 μM are as likely to result from the permissive effect of membrane stabilization as from any requirement for this ion to cross the plasma membrane in order to be effective as a second messenger. There is an added complication for plant cells in that external Ca^{2+}, even at 5 mM, can dramatically affect ethylene production [3] with consequent effects on development. Corriveau and Krul [4] deduced from the inhibitory effect of 10^{-5}M Ag^{2+} on the response of soybean hypocotyls to BA that in this system ethylene mediates cytokinin action, a proposal in need of further investigation by quantitative studies and the use of more stringent controls. The interplay between cytokinins, ethylene and Ca^{2+} is likely to prove difficult to untangle. Cytokinins and external Ca^{2+} can act synergistically to promote ethylene biosynthesis [5] and inhibitory effects of trifluoperazine, a phenothiazine, and W-5, a (comparatively inactive) naphthalene sulphonamide [6] may implicate calmodulin somewhere in this messy relationship.

Inhibition of response by anticalmodulin drugs has contributed to the supposition that cytokinins work via Ca^{2+}[7, 8] but this evidence must be considered tentative at best. At concentrations of 10^{-5}M and upwards these compounds can generate artifacts, behaving as detergents or redox agents, or inhibiting non-calmodulin proteins such as protein kinase C [9]. Anticalmodulin drugs may prove

[1]University of Cambridge, Department of Botany, Botany School, Downing Street, Cambridge CB2 3EA, UK

[2]Agricultural Genetics Company Limited, Unit 154/155, Cambridge Science Park, Milton Road, Cambridge CB4 4GG, UK

more selective at low (10^{-7} to 10^{-5}M) concentrations, but even in this case there is a real danger of artifacts, thanks to the interlocking of Ca^{2+} action and Ca^{2+} transport. Using carrot cell protoplasts loaded with the [Ca^{2+}] indicator quin 2, treatment with 10^{-5}M levels of anticalmodulin drugs was shown to raise cytosolic Ca^{2+} concentrations from 3×10^{-7}M to greater than 10^{-6}M over 15 min [10], probably by inhibiting calmodulin-dependent Ca^{2+}-sequestering activity. A physiological effect of the drug, then, could be due either to the increase in the level of cytosolic Ca^{2+} or to direct inhibition of a separate calmodulin-dependent process, or both. Evidence from the use of quin 2 is not unassailable since it could find its way into Ca^{2+}-rich vacuoles [11] and it can show physiological activity in its own right by chelating Ca^{2+} [12].

An alteration in the control of Ca^{2+}-transport by Ca^{2+} may form part of the mode of action of cytokinins. Treatment of wheat roots with 10^{-8}M BA resulted in large increases in the affinity and capacity of microsomal Ca^{2+}-ATPase activity, apparently as a result of the acquisition by Ca^{2+}-pumps of the ability to respond to calmodulin [13]. One of the earliest papers linking cytokinin action with Ca^{2+} (and still amongst the best evidence for this link) was also a study of the effect of growth substance pre-treatment on microsomal Ca^{2+} transport. Kubowicz, Vanderhoef and Hanson [14] showed that as cells in a soybean hypocotyl made progress towards maturity, they altered in their response to treatment with zeatin from stimulation of microsomal Ca^{2+} uptake (dividing cells), to inhibition (expanding cells) and back to stimulation (fully expanded cells). This is a clear indication that there is no universal type of link between cytokinin and Ca^{2+} as 2nd messenger; different types of 'mechanisms' characterize different stages in development. A similar conclusion can be drawn from a recent re-examination [15] of the best authenticated example of a Ca^{2+}-mediated cytokinin response, the asymmetric cell division in caulonema filaments of *Funaria*. The initial cell division was triggered by 10^{-12}M cytokinin and required external Ca^{2+}, but the subsequent swelling to form a bud in response to 10^{-7}M cytokinin took place in Ca^{2+}-free medium or when Ca^{2+} entry was blocked by La^{3+}. The dramatic shift in sensitivity to cytokinin implies a change in receptor activity as well as in the mode of action of cytokinin.

The ability of Ca^{2+}-channel blockers to influence cytokinin-responses has also been used to adduce second messenger status for Ca^{2+}. However, the metal ions La^{3+} and Co^{2+} are inevitably comparatively non-specific, and also affect both the physical properties of the wall and ethylene biosynthesis respectively. Ca^{2+}-channel-blocking drugs such as verapamil and nifedipine could only begin to be trusted if their effect on Ca^{2+} transport in the experimental material had been fully characterized. The discovery that these two Ca^{2+} channel binders induce callose formation in cells of the liverwort *Riella* [16] indicates that they can facilitate Ca^{2+} entry in some tissues. Verapamil and nifedipine induced callose in different, discrete regions of the cell which suggests the possibility of diversity within the Ca^{2+}-channels, an added complication already apparent from the differential sensitivity of Ca^{2+}-channels in tonoplast and endoplasmic reticulum membrane to inositol 1, 4, 5-trisphosphate [17, 18].

As yet there is very little evidence that inositol derivatives are involved in the processes which link the cytokinin signal with its putative Ca^{2+} second messenger. Results which indicate that cytokinins reduce the incorporation of [^{32}P]P_i into phosphorylated forms of phosphatidyl inositol [19] are hard to interpret since the comparatively inactive cytokinin *cis*-zeatin had the highest activity and the non-cytokinin adenine showed activity. Also, the cell suspension used was not cytokinin-dependent. A report that prolonged exposure to cytokinin altered the levels of inositol phosphates in *Funaria* is published only in abstract form [20].

As far as we know a link between cytokinin and protein phosphorylation has only been documented for Chinese cabbage leaf discs [21]. Microsomes from leaves treated with 2×10^{-5}M cytokinin solution had a reduced (by 25 to 50%) ability to use ATP to phosphorylate membrane proteins as compared with that of microsomes from water-treated controls. Without the results from leaves treated with 2×10^{-5}M adenine solution, we cannot tell if this effect is cytokinin-specific. If cytokinins can act via alteration of the cytosolic calcium concentration it is then possible that cytokinin treatment will result in activation of one or more of the Ca^{2+}-regulated protein kinases shown to be present in plant tissues [22, 23].

As an alternative, or an adjunct, to operating via Ca^{2+}, cytokinins could conceivably influence development via polyamines. Increases in polyamine synthesis following cytokinin stimulation are well documented [24], and plant tissues contain polyamine-stimulated protein kinases which phosphorylate a different set of substrate proteins from those processed by Ca^{2+}-stimulated protein kinases [22, 25].

Stimulation of H^+ efflux is an important component of auxin action in many plant tissues [26]. A recent report indicates that active auxins directly stimulate a plasma membrane NADH oxidase from soybean hypocotyls. Enzyme from tissue which had completed elongation was no longer stimulable [27]. NADH oxidase is implicated in an electron transport-driven H^+-efflux across the plasma membrane, a mechanism which co-exists with H^+-pumping ATPases and, for cultured carrot cells, predominates specifically in young cells [28]. Because BA hyperpolarized the cell membrane and stimulated H^+ efflux in squash cotyledons, it was suggested that the stimulation of the primary H^+-pump might be an early event in signal transduction for cytokinins [29]. Working with our line of cytokinin-dependent soybean cell suspension, Dale Sander's group have shown that treatment with 2×10^{-6}M cytokinin solution induces a 13 to 18 mV hyperpolarisation of the cell membrane, completed within 2 min [30]. The effect is most likely due to stimulation of the plasma membrane H^+-pump. However, *cis*-zeatin, a weak cytokinin, and adenine elicited the same effect, and two N^9-substituted cytokinins did not, even though they are both active cytokinins [30]. The results emphasize the importance of using close analogues as inactive controls. The lack of such controls in earlier studies means there is no evidence that the effect of cytokinins on H^+-transport is directly related to signal transduction.

Apart from thinking that there are a large number of *potential* mechanisms of action, it is difficult to conclude anything from these findings because each of them concerns a different response to cytokinin by a different tissue of a different species at a different stage in development in a different environment. Soybean hypocotyl is probably the most intensively researched example. There is evidence to suggest that in the elongating zone activation by auxin involves NADH oxidase and Ca^{2+} transport at least, while cytokinin may operate both via a counter effect on Ca^{2+}-transport and through the agency of ethylene.

So, it seems that not only can the mode of action of cytokinins change as a tissue progresses through different developmental stages, but even at the same developmental stage we may find more than one 'mode of action'. Are they component arms or redundant alternatives, like belt and braces? The challenge is to determine the functional role in any response of each of the different 'modes of action'.

2 Studies of the Mode of Action of Cytokinin in Cytokinin-Dependent Cell Suspension Cultures of Soybean Callus

Soybean callus of cotyledonary origin (cv Acme) is a principal bioassay for cytokinins [31] with an unusually stable, absolute requirement of cell division for cytokinin. We choose to use it in suspension culture in order to obtain synchronous activation of all cells, allowing us to investigate early events.

An unsettling feature of the system is the instability of its characteristics other than cytokinin-dependency. For example, the time taken following transfer to cytokinin-free medium (at a fixed cell density) before mitotic arrest is liable to shift between experiments. A simple explanation for the shifts might be the spontaneous occurrence in the culture of variant types with altered cytokinin metabolism. The spontaneous occurrence of an unstable line with enhanced activity of zeatin side chain cleavage has been described [32]. It seems that the renowned phenotypic instability of callus tissue [33] is accentuated by suspension culture, in which each small clump of cells is a separate entity with its own selective advantages in the fierce competition to contribute cells to the inoculum for the next batch.

However, phenotypic instability can be exploited, in this case to provide a means to subject correlations to repeated testing. Of the biochemical events triggered by cytokinins, some will be necessary for cell division, others not. Only events necessary for cell division will always be sustained in a culture in which cytokinin-dependency is maintained. The cytokinin-induced events not necessary for cell division will be liable to change or disappear with time, in a manner not correlated with the cytokinin-dependency of cell division. Our results from a study of phospholipid head group turnover following cytokinin activation appear to be an example of the correlation-breaking power of the callus system.

In a series of experiments carried out in 1982 and 1983 [34], we followed the incorporation of [^{32}P]Pi into the bulk phospholipids of soybean cells at 15-min intervals following the addition of kinetin and [^{32}P]Pi solution, or the same volume of [^{32}P]Pi solution, to cultures in mitotic arrest because of cytokinin deprivation (Fig. 1). By the first time point, 15 min, 70% of the total phosphate incorporated was in PI and PA which jointly contribute only 14% of the total phosphate in phospholipids (Table 1). Even at 1 h after the addition of [^{32}P]Pi, the proportion of it in PA was higher than the measured cell content of PA justified, suggesting that the phosphorylation of DAG is a continuing process whether cytokinin is present or not.

At the 15-min time point there was always a major difference in the labelling of *both* PI and PA in response to kinetin, a difference which disappeared by 30 min. Of the bulk phospholipids only PI and PA showed this labelling pattern (Fig. 1) and

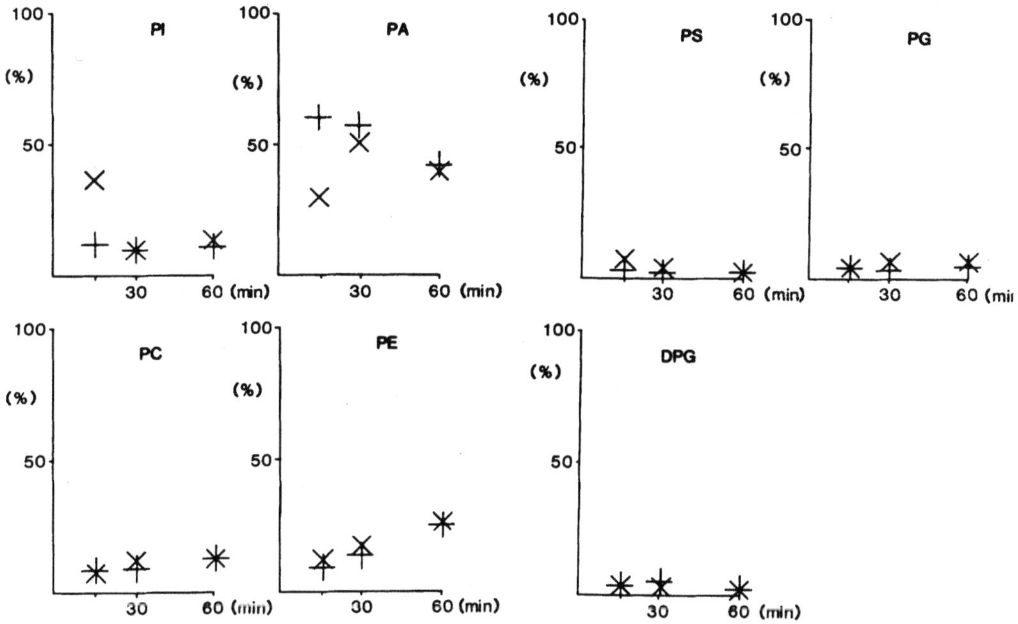

Fig. 1. Effect of kinetin (+) and water (×) on incorporation of [^{32}P]Pi into phospholipids of cytokinin-responsive samples of soybean cells. Cytokinin-responsive samples of soybean cells are suspension cultured cells of soybean (cv Acme) cotyledonary callus brought to mitotic arrest by cytokinin-deprivation. Filter-sterilized kinetin solution (1 ml to 50 ml of cell suspension), or water, was supplied at t = 0. At the same time each sample was supplied with [^{32}P]Pi at 74 kBq ml^{-1}. Cells were killed in hot *iso*-propanol at the times shown on the abscissa. The ordinate represents radioactivity in each phospholipid expressed as a percentage of the total radioactivity recovered from each 2-D TLC plate. The total radioactivity recovered from each plate was: 15 min, water = 28 Bq, kinetin = 11 Bq; 30 min, water = 12 Bq, kinetin = 12 Bq; 60 min, water = 29 Bq, kinetin = 27 Bq

Table 1. Phospholipid composition of cytokinin-responsive samples of suspension-cultured soybean cells[a]

Phospholipid	Abbreviation	Phosphorus content (%)
Phosphatidylinositol	PI	10
Phosphatidic acid	PA	4
Phosphatidylserine	PS	1
Phosphatidylcholine	PC	49
Phosphatidylethanolamine	PE	30
Phosphatidylglycerol	PG	4
Diphosphatidylglycerol	DPG	2

[a] Each value is the mean of measurements from two separate extractions.

only PI and PA are components of the 'PI cycle' which is involved in signal transduction in animal cells [35].

These results prompted us to look for other components of the cycle. In a number of animal cell systems the sustained phase of the cycle's response to hormonal stimuli, (involving the release of DAG from PI and its recycling through PA over a time scale of 20 min) is often preceded by a transient phase of polyphosphoinositide breakdown, peaking at 15 s and completed by 5 min [36]. Phospholipase C-type cleavage of phosphatidylinositol 4,5-bisphosphate ($PI(4,5)P_2$) has been allotted special importance because the inositol 1,4,5-trisphosphate ($Ins(1,4,5)P_3$) that is released mobilises intracellular Ca^{2+}[37]. Cell division of cytokinin-dependent soybean (cv Acme) callus of cotyledonary origin had been shown to be blocked by six different anti-calmodulin drugs [38], suggesting that regulation by elevated levels of cytosolic Ca^{2+} is part of the control of cell division in this tissue. We therefore thought it appropriate to look for cytokinin-stimulation of $Ins(1,4,5)P_3$ release from $PI(4,5)P_2$.

When the acidic lipids of soybean cells heavily pre-labelled with [^{32}P]Pi were resolved by 2-D TLC, no trace of radioactivity could be detected in the position occupied by a sample of authentic $PI(4,5)P_2$ run on the same plate [34]. However [^{32}P]-label *was* found at the location of a phosphatidylinositol 4-phosphate (PI(4)P) standard. This is only a preliminary to the process of proper identification. Co-chromatography of radioactivity, whether from [^{32}P]-Pi or [^{3}H]-inositol, with a phosphoinositide standard in 1-D TLC is not acceptable evidence of identity. It has subsequently been shown that suspension cultured cells of tomato and carrot callus also do not appear to contain detectable levels of $PI(4,5)P_2$ [39].

All our attempts to confirm a role for Ca^{2+} as a second messenger regulating cell division in suspension cultured soybean cells have yielded negative results. In testing the effects of a range of inhibitors we have been careful always to use a range of concentrations and always to check the viability of treated cells. Clearly, whenever inhibitory effects of a compound are accompanied by loss of viability it is impossible to decide whether inhibition is the direct result of a specific interaction between the compound and its target site in the cell or an indirect consequence of general toxicity. Results presented in Fig. 2 indicate that the phenothiazine anticalmodulin drugs chlorpromazine and trifluoperazine did not affect the cell number attained by suspension cultures except at toxic levels. The Ca^{2+}-channel blockers La^{3+}, D600 and verapamil were non-toxic over a wide range of concentrations and did not affect cell division (Fig. 2). Another Ca^{2+}-channel blocking agent, ruthenium red, was also non-inhibitory to cell division at non-toxic levels. The lack of inhibitory effect of Ca^{2+}-channel blockers on soybean cell division can only be used as evidence against a role for *extracellular* Ca^{2+}.

Next we investigated whether the Ca^{2+}-ionophore A23187 in a range of concentrations in combination with a constant (the normal 2mM) level of external Ca^{2+} could substitute for cytokinin and induce cell division in cells mitotically arrested by cytokinin deprivation (Table 2). Fluorescence microscopy showed that A23187 was entering the cells and was not rapidly degraded. Also, the viability of the cells remained high (Table 2). However, no increase in cell number could be detected.

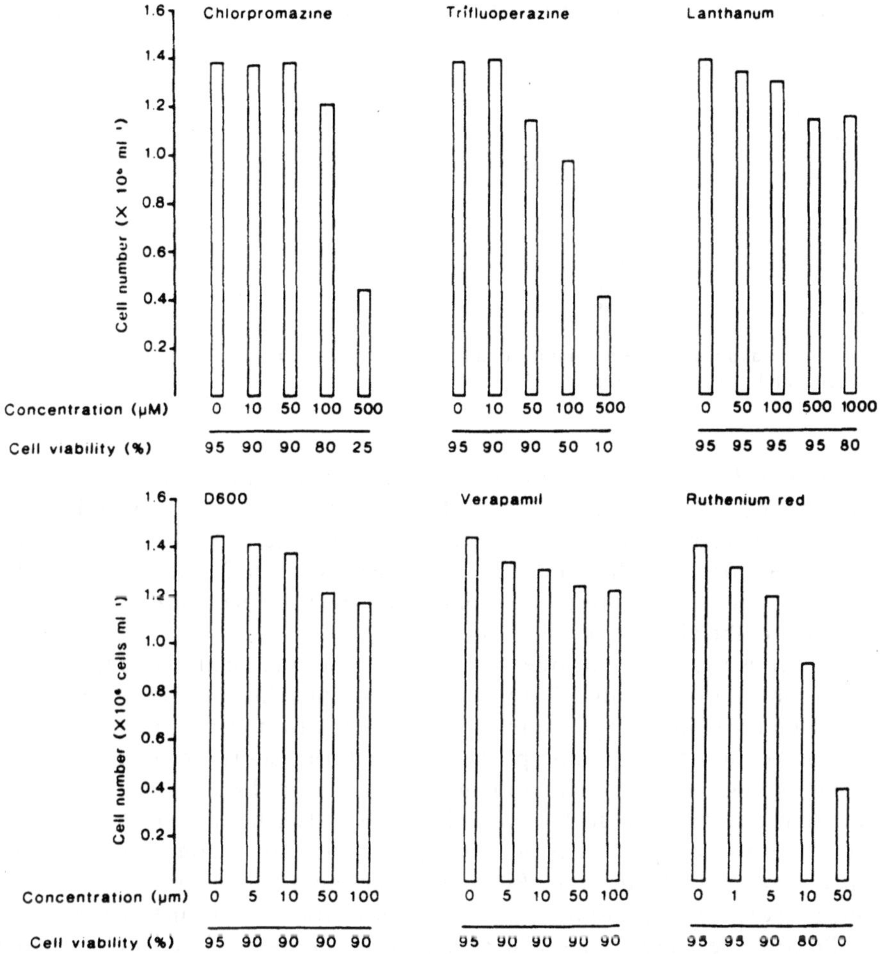

Fig. 2. Effect of anti-calmodulin drugs and Ca^{2+}-channel blockers on cell proliferation in suspension cultures of soybean callus supplied with cytokinin. Cells in the linear phase of the batch growth cycle were suspended at 0.4×10^6 cells/ml. Cell number and viability were estimated after 6 days of culture using methods described in reference [34]. D-600 and verapamil were a generous gift of Knoll A.G., Ludwigshafen am Rhein, FRG

Inhibitors may fail to block cellular processes for any of a number of trivial reasons, e.g. poor transport or high rates of inactivation. We make no claim for these results except that they indicate that anticalmodulin drugs and Ca^{2+}-channel blockers are not universally effective in plant systems. The experiments with ionophore also do not rule out some role for Ca^{2+} in the cell division response to cytokinin. However, the results do suggest that an elevated level of cytosolic Ca^{2+} is not by itself a sufficient signal to replace cytokinin.

In many animal cells the DAG released on phospholipid hydrolysis may be necessary for cell activation in addition to Ins(1,4,5)P_3 [37]. For example, for the

Table 2. Effect of the Ca^{2+}-ionophore A23187 on cell number and cell viability in cytokinin-responsive samples of suspension-cultured soybean callus 6 days after addition

Ionophore concentration[a] (μM)	Cell number (10^6 cells ml^{-1})	Cell viability (%)
0	0.51	95
10	0.51	95
25	0.48	95
50	0.46	95
100	0.39	80

[a] Ionophore was added in the minimum volume of methanol. Cell number and cell viability were estimated as in [34].

initiation of DNA synthesis in hamster fibroblasts by thrombin, release of DAG from PI is necessary but release of Ins(1,4,5)P_3 from PI (4,5)P_2 is not [40]. We have tested the effect of 0.1 μM tetradecanoyl phorbol acetate, a DAG analogue, both by itself and in concert with 10 μM A23187 in the presence of 2 mM Ca^{2+}, on soybean cells in mitotic arrest as a result of cytokinin deprivation. In no case was cell division sustained in the absence of cytokinin.

Collectively, the results offer no support for a functional role of the 'PI cycle' in the control of soybean cell division by cytokinin. In 1987 and 1988 we re-investigated the effects of cytokinin on phospholipid turnover. To increase the level of incorporation of [^{32}P]Pi into phospholipids during short time courses, the level of Pi in both growth and incubation media has been reduced from 2.2 to 0.8 mM, which is in fact the optimum level for cell division. Also, the amount of [^{32}P] supplied to the cells was increased. The phospholipids were analyzed on a 1-D TLC system which resolves the lipids of the PI cycle. In combination these three modifications increased the [^{32}P] in phospholipids from cells fed for only 10 min to a level at which 500 Bq could be loaded on a single track. Figure 3 shows the pattern of phospholipid bands on the TLC plate after analysing lipids from cytokinin-starved soybean cells. Some batches of cells were pre-labelled with [^{3}H]-inositol before adding kinetin or water, others were supplied with [^{32}P]Pi at the same time as kinetin or water was added. All samples were killed rapidly 10 min after adding kinetin or water. The radioactivity recovered from this plate, band for numbered band, is presented as the percentage of the total radioactivity in each track (Table 3). The synthesis of phospholipids from [^{32}P]Pi in 10 min is dominated by lipid co-chromatographing with PI(4)P (45% of the total label recovered in phospholipids), and not with PI(4,5)P_2. Similar results have been reported for suspension cultured cells of tomato callus [39]. In our previous work this lipid, and all more polar lipid species, remained at the origin of the TLC plate with [^{32}P]Pi. The pattern of labelling of the less polar, bulk phospholipids is broadly similar to our earlier findings, except that in this case PI and PA together account for 48% of the labelling in bulk phospholipids after 10 min (still considerably in excess of their contribution to the total phosphorus in these lipids), in both kinetin- and water-treated samples. The pattern of labelling from [^{3}H]-inositol is dominated by PI (20% of the total), the only bulk phospholipid

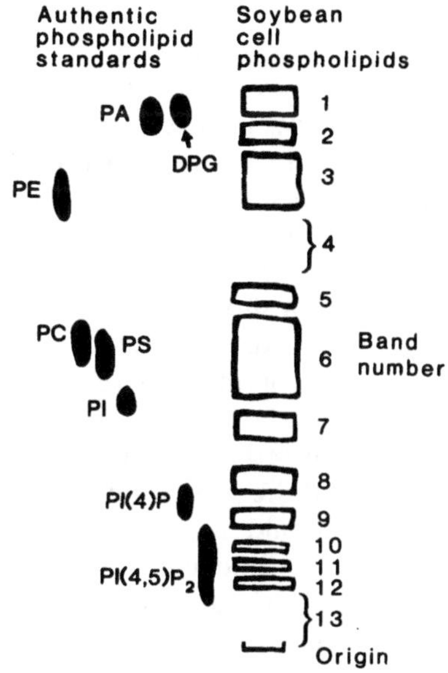

Fig. 3. The pattern of phospholipid bands after 1-D TLC of lipids from cytokinin-responsive soybean cells. Cytokinin-responsive cells were prepared as described in [34], except that the concentration of Pi in all media was 0.8 mM. For labelling from [^{3}H]-inositol, cultures were grown in medium containing 46 kBq ml^{-1} [2-^{3}H]-*myo*-inositol for 7 days before transfer to medium lacking both cytokinin and inositol. For labelling from [^{32}P]Pi, the isotope was added to a final level of 320 kBq ml^{-1} with the kinetin solution (or water in the case of controls) after incubation in cytokinin-free medium. The 1-D TLC system is modified from that of Goppelt and Resch [49]. The distribution of radioactivity between the bands is presented in Table 3

Table 3. Radioactivity in TLC fractions of soybean cell phospholipids as a % of the total label on each track

Band No.	From [^{32}P] added with kinetin/H_2O:		Position of phospholipid markers	From [^{3}H]-inositol supplied in the previous growth phase		Band No.
	+Kinetin[a] (% ± SD)	+H_2O[a] (% ± SD)		+Kinetin[a] (% ± SD)	+H_2O[b] (% ± SD)	
1	8.8 ± 2.3	9.7 ± 0.7	PA	0.4 ± 0.1	0.9 ± 0.6	1
2	2.4 ± 0.1	2.2 ± 0.6		0.3 ± 0.1	0.9 ± 0.6	2
3	3.9 ± 1.1	3.8 ± 0.4	PE	0.6 ± 0.1	1.4 ± 1.3	3
4	1.7 ± 0.4	2.2 ± 0.3		0.9 ± 0.3	2.0 ± 1.1	4
5	1.3 ± 0.2	1.1 ± 0.1		0.6 ± 0.1	1.1 ± 0.1	5
6	4.8 ± 0.8	5.1 ± 0.7	PC/PS	2.5 ± 1.5	1.1 ± 0.3	6
7	4.4 ± 0.8	3.7 ± 0.6	PI	21.1 ± 2.0	16.8 ± 0.8	7
8	6.9 ± 1.0	8.7 ± 2.0	PI(4)P	4.1 ± 0.9	6.7 ± 1.3	8
9	44.1 ± 5.7	45.2 ± 0.5		5.6 ± 0.4	6.5 ± 0.3	9
10	8.5 ± 0.8	9.5 ± 0.5	PI(4.5)P_2	5.6 ± 0.6	6.3 ± 0.2	10
11	2.5 ± 1.5	1.9 ± 0.7		12.4 ± 3.2	13.1 ± 3.1	11
12	2.5 ± 1.5	1.2 ± 0.5		16.8 ± 3.6	19.4 ± 4.7	12
13	7.7 ± 4.3	5.7 ± 0.3		30.4 ± 5.1	24.2 ± 9.8	13

[a] n = 3.
[b] n = 2.
No differences kinetin vs H_2O are statistically significant ($P > 0.05$).

containing inositol. Nevertheless, PI is underlabelled here relative to its enormous chemical preponderance. The phosphoinositide-containing bands 11 and 12 are almost as heavily labelled, but barely detectable chemically.

It follows that the preferential incorporation of [^{32}P]Pi into PI and PA, evidence for the operation of a PI cycle, is confirmed. However, the difference between kinetin and water-treated cells has completely disappeared. The capricious variability of callus tissue has done its deadly experimental work for us and broken the correlation. In the Popperian sense we have made real scientific progress because we have now shown that if there was an effect of kinetin on PI-turnover, then it was gratuitous (like the effect of cytokinin on membrane potential in this system, or on Ca^{2+}-transport in the two end regions of soybean hypocotyls), and not necessary for cell division.

We think that the possibility of gratuitous association between the biochemical events and the subsequent developmental events triggered by plant growth substances should be more widely acknowledged. A correlation is not a causal connection. Biochemical work witn *Catharanthus* cell cultures, linking auxin-stimulation of putative Ins(1,4,5)P_3 release from PI(4,5)P_2 at 1 min with auxin-induced cell division by 24 h [41], remains at the correlation stage. (Results indicating that Ins(1,4,5)P_3 may not be involved in the stimulation of cultured carrot cells by auxin have been reported in an abstract [42].)

In animal cells, GTP-binding proteins, the 'G-proteins', activate the phosphodiesterase which cleaves PI(4,5)P_2 [43]. We have been testing for a functional link between G-proteins and *any* of the phosphoinositide-specific phospholipases in our soybean cells. The cells are pre-labelled with [^{3}H]-inositol, and permeabilized by treatment with digitonin so that phosphoinositide phosphorylation can be maintained by an exogenous ATP-regenerating system. After a 10-min incubation with various test substances the radioactivity released from membranes as water-soluble anions, can be used to monitor phosphoinositide breakdown. Digitonin-permeabilized cells are being used here as an in vitro system to assay an enzyme activity. The results are presented in Table 4. While Ca^{2+} dramatically stimulated the release of radioactivity, the non-hydrolyzable GTP analogue, guanylyl imidophosphate, was completely without effect either by itself or in the presence of added Ca^{2+}. Note that kinetin was also without detectable effect on

Table 4. Label in phosphate esters released from digitonin-permeabilized soybean cells pre-labelled from [^{3}H]-inositol

Time after additions (min)	Additions		Additions: + Kinetin[a] (Bq ± SD)	+ H_2O[a] (Bq ± SD)
0				26.5 ± 2.6
10	No Ca^{2+} or GI added	:	30.3 ± 1.1	30.4 ± 3.5
10	Plus 2 mM Ca^{2+}	:	55.9 ± 2.6	56.3 ± 2.9
10	Plus 100 μM GI	:	28.2 ± 1.1	29.1 ± 3.6
10	Plus 2 mM Ca^{2+} and 100 μM GI	:	53.9 ± 2.7	56.8 ± 2.9

[a]n = 3; GI = guanylyl imidophosphate.

phosphoinositide breakdown, consistent with the other evidence presented here. The membrane-bound PI(4,5)P_2 phosphodiesterase from celery [44] and the enzyme from wheat shoot and roots [45] are both reported to be Ca^{2+}-stimulated but unaffected by GTP-analogues. Using membranes from sycamore cells in suspension culture, a low degree of stimulation of polyphosphoinositide breakdown by GTP and by non-hydrolyzable analogues has been claimed, but the water-soluble compounds released were heterogeneous [46]. If the heterotrimeric G-proteins which potentiate signal transduction in animal cells also occur in plant cells, their function may well be different, and perhaps not directly related to phosphoinositide breakdown.

The mode of action of cytokinin in the control of cell proliferation in cotyledonary callus of soybean remains unknown. Bevan and Northcote [47] showed that the stimulation by kinetin of the increase in numbers of polyribosomes on transfer of these cells to fresh medium was yet another gratuitous effect of cytokinin, not necessary for cell division. They also showed that treatment with cytokinin, as compared with water, led to no detectable change in the relative levels of any mRNAs, and had no detectable effect on bulk protein synthesis or on the synthesis of any particular polypeptides. These results have always pointed to the cytoplasm, rather than the nucleus, as the site of the cytokinin signal-translating machinery. It seems that on cytokinin-starvation, cells can arrest at a number of positions in the mitotic cycle. This suggests that there are a number of different processes in the cell cycle requiring the presence of cytokinin [48]. This may be the reason for the unusual stability of cytokinin-dependence, a feature characteristic of the differentiated state which these cells would otherwise be expected to lose. If there are a large number of cytokinin-surmountable hurdles in the process of cell proliferation for cultured soybean cells, the 'mode of action', when we find it, is unlikely to be simple.

Acknowledgements. We wish to thank Dale Sanders for a pre-print copy of reference [30], and permission to present his results. We acknowledge research funding from the Agricultural and Food Research Council (UK), the Science and Engineering Research Council (UK), and the Gatsby Charitable Foundation.

References

1. Leopold AC (1987) In: Hoad GV, Lenton JR, Jackson MB, Atkin RK (eds) Hormone action in plant development – a critical appraisal. Butterworth London, p 3
2. Kauss H (1987) Annu Rev Plant Physiol 38:47
3. Burns JK, Evensen KB (1986) Physiol Plant 66:609
4. Corriveau JL, Krul WR (1986) J Plant Physiol 126:297
5. Lau OL, Yang SF (1975) Plant Physiol 55:738
6. Yip W-K, Yang SF (1988) Plant Physiol 86 Suppl:113
7. Poovaiah BW, Reddy ASN (1987) Crit Rev Plant Sci 6:47
8. Elliott DC (1986) In: Trewavas AJ (ed) Molecular and cellular aspects of calcium in plant development. Plenum, New York, p 285
9. Roberts DM, Lukas TJ, Watterson DM (1986) Crit Rev Plant Sci 4:311
10. Gilroy S, Hughes WA, Trewavas AJ (1987) FEBS Lett 212:133
11. Cork RJ (1986) Plant Cell Environ 9:157
12. Wolniak SM, Bart KM (1985) Eur J Cell Biol 39:33

13. Oláh Z, Bérczi A, Erdei L (1983) FEBS Lett 154:395
14. Kubowicz BD, Vanderhoef LN, Hanson JB (1982) Plant Physiol 69:187
15. Markmann-Mulisch U, Bopp M (1987) J Plant Physiol 129:155
16. Grotha R (1986) Planta 169:546
17. Schumaker KS, Sze H (1987) J Biol Chem 262:3944
18. Drøbak BK, Ferguson IB (1985) Biochem Biophys Res Commun 130:1241
19. Falkenau C, Heim S, Wagner KG (1987) Plant Sci 50:173
20. Talbot J, Saunders MJ (1986) Plant Physiol 80 Suppl:113
21. Ralph RK, Bullivant S, Wojcik SJ (1976) Biochim Biophys Acta 421:319
22. Ranjeva R, Boudet AM (1987) Annu Rev Plant Physiol 38:73
23. Putnam-Evans CL, Harmon AC, Cormier MJ (1986) In: Trewavas AJ (ed) Molecular and cellular aspects of calcium in plant development. Plenum, New York, p 99
24. Smith TA (1985) Annu Rev Plant Physiol 36:117
25. Datta N, Hardison LK, Roux SJ (1986) Plant Physiol 82:681
26. Cleland RE (1977) Symp Soc Exp Biol 31:101
27. Brightman AO, Barr R, Crane FL, Morré DJ (1988) Plant Physiol 86:1264
28. Barr R, Crane FL (1988) Plant Physiol 88 suppl:79
29. Marré E, Lado P, Ferroni A, Ballarin-Denti A (1974) Plant Sci Lett 2:257
30. Parsons A, Blackford S, Sanders D (1989) Planta 178:215
31. Miller CO (1963) Mod Methods Plant Anal 6:194
32. Horgan R (1987) In: Hoad GV, Lenton JR, Jackson MB, Atkin RK (eds) Hormone action in plant development – a critical appraisal. Butterworth, London, p 119
33. Larkin PJ, Scowcroft WR (1981) Theor Appl Genet 60:197
34. Connett RJA, Hanke DE (1987) Planta 170:161
35. Sekar MC, Hokin LE (1986) J Membr Biol 89:193
36. Griendling KK, Rittenhouse SE, Brock TA, Ekstein LS, Gimbrone MAJr, Alexander RW (1986) J Biol Chem 261:5901
37. Berridge MJ (1987) Annu Rev Biochem 56:159
38. Elliott DC (1983) Plant Physiol 72:215
39. Drøbak BK, Ferguson IB, Dawson AP, Irvine RF (1988) Plant Physiol 87:217
40. Carney DH, Scott DL, Gordon EA, LaBelle EF (1985) Cell 42:479
41. Ettlinger C, Lehle L (1988) Nature 331:176
42. Boss WF, Chen QY, Dengler LA, Hendrix KW, Rincon M, Wheeler JJ (1988) Plant Physiol 86 Suppl:84
43. Gilman AG (1987) Annu Rev Biochem 56:615
44. McMurray WC, Irvine RF (1988) Biochem J 249:877
45. Melin P-M, Sommarin M, Sandelius AS, Jergil B (1987) FEBS Lett 223:87
46. Dillenschneider M, Hetherington A, Graziana A, Alibert G, Berta P, Haiech J, Ranjeva R (1986) FEBS Lett 208:413
47. Bevan M, Northcote DH (1981) Planta 152:24
48. Wang TL, Everett NP, Gould AR, Street HE (1981) Protoplasma 106:23
49. Goppelt M, Resch K (1984) Anal Biochem 140:152

A Functional Cytokinin-Binding Protein in Photochemical Reactions of Chloroplast

H. HUANG[1], S.Y. YANG[2], and Y.W. TANG[2]

1 Introduction

Cytokinins have been suggested to play a critical role in several developmental processes in higher plants [5]. It is well documented that increases in RNA, protein and chlorophyll levels can be observed upon treatment with CKs [2, 6, 15], and, therefore, research on possible mechanisms of CK action in relation to CK-binding sites is logical. Although evidence has accumulated that CK-binding proteins can be isolated, and in fact do exist [3, 4, 14], and macromolecular factors that can interact with CKs and with in vitro RNA synthesis systems have been found [17, 18], the specificities of these interactions remain to be established.

^{6}N-Benzyladenine has been reported to accelerate ATP formation during photophosphorylation in broken spinach chloroplasts [9]. This result implies that CKs may not only act directly at the level of transcription or translation [1, 13, 19], they may also act through regulation of the permeability of photosynthethic system membranes. This would be of importance for photophosphorylation in chloroplasts, and could lead to control over a great many metabolic processes, including transcription and translation. The existence of CK-binding proteins in chloroplasts has been reported in our previous work [7], and several properties of the protein have been determined [8]. The purpose of the investigations reported herein was to isolate the chloroplast CK-binding protein (CCKBP) from chloroplast membrane systems, and to observe whether the CCKBP has a physiological function in photochemical reactions of chloroplasts.

2 Materials and Methods

Chloroplasts from 6-day-old wheat seedlings were used. The culture of wheat seedlings, the isolation of chloroplasts and the determination of CCKBP in chloroplasts were performed as described earlier [8]. The ATP content in detached wheat leaves was measured according to [12], and in vitro photophorphorylation was assayed according to [16]. Delayed light emission (DLE) of chloroplasts was measured as described by [20].

[1] Department of Molecular Biology, Wellman 8, MGH, Boston, MA 02114, USA
[2] Shanghai Institute of Plant Physiology, Academia Sinica, 300 Fonglin Road, Shanghai, People's Republic of China 200032

3 Results and Discussion

3.1 ^{6}N-Benzyladenine-Binding with Soluble CCKBP

The CCKBP of wheat chloroplasts has a high affinity for BA [8]. Since the binding reaction occurs with broken chloroplasts, the CCKBP may exist on the surface of thylakoid membrane. When NaCl (0.5 to 2 M) was used to treat broken chloroplasts, the BA-binding activity to chloroplasts was reduced significantly (Fig. 1). We, thus, conclude that salt in higher concentrations can release the CCKBP from the surface of thylakoid membrane.

A determination of the BA-binding reaction with soluble CCKBP was done after centrifugation of salt-treated chloroplast suspensions, and desalination of the supernatant. Thus, the CCKBP in a soluble state can also bind to BA (Fig. 2A). The BA-binding activity is quite different in various fractions obtained through $(NH_4)_2SO_4$ fractionation. In fact, the BA-binding activity is highly concentrated on the precipitated fraction obtained with 30% saturated $(NH_4)_2SO_4$. This was calculated by means of radioactivity counts in samples, minus those in BSA under the same exprimental conditions (Fig. 2B). Proteins obtained from the 30–80% saturated $(NH_4)_2SO_4$ precipitates would hardly bind BA, even though this category of protein is quite abundant (Fig. 2C). Using the gel filtration assay [14] it is also apparent that the BA-binding reaction occurs primarily with the fraction obtained with 30% saturated $(NH_4)_2SO_4$ (Fig. 3). Thus, the fact that the released CCKBP still retains the ability to bind BA after NaCl treatment allows us to study its physiological functions by using the soluble CCKBP in in vitro systems.

3.2 Functions of CCKBP in DLE of Chloroplasts

A short-time treatment with BA has been reported to have a direct action on stimulating photophosphorylation of spinach chloroplasts [9]. Chloroplasts from wheat seedlings were also observed to respond to BA in photophosphorylation (Table 1). Also, with detached wheat leaves BA caused an increase of ATP

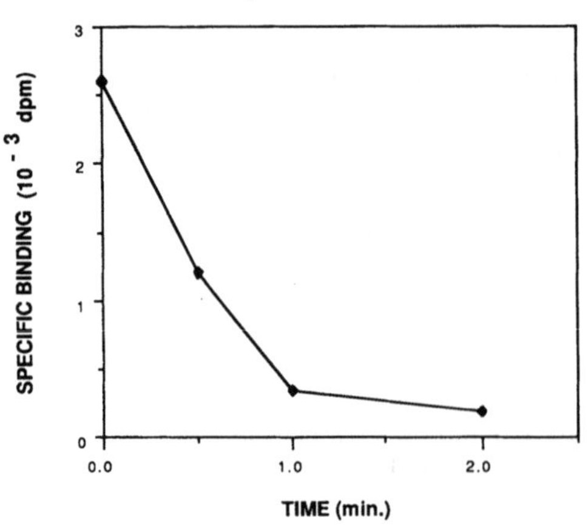

Fig. 1. Effect of pretreatment with NaCl of varying molar strength on specific BA-binding of the CCKBP

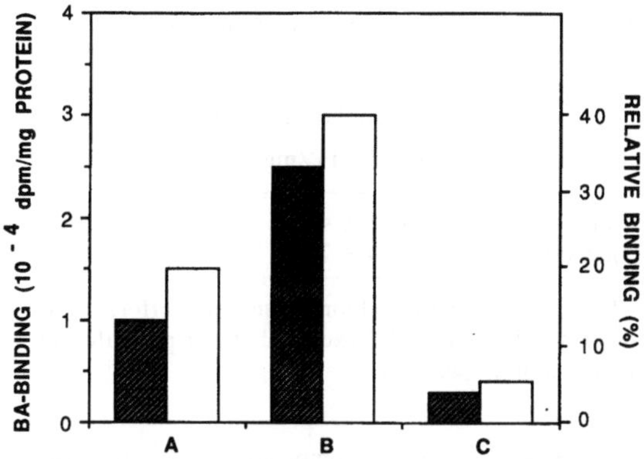

Fig. 2. BA-binding activity of the soluble CCKBP. Isolated chloroplasts were suspended in 2 M NaCl with 0.05 M Tris-HCL buffer (pH 7.6). The suspension was incubated in an ice bath for 15 min followed by centrifugation at 23 000×g for 10 min. The supernatant was desalted with Sephadex G-25. The proteins eluted by Tris buffer were subjected to $(NH_4)_2SO_4$ fractionation. The pellets were suspended in Tris buffer and then dialyzed against Tris buffer at 4°C for 4 h. Usually a 1 ml sample (ca. 500 μg protein) was added to an Eppendorf tube containing 0.12 μCi [^{3}H]-BA. After 30 min at 4°C the mixtures were dialyzed for 8 h (at 4°C) against Tris buffer containing 10 mM NaCl and 0.5% charcoal powder. The radioactivity in each sample was counted with 0.2 ml dialyzed protein solution, plus 7 ml of liquid scintillation solution

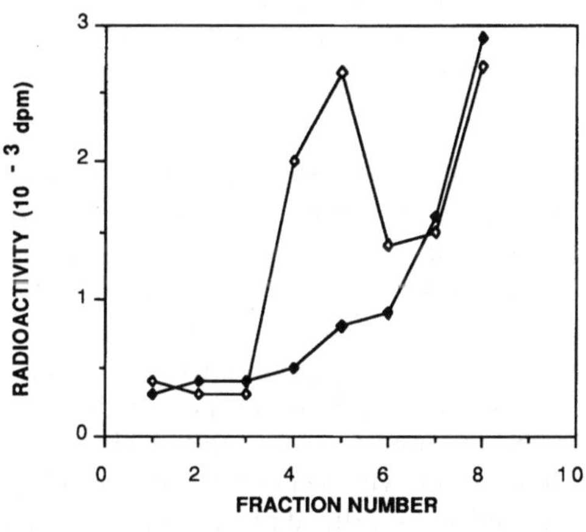

Fig. 3. Gel filtration assay of the CCKBP. Protein used in gel filtration assay came from 30% saturated $(NH_4)_2SO_4$ precipitates (◊——◊). Reaction medium contained 0.44 μCi [^{3}H]-BA and 65 μg desalted proteins in 100 μl Tris buffer (as in Fig. 2). After incubation in an ice bath for 15 min the mixtures were loaded on a Sephadex G-25 column (10×0.7 cm) and eluted with Tris buffer. Aliquots of 0.2 ml were collected and counted for radioactivity as in Fig. 2. Control (♦——♦) was the same protein heated for 10 min at 100°C, and then mixed with 0.44 μCi [^{3}H]-BA before gel filtration

Table 1. Effect of BA on photophosphorylation of wheat chloroplasts[a]

Treatment	Photophosphorylation μmol ATP/mg chl h	%
Control	111.8 ± 4.0	100
3 μM BA	141.4 ± 7.7	126

[a] Preparation of wheat chloroplasts was performed as described by Huang [9] except that the pH value of isolation buffer was 7.6.

Table 2. Effect of BA on ATP formation in detached wheat leaves[a]

Treatment	ATP nmol/g fresh WT	%
Control	11.9 ± 1.5	100
20 μm BA	19.4 ± 2.3	163

[a] Leaves came from 5-day-old wheat seedlings and were incubated with or without BA in dark for 6 h, then illuminated at 5000 lx white light from fluorescent tubes for 30 min. ATP was extracted with 14% perchloric acid. The value of ATP was calculated by: (ATP content in illuminated wheat leaves − ATP content in dark-incubated leaves).

formation when dark-incubated leaves were illuminated (Table 2). To understand the relationship between photophosphorylation processes and CCKBP, we tested the intensity of DLE with chloroplasts from leaves of wheat seedlings.

The DLE of chloroplasts, since its discovery, has been considered to originate from the reversal of the primary photochemical events in photosystem II [10]. The time course of changes in DLE, measured at the millisecond level, consists of at least two components. One component develops slowly during illumination. It is related to the amount of accumulated high energy intermediate or state of photophosphorylation. Another component develops rapidly, and is related to the amount of reduced and oxidized products formed during illumination by photosystem II [11]. When chloroplast suspensions contained 3 μM BA (Fig. 4B), DLE intensity of the slow phase increased in comparison to the control (Fig. 4A). However, salt-washed chloroplasts, which lost CCKBP, did not show this slow-phase increased DLE intensity, even when the suspension contained BA (Figs. 4C and 4D). Neither the CCKBP control solution (Fig. 4G), nor the CCKPB solution containing 3 μM BA (Fig. 4H) had light emissions in either fast and slow phases. But, if the suspension of salt-washed chloroplasts was mixed with the CCKBP solution, then the slow phase light emission reappeared. In this latter case the chloroplast suspension containing 3 μM BA (Fig. 4F) also exhibited stronger DLE intensity in slow phase than did the control suspension (Fig. 4E, no BA). This means that the increase of photophosphorylation level of the chloroplast by BA may depend on the presence

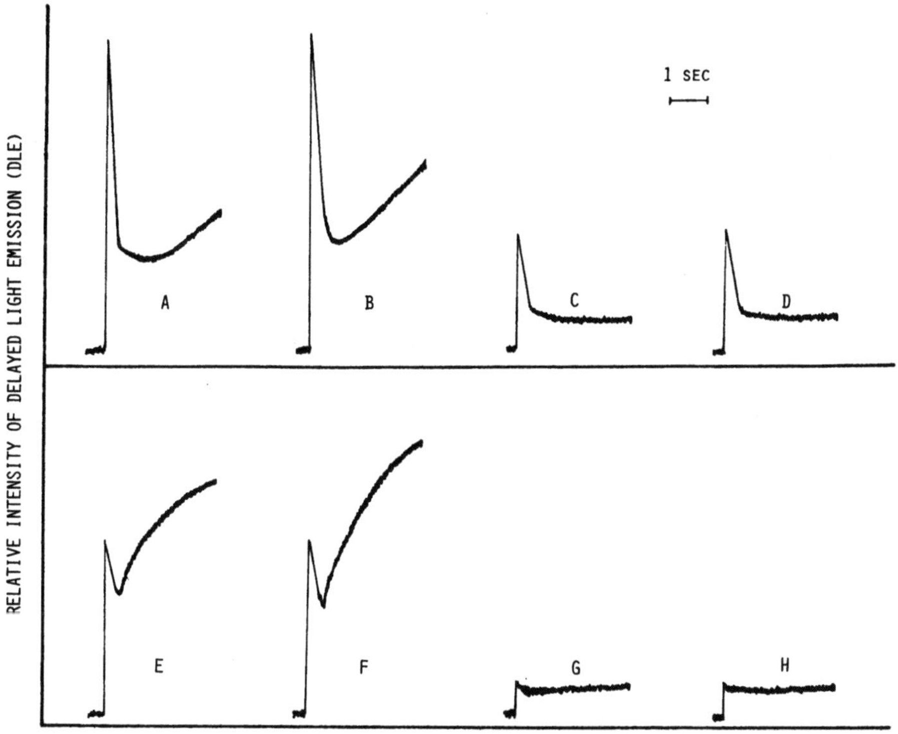

Fig. 4. Effects of the CCKBP and BA on delayed light emission. The procedures for chloroplast preparation were the same as noted in Table 1 and the extraction of the CCKBP was as noted in Fig. 2. Prepared chloroplasts were suspended in Tris buffer containing 10 mM NaCl, and then divided into aliquots (1 ml), to which 10 μl BA or water was added. After dark incubation in an ice bath for 20 min, 0.1 ml of the chloroplast suspension (ca. 0.18 mg chlorophyll per ml), plus 1 ml Tris buffer or Tris buffer containing the extracted crude CCKBP, was used for the DLE assay. The final BA concentration in these mixtures was 3 μM. Excitation light was 10^6 erg $cm^{-2}s^{-1}$ when the sector was stopped. The intensity of DLE at 1.8 ms, after the cessation of 1 ms flash excitation, was measured as described in [20]. *A* control; *B* 3 μM of BA; *C* salt-washed chloroplasts with 3 μM of BA; *D* salt-washed chloroplasts without BA; *E* suspension of salt-washed chloroplasts mixed with CCKBP, but with no BA; *F* Suspension of salt-washed chloroplasts mixed with CCKBP plus 3 μM of BA; *G* CCKBP control solution; *H* CCKBP solution +3 μM of BA

of the CCKBP, and therefore the CCKBP can be considered as a physiologically functional hormone receptor. This receptor, after binding BA, may act through reducting permeability of the membranes of the photosynthetic apparatus. As a result of this reduced permeability, the formation of a high.energy state is promoted in chloroplasts, thereby leading to an increase in photophosphorylation.

4 Conclusions

Use of higher concentrations of NaCl can release the CCKBP from the membranes of the photosynthetic apparatus. The protein, even in a soluble state, still retains its ability to bind BA. The additional increase of DLE intensity in the slowly developed phase, which is closely related to photophosphorylation in chloroplasts, requires the presence of the CCKBP. This protein may thus be a receptor for the CK class of plant hormone.

Acknowledgements. We are grateful to the National Natural Science Foundation of China for the grant (No. 3861067) that supported this research. We are also grateful to Dr. D.Y. Li for his help in the measurement of the delayed light emission of chloroplasts.

References

1. Chen C-M, Ertl J, Yang M-S, Chang C-C (1987) Plant Sci 52:169
2. Chen C-M, Leisner SM (1985) Plant Physiol 77:99
3. Chen C-M, Melitz DK, Petschow B, Eckert RL (1980) Eur J Biochem 108:379
4. Erion JL, Fox JE (1981) Plant Physiol 67:156
5. Flores S, Tobin EM (1986) Planta 168:340
6. Haru K, Naito K, Suzuki H (1982) Physiol Plant 55:247
7. Huang H, Lu JL, Jian ZY, Tang YW (1987) In: Klambt D (ed) Plant hormone receptors. Springer, Berlin Heidelberg New York Tokyo, p 185
8. Huang H, Tang YW (1988) Acta Biol Exp Sin 21:153
9. Huang ZH, Wei JM (1984) Acta Phytophysiol Sin 10:161
10. Itoh S (1980) Plant Cell Physiol 21:873
11. Itoh S, Murata N, Takamia A (1971) Biochim Biophys Acta 245:109
12. Larsson CM, Olsson T (1979) Plant Cell Physiol 20:145
13. Lerbs S, Lerbs W, Klyachko NL, Romanko EG, Kulaeva On, Wallgiehn R, Parthier B (1984) Planta 162:289
14. Moore III, FH (1979) Plant Physiol 64:594
15. Naito K, Nagumo S, Faruya K, Suzuki H (1981) Physiol Plant 52:343
16. Nielsen SO, Lehninger AL (1955) J Biol Chem 215:555
17. Romanko EG, Selivankina SYu, Moshkov IE, Novikova GV (1986) Sov Plant Physiol 33:823
18. Selivankina SYu, Romanko EG, Moshkov IE, Kharchenko VI, Kulaeva ON (1985) Fiziol Rest (MOSC) 32:506
19. Teyssendier de la serve B, Axelas M, Peaud-Lenoel C (1985) Plant Mol Biol 5:155
20. Yu SG, Xie QD (1983) Acta Phytophysiol Sin 9:270

Structure-Activity Relationship Studies and Development of *s*-Triazine and Carbamate Anticytokinins

H. IWAMURA[1]

1 Development of Adenylate Anticytokinins

The importance of cytokinin antagonists for studies on the mode of action of cytokinins was postulated soon after the discovery of kinetin [16]. Earlier searches had also been made among purine analogues known as metabolic inhibitors. Reportedly examined were 2,6-diaminopurine, 6-mercaptopurine, and even puromycin [3]. Their inhibitory activity was, however, very weak and no specific antagonistic nature could be found. About a decade later anticytokinin activity of 7-substituted-3-methylpyrazolo[4,3-d]pyrimidines (*1*), a 9-deazaadenine was reported in the tobacco callus assay [5]. Anticytokinin activity has also been noted for 7-deazaadenine structures. The first ones prepared were 4-substituted 7-(β-D-ribofuranosyl)pyrrolo[2,3-d]pyrimidines (*2*) [7, 8], then 4-substituted-2-methylthio- (*3*) [19] and 4-substituted-2-methylpyrrolo[2,3-d]pyrimidines (*4*) [9]. Furthermore, by structural analogy with these compounds, 4-substituted-2-methylthiopyrido[2,3-d)pyrimidines (*5*), another series of anticytokinins having a fused 6–6 membered ring system have also been developed [10]. These compounds are listed in Fig. 1.

2 Mapping of the Cytokinin Receptor

To find the mode of action of anticytokinins at the site of action, quantitative structure-activity relationship studies have been done [10, 11, 12]. Earlier, a similar analysis was made for N^6-substituted adenines and chemically different DPU cytokinins, revealing how both agonists interact at a common active site, or how both structures correspond to each other when they fit the site, i.e. the cytokinin receptor [6]. A key for the analyses has been the definition of steric parameters suitable for the purpose, and the elucidation and comparison of molecular shapes important for the expression of activity. Figure 2 shows them schematically; L is the length in the stable, staggered conformation of a substituent along the connecting axis (L-axis) to the rest of molecule, W_{max} is the maximum width in the direction perpendicular to the L-axis, T_r is the thickness in the vertical direction to the W_{max} of the right-hand side of the substituent, and T_l is that of the left-hand side. Accordingly, these steric parameters are a kind of vector.

The correlation of the steric parameters with the activity of both the agonist and antagonists has been examined by regression analysis, together with the electronic

[1] Department of Agricultural Chemistry, Faculty of Agriculture, Kyoto University, Kyoto 606, Japan

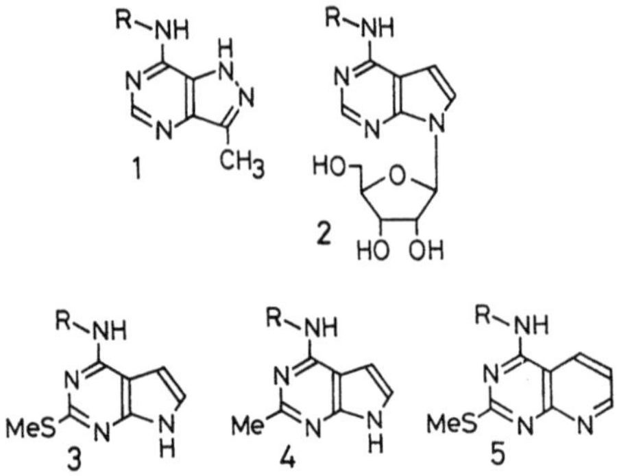

1 2 3 4 5

Fig. 1. Structures of adenylate anticytokinins

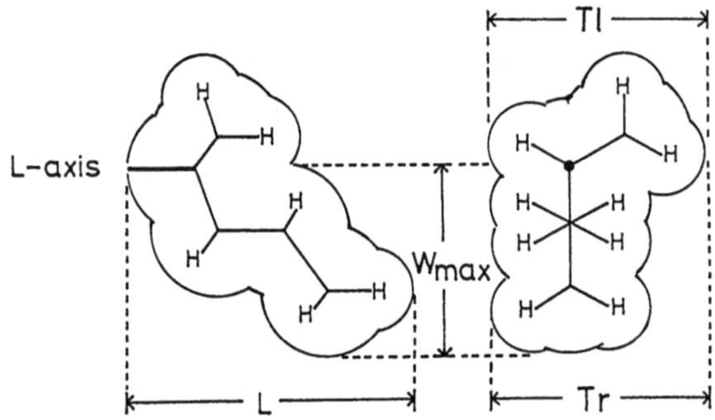

Fig. 2. Schematic representation of the steric parameters. The substituent used as model is l-methylbutyl

and hydrophobic parameters. The significant parameters found in this manner are considered to reflect their importance in binding with the receptor.

Figure 3 schematically shows the combined results for both cytokinins and anticytokinins [6, 10, 12, 18]. The solid lines with a fringe represent the steric interaction site or a spatial wall suggested by the significance of the width and length parameters, and the ovals are those suggested by the significance of the thickness parameters and considered to be located above or below the page plane.

The line suffixed by W_{max} in Fig. 3A was suggested by its significance in the analysis of adenines and drawn in consideration of its direction. L_o and L_p are the lengths of ortho and para groups of ureas respectively. The shaded circle show the region where the hydrophobicity of the molecule is thought to be important for activity. This was drawn according to the fact that the higher the hydrophobicity of meta substituents of DPUs, the more the potency. The electron withdrawal from the NH bridge of both adenylate and urea compounds was also shown to enhance activity. This is considered to suggest an H-bonding interaction between the imino hydrogen atom and a basic group: B of the receptor. Since the W_{max} parameter was also significant in the analyses of anticytokinins, they were accommodated to the model so that their side chain faces the W_{max} wall (Fig. 3B and C). Another fringed line and the ovals were then drawn according to the significance of their length and thickness parameters.

As shown in Fig. 3A, agonistic adenine and urea structures are shown to have a general resemblance.

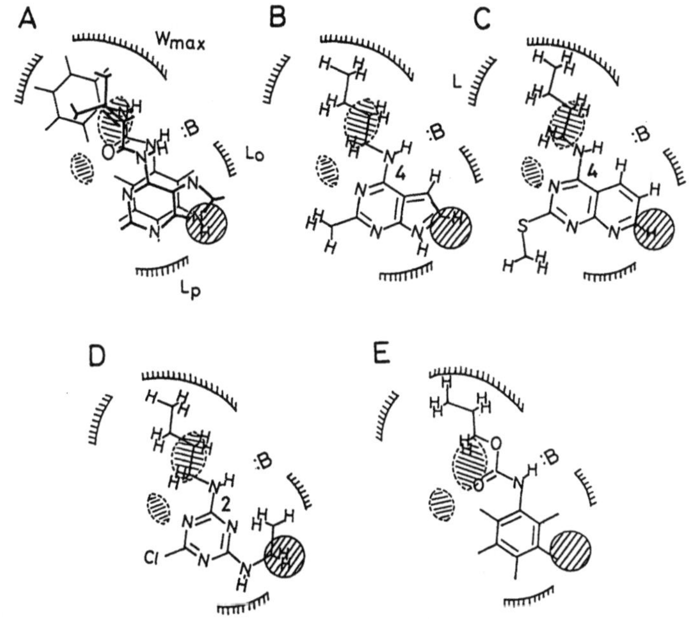

Fig. 3A-E. Schematic substrate-receptor complex of cytokinin-agonistic and antagonistic compounds. Models are for N[6]-substituted adenines and N,N'-diphenylureas (**A**); 4-substituted-2-methylpyrrolo[2,3-d]pyrimidines (**B**); 4-substituted-2-methylthiopyrido[2,3-d]pyrimidines (**C**); N[2]-substituted 2-amino-4-chloro-6-ethylamino-*s*-triazines (**D**); and N-phenylcarbamates (**E**). In **A**, the ends of the bars of the structures represent hydrogen atoms, and the double bonds of the aromatic rings are omitted. **C** and **D** were reproduced from previous papers [12, 18] with permission from the American Chemical Society

3 Development of *s*-Triazine Anticytokinins

The anticytokinins hitherto developed were adenylate compounds. Anticytokinins with a structural resemblance to DPUs, or a non-adenylate structure, were not yet known.

It has been shown in Fig. 3A that DPUs can bind to the receptor in a similar fashion to N[6]-substituted adenines, showing cytokinin activity. By analogy with this, non-adenylate compounds that can fit the receptor similarly to the adenylate anticytokinins could be found. We selected from among all *s*-triazine structures the non-adenylate candidates. We hoped they would bind to the receptor as shown schematically in Fig. 3D, N[2]-substituents at the site where the N[4]-group of adenylate anticytokinins comes on and a 2-ethylamino group at the site the pyrrole or pyridine moiety occupies. The N[2]-hydrogen atom could interact with the basic group: B.

Representatives of the *s*-triazines thus prepared are summarized in Table 1, together with the activity data I_{50}, the molar concentration at which 50% of the tobacco (Wisconsin No. 38) callus growth is obtained on the medium with 0.05×10^{-6} M kinetin but without test compounds [18]. Most potent were *n*-butyl,

Table 1. Anticytokinin activity of N^2-substituted 2-amino-4-chloro-6-ethylamino-*s*-triazines

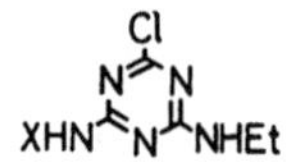

X	I_{50} (10^{-6} M)[a]	X	I_{50} (10^{-6} M)
Et	8.91	*c*-Pr	1.29
n-Pr	5.15	*c*-Bu	0.29
i-Pr	1.10	*c*-Hx	0.47
n-Bu	0.44	*c*-Pr-CH_2	0.47
i-Bu	1.05	*c*-Hx-CH_2	4.79
n-Pent	5.13		

[a] Concentration at which 50% of tobacco callus grown on the medium with 0.05×10^{-6} M kinetin but without anticytokinin is obtained. The callus was derived from *Nicotiana tabacum* var. Wisconsin No. 38 and the medium used is that specified by Linsmaier and Skoog [15].

c-butyl, *c*-hexyl, and *c*-propylmethyl derivatives, the I_{50} being $0.3–0.5 \times 10^{-6}$ M. The compounds with a larger or smaller substituent had weaker activity, suggesting the participation of steric factors in fitting to the receptor.

To confirm their competitive, anticytokinin nature, we used the method of Lineweaver and Burk [14]. The results of the treatment on the *c*-hexyl compound, one of the most active members of the class, is shown in Fig. 4, where the reciprocal of the growth response was plotted against the reciprocal of the concentration of added kinetin. That the resultant set of straight lines has a common intercept suggests that the *s*-triazines share the site of action with cytokinins.

The activity of the *s*-triazine anticytokinins was as high as that of pyrido-[2,3-d]pyrimidines (*5*) but a few times less than that of pyrrolo[2,3-d]pyrimidines (*4*), the most potent class of anticytokinins known so far. The optimization of the structure is under way as well as the examinations of their effects on plant systems other than tobacco.

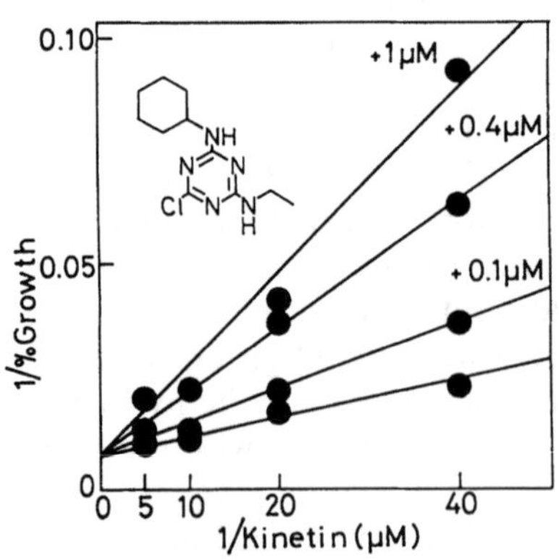

Fig. 4. The reciprocal of the growth rate of tobacco callus plotted as a function of the reciprocal of the concentration of kinetin alone (*bottom line*) and in the presence of N^2-cyclohexyl-2-amino-4-chloro-6-ethylamino-*s*-triazine

4 Development of Carbamate Anticytokinins

Pertinently substituted *s*-triazines are known to inhibit PS II electron flow in chloroplasts and are herbicidal [20], and also phenylurea and carbamate herbicides are known to act at the same site as that of *s*-triazines [17]. In other words, they are bioisosters to each other with respect to the binding to the action site, or the bioisosterism between them is fairly high, irrespective of the apparently different structures. Moreover, phenylureas are known to have cytokinin activity [4, 6]. Thus, the carbamates are expected to behave bioisosterically to phenylureas or *s*-triazines with respect to the interaction with the cytokinion receptor.

Compounds thus prepared retarded the callus growth caused by kinetin, and some of the results are shown in Table 2 (from Shimizu, Iwamura, and Fujita 1988, unpublished). The competitive, antagonistic nature was confirmed as above kinetically (data not shown). Both *O*-alkyl and *O*-phenyl derivatives were active, which is analogous to the fact that compounds having both alkyl and phenyl substituents at the exocyclic amine have activity in the adenylate series. Moreover, compounds having a hydrophobic group like Cl at the meta position of the aniline moiety tended to be higher in activity, and this phenomenon resembles that observed for urea cytokinins [4, 6]. According to these facts, the aniline moiety is considered to correspond to the heterocyclic moiety of the adenylate compounds, the meta substituent coming on the hydrophobic region. The situation is shown in Fig. 3E, where the NH group is more suitably located for interaction with the basic function: B, than that in the compound accommodated upside down.

Table 2. Anticytokinin activity of N-phenylcarbamates

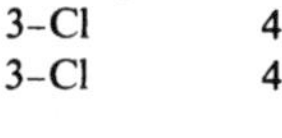

X	R	I_{50} (M)[a]	X	R	I_{50} (M)
3-Cl	*n*-Pr	2.88×10^{-6}	3,4-Cl_2	3-F-C_6H_4	4.26×10^{-6}
4-Cl	*n*-Pr	2.00×10^{-5}	3,4-Cl_2	4-F-C_6H_4	1.31×10^{-6}
2,5-Cl_2	*n*-Pr	5.37×10^{-6}	3,4-Cl_2	3-Me-C_6H_4	5.13×10^{-6}
3,4,5-Cl_3	*n*-Pr	1.78×10^{-6}	3-Cl	4-F-C_6H_4	3.80×10^{-6}
3,4-Cl_2	*n*-Pr	1.48×10^{-5}	3-Cl	4-Cl-C_6H_4	1.00×10^{-6}
3,4-Cl_2	*n*-Pr-CH_2	1.15×10^{-5}			

[a]See footnote a of Table 1.

5 Flower Induction by *s*-Triazine and Carbamate Derivatives in Seedlings of Asparagus Plants

Recently Abe and Kameya [1] have found a novel flower-inducing activity of atrazine, a *s*-triazine herbicide, in seedlings of *Asparagus officinalis* L. (cv Mary Washington 500W). Flowering in asparagus occurs normally two to three years after seeding. However, it occurs after 25 days when the seeds are soaked for 12 days in $200–400 \times 10^{-6}$ M solution of atrazine before planting in soil. The percentage of flowering plants is 20–40%, but they usually die within two months because of the herbicidal action of the compound. This fact prompted us to test our *s*-triazine and carbamate compounds because of their structural congeniality, and also because cytokinins are thought to be involved in flower organ development and sex expression of asparagus plants [13].

Test results of representative compounds are summarized in Table 3. Among *s*-triazines, 2-chloro-4-cyclohexylamino-6-ethylamino-*s*-triazine gave the highest rate of flowering, about 50% at 200×10^{-6} M, without causing death [1]. Conspicuous were the results of carbamates. The flowering brought about by use of *n*-propyl N-(*p*-chloro and bromophenyl)carbamates attains a 90–95% level, and the plants do not die [21]. The low toxicity of the compounds can be attributed to their substituent structures, especially those of 4- and *O*-substituents in *s*-triazines and carbamates respectively, which deviate from those substituent structures which endow the molecules with a high herbicidal activity.

Table 3. Effects of *s*-triazines and N-phenylcarbamates on flowering in seedlings of *Asparagus officinalis* L. cv Mary Washington 500W[a]

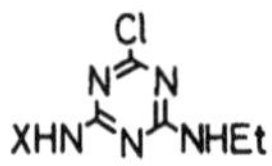

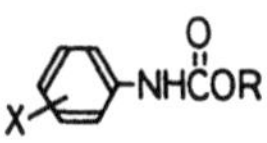

X	Plants with flowers, % (conc. 10^{-4} M)	X	R	Plants with flowers, % (conc. 10^{-4} M)
i-Pr (atrazine)	20 (2)	4-Cl	*n*-Pr	95 (2)
n-Hx	11 (2)	4-Br	*n*-Pr	94 (4)
$MeO(CH_2)_2$	11 (2)	3,4-Cl_2	*n*-Pr	84 (4)
c-Bu	32 (2)	3,4-Cl_2	*n*-Pr	92 (4)
c-Hx	45 (2)	3,4-Cl_2	*c*-Bu	90 (4)
C_6H_5	20 (2)	3,4-Me_2	*n*-Pr	87 (4)

[a]Seeds were soaked in an aqueous solution containing a test compound and less than 0.5% of dimethylsulfoxide. After 12 days, the seedlings were washed with water, planted in Vermiculite, and grown for 13 days at 25°C under a 12-h photoperiod. Flowering rate was expressed by (No. of plants with flowers/No. of plants emerged from Vermiculite) × 100.

Flowering usually occurs only once at the top of the plants (Fig. 5). The flowers are usually normal and fertile. Spraying the plants is not effective, suggesting that the effect is triggered at a very early stage during or after germination. Asparagus is a dioecious species, and male plants are preferred for commercial production because of their greater yield, vigor, and longevity. It is, however, impossible to distinguish the sex until each plant flowers. Thus, our compounds may be useful in determining the sex, and thereby allowing for selection of the more vigorous male plants at quite an early stage.

Fig. 5. Seedlings having flowers of *Asparagus officinalis* L. cv Mary Washington 500W. Treatment was done with *n*-propyl N-(3,4-dichlorophenyl)carbamate

References

1. Abe T, Kameya T (1986) Planta 169:189
2. Abe T, Shimizu R, Iwamura H, Kameya T (1987) Physiol Plant 70:228
3. Blaydes DF (1966) Physiol Plant 19:748
4. Bruce MI (1966) Proc R Soc Lond Ser B 165:245
5. Hecht SM, Bock RM, Schmitz RY, Skoog F, Leonard NJ (1971) Proc Natl Acad Sci USA 68:2608
6. Iwamura H, Fujita T, Koyama S, Koshimizu K, Kumazawa Z (1980) Phytochemistry 19:1309
7. Iwamura H, Ito T, Kumazawa Z, Ogawa Y (1974) Biochem Biophys Res Commun 57:412
8. Iwamura H, Ito T, Kumazawa Z, Ogawa Y (1975) Phytochemistry 14:2317
9. Iwamura H, Masuda N, Koshimizu K, Matsubara S (1979) Phytochemistry 18:217
10. Iwamura H, Masuda N, Koshimizu K, Matsubara S (1983) J Med Chem 26:838
11. Iwamura H, Murakami S, Koga J, Matsubara S, Koshimizu K (1979) Phytochemistry 18:1265
12. Iwamura H, Murakami S, Koshimizu K, Matsubara S (1985) J Med Chem 28:577
13. Lazarte JE, Garrison SA (1980) J Am Sco Hortic Sci 105:691
14. Lineweaver H, Burk D (1934) J Am Chem Soc 56:658
15. Linsmaier EM, Skoog F (1965) Physiol Plant 18:100
16. Miller CO, Skoog F, von Saltza MH, Strong FM (1955) J Am Chem Soc 77:1329
17. Mitsutake K, Iwamura H, Shimizu R, Fujita T (1986) J Agric Food Chem 34:725
18. Shimizu R, Iwamura H, Matsubara S, Fujita T (1989) J Agric Food Chem 37:236
19. Skoog F, Schmitz RY, Hecht SM, Frye FB (1975) Proc Natl Acad Sci USA 72:3508
20. Tischer W, Strotmann H (1977) Biochem Biophys Acta 460:113
21. Yanosaka K, Iwamura H, Fujita T (1989) Z Naturforsch 44c:226

Construction of a System for Regulated Alterations of Endogenous Cytokinins

J.I. MEDFORD[1], J.A. WINTER[2], and H.J. KLEE[3]

1 Introduction

Cytokinins are a class of phytohormones that are believed to regulate development at various times and places [6, 9, 17]. Specific aspects of development that are affected by cytokinins include shoot formation, release of axillary buds, partial floral evocation and inhibition of root elongation. However, there are some instances where cytokinins seemingly have little or no effect on these and other processes, thus producing puzzling discrepancies. These discrepancies are significant, because in order to understand how cytokinins function we need to know which processes are not effected by cytokinins as well as those that are effected.

A unique approach toward understanding the function of phytohormones comes from technological advances in plant transformation. Previously, Klee et al. [10] developed transgenic petunia plants where the endogenous IAA (indole acetic acid) levels were altered using the auxin biosynthetic genes from *Agrobacterium tumefaciens*. *Agrobacterium* also encodes a gene for cytokinin biosynthesis, isopentenyl transferase (*ipt*). This enzyme catalyzes the condensation of adenosine monophosphate and isopentenyl pyrophosphate, giving isopentenyl AMP, a putative precursor to further cytokinin biosynthesis [13].

We are interested in extending our studies on the role of phytohormones in plant development by altering the endogenous cytokinin levels. In order to do this, a chimeric gene was constructed that fused the heat inducible maize hsp70 promoter to the *Agrobacterium ipt* coding region.

2 Results

We designed a system where the *ipt* gene from *Agrobacterium* can be regulated by heat. In order to do this, site-specific mutagenesis [11] was performed on a 700 bp fragment containing the maize hsp70 promoter [15] creating a BglII site at the 3′ end of the hs non-translated leader. This mutagenized promoter, containing transcriptional and translational regulatory signals, was fused to a 750 bp fragment of pTiT37 [7] containing the *ipt* coding region (Fig. 1). The chimeric gene (hs-*ipt*) was

[1] Department of Biology, College of Science, Pennsylvania State University, University Park, PA 16802, USA

[2] Dept. of Biochemistry, Univ. of Missouri, Columbia, MO 65221, USA

[3] Plant Molecular Biology Group, Monsanto Company, 700 Chesterfield Village Parkway, St. Louis, MO 63198, USA

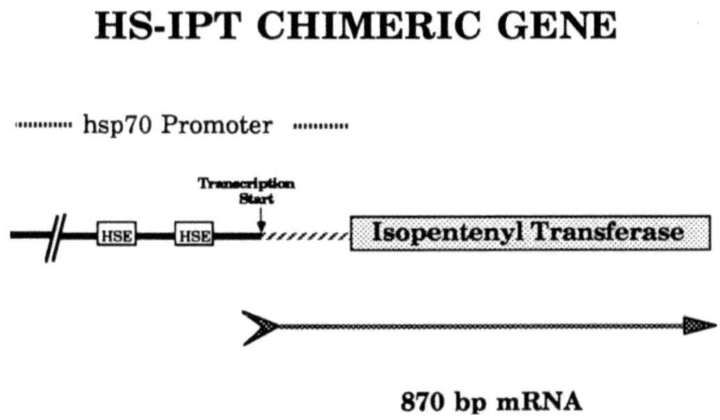

Fig. 1. Chimeric heat shock-*ipt* gene. The entire hsp70 promoter containing two heat shock elements (HSE) and 120 bp of the hs non-translated leader (////////) was ligated to an the *ipt* coding region from pTiT37 [7]. The transcription product from the chimeric gene is an 870 bp mRNA (*arrow*). The chimeric gene was inserted into plant transformation vectors and transferred to plants [12]

inserted into plant transformation vectors [13] and introduced into petunia and *Arabidopsis* plants. Transgenic progeny were identified by Southern blot analysis and antibiotic resistance (data not shown).

In order to verify the functionality of the chimeric hs-*ipt* gene, plants were analyzed for induction of *ipt* mRNA. Leaves from four independent transgenic petunias were excised and incubated in a water bath in vitro at 42 °C for 1 h. Figure 2 shows that the heat shock induced *ipt* mRNA in all four plants. No detectable *ipt* mRNA was found at the control temperature (25 °C) in the four independent plants.

Since we wanted to use the hs-*ipt* gene to induce alterations in cytokinins during various phases of development, conditions that allowed both in vivo heat shock induction and plants survival were defined. Plants were heat shocked by a linear elevation of the temperature over a 15-min period with relative humidity at 70%. The temperature was maintained for 2 h followed by a 15-min linear decline. Both petunia and *Arabidopsis* plants survived the heat shock at all temperatures tested. The effectiveness of the in vivo hs was determined by analysis of endogenous hsp70 induction. Northern blot analysis was done using a petunia hsp70 cDNA clone (Winter, unpubl.) under conditions allowing both cross hybridization between species and to various hsp70 family members and cognates. Even at the control

Fig. 2. Analysis of hs-*ipt* gene expression in response to temperature elevation. Leaves were excised from four independent transgenic petunias (*A*-*D*) and incubated at 25 °C or at 42 °C for 2 h. Total RNA was extracted immediately thereafter and induction of the hs-*ipt* gene analyzed by Northern blotting. Blots were probed with the 750 bp *ipt* coding region

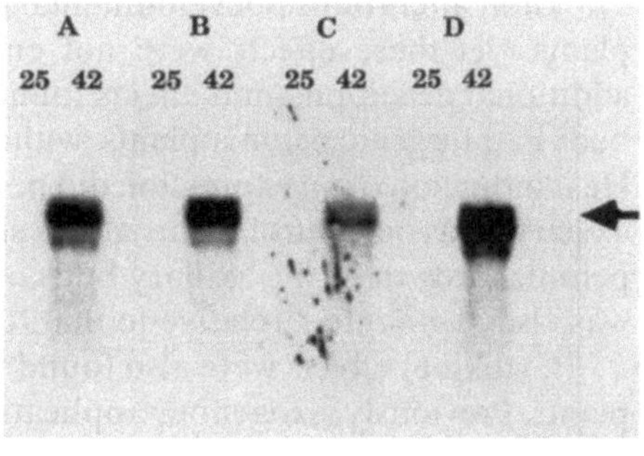

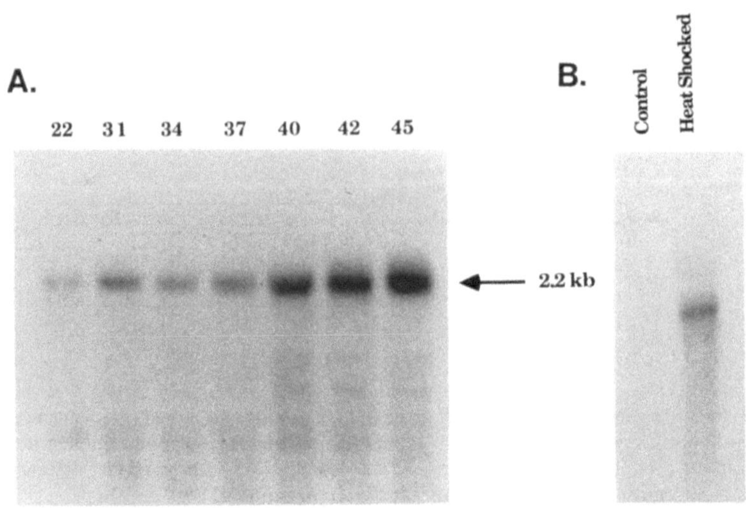

Fig. 3A,B. In vivo induction of hsp70 and *ipt* mRNA. **A** Induction of endogenous hsp70 mRNA. *Arabidopsis* plants (3–4 weeks old) were heat shocked at the various temperatures indicated in a chamber as described in the text. Four h after heat shock, plant material was harvested and total RNA isolated. Induction of the endogenous hsp70 genes was determined by hybridization of a northern blot to a cloned petunia hsp70 gene. The blot was hybridized under conditions to allow cross reactivity between the two species and members of the hsp70 gene family. **B** Induction of *ipt* mRNA. Transgenic *Arabidopsis* plants were hs in vivo at 42°C and RNA isolated 4 h afterwards. The *ipt* mRNA could not be detected at the control temperature

temperature (22°C), hsp70 mRNA is detectable (Fig. 3A). Nonetheless, the temperature elevation in vivo results in accumulation of hsp70 mRNA. Densitometer scanning of the autoradiogram indicated that heat shock at 37°C produced a 2.3-fold induction, and at the highest temperature, 45°C, hsp70 was induced 8.5-fold. As an additional assessment of the system, transgenic plants were analyzed for induction of the *ipt* mRNA. Figure 3B shows that in vivo induction at 42°C also results in the accumulation of an 870 by *ipt* mRNA.

In previous studies using crown gall tumors incited by *A. tumefaciens*, accumulation of *ipt* mRNA resulted in a corresponding accumulation of cytokinins [2]. Since *ipt* mRNA accumulated in plants containing the hs-*ipt* gene, the transgenic plants were examined (both with and without heat induction) for effects of cytokinin alterations. Developmental effects were found in the non-heat shocked plants, yet these effects were not enhanced by heat induction, nor were any additional developmental effects found following hs. Figure 4 shows that axillary buds in transgenic petunia plants, without heat shock, are released from dormancy. Heat induction of *ipt* expression did not enhance axillary bud growth. Comparable axillary buds on control plants at this stage are small and inconspicuous. Similar to petunia, growth of the axillary buds of nonstressed transgenic *Arabidopsis* plants was also augmented, relative to that found in control plants.

Cytokinin effects were also found in the root system of transgenic *Arabidopsis* plants. Previously, exogenous application of kinetin to young corn roots resulted in

Fig. 4. Release of axillary buds in uninduced transgenic petunia plants. Uppermost nodes of a hs-*ipt* transgenic petunia plant showing growth of axillary buds

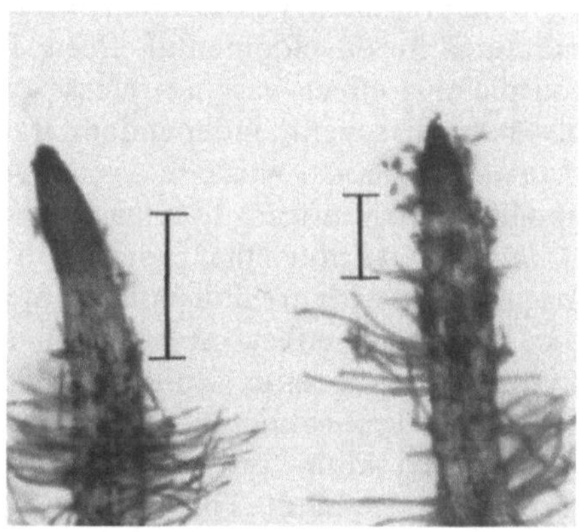

Fig. 5. Morphology of uninduced transgenic *Arabidopsis* root tips. Root tips from transgenic (*right*) and control plants (*left*) grown on agar medium for 7 days. In the transgenic plant root hairs emerged closer to the tip, indicating a reduction in the zone of elongation

a disruption of the pattern of growth [8]. In the roots of transgenic plants 7 days after germination, the root hairs appeared closer to the tip, indicating a reduction in the size of the zone of elongation. Furthermore, the rate of root growth, as measured in terms of elongation, was only 68% of that found in the transgenic plants.

Previous reports on *Arabidopsis* have shown that cytokinins enchance the onset of flowering [4, 14]. To test the effect of enhanced cytokinin levels on flowering, transgenic *Arabidopsis* plants (a facultative long day plant) were grown in short days (8 h). Two-and-a-half months after the start of the experiment, one set of plants (wild type and transgenic) were hs at 42°C every third day for a period of three weeks. The wild-type and transgenic plants (induced and uninduced) flowered 5.5 months after the start of the expriment: yet there was no significant difference in the onset of flowering in any of the plants.

3 Discussion

We have described preliminary work on construction of a system to alter cytokinin levels in plants. In order to do this, a heat inducible promoter (maize hsp70) was fused to the coding region of the *Agrobacterium ipt* gene. To test the functionality of the system, transgenic petunia leaves were heat shocked at 45 °C and assayed for accumulation of the *ipt* mRNA. In four independent transgenic plants, thermal induction produced a corresponding accumulation of the *ipt* mRNA. Having established that the chimeric gene was inducible, conditions were defined for in vivo induction. At all temperatures tested the plants survived and hsp70 mRNA accumulated after thermal induction. After heat shock at 37°C there was only 2.3-fold induction of the hsp70 mRNA, which is in agreement with a recent study on hsp70 in *Arabidopsis* [18]. The accumulation of hsp70 mRNA at 22 °C probably represents expression of hsp70 family members referred to as cognates or hsc70 [5]. Nonetheless, the signal at 22°C does suggest that there is some non-thermal regulation of plant hsp70s. Nonthermal expression of hs-*ipt* mRNA would account for the fact that cytokinin effects were found in non-heat shocked transgenic plants. Although the uninduced hs-*ipt* mRNA could not be detected, even a low level of *ipt* expression could result in considerable cytokinin production.

As a preliminary assessment of cytokinin alterations the transgenic plants were examined for developmental effects. In uninduced transgenic petunias, the most conspicuous effect was the release of axillary buds from dormancy. This was observed in several independent transgenic petunia plants and in transgenic *Arabidopsis* plants. This effect is in agreement with studies in which exogenously applied cytokinin leads to release of dormant buds [16]. In uninduced transgenic *Arabidopsis*, we noted that root growth was inhibited in the transgenic plants, again in agreement with previous studies [8]. In contrast to the effects on axillary bud and root growth, the effects on flowering in transgenic *Arabidopsis* plants differs from studies with exogenous application of cytokinin. One explanation for this is that in exogenous application the cytokinins were applied directly to the shoot apex, whereas cytokinin production in the transgenic plants presumably occurs throughout the plant. This may be because cytokinin enhancement of flowering through exogenous application is more a consequence of selective nutrient distribution rather than an action of cytokinin on the developmental program of the plant. Alternatively, cytokinins may produce only a partial evocation, which would require more extensive analysis of the shoot apex to ascertain.

The unexpected result of this work is that heat induced increases in the *ipt* mRNA (Figs. 2 and 3), but did not produce effective alterations in plant development. Preliminary analysis of cytokinin levels indicates that even without heat induction, the transgenic plants contain cytokinin levels 8-fold over that of control plants, while cytokinin levels in heat-induced plants is 60-170-fold over that found in control plants (Horgan, unpubl.). One explanation for these results is that the cytokinin levels after heat shock may be beyond the ability of the plant to respond. The supraoptimal cytokinin levels after hs may negate the developmental effects of cytokinins that were found at control temperatures. Our current work is focusing more closely on defining the optimal response times and processes.

Abbreviations

ipt-isopentenyl transferase gene
hs-ipt chimeric heat shock isopentenyl transferase gene
bp-base pairs

References

1. Akiyoshi D, Klee H, Amasino R, Nester EW, Gordon MP (1984) Proc Natl Acad Sci USA 81:5994
2. Akiyoshi DE, Morris RO, Hinz R, Mischke BS, Kosuge T, Garfinkel D, Gordon MP, Nester EW (1983) Proc Natl Acad Sci USA 80:407
3. Barry GF, Rogers SG, Fraley RT, Brand L (1984) Proc Natl Acad Sci USA 81:4776
4. Besnard-Wilbaut C (1981) Physiol Plant 53:205
5. Bond U, Schlesinger MJ (1987) Adv Gen 24:1–30
6. Finkelstein RR, Estelle MA, Martinez-Zapater JM, Somerville CR (1987) In: Goldberg RB, Verma DP (eds) Temporal and Spatial Regulation of Plant Genes. Springer, Berlin Heidelberg New York Tokyo
7. Goldberg SB, Flick JS, Rogers SG (1984) NAR 12:4665
8. Kappler R, Kristen U (1986) Bot Gaz 147:247
9. King P (1988) Trends Genet 4:157
10. Klee HJ, Horsch RB, Hinchee MA, Hein MB, Hoffmann NL (1987) Genes Dev 1:86
11. Kunkel TA (1985) Proc Natl Acad Sci USA 82:488
12. Lloyd A, Barnason A, Rogers SG, Byrne M, Fraley RT, Horsch RB (1986) Science 234:464
13. Medford JI, Horgan R, Klee HJ (1989) The Plant Cell 1:403
14. Michniewicz M, Kamienska A (1965) Naturwissenschaften 52:623
15. Rochester DE, Winter JA, Shah DM (1986) EMBO J 5:451
16. Tamas IA (1987) In: Davies RJ (ed) Plant hormones and their roles in plant growth and development. Nijhoff, Boston, p 393
17. Wareing PF, Phillips IDJ (1981) Growth and differentiation in plants, Pergamon, New York
18. Wu CH, Caspar T, Browse J, Lindquist S, Somerville C (1988) Plant Physiol 88:731

Ethylene Binding Receptors – Is There More Than One?

E.C. SISLER[1]

1 Introduction

In the late 1970s, ethylene binding was reported in tobacco leaves [13, 20]. About the same time a report on compartmentation of ethylene in *Phaseolus vulgarus* cotyledons appeared [10]. Here, too, ethylene appears to be binding. Since then binding has been reported in a large number of plant tissues, and a report has appeared on non-physiological binding [1]. The usual parameters of binding have been determined in attempts to relate ethylene binding with physiological activity. Various criteria have been made for considering an isolated receptor as physiologically important. Trewavas and Jones [24] have presented the following: (1) it should show reversible, specific and saturable binding; (2) its binding dissociation constants should be at or lower than the biologically active concentrations; (3) the number of binding sites should be small; and (4) chemical modification of substrates should lead to equivalent modification in binding affinity and biological activity. For the most part, the data collected to date meet these criteria, and much of the data would suggest a single binding site. Ethylene is the only plant hormone for which binding can be measured in vivo, which adds another dimension to studies by allowing another parameter (diffusion from the site) to be measured.

Recent experiments involving diffusion of bound [^{14}C]ethylene from the binding site in intact tissue point to more than one binding site in the same tissue. Some of the bound ethylene dissociates rapidly, some appears to be intermediate in binding time, and some of the binding appears to dissociate very slowly [3, 4, 5, 6, 7, 10, 13, 15]. All forms of binding appear to be reversible. Purification data also suggest more than one type of site [14, 17]. Further evidence is presented here that there is more than one type of binding, but more research will be necessary to determine the significance of each. Although many of the criteria point to one or more of these components as being the physiological receptor, none has yet been shown to be a receptor.

2 Materials and Methods

Mung bean sprouts (*Vigna radiata* L.) were obtained from commercial sources and used within 2–3 days after harvest. Tobacco leaves (*Nicotiana tabacum* L.) were grown in a greenhouse. Sprouts or extracts were exposed to [^{14}C]ethylene in a

[1]Department of Biochemistry, North Carolina State University, Box 7622, Raleigh, NC 27695–7622, USA

desiccator in the presence or absence of unlabeled ethylene [13]. Exposure of living material was usually 45 min to minimize the effects of endogenous ethylene; exposure of extracts was usually 3 h. After the exposure time, the plant material was removed and placed in a 250-ml container along with a scintillation vial containing 0.2 ml of $Hg(ClO_4)_2$ on glass fiber which was used to collect the labeled ethylene [13, 26]. The containers were placed in an oven at 60°C for approximately 4 h, followed by 8 h at 25°C; after removal of the vials, scintillation fluid was added and the samples were counted. In experiments where the gas phase was to be measured, 20 g of sprouts were blended with 35 ml of water in a 250-ml plastic blender container for 1 min. The container had a hole and a short tube in the bottom. After the blending was complete, a stopper was removed from the hole, the liquid phase was drained into another 250-ml container. A scintillation vial with $Hg(ClO_4)_2$ was attached to the short tube on the blending container to collect the labeled ethylene in the gas phase in the container. A vial with $Hg(ClO_4)_2$ was placed in the container with the liquid phase to collect the labeled ethylene in this phase. The gas phase was collected for 12 h at room temperature and the liquid phase was collected for 12 h at 60°C. Scintillation fluid was then added to the vials and the samples were counted.

3 Results and Discussion

3.1 Evidence Based on Binding Constants

The binding constant K_d is one of the parameters used to characterize hormone receptors. The early measurements indicated that the binding constant K_d for ethylene was close to the K_m for a physiological response. Thus, the K_d for binding in tobacco was 0.27 μl/l and the K_m for a physiological response was 0.3 μl/l. For propylene the constants were 42 μl/l for binding and 50 μl/l for a physiological response [13]. In each case, the values are reasonably close. When experiments were initially done, it was not realized that there probably are two or more components to the binding, and the values represent mainly the short-time binding component. The long-time binding component(s) is a much smaller portion of the total, but some ethylene from the longer binding component would have been collected. Whether or not both of the components have the same K_d has not been determined. Since the original values were taken, many more values have been compiled. These values range from 0.09 μl/L to 0.3 μ/l [18]. Recently a much higher binding constant has been reported for pea tissue [12]. This binding component has a much higher K_d, and the time of association is much longer than is usually observed, making its significance difficult to interpret. High amounts of ethylene do exist in tissues, such as apples, with no apparent role, and perhaps a higher K_d binding site does play some as yet unknown role. Studies on binding of compounds other than ethylene, some giving a response and some blocking the response, have been made [2, 8, 16, 19], but no large differences between the different preparations have been noted.

3.2 Diffusion of Ethylene from the Binding Site in Mung Bean Sprouts

In experiments where ethylene has been allowed to bind to mung bean sprouts, and the unbound ethylene has been subsequently removed, it is possible to follow the diffusion of the labeled ethylene from the tissue. Ethylene not bound to the site rapidly diffuses out. Bound ethylene diffuses out of mung bean sprouts in a complex manner (Fig. 1). A large component (approximately 65%) diffuses out rapidly with a half-life of about 6 min. Another component (approximately 12%) diffuses out with a half-life of about 2 h, and still another component (approximately 23%) diffuses out with a half-life of about 50 h. Are these components different receptors? At the present time the difference in diffusion rates is the only evidence that they are different, but such large differences in diffusion rate, together with differences in behavior during purification could suggest different roles.

3.3 Diffusion of Ethylene from the Binding Site in Extracts of Mung Bean Sprouts

When extracts of mung bean sprouts are exposed to [^{14}C[ethylene binding occurs; however, on an equivalent weight basis, the amount of binding is much less than with intact bean plants (Fig. 1). After binding, the ethylene diffuses away. One component has a half-life of approximately 1h, the other component has a half-life of approximately 50 h. There does not appear to be a very short half-life component (i.e., a few minutes) as occurs in intact germinant plants. This raises the question of whether the short half-life diffusing component observed in vivo binds ethylene in vitro. Either such binding must be much reduced, or is totally absent from the extract.

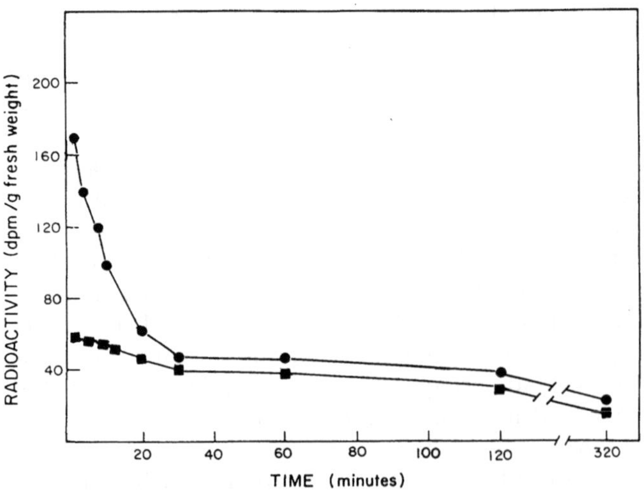

Fig. 1. Diffusion of bound ethylene from intact mung bean sprouts (●), and diffusion from a mung bean sprout extract (■) prepared from an equivalent amount of mung beans. Ten-gram samples of sprouts were used and values were corrected for unbound ethylene; SD = 5

3.4 Ethylene in the Gaseous Phase after Binding and Blending Mung Bean Sprouts

After [^{14}C]ethylene is bound and removed from the source of unbound ethylene, blending of the sprouts causes the release of a substantial amount of bound ethylene very rapidly. Figure 2A shows the amount of [^{14}C]ethylene released into the gaseous phase after blending for various time intervals up to 20 min. Much of the ethylene is released during the first minute. Since all of the samples were started at the same time, the amount of ethylene in the gas phase plus liquid phase should total the amount in the sprouts at 1 min. Figure 2B shows the amount of ethylene released into the gas phase after allowing diffusion to occur for the specified time, then

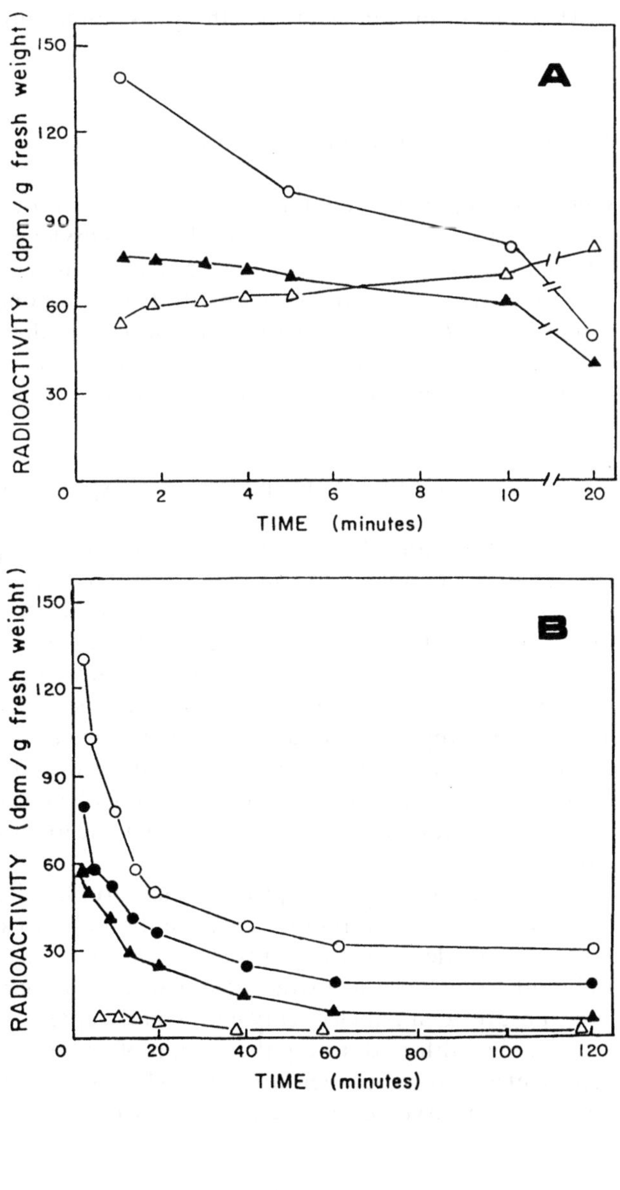

Fig. 2. A Ethylene from intact mung bean sprouts (○), and in liquid phase (▲) and gas phase (△) after blending in water. Blending was started at zero time for the number of minutes indicated. Values have been corrected for unbound ethylene; SD = 6. **B** Ethylene from intact mung bean sprouts (○), and in liquid phase (●) and gas phase (▲) after blending in water. Blending was for 1 min after they were aired for the indicated time. △ is the amount of ethylene released per minute from intact sprouts after airing for the indicated time. Values have been corrected for unbound ethylene; SD = 4

blending for 1 min. The amount released in a 1-min interval appears to be much more than would be released during the same time interval from intact germinant plants (compare Figures 2A and 2B). The amount of ethylene in the gas phase plus liquid phase should total the amount in the sprouts at each point. It is not likely that this represents ethylene that was previously released and was simply trapped in the tissue. Unbound ethylene diffuses from the tissue with a half-life of approximately 22 s. It would thus appear that blending and disrupting of the tissue causes an abrupt release of a portion of the ethylene from the tissue. This appears to correspond to an amount of ethylene that possibly comes from the short-time binding component. A portion of the ethylene is not rapidly released, and this corresponds approximately to the longer binding component(s).

A study needs to be undertaken to determine whether the rapid release of ethylene can be prevented by such things as buffers, protease inhibitors, or osmotic regulators.

3.5 Diffusion of Ethylene from the Binding Site of Extracts of Mung Bean Seeds

Diffusion of ethylene from mung bean seeds occurs over a long period of time (Fig. 3A). The time for one-half of the bound ethylene to diffuse from the site is approximately 50 h and does not suggest more than one phase or, if there is more than one, the others are very small. This component resembles one studied extensively in *Phaseolus vulgaris*. It has been solubilized and partially purified [22, 23] and some of its properties have recently been summarized [9, 21]. Further studies will be required to determine if the two are identical.

3.6 Diffusion of Ethylene from Sites after Exposure to Unlabeled Ethylene

If extracts from mung bean seeds are exposed to 10 μl/l of ethylene for 12 h and then subjected to vacuum for 20 min to remove unbound ethylene, most of the binding sites are occupied by unlabeled ethylene. Exposure to [^{14}C]ethylene for 1 h reveals only a very small amount of binding of labeled ethylene (Fig. 3A). This amount of binding represents the number of sites not occupied by unlabeled ethylene and any binding of [^{14}C]-ethylene due to an exchange between bound unlabeled ethylene and labeled ethylene. The unoccupied sites would represent the amount of unlabeled ethylene that had diffused away during the 80 min (1 h treatment time plus 20 min vacuum time). The amount of bound labeled ethylene (Fig. 3A) does seem to be a little greater than would be expected due to diffusion of unlabeled ethylene from the site.

Exposure of mung bean sprout extracts to unlabeled ethylene for 12 h, applying vacuum for 20 min to remove excess unbound ethylene, and then treating with unlabeled ethylene for 1 h results in a reduced amount of binding. The results are different from those in extracts from seeds. The labeled ethylene bound after preexposure to unlabeled ethylene would represent binding sites from which ethylene had diffused during the 80 min of vacuum and treatment. It could also represent exchange between labeled ethylene and bound unlabeled ethylene. If the values (lower curve) represented only sites which had become vacant during 80 min,

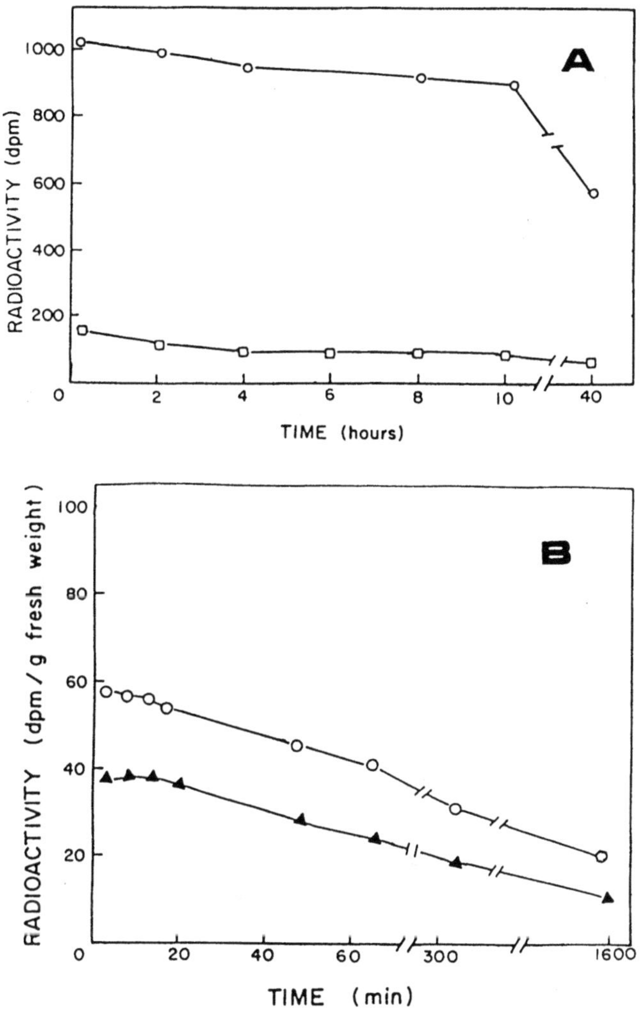

Fig. 3. **A** Bound ethylene remaining in an extract of mung bean seeds. Each point represents an equal amount of extract. The *upper line* (○) was without pretreatment with unlabeled ethylene. The *lower line* (□) represents equivalent samples pretreated 12 h with 10 μl/l of unlabeled ethylene. Samples have been corrected for unbound ethylene; SD = 8. **B** Diffusion of bound ethylene from the mung bean sprout extract. *Upper line* (○) without pretreatment with unlabeled ethylene. *Lower line* (▲) represents pretreatment for 12 h with 10 μl/l unlabeled ethylene. Values are corrected for unbound ethylene; SD = 5

the shape of the diffusion curve should appear much different from the curve obtained in the absence of unlabeled ethylene (upper curve). It should be almost devoid of long-time binding component because that component should be nearly saturated with unlabeled ethylene. This is not the case. There is much more long-time binding component than would be expected. Further, the kinetics of diffusion of the more recently bound component are essentially the same as those of the component before treatment with unlabeled ethylene. Both a medium- and

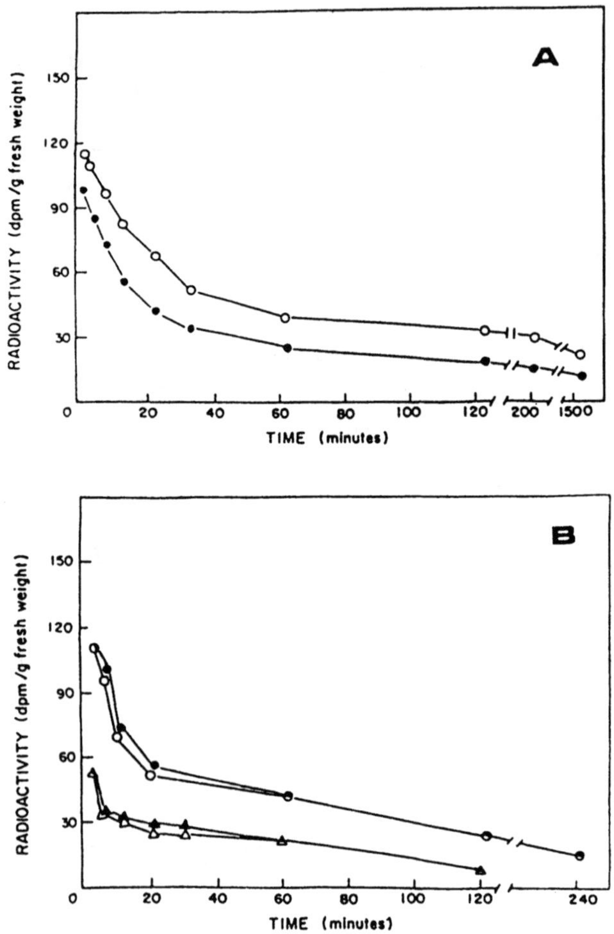

Fig. 4. A Diffusion of bound ethylene from intact mung bean sprouts. *Upper line* without pretreatment with unlabeled ethylene; *lower line* diffusion after pretreatment with unlabeled ethylene (10 μl/l) prior to exposure to [^{14}C]ethylene. Values are corrected for unbound ethylene; SD = 7. **B** Release of bound ethylene in the presence and absence of 10.000 μl/l of unlabeled ethylene. Ethylene present (●) and absent (○) in mung bean sprouts. Ethylene present (▲) and absent (△) in tobacco leaves

long-time binding component are present (Fig. 3B). This may mean the medium-time and long-time binding components are linked, but it is not clear how.

The kinetics of diffusion from intact mung bean sprouts resemble those from extracts of sprouts. A large portion of the long-time binding component remains (Fig. 4A) after pretreatment with unlabeled ethylene. Again, this may represent exchange between bound unlabeled ethylene and unbound [^{14}C] labeled ethylene.

3.7 Test for Cooperative Release or Exchange of Bound Ethylene

In some situations more than one molecule of ligand may bind to the receptor. Usually they bind with different kinetics. Dissociation could also be linked and the presence of one bound ethylene could alter the dissociation of another. One test would be to see if the presence of saturating amounts of ethylene added after [^{14}C] ethylene was bound would alter dissociation from the sites. This was tested using a large amount of ethylene, but no difference was observed (Fig. 4B). Although this does not prove that more than one molecule does not bind to a single component, it suggests that the presence of one molecule does not influence the rate of diffusion from the site of another.

4 Conclusions

The evidence obtained so far would indicate that more than one type of ethylene-binding site is present in most plants. In mung beans there are at least two and perhaps three or more. There is no way to tell at present if any of these is the "physiological" receptor. They may all be, and should be considered so until shown to be otherwise. However, it is of course possible that none may be. The short-time binding component seems to be the most likely prospect for rapid ethylene responses [25]. This component does not appear to bind ethylene in vitro, which poses a serious problem. It may be necessary to use some technique such as photoaffinity labeling to localize it in vivo and identify it in vitro. If a label can be attached in vivo and followed in vitro, it may be possible to show the relation of this component to other binding components and to the physiological receptor. Experiments along these lines are being conducted [11], and may eventually reveal a relation to the different binding components. In the meantime, efforts should be made to purify and study those components that can be measured so that this complex situation can be better understood.

References

1. Abeles FB (1984) Plant Physiol 74:525
2. Beggs MJ, Sisler EC (1986) Plant Growth Regul 4:13
3. Bengochea T, Dodds JH, Evans DE, Jerie PH, Niepel B, Sharri AR, Hall MA (1980) Planta 148:397
4. Evans DE, Bengochea T, Cairns AR, Dodds JH, Hall MA (1982) Plant Cell Environ 5:101
5. Evans DE, Smith AR, Taylor JE, Hall MA (1984) Plant Growth Regul 2:187
6. Goren R, Sisler EC (1984) Tob Sci 28:110
7. Goren R, Sisler EC (1986) Plant Growth Regul 4:43
8. Hall MA, Smith AR, Thomas CJR, Howarth CJ (1984) In: Fuchs Y, Chalutz E (eds) Binding sites for ethylene. Nijhoff/Junk, The Hague, p 55
9. Hall MA, Howarth CJ, Robertson D, Sanders IO, Smith AR, Smith PG, Starling RJ, Tang Z-D, Thomas CJR, Williams RAN (1987) In: Molecular biology of plant growth control. Liss, New York, p 335
10. Jerie PH, Shaari AR, Hall MA (1979) Planta 144:503
11. Quinn JM, Yang SF (1988) Plant Physiol 86:(Suppl) p 113
12. Sanders IO, Robertson DR, Smith AR, Hall MA (1987) In: Klambt D (ed) Plant hormone receptors. Springer, Berlin Heidelberg New York Tokyo, p 289
13. Sisler EC (1979) Plant Physiol 64:538
14. Sisler EC (1980) Plant Physiol 66:404
15. Sisler EC (1982) J Plant Growth Regul 1:219
16. Sisler EC (1982b) J Plant Growth Regul 1:211
17. Sisler EC (1987) In: Klambt D (ed) Plant hormone receptors. Springer, Berlin Heidelberg New York Tokyo, pp 297
18. Sisler EC, Wood C (1987) In: Klambt D (ed) Plant hormone receptors. Springer, Berlin Heidelberg New York Tokyo, p 239

19. Sisler EC, Wood C (1988) Plant Growth Regul 7:181
20. Sisler EC, Wylie PA (1978) Plant Physiol 61:(Suppl) p 131
21. Smith AR, Hall MA (1985) In: Roberts JA, Tucker GA (eds) Ethylene and plant development. Butterworth, London, p 101
22. Thomas CJR, Smith AR, Hall MA (1984) Planta 160:474
23. Thomas CJR, Smith AR, Hall MA (1985) Planta 164:272
24. Trewavas AJ, Jones AM (1981) What's New in Plant Physiol 12:5
25. Warner IN, Leopold AC (1971) Biochem Biophys Res Commun 44:989
26. Young RE, Pratt HK, Biale JB (1952) Anal Chem 24:551

III Hormones and Calcium

Calcium and Second Messengers in Hormonal Regulation

A.C. LEOPOLD[1]

1 Introduction

From my point of view, there are three major ambiguities in the present understanding of the plant hormone field: (1) There is a frustrating lack of specificity in the functions regulated by the various hormones; that is, nearly any major developmental function in plant growth and development can be altered by any of the five plant hormones. This confusing overlap in regulation has been recognized for more than a decade [14]; (2) In many instances there is no apparent correlation between the amount of a given hormone present in plants and the activity or function which the hormone is presumed to regulate. Thus, a hormone which is presumed to be an inhibitor of growth, such as ABA, is often found in greatest abundance in tissues which are growing most rapidly. The report of high ABA levels in rapidly growing rice plants is a vivid example (Takahashi, this volume); and (3) After five decades of research on plant hormones, we still are ignorant of the mechanism of action of any one of them.

It is my plan to review in a general way the principal components of hormonal regulatory systems as we perceive them today, with emphasis on calcium and other second messengers, and then to return to the three major ambiguities which we face, seeking possible means of dealing with them through an expanding knowledge of mechanisms.

2 Agonists

Of course, the most obvious agonists of regulatory functions in plants are the five well-known classes of plant hormone, IAA, GAs, CKs, ABA and ethylene. In addition, there are several newer candidates including the brassinosteroids, jasmonate and the polyamines. The hormonal status of these newer entries is still somewhat uncertain.

Because of the widespread instances in which phytochrome appears to regulate growth and development, either through interactions with the hormones, or independently, I suggest that we should consider it as a major agonist along with the hormones.

[1]Boyce Thompson Institute for Plant Research, Tower Road, Ithaca, NY 14853, USA

3 Receptors

The classic model of the action of a hormone or other agonist as a regulator of development involves first its attachment to a binding site or receptors. Among the animal hormones other than steroids, the receptor or binding sites are characteristically on the plasmalemma, and serve to transmit the agonist signal from the outside to the inside of the cell. Steroids apparently find their receptors in the nucleus and there they interact with the chromosomes. It is not infrequent for a hormone to have multiple binding sites; in some instances the sites may each be specific to a given hormonal function, and in others the binding sites may constitute pools of bound hormone or even sites of hormonal metabolism. Of course, the possibility always exists that a hormone binding site has no regulatory function.

Binding sites for plant hormones are often located in membranes, although numerous cytoplasmic binding sites have been reported [26, 31]. The brassinosteroids, by analogy, possibly bind in the nucleus, although this has not been determined. What role the cytoplasmic binding sites might play is entirely obscure.

In animal hormone systems, changes in responsiveness often occur as a consequence of changes in the abundance or the affinity of receptors. The commonest means of lowering responsiveness to a given hormone is by reducing the abundance of receptor sites, a phenomenon called down-regulation. This may involve the transfer of receptors into the cytoplasm where they will be degraded. Alternatively, responsiveness to a hormone may be achieved by up-regulation, or the proliferation of new binding sites. Changes in responsiveness may also occur through changes in the affinity of the receptor, most often achieved by phosphorylation of the receptor.

While the dynamics of hormone receptors in plants is poorly understood, there are several rather hopeful experiments that have been reported. With regard to the possible correlations between abundance of receptors and hormonal responsiveness, some reports are appearing in the literature indicating that a correlation of this sort exists [11, 28]. In view of the abundance of receptors in different fractions of tissues, one may entertain the uneasy possibility that each of the various receptors might be involved in distinctive regulatory functions of the hormone. The first relevant evidence has recently appeared, and this indicates that a mutant which is deficient in an ethylene receptor loses each of the characteristic responses to ethylene, thus suggesting that a single receptor mediates the broad array of responses to this hormone [2].

4 Second Messengers

The binding of an agonist to a receptor will ordinarily activate a second messenger, or a chemical signal within the cytoplasm. In plants, the best known second messengers are Ca^{++} and the phosphatidylinositol messenger system (PI).

It has been known since 1974 that supplementary additions of Ca^{++} could have large effects on the regulatory activity of each of the five plant hormones [15]. This observation has been expanded recently by the finding of a general sensitivity of each of the hormone actions to calmodulin inhibitors [8, 12]. The range of Ca^{++}

effects is quite extraordinary, including effects on growth, differentiation, transport, hormone binding, mitosis, streaming and shock reactions. Surely Ca^{++} is a central participant in regulatory functions in plants, playing a major role in the hormone and phytochrome regulatory effects, as well as in effects which are not identifiable with specific agonists. Ca^{++} has the capability of directly modifying specific enzymes [see review of Marme, 16]. While many Ca^{++} effects are mediated by its attachment to calmodulin, certainly some of its effects are produced by the cation itself, including effects associated with Ca^{++} binding to mebranes or to cytoskeletal components [5, 27].

In animal systems, a precise control of cytoplasmic Ca^{++} is obtained by the PI, in which an agonist attachment to a receptor on the membrane is translated into a release of inositol trisphosphate and diacylglycerol, the former stimulating Ca^{++} release and the latter stimulating protein kinases [21]. The attractive possibility that the PI system operates in response to plant hormones, or to phytochrome-regulated actions has received much attention recently. Labeling experiments indicate the formation of polyphosphorylated PI in tomato tissues and the possible formation of the inositol trisphosphate [7]. Evidence indicating PI participation in plant hormone actions has been reported for auxin effects [13, 17, 37], ABA effects [22] and GA effects [19], as well as for phytochrome effects [6, 10, 18, 23]. However, the changes in inositol metabolites reported in these experiments are small and transient, and there is some reason for concern about the method of quantification [3, 25]. So, the involvement of this second messenger remains a most interesting and reasonable possibility in hormonal regulation in plants, but definitive experimental evidence is not yet at hand.

5 Protein Kinase

The release of a second messenger in the cytoplasm can bring about a physiological response through the activation of a protein kinase, the phosphorylation of enzymes being a means of increasing or decreasing specific enzymic activities. In animal systems, more than a hundred protein kinases have been identified, some of which require Ca^{++} plus calmodulin, others requiring Ca^{2+} alone, and still others being apparently independent of Ca^{2+} [9].

In plants, reports that protein kinase activities are associated with regulatory signals are appearing [24], and the identification of protein kinases is just emerging [32, 36]. The phosphorylation of phytochrome itself may be a component of regulatory systems [34].

Collectively, these various lines of evidence suggest that in plants, regulatory actions by hormones and phytochrome may involve attachment to a receptor, activation of second messengers (Ca^{++}, and possibly PI), and subsequent activation of protein kinases, which then serve to regulate specific enzymes bringing about the regulatory action.

6 Speculation

For the five classes of hormones, binding sites which may serve as receptors have been identified, though the participation in regulatory actions have not been established for any of them. Second messenger functions for Ca^{++} have been established for a wide range of actions by hormones and phytochrome, often associated with a requirement for calmodulin. Tentative evidence now exists for the participation of the PI second messenger system, linking the hormone receptor site to the Ca^{++} and protein kinase regulated functions in the cytoplasm. And finally, evidence is emerging for the existence of protein kinases, which may serve as switching mechanisms in regulatory actions. This outline of possible regulatory mechanisms in plant remains speculative in part, but has the attractions of being consistent with animal regulatory systems, and offering a coherent organization for hormonal functions.

With this outline in mind, what can be said in response to the lack of specificity with which developmental functions respond to hormones? If each of the hormones and phytochrome bring about regulatory actions through the Ca^{++} second messenger system, then the specificity of their actions should perhaps be presumed to be a consequence of specific protein kinases activated by Ca^{2+}. It is not easy to conceive how hormonal specificity can be achieved through such a selective action by a common second messenger. In the past it has been easy to think of cells or tissues showing specific responses to individual hormones having differences in receptivity, which translates into differences in functional receptor sites. But the model of hormonal function we have just described would have each hormone "turning on" the same second messenger in the cell, regardless of receptors. I find it puzzling to think of such a commonality of second messenger between hormones.

Another difficulty concerning hormonal regulation is the frequency of poor correlations between the amounts of a given hormone and the function which is alleged to be regulated. The easy response to this question is to assume that there are differences in sensitivity of cells or tissues to a given hormone [30], and that sensitivity is a reflection of the abundance of hormone receptor sites. Such an explanation would have real appeal. But, if that were the answer to the question of sensitivity, then one should not find the largest amounts of specific binding sites for given hormones in non-responsive tissues. However, as an illustration, we might note the isolation of abundant ethylene binding sites from non-responsive seed tissues [1]. In the plant literature there has been a singular lack of positive correlations between abundance of receptors and responsiveness to hormones. The first evidence of such a correlation is just emerging [cf. 2, 11]. There is of course abundant evidence for changes in sensitivity of tissues to hormones as a function of developmental changes [35], and even with environmental signals [4]. But the most obvious possibility of accounting for sensitivity changes, that is through modulation of receptor sites, is lacking. Another possibility that should be considered is that changes in sensitivity may reflect differences in availability of second messengers (Ca^{2+}, PI?) or of appropriate protein kinases.

Perhaps the most frustrating aspect of hormonal physiology is that after 50 years of research on plant hormones, the mechanisms of action remain essentially obscure. Progress in this aspect of regulatory physiology probably hinges first on the clarification of receptor sites which function in hormonal action. Receptor sites may be coupled to second messengers, as outlined above, but there is an interesting possibility that the receptor may in some instances be a direct regulator of ion channels. Preliminary evidence for this possibility has been suggested for animal systems [20, 27] and for plants as well [33]. A variant of such a model of hormonal action could be the alteration of Ca^{2+} binding to the membrane by a hormone [5], thus forming a direct linkage between the hormone/receptor complex and the second messenger. The pervasive evidence that plant hormones lead to changes in the nucleic acid-directed protein synthesis has until now found no linkage to the various proposed mechanisms of hormone action.

In conclusion, we are at a rather frustrating state in hormonal physiology of plants. We have inherited an attractive mechanistic scheme from animal biochemists, and have thus expanded our vision to include a sequence of receptor sites, second messengers and protein kinases, but we do not find coherence between this new model and the three major ambiguities of regulatory plant physiology.

References

1. Bengochea T, Acaster MA, Dodds JH, Evans DE, Jerie PH, Hall MA (1980) Planta 148:407
2. Bleeker AB, Estelle MA, Somerville C, Kende H (1988) Science 241:1086
3. Boss WF, Chen Q, Dengler LA, Hendrix KW, Rincon M, Wheeler JJ (1988) Plant Physiol 86(S):502
4. Brock TG, Kaufman PB (1988) Plant Physiol 87:130
5. Buckhout TJ, Young KA, Low PS, Morre DJ (1981) Plant Physiol 68:512
6. Cote GG, Morse MJ, Crain RC, Satter RL (1987) Plant Physiol 83(S):820
7. Drobak BK, Fergusen IB, Dawson AP, Irvine RF (1988) Plant Physiol 87:217
8. Elliott DC, Batchelor SM, Cassar RA, Marinos NG (1983) Plant Physiol 72:219
9. Hanks SK, Quinn AM, Hunter T (1988) Science 241:42
10. Hartmann E, Pfaffmann, Weber M (1988) In: Proc 1st internant workshop on second messengers and phosphoinositides in plants. Lafayette, IN, Abst A-18
11. Jones AM, Venis MA (1988) IPGSA Abst 16
12. Kelly GJ (1984) TIBS 9:4
13. Lehle L, Ettlinger C (1988) In: First internat workshop on second messengers in plants. Lafayette, IN, Abstr A-23
14. Leopold AC (1972) In: Kaldewey H, Vardar Y (eds) Hormonal regulation in plant growth and development. Verlag Chemie, Weinheim, p 245
15. Leopold AC, Poovaiah BW, dela Fuente RK, Williams RJ (1974) In: Masuda Y (ed) Plant growth substances, 1973. Hirokawa, Tokyo, p 780
16. Marme D (1982) What's new in plant physiol 13:37
17. Morre DJ, Drobes B, Jaffmann HP, Hartmann E (1988) Plant Physiol 86(S):689
18. Morse MJ, Crain RC, Satter RL (1987) Plant Physiol 83:640
19. Murthy P, Renders J (1988) IPGSA Abst 357
20. Neer EJ, Clapham DE (1988) Nature 333:129
21. Nishikuza Y (1986) Science 233:305
22. Perdue DO, Leopold AC (1987) Plant Physiol 83(S):615
23. Poovaiah BW, Reddy ASN (1987) Crit Rev Plant Sci 6:47
24. Raghothama KG, Reddy ASN, Friedmann M, Poovaiah BW (1987) Plant Physiol 83:1008
25. Rincon M, Boss WF (1988) In: Proc 1st internat workshop on second messengers and phosphoinositides in plants. Lafayette, IN, Abst A-36
26. Rubery PH (1981) Annu Rev Plant Physiol 32:596
27. Shlatz L, Marinetti GV (1972) Science 176:175
28. Smith AR, Bell MH, Connern C, Sanders IO, Williams RAN, Wood C, Hall MA (1988) IPGSA Abst 15

29. Takahashi K, Mazaredo AM, Aguilar A, Vergara BS (1988) IPGRA Abst 420
30. Trewavas A (1986) Aust J Plant Physiol 13:447
31. Venis M (1985) Hormone binding sites in plants. Longman, New York
32. Verkey S, Gaiser JC, Lomax T (1988) IPGSA Abst 424
33. Wilson KJ, Schauf CL, Stillwell W (1988) Am J Bot 75(S):127
34. Wong YS, Cheng HC, Walsh DA, Logarias JC (1986) J Biol Chem 261:12089
35. Wright STC (1966) J Exp Bot 17:165
36. Yang YP, Randall DD (1988) Plant Physiol 86(S):136
37. Zbell B, Walter-Back C, Bucher H (1988) IPGSA Abst 13

The Role of Calcium in the Response of Roots to Auxin and Gravity

M.L. Evans, C.L. Stinemetz, L.M. Young, and W.M. Fondren[1]

1 Introduction

There is growing interest in the possibility that calcium (Ca) serves as a second messenger in the response of plant cells to hormones and environmental stimuli [5, 11, 22]. Some studies indicate that cytoplasmic Ca levels in plant cells may be controlled by the inositol trisphosphate pathway in much the same manner as in animal cells [5]. Schumaker and Sze [24] reported that IP_3 (inositol 1,4,5-trisphosphate) induces Ca release from vacuolar membrane vesicles and that this effect is blocked by TMB-8 [8-(N,N-dimethylamino)-octyl 3,4,5-trimethoxybenzoate-HCl], an antagonist of intracellular Ca release (also see references cited in [5]). Recent evidence indicates that hormonal promotion of cell elongation and perhaps cell wall biosynthesis may be mediated by changes in cytoplasmic Ca levels. Some of these effects may depend upon activation of calmodulin. (CaM) by Ca since CaM antagonists interfere with auxin-induced cell elongation [3, 23]. Brummell and Maclachlan [1] have found that TMB-8 inhibits both auxin-induced growth and auxin-induced wall biosynthesis in pea stem segments and they noted that Ca could restore auxin-induced enhancement of wall biosynthesis in segments treated with TMB-8.

The stimulus-response coupling mechanism associated with the response of roots to gravity has been the subject of intense investigation in recent years. This response system is especially intriguing since (1) thereis spatial separation between the sensing cells (columella cells of the root cap) and the responding cells (elongation zone), (2) the response appears to require development of an asymmetry of hormone concentration or sensitivity across the elongation zone, (3) the response appears to require or at least be modified by Ca (see below), and (4) in roots of some cultivars of maize the orthogravitropic response requires light.

This report considers recent evidence for the involvement of Ca and calmodulin in root gravitropism and discusses potential interactions between Ca, growth regulators, and light in the gravitropic response.

2 Materials and Methods

2.1 Plant Material

Experiments described from this laboratory were done using 3-day-old light-grown or dark-grown seedlings of the maize cultivar Merit. The seedlings were raised as described in [20].

[1] Department of Botany, Ohio State University, Columbus, Ohio, USA

2.2 Auxin Transport Experiments

Experiments on the transport of ^{3}H-IAA (833 GBq/nmol, Amersham, Arlington Heights, IL, USA) were done using the agar donor/receiver method essentially as described in [8]. For measurement of lateral transport of labeled auxin across the root cap the same method was used with the donor and receiver blocks placed on opposite sides of the cap.

2.3 Calcium Transport Experiments

Calcium movement across the caps of gravistimulated roots was measured using $^{45}CaCl_2$ as described by Lee et al. [13].

2.4 Plasmolysis and Microscopy of Root Caps

The tips (1–2 mm) of vertically-oriented roots were immersed in 0.5 M mannitol for 15 min and then in distilled water for 30 min. Controls were immersed in distilled water for 45 min. The caps were fixed and prepared for light microscopy as described by Moore et al. [19].

3 Results and Discussion

3.1 Evidence for a Role for Calcium in Root Gravitropism

Evidence indicating a role for Ca in root gravitropism includes the observations that: (1) Application of high levels of EDTA to the cap causes maize roots to lose gravitropic responsiveness [14], (2) gravistimulation induces polar movement of Ca to the lower side of the root cap [13], and (3) maize roots curve toward the high side of Ca gradients applied across the root cap [10]. Although these findings are indicative of a role for Ca, they do not allow us to determine whether Ca plays a primary role or a supporting role in the gravitropic response mechanism. We have studied the nature of gravi-induced polar Ca movement across maize root tips and we find [13, 15] that (1) it is sensitive to metabolic inhibitors, (2) it requires an intact root cap, and (3) it is inhibited by auxin transport inhibitors. In an attempt to gain information on the pathway of lateral Ca movement in caps of gravistimulated roots, we compared $^{45}Ca^{2+}$ movement across the caps of control seedlings vs seedlings in which the cap cells had been strongly plasmolyzed (Fig. 1) and rehydrated prior to gravistimulation. We found that plasmolyzed/rehydrated roots showed strong gravitropism but little or no polar Ca movement (Table 1) as measured by Ca movement into externally applied agar receivers. These results indicate that Ca asymmetry may not be essential for the gravitropic response. However, since our earlier studies indicate that Ca asymmetry is important, we need to determine whether or not an internal Ca gradient develops across the caps of plasmolyzed/deplasmolyzed roots. Leopold and coworkers [21] have found that

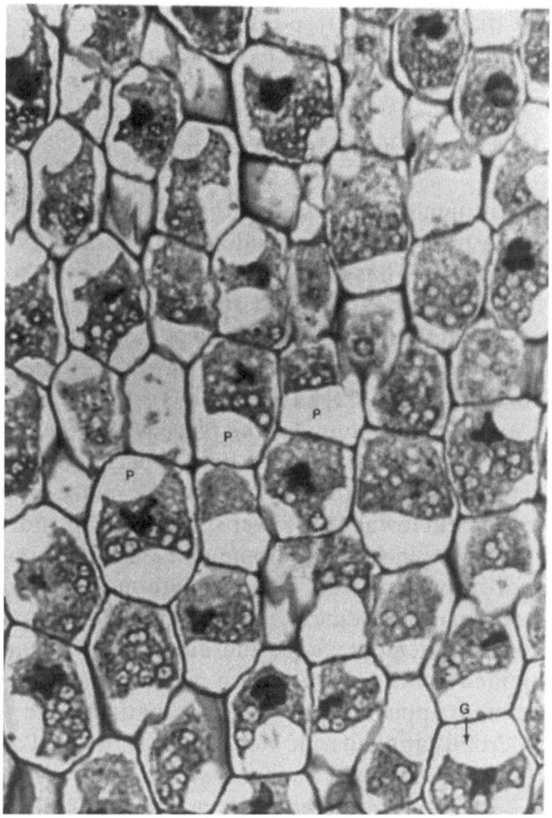

Fig. 1. Radial section of the cap of a primary root of *Zea mays* after submersion of the apex in 0.5 M mannitol for 15 min. *P* plasmolyzed cell. *Arrow* represents gravity vector during plasmolysis and fixing. ×670

Table 1. Effect of root tip plasmolysis/deplasmolysis on gravitropism and gravi-induced lateral movement of calcium

Treatment	Polarity[a]	Curvature (deg)[b]
Control	1.5	48 ± 6
Plasmolyzed	1.1	42 ± 7

[a]Receiver blocks were applied either to the top or bottom of the cap of a horizontally-oriented root with a donor containing $^{45}Ca^{2+}$ applied to the opposite side. Polarity = cpm lower receiver/cpm upper receiver after 90 min.
[b]Curvature ± SEM 90 min after gravistimulation.

the diagravitropic response of roots of dark-grown seedlings of the Merit cultivar of maize is not affected by elution of Ca from the roots.

3.2 Calcium and Auxin Transport

Dela Fuente [2] has shown that tissues depleted of Ca lose the capacity for polar auxin transport. We considered the possibility that the importance of Ca to root gravitropism has to do with maintenance of auxin transport or to a role for Ca in determining the preferred direction of polar auxin movement. In a recent study [9] we found that application of Ca to the caps of maize roots enhanced the movement of labeled IAA from the cap toward the elongation zone. EGTA had the opposite effect. Konings [12] has presented evidence that the root cap is not only the site of graviperception but also the place where the initial lateral redistribution of auxin occurs. We tested this idea by comparing upward vs downward auxin movement across the caps of gravistimulated maize roots. We found preferential auxin movement toward the lower side of the cap and noted a strong correlation between the time course of development of auxin transport polarity and the development of gravitropic curvature (Fig. 2). Although curvature began before we could detect polar auxin movement, this may be a consequence of the method used to measure auxin movement. Since the auxin was collected and measured in agar blocks applied to the root surface, it is likely that the auxin gradient develops in the cap before it appears externally. Alternatively, the appearance of auxin asymmetry in externally applied agar blocks may be delayed by removal of auxin from the cap by the basipetal auxin transport system.

In gravistimulated roots pretreated with EGTA, there was little or no polar auxin movement across the caps. Polar auxin movement was restored by treatment with Ca prior to gravistimulation (Table 2). These results are in agreement with Koning's conclusion that the cap is the site of initial auxin redistribution, and they suggest that Ca is important to the auxin distribution mechanism.

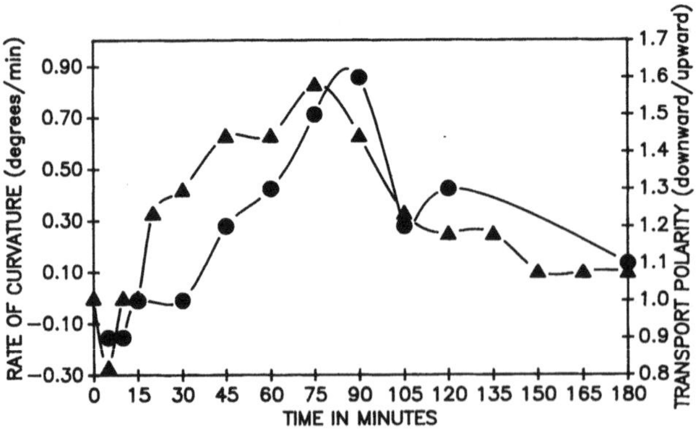

Fig. 2. Correlation between rate of curvature (*triangles*) and magnitude of downward auxin transport polarity across the root cap (*circles*). Rate of curvature was estimated from plots of curvature (measured from projected negatives) vs time

Table 2. Effects of EGTA and calcium on gravi-induced polar auxin movement across the caps of maize roots

Treatment	Polarity[a]
Control	1.6
EGTA[b]	1.1
EGTA, H_2O[c]	1.3
EGTA, calcium[c]	2.0

[a]Donors containing 50 000 cpm ^{3}H-IAA were applied either to the top or bottom of the root cap with receiver on the opposite side. Polarity = cpm lower receiver/cpm upper receiver after 90 min gravistimulation.
[b]Tips of the roots were submerged in aerated 2 mM EGTA for 30 min.
[c]Following EGTA treatment tips of the roots were submerged in aerated water or 5 mM $CaCl_2$ for 30 min prior to gravistimulation.

3.3 Interactions of Calcium and Light in the Control of Gravitropic Sensitivity

Merit is among those cultivars of maize in which the primary root of the seedling requires light for normal positive orthogravitropism. When gravistimulated in the dark the roots are diagravitropic, i.e. they tend to assume an angle about 90° to the gravity vector. Upon illumination the roots become positively orthogravitropic, i.e. they curve downward. The response appears to be a red-light response mediated by phytochrome in the root cap [7]. Although the basis of the light effect is unknown, there are indications that it may involve changes in cytoplasmic Ca and CaM levels. Calmodulin levels are high in root caps compared with adjacent tissue [17, 25]. However, in dark-grown roots of Merit, the CaM activity in the root cap is low. Upon illumination the CaM activity rises dramatically, and this increase precedes the development of normal positive gravitropism [25]. These observations indicate that CaM may play a role in transduction of the gravitropic signal. Leopold and coworkers [16, 21] have tested a number of factors for their ability to substitute for red light in the induction of orthogravitropism in roots of Merit. They found that both ABA and the Ca ionophore, A23187, were effective. In roots treated with EDTA, ABA only partially substituted for red light. However, adding supplemental Ca restored the full effect of ABA (Table 3). Friedmann and Poovaiah (cited in [22]) obtained similar results using EGTA and A23187 to deplete roots of Ca. They found that light failed to induce gravitropic competency in dark-grown roots depleted of Ca. Adding Ca restored the ability of light to induce gravitropic competency. Leopold and coworkers suggest that red light [21] and ABA [16] induce the capacity for orthogravitropism by increasing cytoplasmic Ca levels. In this regard it may be significant that light increases the concentration of ABA in the elongation zone [6] since we have found that the sensitivity of roots to auxin is affected by Ca (see below).

Table 3. Effects of calcium and EDTA on the ability of ABA to induce graviresponsiveness in dark-grown roots of Merit[a]

Treatment[b]	Dark curvature (deg)[c]
Buffer	4
ABA	65
ABA + EDTA	30
ABA + EDTA + Ca	65

[a] Data from AC Leopold and A LaFavre, by permission.
[b] Terminal 1–2 mm immersed in MES buffer, pH 6.0 for 2 h prior to gravistimulation. EDTA, Ca (1 mM). ABA (10 μM).
[c] Curvature 4 h after placing horizontally. LSD = 6.1.

3.4 Effects of Calcium on Sensitivity to Auxin

Considerable evidence indicates that auxin mediates the downward curvature of gravitropically-responding roots [18]. In view of the evidence that factors influencing gravitropic sensitivity do so by enhancing cytoplasmic Ca levels, it is important to consider the influence of Ca on auxin action in roots. Hasenstein and Evans [8] compared the auxin sensitivity of roots of maize seedlings depleted of Ca with that of seedlings receiving supplemental Ca. They found that auxin severely inhibited the growth of high-calcium roots at concentrations that only slightly inhibited low-calcium roots. This indicates that cytoplasmic Ca levels (or the size of internal Ca pools that can release Ca to the cytoplasm) are important to the action of auxin on cell elongation in roots. Thus, Ca appears to influence both auxin transport and auxin sensitivity in roots. Figure 3 shows a diagrammatic representation of some possible interactions between light, Ca, ABA, and IAA in the control of root gravitropism.

4 Conclusions

There are numerous potential modes of interaction of light, Ca, and auxin in the growth and gravitropic response of maize roots. Calcium enhances basipetal auxin movement as well as polar auxin movement across the caps of gravistimulated roots. Ca also increases the inhibitory effect of auxin on root growth. Light increases ABA levels in the elongation zone, and there is indirect evidence that ABA may increase cytoplasmic Ca levels. Light also increases the activity of CaM in the maize root cap. Viewed as a whole the data indicate that light-induced changes in gravitropic sensitivity as well as sensitivity to auxin may be mediated by light-induced changes in cytoplasmic Ca levels.

Acknowledgements. Some of the work reported here was supported by National Science Foundation grant DMB 8608673 and by grant NAGW-297 from the National Aeronautics and Space Administration. C.L. Stinemetz was supported by a fellowship from the NASA Graduate Student Researchers Program.

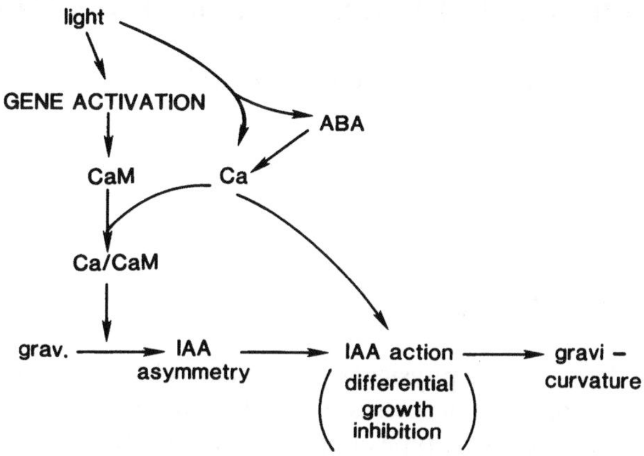

Fig. 3. Proposed involvement of calcium in the photoregulation of positive orthogravitropism in roots of maize. Light induces the synthesis of CaM and elevates the level of ABA in the root. Light also causes influx of extracellular calcium and this effect may be mediated by ABA. Ca/CaM (or perhaps free calcium) is necessary for gravi-induced auxin redistribution. Elevated cytoplasmic Ca also potentiates the growth-suppressing action of auxin. ABA can substitute for light in some cases (perhaps depending on the CaM status of the cap). However, ABA alone is not sufficient for induction of positive orthogravitropic competency since we find (Hasenstein and Evans, unpublished) seed lots of some light-requiring cultivars of maize in which light will induce positive orthogravitropism but ABA will not

References

1. Brummell DA, Maclachlan GA (1987) In: Cosgrove DJ, Knievel DP (eds) Physiology of cell expansion during plant growth. Am Soc Plant Physiol, Rockville, Maryland, USA, p 298
2. Dela Fuente RK (1984) Plant Physiol (Bethesda) 76:342
3. Elliot DC, Batchelor SM, Cassar RA, Marinos NG (1983) Plant Physiol (Bethesda) 72:219
4. Evans ML (1986) Plant Physiol (Bethesda) 58:599
5. Evans ML, Hasenstein KH (1987) In: Cosgrove DJ, Knievel DP (eds) Physiology of cell expansion during plant growth. Am Soc Plant Physiol, Rockville, Maryland, USA, p 202
6. Feldman LJ, Arroyave NJ, Sun PS (1985) Planta 166:483
7. Feldman LJ, Briggs WR (1987) Plant Physiol 83:241
8. Hasenstein KH, Evans ML (1986) Plant Physiol (Bethesda) 81:439
9. Hasenstein KH, Evans ML (1988) Plant Physiol (Bethesda) 86:890
10. Hasenstein KH, Evans ML, Stinemetz CL, Moore R, Fondren WM, Koon EC, Higby MA, Smucker AJM (1988) Plant Physiol (Bethesda) 86:885
11. Hepler PK, Wayne RO (1985) Annu Rev Plant Physiol 36:397
12. Konings H (1968) Acta Bot Neerl 17:203
13. Lee JS, Mulkey TJ, Evans ML (1983) Plant Physiol (Bethesda) 73:874
14. Lee JS, Mulkey TJ, Evans ML (1983) Science 220:1375
15. Lee JS, Mulkey TJ, Evans ML (1984) Planta 160:536
16. Leopold AC, LaFavre AK (1987) Plant Physiol (Bethesda) 89:875
17. Lin CT, Sun D, Song GX, Wu JY (1986) J Histochem Cytochem 34:561
18. Moore R, Evans ML (1986) Am J Bot 73:574
19. Moore R, Fondren WM, Marcum H (1987) Am J Bot 74:329
20. Mulkey TJ, Kuzmanoff KM, Evans ML (1981) Planta 152:239
21. Perdue DO, LaFavre AK, Leopold AC (1988) Plant Physiol (Bethesda) 86:1276
22. Poovaiah BW, McFadden JJ, Reddy ASN (1987) Physiol Plant 71:401
23. Raghothama KG, Mizrahi Y, Poovaiah BW (1985) Plant Physiol (Bethesda) 79:28
24. Schumaker KS, Sze H (1987) J Biol Chem 262:3944
25. Stinemetz CL, Kuzmanoff KM, Evans ML, Jarrett HW (1987) Plant Physiol (Bethesda) 84:1337

The Role of Calcium in Stimulus-Response Coupling

B.W. Poovaiah and A.S.N. Reddy[1]

1 Introduction

Chemical and physical signals such as hormones, light and gravity control diverse physiological processes in plants. Voluminous descriptive data is available in the literature on the type of physiological responses elicited by these signals, especially by hormones. However, biochemical and molecular mechanism(s) by which these signals control diverse processes are poorly understood. In recent years, signal transduction has become an area of intense investigation. Recent developments in recombinant DNA technology and recognition of calcium as a messenger have opened new avenues to investigate the biochemical and molecular mechanisms involved in signal transduction.

The biochemical events involved in transduction of extracellular signals into intracellular events has been extensively studied in animal systems [1, 32]. These studies have established an important role for cAMP, calcium, inositol 1,4,5 trisphosphate (possibly other inositol polyphosphates) and diacylglycerol as messengers in transmitting the signal received on the surface of an individual cell to the metabolic machinery. A transient increase in cytosolic calcium through inositol phospholipid turnover and regulation of protein phosphorylation by calcium-calmodulin-dependent and calcium-phospholipid-dependent protein kinases are considered to be the primary events in the transduction of some signals. In addition, regulation of gene expression at the transcriptional level through specific receptors has been shown to be the mechanism of action of some animal hormones such as steroids [4]. Until recently, very little has been known in plants about messenger molecules involved in signal transduction and the regulation of gene expression by hormones and environmental signals. Although there is evidence for the presence of cAMP in plants, no evidence has been obtained for its physiological role as a messenger [3, 38]. Investigations during the last decade indicate the potential role for calcium as a messenger in signal transduction in plants [18, 38, 55]. Calcium has been shown to be involved in various signal-induced processes. This, coupled with the discovery of calmodulin, calcium-calmodulin-regulated enzymes including protein kinases and the components of the phosphoinositide pathway, indicates that calcium could play a similar role in plants as in animals [38]. With the aid of recent sophisticated recombinant DNA techniques, regulation of gene expression by transcription by plant hormones has been demonstrated [16, 62].

[1]Laboratory of Plant Molecular Biology and Physiology, Department of Horticulture and Land. Arch., Washington State University, Pullman, WA 99164-6414, USA

In this article, we will summarize the role of calcium in the transduction of light and hormone signals. In doing so, we will be focusing on light-regulated gravitropism in corn roots and auxin-induced cell elongation in pea epicotyl segments.

2 Light-Regulated Gravitropism

Most roots show positive gravitropic response in either dark or light. However, the roots of some varieties of plants (e.g., *Zea mays* L., cv Merit and *Zea mays* L., cv Golden Cross Bantam 70) respond positively to gravity once they have been exposed to light [13, 58, 61]. The effectiveness of various wavelengths of light in promoting positive gravitropism has been investigated by several researchers [12, 26, 58, 61]. In Merit corn a brief red light treatment of the root cap is sufficient to induce positive gravitropism, also known as orthogravitropism [12, 26, 35, 55, 58]. The red light induction of positive gravitropism is mediated by phytochrome [12, 58]. Hence, roots of these varieties of corn serve as an excellent experimental system to study the biochemical changes triggered by signals such as light and gravity. Recent investigations from various laboratories indicate a potential role for calcium as a messenger in signal transduction in plants. We have investigated the role of calcium and the rapid events involved in light-signal transduction using roots of Merit corn.

2.1 Calcium and Gravitropism in Merit Corn Roots

The importance of calcium in light-independent gravitropism in roots has been known for several years [23, 42]. Manipulation of calcium gradients in the root tip affect gravitropic response [23]. Loss of gravisensitivity by calcium depletion and its restoration by calcium has been shown in normal corn roots [23].

Merit corn roots that are depleted of calcium by calcium chelator and ionophore prior to light treatment do not show positive gravitropic curvature. However, replenishment of calcium to depleted roots restores light-regulated positive gravitropic response [49]. Treatments that cause an influx of calcium can substitute for light causing positive gravitropism in the dark [35]. Furthermore, verapamil, a calcium channel blocker, can inhibit light-regulated gravitropism. These results indicate an important role for calcium in light-regulated gravitropism.

2.2 Possible Role of Phosphoinositides in Light-Regulated Gravitropism

A component of the gravitropic response in both normal and Merit corn roots is believed to be an increase in cytosolic calcium in the columella cells of the root cap [41, 55]. The mechanism(s) by which light and gravity signals could bring about changes in cytosolic calcium is under intense investigation. Gravity-induced displacement of amyloplasts in the columella cells of the root cap has been proposed to induce localized changes in cytosolic calcium [41, 55]. Starchless mutants of *Arabidopsis* show gravicurvature as do controls containing amyloplasts, indicating that sedimentation of amyloplasts may not be a prerequisite for gravity response [for review see 42].

Turnover of inositolphospholipids has been shown to play a role in raising cytosolic calcium in response to external signals in animals [1, 32]. In animal cells, signals such as light, hormones, growth factors and neurotransmitters can induce the hydrolysis of phosphatidylinositol 4,5 bisphosphate by activating phospholipase C through G-proteins. This results in the production of inositol 1,4,5-trisphosphate and diacylglycerol. IP_3 is known to induce calcium release from intracellular stores whereas DG is shown to activate protein kinase C. A schematic model illustrating signal-induced turnover of inositol phospholipids and the role of turnover products is shown in Fig. 1. During the last couple of years several labs have been actively investigating the existence of this pathway and its importance in signal transduction in plants. Results of these investigations indicate the presence of various components of this pathway supporting its possible role in signal transduction [28, 31, 39, 43].

Since calcium is found to have an important role in the light-regulated gravity response [40], we investigated the involvement of phosphoinositides in light-signal transduction. Preliminary results indicate that exposure of dark-grown roots to

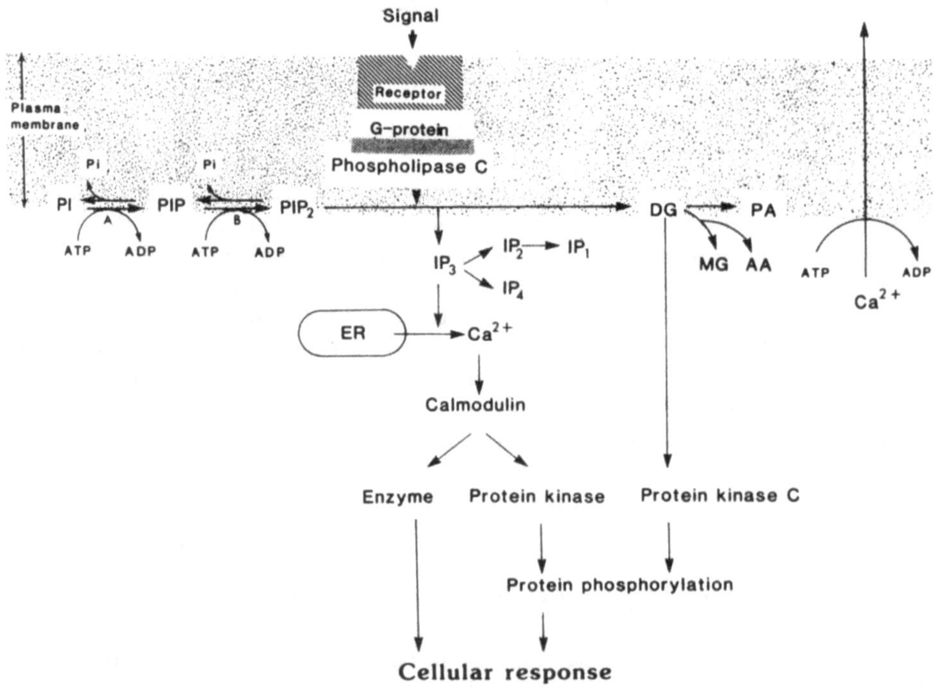

Fig. 1. Stimulus-induced turnover of phosphatidylinositol 4,5-bisphosphate (PIP_2) and the role of turnover products in signal transduction. PI, phosphatidylinositol; PIP, phosphatidylinositol 4-phosphate; PIP_2, phosphatidylinositol 4,5 bisphosphate; IP_3, inositol trisphosphate; IP_4, inositol tetrakisphosphate; IP_2 inositol bisphosphate; IP_1, inositol monophosphate; ER, endoplasmic reticulum; DG, diacylglycerol; MG, monoglyceride; AA, arachidonic acid; PA, phosphatidic acid. [Adapted from 38]

light may increase inositol phosphate levels [49]. Similar results were reported by Leopold's group [36]. Light stimulated hydrolysis of PIP_2 and an associated increase in the levels of IP_3 and DG have been shown in the pulvini of *Samanea saman* which exhibit light-regulated rhythmic movement [31]. 5-hydroxytryptamine, an animal hormone, is known to promote PIP_2 hydrolysis [24]. Treatment of dark-grown roots with 5-HT induced positive gravitropic curvature in the dark, and we have also obtained some evidence that 5-HT treatment of roots results in higher levels of inositol phosphates [49]. Perdue et al. have shown that the compounds, such as 5-HT and deoxycholate, that are capable of triggering PIP_2 hydrolysis in some animal systems can substitute for light in inducing gravitropism in roots [35]. Furthermore, lithium, an inhibitor of the phosphoinositide pathway, has been shown to inhibit light-induced gravitropism. These results suggest that light could promote the hydrolysis of PIP_2 and produce IP_3 and DG. The IP_3 thus released could raise cytosolic calcium from intracellular calcium stores which in turn could activate calcium-calmodulin-dependent enzymes including protein kinases. Release of calcium from microsomes and vacuoles by IP_3 has been shown in plants [9, 46, 47, 57].

Signal-induced hydrolysis of PIP_2 is achieved in animal systems through the activation of phospholipase C, and a specific class of GTP binding proteins known as G-proteins are believed to be involved in the activation of phospholipase C [22]. Recently, we obtained evidence for the presence of specific GTP binding proteins using [α-^{32}P]GTP in the membranes isolated from Merit corn root tips [41]. Three polypeptides have been found to bind GTP (Fig. 2A). Binding of labeled GTP to proteins is competed by cold GTP, GTP(γS) or GDP whereas ATP is found to be ineffective in competing for GTP binding (Fig. 2B,C). Preliminary evidence indicates that GTP and GTP analogs may release IP_3 from membranes isolated from Merit corn root tips [36]. Proteins that bind to GTP have been reported in other plant systems as well [10, 17]. However, it is not clear at the present time whether these proteins are similar to G-proteins. Furthermore, there is no evidence for the role of plant GTP binding proteins in activating phospholipase 'C'. An enzyme that

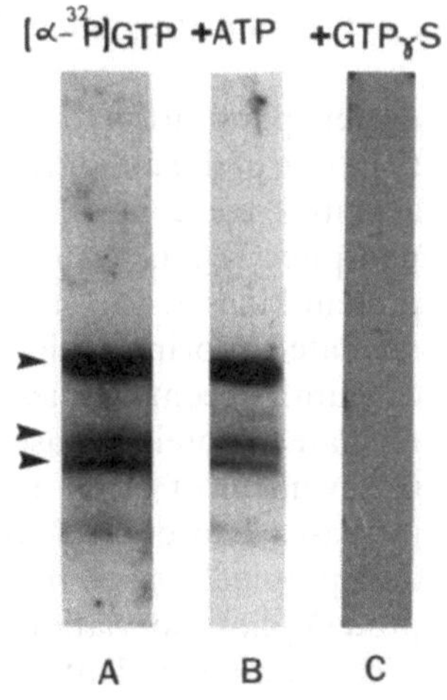

Fig. 2. A Binding of [α-^{32}P]GTP to membrane proteins isolated from Merit corn root tips (0.5 mm). Membrane proteins were isolated, separated on 12% SDS-PAGE and electrophoretically transferred to nitrocellulose filters. Binding of [α-^{32}P]GTP to proteins on nitrocellulose blot was determined as described earlier [10, 22]. Autoradiograph shows the proteins (29, 27, and 24 kDa) that bind to labelled GTP. **B** and **C** Effect of cold ATP and GTP [γS] on the binding of [α-^{32}P]GTP to corn membrane proteins. Nitrocellulose blots were prepared as described above and incubated with 10 mM ATP (**B**) or GTP[γS] (**C**) for 10 min following which [α-^{32}P]GTP was added and further incubated for 60 min [39]

catalyzes the hydrolysis of PIP_2 is found to be present in the plasmamembranes prepared from Merit corn root tips [Friedmann and Poovaiah, unpubl. results]. Since the hydrolysis products are not identified, we do not know at this time whether the observed activity is due to phospholipase 'C' or other phospholipase. The results presented here suggest a possible role for inositol phospholipid turnover and PIP_2 turnover products. However, more information is clearly needed to further ascertain the role of the PI pathway in signal transduction.

2.3 Involvement of Calmodulin in Gravitropism

Calmodulin, a highly conserved calcium binding protein in eukaryotes, mediates calcium action in many plant and animal cells [38, 54]. Calmodulin inhibitors have been shown to inhibit gravitropism without inhibiting growth [2, 55]. Calmodulin antagonists have been shown to inhibit polar transport of calcium in roots indicating that establishment of a calcium gradient may be a calmodulin-dependent process [41, 55]. Light-induced gravitropism is found to be inhibited by calmodulin antagonists such as calmidazolium and compound 48/80. Root tips are found to contain high amounts of calmodulin [45]. A recent report indicates an increase in the level of calmodulin and its activity in Merit corn roots that are exposed to light [59]. We have recently isolated a calmodulin cDNA (pPCM-l) from potato [20]. The nucleotide sequence of potato calmodulin is shown in Fig. 3. The 922 base pair sequence contains the entire coding region and 80 nucleotides of 5′ and 376 nucleotides of 3′ untranslated region. The deduced amino acid sequence of potato calmodulin showed an overall homology of 98% with spinach calmodulin [25]. Using potato calmodulin cDNA as a probe we studied the effect of light on calmodulin mRNA level in dark- and light-treated Merit corn roots. As shown in Fig. 4, light-treated Merit corn roots contained an increased level of calmodulin RNA as compared to the dark control. All these reports suggest a role for calmodulin in positive gravitropism.

2.4 Calcium-Regulated Phosphorylation

Protein phosphorylation, one of the post-translational modifications, is considered to be one of the major regulatory mechanisms by which various cellular metabolic activities are controlled in animals [7]. Several key enzymes undergo protein phosphorylation, resulting in the alteration of their properties. The activity of protein kinases, the enzymes that catalyze protein phosphorylation reactions, is regulated by primary signals through messengers such as phospholipids, cAMP and calcium. Recent studies established the presence of calcium and calmodulin-regulated protein kinases in plants [19, 37, 38, 64, 65]. Since the results suggest involvement of calcium in light-regulated gravitropism, it is possible that light increases cytosolic calcium, thereby activating calcium- and calmodulin-dependent protein kinases. If this is so, one would expect changes in calcium-dependent protein phosphorylation in vivo when roots are exposed to light. To study the role of calcium-dependent protein phosphorylation in light-dependent gravitropism, in

```
        10                  30                  50                  70                  90
TTTTGAGGAGAGAGATAAAACTCAAATCACATACATATTAGTACTTAACTGTACAGCTGTTTCTAGGAGTACGAAAAAAAATGGCAGAGCAGCTGACGGA
                                                                                 MetAlaGluGlnLeuThrGl

        110                 130                 150                 170                 190
GGAGCAGATCGCCGAGTTCAAGGAAGCTTTTAGCCTTTTCGACAAGGATGGCGATGGCTGTATTACTACCAAGGAGTTGGGAACAGTGATGAGATCACTT
uGluGlnIleAlaGluPheLysGluAlaPheSerLeuPheAspLysAspGlyAspGlyCysIleThrThrLysGluLeuGlyThrValMetArgSerLeu

        210                 230                 250                 270                 290
GGTCAGAATCCCACTGAAGCTGAACTACAGGATATGATCAGTGAAGCTGATGCTGATCAGAATGGAACCATTGATTTTCCAGAGTTCTTGAATCTGATGG
GlyGlnAsnProThrGluAlaGluLeuGlnAspMetIleSerGluAlaAspAlaAspGlnAsnGlyThrIleAspPheProGluPheLeuAsnLeuMetA

        310                 330                 350                 370                 390
CACGTAAGATGAAGGACACTGATTCTGAGGAGGAACTCAAAGAGGCTTTCAAGGTTTTCGATAAAGATCAGAATGGCTTTATTTCTGCAGCTGAGCTTCG
laArgLysMetLysAspThrAspSerGluGluGluLeuLysGluAlaPheLysValPheAspLysAspGlnAsnGlyPheIleSerAlaAlaGluLeuAr

        410                 430                 450                 470                 490
TCATGTAATGACAAACCTTGGAGAGAAGCTGACTGATGAAGAGGTGGATGAGATGATCCGAGAGGCAGATATTGATGGTGATGGGCAAGTTAATTATGAG
gHisValMetThrAsnLeuGlyGluLysLeuThrAspGluGluValAspGluMetIleArgGluAlaAspIleAspGlyAspGlyGlnValAsnTyrGlu

        510                 530                 550                 570                 590
GAGTTTGTCCGTATGATGCTTGCCAAGTGATGGCTTAAGATTCTCTTAGCTACTGTGTAATTTATGATGGCCGCTTAGTTACTACAACTTCTAGCTGGCA
GluPheValArgMetMetLeuAlaLysEnd

        610                 630                 650                 670                 690
GTTATATTCTGTTCTGTTAAGACAAACAAATGTGTCGTATGGTTTTACTAGCATCTAGACTCCTTTCAGTTTTATGTTAACTTATGGACTACAGGTGTAT

        710                 730                 750                 770                 790
GCTGCTTTAGTCCCTTGCCCATCCAGAGGGGAAAAGAAAAGGAGAAATTAGATAGTTCATTCGTAATATCTTGTTAGTTGCCTTGGTTTCTTTCCAGGCT

        810                 830                 850                 870                 890
ACAGTCTGTATGTGTTTCACTGTCTCTAACTGTTATATTTCCCGTTTCTGGTTTTCGCTTTTATCACTGATTAATGGGAAGGTCAAATGCAAGGTGATTA

        910
TTTCCTAAAAAAAAAAAAAAAAA
```

Fig. 3. Nucleotide sequence of potato calmodulin cDNA clone pPCM-1. The numbers indicate nucleotide positions. The initiation codon, termination codon and the possible polyadenylation signals are *underlined*. The amino acid sequence derived from nucleotide sequence is shown below the nucleotide sequence [20]

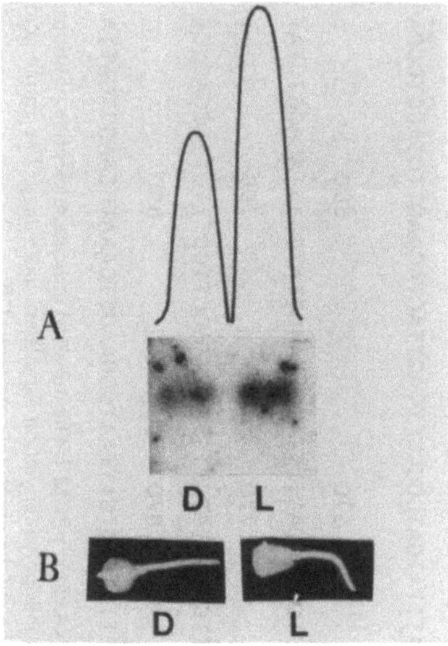

Fig. 4. **A** The level of calmodulin mRNA in Merit corn root tips that were left in dark (*D*) or exposed to light (*L*). Seedlings (48-h-old) were kept either in dark or exposed to light for 40 min. Then the root tips (2 mm) were excised, frozen immediately in liquid N_2 and used for poly$(A)^+$ isolation. Two μg of RNA was probed with radiolabeled pPCM-1. The autoradiogram and densitometer scanning results are presented on the lower and upper part of the figure, respectively. **B** Light-regulated positive gravitropism in Merit corn roots. Corn seedlings (48-h-old) were kept dark (*D*) or exposed to light for 10 min and left in the dark (*L*) for 6 h [20]

vivo protein phosphorylation studies were performed in dark-grown and light-treated roots by manipulating tissue calcium levels. Seven-minute light treatment of dark-grown roots promoted phosphorylation of specific polypeptides of M_r 48 000, 92 000 and 94 000 (Fig. 5). Depletion of calcium in root tips by EGTA and A23187 prior to light treatment significantly decreased light-induced promotion of phosphorylation of these polypeptides (Fig. 5) [27]. Replenishment of calcium to depleted root tips restored the light effect on protein phosphorylation. Furthermore, light-induced changes in protein phosphorylation could be observed within one min [40]. Exposure of root bases to light did not show any specific enhancement of protein phosphorylation. Interestingly, the changes in phosphorylation were observed only in the root tips, which are considered to be the site of light and gravity perception. These results provide evidence for light-induced rapid changes in protein phosphorylation and these changes are influenced by calcium. Furthermore, these results suggest the specificity of the changes in protein phosphorylation in light responsive tips.

Our results suggest that the transduction of the light signal could occur by a signal pathway that involves turnover of inositol phospholipids, calcium and calcium-dependent protein phosphorylation.

3 The Role of Calcium in Auxin Action

Auxins control many aspects of plant growth and development. Physiological processes such as cell elongation, cell division, differentiation, tropisms, rooting, fruit development and abscission are influenced by auxin [16, 50, 62]. However, the mechanism by which this hormone regulates diverse physiological processes is

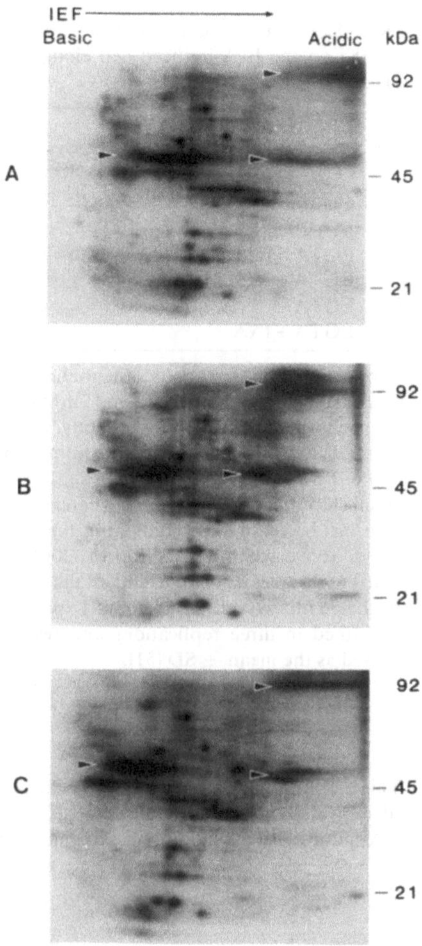

Fig. 5A-C. Rapid changes in protein phosphorylation in roots of Merit corn. Apical segments of dark-grown roots were preloaded with ^{32}P for 1 h, then washed in buffer. **A** Dark controls were left in buffer for 15 min; **B** for light treatment, roots were exposed to light for 7 min after 8 min dark incubation; **C** light treatment was the same as in B, but EGTA + A23187 was present for 15 min. Proteins were extracted and separated by two-dimensional gel electrophoresis as described earlier [45]. *Arrows* indicate the phosphoproteins that are affected by light [27]

unclear. Numerous studies indicate possible involvement of calcium in auxin action [6, 8, 15, 21, 38]. In our laboratory we have been investigating the role of calcium in auxin-induced cell elongation and auxin-induced delay of abscission, using pea epicotyl segments and bean explants, respectively.

Treatment of pea epicotyl segments with EGTA or EGTA with the calcium ionophore A23187 in the presence of IAA inhibited auxin-induced cell elongation as compared to untreated tissue (Table 1) [51]. Neither of these treatments reduced elongation in the absence of auxin. The calcium channel blocker D-600 and the calcium chelator chlorotetracycline reduced the auxin-induced growth (Table 2). These results suggest that calcium has an essential role in auxin-induced cell elongation. We have also reported a possible role for calcium in auxin-induced delay of abscission in bean pulvinar explants [44]. Recent evidence suggests that auxin may induce rapid (within minutes) hydrolysis of inositol phospholipids

Table 1. Effect of EGTA and calcium ionophore A23187 on the IAA-induced cell elongation in pea epicotyl tissue

	Treatment	$^{\Delta}$Elongation
		mm
1	Control	12.7 ± 0.5
2	IAA 20 μM	28.7 ± 0.5
3	EGTA 5 mM + A23187	14.7 ± 0.9
4	EGTA + A23187 + IAA	15.7 ± 0.5
5	EGTA	16.3 ± 1.2
6	EGTA + IAA	20.3 ± 0.6

$^{\Delta}$Elongation = L_{final}-$L_{initial}$. Incubation conditions: All samples were treated for 1 h in 15 mM sucrose + 50 μg/ml chloramphenicol. Subsequently, samples 1 and 2 were incubated in incubation medium alone (1 mM citrate, 1 mM PIPES, 15 mM sucrose, 1 mM KCl, 50 μg/ml chloramphenicol pH 6.0) whereas samples 3–6 were incubated in I.M. containing the components shown above without IAA for 1h. 20μM IAA was added to samples 2, 4 and 6, and all the samples were incubated for two additional hours. Experiments were performed in three replications and results are expressed as the mean ± SD [51].

Table 2. Effect of calcium channel blocker D-600 and calcium chelator chlorotetracycline (CTC) on the IAA-induced cell elongation in pea epicotyl tissue

	Treatment	$^{\Delta}$Elongation (mm)
1	Control	10.7 ± 0.5
2	IAA 20 μM	26.0 ± 0.8
3	D-600 5 μM	9.6 ± 0.9
4	D-600 + IAA	15.7 ± 0.5
5	CTC 500 μM	10.0 ± 0.8
	CTC + IAA	16.3 ± 0.5

$^{\Delta}$ Elongation = L_{final}-$L_{initial}$. D-600:-α-Isopropyl-α- [(N-methyl- N-homoveratryl) -α-aminopropyl]-3,4,5-trimethoxyphenylacetonitrile hydrochloride. Incubation conditions: All samples were treated for 1.5 h in 15 mM sucrose + 50 μg/ml chloramphenicol and then for 0.5 h in incubation medium alone (1 mM citrate, 1 mM PIPES, 15 mM sucrose, 1 mM KCl, 50 μg/ml chloramphenicol pH 6.0) and for an additional 10 min in I.M. containing the components shown above without IAA. 20 μM IAA was added to samples 2, 4 and 6 for two additional hours. Experiments were performed in three replications and results are expressed as the mean ± SD [51].

resulting in the production of inositol trisphosphate [11, 66] which is believed to increase cytosolic calcium by releasing calcium from intracellular stores. Based on these results, it is proposed that alteration of cytosolic calcium level is a contributing event in the mechanism of auxin action.

3.1 Auxins and Protein Phosphorylation

It is well known that protein phosphorylation, regulated by messengers such as calcium, plays a key role in signal transduction in animals. To test the effect of auxin on protein phosphorylation in vivo, protein phosphorylation studies were performed in pea epicotyl segments. As shown in Fig. 6, auxin appears to affect phosphorylation of specific polypeptides without affecting ^{32}P uptake [48]. Phosphorylation of the 23 000, 82 000, 105 000 and 110 000 molecular weight polypeptides was markedly reduced in the presence of auxin. In addition, a slight promotion in phosphorylation of the 19 000, 24 000 and 28 000 molecular weight polypeptides was observed. We have also shown the possible involvement of calmodulin in auxin-regulated protein phosphorylation. Murray and Key [30] have reported an auxin-induced phosphorylation of soybean nuclear protein. Auxin-induced phosphorylation of ribosomal protein and its implication in translational control has been reported recently [56].

3.2 Auxins and Gene Expression

Studies using recombinant DNA techniques have shown rapid induction of specific mRNAs by auxin in epicotyl, hypocotyl and coleoptiles of various plants [16, 63]. Experimental evidence obtained from mammalian cells implicates calcium in the regulation of gene expression at the level of transcription and translation [5, 38, 60]. A possible role for calcium in the regulation of gene expression in plants is proposed [14, 43]. Since auxin regulates gene expression and calcium appears to be involved in auxin action, it is interesting to know whether calcium has any role in auxin-regulated gene expression. Treatment of pea epicotyl segments with calcium chelators, calcium ionophore and calcium channel blockers did not have any effect on the steady state level of mRNAs corresponding to pIAA4/5, pIAA6 cDNA clones [51]. The possible role of calcium in the translation of these mRNAs or in the expression of other auxin-regulated genes is not ruled out from our studies.

Strawberry fruit growth is regulated by auxin [29, 33, 34]. We have recently isolated several cDNA clones corresponding to auxin-induced and auxin-repressed mRNAs of developing strawberry fruit [52, 53]. Since calmodulin is implicated to play a role in auxin-induced processes, we have investigated the effect of auxin on calmodulin expression using potato calmodulin cDNA. Northern analysis of RNA isolated from auxin-treated and auxin-deprived fruits showed that auxin treatment increases the level of calmodulin mRNA [20] (Fig. 7). This data suggests than an auxin-induced increase in calmodulin expression could be one of the molecular events in auxin-induced strawberry fruit growth.

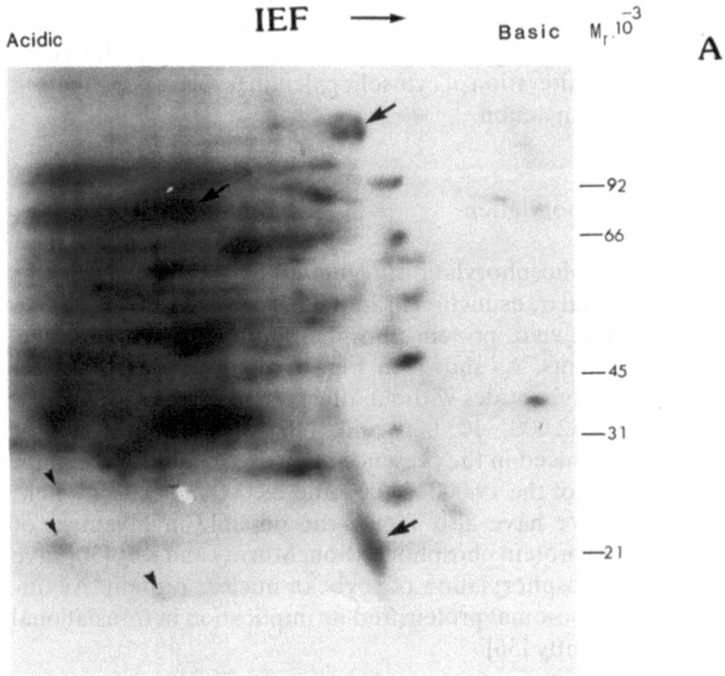
IEF
Acidic
Basic
$M_r \cdot 10^{-3}$
A
92
66
45
31
21

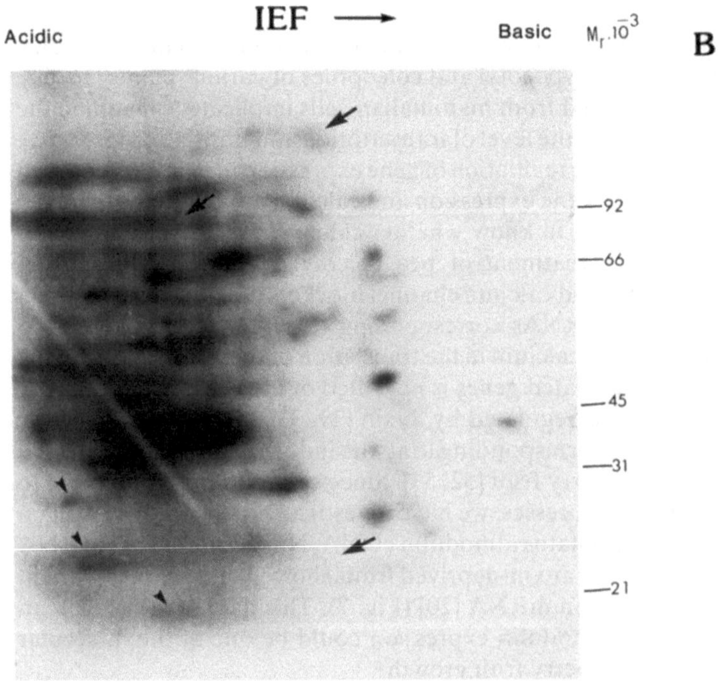
IEF
Acidic
Basic
$M_r \cdot 10^{-3}$
B
92
66
45
31
21

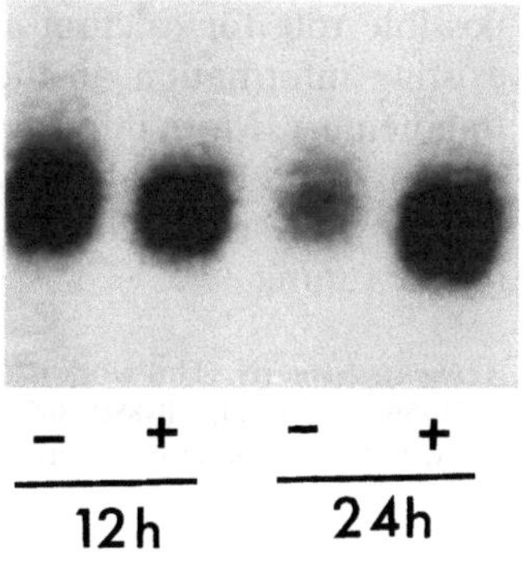

Fig. 7. Hybridization of RNA from auxin-treated and depleted fruits to plant calmodulin cDNA clone (pPCM-l). Seven-day-old pollinated fruits were deachened and treated with (+) or without (−) auxin for different time periods. Thirty μg of total RNA was probed with pPCM-l [20]

In Fig. 8, we have illustrated the proposed biochemical and molecular events that could occur during signal-induced response. Depending on the signal, and the tissue, all or part of the events might occur.

4 Conclusion

Recent developments in recombinant DNA techniques and recognition of the potential role for calcium in signal transduction have opened new approaches to understand the mechanisms involved in signal transduction. The progress with these new approaches is very encouraging. Evidence in the literature suggests a

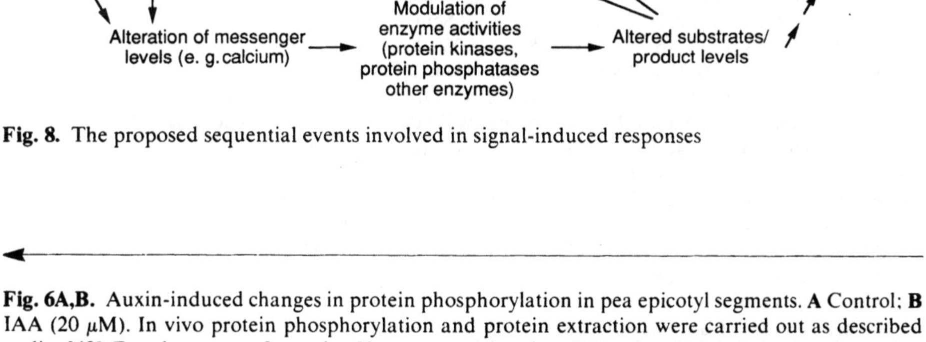

Fig. 8. The proposed sequential events involved in signal-induced responses

Fig. 6A,B. Auxin-induced changes in protein phosphorylation in pea epicotyl segments. **A** Control; **B** IAA (20 μM). In vivo protein phosphorylation and protein extraction were carried out as described earlier [48]. Equal amount of protein (50 μg) was analyzed on IEF gels (pH 3.5–8) for first dimension followed by second dimension on 8% to 16% linear SDS-PAGE. Dried gels were exposed to X-ray films at -70°C in the presence of intensifying screen. *Arrows* indicate the phosphoproteins that are affected by auxin [48] possible role for calcium as a messenger in signal transduction. However, the existing information on the role of phosphoinositides in signal transduction is fragmentary. More information is needed before one can make firm conclusions. Proving the role of calcium and determining the identity and function of proteins whose synthesis is regulated by signals such as hormones will be an important task for the future.

Acknowledgements. This work has been supported by grants from the National Science Foundation DCB-8615978 and DCB-8801860 and National Aeronautics and Space Administration NAG-10-0032. Contributions of other members of our research team are gratefully acknowledged.

References

1. Berridge MJ, Irvine RF (1984) Nature 312:315
2. Bjorkman T, Leopold AC (1987) Plant Physiol (Bethesda) 84:847
3. Brown EG, Newton RP (1981) Phytochemistry 20:2453
4. Cato ACB, Geisse S, Wenz M, Westphal HM, Beato M (1984) EMBO J 3:2771
5. Chin K-H, Cade C, Brostrom CO, Galuska EM, Brostrom MA (1987) J Biol Chem 262:16509
6. Cleland RE, Rayle DL (1977) Plant Physiol (Bethesda) 60:709
7. Cohen P (1985) Eur J Biochem 151:439
8. Cunninghame ME, Hall JL (1986) Protoplasma 133:149
9. Drobak BK, Ferguson IB (1985) Biochem Biophys Res Commun 130:1241
10. Drobak BK, Allan EF, Comerford JG, Roberts K, Dawson AP (1988) Biochem Biophys Res Commun 150:899
11. Ettlinger C, Lehle L (1988) Nature 331:176
12. Feldman LJ, Briggs WR (1987) Plant Physiol (Bethesda) 83:241
13. Feldman LJ, Gildow V (1984) Plant Physiol (Bethesda) 74:284
14. Fluhr R, Kuhlemeier C, Nagy F, Chua N-H (1986) Science 232:1106
15. Griffling LR, Ray PM (1979) Plant Physiol (Bethesda) Suppl 63:283
16. Guilfoyle TJ (1986) CRC Crit Rev Plant Sci 4:247
17. Hasunuma K, Furukawa K, Tomita K, Mukai C, Nakamura T (1987) Biochem Biophys Res Commun 148:133
18. Hepler PK, Wayne RO (1985) Annu Rev Plant Physiol 36:397
19. Hetherington AM, Trewavas A (1982) FEBS Lett 145:67
20. Jena PK, Reddy ASN, Poovaiah BW (1989) Proc Natl Acad Sci USA 86:3644
21. Kubowicz BD, Vanderhoef LN, Hanson JB (1982) Plant Physiol (Bethesda) 69:187
22. Lapetina EG, Reep BR (1987) Proc Natl Acad Sci USA 84:2261
23. Lee JS, Mulkey TJ, Evans ML (1983) Science 220:1375
24. Litosch C, Wallis JN, Fain JN (1985) J Biol Chem 260:5464
25. Lukas TJ, Iverson DB, Schleicher M, Watterson DM (1984) Plant Physiol 75:788
26. Mandoli DF, Tepperman J, Huala E, Briggs WR (1984) Plant Physiol (Bethesda) 75:359
27. McFadden JJ, Poovaiah BW (1988) Plant Physiol (Bethesda) 86:332
28. McMurray WC, Irvine RF (1988) Biochem J 249:877
29. Mudge KW, Narayanan KR, Poovaiah BW (1981) J Am Soc Hortic Sci 106:80
30. Murray MG, Key JL (1978) Plant Physiol (Bethesda) 61:190
31. Morse MJ, Crain RC, Satter RL (1987) Proc Natl Acad Sci USA 84:7075
32. Nishizuka Y (1986) Science 233:305
33. Nitsch JP (1950) Am J Bot 37:211
34. Nitsch JP (1955) Plant Physiol 30:33
35. Perdue DO, LaFarve AK, Leopold AC (1988) Plant Physiol (Bethesda) 86:1276
36. Perdue DO, Leopold AC (1988) Plant Physiol (Bethesda) Suppl 86:68
37. Polya GM, Davies JR (1982) FEBS Lett 150:167
38. Poovaiah BW, Reddy ASN (1987) CRC Crit Rev Plant Sci 6:47

39. Poovaiah BW, Reddy ASN In: Boss W, Loewus F, Morre J (eds) Inositide Metabolism in Plants. Liss, New York (in press)
40. Poovaiah BW, Reddy ASN In: Proc Intl Congress Plant Physiol. Cambridge Univ Press (in press)
41. Poovaiah BW, Reddy ASN In: Waisel Y, Eshel A, Kafkafi U (eds) Plant roots. Dekker, New York (in press)
42. Poovaiah BW, McFadden JJ, Reddy ASN (1987) Physiol Plant 71:401
43. Poovaiah BW, Reddy ASN, McFadden JJ (1987) Physiol Plant 69:569
44. Poovaiah BW, Friedmann M, Reddy ASN, Rhee JK (1988) Physiol Plant 73:354
45. Raghothama KG, Reddy ASN, Friedmann M, Poovaiah BW (1987) Plant Physiol (Bethesda) 83:1008
46. Ranjeva R, Carrasco A, Boudet AM (1988) FEBS Lett 230:137
47. Reddy ASN, Poovaiah BW (1987) J Biochem 101:569
48. Reddy ASN, Chengappa S, Poovaiah BW (1987) Biochem Biophys Res Commun 144:944
49. Reddy ASN, McFadden JJ, Friedmann M, Poovaiah BW (1987) Biochem Biophys Res Commun 149:334
50. Reddy ASN, Friedmann M, Poovaiah BW (1988) Plant Cell Physiol 29:179
51. Reddy ASN, Koshiba T, Theologis A, Poovaiah BW (1988) Plant Cell Physiol 29:1165
52. Reddy ASN, Poovaiah BW (1990) Plant Mol Biol 14:127
53. Reddy ASN, Jena PK, Mukherjee SK, Poovaiah BW (1990) Plant Mol Biol 14:643
54. Roberts DM, Lukas TJ, Watterson DM (1986) CRC Crit Rev Plant Sci 4:311
55. Roux SJ, Serlin BS (1987) CRC Crit Rev Plant Sci 5:205
56. Sanchez-de-Jimenez E, Perez A, Perez L (1988) Plant Physiol (Bethesda) Suppl 86:929
57. Schumaker DS, Sze H (1987) J Biol Chem 262:3944
58. Shen-Miller J (1978) Plant Cell Physiol 19:445
59. Stinemetz CL, Kuzmanoff KM, Evans ML, Jarrett HW (1987) Plant Physiol 84:1337
60. Stratowa C, Rutter WJ (1986) Proc Natl Acad Sci USA 83:4292
61. Suzuki T, Fujii T (1978) Planta 142:275
62. Theologis A (1986) Annu Rev Plant Physiol 37:407
63. Theologis A, Huynh TV, Davis RW (1985) J Mol Biol 183:53
64. Veluthambi K, Poovaiah BW (1984) Science 223:167
65. Veluthambi K, Poovaiah BW (1986) Plant Physiol (Bethesda) 81:836
66. Zbell B, Walter C (1987) In: Klambt D (ed) Plant hormone receptors. Springer, Berlin Heidelberg New York Tokyo, p 141

IV Hormone Synthesis and Metabolism

Recent Studies of the Metabolism of Abscisic Acid

J.A.D. ZEEVAART[1], D.A. GAGE[2], and R.A. CREELMAN[3]

1 Introduction

Abscisic acid is the only one of the five major plant growth substances of which the biosynthetic pathway in higher plants remains unknown. In fungi, there is good evidence that the direct pathway from MVA via farnesyl pyrophosphate operates, with α- or γ-ionylidene derivatives as intermediates in the later stages of the pathway [26]. On the other hand, there is considerable evidence that in higher plants ABA is synthesized via the indirect pathway in which ABA is a breakdown product of a larger precursor molecule, probably a carotenoid [26]. Results of $^{18}O_2$ labeling studies of ABA produced in water-stressed leaves strongly favor the indirect pathway. In this paper we report on recent investigations of ^{18}O incorporation into ABA from stressed as well as from non-stressed tissues.

Levels of ABA in a tissue are determined not only by biosynthesis, but by breakdown as well. While the catabolic pathway of ABA is known, the regulation of the conversion of ABA to PA by mechanisms other than the water status of the tissue has not been reported. Herein, we present data that this key step in ABA inactivation is inhibited by the growth retardant tetcyclacis (Tcy).

2 Incorporation of ^{18}O into Abscisic Acid

In the initial experiments with stressed *Xanthium* leaves incubated with $^{18}O_2$, one atom of ^{18}O was present in the carboxyl group of ABA [9]. In this early work, isotope enrichment was determined by MS with electron impact (EI) ionization. Since the molecular ion of Me-ABA is an ion of very low abundance, this method is not sufficiently sensitive to detect small amounts of isotope enrichment. Negative chemical ionization (NCI)-MS has the advantage that it is very sensitive for electrophilic compounds, such as Me-ABA. Moreover, the molecular ion $[M]^{-}$ is the base peak in the spectrum, although a few fragment ions are present. Initially, this required the positions of the ^{18}O atoms within Me-ABA to be determined by analysis of the EI-MS fragmentation pattern. In addition, NCI MS-MS was also used to localize labeled oxygen atoms [8]. With the recent complete interpretation

[1] MSU-DOE Plant Research Laboratory, Michigan State University, East Lansing, MI 48824, USA
[2] MSU-NIH Mass Spectrometry Facility, Department of Biochemistry, Michigan State University, East Lansing, MI 48824, USA
[3] Department of Biochemistry and Biophysics, Texas A&M University, College Station, TX 77843, USA

of the fragmentation pattern of Me-ABA by NCI-MS [17], ^{18}O atoms can be readily assigned to their various positions within Me-ABA.

Using NCI-MS, it was shown with long-term labeling experiments [8] that in stressed *Xanthium* leaves three of the four oxygen atoms were derived from molecular oxygen. One ^{18}O atom was rapidly incorporated in the carboxyl group, while much smaller amounts of ^{18}O were present in the ring oxygens of ABA. The unequal enrichment with ^{18}O at the various positions of the ABA molecule contrasts with the equal enrichment in a sesquiterpenoid synthesized via the direct pathway, the phytoalexin rishitin [5]. The fourth oxygen atom of ABA, in the carboxyl group, was derived from water [8]. These results indicate that stress-induced ABA was derived from a large precursor pool which already contains two of the four oxygens present in ABA. As this primary precursor is depleted over time, other precursors containing fewer oxygen atoms feed into the biosynthetic pathway [8]. The high degree of isotope enrichment in the carboxyl group suggests oxidative cleavage of a larger molecule, probably yielding an aldehyde. This intermediate could be further oxidized by a dehydrogenase and incorporate an oxygen atom from water into the carboxyl group of ABA. This idea is supported by the recent finding that ABA-aldehyde is the immediate precursor of ABA in higher plants [20,22].

The work with *Xanthium* leaves has been extended to other species, such as *Hordeum vulgare, Nicotiana tabacum, Persea americana*, and *Phaseolus vulgaris*. In each species most ^{18}O was incorporated into the carboxyl group of ABA, whereas isotopic enrichment was much less in the oxygen atoms of the ring. In both *Hordeum* and *Phaseolus* the 1′-hydroxyl group contained more ^{18}O than it did in ABA from *Xanthium*. This may be due to a smaller primary precursor pool in the former plants than in the latter [15]. Thus, although there are minor variations between different species, it is clear that in all stressed leaves ^{18}O is most rapidly incorporated into the carboxyl group of ABA, which adds further support to the hypothesis that ABA originates by oxidative cleavage of a precursor with more than 15 carbon atoms.

Similar ^{18}O experiments have been conducted with unripe avocado fruits. In Fig. 1, an NCI mass spectrum of Me-ABA from avocado mesocarp incubated in $^{18}O_2$ can be compared with that from ^{18}O-labeled Me-ABA from stressed *Phaseolus* leaves. It is clear that the pattern of ^{18}O incorporation into ABA was very similar in the two tissues. In each case, the $[M]^-$ ion was shifted from m/z 278 to 280, indicating high ^{18}O enrichment in one oxygen atom. The position of this ^{18}O atom is in the carboxyl group in both tissues, as is evident from the shift of the ion at m/z 141 to 143. This ion represents the side chain of Me-ABA [17]. In the doubly labeled Me-ABA (M^- at m/z 282), the second ^{18}O was located in the tertiary hydroxyl group. This conclusion is based on the relative abundances of the ions m/z 280 and 282 compared to m/z 262 and 264. The pair of ions at lower mass is formed by the loss of water from the 1′-hydroxyl group [17]. Loss of 18 amu from m/z 280 gives 262. Likewise, loss of 1′-^{18}OH from m/z 282 gives 262. The latter two losses were confirmed by NCI MS-MS. The fragment at m/z 264 is derived from m/z 284 by loss of 1′-^{18}OH. Thus, based on the ^{18}O labeling pattern of ABA from unripe avocado mesocarp and stressed leaves, the biosynthetic pathway of ABA appears to be the same in the two tissues.

At different stages of ripening avocado fruits displayed different ^{18}O labeling patterns. In some samples, the base peak was shifted from m/z 278 to 282, indicating

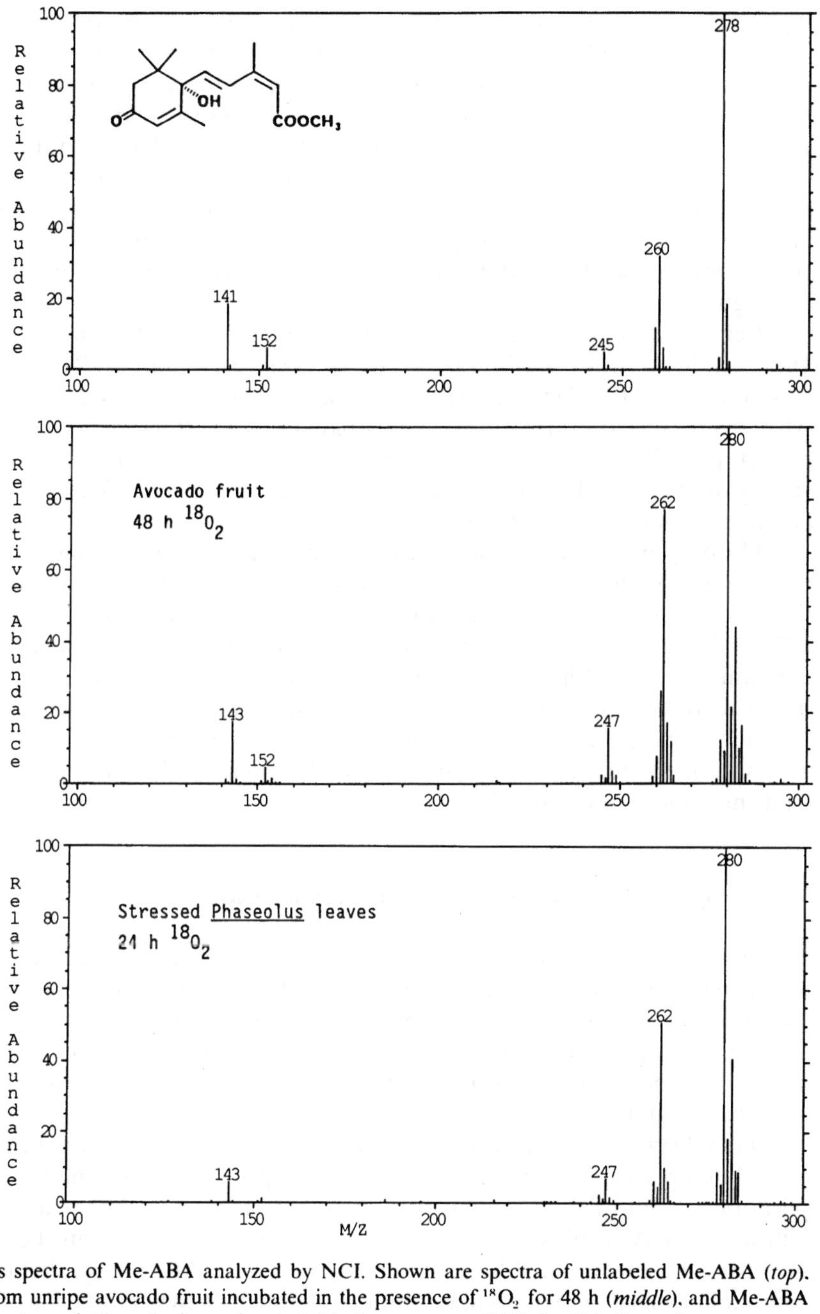

Fig. 1. Mass spectra of Me-ABA analyzed by NCI. Shown are spectra of unlabeled Me-ABA (*top*). Me-ABA from unripe avocado fruit incubated in the presence of $^{18}O_2$ for 48 h (*middle*). and Me-ABA isolated from stressed *Phaseolus* leaves incubated in the presence of $^{18}O_2$ for 24 h (*bottom*)

high ^{18}O incorporation into two positions, with a substantial amount of triply labeled ABA (m/z 284), and relatively little singly labeled ABA (m/z 280) formed. Analysis of such spectra indicated that of the total ABA labeled, the carboxyl group was always the most enriched in ^{18}O, with the tertiary hydroxyl a close second. The third ^{18}O atom, in the 4′-keto position, was always considerably less abundant than the one in the 1′-hydroxyl. Assuming that ^{18}O in the 4′-keto position exchanged to some extent with water [8], both ring oxygens would be highly labeled. Thus, the fruit stage at which this labeling pattern was observed presumably represents the time when the primary precursor pool was depleted, but biochemical machinery for ABA biosynthesis was still highly active.

Very little incorporation of ^{18}O was observed in turgid *Xanthium* leaves [8]. However, in recent work with turgid *Phaseolus* leaves, significant ^{18}O incorporation was obtained with a labeling pattern similar to that in stressed leaves. Likewise, ABA from excised maize embryos incubated in $^{18}O_2$ showed the same labeling pattern [11]. Thus, the ABA biosynthetic pathway appears to be the same in a variety of tissues, although synthesis is probably controlled differently in stressed and non-stressed leaves, and possibly in other tissues.

The only exception encountered to the rule that ^{18}O is always most abundantly incorporated in the carboxyl group of ABA was observed with apple fruit tissue. In this case, the highest ^{18}O enrichment was in the tertiary hydroxyl group. Clearly, if ABA is derived from a larger precursor molecule by oxidative cleavage, then ^{18}O enrichment in the 1′-hydroxyl group can at most be equal to, but not higher than in the carboxyl group. One possible explanation would be post-cleavage incorporation of ^{18}O at the 1′-position, but such a mechanism would rule out xanthoxin as an intermediate in ABA biosynthesis [18]. Obviously, ABA biosynthesis in apple fruits needs to be explored further.

3 Effect of Mevinolin on Abscisic Acid Accumulation

Mevinolin has been described [1] as a potent and specific inhibitor of the enzyme 3-hydroxy-3-methylglutaryl coenzyme A (HMG-CoA) reductase which catalyzes the formation of MVA, the first committed step in isoprenoid biosynthesis. Assuming that ABA is synthesized from a larger precursor, one would expect mevinolin to be without effect on short-term ABA biosynthesis. However, as shown in Table 1, mevinolin strongly inhibited ABA accumulation in water-stressed *Xanthium* leaves, 50% inhibition being caused by approximately 0.4 mM mevinolin. Attempts to reverse the inhibition with prior or simultaneous MVA applications failed (Table 2). This indicates that HMG-CoA reductase is not the critical step in ABA biosynthesis that is blocked by mevinolin. It must be assumed, therefore, that mevinolin inhibits other reactions in the ABA biosynthetic pathway. Although reversal of mevinolin inhibition of growth by MVA has been observed in plants [3], failure to obtain reversal has been observed with the accumulation of sesquiterpenoid phytoalexins in potato [21] and tobacco [6].

Table 1. Effect of mevinolin on abscisic acid accumulation in water-stressed leaves of *Xanthium strumarium*[a]

Treatment	Mevinolin mM[b]	ABA $\mu g\ g^{-1}$ dry wt.
Turgid	0	1.4
Stressed	0	26.5
Stressed	0.1	21.8
Stressed	0.3	18.0
Stressed	0.5	8.3
Stressed	0.8	5.3

[a] Leaves were harvested after 6 h stress (12% loss of fresh wt.).
[b] Mevinolin was hydrolyzed according to [13] to give the K-salt, and fed via the petioles to excised leaves prior to stress. The concentrations were estimated on the basis of amount taken up and water content of leaves.

Table 2. Effects of mevinolin and mevalonic acid on abscisic acid accumulation in water-stressed leaves of *Xanthium strumarium*

Treatment	Mevinolin[a] mM	Mevalonic acid[a] mM	ABA $\mu g\ g^{-1}$ dry wt.	
			Expt. 1	Expt. 2
Turgid	0	0	1.9	2.8
Stressed	0	0	23.4	28.4
Stressed	0.5	0	7.8	8.1
Stressed	0	1	23.4	33.3
Stressed	0.5	1	5.1	8.8

[a] Mevalonic acid and mevinolin were fed as their K-salts via the petioles to excised leaves prior to stress. Further details as in Table 1.

4 Effect of Tetcyclacis on Abscisic Acid Catabolism

The growth retardant Tcy inhibits the oxidation of *ent*-kaurene to *ent*-kaurenoic acid [19]. The three oxidative steps involved are catalyzed by Cyt P-450 monooxygenases. Since 8′-hydroxylation of ABA which gives rise to PA also involves a Cyt P-450 dependent step [12], it was of interest to see if this reaction could be inhibited by Tcy. Excised *Xanthium* leaves were allowed to take up a solution of Tcy before they were wilted to induce rapid ABA accumulation. After a wilting period of 5 h, the leaves were rehydrated, which resulted in a rapid decrease in ABA with a concomitant increase in PA [24]. The leaves were analyzed for ABA, PA, and ABA-GE at various times before, during, and after wilting. The results in Fig. 2 indicate that Tcy had no effect on the accumulation of stress-induced ABA. However, when the wilted leaves treated with Tcy were rehydrated, ABA levels

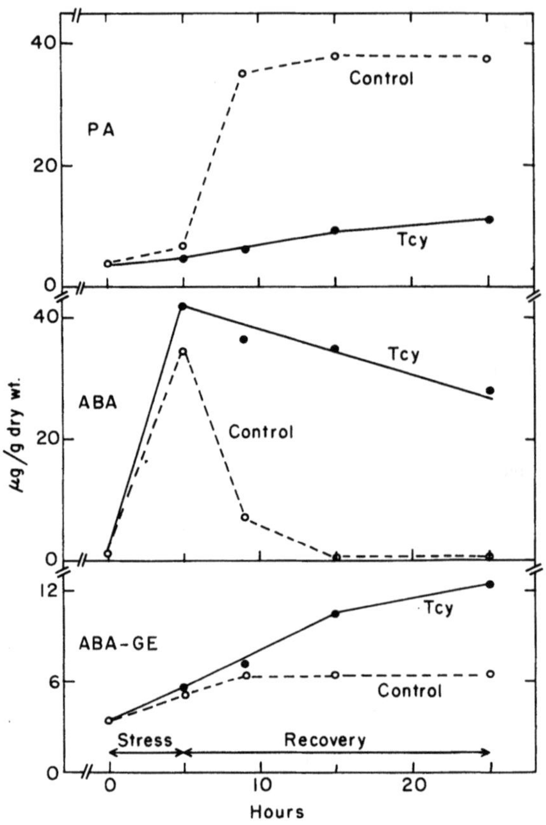

Fig. 2. Inhibition of the conversion of ABA to PA by Tcy following rehydration of water-stressed *Xanthium strumarium* leaves. Three ml of an aqueous solution of 5×10^{-5} M Tcy were fed to excised *Xanthium* leaves via the petioles prior to stress. Recovery was initiated by submerging the wilted leaves in distilled water for 5 min

decreased very slowly in comparison with those in control leaves, and very little PA accumulated. More ABA-GE was formed in Tcy-treated than in control leaves, presumably due to the continued high level of free ABA, the substrate for the conjugation reaction. The inhibitory effect of Tcy on the conversion of ABA to PA was also observed in stressed and subsequently rehydrated leaves taken from *Xanthium* plants treated via the soil with Tcy. Furthermore, Tcy also inhibited the catabolism of radioactive ABA. In the case of (+)-ABA, 64% was converted to PA and 14% to ABA-GE in control *Xanthium* leaves, while in Tcy-treated leaves these percentages were 24 and 41, respectively, with 35% of the radioactivity remaining unmetabolized. When (—)-ABA was fed to *Xanthium* leaves, Tcy severely inhibited formation of 7′-hydroxy (—)-*R*-ABA [25]. Increased levels of ABA in plants treated with Tcy and other growth retardants with a similar mode of action have been reported [2, 16; Rademacher, personal communication], as well as responses in treated plants that are reminiscent of increased ABA levels [2, 4]. It has been suggested that inhibition of GA biosynthesis by Tcy could result in increased substrate availability for the ABA biosynthetic pathway [4]. However, on the basis of our results it is more likely that the increased ABA levels in Tcy-treated plants are due to decreased catabolism of ABA to PA. The mechanism of Tcy action, at least in short-term experiments, is probably via inhibition of the Cyt P-450 monooxygenase involved in ABA oxidation.

5 Comparison of the Catabolism of (−)- and (+)-ABA

We have reported that in *Xanthium* leaves (−)-ABA is converted to (−)-ABA-GE and 7′-hydroxy (−)-*R*-ABA [25]. Formation of the latter compound has also been observed in *Hordeum* leaves [7]. Recently, 7′-hydroxy ABA [assigned as (+)-*S*, although no optical rotation was given] has been reported as an endogenous compound in *Vicia faba* [14]. To verify this finding, we have investigated the catabolism of radioactive (+)- and (−)-ABA in this species. After feeding (+)-ABA, the two major catabolites were PA and DPA, but no radioactive 7′-hydroxy ABA was detected. In the case of (−)-ABA metabolism, approximately 5% was converted to 7′-hydroxy (−)-ABA, whereas in a parallel experiment with *Xanthium* leaves 31% of the radioactivity was present in this catabolite [Cf. 25].

A possible explanation for the occurrence of 7′-hydroxy ABA in *V. faba* [14] is as follows (Milborrow, personal communication). The 1′,4′-*cis*-diol of ABA has been identified in immature seeds of *V. faba* [10]. In acidic conditions this compound is epimerized to the 1′,4′-*trans*-diol of (−)-ABA [23]. Thus, endogenous 7′-hydroxy ABA could be produced in *V. faba* via the sequence: 1′-4′-*cis*-diol of ABA → 1′,4′-*trans*-diol of (−)-ABA → (−)-ABA → 7′-hydroxy (−)-ABA. Clearly, to determine the biosynthetic origin of 7′-hydroxy ABA present in *V. faba* [14], it is essential to establish its absolute configuration.

6 Conclusions

Evidence obtained with ^{18}O labeling of ABA indicates that the biosynthetic pathway is probably the same in stressed and non-stressed leaves, as well as in other tissues. This means that in all cases ABA is derived by oxidative cleavage from a precursor that has more than 15 carbon atoms. However, results with apple fruit tissue suggest that variations on this scheme may exist.

Mevinolin inhibits the accumulation of stress-induced ABA in leaves, but the inhibition cannot be overcome by MVA. This suggests that mevinolin also interferes with other steps in the ABA biosynthetic pathway.

The growth retardant Tcy is a potent inhibitor of the conversion of ABA to PA which results in accumulation of ABA and ABA-GE.

Metabolic studies with both enantiomers of ABA do not support a recent report that 7′-hydroxy (+)-*S*-ABA occurs naturally in *Vicia faba*.

Acknowledgments. Work supported by the United States Department of Energy under Contract DE-AC02-76ERO-1338, National Science Foundation grants PCM83-14321 and DMB-8703847, and National Institutes of Health grant DRR 00480. Dr. T.L. Davenport, University of Florida TREC, Homestead, FL, kindly supplied the avocado fruits.

References

1. Alberts AW, Chen J, Kuron G, Hunt V, Huff J, Hoffman C, Rothrock J, Lopez M, Joshua H, Harris E, Patchett A, Monaghan R, Currie S, Stapley E, Albers-Schonberg G, Hensens O, Hirshfield J, Hoogsteen K, Liesch J, Springer J (1980) Proc Natl Acad Sci USA 77:3957
2. Asare-Boamah NK, Hofstra G, Fletcher RA, Dumbroff EB (1986) Plant Cell Physiol 27:383
3. Bach TJ, Lichtenthaler HK (1983) Physiol Plant 59:50
4. Bonham-Smith PC, Kapoor M, Bewley JD (1988) Physiol Plant 73:27
5. Brindle PA, Coolbear T, Kuhn PJ, Threlfall DR (1985) Phytochemistry 24:1219
6. Chappell J, Nable R (1987) Plant Physiol (Bethesda) 85:469
7. Cowan AK, Railton ID (1987) Plant Physiol (Bethesda) 84:157
8. Creelman RA, Gage DA, Stults JT, Zeevaart JAD (1987) Plant Physiol (Bethesda) 85:726
9. Creelman RA, Zeevaart JAD (1984) Plant Physiol (Bethesda) 75:166
10. Dathe W, Sembdner G (1982) Phytochemistry 21:1798
11. Gage DA, Fong F, Zeevaart JAD (1989) Plant Physiol (Bethesda) 89:1039
12. Gillard DF, Walton DC (1976) Plant Physiol (Bethesda) 58:790
13. Kita T, Brown MS, Goldstein JL (1980) J Clin Invest 66:1094
14. Lehmann H, Schwenen L (1988) Phytochemistry 27:677
15. Li Y, Walton DC (1987) Plant Physiol (Bethesda) 85:910
16. Lürssen K (1987) Pestic Sci 21:310
17. Netting AG, Milborrow BV, Vaughan GT, Lidgard RO (1988) Biomed Environ Mass Spectrom 15:375
18. Parry AD, Neill SJ, Horgan R (1988) Planta 173:397
19. Rademacher W, Fritsch H, Graebe JE, Sauter H, Jung J (1987) Pestic Sci 21:241
20. Sindhu RK, Walton DC (1988) Plant Physiol (Bethesda) 88:178
21. Stermer BA, Bostock RM (1987) Plant Physiol (Bethesda) 84:404
22. Taylor IB, Linforth RST, Al-Naieb RJ, Bowman WR, Marples BA (1988) Plant Cell Environ 11:739
23. Vaughan GT, Milborrow BV (1988) Phytochemistry 27:339
24. Zeevaart JAD (1980) Plant Physiol (Bethesda) 66:672
25. Zeevaart JAD, Boyer GL, Cornish K, Creelman RA (1985) In: Bopp M (ed) Plant growth substances 1985. Springer, Berlin Heidelberg New York Tokyo, p 101
26. Zeevaart JAD, Creelman RA (1988) Annu Rev Plant Physiol Plant Mol Biol 39:439

Recent Investigations of the Biochemistry of Abscisic Acid

B.V. MILBORROW[1]

1 The Diols of ABA: A Cautionary Tale

Over the last twenty-five years ABA has displayed a continuing series of unexpected, quirkish properties. As a quick review one can mention:

1. Natural (+)-*S*-ABA melts at 161°, Synthetic (+)-*R,S* at 190.
2. Homoallylic conjugation is responsible for the apparent inversion of the ORD spectrum when the 4′-double bond is saturated [24].
3. ABA is the only compound that does not obey Mills' rule [21].
4. The 1′-OH cannot be acetylated but plants can form the 1′-O-β-D-glucoside [8].
5. The 1′-O-glucoside rearranges spontaneously to ABA glucose ester [8].
6. The Electron Impact Mass Spectrum is totally different from the Chemical Ionization Mass Spectrum [8].
7. The ketal of ABA, but not that of the 2-*trans* isomer, hydrolyzes spontaneously to ABA (Abrams and Milborrow, unpubl.).
8. ABA can be made by a Wittig reaction on a ketone whereas normally only aldehydes react [20].
9. The ^{1}H NMR signal of the 5′-*pro-R* hydrogen atom is upfield of its 5′-*pro-S* partner in ABA and the 1′,4′-*cis*-diol but downfield in the 1′,4′-*trans*-diol [11].
10. In the ^{13}C NMR spectrum of ABA the signals of the *geminal* 8′ and 9′ methyl groups are downfield from those of the C-6 and C-7′ methyl groups and that of the axial, C-8′-*pro-S* is upfield from that of the equatorial C-9′-*pro-R*. In the ^{1}H spectrum the signals of the *geminal* C-8′ and C-9′ methyl groups are upfield of those of C-6 and C-7′ methyl groups and the axial C-8′-*pro-S* is downfield from that of the equatorial, C-9′-*pro-R* [11].
11. In spite of being one of the most optically active compounds known, both (+)-*S*-and (−)-*R*-ABA inhibit growth almost equally.
12. (+)-*S*-ABA causes stomata to close, (−)-*R*-ABA has little effect.

These are the some of the credentials that establish the idiosyncratic quiddity of ABA and demand that great care is taken in the rigorous scrutiny of conclusions drawn from experimental results.

A further example of the unexpected behaviour comes from work done with Vaughan [22, 23] on the 1′,4′-diols of ABA. We examined their physicochemical

[1] School of Biochemistry, University of New South Wales, P.O. Box 1, Kensington, New South Wales 2033, Australia.

oxidation and found that, not only were they oxidised to ABA at pHs below about 5 (Fig. 1), but they were interconverted.

Furthermore the (+)-*trans*-diol is converted into the (+)-*cis*-diol but it is also converted into the (—)-*cis*-diol, and vice versa. The reaction is believed to proceed as shown in Fig. 2.

Thus epimerization occurs at C-1′ and C-4′ and when these reactions were carried out in [^{18}O]H_2O, and the compounds analyzed by mass spectrometry, it was found that ^{18}O was present in both hydroxyl groups, even of molecules that had not apparently suffered epimerization, although exchange at C-4′ exceeded that at C-1′ (Table 1).

The oxygen atom of the 4′-ketone group of ABA exchanges with the medium but the oxygen atom of the 1′-hydroxyl group is completely stable. Consequently precursors of ABA with a ring bearing two hydroxyl groups may have exchanged one or both of their oxygen atoms during biosynthesis. The relative enrichments of ^{18}O from $^{18}O_2$ at C-1′ and C-4′ of ABA should, therefore, be interpreted with care, apparent incorporation may have been modified by loss of ^{18}O by exchange processes.

The epimerization of the hydroxyl groups of the *trans*-, and particularly the *cis*-diol, may account for the contradictory reports on the occurrence of 7′-hydroxy ABA or nigellic acid as a natural or an unnatural metabolite. After the compound was characterized [5], Boyer and Zeevaart [1] found that it was formed solely from the unnatural (—)-*R* enantiomer of (+)-*RS* [^{14}C]ABA. More recently, Lehmann and Schwenen [6] found that it occurred naturally and concluded, therefore, that it was a derivative of (+)-*S*-ABA. This is not compatible with the detailed mechanism proposed for the formation of nigellic acid [12], but before the suggestion is

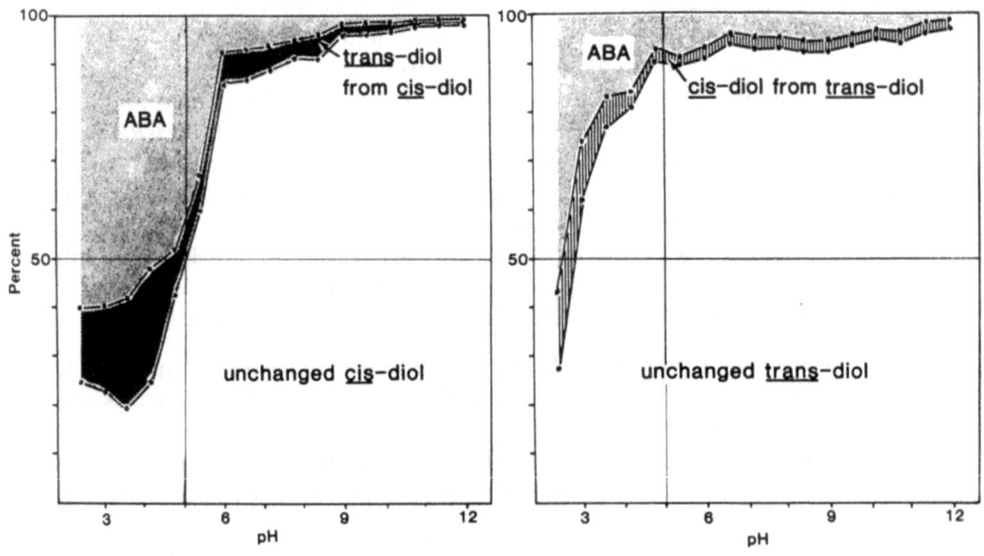

Fig. 1. Effect of pH on the stability of the [2-^{14}C]1′,4′-diols of ABA. Aqueous solutions of each diol were buffered from pH 2.4 to 12, extracted into ether after 72 h and separated by silica gel tlc. The amounts of ABA and the other isomer were then determined by scintillation spectrometry

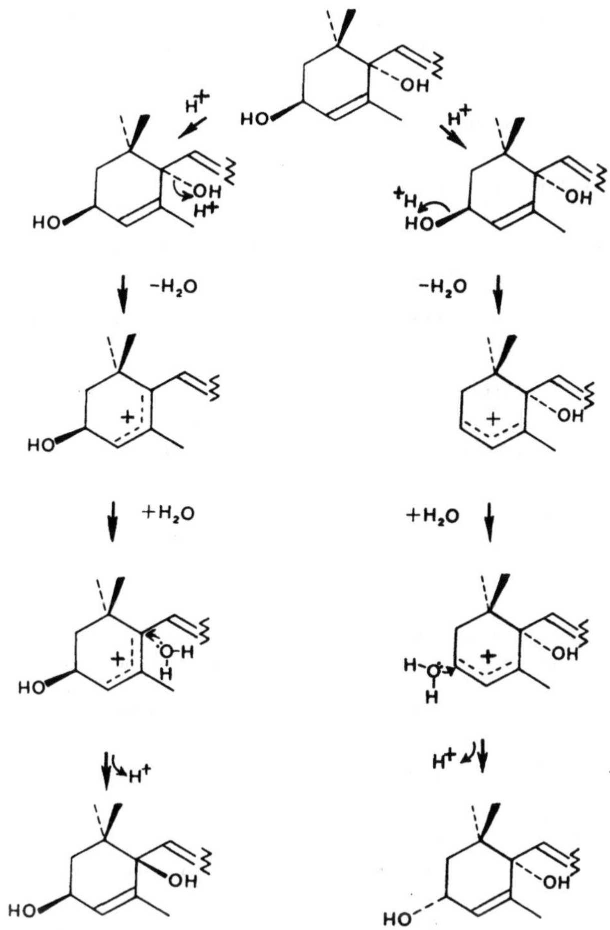

Fig. 2. The interconversion of the diols in acid. The observed loss of [18]0 from the 1′ and 4′ hydroxyl groups is compatible with mechanism proposed where protonation of a hydroxyl group, followed by the elimination of water produces a carbocation. Attack from either face reforms a diol after deprotonation

abandoned it is possible to account for its natural occurrence as a metabolite of (—)-*R*-ABA.

It is noteworthy that the plant species chosen by Lehmann and Schwenen: broad bean (*Vicia faba*), is the same one from which Dathe and Sembdner [3] isolated the *cis*-diol of ABA. We have now shown that the (+)-*cis*-diol can epimerize to the (—)-*cis*-diol and to the (—)-*trans*-diol [23]. Thus, if (—)-*cis*-diol and (—)-*trans*-diol formed spontaneously in vivo, were to be oxidised physicochemically or enzymically then (—)-*R*-ABA would be present endogenously. Its hydroxylation to (—)-7′-OH ABA could then reconcile Lehmann and Schwenen's (b) observations with those of Boyer and Zeevaart [1]. The diols are relatively abundant in pea shoots, less so in avocado fruit and indetectable in tomato shoots (Table 2) [19].

The amounts of (—)-*cis* and (—)-*trans* diol formed by epimerization (Fig. 1) would be expected to comprise a small part of the total diols and any conversion into ABA would produce a very minor proportion of the total, free ABA. It would be

Table 1. Exchange of the oxygen atoms at C-1′ and C-4′ of the 1′,4′-*cis*- and 1′,4′-*trans* diols with and without epimerization. The Me esters were incubated for 72 h at pH 3.5 and then analyzed by mass spectrometry

Starting material in $[^{16}O]H_2O$	*After incubation* $\%^{18}O$	$\%^{18}O$
Me 1′-$[^{18}O]$*trans*-diol ($41\%\ ^{18}O = 100\%$)	*trans*-diol 39.6% (96.5% of initial)	*cis* diol 30.2% (73.7% of initial)
Me 1′-$[^{18}O]$*cis*-diol ($40.3\%\ ^{18}O = 100\%$)	37.6(93.3% of initial)	35.8% (88.86% of initial)

No loss of deuterium occurred from C-4′ of the *cis*- or *trans*-diols during incubation in $[^1H]H_2O$.

Starting material in $[^{18}O]H_2O$ (97.2 atoms %)	*After incubation* $\%\ ^{18}O$	$\%\ ^{18}O$
	trans-diol	*cis*-diol
Me 4′-$[^{16}O]$*trans*-diol	1′-$[^{18}O]$23.6% 4′-$[^{18}O]$1.5%	*cis*-diol destroyed
Me 4′-$[^{16}O]$*cis*-diol	1′-$[^{18}O]$11.7% 4′-$[^{18}O]$26.3%	1′-$[^{18}O]$0% 4′-$[^{18}O]$ 84%

Table 2. A Contents of ABA, *cis*- and *trans*-diols (μg/kg connected for losses during extraction). **B** RS-$[2^{14}C]$ ABA fed. dpm present in the free diols after 24 h

A.	ABA	*trans*-diol	*cis*-diol
Pea shoots	40 (22 conjugated)	5.2	Not detected
Avocado fruit	8450 (1340 conjugated)	93 (6.7 conjugated)	14.5
Tomato		Not detected	Not detected

B.	(+)-1′-*S*-*trans*	(−)-1′-*R*-*trans*	(+)-1′-*S*-*cis*	(−)-1′-*R*-*cis*
Avocado fruit	3636	6171	389	2293
Broad bean shoots	837	3790	300	541
Pea shoots	846	1055	Not detected	Not detected

impossible to detect this small proportion by measurement of specific rotation. The combination of a chromatographic method of resolving ABA [16] with a new negative ion, chemical ionization, gas chromatographic/mass spectrometic method for determining ABA as a pentafluorobenzyl ester [15] allows the presence of traces of (−)-ABA to be detected. It must be pointed out that any free (−)-ABA could have arisen as a physico-chemically induced artefact by epimerization and oxidation of the *cis*-diol during the isolation procedure. However, if (−)-ABA were present in the plants as a normal constituent then a large part of it would be converted into ABA glucose ester. The presence of (–)-ABA in the hydrolyzed glucose ester fraction must be sought.

The glucose esters and 4′-O-β-D-glucopyranosides of the diols have also been found in peas, so the diols may comprise a minor, alternative pathway of inactivation of ABA. Surprisingly, the growth of pea plants in an anaerobic atmosphere in darkness did not cause an increase in the amount of diols (Fig. 3).

ABA is inactivated by hydroxylation of the 8′-methyl group followed by cyclization to phaseic acid. This in turn is reduced to dihydrophaseic acid (DPA) and traces of *epi*-DPA. All of these compounds and the diols are glucosylated at various sites: esters on carboxyl groups, glucosides on 1′- and 4′-hydroxyl groups. The only other derivative groups are the hydroxy methyl glutaryl residue at C-8′ of the unstable 8′-hydroxy ABA [4] and the maltosyl ester of ABA. The differences between plant species appear to lie in the relative abundances of the different compounds, rather than which ones are formed. For example, the major metabolite of tomato and most plants is the 4-O-β-D-glucopyranoside of DPA, while phaseic acid glucose ester is formed in mango leaves, almost to the exclusion of DPA [9, 2].

2 Glucosyl Conjugates of Diols

The main metabolites formed by tomato shoots from large quantities of *RS*-[2-^{14}C]*trans*-diol (134 μg/g fresh weight) were the glucose esters and 4′-O-glucosides of the 1′-*R* and 1′-*S* enantiomers. When smaller quantities of 1′-*S*-*trans*-diol (3.6 μg/g fresh weight) were supplied it was almost all converted into DPAGS, presumably after oxidation of the 4′-hydroxyl group to produce ABA [12]. A small quantity of 1′-*R*-*trans*-diol (3.1 pg/g) was converted into its glucose ester, glucoside and metabolites of ABA. A small amount of DPAGS detected (normally derived entirely from *S*- ABA) could have arisen from *S*-[^{14}C]ABA formed by epimerization at C-1′ of the 1′-*R*-diol.

The diol 4′-O-β-D-glucopyranosides and β-D-glucosyl esters have been characterized as naturally occurring metabolites in peas and avocadoes but the quantities present were about 10%, or less, of the free diols. In these tissues, at least, (Table 1) the diol pathway for inactivation of ABA is a minor one. Against this must be weighed the instability of the diols. Great care has to be taken to prevent oxidation to ABA and conventional extraction procedures, monitored by adding racemic G-^{3}H labelled material of high specific activity (33.2 mCi/μmole), disclosed between 2 and 12% recovery for the *trans*- and 0% for the *cis*-diol. It is probable therefore, that the presence of the diols has been overlooked frequently in the past.

The diols and DPA have three, and ABA and PA have two, sites at which glucosylation can occur. However, so far no metabolite has been detected with a glucosyl residue at more than one position. This may be trivial or it may be indicative of the mechanism of conjugation's being in a membrane at an interface to that the polar metabolite is passed through the membrane as it is formed and then released into a vacuolar or golgi vesicle. The polarity of the conjugates debar them from reentering the cytosol and thereby prevents their gaining another sugar residue at a second site.

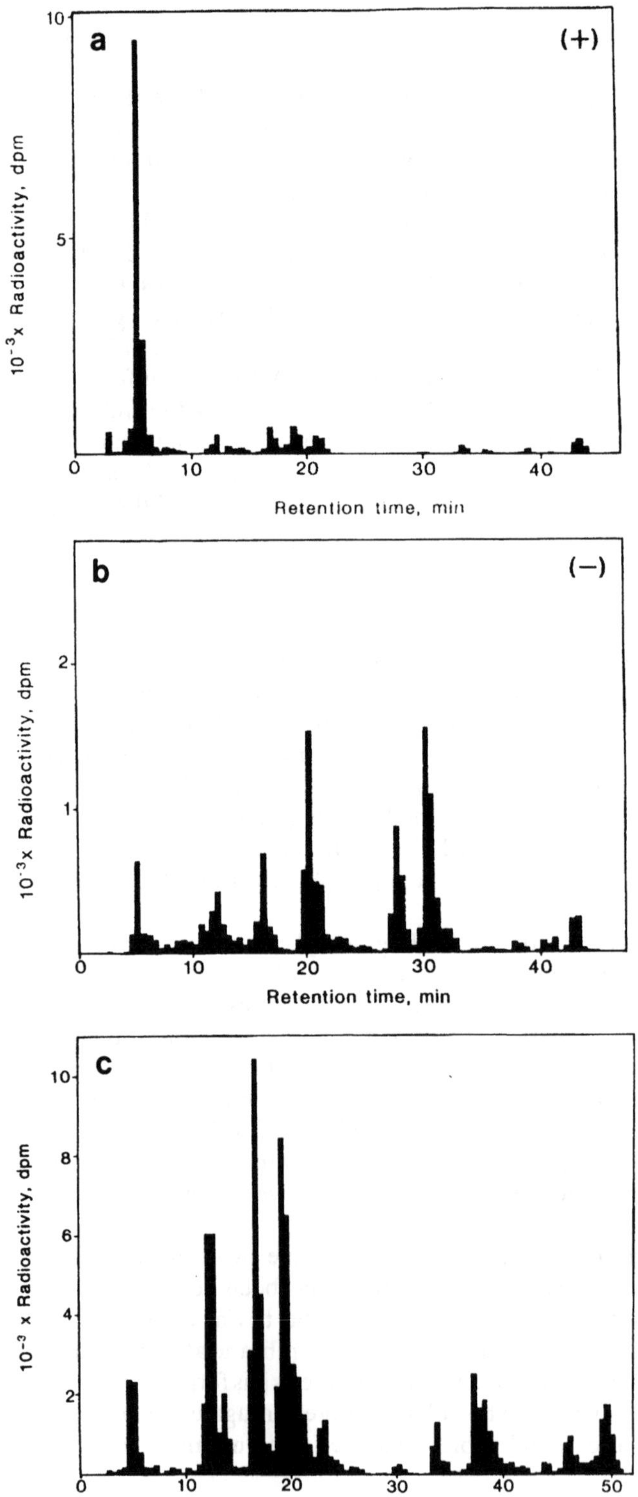

Fig. 3. Metabolites of [2-^{14}C] *trans*-diol fed to tomato shoots. **3a** *RS-trans* diol (134 μ/g to 56 g shoots); **3b** 1′-*S-trans* diol (3.96 μg/g to 5.6 g shoots); **3c** 1′-*R-trans*-diol (3.1 μg/g to 6.0 g shoots). Reversed-phase HPLC C_{18} column, ethanol /0.2% aqueous acetic acid (3:17 v/v) for 22 min then a linear gradient to ethanol /0.2% aqueous acetic acid (1:3 v/v) over 15 min. DPAGS, 6 min; 1′*S-trans*-diol, 12 min; 1′-*R-trans*-diol, 14 min; 1′-*R-trans*-diol-4′-glucoside, 17 min; 1′-*S-trans*-diol-4′-glucoside, 20 min; ABAGE, 21 min; ABAGS, 28 min; 7′-OHABA, 31 min; *trans*-diol, 34 min; ABA, 38 min. NB Conjugation of the *R* and *S-trans* diol with D glucose forms compounds which are no longer enantiomeric and so become separable chromatographically

3 Formation of Phaseic Acid

The discovery of the structure of phaseic acid (PA) began with the isolation of "metabolite C" (8'-OH ABA) from tomato shoots fed with [^{14}C]ABA. The needle-like crystals melted at 195 but immediately re-crystallized as hexagonal plates which then melted at 207°. The strong (+) Cotton effect of 8'-HO ABA, similar to that of (+)-ABA, became a weak negative one. Metabolite C had become phaseic acid. The 9 mg of needle-like crystals of [8'-OH ABA] were methylated with diazomethane and the Me ester was analyzed by NMR spectrometry [10].

The structure of PA was determined and the rearrangement was deduced to be the cyclization reaction: a nucleophilic attack of the 8' hydroxy methyl group on C-2'. There are two possible causes: a base in the diazomethane as an aerosol of KOH was swept over during the formation of diazomethane, or (b) acid as ^{2}HCl can be formed by the breakdown of C^2HCl_3 which was used as an NMR solvent. Thus, 8'-hydroxy ABA can be rearranged physicochemically to PA. PA occurs in vivo, but it is not known whether its formation is physico-chemical or catalyzed by an enzyme.

If the cyclization were physico-chemical then the addition of H^+ to C-3' could be expected to occur almost equally from the α- or β-faces of the ring (Fig. 4). If the reaction were catalyzed enzymically then the addition of H^+ to C-3' of 8'-OH ABA could be expected to be stereospecific. This can be tested by feeding $^2H_6$ABA and isolating the dihydrophaseic acid glucoside (DPAGS). The experiment was done with tomato shoots where only the (+)-S-ABA is oxidised to PA and DPA. Not only is DPAGS the most abundant metabolite but, more importantly, the new H atom at C-4' is of known absolute configuration. Five hundred MHz high field ^{1}H NMR spectrometry can determine the coupling constants with the 4 H atoms at C-3' and C-5' and thereby establishes which gives rise to which signal. Deuterium was present in the 3'-*pro-S*, 5'-*pro-S* and 5'-*pro*-R positions of the DPAGS while the signal of the 3'-*pro-R* hydrogen atom was not diminished (Fig. 5). The C-3'-*pro*-R position, therefore, contained no deuterium from ABA and originated from ^{1}H atoms derived from the medium. A further check on the validity of the result was obtained by carefully exchanging PA in ^2H_2O at pH 10.55 and monitoring the replacement of the ^{1}H atoms at C-3' and C-5'.

The cyclization of 8'-OH ABA proceeds through a carbanion at C-3' which then gains a proton from the medium. Similarly, the enolization of the 4'-ketone also proceeds via a carbanion at C-3'.

If the exchange reaction were strongly sterically directed by the conformation of the ring then the apparent selectivity of the cyclization could be attributed to physicochemical features of the mechanisms. However, the opposite result was obtained. The 3'-*pro-R* of DPA is derived from the medium while the H atom of PA that becomes the 3'-*pro*(*S*) of DPA exchanges with the medium. Consequently, the specific addition of H^+ from the medium to C-3' of 8'-OH ABA cannot be accounted for by physicochemical processes and the cyclization reaction, therefore, must be enzyme catalyzed [17]. This result does nothing to explain why on one occasion we were able to isolate 9 mg of 8'-OH ABA while in every subsequent experiment, with all solutions buffered close to neutrality, at 4°C and in near darkness, with redistilled solvents and antioxidants present, only PA was isolated.

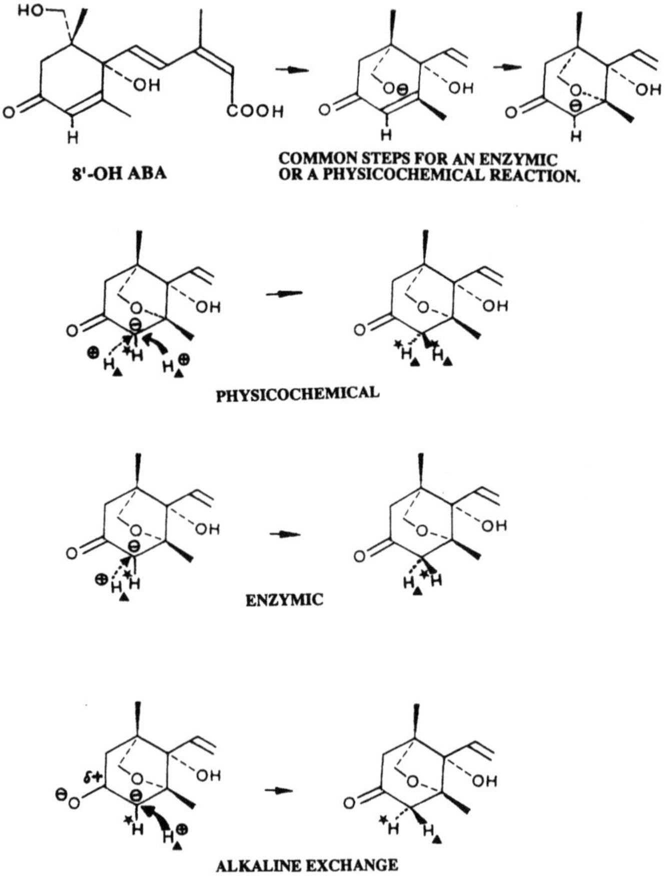

Fig. 4. Representation of the steric constraints on the intermediates during the cyclization of 7'-HO ABA to form phaseic acid. The reaction proceeds via a carbanion (1) at C-3'. If the reaction were entirely physicochemical then the H^+ taken up from the medium would be expected in both positions at C-3'. (2) If the reaction were catalyzed enzymically then the H^+ taken up would be expected to be restricted to just one position. (3) Base catalyzed exchange of PA also goes through a carbanion intermediate but a majority of the H^+ taken up from the medium occupies the opposite position in comparison with the third line

3.1 Patterns of Metabolites

This leads to a consideration of the differences in the patterns of metabolites formed from [^{14}C]ABA. In 1983 [9] histograms of the distribution of labelled metabolites formed by four plant species showed that the patterns differed considerably [11], but recent work has shown that it is the ratios between different metabolites that vary, rather than the variety of compounds.

For example, pea seedlings reduce more ABA to the *trans*-diol than other species examined. Avocado fruit mesocarp and broad bean shoots reduce (+)-[^{14}C]ABA and (–)-[^{14}C]ABA to their *trans*-diols and smaller amounts of (+)- and

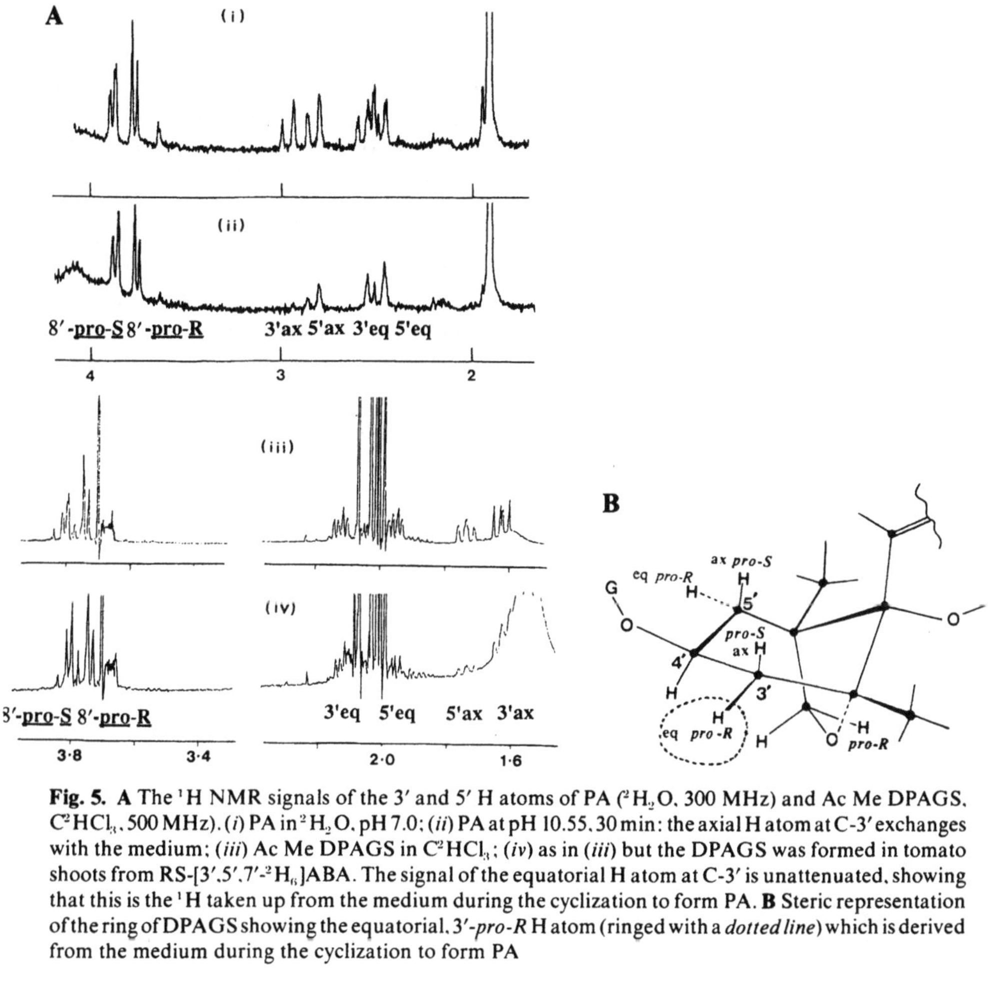

Fig. 5. A The ^{1}H NMR signals of the 3′ and 5′ H atoms of PA (^2H_2O, 300 MHz) and Ac Me DPAGS, C^2HCl_3, 500 MHz). (*i*) PA in ^2H_2O, pH 7.0; (*ii*) PA at pH 10.55, 30 min: the axial H atom at C-3′ exchanges with the medium; (*iii*) Ac Me DPAGS in C^2HCl_3; (*iv*) as in (*iii*) but the DPAGS was formed in tomato shoots from RS-[3′,5′,7′-$^2H_6$]ABA. The signal of the equatorial H atom at C-3′ is unattenuated, showing that this is the ^{1}H taken up from the medium during the cyclization to form PA. **B** Steric representation of the ring of DPAGS showing the equatorial, 3′-*pro-R* H atom (ringed with a *dotted line*) which is derived from the medium during the cyclization to form PA

(−)-*cis* diols were also formed. However, no *cis*-diol could be detected in pea shoots and neither diol could be detected in tomato shoots (Table 2). Dihydrophaseic acid 4′-O-β-D-glucopyranoside is the major metabolite of ABA in long term experiments in tomato and several other species examined, while in mango leaves it is almost indetectable and PA glucose ester is abundant [9]. The 2-*trans* isomers of ABA and all the metabolites can be formed by photolytic isomerization in vivo and are metabolized along the same pathways as the normal, 2-*cis* compounds. Thus, (+)-*S*-2-*trans*-ABA is rapidly converted into 2-*t*-PA and on to 2-*t*-DPA and 2-*t*-DPAGS, while (−)-*R*-2-*trans*-ABA becomes the isomeric glucose ester and 1′-O-β-D-glucopyranoside. The oxidation of ABA to PA and the reduction of PA to DPA or *epi* DPA and ABA to the diols, coupled with the formation of glucose esters on carboxyl groups and glucosides on hydroxyl groups can account for all the metabolites known at present, with two minor exceptions.

4 Deuterium in Xanthoxin, ODA and ABA

One of the surprises we have uncovered with the biosynthesis of ABA came from attempts to measure its rate of turnover by replacing a proportion of the water in tomato plants with ^2H_2O and analyzing the ABA for incorporation of ^2H [17]. Surprisingly, virtually no ^2H was detectable in ABA for up to 3 days.

After about 6 days up to 32% of the ABA was found to be labelled with ^2H atoms (Fig. 6).The pattern suggests that one deuterium is added during cyclization (at C-5′). Evidence to support this was obtained by dissolving the ABA synthesized from mevalonate by an avocado fruit system in which 30% of the water ^2H_2O, in strongly alkaline water (C-3′, C-5′ and 7′ exchange their hydrogen atoms with the medium). The single ^2H was removed, thereby establishing that it was present at one of these sites (C-5′).

The long delay in the appearance of ^2H in ABA is taken to indicate a large pool of precursors and a slow turnover in turgid tomato plants. Most of the precursors

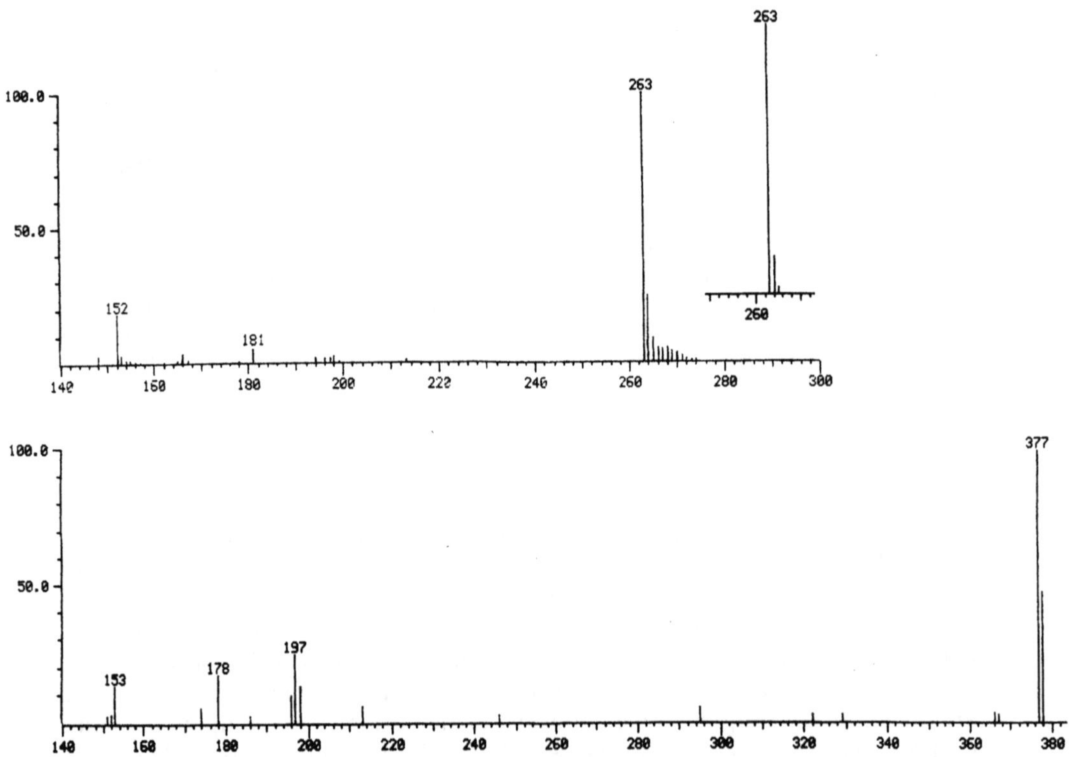

Fig. 6. Methane chemical ionization negative ion mass spectra of pentafluoro benzyl ABA (*upper* with insert of unlabelled material) and dipentafluoro benzyl ODA (*lower*). The compounds were isolated from tomato shoots supplied with 40% ^2H_2O for six days, then severely wilted. M/z-263⁻ = molecular ion − 1 mass unit after the loss of the pentafluorobenzyl residue +H.M/z 377⁻ = molecular ion of dipentafluorobenzyl ODA-1 mass unit after the loss of one pentafluorobenzyl residue +H. *Ordinate*, relative intensity; *abscissa*, mass numbers

measure the rate of turnover of the endogenous ABA by adding a small amount of [^{14}C]labelled ABA to tomato shoots via the sapstream and that it would equilibrate with the endogenous material. It has subsequently been shown by Zeevaart [25], using $^{18}O_2$ labelling, that the endogenous ABA turns over very slowly while added ABA turns over more rapidly. However, the large size of the precursor pool requires that meaningful results can be obtained only when a series of measurements are extended over about a week. The predominance of ABA with one ^2H atom suggests that it is unlikely that ABA is derived from the carotenoid pool. Whether or not a small fractional subpool of carotenoids does give rise to ABA cannot be excluded. Wilting the tomato plants caused a lowering of the specific ^2H atoms % of the ABA. This is interpreted as the mobilization of a larger pool comprising unlabelled, previously synthesized precursor (Table 3).

The favoured candidate as a close precursor for ABA has been xanthoxin. It is present in leaves, is readily converted into ABA and can be formed photolytically and enzymically from violaxanthin. However, ABA was labelled 20 to 30% with from 3 to 14 ^2H atoms in tomato plants supplied with ^2H_2O, whereas the xanthoxin was not labelled with more than one ^2H atom. Xanthoxin, therefore, cannot be a precursor of ABA [17].

Recent work by Taylor and his colleagues [7], has identified a compound present in ABA-less mutant tomato plants as 2,4-octadienedioic acid (ODA) and suggested that it is the C_{10} central residue of a C_{40} carotenoid from which two C_{15} (ABA precursor) residues have been removed. When the same ^2H_2O experiment was carried out with normal tomato plants, which were then wilted severely, the ABA became labelled with up to 11 deuterium atoms but the ODA was unlabelled except that 20% of the molecules contained just one deuterium atom. This is compatible with the ODA's being formed from an unlabelled carotenoid during the course of the experiment and one of the three doublebonds being saturated by NAD H $^2H^+$ (Fig. 7).

Thus, ODA cannot be formed from the same molecules as ABA; it cannot be a by-product of ABA biosynthesis. ODA may be formed from carotenoids in response to stress caused by severe wilting in normal tomato plants and the stress of semi-permanent wilting condition of some tomato mutants [14].

We calculated the total pool size of all the precursors to be some 30 times the size of the pool of free ABA and roughly equal to the amount of violaxanthin, lutein or β-carotene. However, the long half-life of turnover for endogenous ABA, determined by $^{18}O_2$ labelling, would reduce the size of the pool calculated on the

Table 3. Distribution of ^2H in ABA from shoots fed ^2H_2O for 6 days and then extracted, or wilted, then extracted. Measured by negative ion methane chemical ionization mass spectrometry

Numbers of ^{2} H atoms per molecule	*Expt. 1* % of labelled molecules (corrected for natural abundance)		*Expt. 2*	
	Turgid	Wilted	Turgid	Wilted
1	4.4	5.8	8.8	7.4
3 to 14	22.6	14.2	23	14
Total	27	20	32	22

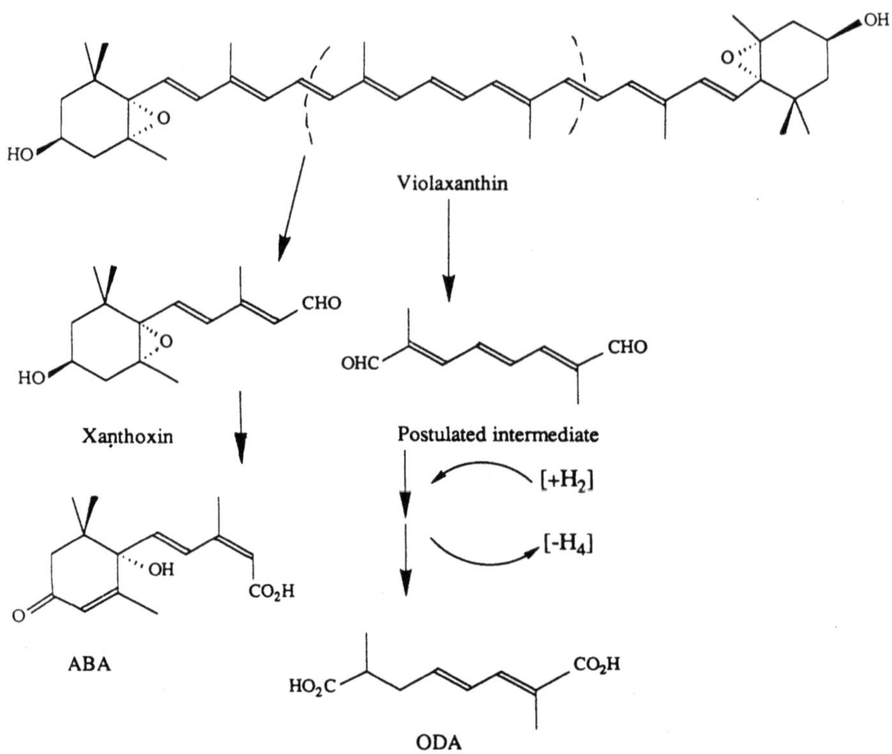

Fig. 7. Oxidative cleavage of violaxanthin or a similar xanthophyll at the two places marked with *broken lines* would form two molecules of xanthoxin and a C_{10} precursor which, by undergoing two oxidations and one reduction, would produce 2,7-dimethyl octa-2,4-dienedioic acid. Xanthoxin is rapidly converted into ABA in plants

basis of our assumptions, which relied on the $t_{1/2}$ measured from the breakdown of added [^{14}C]ABA, to close to the pool size of free ABA or two or three times larger. Thus, the present evidence is compatible with carotenoid being a precursor of ABA. Conversely, if carotenoids do give rise to ABA then they, as precursors, would have to be deuteriated. The carotenoids examined were violaxanthin and lutein epoxide. They were exposed to light to cleave off xanthoxin which was then reacted with pentafluorobenzylhydroxylamine to form xanthoxin pentafluorobenzyloxime. This was subjected to GC/MS negative ion, chemical ionization (NICI) (which virtually gives parent ions only) and good spectra are obtained at the picogram level or below. The MS of the xanthoxin derivative formed from the xanthophylls showed that none contained more than one deuterium, but between 13 and 30% of the ABA isolated from the same tissue was labelled with from 3 to 14 ^{2}H atoms [18].

The importance of choosing the appropriate conditions for analyzing substances of interest is well demonstrated in the analysis of xanthoxin. The violaxanthin and lutein epoxide in plant extracts are cleaved by light, oxygen and water to give a mixture of the *cis* and *trans* isomers of xanthoxin. If care is taken to keep the extracts in darkness, exclude oxygen, add antioxidants, keep the extracts cool and dry and use ammonia rinsed glassware, then the xanthoxin concentrations measured are extremely low and only the 2-*trans* isomer is present [18].

Acknowledgements. Thanks are due to Drs. A.M. Duffield, A.G. Netting, H.M. Nonhebel and G.T. Vaughan, together with Messrs N.J.Carrington and R.D. Willows, who have contributed greatly to the work described here. The investigations were supported by the Australian Research Grants Scheme.

References

1. Boyer GL, Zeevaart JAD (1986) Phytochemistry 25:1103
2. Carrington NJ, Vaughan GT, Milborow BV (1988) Phytochemistry 27:673
3. Dathe W, Sembdner G (1982) Phytochemistry 21:1798
4. Hirai N, Fukui H, Koshimizu K (1978) Phytochemistry 17:1625
5. Lehmann H, Preiss A, Schmidt J (1983) Phytochemistry 22:1277
6. Lehmann H, Schwenen L(1988) Phytochemistry 27:677
7. Linforth RST, Taylor IB, Hedden P (1987) J Exp Bot 38:1734
8. Loveys BR, Milborrow BV (1981) Aust J Plant Physiol 8:571
9. Loveys BR, Milborrow BV (1983) In: Crozier A, Hillman JR (eds) The biosynthesis and metabolism of plant hormones, SEB Symposium Cambridge University Press, p 17
10. Milborrow BV (1969) Chem Commun, p 966
11. Milborrow BV (1984) Biochem J 220:325
12. Milborrow BV (1986) In: Bopp M (ed) Plant growth substances 1985. Springer, Berlin Heidelberg New York Tokyo, p 108
13. Milborrow BV, Carrington NJ, Vaughan GT (1988) Phytochemistry 27:757
14. Milborrow BV, Nonhebel HM, Willows RD (1988) Plant Sci 56:49
15. Netting AG, Milborrow BV (1988) Biomed Environ Mass Spectrom 17:281
16. Nonhebel HM (1987) J Chromatogr 402:374
17. Nonhebel HM, Milborrow BV (1986) J Exp Bot 37:1533
18. Nonhebel HM, Milborrow BV (1987) J Exp Bot 38:980
19. Okamoto M, Hirai N, Koshimizu K (1987) Phytochemistry 26:1269
20. Roberts DC, Heckman RA, Hege BP, Bellin SA (1968) J Org Chem 33:3566
21. Ryback G (1972) Chem Commun, p 1190
22. Vaughan GT (1986) Ph D Thesis, University of New South Wales, Australia
23. Vaughan GT, Milborrow BV (1988) Phytochemistry 27:339
24. Weiss G, Koreeda M, Nakanishi K (1973) Chem Commun p 565
25. Zeevaart JAD (1986) Michigan State University/Dept. of Energy, Annu Rep 1986, p 134

Tissue-Specific Metabolism of both Abscisic Acid and Gibberellins in Legume Seeds

M.L. Brenner, S.K. Stombaugh, and P.R. Birnberg[1]

1 Introduction

The maternal plant can contribute a significant portion of the ABA that is found in developing soybean seeds [2, 10]. This is especially the case during mid pod-fill [6]. In source leaves of sugar beet plants, the major metabolic products of ABA are DPA and DPA-G [3], while in sugar-beet roots (sink tissue) the major products of ABA are PA and an unidentified compound. This observation of tissue-specific metabolism of ABA possibly relating to regulation of partitioning of photoassimilates raises an important question. Is this differential metabolism a common phenomenon and does it have important regulatory function? Earlier, Dashek et al. [4] reported tissue-specific metabolism of ABA occurring in barley grains. Barley aleurone tissue metabolizes ABA to PA, DPA and other more polar metabolites, while ABA is not metabolized in endosperm tissue.

2 Differential Metabolism of ABA

Recognizing that substances transported to developing seeds are unloaded in the seed coat and move to the embryo apoplastically [14], we thought it would be appropriate to determine if metabolism of ABA differed at the point of phloem unloading, the seed coat, and the point of uptake, the embryo.

Isolated soybean (*Glycine max* Merr L. cv Clay) seed-coat halves and cotyledons were treated with [^{3}H](S)ABA (0.38 μCi, 69 μCi/nmol). The [^{3}H](S)ABA was recovered from [^{3}H](R/S)ABA (Amersham) using immunoaffinity chromatography with monoclonal antibodies bonded to Affigel-10 (Bio-Rad). The [^{3}H](S)ABA was applied in 20 μl of 40 mM MES buffer (pH 6.0) to the adaxial surface of the cotyledons or the interior surface of seed coats. Both seed parts were placed on moistened filter paper in Petri dishes and incubated in the dark for the duration of the experiment.

At the end of the metabolism period, each piece was frozen on solid CO_2. The tissue was extracted following the procedures previously described [11]. The sample was resuspended in 0.1 N acetic acid, filtered and applied to a reversed phase C18 HPLC column eluted with a gradient of 0.1 N acetic acid to 0.1 N acetic acid in acetonitrile. The occurrence of [^{3}H]-labeled peaks were monitored with an on-line radioactivity detector (Packard Trace 7140).

[1]Department of Horticultural Science and Landscape Architecture, University of Minnesota, St. Paul, MN 55108, USA

Identification of the respective peaks was accomplished by adding samples of [^{3}H]-labeled peaks to a bulk extract of seed coats and cotyledons. The extract was purified through three sequential HPLC purification steps based on the occurrence of labeled material. The purified peaks, methylated before the last HPLC step, were acetylated as necessary for GC-MS (Carlo Erba/Kratos M-25). Methyl-PA, methyl-DPA and acetylated-methyl DPA-G were identified by their total ion scan mass spectra.

Seed coats efficiently metabolized [^{3}H](S)ABA to [^{3}H]PA, and another product (peak G) which we have yet to identify (Fig. 1a). PA was steadily formed during the first 12 h and then at a slower rate for the next 12 h of the experimental period. Only minimal amounts of DPA were formed in the seed coat tissue during this experimental period. In contrast, PA was the major product in cotyledons for only the first 3 h after which its level stabilized, while DPA and DPA-G were efficiently produced (Fig. 1b). Although the data are not shown, we also found a number of minor metabolites of [^{3}H](S)ABA in both seed parts, especially the seed coats.

The significance of the differential metabolism of ABA by the two seed parts is not clear. While there is no question that ABA has regulatory activity, there are only a few reports that PA has biological activity [8, 12]. Soybean seed coat tissue efficiently oxidizes ABA to PA, but seems relatively inefficient in metabolizing PA. The seed coat's capacity to efficiently metabolize ABA, but not PA, could be an indication that ABA be an active agent in this tissue and while PA is not. Ross et al. [9] have shown that ABA rapidly (< 10 min) stimulates phloem unloading from pea seed coats still attached to the plant.

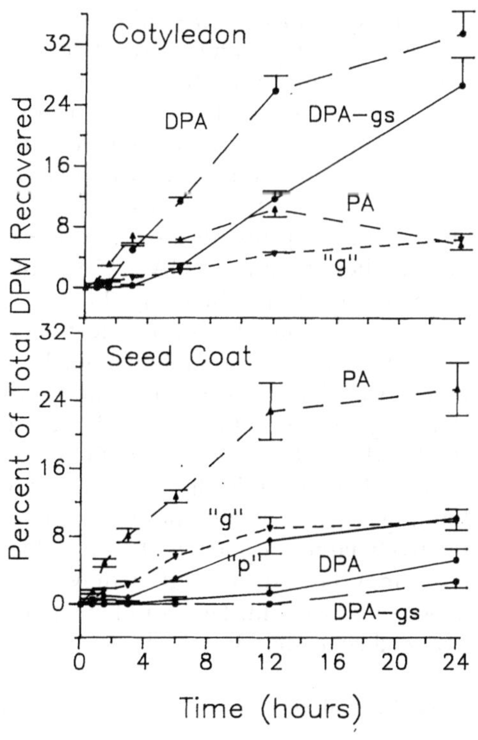

Fig. 1. Metabolic fate of [^{3}H](S)ABA applied to isolated soybean seed coats and cotyledons. Products "*g*" and "*p*" represent unidentified peaks recovered from a reversed phase C18 HPLC column. The Rt of *g* = 12.8 min, *p* = 16 min, *PA* = 25 min, *DPA* = 19 min, and *DPA*-gs (DPA-glucoside) = 14.8 min

3 Differential Metabolism of GA_{12} ald

The contrasting metabolic fates of ABA in seed coats and cotyledons of soybean made us wonder if GAs, which often interact with ABA in controlling development, might also have tissue-specific metabolic fates. Sponsel [13] has already demonstrated that GA_{20} and GA_{29} are metabolized differently in cotyledons and seed coats of pea (*Pisum sativum* L.) – GA_{20} is oxidized (to GA_{29}) much more rapidly in cotyledons, while G_{29} is oxidized (to GA_{29}catabolite) much more rapidly in seed coats. A related, and perhaps more fundamental question is whether the universal GA precursor, GA_{12}ald has different metabolic fates in seed coats and cotyledons of legumes. Because of their demonstrated value as a species for studying GA metabolism [5, 7, 13], we used peas to obtain data bearing on this question.

The protocol for these experiments closely followed that described above for ABA. Seed at the half-filled stage (ca. 210 mg fw/seed) of peas (genotype G2) were used. Enzymatically synthesized [1] [^{14}C]GA_{12}aldehyde – ca. 200 Ci/mole, 200 000 dpm – was applied in 10 μl of 95% ethanol to each seed part. Metabolites were extracted and separated on reversed-phase C_{18} HPLC as described by Maki et al. [7]. Identifications of peaks involved GC-MS of purified, radio-labeled compounds and are based on work by Maki et al. [7] and by Yu-Xian Zhu and Peter Davies [this meeting, abstract #378; and personal communication].

The separations of the metabolites obtained are illustrated for typical experiments in Figs. 2a-2d. The metabolic fates of GA_{12}ald in seed coats and in cotyledons differed in two ways: (1) Metabolism was much more rapid in cotyledons. (2) Most of the GA_{12}ald in cotyledons was directed into all the GAs on the 13-hydroxy pathway (GA_{12}ald $\rightarrow\rightarrow$ GA_{53} $\rightarrow$ GA_{44} $\rightarrow$ GA_{19} $\rightarrow$ GA_{20} $\rightarrow$ GA_{29} $\rightarrow$ GA_{29}catabolite) with the exception of the last GA_{29}catabolite. These observations conform to those reported by Maki et al. [7] that GA_{12}ald was metabolized to GA_{53}, GA_{44}, GA_{19}, and GA_{20}; and by Sponsel [13] that cotyledons converted GA_{20} to GA_{29} but not all the way to GA_{29}catabolite. In seed coats, on the other hand, GA_{12}ald was first converted primarily to the material eluting at 26.3′ one, which is mostly a saccharide ester of GA_{12}ald, some of another conjugate, (the material eluting at 17.7′, probably a saccharide ester of GA_{53}ald), and also GA_{29}catabolite[2] plus smaller amounts of GA_{19}, GA_{20}, and GA_{44}. By 24 h, GA_{29}catabolite became the predominant product (Fig. 2d).

Thus, GA metabolism in a legume seed is also tissue-specific. In cotyledons, metabolism is more rapid than in seed coats and proceeds via the early 13-hydroxy pathway to GA_{20}. In seed coats, most is converted to sugar conjugates, and the rest is rapidly converted to GA_{29}catabolite; some of the sugar conjugates appear to be stable, while others may release GAs to (re)join the 13-hydroxy pathway (unpubl. results). It is possible that the seed coat's enzymatic-oxygenase capacity is rate limiting, or that it has more efficient conjugating enzymes. Alternatively, the relatively large proportion of GA_{12}ald and GA_{53}ald conjugates recovered in seed coats may be indicative that they were formed as a means to "detoxify" the exogenous GA_{12}ald. This could mean that GA_{12}ald occurs at much lower levels in pea seed coats than in cotyledons.

[2] Note: The peak identified as GA_{29}catabolite is a conjugate of an unknown GA (GA_x).

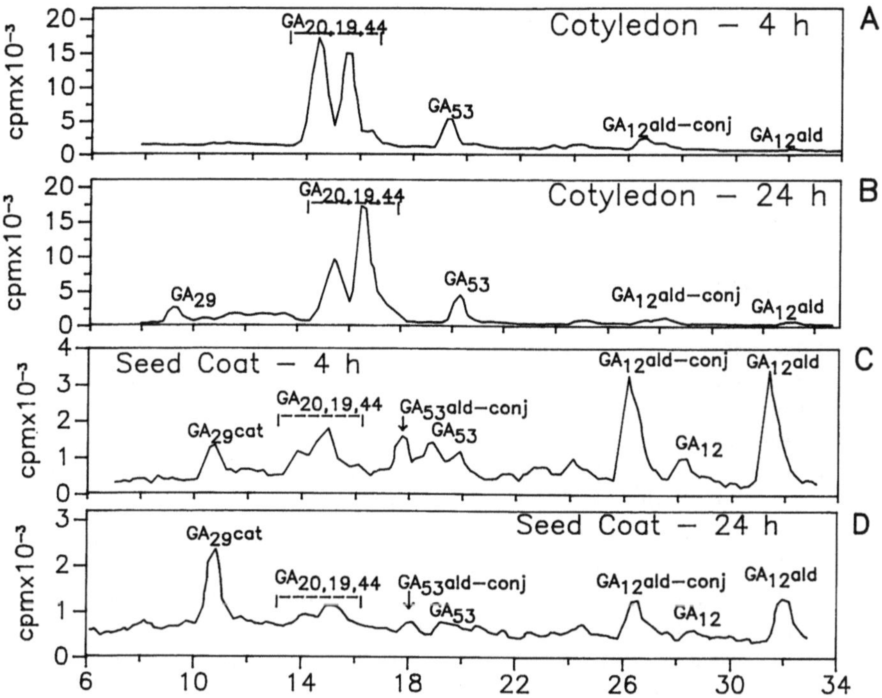

Fig. 2. HPLC separation of the metabolic products of [^{14}C]-GA_{12}ald applied to isolated pea seed coats and cotyledons. Note: The peak labeled as GA_{29}cat should be labeled GA_x-conjugate

Acknowledgements. Supported in part by the United States Department of Agriculture under grant 87-CRCR-1-2467 from the Competitive Research Grants Office; and NSF/DMB-8607749-01 from the National Science Foundation; Contribution from the University of Minnesota Agricultural Experiment Station, St. Paul, MN 55108, Paper No. 17,160 of the Proceeding Series.

References

1. Birnberg PR, Maki SL, Brenner ML, Davis GC, Carnes MG (1986) Anal Biochem 153:1
2. Brenner ML, Hein MB, Schussler J, Daie J, Brun WA (1982) In: Wareing PF (ed) Plant growth substances 1982. Academic Press, New York, p 343
3. Daie J, Wyse R, Hein MB, Brenner ML (1984) Plant Physiol 74:810
4. Dashek WV, Singh BN, Walton DC (1979) Plant Physiol 64:43
5. Davies PJ, Birnberg PR, Maki SL, Brenner ML (1986) Plant Physiol 81:991
6. Hein MB, Brenner ML, Brun WA (1984) Plant Physiol 76:955
7. Maki SL, Brenner ML, Birnberg PR, Davies PJ, Krick TP (1986) Plant Physiol 81:984
8. Nolan RC, Ho T-HD (1988) Plant Physiol 88:588
9. Ross GS, Minchin PEH, McWha JA (1987) J Plant Physiol 129:435
10. Schussler JR (1986) PhD thesis, University of Minnesota, St Paul, MN
11. Schussler JR, Brenner ML, Brun WA (1984) Plant Physiol 76:301
12. Sharkey TD, Raschke K (1980) Plant Physiol 65:291
13. Sponsel VM (1983) Planta 159:454
14. Thorne JH (1985) Annu Rev Plant Physiol 36:317

An Overview of Cytokinin Biosynthesis

L.M.S. PALNI[1,2], S.K. NANDI[1,2], S. SINGH, and D.S. LETHAM[2]

1 Introduction

A critical element in understanding the physiological role of cytokinins is the knowledge of how plants control their cytokinin levels. Therefore, we need to know the pathway(s) and site(s) of cytokinin synthesis, the metabolic fate of cytokinins, and the biochemical mechanisms controlling their biosynthesis and degradation.

Although roots, particularly the root apices, are a major site of cytokinin biosynthesis in plants, there is evidence now that other meristematic tissues and organs including the cambium, developing buds, seeds and fruits and the embryonic axis of germinating seed have the ability to synthesize cytokinins under optimal growth conditions [6, 14]. Recently, stem and leaves have also been shown to be additional sites of cytokinin production [6]. The view that root-produced cytokinins move in the xylem to the shoot to participate in the control of development and senescence is widely accepted [15]. The major question that remains to be clarified is under what conditions the observed cytokinin activity in other plant parts is derived solely from the roots, and when, and to what degree it is derived by synthesis in situ.

Experiments demonstrating incorporation of radioactive precursors, commonly adenine, into cytokinins in plant tissues have been used to examine the de novo pathway of cytokinin biosynthesis [15, 16, 22 and references therein]. Some of these in vivo labelling experiments have highlighted the importance of cytokinin nucleotides as the primary products of biosynthesis [19]. Further confirmation for this proposal was provided by the in vitro work which demonstrated that a cell free preparation from a slime mold catalyzed the formation of iPA 5′-monophosphate from 5′-AMP and IPP [26]. Neither adenine nor adenosine acted as a substrate for the enzyme, Δ^2-IPP: AMP-Δ^2-isopentenyl transferase, a prenyl transferase commonly termed cytokinin synthase. The enzyme has subsequently been purified from a number of higher plant sources [7, 11]. In addition it has been established that cytokinin (and auxin) production in crown gall tissues is under the control of T-DNA, and genes have been identified which directly code for the production of these phytohormones [17].

In this paper we have examined [^{3}H]-MVA incorporation into cytokinins in crown gall tissues. In addition, results are described which provide evidence of [^{14}C]-adenine incorporation into cytokinins by germinating seeds and young leaves under normal physiological conditions.

[1] Plant Cell Biology Group, Research School of Biological Sciences, The Australian National University, Canberra City, ACT 2601, Australia
[2] Present address: C.S.I.R. Complex, PALAMPUR, H.P., India – 176061

2 Cytokinin Biosynthesis in Crown-Gall Tissues

Crown-gall tissues have proved very useful for biosynthetic studies because the cultures can be maintained indefinitely on defined medium under aseptic conditions. The endogenous cytokinins have been unequivocally identified, and their levels are sufficiently high to permit feeding of precursors and intermediates in amounts which are physiological, yet high enough to permit identification of products by reliable methods. Thus crown gall tissues of *Vinca rosea* and *Datura innoxia* were utilized for incorporation of labelled adenine into cytokinins [19, 20]. Recently, incorporation of [$^{15}N_5$]-adenine into endogenous cytokinins by *Datura* crown gall tissues has been demonstrated using unambiguous methods [22].

MVA would appear to be a better precursor than adenine and its derivatives for studies of cytokinin biosynthesis because of its reduced conversion into undesirable basic compounds [15], and the fact that its incorporation into isopentenyl groups of tRNA cytokinins has been established [15]. However, there are only a few reports of MVA's incorporation into free cytokinins [2, 5], and unfortunately the incorporation was extremely low, preventing proper characterization of the labelled cytokinins. Furthermore, MVA incorporation into cytokinins by *V. rosea* crown gall tissues could not be detected under conditions which result in maximal incorporation of [^{14}C]-adenine into cytokinins [19]. The uptake of MVA (a mixture of optical isomers) by cell cultures is often very poor. Furthermore the radioactivity taken up is possibly swamped by a large endogenous pool of MVA. This severely limits the chance of detecting MVA's incorporation into cytokinins.

3-Hydroxy-3-methylglutaryl coenzyme A (HMG-CoA) reductase (EC 1.1.1.34) catalyzes the biosynthesis of MVA, and is generally accepted to be a rate-limiting enzyme in isoprenoid biosynthesis. Two novel compounds of fungal origin, ML-236B (compactin) and monacolin K (mevinolin, an analogue of compactin) are potent inhibitors of this enzyme [8]. Although the effects of these inhibitors on mammalian isoprenoid synthesis and cholesterol metabolism have been well established [4], there are only a few reports indicating inhibition of plant growth and sterol synthesis by compactin [9, 23] and mevinolin [1]. In this report we have used compactin to demonstrate incorporation of [^{3}H]-MVA into free cytokinins. Compactin probably accomplishes this by lowering endogenous MVA levels, thereby increasing the specific radioactivity of MVA pool.

Compactin (a gift of Dr. A. Endo, Tokyo Noko University, Japan) was converted to the acid form [3] and stored as 1 mM solution in multiple aliquots at -20°C until used. Initial experiments were carried out with a shooty line of tobacco (*Nicotiana tabacum* cv Xanthi) crown gall tissue in which [9R]Z was the major endogenous cytokinin. The tissue (10 g) was incubated in liquid medium (10 ml) for 5 h in the absence or presence of compactin (5 μM) on a shaker (80 rpm, 22°C, under fluorescent light). At the end of this preincubation period 4.5 μCi of DL-[2-^{3}H]-MVA lactone (1.28 Ci/mmol; Amersham, UK) was added to each flask. Following further incubation for 10 h the tissues were removed, extracted and purified as previously reported [19]. The uptake of [^{3}H]-MVA was 59 and 56.7% by control and treated tissues, respectively. In both cases about 56% of radioactivity taken up was extracted, and more than 90% of extracted radioactivity was not retained on cellulose phosphate columns. The basic fractions and the nucletoide-derived, butanol-soluble fractions were initially analyzed by 2D-TLC (Silica gel; 1st dim.: n-BuOH:HOAc:H_2O, 12:3:5, 2nd dim.: n-BuOH:NH_3:H_2O, 6:1:2, upper phase) following the addition of unlabelled cytokinin standards. Only the nucleotide-derived fraction from compactin-treated tissue indicated incorporation of [^{3}H] into [9R]Z. The remainder of this fraction was then fractionated on a Sephadex LH-20 column (elution with 35% ethanol) which indicated that all the [^{3}H] (applied to the column) was present in the elution volume of [9R]Z/(diH) [9R]Z (Fig. 1A). This result looked very promising. However, HPLC of appropriately pooled fractions showed that most of the radioactivity eluted very early, and only about 8% was found at the Rt of [9R]Z (Fig. 1B).

Unfortunately, this experiment could not be repeated due to the loss of the tobacco crown gall line by contamination of stock cultures. Further experiments were therefore carried out with *D. innoxia* crown gall tissue. The details of incubation conditions were exactly as described in a previous study of [U-^{14}C]-adenine incorporation into cytokinins by tissues of different ages [22]. The tissues (2 g) were preincubated with compactin (0, 5, 20, 100 μM) for 4 h after which 38 μCi of [^{3}H]-MVA lactone was added to each dish. The tissues were extracted and analyzed following further incubation for 8 h. Two additional treatments were included in which compactin (20 μM) was present only during preincubation or during incubation with [^{3}H]-MVA (Table 1). The extracts were first chemically treated with sodium periodate [19] to convert cytokinin ribosides and nucleotides

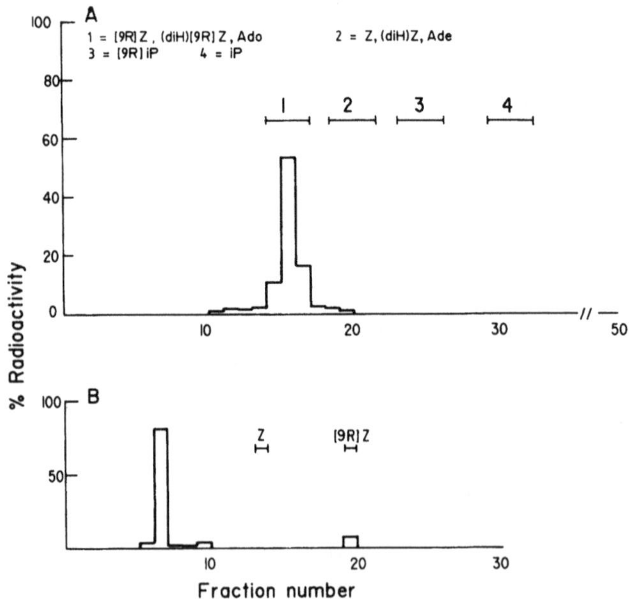

Fig. 1A,B. [^{3}H]-Mevalonic acid incorporation into cytokinins by tobacco crown gall tissue. **A** Sephadex LH-20 column chromatography of the nucleotide-derived, BuOH-soluble fraction from compactin (5 μM)-treated tissue; **B** HPLC of fraction 1 obtained from the analysis shown above (column: μBondapak C_{18}, gradient elution with 10–50% MeOH (containing 1% HOAc) in 30 min, 3 ml/min)

Table 1. Incorporation of DL-[2-^{3}H]-mevalonic acid lactone into putative cytokinins by *Datura innoxia* crown-gall tissue with and without the enzyme inhibitor compactin[a]

Compactin concentration (μM) Pre-incubation 4 h	Compactin concentration (μM) Incubation 8 h	Uptake As % of [^{3}H] supplied	Radioactivity incorporation into cytokinins: Isopentenyladenine type	Radioactivity incorporation into cytokinins: Zeatin type
			As % of [^{3}H] applied to 2D-TLC; values in round brackets are as % of [^{3}H] extracted from tissue	
0	0	29.8	0.66 (0.0086)	1.19 (0.0155)
5	5	29.9	0.96 (0.0059)	2.19 (0.0136)
20	20	31.2	0.73 (0.0055)	1.11 (0.0084)
100	100	28.7	1.45 (0.0105)	31.42 (0.2330)[b]
20	0	29.2	0.99 (0.0090)	1.68 (0.0152)
0	20	38.0	1.30 (0.0076)	0.95 (0.0056)

[a] Three-week-old tissue (2 g fw) was used. [^{3}H]-mevalonic acid lactone was added at the end of 4 h pre-incubation period.
[b] Radioactivity incorporation into this fraction was confirmed by further analysis (see text).

into corresponding bases, and then sequentially purified by cellulose phosphate chromatography and *n*-BuOH partition. Small aliquots were then analyzed by 2D-TLC after the addition of appropriate cytokinin standards. The results are shown in Table 1.

Appreciable increase in [^{3}H]-incorporation was found only in the case of tissue which had been treated continuously with 100 μM compactin, and the increase was limited to Z-type cytokinins. Further analyses were therefore concentrated only on the extract from this treatment. The extract was subjected to preparative 1D-TLC (Silica gel; *n*BuOH:HOAc:H_2O, 12:3:5) and the zone (R_f 0.48–0.53) corresponding to co-chromatographing Z/(diH)Z/[9R]Z/(diH) [9R]Z was removed and eluted (60% EtOH containing 4% HOAc). The eluate was filtered, repeatedly evaporated after the addition of MeOH to remove traces of HOAc, and finally dissolved in 2 ml of 70% EtOH. This contained only about 2.5 kdpm. Further analysis to confirm [^{3}H]-MVA incorporation into Z (may also include its riboside and 5′-phosphate derivatives) was carried out as follows: A portion was analyzed by HPLC (Novapak C_{18}, 5μ, 8 × 100 mm; 40% MeOH containing 1% HOAc; 3 ml/min) which showed coincidence of [^{3}H]- with Z, but not with (diH)Z (Fig. 2A). The remaining sample was divided into two portions and one portion was dried and converted to its *O*-acetyl derivative[3]. The derivatized material and underivatized sample was then analyzed by 1D-TLC (Silica gel: $CHCl_3$: MeOH, 9:1, with a trace of NH_3). This system resolved (*cis*)Z from Z (Fig. 2C) indicating incorporation into Z only, which was further confirmed by co-chromatography of the derivatized metabolite with authentic acetyl-Z (Fig. 2B).

[3] *O*-Acetyl derivatives of cytokinin bases can be prepared by reacting thoroughly dried sample with pyridine – acetic anhydride (2:1) at 26°C for 24 h. The mixture is then dried under vacuum and the derivative redissolved in MeOH or $CHCl_3$ for chromatography [24].

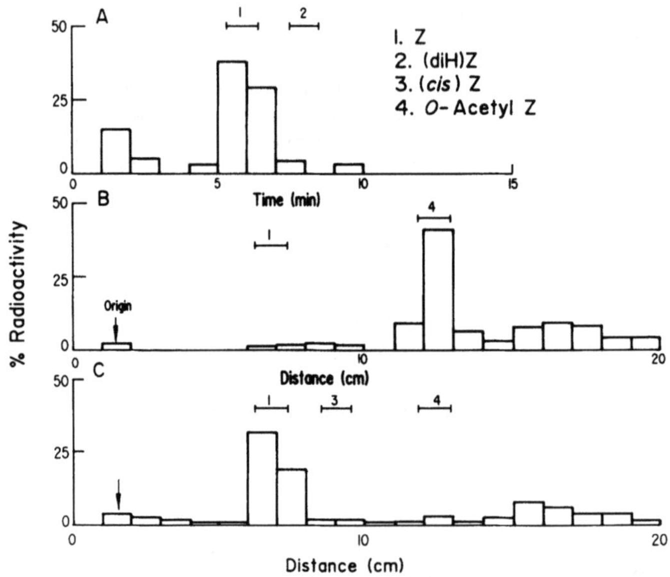

Fig. 2A-C. [^{3}H]-Mevalonic acid incorporation into cytokinins by *Datura* crown gall tissue treated with 100 μM compactin. The sample was purified by preparative TLC and a fraction which would contain Z/(diH)Z/[9R]Z/(diH)[9R]Z was removed, eluted and further analyzed by HPLC (**A**) and TLC (**B** and **C**). Details are given in the text

3 Cytokinin Biosynthesis by Lupin Seeds During Early Stages of Germination

In view of low cytokinin levels in dry seeds [27] and their rapid metabolism following imbibition [10, 13], germinating seeds may have the capacity to synthesize cytokinins. A marked increase in cytokinin activity during seed germination has been noted [14]. Furthermore, the reported axial control of reserve mobilization in seeds and its replacement by cytokinins [12] also indicates that germinating seeds, particularly the embryonic axes, synthesize cytokinins. Recently, the embryonic axis was also shown to control expansion and chlorophyll formation in cotyledons of yellow lupin, and cytokinins appeared to constitute an important part of the stimulus emanating from the axis [18]. These facts led us to investigate cytokinin biosynthesis in germinating lupin seeds [18].

Partially de-coated (about 20% of testa removed) seeds of yellow lupin (*Lupinus luteus* L. cv Weiko III) were surface sterilized for 5 min, washed thoroughly, and allowed to imbibe water for 2 h at 22°C in the dark. The seeds were then fully de-coated and incubated in water containing 42 μCi of [U-^{14}C]-adenine (296 mCi/mmol; Amersham, UK) for 4 h (22°C, dark). Following exposure to [^{14}C]-adenine, the seeds were washed with water and further incubated for 6 h (22°C, dark). The cotyledons (425 g) and embryonic axes (embryos, 13.5 g) were then separated and extracted individually using solvents known to inactivate phosphatases [19]. The extracts were purified on cellulose phosphate columns, and the nucleotides were degraded either enzymatically (to ribosides) with alkaline phosphatase, or chemically (to bases) following further purification on DEAE-cellulose.

The basic fractions (which would contain bases, ribosides and glucosides) and the nucleotide-derived, BuOH-soluble fractions from the embryo and cotyledon extracts were initially fractionated on Sephadex LH-20 columns. Appropriate fractions were then analyzed by 2D-TLC on silica gel plates. This analysis indicated [^{14}C]-incorporation into a number of cytokinins (Table 2). All of the embryo-derived Z/(diH)Z fraction was used up in this analysis, however, further analysis by HPLC of this fraction from cotyledons failed to confirm incorporation into Z or (diH)Z. The corresponding ribosides (both from the embryo and cotyledon fractions, and following enzymatic hydrolysis of nucleotides from cotyledons) were further analyzed. After the addition of small amounts of [^{3}H]-(diH) [9R]Z these three fractions were extracted with *n*-BuOH and the BuOH-soluble materials were subjected to preparatives TLC (Silica gel; *n*-BuOH:HOAc:H_2O, 12:3:5). The zone containing the [^{3}H] marker, and co-chromatographing with [9R]Z/(diH) [9R]Z/Z/(diH)Z, was eluted in each case. This step removed bulk of [^{14}C] radioactivity due to adenine and adenosine. The TLC eluates were purified by HPLC (μBondapak C_{18}, 7.8 × 300 mm, Waters; elution with 20% aqueous MeOH containing 1% HOAc at 3 ml/min). All the [^{3}H] and most of the [^{14}C] eluted at the R_t of Co-chromatographing [9R]Z and (diH) [9R]Z. These fractions were collected and small portions analyzed as follows:

1. 1D-TLC (silica gel; $CHCl_3$:MeOH, 9:1 with a trace of NH_3) which resolved (*cis*) [9R]Z from [9R]Z/(diH) [9R]Z. Both [^{3}H] and [^{14}C] radioactivity co-chromatographed with [9R]Z/(diH) [9R]Z, but not with (*cis*) [9R]Z.
2. Following treatment with 0.01% aqueous solution of $KMnO_4$ the samples were analyzed by 1D-TLC (Silica gel; *n*-BuOH:HOAc:H_2O, 12:3:5). Both [^{3}H] and [^{14}C] radioactivity co-chromatographed with treated (diH) [9R]Z. This treatment (oxidation) changes the R_f of [9R]Z, but that of (diH) [9R]Z remains unchanged.

Table 2. [^{14}C]-Adenine incorporation into putative cytokinins by intact lupin seed and isolated embryos and cotyledons

[^{14}C]-Adenine supplied to	Fraction	Radioactivity incorporation into putative cytokinins (as % of [^{14}C] supplied)		
		Z/ (diH)Z	[9R]Z/ (diH) [9R]Z	[9R]Z/ (diH) [9R]Z[a]
Intact seed	Embryo	0.00021 (0.000016)[b];	0.011 (0.00081)[c]	ND
	Cotyledon	0.00810 (0.000019)[d]	0.054 (0.00013)[c]	0.08 (0.00019)[c]
Isolated organs	Embryo	ND	0.064 (0.16)[c]	ND
	Cotyledon	ND	0.061 (0.032)[d]	ND

The values in parentheses represent [^{14}C]-adenine incorporation on per g fw basis. ND = Not detected.
[a] Derived from enzymatic hydrolysis of corresponding nucleotides.
[b] Sample was not subjected to further analysis.
[c] Further analyses confirmed [^{14}C] incorporation into (diH) [9R]Z only.
[d] Further analyses failed to confirm [^{14}C] incorporation into [9R]Z or (diH) [9R]Z.

3. The isopropylidene derivatives[4] were prepared and analyzed by 1D-TLC (Silica gel; $CHCl_3$:MeOH, 9:1 with a trace of NH_3). [^{3}H] and [^{14}C] radioactivity was associated with the authentic derivative of (diH) [9R]Z.

This is good evidence for [^{14}C] incorporation into (diH) [9R]Z by embryos and cotyledons, and into (diH) [9R]Z (derived from its 5′-nucleotides) by cotyledons. The nucleotide fraction was also degraded chemically, and [^{14}C] was shown to be associated with (diH)Z.

To further define the site of cytokinin synthesis in germinating seeds, embryonic axes (0.4 g) and cotyledons (2 g) were carefully excised from partially imbibed (2 h) lupin seeds, and incubated in Petri dishes with [U-^{14}C]-adenine in HEPES buffer (embryos: 11 μCi in 0.6 ml buffer; cotyledons: 15 μCi in 2 ml buffer; pH 7) on a shaker (80 rpm; 4 h, 22°C, dark). Following 4-h exposure to [^{14}C]-adenine these were washed with buffer and further incubated for 6 h. The cotyledons and embryos were then extracted and purified as before. The results of 2D-TLC analyses are shown in Table 2. [^{14}C] incorporation into (diH) [9R]Z by embryos was confirmed by further analyses. However, similar analyses indicated a lack of [^{14}C] incorporation into (diH) [9R]Z by isolated cotyledons.

These results of [^{14}C]-adenine incubation studies directly demonstrate cytokinin synthesis by germinating seeds, and further indicate that possibly only the embryonic axes have the capacity to synthesize cytokinins, which are then translocated to the cotyledons, apparently accumulating therein to evoke physiological response. This is reflected in experiments with the intact seed where [^{14}C] incorporation into cytokinins (per g fw) was considerably higher in embryos than in the cotyledons (Table 2), while incorporation per organ is greater for the cotyledons [18]. Translocation experiments carried out with selective application of [^{3}H]-(diH) [9R]Z to embryos or cotyledons also indicate a polar movement of cytokinins from the embryonic axes to the cotyledons. The incorporation of [^{14}C]-adenine into (diH) [9R]Z is particularly interesting because (diH)Z-type cytokinins predominate in lupin seed.

4 Cytokinin Biosynthesis in Tobacco Leaves

This investigation was based on studies of sequential leaf senescence in tobacco (*Nicotiana rustica*) plants where leaf cytokinin levels are important in control of sequential leaf senescence [24]. The basal yellow leaves had much reduced levels of cytokinins (22 ng/g fw) in comparison to green, fully expanded leaves (65 ng/g fw) and very green, expanding leaves (57 ng/g fw) near the apex; Z and [9R]Z were major cytokinins [24]. This difference in cytokinin levels could not be attributed to any differences in the translocation and metabolism of xylem-supplied [^{3}H]-[9R]Z

[4] 2′, 3′-O-Isopropylidene ethers of cytokinin ribosides can be made by dissolving sample in 2, 2-dimethoxypropane (30 μl, Fluka) to which conc. HCl (1 μl) is added. Samples are allowed to stand for 5 min at 30°C, then rapidly evaporated under a stream of N_2. These can be further converted to *tert*-butyl-dimethylsilyl derivatives if required [10].

and [^{3}H]-(diH) [9R]Z. Thus, the differing cytokinin levels in leaves of various maturity may perhaps be explained in terms of differences in cytokinin biosynthetic capacity.

Tobacco leaves of differing maturity were then supplied with [^{14}C]-adenine in nutrient solution via the petiole. The putative cytokinin fractions were sequentially purified by cellulose phosphate chromatography, *n*-BuOH partitioning, cellulose TLC, silica gel TLC and C_{18} reversed phase HPLC. The analyses indicated [^{14}C] incorporation into Z and from HPLC results, the [^{14}C] attributable to Z was 0.00253, 0.00004 and 0.00002% of supplied radioactivity in very green, green and slightly yellow leaves, respectively. However, when the putative [^{14}C]-Z fraction was derivatized (9-chlorocyanoethyl derivative[5] and *O*-acetyl derivative) only the metabolite from the very green, expanding leaves still co-chromatographed with the authentic derivatives of Z. This did not occur for the derivatized fractions from older leaves. Thus, higher levels of cytokinin in upper green leaves may be due to their synthesis in situ. It follows that sequential leaf senescence in tobacco may result, at least in part, from loss of ability to synthesize cytokinins within the lower leaves.

5 Concluding Remarks

Adenine and its derivatives are subject to a rapid and very diverse metabolism in plants. However, biosynthesis of free cytokinins represents only a very minute pathway in this metabolic complex. Hence, isolation and purification of extremely small amounts of labelled putative cytokinins from plant tissues is technically very difficult. Rigorous characterization of cytokinin metabolites by high resolution chromatographic analysis is therefore essential, but is fraught with problems. A number of quite stable derivatives can be prepared for further characterization of metabolites by HPLC and TLC. These include *tert*-butyldimethylsilyl [20] and permethyl derivatives [21], in addition to the isopropylidene, cyanoethyl and *O*-acetyl derivatives described herein. Enzymes like xanthine oxidase, nucleoside phosphorylase and 5′-nucleotidase, which show selectivity, can also be appropriately used to further confirm precursor incorporation into cytokinins. With the extremely low precursor incorporation usually observed it is rarely possible to purify labelled cytokinins to constant specific radioactivity. Although the use of immunoaffinity chromatography as a selective purification step may hold promise, its application in cytokinin biosynthetic studies with plant tissue remains to be demonstrated. Furthermore, it is hoped that the use of antibodies against cytokinin synthase, preferably of plant origin, or from the *ipt* gene product if it is sufficiently similar to plant enzyme, may be helpful in localization of biosynthetic sites in plants at both histological and intracellular levels.

[5] Cyanoethyl derivatives of cytokinin bases can be prepared by reacting thoroughly dried samples in a reacti vial with anhydrous *n*-BuOH (150 μl), acrylonitrile or 2-chloroacrylonitrile (30 μl) and triethylamine (10 μl). The mixture is kept at 80°C for 0.5 h and then quickly evaporated. A few drops of $CHCl_3$ are added and the sample dried again [24].

References

1. Bach T, Lichtenthaler S (1983) Physiol Plant 59:50
2. Barnes MF, Tien CL, Gray JS (1980) Phytochemistry 19:409
3. Braithwaite AW, Palni LMS (1986) Eur J Cell Biol 41:121
4. Brown MS, Goldstein JL (1980) J Lipid Res 21:505
5. Burrows WJ, Fuell KJ (1981) In: Guern J, Peaud-Lenoel C (eds) Metabolism and molecular activities of cytokinins. Springer, Berlin Heidelberg New York, p 352
6. Chen C-M, Ertl JR, Leisner SM, Chang C-C (1985) Plant Physiol 78:510
7. Chen C-M, Melitz DK (1979) FEBS Lett 107:15
8. Endo A (1981) Methods Enzymol 72:56
9. Hata S, Takagishi H, Egawa Y, Ota Y (1986) Plant Growth Regul 4:335
10. Hocart CH (1985) Ph D Thesis, Australian Natl University, Canberra, p 253
11. Hommes NG, Akiyoshi DE, Morris R (1985) Methods Enzymol 110:340
12. Ilan I, Gepstein S (1980/81) Isr J Bot 29:193
13. Knypl JS, Letham DS, Palni LMS (1985) Biol Plant 27:188
14. Letham DS (1978) In: Letham DS, Goodwin PB, Higgins TJV (eds) Phytohormones and related compounds: a comprehensive treatise, vol 1. Elsevier/North Holland Biomedical, Amsterdam, p 641
15. Letham DS, Palni LMS (1983) Annu Rev Plant Physiol 34:163
16. McGaw BA, Scott IM, Horgan R (1984) SEB Semin Ser 24:105
17. Morris R (1986) Annu Rev Plant Physiol 37:509
18. Nandi SK (1988) Ph D Thesis, Australian Natl University, Canberra, p 302
19. Palni LMS, Horgan R, Darrall NM, Stuchbury T, Wareing PF (1983) Planta 159:50
20. Palni LMS, Tay SAB, Nandi SK, Pianca DJ, de Klerk GJM, Wong OC, Letham DS, MacLeod JK (1985) Biol Plant 27:195
21. Palni LMS, Tay SAB, MacLeod JK (1986) In: Linskens HF, Jackson JF (eds) Modern methods of plant analysis, new series, vol 3. Springer, Berlin Heidelberg New York Tokyo, p 304
22. Palni LMS, Tay SAB, MacLeod JK (1987) Plant Physiol 84:1158
23. Ryder NS, John Goad L (1980) Biochim Biophys Acta 619:424
24. Singh S (1988) Ph D Thesis, Australian National University, Canberra, p 221
25. Stuchbury T, Palni LMS, Horgan R, Wareing PF (1979) Planta 147:97
26. Taya Y, Tanaka Y, Nishimura S (1979) Nature 271:545
27. Van Staden J (1983) Physiol Plant 58:340

Genetic Differences in the Enzymatic Regulation of Zeatin Metabolism in *Phaseolus* Embryos

D.W.S. Mok[1], M.C. Mok[1], G. Shaw[2], S.C. Dixon[1], and R.C. Martin[1]

1 Introduction

Hormones play a significant role in all phases of growth and development of higher plants. Although the site(s) of action of hormones are largely unknown, it can be postulated that rather precise regulatory mechanisms must exist to maintain a critical hormone balance in order for controlled development to occur. We have focused our research on cytokinin metabolism in *Phaseolus* with the objective of identifying genetic mechanisms regulating cytokinin metabolism and biosynthesis. A systematic approach was taken by screening for genetic variations of interest, followed by genetic and biochemical characterizations. This approach has been successfully applied to callus systems and a number of reports describing inter- and intra-specific differences have been published [1, 16, 17, 18, 19, 20, 21, 23]. In this paper we summarize recent findings concerning the genetic differences in zeatin (Z) metabolism occurring in immature *Phaseolus* embryos. The studies described here have led to the identification of a new Z metabolic pathway and the discovery of differential expression of cytokinin-specific enzymes in *Phaseolus*.

2 Zeatin Metabolism in Higher Plants

Naturally occurring cytokinins are structurally related to Z, although a few exceptions are known [3, 10]. The spectrum of endogenous cytokinins and the metabolites formed from exogenous feeding experiments are represented by the following conversions: (1) formation of nucleosides and nucleotides [29, 33]; (2) N-glucosylation at the 9-, 7- or 3-position of the purine ring [4, 9, 26]; (3) attachment of the amino acid alanine to the 9-position of the purine ring to form lupinic acid [6, 30]; (4) O-glucosylation [7, 12, 25, 28], (5) O-xylosylation [11, 24, 35] and (6) reduction [27, 30, 33] of the side chain of Z; and (7) degradation [2, 19, 26, 34, 36] of the N^6-side chain. Summaries of earlier findings have been presented in several recent review articles [13, 14, 15]. The studies of O-xylosylation and side chain reduction of Z in *Phaseolus* and the genetic differences in the enzymatic reactions will be described in more detail below.

[1]Dept. of Horticulture, Oregon State University, 2042 Cordley Hall, Corvallis, OR 97331-2911, USA
[2]School of Chemistry, University of Bradford, Bradford, West Yorkshire, BD7, 1DP, UK

3 Discovery of a New Metabolite, O-Xylosylzeatin ((OX)Z), in Phaseolus Embryos

As callus tissues of *Phaseolus* displayed large differences in responses to and requirements for cytokinins [1, 16–21], we expected that significant genetic variations in cytokinin metabolism would also occur in organized tissues. Such differences should be reflected by qualitative or substantial quantitative differences in the array of metabolites recovered after incubation with radiolabeled cytokinins. Based on this assumption, the metabolism of [^{14}C]-Z was determined in immature embryos of *P. vulgaris* and *P. lunatus* [10]. The most interesting finding was that two novel metabolites, designated as Met I and Met II, occurred only in *P. vulgaris* embryos, whereas rapid conversion to O-glucosylzeatin ((OG)Z) was observed only in *P. lunatus*.

Degradation tests provided the initial information concerning the possible structure of Met I. The compound could be partially degraded by β-glucosidase and acid hydrolysis with trifluoroacetic acid, yielding Z. Combined treatment with periodate and cyclohexamide (which opens and cleaves ring structures of glycosyl moieties) resulted in the formation of Z. Finally, permanganate treatment (removal of the N^6-side chain) of Met I gave rise to adenine. These results indicated that the structural modification of Met I was located on the side chain of Z and the additional moiety attached was likely a glycosyl ring. Similar enzymatic and chemical degradation tests of Met II suggested that it was the ribonucleoside of Met I. Subsequently, structural analyses of the permethylated derivative of Met I were performed by GC-MS which revealed that a five-carbon glycosyl moiety was added to Z [11].

Further confirmation of the structures of the novel Z metabolites was obtained by two means: (1) determination of the pentose donor in the formation of Met I in vitro, and (2) chemical synthesis of the compound based on information obtained from structural analyses and enzymatic studies. Of all the potential pentose donors tested, only UDP-xylose (UDPX) was able to react with labeled Z in the presence of enzyme extracts, to form a product with identical chromatographic and chemical properties as Met I produced in vivo [35]. This observation indicated that Met I was an O-xylosyl derivative of Z, which was in agreement with the partial sensitivity of Met I to β-glucosidase treatment as well as the GC-MS results indicating the presence of a five-carbon moiety. The unambiguous confirmation was obtained by chemical synthesis of O-β-D-xylopyranosylzeatin (Fig. 1) which had identical chemical, chromatographic and GC-MS properties as Met I [32].

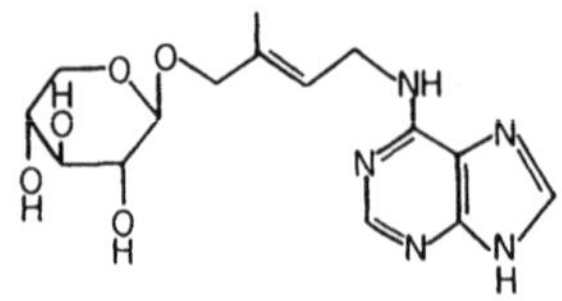

Fig. 1. Structure of O-β-D-xylopyranosylzeatin (OX)Z

4 The Occurrence of O-xylosylzeatin in *Phaseolus* Species

Even though the conversion of Z to (OX)Z had been clearly demonstrated, the natural occurrence of (OX)Z needed to be confirmed. Therefore, extracts were obtained from young embryos of *P. vulgaris* and purified by a combination of Sephadex, DEAE and reversed phase HPLC columns. GC-MS analyses of the purified extracts resulted in the detection of (OX)Z and its ribonucleoside as well as O-xylosyldihydrozeatin ((OX) (diH)Z). Thus the O-xylosyl derivatives of Z are naturally occurring compounds. Incubation of embryos of two other species, *P. coccineus* and *P. acutifolius*, with radiolabeled Z also led to the formation of (OX)Z and its dihydro-derivative [24]. These results indicate that O-xylosylation occurs in three of the four *Phaseolus* species examined thus far, *P. lunatus* being the exception. The interspecific differences are further elaborated on in some of the following sections.

5 The Biological Significance of O-xylosylation of Zeatin

The occurrence of O-xylosyl derivatives of Z and (diH)Z has only been detected in immature embryos of certain species [24]. The implications of selective expression of O-xylosylation with regard to embryo development are unknown. However, it may be speculated that some biological function must be associated with a metabolic activity which accounts for conversion of up to 80% of the exogenously supplied Z. As an initial step, the biological activities of (OX)Z and Z were compared in *P. vulgaris* and *P. lunatus* callus bioassays [22]. In *P. vulgaris* tissues, (OX)Z was 100 times more active than Z, while in *P. lunatus* the compound was slightly less active than Z. Since (OX)Z was equally active in both species, the difference was due to the much lower activity of Z in *P. vulgaris*. If (OX)Z serves merely a storage role as hypothesized for (OG)Z, conversion of (OX)Z back to Z should be accompanied by a reduction of activity in *P. vulgaris* tissues as compared with *P. lunatus*. Since a reduction in activity was not observed and very little conversion to Z was detected in short term incubation studies [22], (OX)Z itself may be biologically active. It is also possible that the compound is compartmentalized thus eluding factors which usually render Z nearly inactive, or is transported closer to the site(s) of cytokinin action.

6 O-Xylosylation vs O-Glucosylation and Characterization of Respective Enzymes

As (OX)Z and (OG)Z occur in *P. vulgaris* and *P. lunatus* embryos respectively, we were interested in determining if the variation is related to distinct enzymatic reactions or differential availability of glycosyl donors. Therefore, the enzyme catalyzing the formation of (OX)Z was isolated from *P. vulgaris* embryos and partially purified using gel filtration and anion exchange HPLC [35]. The enzyme (UDP-xylose:Z-xylosyltransferase, E. C. 2.4.2.--) was highly specific, recognizing Z and (diH)Z but not *cis*-Z or ribosylzeatin ((9R)Z). The affinity of the xylosyltransferase was also very high with Kms of 2 and 10 μM for Z and (diH)Z respectively. Enzyme preparations at various stages of purification did not catalyze the formation of (OG)Z from Z and UDP-glucose (UDPG).

(OG)Z [11] was detected previously in *Lupinus* [6], *Vinca rosea* [25, 31] and *Populus* [7], but the enzymatic reactions have not been examined. With the intention of confirming a genetic difference in O-glycosylation between *P. vulgaris* and *P. lunatus* at the enzyme level, we have initiated experiments to study the enzyme mediating the formation of (OG)Z in *P. lunatus*. The isolation procedures include precipitation with ammonium sulfate and chromatography using anion exchange and affinity HPLC columns. The enzyme was purified approximately 650-fold (Table 1). Increased purity of the enzyme (to over 2500-fold) was obtained recently [5]. The enzyme catalyzes the formation of (OG)Z from Z and UDPG. However, no N-glucosides of Z could be detected, indicating that the enzyme is different from the previously reported N-glucosylation enzymes [4, 8, 9]. Identical extraction methods consistently failed to detect O-glucosylation enzyme activity in embryos of *P. vulgaris*.

Table 1. Purification of UDPG:Z-glucosyltransferase from *P. lunatus* embryos

	Enrichment *(fold)*	Recovery %
Crude extract	1	100
$(NH_4)_2SO_4$ precipitation	9	100
Ax-300, HPLC	246	80
Affinity HPLC	650	37

It was somewhat unexpected that highly purified O-glucosylation enzyme preparations isolated from *P. lunatus* could mediate the formation of (OX)Z (in the presence of Z and UDPX), but with lower activity than O-xylosyltransferase of *P. vulgaris*. Thus far, the two O-glycosylation activities in the *P. lunatus* extracts are inseparable by affinity and anion exchange chromatography. Competition experiments (Fig. 2) with both substrates, UDPG and UDPX, at equal concentrations resulted in the formation of predominantly (OG)Z, suggesting that there may be a single enzyme in *P. lunatus* with a preference for UDPG as the glycosyl donor.

The results obtained thus far indicate that two distinct O-glycosylation enzymes operate in different species. The O-xylosyltransferase isolated from *P. vulgaris* reacts only with UDPX whereas the O-glucosylation enzyme of *P. lunatus* seems to have a broader affinity for glycosyl moieties. Further characterization of these two enzymes should provide an insight to the genetic basis of evolutionary divergence of cytokinin metabolic enzymes.

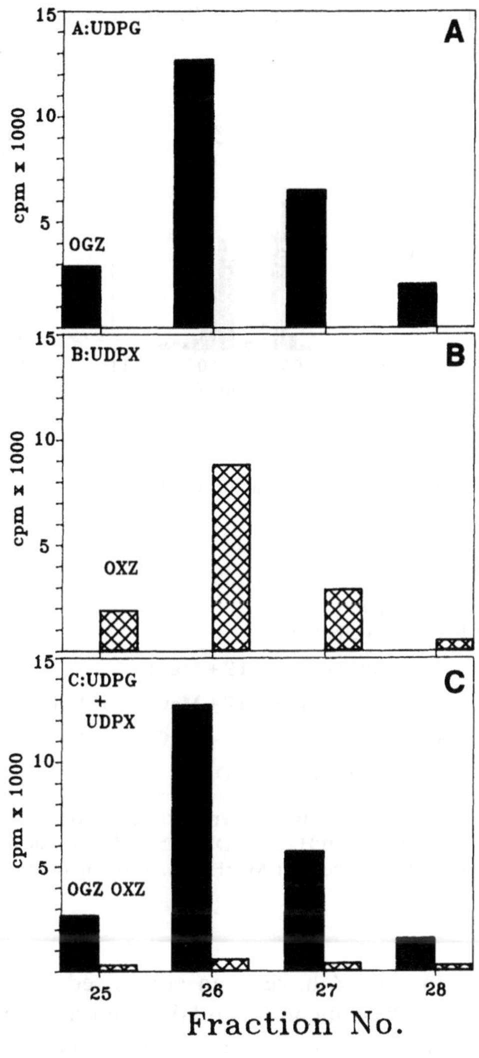

Fig. 2A-C. Substrate specificity of *UDPG*:Z-glucosyltransferase isolated from *P. lunatus* embryos and purified by anion exchange HPLC. **A** UDPG plus Z; **B** UDPX plus Z; **C** UDPG and UDPX plus Z. [Each assay contained enzyme extracted from 100 mg of embryos. Reaction condition: UDPG and UDPX (2 mM), ATP (0.3 mM), $MgCl_2$ (30 mM), [^{14}C]-Z (33 000 cpm) in 0.08 M Tris-HCl buffer, pH 8.0. 1-h incubation at 27°C. *OGZ* O-glucosylzeatin; *OXZ* O-xylosylzeatin]

7 Genetic Differences in the Reduction of Zeatin

Embryos of *P. coccineus* and *P. vulgaris* convert Z to (OX)Z and (OX) (diH)Z [24], and both compounds have been shown to occur naturally in the latter. The reduced forms of O-xylosyl derivatives were not detected in *P. lunatus* embryos. To determine the pathway leading to the reduced conjugate, the metabolites of [^{14}C]-Z and [^{14}C]-(OX)Z were examined after incubation with enzyme extracts obtained from *P. coccineus* embryos. (OX)Z was not converted to its dihydro derivative whereas Z was converted to (diH)Z under the appropriate reaction conditions (Fig. 3). Therefore, the reduction of the side chain seems to preceed O-xylosylation in *P. coccineus* embryos. Preliminary experiments to characterize the enzyme converting

Reduction of zeatin to dihydrozeatin

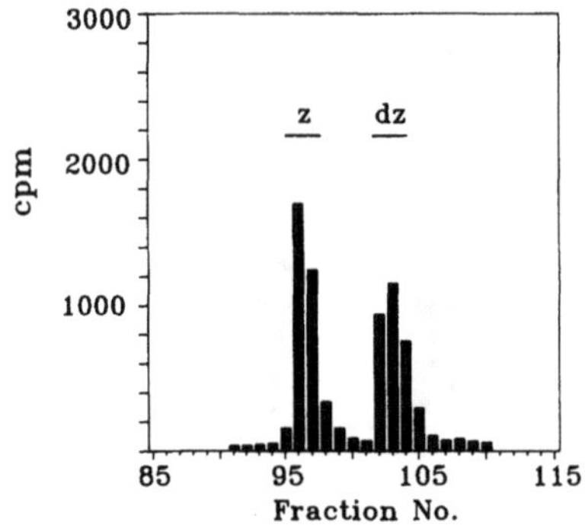

Fig. 3. HPLC profile of reaction products after incubation of Z with Z reductase. [The assay contained enzyme extracted from 100 mg of embryos. Reaction condition: NADPH (0.25 mM), [^{14}C]-Z (35 000 cpm), incubation at 27°C for 1 h. Products were chromatographed on a reversed phase C_{18} column with 0.2 M triethylamine buffer at pH 4.8 as the mobile phase. A gradient of 5 to 50% methanol over 90 min was used to elute Z and its derivatives. The flow rate was 1 ml/min and 0.5 ml fractions were collected]

Table 2. Effects of co-factors on the conversion of Z to (diH) Z

	Distribution of radioactivity (% of Z + (diH)Z)[a]	
	Z	*(diH)Z*
Co-factors		
1. Phosphate buffer	100	0
2. Phosphate buffer + ATP + $MgCl_2$	100	0
3. Phosphate buffer + ATP + $MgCl_2$ + NADH	100	0
4. Phosphate buffer + ATP + $MgCl_2$ + NADPH	54	46
5. Phosphate buffer + NADPH	0	100

[a] Each assay contained enzyme extracted from 250 mg of embryos. Conditions: ATP (0.5 mM), $MgCl_2$ (0.05 M), NADPH or NADH (0.5 mM), enzyme (100 μl), reaction volume, 200 μl by the addition of phosphate buffer (0.1 M, pH 7.3). Reaction time: 1 h at 27°C.

Z to (diH)Z have been conducted (Plant Physiol, submitted). The reaction is NADPH-dependent and does not require the presence of ATP and cations (Table 2). The enzyme has been partially purified using ammonium sulfate precipitation, anion exchange and affinity column chromatography. The affinity of the enzyme for i^{6}Ade (N^6-(Δ^2-isopentenyl)adenine), the unhydroxylated counterpart of Z, is negligible. Preliminary estimates of Z reductase activities varied substantially between species, with high activity in *P. coccineus* and *P. vulgaris* embryos but only marginally detectable activity in *P. lunatus*.

8 Conclusion

It was somewhat unexpected to find two distinct O-glycosylation enzymes in *Phaseolus* species. Although (OG)Z has been detected in many species, (OX)Z has not been identified in any genera other than *Phaselous*. Based on the biological activity and the metabolism of (OX)Z in different bioassay systems, it may be postulated that O-xylosylation is related to other important functions and does not serve a mere storage role. It appears that the pathways which occur in *P. vulgaris* represent processes converting Z to compounds (i.e. (OX)Z and (diH)Z) which are more active than Z in the *P. vulgaris* bioassay. Furthermore, these processes are regulated by very specific enzymes with high substrate specificity and affinity. There may be a relationship between these conversions and the occurrence of cytokinin oxidases which are active in this species.

Unlike some other cytokinin metabolic enzymes with affinity for cytokinins and other purines, the three *Phaseolus* enzymes described here appear to recognize only *trans*-Z (and (diH)Z in the case of the O-xylosyltransferase). Thus, these enzymes are capable of distinguishing between the *trans*- and *cis*-isomers of Z as well as the free base and ribonucleoside forms. The fact that similar metabolic conversions of *cis*-Z do not occur underscores the importance of the enzymes in regulating cytokinin activity in relation to plant growth, which also involves only the *trans*-isomer of Z. Moreover, the inability of the enzymes to recognize structures other than the free base supports the central role of cytokinin free base as opposed to the ribonucleoside and ribonucleotide forms.

The interspecific variations in enzyme activities may indicate regulatory mechanisms effecting selective expression of specific enzymes. We intend to determine the genetic basis of divergence in Z O-glycosylation, whether it lies in the sequence of the structural genes, the transcriptional, translational and/or post-translational levels. The similarities in function and chromatographic properties indicate at least a close relationship between the two enzymes. Similar molecular analyses can be applied to elucidate the mechanisms controlling the reductase activity. As more refined methods of detecting gene expression, such as reaction with specific antibodies, can be adapted for these cytokinin enzymes, their role in regulating cytokinin levels during plant development can be further assessed.

Acknowledgements. The research is supported by grants from USDA/CRGO (86-CRCR-1-1998) and NSF (US-UK Coop. Sci. INT-8513026). This is technical paper No. 8525 of the Oregon Agricultural Experiment Station.

References

1. Capelle SC, Mok DWS, Kirchner SC, Mok MC (1983) Plant Physiol (Bethesda) 73:796
2. Chatfield JM, Armstrong DJ (1987) Plant Physiol (Bethesda) 84:726
3. Chaves das Neves HJ, Pais MS (1980) Tetrahedron Lett 21:4387
4. Cowley DE, Duke CC, Liepa AJ, MacLeod JK, Letham DS (1978) Aust J Chem 31:1095
5. Dixon SC, Martin RC, Mok MC, Shaw G, Mok DWS (1989) Plant Physiol (Bethesda) 90:1316
6. Duke CC, MacLeod JK, Summons RE, Letham DS, Parker CW (1978) Aust J Chem 31:1291
7. Duke CC, Letham DS, Parker CW, MacLeod JK, Summons RE (1979) Phytochemistry 18:819
8. Fox JE, Sood CK, Buckwalter B, McChesney JD (1971) Plant Physiol (Bethesda) 47:275
9. Fox JE, Cornette J, Deleuze G, Dyson W, Giersak C, Niu P, Zapata J, McChesney J (1973) Plant Physiol (Bethesda) 52:627
10. Horgan R, Hewitt EW, Morgan JM, Purse J, Wareing PF (1975) Phytochemistry 14:1005
11. Lee YH, Mok MC, Mok DWS, Griffin DA, Shaw G (1985) Plant Physiol (Bethesda) 77:635
12. Letham DS, Parker CW, Duke CC, Summons RE, MacLeod JK (1977) Ann Bot 41:261
13. Letham DS, Tao GQ, Parker CW (1982) In: Wareing PF (ed) Plant growth substances 1982. Proc 11th Int Conf Plant Growth Subst, Academic Press New York, p 143

14. Letham DS, Palni LMS (1983) Annu Rev Plant Physiol 34:163
15. Letham DS, Parker CW, Zhang R, Singh S, Upadhyaya MN, Dart PJ, Palni LMS (1989) Plant growth substances 1988 Springer-Verlag, Berlin (This volume)
16. Mok MC, Mok DWS, Armstrong DJ (1978) Plant Physiol (Bethesda) 61:72
17. Mok MC, Kim SG, Armstrong DJ, Mok DWS (1979) Proc Natl Acad Sci (USA) 76:3880
18. Mok MC, Mok DWS, Armstrong DJ, Rabakoarihanta A, Kim SG (1980) Genetics 94:675
19. Mok MC, Mok DWS, Dixon SC, Armstrong DJ, Shaw G (1982) Plant Physiol (Bethesda) 70:173
20. Mok MC, Mok DWS, Armstrong DJ, Shudo K, Isogai Y, Okamoto T (1982) Phytochemistry 21:1509
21. Mok MC, Mok DWS (1985) Physiol Plant 65:427
22. Mok MC, Mok DWS, Marsden KE, Shaw G (1987) J Plant Physiol 130:423
23. Mok MC, Mok DWS, Turner JE, Mujer CV (1987) HortScience 22:1194
24. Mok DWS, Mok MC (1987) Plant Physiol (Bethesda) 84:596
25. Morris RO (1977) Plant Physiol (Bethesda) 59:1029
26. Paces V, Kaminek M (1976) Nucleic Acids Res 3:2309
27. Palmer MV, Scott IM, Horgan R (1981) Plant Sci Lett 22:187
28. Parker CW, Wilson MM, Letham DS, Cowley DE, MacLeod JK (1973) Biochem Biophys Res Commun 55:1370
29. Parker CW, Letham DS (1974) Planta 115:337
30. Parker CW, Letham DS, Gollnow BI, Summons RE, Duke CC, MacLeod JK (1978) Planta 142:239
31. Scott IM, Martin GC, Horgan R, Heald JK (1982) Planta 154:273
32. Shaw G, Mok MC, Mok DWS (1987) J Chem Res (S) 109
33. Sondheimer E, Tzou D (1971) Plant Physiol (Bethesda) 47:516
34. Terrine C, Laloue M (1980) Plant Physiol (Bethesda) 65:1090
35. Turner JE, Mok DWS, Mok MC, Shaw G (1987) Proc Natl Acad Sci (USA) 84:3714
36. Whitty CD, Hall RH (1974) Can J Biochem 52:789

Xylem-Translocated Cytokinin: Metabolism and Function

D.S. Letham, C.W. Parker, R. Zhang, S. Singh, M.N. Upadhyaya, P.J. Dart, and L.M.S. Palni[1]

1 Introduction

The concept that cytokinin synthesized in the root moves via the xylem to the shoot to regulate some aspects of development, and especially leaf senescence, is often proposed [2]. However, studies of the distribution and metabolism of radioactive cytokinins supplied to the xylem or root have been invariably inadequate for one or more of the following reasons: (1) cytokinins endogenous to the xylem of the species under study were not supplied and often synthetic (unnatural) cytokinins were used; (2) unphysiological and even toxic levels (e.g. mM) were given; (3) the radioactive metabolites in the shoot were not identified and attempts to do so were based on inadequately prepared extracts, and employed crude fractionation procedures and chromatography without appropriate markers. A study of the metabolism of xylem cytokinins in blue lupin, which avoids these problems, has been reported [2] and further results are described below. A similar study, in relation to sequential leaf senescence of tobacco, is also outlined. In order to evaluate the role of xylem cytokinins in shoot development, it is necessary to develop more natural methods which would allow manipulation of the endogenous cytokinin level in the transpiration stream. This has been achieved using *Rhizobium* mutants and cytokinin levels have been related to shoot development as discussed below.

2 Translocation and Metabolism of Xylem Cytokinins in Lupin

In earlier work, [^{3}H]-[9R]Z[2] was supplied to the transpiration stream of both de-rooted and intact blue lupin (*Lupinus angustifolius* L.) plants [2]. Features of this work included: (1) the very low proportion of [9R]Z which moved to the seed, although [9R]Z reached the pod walls in significant amounts and tended to be conserved therein; (2) the direct lateral movement of [9R]Z and/or related cytokinins from xylem to bark (all tissues outside secondary xylem); (3) the detection of a metabolite, or closely related metabolites, of unknown nucleotide-like structure in pod walls, stem bark and developing lateral shoots. For convenience, the metabolites were all designated U-NT (meaning unknown nucleotides), because all

[1] Research School of Biological Sciences, The Australian National University, P.O. Box 475, Canberra City, ACT, 2601, Australia

[2] Abbreviations for cytokinins follow the system in Annu Rev Plant Physiol 34:167 (1983)

exhibited the same chromatographic characteristics during TLC. Further work concerning each of the above three areas is outlined below.

The radioactive compounds, derived from xylem-supplied [^{3}H]-[9R]Z and [^{3}H]-(diH)[9R]Z, found in seed (20 days after flowering) were characterized chromatographically. The total solvent-extractable radioactivity in seed coats was over 10 times that in embryos, and the seed coats contained [^{3}H]-labelled cytokinins, namely, (diH)[9R]Z and (OG diH)Z derived from xylem-supplied [9R]Z, and (diH)[9R]Z, (OG diH)Z and (diH)Z nucleotide when (diH)[9R]Z was supplied. These compounds accounted for about 20% of the extracted radioactivity. However, no radioactive cytokinins could be unequivocally identified in embryos when [^{3}H]-[9R]Z was supplied to the xylem. When [^{3}H]-(diH)[9R]Z was supplied, radioactive (diH)[9R]Z, (OG diH)Z and (diH)Z nucleotide were identified in embryos, but at greatly reduced levels relative to seed coats. Thus, xylem cytokinins appear to make some contribution to the seed cytokinin pool, especially that of the seed coats. While the seeds are clearly not a site to which xylem cytokinins move preferentially, an important unanswered question is the following: could the low proportion of xylem cytokinins which reach the seeds over a prolonged period, and are conserved therein, account for the high endogenous cytokinin level found in the seed? Relevant calculations were based on the following: (1) the maximum transpiration rate for the blue lupin plant; (2) the period of seed development prior to the time of our determination of endogenous seed cytokinin level; (3) the total cytokinin level in xylem sap (28 pmol ml^{-1}); (4) the percentage of [^{3}H] supplied to the xylem as [^{3}H]-[9R]Z and -(diH)[9R]Z which was in embryos and seed coats in extractable form; (5) the proportion of radioactivity due to actual cytokinins; (6) the weight of seed tissue per plant. Assuming the seed cytokinin is entirely xylem derived, the calculated cytokinin contents of the lupin seed embryo and seed coats should be 0.88 and 10.8 pmol g^{-1}, respectively. This is less than 1.1% of the observed contents of 219 and 1041 pmol g^{-1} respectively. In fact, for several reasons, the calculated values are probably overestimates.

In blue lupin, xylem cytokinins which do reach the seed are largely confined to the seed coats. A similar but less marked situation occurs in soybean [5, 7]. The seed coats of legume seeds thus appear to shield the embryos from even the very low amount and proportion of xylem cytokinins which reach the seed. However, the common cytokinin breakdown product, adenosine, appears to move readily from the xylem to embryo [5]. It is thus reasonable to conclude that xylem-derived cytokinins have little or no direct control over embryo development in legume seeds, which may be autonomous with respect to cytokinins.

The U-NT metabolites in bark and podwalls were transient – that is they were present only during [9R]Z uptake and immediately afterwards [2]. When (diH)[9R]Z was supplied to de-rooted lupin plants via the transpiration stream, a transient polar metabolite was again detected in the bark tissues. This metabolite exhibited normal phase TLC properties identical to those of U-NT; both compounds exhibited R_f values which were greater than those of Z nucleotide and similar to those of iP nucleotide. Hence this metabolite of (diH)[9R]Z, termed DU-NT henceforth, and U-NT were presumably analogous.

Initial studies of the structure of DU-NT utilized the [^{3}H]-labelled metabolite and degradation products were characterized chromatographically. Treatment of

DU-NT with alkaline phosphatase yielded a [^{3}H]-labelled product which chromatographed between (diH)[9R]Z and [9R]iP on TLC (silica gel, butanol-acetic acid-water 12:3:5). Degradation with periodate and cyclohexylamine, a procedure which converts riboside 5′-phosphates, but not 2′- or 3′-phosphates, into bases [6], gave a derivative which co-chromatographed with (diH)Z. Hence DU-NT behaved as a riboside 5′-phosphate. When dephosphorylated DU-NT was treated with esterase, the main product co-chromatographed with (diH)[9R]Z. Hence DU-NT appeared to have a carboxylic ester group. If a (diH)[9R]Z moiety was indeed present in DU-NT, the ester linkage must involve the oxygen of the N^6-sidechain, since the ribose 2′- and 3′-hydroxyls were free to react with periodate. When supplied to a lupin shoot explant, the dephosphorylated form of DU-NT was completely converted to compounds which co-chromatographed with (diH)Z and (OG diH)Z. Hence, DU-NT appeared to contain an ester group which is cleaved in plant tissues. This could explain the apparent transient nature of the metabolite and the related [^{3}H]-labelled metabolite, U-NT. Dephosphorylated DU-NT (henceforth termed DU-Ri; U-Ri denotes the corresponding metabolite of [9R]Z) was hydrolyzed by 1 N ammonia to (diH)[9R]Z and would therefore be degraded during normal cellulose phosphate column purification. TLC studies indicate that DU-Ri and U-Ri are formed in bark as metabolites of (diH)[9R]Z and [9R]Z respectively (see below).

CI-MS indicated the molecular weight of DU-Ri was 395. Hence, DU-Ri appeared to be the 9-riboside of *O*-acetyldihydrozeatin and the EI mass spectrum of tri-TMS DU-Ri (intense peak at m/z 292) supported this suggestion. Therefore, this acetyl riboside was synthesized and found to co-chromatograph with DU-Ri in diverse systems including GC. The synthetic and natural ribosides exhibited the same EI mass spectrum. Accordingly, DU-NT was assigned the structure *O*-acetyl-9-β-D-ribofuranosyldihydrozeatin 5′-monophosphate and this was confirmed by comparison with authentic synthetic nucleotide. The [9R]Z metabolites U-NT and U-Ri, which correspond chromatographically to DU-NT and DU-Ri, have not been characterized by MS, but are likely to be the unsaturated analogues of DU-NT and DU-Ri, respectively.

In blue lupin, a rapid lateral movement of [^{3}H]-[9R]Z and [^{3}H]-(diH)[9R]Z from xylem to bark was observed using girdled stems with leaves removed. This movement was somewhat more pronounced in the two main lateral stems which develop just below the primary inflorescence. Thus, when [^{3}H]-[9R]Z (2 μM) was supplied to derooted plants via the xylem stream, for 1.5 h followed by a water "chase" for 3.5 h, 24% of the total [^{3}H] extracted from the stem was derived from the bark. Kinetic studies of metabolites in xylem plus pith and in bark following [^{3}H]-[9R]Z and [^{3}H]-(diH)[9R]Z uptake through the xylem of de-rooted lupin plants, together with studies of the metabolism of these ribosides in excised stem tissues, indicated that: (1) the supplied ribosides were translocated to bark per se and possibly as the nucleotides; (2) U-Ri and DU-Ri were formed exclusively in the bark and were probably the precursors of U-NT and DU-NT, respectively; (3) ribose cleavage and sidechain reduction were largely confined to bark.

U-NT and DU-NT were detected only in tissues where the transpirational flux is low, relative to that in laminae. Hence, these nucleotides may be involved in cytokinin uptake and movement through tissue where active transport from cell to cell is necessary. They are transient metabolites and may represent stabilized nucleotides resistant to cytokinin oxidase which can release free active cytokinins by enzymic action after translocation.

3 Xylem Cytokinins and the *Rhizobium*-induced Leaf Curl Syndrome in Pigeonpea

The Physiology and Genetics of the Syndrome. The *Rhizobium* strain IC3342 nodulates *Cajanus cajan* (L.) Millsp. (pigeonpea) as well as inducing a systemic response which results in abnormal shoot development. The symptoms start 25–30 days after sowing and inoculation. There is a typical tip bending, followed by hyponasty, curling of leaves, release from apical dominance and proliferation of lateral buds. Internodal length is reduced. Effective nodulation is essential for induction of the leaf curl syndrome by strain IC3342. Five unlinked genetic regions of strain IC3342 involved in the induction of the syndrome were identified; three are on a megaplasmid (Sym plasmid) containing *nif* and *nod* genes, and two are on the chromosome. Some of the genes involved are required for both N_2 fixation and leaf curl induction. Of these five genetic regions identified by Tn5 mutagenesis, one is of particular interest. This region (11 K bases; defined by studies with cosmids containing overlapping sequences), which is located on the Sym plasmid and is designated *lcr* (leaf curl response), has the capacity to confer leaf curl-inducing ability to a very closely related but non-leaf curl strain (ANU240). Hybridization studies with strain ANU240 indicated the existence of sequences homologous to each of the other four identified genetic regions from strain IC3342 which are involved in leaf-curling, but not to the *lcr* region. It is the *lcr* region that contains genes involved in the curl induction pathway which interact with the other genetic loci.

Characterization of the Principle Which Induces the Leaf Curl Syndrome. Approach grafting experiments indicated that a leaf curl-inducing principle was produced in the roots/nodules, translocated to the growing shoots through the xylem, and that a continuous supply was essential for manifestation of the symptoms.

Feeding experiments were conducted with nodule extract, xylem exudate (sap) and leaf extract. Feeding of the sap was carried out by a wick method [2], while the extracts were fed via the growth medium. Xylem sap and extracts obtained from normal healthy plants were also fed, and these served as controls. Initial symptoms of the leaf curling syndrome (tip-bending) were observed in plants fed by the wick method with a 10-fold-concentrated xylem sap collected from leaf curl plants. Tip-bending was also observed in plants fed through the rooting medium with leaf-extract from leaf curl plants, but nodule extracts failed to produce bending. This confirmed that the curl syndrome was induced by a xylem translocated factor and this factor appeared to accumulate in the leaves.

BAP riboside, a synthetic cytokinin, supplied via the root system induced some effects in the shoot which are characteristic of the leaf curl syndrome, especially release of lateral buds from apical dominance and hyponasty.

Because of the above results, the cytokinin levels in IC3342-inoculated plants and normal plants were compared. *Amaranthus* betacyanin bioassays indicated

Table 1. Cytokinin levels in the xylem exudate of pigeonpea plants inoculated with three different strains of *Rhizobium*[a]

Inoculum	Cytokinin levels (ng ml^{-1})			
	Z	[9R]Z	(diH)Z	(diH) [9R]Z
IC3342	1.81	10.50	2.44	8.92
Tn5 mutant (*lcr*) of IC3342	1.95	1.68	0.96	1.89
None (nitrate supplied)	0.06	0.58	0.23	0.66
Normal *Rhizobium* strain[b]	0.13	1.23	0.82	1.34
LSD (P = 0.01)		4.35	1.47	4.35

[a] Exudate was collected from 34- to 38-day-old plants for 1 h following stem excision 3 cm above the level of the potting mixture (sand-vermiculite). Exudate was purified by chromatography on cellulose phosphate and by HPLC prior to RIA [2]. Values for Z did not differ significantly.
[b] Cowpea group *Rhizobium* (a wild-type normal strain, IHP100 from ICRISAT).

that cytokinin levels in xylem sap and leaf extracts of curl plants were greater than those of normal plants. In a replicated experiment, the cytokinin levels in xylem exudate of leaf curl and normal plants were then determined by RIA after purification of the sap samples (Table 1). Sap from uninoculated nitrate-fed control plants was also assessed. The principal cytokinins in the xylem exudate of normal nodulated plants and leaf curl plants were [9R]Z and (diH)[9R]Z. However the riboside level in the exudate of the latter plants was about 8 times that of the former, which was similar to the level of uninoculated nitrate-supplied plants. A 10-fold increase in BAP riboside concentration supplied to roots of pigeonpea markedly affected shoot development; an increase from 1 to 10 nM induced hyponasty, while increase from 1 to 10 μM caused lateral bud proliferation. Hence, the 8-fold increase in riboside concentration noted in Table 1 is probably of regulatory significance. It is particularly significant that plants inoculated with a mutant of IC3342 (Tn5 insertion in *lcr* locus), which did not evoke the leaf curl syndrome, contained the riboside levels of normal plants. Hence the *lcr* region of IC3342 contains genes involved in the overproduction of xylem cytokinins. [9R]Z and (diH) [9R]Z must be regarded as the transmitted signal which evokes the curl syndrome, or as important components of the induction signal. Since the Z level in IC3342 sterile culture filtrate (2. 19 μg l^{-1}) was 25 times that in the closely related normal strain ANU240, genes in the *lcr* region presumably enhance or induce cytokinin production by the bacteria. A logical suggestion would be that a cytokinin biosynthetic gene is located at this site.

Hence, a 3.3 K base region centred on the Tn5 insertion site in the *lcr* locus was sequenced and this revealed the presence of five open reading frames (ORFs) capable of encoding proteins. These are designated ORFs 1–5. ORFs 1 to 3 are on one strand, while 4 and 5 are located on the complementary strand of the sequence. The sequence of ORF2 shows strong homology to that of the *E. coli* regulatory gene *ompR*, while ORF4 exhibits strong homology to the regulatory gene *fnr* from *E. coli*. Hence, the gene products encoded by these two ORFs may well exert a regulatory role. The observed overlap of ORF2 with ORF3, and ORF4 with ORF5 suggests translational coupling of these gene pairs. Since the Tn5 insertion is located in the

N-terminal region of ORF3, and the C-terminal region of ORF4, transcription and translation of these two ORFs would be affected. Furthermore, Tn5 insertion may also exert polar effects on the gene encoded by ORF5. Hence, ORF3, ORF4 and ORF5 are all potentially involved in leaf curling and cytokinin overproduction.

However, total genomic DNA from IC3342 and cosmid clones containing wild type sequences of each mutant locus (including *lcr*) were probed with a sequence of the cloned cytokinin biosynthetic gene (*ipt*) from *A. tumefaciens*. Hybridization, even under conditions of low stringency, revealed no homology with the *ipt* gene probe. Sequence comparison failed to detect homology of any of the ORFs of the *lcr* region to the known genes involved in cytokinin biosynthesis, i.e. *ipt* and *tzs* from *A. tumefaciens* [4], and *ptz* from *Pseudomonas savastanoi* [4].

The following possibilities at the molecular level merit consideration: (1) the extent of sequence homology between cytokinin biosynthetic genes from the strain IC3342 and *Agrobacterium* is not sufficient to be detected by hybridization; (2) this *Rhizobium* strain has a different biosynthetic pathway for the production of cytokinins compared to that of *Agrobacterium*; (3) IC3342 may contain a gene not present in ANU240, which is involved in conversion of iP-type cytokinins to Z-type cytokinins. These molecular aspects require clarification. However, the IC3342 nodulated pigeonpea plant constitutes a novel intact plant system, one whereby the role of *endogenous* xylem cytokinins in apical dominance, lateral shoot proliferation and stem elongation can be readily studied. Cytokinin appears to play a key role in regulation of the first two phenomena [3] and evidence, both for [9] and against [8] their control by root-produced cytokinin, has been presented based on studies with de-rooted plants. However, in the studies with pigeonpea, endogenous cytokinin level in the xylem of intact plants was related to lateral bud development in the shoot. The work also showed that nodule-produced cytokinins are translocated via the xylem.

4 Translocation of Cytokinins in the Xylem of Tobacco in Relation to Sequential Leaf Senescence

Four types of evidence that endogenous cytokinin levels are involved in control of sequential leaf senescence of tobacco were obtained. First, the cytokinins normally present in tobacco leaves, when applied exogenously to intact leaves, as well as to excised leaves and leaf discs, markedly delay senescence. Secondly, the levels of cytokinin bases (the probable active forms of cytokinin) decline as senescence occurs; these differences in cytokinin level are similar to those caused when cytokinins are applied to leaves in concentrations which retard senescence. Thirdly, NH_4^+ ions were found to delay senescence and they also resulted in elevated endogenous cytokinin base levels. Fourthly, treatments which would be expected to elevate cytokinin levels (detopping, optimal nutrition) reduce the response to exogenous cytokinin. Endogenous cytokinin levels in leaves were determined by both RIA and GC-MS. Quantification of Z and (diH)Z by GC-MS was greatly facilitated by the use of O-t-butyldimethylsilyl 9-pentafluorobenzyl derivatives. The purified leaf extract was first derivatized to yield O-t-butyldimethylsilyl-Z and -(diH)Z[1] and was then rederivatized with pentafluorobenzyl bromide [10] to give the disubstituted compounds. These are very stable, can be further purified by

HPLC, and separate well during GC. The derivatives give intense negative ion CI spectra in which the ion current is concentrated in one fragment ion derived by benzyl cleavage. The ion current, in the form of this negative ion, was 80 times that in the most intense positive ion (MH^+).

The differing levels of cytokinins in tobacco leaves of varying age could be due to one or a combination of the following factors: (1) differential translocation of xylem cytokinins; (2) differential metabolism of xylem cytokinins; (3) differential retention of xylem cytokinins; and (4) differential biosynthesis of cytokinins in situ in the leaves. These four factors were assessed. By RIA, the principal cytokinins in tobacco xylem exudate were identified as Z, (diH)Z, [9R]Z and (diH)[9R]Z. The distribution and metabolism of the two ribosides in the tobacco shoot were determined after supply via the xylem. The major metabolites of [9R]Z in tobacco leaf laminae were adenine, adenosine and AMP, while the principal metabolite of (diH)[9R]Z was the 7-glucoside of (diH)Z. However, expanded pre-senescent and early senescent laminae did not differ in their cytokinin metabolism, while small expanding laminae showed a higher rate of metabolism. There was no differential distribution of the ribosides to these laminae, and no differential retention in leaves of differing maturity.

Upper small green leaves of tobacco were found to incorporate [^{14}C]adenine into zeatin but more mature leaves were less active or inactive in synthesizing cytokinins (for further discussion see Palni et al., this volume). Thus sequential leaf senescence in tobacco may result from loss of ability to synthesize cytokinins within the leaf after full expansion.

To summarize, cytokinin translocation via the xylem has been studied in three systems, namely, blue lupin, pigeonpea and tobacco. These provide insight into seed development and new metabolites, lateral shoot development, and leaf senescence, respectively. The view that xylem cytokinin level greatly influences the cytokinin status of the shoot and thus affects aspects of shoot development and also senescence is widely accepted although it still lacks adequate validation. The results with pigeonpea and work with soybean explants (see Noodén and et al., this Vol.) indeed support the concept of a regulatory role for xylem cytokinin. However, the tobacco system clearly shows that cytokinin level in leaves is governed by additional mechanisms such as biosynthesis in the leaf per se.

References

1. Hocart CH, Wong OC, Letham DS, Tay SAB, MacLeod JK (1986) Anal Biochem 153:85
2. Jameson PE, Letham DS, Zhang R, Parker CW, Badenoch-Jones J (1987) Aust J Plant Physiol 14:695
3. Martin GC (1987) HortScience 22:824
4. Morris RO (1986) Annu Rev Plant Physiol 37:509
5. Noodén LD, Letham DS (1984) J Plant Growth Regul 2:265
6. Parker CW, Letham DS, Gollnow BI, Summons RE, Duke CC, MacLeod JK (1978) Planta 142:239
7. Singh S, Letham DS, Jameson PE, Zhang R, Parker CW, Badenoch-Jones J, Noodén LD (1988) Plant Physiol 88:788
8. Wang TL, Wareing PF (1979) New Phytol 82:19
9. Woolley DJ, Wareing PF (1972) New Phytol 71:1015
10. Zhang R, Letham DS, Wong OC, Noodén LD, Parker CW (1987) Plant Physiol 83:334

Cytokinin Oxidase and the Degradative Metabolism of Cytokinins

R. HORGAN[1], L.R. BURCH[1], and L.M.S. PALNI[2]

1 Introduction

Over the last 16 years considerable knowledge has accumulated about the metabolism of naturally occurring and synthetic cytokinins. In the case of naturally occurring cytokinins, many of the metabolites identified after the external application of a cytokinin to a variety of plant tissues have been subsequently identified as endogenous compounds. Thus, a comprehensive picture now exists of the metabolic reactions that may be involved in the regulation of cytokinin levels in plants. A basic biochemical model for the control of cytokinin levels would envisage a dynamic balance between biosynthesis, the anabolic reactions leading to the various cytokinin conjugates and the catabolic reactions leading to loss of cytokinin activity. Whilst it is still unclear as to the precise function of cytokinin conjugation, i.e. to what extent it leads to the formation of storage, transported or inactivated forms of the cytokinins; the degradative metabolism of cytokinins via N^6 side chain cleavage is clearly a process whereby plant tissues may regulate the levels of biologically active cytokinins.

2 Biochemistry of Cytokinin Degradation

Many studies of cytokinin metabolism have revealed that externally applied cytokinins are broken down to common purinyl compounds. In the case of cytokinins possessing a Δ^2 double bond in the side chain the initial products of breakdown are N^6 amino purines. There has been little research into the breakdown of cytokinins which lack a Δ^2 double bond. The N^6 side chains of such compounds are more resistant to cleavage and the initial products of breakdown appear to be 6-oxo purines.

Paces et al. [12] first described an enzyme activity from cultured tobacco tissue which converted 8-^{14}C[9R]iP to adenosine. A similar enzyme was partially purified from *Zea mays* kernels by Whitty and Hall [14] and named "cytokinin oxidase". These workers characterized the enzyme as having a M_r of about 88 000, a requirement for molecular oxygen and the presence of a Δ^2 double bond in the side chain of the substrate. Subsequenly it was shown [1] that the enzyme produced

---

[1]Department of Botany and Microbiology, The University College of Wales, Aberystwyth SY23 3DA, UK

[2]Research School of Biological Sciences, The Australian National University, P.O. Box 475, Canberra City, Australia

adenine and 3-methyl-2-butenal when iP was the substrate and postulated that these products arose via oxidation of the N^6-C-1 bond to give an imine intermediate which was then cleaved by hydrolysis. This suggestion has been confirmed by Laloue and Fox [6] who demonstrated the formation of 6-(Δ^2 isopentenylimino)purine riboside as an intermediate in the degradation of [9R]iP by a partially purified cytokinin oxidase preparation from wheat germ. They also found that the enzyme was strongly inhibited by the cytokinin active ureas, DPU and N-(2-chloro-4-pyridyl)-N-phenyl urea.

Cytokinin oxidase has been partially purified from *Phaseolus vulgaris* callus tissue [3]. The enzyme activity could be induced by external application of cytokinin and the cytokinin active urea derivative, Thidiazuron [3]. Thidiazuron also inhibited the activity of cytokinin oxidase. It was also shown [4] that the in vitro activity of the enzyme is enhanced when assayed in the presence of copper-imidazole complexes.

The substrate specificity of partially purified cytokinin oxidases from *Zea mays* kernels and *Vinca rosea* crown gall tissue was investigated [7]. Both enzymes were only active on substrates with a Δ^2 double bond in the side chain and were essentially unaffected by substituents on the purine ring. However, the presence of bulky substituents on the side chain, as in (OG)Z, made the cytokinin resistant to both oxidases. M_r of the corn enzyme was estimated to be 94 000 by gel filtration which was close to the 88 000 value of [14]. The *Vinca* enzyme, although kinetically similar to the *Zea* enzyme, appeared to have a much lower M_r of 25 000.

From the above studies it can be concluded that plant tissues contain a specific enzyme for the degradation of naturally occurring cytokinins having a Δ^2 isopentenyl side chain. Mechanistically it would appear to be a copper-dependent amine oxidase. From the widespread occurrence of adenine and related compounds as metabolites of externally applied Δ^2 isopentenyl cytokinins [8], it may be concluded that cytokinin oxidase is present in a large number of plant species.

3 Physiology of Cytokinin Degradation

Given the widespread occurrence of cytokinin oxidase it is pertinent to ask what is the physiological significance of this enzyme? Several studies have shown that the response of tissue cultures to applied cytokinins may in some instances depend upon the level of cytokinin oxidase. Thus, tissue cultures of *Phaseolus vulgaris* grow better on (diH)Z and its riboside, which are not degraded by cytokinin oxidase, than on Z and [9R]Z [10]. The reverse is true for *P. lunatus* tissue cultures [10]. This is probably due to a greater rate of breakdown of 8-^{14}C-[9R]iP by *P. vulgaris* callus [11]. This presumably involves cytokinin oxidase to which (diH)Z and its riboside are resistant. Horgan [5] has shown that the response of soybean cell suspension cultures to Z can be inversely correlated with their ability to break down this cytokinin. Two strains of soybean cell suspension cultures which responded identically to BA, differed markedly in their ability to grow on Z. When supplied with 8-^{14}C-Z the slower growing strain exhibited a significantly greater capacity to degrade 8-^{14}C-Z to adenine and related metabolites. This result suggests that the sensitivity of the cultures to Z was governed by cytokinin oxidase levels.

Whilst the above findings are interesting with regard to tissue responses to externally applied cytokinins the most important question to address with regard to cytokinin oxidase is to what extent may it regulate the level of endogenous cytokinins? Some evidence for a role for cytokinin oxidase comes from recent studies on tobacco crown gall tissues. Crown gall tissues carrying insertionally inactivated auxin genes (*tms* genes) on the T-DNA exhibit extremely elevated levels of cytokinins. The most significant feature of the cytokinin content of these tissues is a greatly elevated level of [7G]Z (up to 100 times that of normal transformed tissue) [9]. A possible explanation of this phenomenon is that cytokinin oxidase mediated degradation is reduced in tms mutant tissues as a consequence of reduced auxin levels. Palni et al. [13] have shown that tobacco pith when grown on increasing levels of NAA exhibits an increased capacity to metabolise ^{3}H-[9R]Z to adenosine and related compounds. They also observed up to 30% stimulation of activity of a semi-purified cytokinin oxidase preparation from *Zea mays* kernels by 5 10^{-5} M NAA. The full significance of these findings is unknown, but they strongly suggest that the cytokinin oxidase activity of a tissue may play a significant role in regulating its level of certain cytokinins. Indeed the inhibitory effects of certain phenyl ureas [3] on cytokinin oxidase activity may be related to their cell division promoting activity.

4 Purification of Cytokinin Oxidase from *Zea mays* Kernels

Zea mays kernels were chosen as a source of cytokinin oxidase because published work [1, 7, 14] and our own prior experience indicated that they were a relatively rich source of the enzyme. Although activity could not be detected in a crude buffer extract it was present in a 60% NH_4SO_4 precipitate. The purification procedures of this precipitate are shown in Figs. 1–3. Cytokinin oxidase activity in the column

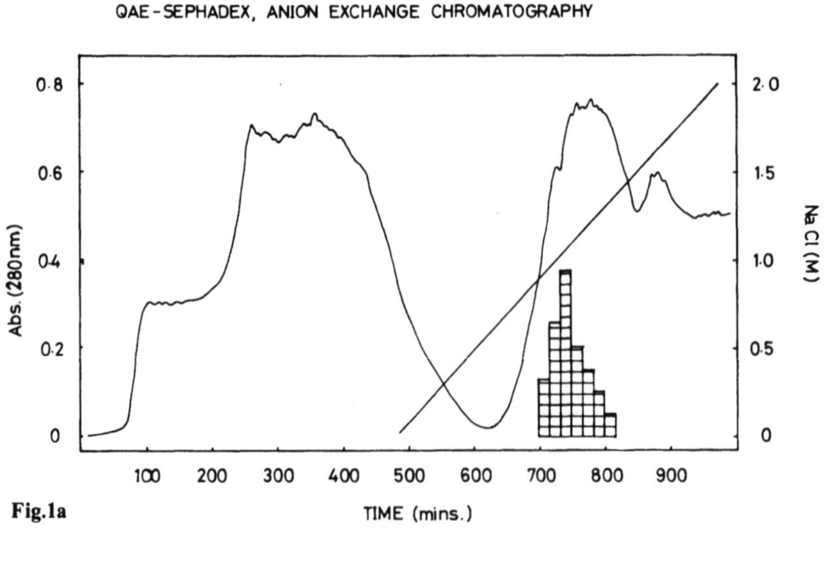

Fig. 1a

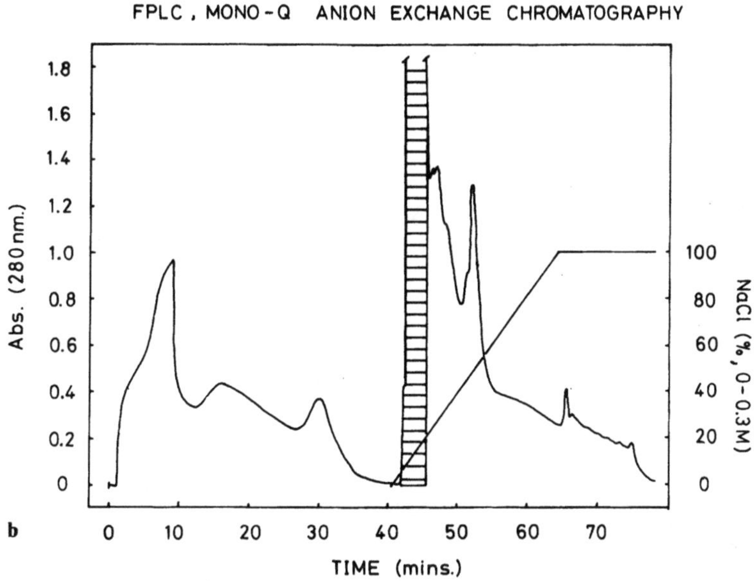

b

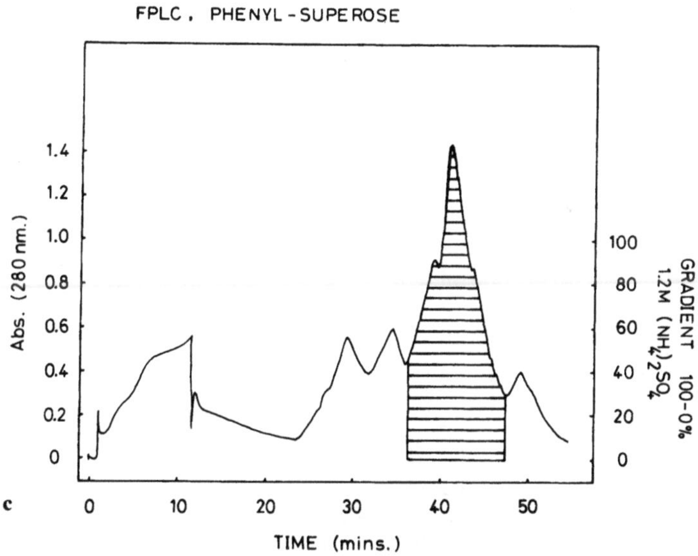

c

Fig. 1a-c. UV absorbance and cytokinin oxidase activity elution profiles for the initial stages of purification of cytokinin oxidase from *Zea mays* kernels. The FPLC was carried out on a Pharmacia FPLC system. The *shaded areas* represent the portions of the eluates with significant oxidase activity which were used for subsequent steps

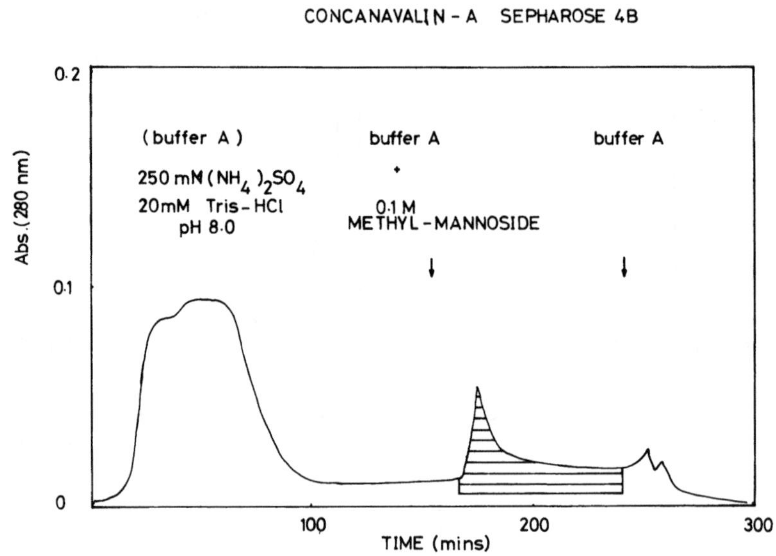

Fig. 2. Concanavalin A-Sepharose chromatography of FPLC Phenyl Superose fraction shown in Fig. 1. The *shaded portion* represents cytokinin oxidase activity which was collected in 12 4 ml fractions

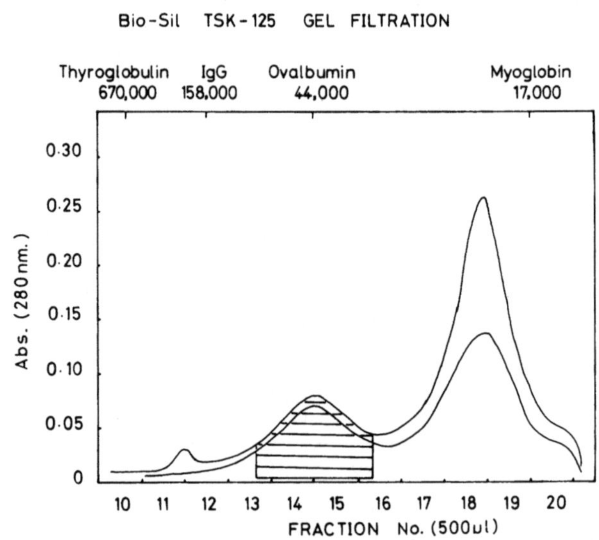

Fig. 3. Elution profile of bulked cytokinin active fractions from a con A-Sepharose column after chromatography on a Bio Sil TSK-125 column. Cytokinin oxidase activity was located exclusively in fractions *13, 14, 15* and *16* and was proportional to the intensity of staining of the single protein band seen in PAGE analysis of these fractions

eluates was determined by measuring the conversion of 8-^{14}C-iP to adenine using reversed phase HPLC-RC. Although significant removal of the bulk of interfering proteins was achieved in all steps, the extent of purification was not always reflected in the specific activity values, presumably due to two factors. First, enzyme activity was often assayed, for convenience, under non-optimal conditions. Secondly, there was considerable loss of activity, for as yet undetermined reasons, throughout the procedure. Since we wished to obtain pure protein for immunological purposes the purification was monitored throughout by SDS-PAGE.

Although the PAGE analyses up to the con A-sepharose stage clearly indicated the removal of major contaminating proteins they were not interpretable with regard to the association of a specific band with cytokinin oxidase. However, the methyl mannoside eluate of the con A-Sepharose column when examined by PAGE with silver staining, revealed the presence of five proteins with estimated M_rs of 78 000, 32 000, 29 000, 18 000 and 14 000. The very high degree of purification achieved on con A-Sepharose is in agreement with the findings of Chatfield [2] who first identified cytokinin oxidase as a glycoprotein and utilised con A-Sepharose as a medium for its purification. The cytokinin oxidase activity from con A-Sepharose was spread over a considerable volume of the methyl mannoside eluate (Fig. 2). By monitoring successive fractions by PAGE it was possible to show that only four of these proteins (M_rs 78 000, 29 000, 18 000 and 14 000) correlated with cytokinin oxidase activity. Attempts to recover enzyme activity from the gels were unsuccessful but the relatively large molecular weight differences between the proteins enabled their separation to be achieved by high resolution gel filtration. Figure 3 shows the UV elution profile and cytokinin oxidase activity of the bulked methyl mannoside eluate of the con A-Sepharose column when chromatographed on a Bio Sil TSK-125 column. Cytokinin oxidase activity was associated only with a peak eluting at an approximate M_r of 44 000. PAGE analysis of 4 successive fractions collected across the peak revealed that the enzymic activity was associated with the presence of a single protein of M_r 78 000. The homogeneity of the bulked active fractions from the TSK-125 column was assessed by 2D IEF/PAGE. The results of this analysis, shown in Fig. 4, indicated that the active fractions from the TSK-125 column contained a single protein and that this protein was most probably cytokinin oxidase. Approximately 100 μg of purified protein were obtained from 1.7 kg of *Zea mays* kernels. No efforts were made to optimize yields and the overall recovery was 0.7%.

The purified protein was used to raise polyclonal antibodies in a rabbit. The purified IgG showed significantly enhanced immunoreactivity towards a purified cytokinin oxidase preparation, when compared to non-immune control serum, by dot blotting using a gold labelled goat anti-rabbit second antibody procedure with silver enhancement. Most significantly, the antibody preparation was able to precipitate 80% of the cytokinin oxidase activity of a highly active, partially purified preparation from *Zea mays* kernels in the presence of fixed *Staphylococcus aureus* cells. No precipitation occurred with non-immune serum.

Although raised to native protein the cytokinin oxidase antibodies appear to recognize SDS denatured protein on PAGE gels and to cross-react with a similar protein in partially purified cytokinin oxidase preparations from dry wheat seeds. Figure 5 shows the results of a Western blot analysis of a purified cytokinin oxidase

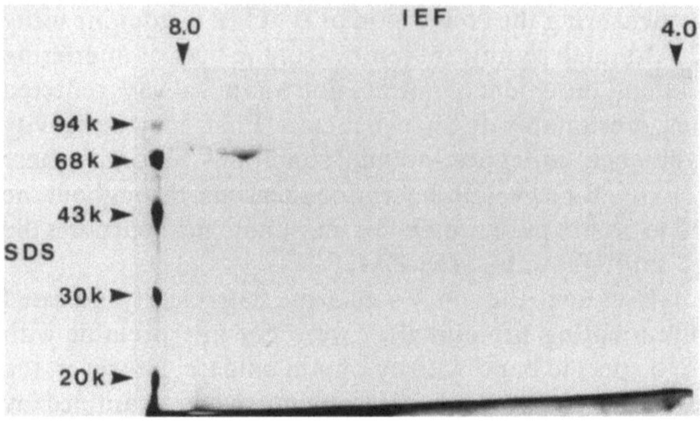

Fig. 4. 2D IEF/SDS-PAGE analysis of the bulked TSK-125 fractions *13, 14, 15* and *16* from Fig. 3. Approximately 3 μg of protein was loaded on the IEF gel

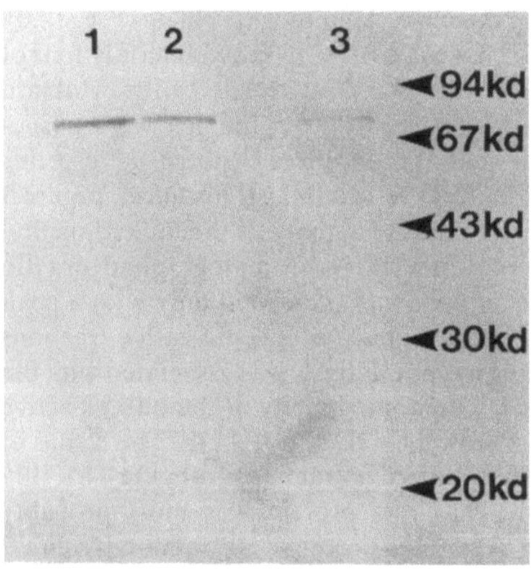

Fig. 5. Immunodetection of cytokinin oxidase from *Zea mays* (lanes *1* and *2*) and wheat seeds (lane *3*) on a Western blot, using antibodies to *Zea* oxidase. Rabbit IgG was detected by gold-labelled goat anti-rabbit second antibody with silver enhancement

preparation from *Zea mays* (lanes 1 and 2) and a partially purified preparation from dry wheat seeds (lane 3). It can be seen that the wheat preparation contains a protein with the same electrophoretic mobility as *Zea* cytokinin oxidase which cross-reacts with the antibodies to this enzyme. We therefore conclude that the cytokinin oxidase of wheat germ studied by Laloue and Fox [6] is very similar to our *Zea mays* enzyme.

5 Conclusions

Cytokinin oxidase appears to be a highly specific enzyme which can degrade a large number of naturally occurring cytokinins. Its activity in plant tissues can clearly influence their responses to externally applied Δ^2 isopentenyl type cytokinins. The role of cytokinin oxidase in regulating endogenous cytokinin levels is still unclear. Extremely elevated levels of cytokinins in tms mutants of tobacco crown gall tissue appear may be caused by a reduction in cytokinin oxidase activity. The cytokinin activity of DPU, N-(2-chloro-4-pyridyl)-N-phenyl-urea and Thidiazuron may be related to their ability to inhibit the activity of cytokinin oxidase, thus increasing the level of endogenous cytokinins. Cytokinin oxidase has been purified to apparent homogeneity from *Zea mays* kernels. Polyclonal antibodies to this enzyme cross react with a similar protein in partially purified cytokinin oxidase preparations from dried wheat seeds. The availability of pure cytokinin oxidase, and antibodies to it, should greatly facilitate research into its biochemical and physiological role.

Abbreviations

[9R]iP	Isopentenyl adenosine
M_r	Molecular weight
iP	Isopentenyl adenine
DPU	Diphenylurea
[OG]Z	O-glucosyl zeatin
[diH]Z	Dihydrozeatin
Z	Zeatin
[9R]Z	Zeatin riboside
BA	Benzyl adenine
[7G]Z	Zeatin-7-glucoside
NAA	Napthaleneacetic acid
SDS	Sodium docedyl sulphate
PAGE	Polyacrylamide gel electrophoresis
IEF	Isoelectric focusing
Con A-Sepharose	Concanavalin A Sepharose 4B (Sigma)
FPLC	Fast protein liquid chromatography (Pharmacia)
HPLC-RC	High performance liquid chromatography with online radioactivity monitoring

Acknowledgments. We wish to thank Mrs. Averil Rees for expert technical assistance and the AFRC for a postdoctoral fellowship to LRB.

References

1. Brownlee GB, Hall RH, Whitty CD (1975) Can J Biochem 53:37–41
2. Chatfield JM (1986) PhD Thesis, Oregon State University
3. Chatfield JM, Armstrong DJ (1986) Plant Physiol 80:493–499
4. Chatfield JM, Armstrong DJ (1987) Plant Physiol 84:726–731
5. Horgan R (1987) In: Hoad GV, Lenton JR, Jackson MB, Atkin RK (eds) Hormone action in plant development. Butterworth, London, p 119
6. Laloue M, Fox JE (1985) Abstracts of 12th international conference on plant growth substances. Bopp M, Knoop B, Rademacher W (eds) Heidelberg, p 23

7. McGaw BA, Horgan R (1983) Planta 159:30–37
8. McGaw BA, Scott IM, Horgan R (1984) In: Crozier A, Hillman JR (eds) The biosynthesis and metabolism of plant hormones. Cambridge University Press, p 105
9. McGaw BA, Horgan R, Heald JK, Wullems GJ, Schilperoort RA (1988) Planta 176:230–234
10. Mok MC, Mok DWS, Armstrong DJ (1978) Plant Physiol 61:72–75
11. Mok MC, Mok DWS, Dixon SC, Armstrong DJ, Shaw G (1982) Plant Physiol 70:173–178
12. Paces V, Werstiuk E, Hall RH (1971) Plant Physiol 48:775–778
13. Palni LMS, Burch LR, Horgan R (1988) Planta 174:231–234
14. Whitty CD, Hall RH (1974) Can J Biochem 52:781–799

Metabolic Aspects of Ethylene Biosynthesis

S.F. Yang, W.K. Yip, S. Satoh, J.H. Miyazaki, X. Jiao, Y. Liu, L.Y. Su, and G.D. Peiser[1]

1 Introduction

The biosynthesis of ethylene has been reviewed recently [9, 33]. In this paper we describe research progress that has occurred since then with regard to (a) ACC synthase, (b) ACC oxidase, (c) ACC N-malonyltransferase, and (d) the methionine cycle, as they related to ethylene biosynthesis.

2 ACC Synthase

The gaseous plant hormone ethylene is biosynthesized in plant tissue by the following sequence: methionine → SAM → ACC → ethylene [33]. In plant tissues ACC synthase which catalyzes the conversion of SAM to ACC and MTA, is generally the rate-controlling enzyme in the pathway. Soon after Adams and Yang [2] elucidated that ACC derived from SAM served as the immediate precursor of ethylene, Boller et al. [5] and Yu et al. [37] demonstrated ACC synthase (EC 4.4.1.14) activity from tomato fruit. As predicted from in vivo study, this enzyme requires pyridoxal phosphate.

Since ACC synthase exists in very low concentration in plant tissues, and is unstable, progress in the purification of this enzyme has been slow. Using a combination of various purification procedures, Bleecker et al. [4], have purified the enzyme 6500-fold from wounded tomato pericarp. SDS-PAGE revealed a molecular weight of 50 kD. They prepared monoclonal antibodies against ACC synthase and demonstrated that these antibodies recognized the native enzyme. When the antibody was linked to a Sepharose-4B gel, this matrix was shown to be effective in isolating ACC synthase from a crude enzyme preparation. The enzyme eluted from the immunoaffinity gel was, however, denatured and contaminated with IgG proteins derived from the antibodies. They have estimated the specific activity of the pure enzyme at about 7 μmol/min-mg protein. Nakajima et al. [20] have purified the enzyme to homogeneity from sliced and aged mesocarp of winter squash fruits. The enzyme had a molecular weight of 45–50 kD by SDS-PAGE; the specific activity was estimated to be 2.4 μmol/min-mg protein. On the other hand Tsai et al. [31] purified ACC synthase from hormone-treated etiolated mungbean hypocotyls. The molecular weight of the native enzyme was 125 kD, which consisted of two subunits of 65 kD. The specific activity was 0.35 μmol/min-mg

[1] Vegetable Crops Department, Mann Laboratory, University of California, Davis, CA 95616, USA

protein, a value much lower than that for tomato or winter squash. Thus, the enzymes from these tissues differed in final specific activity and in molecular weight.

The stereochemical courses of the reaction catalyzed by ACC synthase have been studied in several laboratories. There are two chiral centers on C-2 and S in the methionine moiety of SAM. Khani-Oskouee et al. [10] have shown that only the natural (–)-L-SAM serves as the ACC synthase substrate. Ramalingam et al. [25] synthesized (±)-S-adenosyl[3,4-$^2H_2$]methionine with the deuteriums in either *cis* or *trans* configuration, and demonstrated that *trans*-SAM gave *trans*-ACC and *cis*-SAM gave *cis*-ACC by the action of ACC synthase. Their data indicate that the enzymatic ring formation involves an inversion at the C-4 center (γ-position) of the methionine moiety of SAM. On the other hand, Wiesendanger et al. [32] synthesized S-adenoysl-L-[4-$^2H_2$]methionine as the substrate of ACC synthase, and determined that the deuterium was on the pro-(S) methylene group of ACC. Since the absolute configuration of the α-carbon of methionine in SAM is (S), and that of C-1 of ACC is (R), they concluded that the enzymatic conversion of SAM to ACC involves an inversion at the α-carbon of the methionine moiety. These findings are depicted in Fig. 1.

An important characteristic of ACC synthase in plant tissues is its lability. It has been well recognized that when ethylene production in plant tissues is induced by IAA treatment or by various stresses [33], ethylene production declines rapidly following induction, and this decline is accompanied by a corresponding decline in ACC content. Two mechanisms are responsible for this rapid decline in ethylene production: one is the conjugation of ACC into MACC catalyzed by ACC malonyltransferase resulting in a reduced ACC level in the tissue, and the other is the inactivation of ACC synthase. The apparent half-life of ACC synthase in wounded green tomato pericarp and in IAA-treated mungbean hypocotyls has been estimated to be 30–50 min, based on the decay kinetics of the enzyme activity extracted from induced tissues in the presence of cycloheximide which blocks the new synthesis of the enzyme [1, 36].

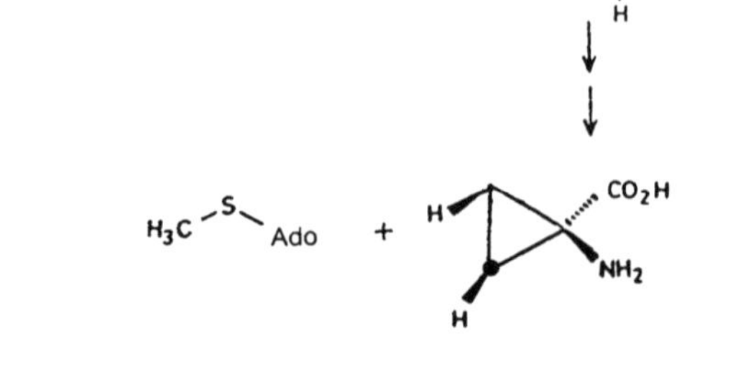

Fig. 1. Reaction mechanism and stereochemical course of the conversion of SAM to ACC catalyzed by the pyridoxal phosphate (PLP)-dependent ACC synthase. C-3 (●) of the L-methionine portion of SAM becomes the pro-(R) methylene group of ACC, whereas the two hydrogens attached to C-3,4 of the methionine moiety in the cis-configuration yield cis-ACC

Satoh and Esashi [26] observed that ACC synthase isolated from mungbean hypocotyls was inactivated in vitro by its substrate SAM during its catalytic reaction. Since the half-life of ACC synthase in vitro was similar to that reported in the tissue, they suggested that the SAM-induced inactivation is responsible for the rapid inactivation of the enzyme found in the tissue. Recently, Satoh and Yang [27] demonstrated that when a partially purified ACC synthase preparation isolated from wounded tomato pericarp tissue was incubated with S-adenosyl-L-[3,4-^{14}C]methionine and the resulting proteins were analyzed by SDS-PAGE, only one radioactive protein band was observed (Fig. 2). This protein was judged to be ACC synthase, based on the observations that its molecular weight was 50 kD and that it was specifically bound to a monoclonal antibody against ACC synthase prepared by Bleecker et al. [4]. S-Adenosyl-L-[carboxyl-^{14}C]methionine, but not S-adenosyl-L-[methyl-^{14}C]methionine, was found to radiolabel ACC synthase (S. Satoh, unpublished results). These results suggest that the SAM-induced inactivation of ACC synthase involves a covalent linkage of a fragment of the SAM molecule, probably 2-aminobutyric acid (ABA), into the active site of ACC synthase. A possible relationship between the catalytic reaction and the enzyme-activated inactivation of ACC synthase is depicted below:

$$\mathrm{E + SAM} \rightleftharpoons \mathrm{E{\cdot}SAM} \begin{cases} \xrightarrow{k_1} \mathrm{E + ACC + MTA} \\ \xrightarrow{k_2} \mathrm{E\text{-}ABA + MTA} \end{cases}$$

The turnover number (k_1) of ACC synthase for its catalytic reaction was estimated to be about 300 min^{-1} [4], whereas k_2 for the inactivation reaction was estimated to be 0.01 min^{-1} based on its half-life of 66 min. Thus, the ratio of k_1/k_2 is estimated to be 30 000.

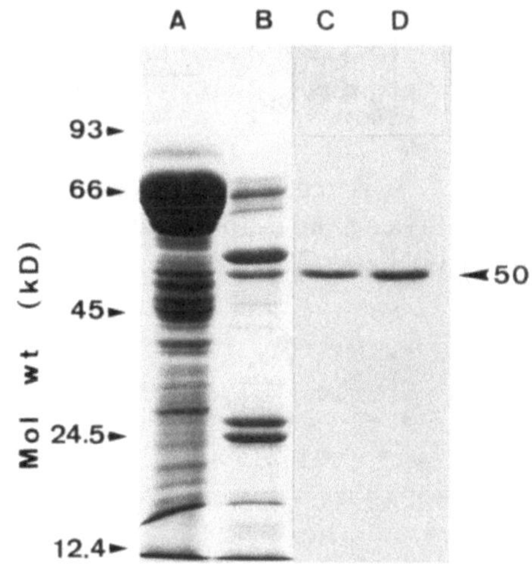

Fig. 2. SDS-PAGE analysis of [^{14}C]SAM-treated ACC synthase preparations. A sample of ACC synthase preparation was incubated at 30° C for 6 h with S-adenosyl-L-[3,4-^{14}C]methionine and the resulting proteins were analyzed either before (lanes *A* and *C*) or after (lanes *B* and *D*) purification by the immunoaffinity gel against ACC synthase [4]. At least two of the bands eluted from the immunoaffinity gel (lane *B*) are the antibody IgG proteins. Lanes *A* and *B* are Coomassie blue stain, and lanes *C* and *D* are the fluorograms of lanes *A* and *B* respectively. The 50 kD *arrow* indicates the position of ACC synthase protein. From Satoh and Yang [27]

3 ACC Oxidase

The final step in the biosynthesis of ethylene is catalyzed by ACC oxidase. Application of ACC to most plant tissues results in a marked increase in ethylene production, indicating that ACC oxidase in these tissues is constitutive and not rate-limiting. Although ACC-dependent ethylene production is readily demonstrated in intact tissues, in vitro ACC oxidase activity has not been demonstrated independent of intact cellular material. Although intact protoplasts and vacuoles retained the characteristics of tissue ACC oxidase activity [9], Porter and John [24] reported that their preparations retained less than 5% of the activity associated with the parent tissue. They proposed that full ACC oxidase activity requires tissue integrity in addition to the previously noted cell membrane integrity.

Although ACC oxidase remains to be isolated and characterized, some important information about ACC oxidase has been obtained using plant tissues. The dependence of ethylene production on ACC and oxygen concentrations has been studied in a number of plant tissues. The K_m value for O_2 varied greatly depending on the internal ACC content. When ACC levels in the tissue were low (below its K_m value), the concentration of O_2 giving half-maximal ethylene production rate ($[S]_{0.5}$) ranged between 5 and 7%, and was similar among different tissues. As the concentration of ACC was increased above its K_m value, $[S]_{0.5}$ for O_2 decreased markedly. In contrast, the K_m value for ACC was little dependent on O_2 concentration, but varied greatly among different plant tissues, ranging from 8 uM in apple tissue to 120 uM in etiolated wheat leaves. Such a great variation was thought to be due to different compartmentation of ACC within the cells in various tissues [34]. Since ACC oxidase is a bi-substrate enzyme, the $[S]_{0.5}$ values for O_2 can vary as the ACC concentration changes and vice versa. The above kinetic data agree with the view that ACC oxidase follows an ordered binding mechanism in which ACC oxidase first binds oxygen and then ACC.

Hiyama et al. [7] reported previously that 1-phenylcyclopropylamine is oxidized by NaOCl or lead tetraacetate to ethylene and benzonitrile via the intermediacy of the nitrenium ion:

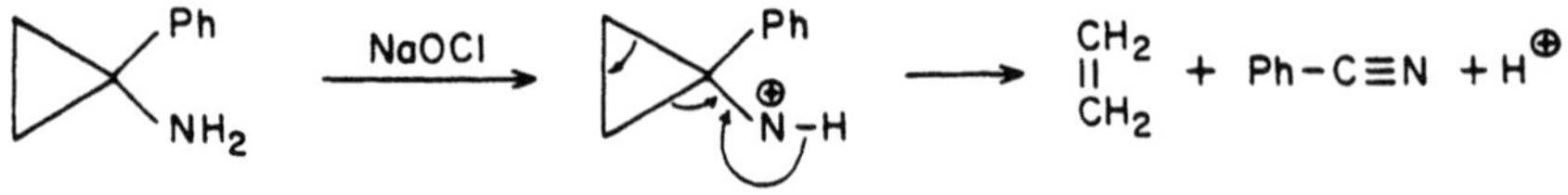

It is reasonable to assume that ACC is similarly oxidized by NaOCl into ethylene and cyanoformic acid, the latter further decomposing spontaneously into HCN and CO_2.

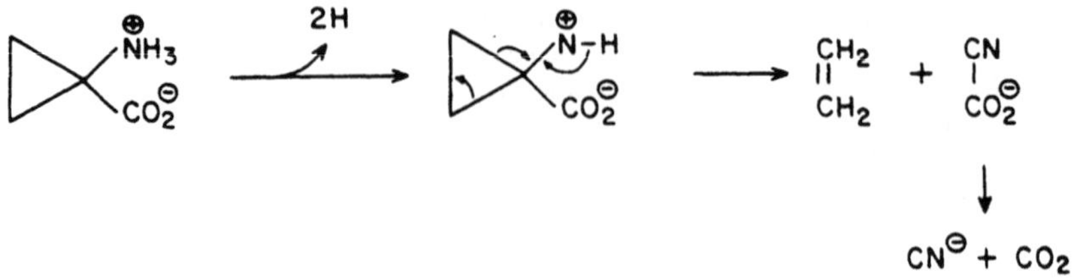

Support for this hypothesis was provided by Peiser et al. [21], who prepared [carboxyl-^{14}C]ACC and [1-^{14}C]ACC, and showed that the carboxyl carbon of ACC is liberated as CO_2, whereas the C-1 of ACC is recovered not as free HCN but as cyanide conjugates, in an amount equivalent to that of the ethylene produced. Their results indicated that ACC is degraded into ethylene, CO_2 and HCN, with HCN being rapidly metabolized to yield *β*-cyanoalanine; *β*-cyanoalanine is then hydrated to asparagine, but in some plants, such as *Vicia sativa*, γ-glutamyl-*β*-cyanoalanine is the main product [21]. This sequence of reactions is illustrated below:

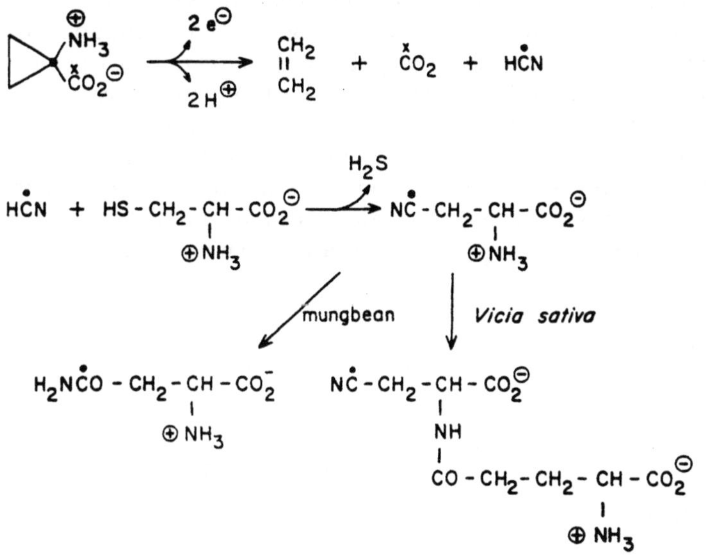

Since no free HCN was detected even in plant tissues which produced ethylene at high rates, Peiser et al. [21] assumed that plants must have ample capacity to metabolize the HCN originating from ACC oxidation. In higher plants the key enzyme to detoxifying HCN is *β*-cyanoalanine synthase (EC 4.4.1.9), which catalyzes the conversion of cysteine and HCN to *β*-cyanoalanine and H_2S. *β*-cyanoalanine synthase is widely distributed in both cyanogenic and non-cyanogenic plants [18]. Since HCN is known to inhibit cytochrome oxidase and triggers cyanide-resistant respiration, it is important to know the steady-state concentration of HCN in plant tissues. By employing an isotope dilution method, Yip and Yang [35] estimated that the steady state concentration of HCN was below 0.2 μM in ripening fruits and in IAA-treated mung-bean hypocotyls, both of which produce ethylene at high rates. The concentration of HCN which results in 50% inhibition of the cytochrome-mediated respiration in plant tissues has been estimated to be 10–20 μM [30]. Based on the early hypothesis of Solomos and Laties [28] that ethylene may induce the cyanide-resistant respiration and that this cyanide-resistant respiration may operate in ripening fruits, Pirrung and Brauman [23] hypothesized that the increased ethylene biosynthesis during fruit ripening may result in an increased HCN level, which in turn inhibits cytochrome oxidase and triggers cyanide-resistant respiration. However, Yip and Yang [35] demonstrated that the HCN level in ripening fruits is too low to cause any significant inhibition of cytochrome oxidase, and support the previous notion of Peiser et al. [21] that plant tissues have ample capacity to detoxify HCN formed during ethylene biosynthesis. Theologis and Laties [30] have established that the respiration in either preclimacteric or climacteric avocado fruits is mediated by the cytochrome respiratory pathway and that there is no involvement of the cyanide-resistant pathway in either tissue. Assuming that a steady-state level of HCN in the plant tissue is maintained between the rate of HCN generation by ACC oxidase and the rate of HCN metabolism by β-cyanoalanine synthase, Yip and Yang [35] calculated that for the plant tissue to maintain HCN levels below the safe level of 1 μM, the tissue β-cyanoalanine synthase activity at saturating HCN concentration should be at least 500 times higher than the tissue ethylene production rate. In ripening fruits [35] and senescing carnation flowers [17] ethylene production increases several hundred-fold, but β-cyanoalanine synthase activity increases only 1- to 2-fold, indicating that the basal level of β-cyanoalanine synthase in these tissues was already very high before the climacteric increase in ethylene production. The relationship between ethylene production rates, β-cyanoalanine synthase activities and tissue HCN levels in ripening apple fruits are shown in Table 1.

Table 1. Changes in ethylene production rate (v_1). extractable β-cyanoalanine synthase activity (V). and tissue cyanide content in ripening apple fruit. Data are from Yip and Yang [35]

Stage	C_2H_4 (v_1)	Cyanoalanine synthase (V)	v_1/V	Tissue [HCN]
	nmol/g-h	nmol/g-h		μM
Unripe	0.01	630	1:63000	<0.1
Ripe	3.3	1650	1: 500	0.2

4 ACC N-Malonyltransferase

Aside from its conversion to ethylene. the other metabolic fate of ACC in plant tissues is its conjugation into MACC, a biologically inactive end product of ACC. Because endogenous levels of ACC can increase during development or in response to stress, it seems logical that the plant would require some means to sequester ACC to prevent overproduction of ethylene. The N-malonylation of ACC serves this purpose by reducing the tissue level of ACC and. consequently. ethylene production.

Studies on the subcellular compartmentation of ACC and MACC revealed that the synthesis of ACC and MACC occurs in the cytoplasm [3, 6]. The MACC synthesized in the cytoplasm is then transported and stored in the vacuole. While MACC in the cytoplasm can be transported outside the cell, vacuolar MACC remains sequestered within the vacuole [6]. Hydrolysis of MACC to ACC was observed in some tissues when a high level of MACC was administered exogenously [8]. However, this process may not play an important role under physiological conditions if MACC is compartmentalized largely in the vacuole and is unavailable for further metabolism.

N-Malonylation of D-amino acids commonly occurs in higher plants. It is thought that the physiological significance of N-malonylation is to inactivate foreign and potentially toxic substances such as D-amino acids or herbicides. Since ACC has no asymmetric carbon, it can be recognized by an enzyme as a D- as well as an L-amino acid. Conceivably, the malonylations of ACC and of D-amino acids can be interrelated. Indeed, various D-amino acids (D-Phe, D-Met, D-Ala) inhibit the malonylation of exogenously administered ACC in mungbean hypocotyl segments, resulting in an increased free ACC concentration and in an increased ethylene production rate; L-enantiomers are, however, ineffective [13]. Reciprocally, ACC or D-phenylalanine greatly inhibits the formation of N-malonyl-D-methionine from exogenously administered D-methionine. These results indicate an intimate relationship between the malonylation of ACC and D-amino acids, and further suggest that both reactions are catalyzed by the same enzyme. Such a conclusion is in agreement with the data obtained with a cell-free system.

ACC N-malonyltransferase, which catalyzes the following reaction, has been isolated and partially purified from mungbean hypocotyls [3, 11, 29].

$$\text{ACC} + \text{Malonyl-CoA} \rightarrow \text{MACC} + \text{CoA}$$

Of particular interest is the intimate relationship between ACC malonyltransferase and D-amino acid malonyltransferase activities. In addition to ACC, AEC, non-polar D-amino acids (D-Met, D-Phe, and D-Ala) and α-aminoisobutyric acid can also be malonylated [11, 29]. Based upon the following observations, D-amino acid malonyltransferase and ACC malonyltransferase are thought to be the same enzyme: (a) the enzyme preparations malonylate both D-amino acids and ACC, (b) the K_m values of those amino acids serving as substrates of malonyltransferase agree with their corresponding K_m values when the same amino acids act as competitive inhibitors of ACC malonyltransferase, and (c) among four stereo-isomers of AEC, those isomers with the D-amino acid configuration are more effective substrates and inhibitors of malonyltransferase than those with the L-configuration (Table 2). The confirmation that a single enzyme carries out both malonylation reactions awaits purification of the enzymes.

Table 2. K_m values of 1-amino-2-ethylcyclopropane-1-carboxylic acid (AEC) isomers serving as substrates of malonyl transferase and K_i values of these isomers serving as competitive inhibitors of the enzyme utilizing ACC as the substrate[a]

AEC isomers	K_m (mM)	K_i (mM)
(1R,2S)	0.2	0.1
(1R,2R)	0.2	0.15
(1S,2R)	1.0	0.8
(1S,2S)	> 10	> 10

[a] It is to be noted that (1R,2S)- and (1R,2R)-AEC, which have an R-configuration at C-1 center as D-amino acids, have lower K_m and K_i values than (1S,2R)- and (1S,2S)-AEC, which have an S-configuration as L-amino acids. Data are from Liu et al. [15].

When excised citrus flavedo tissue [14] tobacco leaf discs [22], or intact green tomato fruit [16] are treated with ethylene, their capability to malonylate ACC is promoted, which is accompanied by an increase in extractable ACC/D-amino acid malonyltransferase activity. In these tissues, therefore, the observed autoinhibition of ethylene production results from an increase in malonyltransferase activity. In these cases ethylene presumably promotes gene expression of ACC malonyltransferase, resulting in reduced level of ACC and thereby reduced ethylene production.

5 Methionine Cycle

While apple fruit maintains a high rate of ethylene production for extended periods, its methionine level is quite low. This observation led to the suggestion that the methionine sulfur released during the conversion of methionine to ethylene must be efficiently recycled to replenish the methionine pool for continued ethylene production. Subsequent work with plant tissues has shown that during ACC synthesis, the SAM sulfur is released as MTA, which undergoes hydrolytic cleavage by MTA nucleosidase to yield MTR; MTR thus formed is efficiently recycled back to form methionine by a salvage pathway in which MTR provides both the methylthio and 2-aminobutyrate portions of methionine [33]. In avocado extracts, Kushad et al. [12] have demonstrated that MTR is converted to 2-keto-4-methylthio-butyrate (KMB), the keto analog of methionine, in the presence of ATP, whereas the conversion of MTR-1-P to KMB is ATP-independent. These in vitro results indicate that MTR is first phosphorylated by MTR kinase to MTR-1-P, which is then metabolized to KMB. Finally KMB is transaminated into methionine by a specific transaminase. Among several potential donors examined, L-glutamine was shown to be the most effective amino donor (19). The sequence of this pathway is shown below:

$$\text{MTA} \rightarrow \text{MTR} \rightarrow \text{MTR-1-P} \rightarrow \text{KMB} \rightarrow \text{methionine.}$$

While MTA nucleosidase and MTR kinase have been purified and characterized, the details of the biochemical conversion of MTR-1-P to KMB remain elusive. Here, the five-carbon ribose moiety of MTR is transformed into the four-carbon 2-ketobutyrate portion of KMB. Thus, one of the five carbons must be released. Recently Miyazaki and Yang [19] have shown that the amounts of labeled HCOOH and methionine derived from methylthio [U-^{14}C] ribose catalyzed by an avocado extract are equivalent on a molar basis, indicating that the conversion involves a loss of formate, presumably from C-1 of MTR. This conversion of MTR to KMB and formate represents a 4-electron oxidation. While a stoichiometric consumption of O_2 and production of formate was demonstrated in rat liver extracts, Miyazaki and Yang [19] were unable to observe such a requirement for molecular oxygen.

The overall result of the methionine cycle is that the 4-carbon moiety of methionine, from which ACC is derived, is ultimately furnished from the ribose moiety of ATP via SAM, while the CH_3S group of methionine is conserved for continued regeneration of methionine. The fate of ATP in relation to ethylene production is shown below, where the numbers refer to the carbon position of the ribose moiety of ATP from which each product is derived.

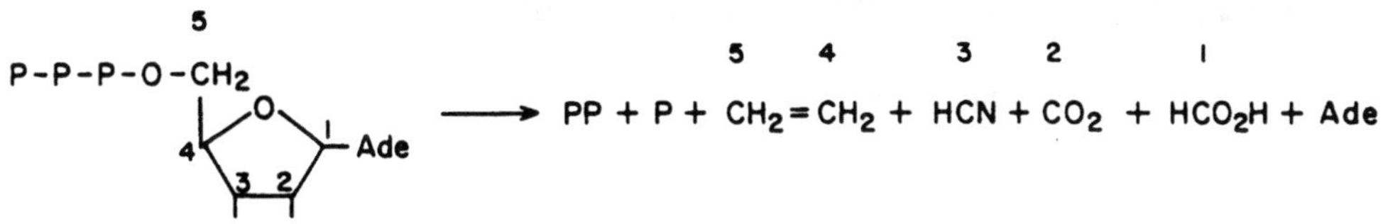

Acknowledgements. Our work cited herein was supported by research grants from the National Science Foundation (PCM-8414971).

References

1. Acaster MA, Kende H (1983) Plant Physiol 72:139
2. Adams DO, Yang SF (1979) Proc Natl Acad Sci USA 76:170
3. Amrhein N, Forreiter C, Kionka C, Skorupka H, Tophof S (1987) In: Schreiber K, Schutte HR, Sembdner G (eds) Conjugated plant hormones. Institute of plant biochemistry. Halle, East Germany, p 102
4. Bleecker AB, Kenyou WH, Somerville SC, Kende H (1986) Proc Natl Acad Sci USA 83:7755
5. Boller T, Herner RC, Kende H (1979) Planta 145:293
6. Bouzayen M, Latche A, Albert G, Pech J-C (1988) Plant Physiol 88:613
7. Hiyama T, Koide H, Nozaki H (1975) Bull Chem Soc Jpn 48:2918
8. Jiao XZ, Philosoph-Hadas S, Su LY, Yang SF (1986) Plant Physiol 81:637
9. Kende H, Bleecker AB, Kenyou WH, Mayne RG (1986) In: Bopp M (ed) Plant growth substances 1985. Springer, Berlin Heidelberg New York Tokyo, p 120
10. Khani-Oskouee S, Jones JP, Woodard RW (1984) Biochem Biophys Res Commun 121:181
11. Kioka C, Amrhein N (1984) Planta 162:226
12. Kushad MM, Richardson DG, Ferro AJ (1983) Plant Physiol 73:257
13. Liu Y, Hoffman NE, Yang SF (1983) Planta 158:437
14. Liu Y, Hoffman NE, Yang SF (1985) Planta 164:565
15. Liu Y, Su LY, Yang SF (1984) Arch Biochem Biophys 235:319
16. Liu Y, Su LY, Yang SF (1985) Plant Physiol 77:891
17. Manning K (1986) Planta 168:61
18. Miller JM, Conn EE (1980) Plant Physiol 65:1199
19. Miyazaki JH, Yang SF (1987) Plant Physiol 84:277
20. Nakajima N, Nakagawa N, Imaseki H (1988) Plant Cell Physiol 29 (in press)
21. Peiser GD, Wang TT, Hoffman NE, Yang SF, Liu HW, Walsh CT (1984) Proc Natl Acad Sci USA 81:3059
22. Philosoph-Hadas S, Meir S, Aharoni N (1985) Physiol Plant 63:431
23. Pirrung MC, Brauman JI (1987) Plant Physiol Biochem 25:55
24. Porter AJ, Borlakoglu JT, John P (1986) J Plant Physiol 125:207
25. Ramalingam K, Lee K, Woodard RW, Bleecker AD, Kende H (1985) Proc Natl Acad Sci USA 82:7820
26. Satoh S, Esashi Y (1986) Plant Cell Physiol 27:285
27. Satoh S, Yang SF (1988) Plant Physiol 88:109
28. Solomos T, Laties GG (1976) Biochem Biophys Res Commun 70:663
29. Su LY, Liu Y, Yang SF (1985) Phytochemistry 24:1141
30. Theologis A, Laties GG (1978) Plant Physiol 62:232
31. Tsai D, Arteca RN, Bachman JM, Philips AT (1988) Arch Biochem Biophys (in press)
32. Wiesendanger R, Martinoni B, Boller T, Arigoni D (1986) J Chem Soc Chem Commun 1986:238
33. Yang SF, Hoffman NE (1984) Annu Rev Plant Physiol 35:155
34. Yip WK, Jiao XZ, Yang SF (1988) Plant Physiol 88:553
35. Yip WK, Yang SF (1988) Plant Physiol 88:473
36. Yoshii H, Imaseki H (1982) Plant Cell Physiol 23:639
37. Yu YB, Adams DO, Yang SF (1979) Arch Biochem Biophys 198:280

Conjugation of Gibberellins in *Zea mays*

G. SCHNEIDER and J. SCHMIDT[1]

1 Introduction

Glucosyl conjugates represent the main group of endogenous GA conjugates in higher plants [19]. In addition, the metabolic formation of GA glucosyl conjugates following feeds of GAs to various plant tissues is a well-established phenomenon. However, the physiological role(s) of GA glucosyl conjugates is (are) still unclear, although certain functions may logically be expected from their structural features, especially when considering the GA-O-glucosides (Fig. 1).

For 2β-hydroxylated biologically inactive GA metabolites [25] including GA_8, GA_{29}, and GA_{34}, the glucosylation leads to a more polar derivative, possibly allowing better transport and compartmentation, such as into the vacuole [7].

For biologically active GAs, such as GA_1, GA_3, GA_4, and GA_{20} the conversion into glucosides results in the loss of activity [22]; their reconversion (deglucosylation) liberates the active aglucones. The reversibility of the glucosylation supports the position that GA-O-glucosides may be involved in GA storage as well as the regulation of the active GA pool.

During the last few years more results have been obtained that support the postulated storage function of GA glucosyl conjugates. Developing seeds of *Prunus armeniaca* [1] and of *Pharbitis nil* [2] were shown to transform $[^3H]GA_5$ into a series of glucosylated GAs. During maturation, caryopses of *Zea mays,* converted the majority of $[^3H]GA_4$ and $[^3H]GA_{20}$ into glucosylated GAs from which free GAs were released during imbibition and early seedling growth [13]. The ratio of free and conjugated GAs during germination was influenced by light [11, 12] and these data represent the first evidence for the involvement of GA conjugation in the homeostatic regulation of GAs.

The significance of GA conjugation as a mechanism for the regulation of GA pools in the plant appears to depend on the taxonomic status of the plant. GA conjugation and 2β-hydroxylation are basically concomitant and competing processes for the deactivation GAs [19, 25]. Thus in *Phaseolus coccineus,* which contains high amounts of GA conjugates, high rates of glucosylation of GA_4 and its metabolites were observed, especially of GA_{34} [27] and of GA_1 [15, 17]. In *Vicia faba* [4], 2β-hydroxylation represents the main route for GA deactivation, although low levels of glucosylating activity toward GA_{20} and GA_{29} is still present [5].

If glucosyl conjugates are involved in the control of GA pools, specific enzymes for both the synthesis and hydrolysis should be present in the plant tissues. The

[1]Institute of Plant Biochemistry, Halle/Saale Academy of Sciences of the German Democratic Republic, Weinberg, 4050 Halle/Saale, GDR

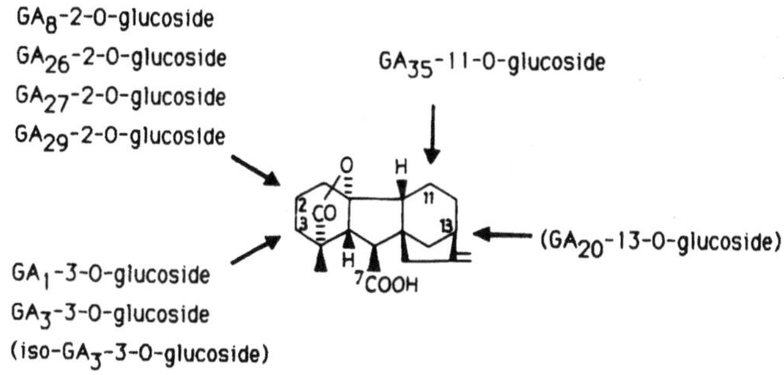

Fig. 1. Orientation of the glucosyl moiety in endogenous gibberellin-O-glucosides

β-glucosidases, which have yet to be isolated, exhibit their highest activity toward 2β-O-glucosides [14, 16]. In contrast, β-glucosyl transferases from pods of *Phaseolus coccineus* favour GA_3 and GA_7 as substrates for 3-O-glucosylation [6, 24].

Improved analytical methods must be developed in order to clarify the physiological role of GA glucosyl conjugates. The majority of investigations to date have relied upon partial identification of GA glucosyl conjugates based on HPLC-RC, bioassay, hydrolysis or other procedures. These are inaccurate methods of quantification and incomplete and equivocal for qualitative identifications.

2 GC-MS of Permethylated GA-O-Glucosides

For studying the role of GA glucosylation in *Zea mays,* we had to find conditions where isomeric 2-O-, 3-O-, and 13-O-glucosides could be separated and identified. This identification is especially important if di- and trihydroxylated GAs like GA_1, GA_8 or GA_{29} are involved. Only GC-MS provides both the required selectivity and sensitivity.

A series of synthetic GA-O-glucosides (GA-O-G) [19, 23] were permethylated according to Rivier et al. [10] and run on a crosslinked methyl silicone capillary column (Table 1). Most of the closely related and isomeric GA-O-G can be separated under these conditions (Table 1). Suitable calibration standards for quantification can be obtained by using deuterated methyl iodide for the derivatization. The average El-spectra of the GA-O-G were measured and compared. The general scheme of their fragmentation is shown in Fig. 2. In terms of distinguishing between ring A/ring C glucosylated GAs, the ions c_1 and c_2 are especially important. The spatial arrangement of the glucosyloxy moiety (axial/equatorial) can be determined from the abundance ratio of the ions *e* and *f*. Thus, on the basis of as few as 8 diagnostic ions (M^+, a,b,c,c_1,c_2,e and f), the location of the glucosyloxy moiety in a given GA-O-G can be determined (Table 2). From the MS data in combination with the KRI, such closely related pairs as GA_1-3-O-G/GA_1-13-O-G (ring A/C isomers), 3-epi-GA_1-3-O-G/GA_{29}-2-O-G (3-O/2-O isomers), and GA_1-3-O-G/3-epi-GA_1-3-O-G (axial/equatorial stereoisomers) can be distinguished [18].

Table 1. Retention (R_t) and Kovats Retention Indices (KRIs) of permethylated gibberellin-O-glucosides (HP 5890/5970B/GC-MS configuration, 25 m × 0.31 mm i.d. crosslinked methyl silicone fused silica column, film thickness 0.17 μm, He 2.5 ml/min, temperature program from 60 °C (1 min) to 260 °C (25 °C/min)

Permeth. Compound	t_R [min]	KRI
GA_1-3-O-glucoside	24.37	3602
GA_1-13-O-glucoside	22.04	3532
3-epi-GA_1-3-O-glucoside	25.36	3625
3-epi-GA_1-13-O-glucoside	24.21	3597
GA_3-3-O-glucoside	21.46	3514
GA_3-13-O-glucoside	21.60	3518
GA_4-3-O-glucoside	19.42	3440
GA_5-13-O-glucoside	18.20	3392
GA_7-3-O-glucoside	18.17	3391
GA_8-2-O-glucoside	26.01	3640
GA_8-13-O-glucoside	24.48	3604
GA_{20}-13-O-glucoside	18.32	3398
GA_{29}-2β-O-glucoside	23.29	3569
GA_{35}-11β-O-glucoside	18.76	3415

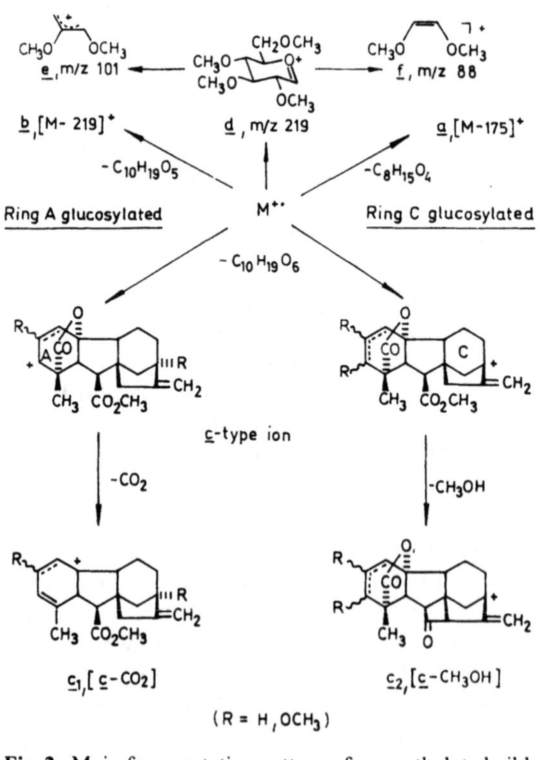

Fig. 2. Main fragmentation pattern of permethylated gibberellin-O-glucosides

Table 2. Relation between structural feature and key ions of El-mass spectra of permethylated gibberellin-O-glucosides

Ion	*Site of glucosylation* Ring A	Ring C
M	+	–
b	+	–
base peak	e or f	c
$c_1 > c_2$	+	
$c_2 > c_1$		+

Stereochemistry of the glucosyloxy moiety equatorial	axial
f > e	e > f
a > 10%	a < 10%

The selected key ions are not only useful for the full scan identification, but can be also utilized to increase the analytical efficiency by SIM. During routine work with $[^2H_2]GA_{20}$-13-O-G the detection limit by full scan was found to be about 10 ng, whereas the SIM mode could detect as little as 0.1 to 1 ng of the GA-O-G.

3 GC-MS Identification of Metabolically Formed GA-O-Glucosides in *Zea mays*

Using GC-MS we were able to confirm the earlier results of the metabolism of GA_{20} in *Zea mays* seedlings [9, 20, 21]. Both GA_{20}-13-O-G and GA_{29}-2-O-G were identified unequivocally. These data demonstrate that *Zea mays* is capable of glucosylating at least the 2-O and 13-O positions of GAs. After feeding GA_{20}-13-O-G to tassels of *Zea mays* we identified GA_1, GA_{29}, GA_{29}-2-O-G and GA_1-13-O-G, in addition to liberated GA_{20}. The identification of GA_1-13-O-G supports the direct transformation of intact GA-O-glucosides. This conclusion was reached because cofeeding excess GA_1 together with $[^3H]GA_{20}$-13-O-G did not influence the amount of $[^3H]GA_1$-13-O-G formed [3]. The proposed GA metabolic pathways, including the glucosylating steps, are summarized in Fig. 3.

We have investigated the role of GA conjugation in the GA deficient dwarf mutants of *Zea mays*. One of these mutants, *d1*, cannot metabolize GA_{20} to GA_1 [26]. When fed to *d1*, GA_{20} was metabolized to a considerably higher amount of GA_{20}-13-O-G than when fed to the wild type (Table 3). The implication of these results is that both conjugation and 2β-hydroxylation [8] are processes that protect the plant from an excess of bioactive dihydroxylated GAs, such as GA_1 and GA_3 [21].

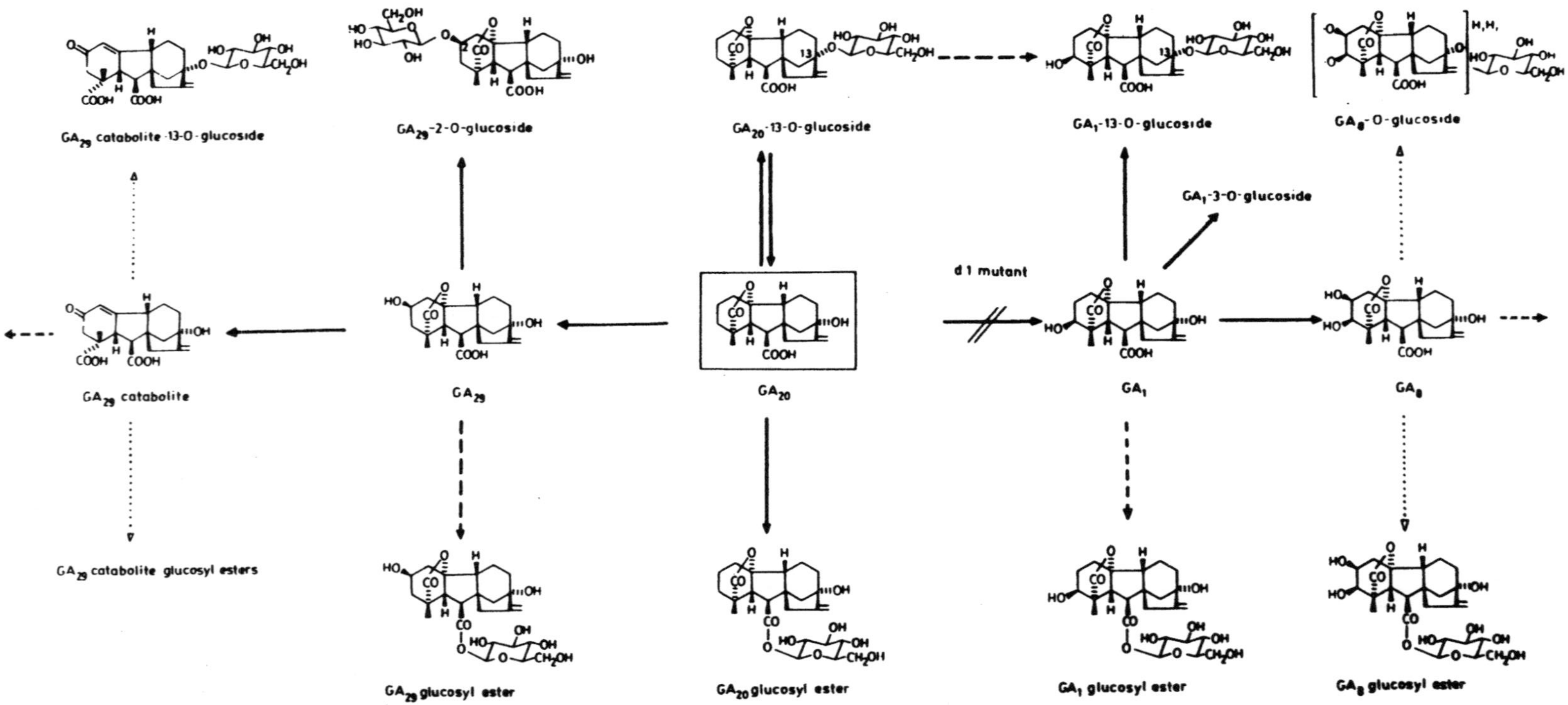

Fig. 3. Proposed scheme for the metabolic conversion of GA_{20} on *Zea mays* (*full lines*: established by GC-MS, HPLC-RC; *broken lines*: some evidence; *dotted lines*: assumption)

Table 3. Quantification of GA_{20}-13-O-glucoside formed from GA_{20} (2 mg) fed to seedlings of normal plants and the *dl* mutant of *Zea mays*[a]

	Recovered Amount (μg) HPLC	Activity (Bq)	Spec. activ. (Bq/μmol)	Total (μg)	Metabolically formed percentage (%)
Normal plants	190	2.67×10^4	6.79×10^4	297	3.6
dl mutant	1300	2.56×10^4	9.53×10^3	2127	25.9

[a] 0.5 μg [^{3}H]GA_{20}-13-O-glucoside added to the extract (4.2×10^4 Bq).

4 Glucosylation of GAs in Seedlings of the *dl* Mutant of *Zea mays*

In order to investigate the specificity of GA glucosylating enzymes in the *dl* mutant, a series of 3- and 13-hydroxylated GAs and 3,13-dihydroxylated GAs were applied to 7-day-old seedlings for 48 h. The experiments were performed with higher than physiologically relevant GA concentrations. The purified and permethylated GA glucoside fractions were subjected to GC-MS for the identification and approximate quantification of GA-O-Gs which were formed from exogenously applied GAs. The 13-hydroxy GA, GA_5 was converted into the corresponding 13-O-glucoside, thus confirming the 13-O-glucosylating capability of the *dl* system. 3-hydroxy GAs like GA_4 were also converted to a small but significant amount of GA_4-3-O-glucoside. The amount of the 3-O-G was greater when GA_7 was fed. Based on these results, feeds of 3,13-dihydroxy GAs should give both the 3-O-G and the 13-O-G. The ratio of the two should reflect the specificity of the glucosylating enzyme(s). In fact, GA_1 was converted into both GA_1-3-O-G and GA_1-13-O-G, but 13-O-glucosylation was the favoured process (GA_1-3-O-G:GA_1-13-O-G = 1:2). In contrast, feeding GA_3 led preferentially to GA_3-3-O-G and only small amounts of GA_3-13-O-G (GA_3-3-O-G:GA_3-13-O-G = 9:1).

5 Results and Conclusions

1. GC-MS of permethylated GA-O-G is an efficient method for their separation and identification.
2. Feeding GA_{20} to seedling of *Zea mays* yielded GA_{20}-13-O-G and GA_{29}-2-O-G (identified by GC-MS).
3. Feeding GA_{20}-13-O-G to tassels of *Zea mays* yielded GA_1-13-O-G and GA_{29}-2-O-G as well as the free GA_{20}, GA_1 and GA_{29} (identified by GC-MS).
4. The glucosylation rate of GA_{20} was higher in seedlings of the *dl* dwarf mutant of *Zea mays* than in the normal (wild type).
5. Seedlings of the *dl* mutant of *Zea mays* transform GA_5 into GA_5-13-O-G as well as GA_4 and GA_7 into the corresponding 3-O-G.
6. Feeding GA_1 and GA_3 to seedlings of the *dl* mutant of *Zea mays* yielded both the 3-O-G and the 13-O-G. For GA_1, 13-O-glucosylation is the major process, but for GA_3, 3-O-glucosylation is the major process.

Acknowledgements. The authors are highly indebted to Prof. O. Junttila and Dr. E. Jensen, Institute of Biology and Geology, University of Tromsö (Norway) for the opportunity to measure the GC-MS data. We thank Dr. R.C. Spray, Dept. Biology, University of California Los Angeles for his help in preparing the manuscript.

References

1. de Bottini G, Bottini R, Koshioka M, Pharis RP, Coombe BG (1987) Plant Physiol 83:137
2. Koshioka M, Pharis RP, King RW, Murofushi N, Durley RC (1985) Phytochemistry 24:663
3. Lattke P (1986) Untersuchungen zum Stoffwechsel von GA_{20} and GA_{20}-13-O-glucoside in *Vicia faba* L. Dissert Thesis, Halle (Saale), GDR
4. Lattke P, Schliemann W, Schneider G (1987) Biochem Physiol Pflanz 182:385
5. Lattke P, Schneider G (1985) J Plant Growth Regul 4:71
6. Lehmann H, Sembdner G (1986) In: Purohit (ed) Hormonal regulation of plant growth and development, vol 3. Bikaner Agro Botanical Publishers, Bikaner, India, p 245
7. O'Neill SD, Keith B, Rappaport L (1986) Plant Physiol 80:812
8. Phinney BO (1989) In: Pharis RP, Rood SB (eds) Plant growth substances 1989, Springer Berlin Heidelberg New York Tokyo, this vol.
9. Phinney BO, Schneider G (1987) In: Schreiber K, Schütte HR, Sembdner G (eds) Conjugated plant hormones-structure, metabolism and function. VEB Deutscher Verlag der Wissenschaften Berlin, GDR, p 167
10. Rivier L, Gaskin P, Albone KS, MacMillan J (1981) Phytochemistry 20:687
11. Rood SB, Beale FD, Pharis RP (1986) Plant Physiol 80:448
12. Rood SB, Pharis RP (1987) In: Schreiber K, Schütte HR, Sembdner G (eds) Conjugated plant hormones-structure, metabolism and function. VEB Deutscher Verlag der Wissenschaften, Berlin, GDR, p 183
13. Rood SB, Pharis RP, Koshioka M (1983) Plant Physiol 73:340
14. Schliemann W (1984) Plant Physiol 116:123
15. Schliemann W (1987) Biochem Physiol Pflanz 182:153
16. Schliemann W (1988) Phytochemistry 27:689
17. Schliemann W, Schneider G (1989) Plant Growth Regul 8:85
18. Schmidt J, Schneider G, Jensen E (1988) Biomed Environ Mass Spectrum 17:7
19. Schneider G (1983) In: Crozier A (ed) The biochemistry and physiology of gibberellins, vol 1. Praeger Scientific, New York, p 389
20. Schneider G (1987) In: Schreiber K, Schütte HR, Sembdner G (eds) Conjugated plant hormones-structure, metabolism and function. VEB Deutscher Verlag der Wissenschaften, Berlin, GDR, p 158
21. Schneider G, Schmidt J, Phinney BO (1987) J Plant Growth Regul 5:217
22. Schneider G, Sembdner G, Phinney BO (1984) J Plant Growth Regul 3:297
23. Schneider G, Sembdner G, Schreiber K, Phinney BO (1989) Tetrahedron 45:1355
24. Sembdner G, Knöfel HD, Schwarzkopf E, Liebisch HW (1985) Biol Plant 27:231
25. Sponsel V (1983) In: Crozier A (ed) The biochemistry and physiology of gibberellins, vol 1. Praeger Scientific, New York, p 151
26. Spray CR, Phinney BO, Gaskin P, Gilmour SJ, MacMillan J (1984) Planta 160:464
27. Turnbull CGN, Crozier A, Schneider G (1986) Phytochemistry 25:1823

Metabolism of Gibberellins A_{20} and A_9 in Plants: Pathways and Enzymology

J. MacMillan[1]

1 Introduction

At the previous (12th) International Conference on Plant Growth Substances we reported on the design of GA-derivatives to probe for GAs and GA-receptors in plants [1]. The application of such derivatives to the study of GA-receptor(s) in aleurone cells is described elsewhere in these Proceedings by Hooley et al. [10]. The preparation of monoclonal antibodies with high epitope specificity for GAs from the GA-protein conjugates, described by Beale et al. [1], has been published by Knox et al. [14, 15].

This report describes unpublished work, including collaborative studies with Prof. B.O. Phinney and his colleagues at UCLA and with Dr. J.B. Reid and colleagues at the University of Tasmania. These studies, which will be documented elsewhere in full, are concerned with the early 13-hydroxylation pathway – the metabolism of GA_{20} and the related enzymology; and the non-13-hydroxylation pathway – the metabolism of GA_9. An attempt is made to draw together the results from which the following conclusions are derived.

1. The mutations in the *d1*, *d2*, *d3* and *d5* mutants of *Zea mays* and the *le* mutant of *Pisum sativum* are "leaky" and the "leakiness" may be the result of altered enzymes, not of lower levels of unaltered enzymes.
2. Stem elongation in *Pisum sativum* correlates with GA_{20} 3β-hydroxylation.
3. In plants, GA_{20} 3β-hydroxylase(s) may catalyze the conversion of GA_{20} to GA_5, as well as the conversion of GA_{20} to GA_1.
4. Gibberellin A_3 is an endogenous GA in shoots of *Zea mays* and is formed from GA_{20} *via* GA_5.
5. Enzyme preparations from seeds of *Marah macrocarpus* and *Malus domestica* convert GA_9 to GA_7 via 2,3-dehydroGA_9.
6. In higher plants the biosynthetic origin of GA_3 and GA_7 is different from that in *Gibberella fujikuroi*.

A composite pathway, summarizing the results presented in this report, is shown for the early 13-hydroxylation pathway from GA_{20} in Fig. 1 and for the non-13-hydroxylation pathway from GA_9 in Fig. 2.

[1] School of Chemistry, The University of Bristol, Cantock's Close, Bristol BS8 1TS, UK

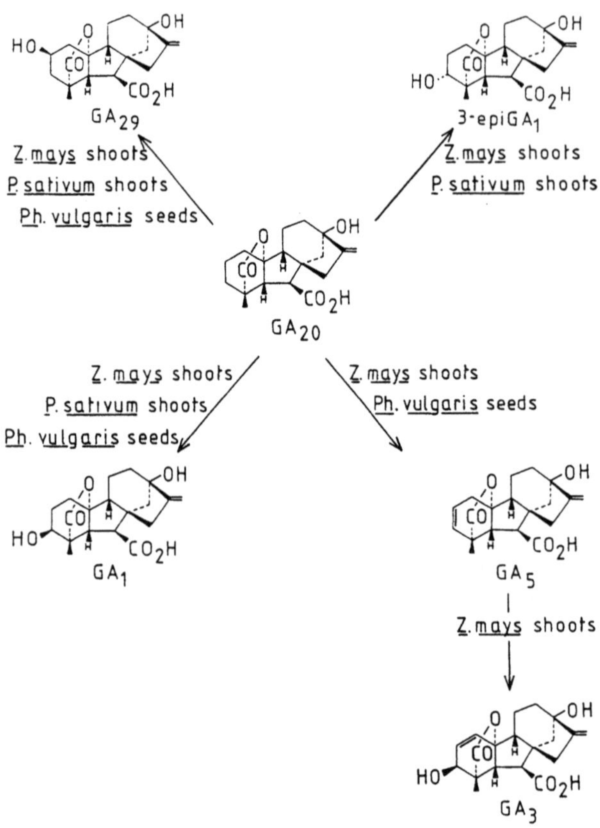

Fig. 1. Composite pathways from GA_{20}

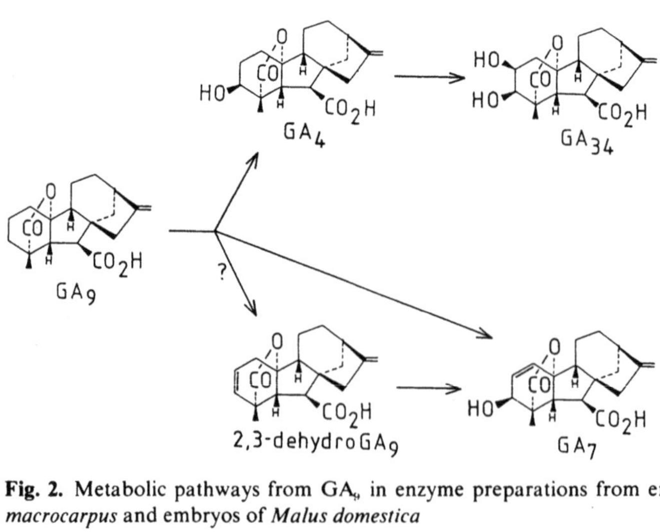

Fig. 2. Metabolic pathways from GA_9 in enzyme preparations from embryo-endosperm of *Marah macrocarpus* and embryos of *Malus domestica*

2 Results

2.1 The Mutant Genes, *d1*, *d2*, *d3* and *d5* of *Zea mays* are "Leaky" [22]

Gibberellins A_{12}, A_{53}, A_{44}, A_{19}, A_{17}, A_{20}, A_{29}, A_1 and A_8 have been identified in extracts of vegetative tissue of normal maize by full scan GC-MS and KRI data. Seven of them (GA_{53}, GA_{44}, GA_{19}, GA_{20}, GA_{29}, GA_1 and GA_8) have been quantified using [^{13}C]-, [^{2}H]- and [^{14}C]-labelled internal standards and GC-SIM, in extracts of normal, *d1*, *d2*, *d3* and *d5* seedlings. The levels of these GAs in *d2*, *d3* and *d5* were less than 10% of those in normal seedlings, consistent with these mutants being blocked for GA-biosynthesis before GA_{53} in an early 13-hydroxylation pathway. In the case of *d1* the levels of GA_1 were less than 2% of those in normal seedlings, whereas the levels of GA_{20} and GA_{29} were more than 10 times those in normals; these results are consistent with the *d1* mutation operating in the conversion of GA_{20} to GA_1. These data show that *d1*, *d2*, *d3* and *d5* are leaky mutants (see also the following Sect. 2.2).

2.2 Internode Extension in *Pisum sativum* Correlates with GA_{20} 3β-hydroxylation [23]

The "leakiness" of the *le* mutation in *Pisum sativum* has previously been demonstrated by Ingram et al. [11]. The authors fed [17-^{13}C, $^{3}H_2$]GA_{20} to the uppermost fully expanded leaves of 15-day-old seedlings of *le*^{d} (short dwarf phenotype), *le* (intermediate) and *Le* (normal). They showed that $\log_{10}$ of the levels of the GA_8 produced from the fed GA_{20} (as determined by HPLC-RC) was linear with respect to the internode extension for each of the three phenotypes. In these experiments the low levels of GA_1 in the dwarf phenotypes could not be determined in the presence of the much higher levels of 3-epiGA_1. Graebe [9] has pointed out that these results could also be interpreted as showing that internode extension is directly related to the metabolism of GA_1 to GA_8. However, this alternative explanation has now been excluded by the following results: [17-^{13}C, $^{3}H_2$]Gibberellin A_{20} was fed to the three genotypes and the three metabolites, [17-^{13}C, $^{3}H_2$)-labelled GA_1, GA_8 and 3-epiGA_1, were separated by HPLC-RC. The [^{13}C] content in each was determined by GC-SIM at the correct R_t before and after dilution with appropriate amounts of unlabelled standards. From these [^{13}C] data, calculations were made of the amounts of unlabelled (endogenous) and labelled (metabolites) GA_1, GA_8 and 3-epiGA_1 present in the tissue above the treated node in each phenotype. The $\log_{10}$ of the endogenous levels of GA_1 (and GA_8) were linearly related to the internode extension in the *le*^{d}; *le* and *Le* seedlings.

Furthermore, although the levels of 3-epiGA_1, both endogenous and [^{13}C]-labelled, decreased in the order *Le, le, le*^{d}, the ratio of 3-epiGA_1 to GA_1 increased in that order. The observed change in product ratio would suggest that the *le* mutation results in an altered enzyme. If GA_1 and 3-epiGA_1 are indeed common products of the GA_{20} 3β-hydroxylase the mutations at the *le* locus may alter the stereospecificity of the enzyme, favouring formation of 3-epiGA_1 over GA_1.

2.3 The GA_{20} 3β-Hydroxylase from Seeds of Phaseolus vulgaris [24, 25]

The GA_{20} 3β-hydroxylase in apical regions of *Pisum sativum* is of particular interest, since expression of the *Le* gene for this enzyme may be controlled by different tissues [13, 20] and by light [5, 8]. Cell-free preparations catalyzing the conversion of GA_{20} to GA_1 have been obtained from shoots of normal seedlings of *Pisum sativum* and characterization of the enzyme is in progress.

More detailed information has, however, been obtained for a partially purified GA_{20} 3β-hydroxylase, obtained from 21-day-old seeds of *Phaseolus vulgaris*. This enzyme (MR 45 kD) is very hydrophobic and acid-labile. Using [1β,2β,3β-$^3H_3$]GA_{20} as substrate and TOH-release as an assay, the enzyme activity is dependent on Fe^{2+}, 2-oxoglutarate and ascorbate and has a Km 0.95 μM and a Vmax 9.55 nmole h^{-1} mg^{-1} with an apparent kinetic isotope effect of 3.5. Using [17-^{13}C,$^3H_2$]GA_{20} as substrate and product analyses by HPLC-RC and GC-MS, the products are GA_1, GA_5 and GA_{29}, formed at the respective rates of 31, 10 and 1.75 pmol h^{-1} mg^{-1}. These rates were unchanged under a range of incubation conditions. The stereochemistry of the reactions has been determined using [2α-$^2H_1$], [2β-$^2H_1$], [3α-$^2H_1$] and [3β-$^2H_1$]-labelled GA_{20} as substrates, and [17-^{13}C, $^3H_2$]GA_{20} as a control to determine the levels of endogenous GAs. Gibberellin A_1 is formed from GA_{20} by loss of the 3β-hydrogen (retention of configuration) and retention of the 2α- and 2β-hydrogens, whereas GA_5 is formed from GA_{20} by loss of the 2β- and 3β-hydrogens. Thus GA_1 is not formed via GA_5. Indeed there is evidence that GA_1 and GA_5 are formed from GA_{20} by the same enzyme. Firstly (as noted above), the rates of formation of the products are unchanged over a range of incubation conditions. Secondly, the ratio of GA_1 to GA_5 formed [2β-$^2H_1$]GA_{20} is increased by a factor of more than 5 over that formed from [17-^{13}C, $^3H_2$]GA_{20}, indicating that the rate of formation of GA_1 is increased as the rate of formation of GA_5 is decreased. The decreased rate of formation of GA_5 is assigned to the kinetic isotope effect on the breaking of the 2β-^2H bond. Thirdly, a monoclonal antibody raised against the partially purified enzyme inhibits the formation of both GA_1 and GA_5 to the same extent and the inhibition is linearly related to the ratio of the enzyme and antibody.

The possibility that the same enzyme catalyzes the conversion of GA_{20} to GA_1 and GA_{20} to GA_5 may also be indicated by the results, as described in the following section.

2.4 Gibberellin A_3 Occurs in Vegetative Shoots of Zea mays and is Formed from GA_{20} via GA_5 [26, 27]

Quantitative studies, (described in Section 2.1), revealed the presence of trace amounts (< 0.1 ng 100 g^{-1} fresh weight) of GA_5 in vegetative shoots of normal (+/*d5*), *d2*, *d3* and *d5* seedlings of maize. In *d1* seedlings in which GA_1 levels are very low, GA_5 has not been detected. Similarly in shoots of homozygous (D8/D8), heterozygous (D8/+) dominant dwarf and normal (+/+), in which the amounts of GA_1 are high, the levels of GA_5 are also higher (5–10 ng 100 g^{-1} f.wt.). Thus, the high level of GA_1 is correlated with the presence of GA_5. Gibberellin A_3 was also identified in extracts of the vegetative shoots of (+/+), (+/D8) and (D8/D8) seedlings by full scan GC-MS and KRI data. In these seedlings [17-^{13}C, $^3H_2$]GA_{20} was metabolized to [^{13}C]-labelled GA_{29}, GA_1, GA_5 and 3-epiGA_1, and [17-^{13}C, $^2H_2$]GA_5 was metabolized to [^{13}C]GA_3, identified by full scan GC-MS and KRI data. Thus GA_3 is biosynthesized in maize seedlings from GA_{20} via GA_5.

These results are consistent with the possibility (Sect. 2.3) that GA_{20} is converted into GA_1 and GA_5 by the same enzyme.

2.5 The Biogenesis of GA_7 in Higher Plants is Parallel to that of GA_3 [28]

The establishment of the biogenetic origin of GA_3 from GA_{20} via GA_5 in vegetative shoots of maize provided a vital clue to the resolution of a long standing problem of the origin of GA_7 in seeds of *Marah macrocarpus* and *Malus domestica*.

Gibberellin A_4 is metabolized to GA_3 via GA_7 in cultures of *Gibberella fujikuroi* with loss of the 1α- and 2α-hydrogens [7, 18]. This is consistent with the fact that 1α- and 2α-hydroxylated GAs predominate in the fungus. However, in higher plants 1β- and 2β-hydroxylated GAs predominate and this author predicted (unpublished) that GA_7 was formed in higher plants from GA_4 with the loss of the 1β- and 2β-hydrogens. Since GA_4 and GA_7 were known to occur in seeds of *Marah* [2] and *Malus* (Hedden and Gaskin, unpublished data), enzyme systems from these seeds have been used in this laboratory to study the stereochemistry of the formation of GA_7 from GA_4. However, over four successive years all attempts to observe the conversion of GA_4 to GA_7 were unsuccessful and the study was discontinued until the discovery that GA_3 was formed from GA_{20} via GA_5 in shoots of maize. It then became evident that the most likely precursors of GA_7 in plants were GA_9 (i.e. 13-deoxyGA_{20}) and 2,3-dehydroGA_9 (i.e. 13-deoxyGA_5). This has been shown to be correct. The main results are summarized in Fig. 2. All products were identified by full scan GC-MS.

In enzyme preparations from *Marah* embryo-endosperm, 2,3-dehydro-[17-$^2H_2$]GA_9 was converted into [17-$^2H_2$]GA_7 in very high yield. [17-$^2H_2$]Gibberellin A_9 was metabolized to [17-$^2H_2$]-labelled GA_4, GA_7 and GA_{34} and [17-^{13}C, $^3H_2$]GA_4 gave [17-^{13}C]GA_{34}. Interestingly, [17-^{13}C, $^3H_2$]GA_{20} was metabolized to [17-^{13}C]-labelled GA_1, GA_{60}, GA_{29} and GA_8 but [17-^{13}C]GA_5 and [17-^{13}C]GA_3 were not detected. However [17-^{13}C, $^3H_2$]GA_5 was almost quantitatively converted to [17-^{13}C]GA_3; traces of 1β-hydroxy-[^{13}C]GA_5 and [^{13}C]GA_6 were also detected. 2,3-Dehydro-[1β,3-$^2H_2$]GA_9 and [1β,3-$^2H_2$] GA_5 were respectively metabolized to [3α-$^2H_1$]GA_7 and [3α-$^2H_1$]GA_3 and to the corresponding [1β,3α-$^2H_2$] epoxides. The loss of label from the 1β-position in these substrates and the loss of label from [2β-$^2H_1$]GA_{20} in the formation of GA_5 from the *Ph. vulgaris* enzyme shows that the overall formation of the 1,2-double bond in GA_3 (and probably GA_7) involves the loss of the 1β- and 2β-hydrogens but not by direct dehydrogenation as originally postulated.

In the enzyme preparation from *Malus* embryos, 2,3-dehydro-[17-$^2H_2$]GA_9 was converted to [17-$^2H_2$]GA_7 and [17-$^2H_2$]GA_9 was metabolized to both [17-^{13}C]GA_4 and [17-^{13}C]GA_7. In contrast to *Marah*, the *Malus* preparation converted

$[17\text{-}^{13}C, {}^{3}H_2]GA_5$ to $[17\text{-}^{13}C]GA_3$ in about 0.1% yield and a trace (0.02%) of 1β-hydroxy-$[17\text{-}^{13}C]GA_5$. No detectable products were observed from $[17\text{-}^{13}C, {}^{3}H_2]GA_{20}$ in the *Malus* system.

2,3-dehydroGA_9 was not detected as a natural product in seed of *Marah macrocarpus* and *Malus domestica* or as a metabolite of GA_9 in the enzyme preparations. Presumably 2,3-dehydroGA_9 is rapidly converted to GA_7.

3 General Comments

As discussed in Section 2, four separate metabolites (GA_{29}, GA_1, GA_5 and 3-epiGA_1) of GA_{20} have been identified (Fig. 1). Thus, GA_{20} may occupy a central position late in the early 13-hydroxylation pathway of GA-biosynthesis. The presence of separate GA_{20} 2β- and 3β-hydroxylases is indicated by studies on purified 2β-hydroxylases from seeds [21] and epicotyls (VA Smith and J MacMillan, unpublished) of *Pisum sativum*. This is supported by genetic and metabolic evidence from the *d*1 mutant of maize and the *le* mutant of pea, in which the conversion of GA_{20} to GA_1 is impaired and the conversion of GA_{20} to GA_{29} is enhanced [Section 2.1 and Ref. 12]. The 3β-hydroxylase(s) which convert GA_{20} to GA_1 may also catalyze the conversion of GA_{20} to both GA_5 and 3-epiGA_1. However, it cannot be assumed that the GA_{20} 3β-hydroxylase(s) are the same in all plant tissues. For example, the partially purified GA_{20} 3β-hydroxylase from cotyledons of *Phaseolus vulgaris* catalyzes the formation of both GA_1 and GA_5 from GA_{20}, but not 3-epiGA_1 from GA_{20}. In *le*^{d}, *le* and *Le* seedlings of *Pisum*, GA_1 and 3-epiGA_1 are formed from GA_{20}, but GA_5 has yet to be detected as a metabolite of GA_{20}. In shoots of *Zea mays*, GA_1 and GA_5 have been shown to be metabolites of GA_{20}.

The identification of GA_3 in vegetative shoots of maize and the establishment of its biogenesis from GA_{20} by $[^{13}C]$-labelling and GC-MS, firmly establishes that GA_3 is an endogenous plant gibberellin. Previously identifications of GA_3 have been uncertain because of the possibility of contamination of plant extracts by the ubiquitous use of GA_3 as a reference compound. Previously the formation of $[^{3}H]GA_3$ from $[^{3}H]GA_5$ has been reported on the basis of GC-RC [6] or GC-SIM [3, 4, 16, 17] in several plant tissues in which either GA_5 and GA_3 were known to occur [6, 16] or only GA_5 had been identified [3, 4, 17].

The occurrence of GA_3 in maize shoots means that GA_3 must be included with GA_1 as a native GA responsible for stem elongation [19], although GA_1 predominates. Both GA_1 and GA_3 (through GA_5) may be formed by the same enzyme and stem elongation may be regulated through the GA_{20} 3β-hydroxylase.

In the non-13-hydroxylation pathway, GA_9 is analogous to GA_{20} in the early 13-hydroxylation pathway (Fig. 1). The pathway from GA_9 to GA_7 in the *Marah* and *Malus* systems, and the pathway from GA_{20} to GA_3 in maize is different from that in the fungus. This difference has phyllogenetic significance.

Acknowledgements. The permission of all colleagues to quote their unpublished results is gratefully acknowledged.

References

1. Beale MH, Hooley R, MacMillan J (1986) In: Bopp M (ed) Plant growth substances 1985. Springer, Berlin Heidelberg New York Tokyo, p 65
2. Beeley LJ, Gaskin P, MacMillan J (1975) Phytochemistry 14:779
3. de Bottini G, Bottini R, Koshioka M, Pharis RP, Coombe DG (1987) Plant Physiol 83:137
4. de Bottini G, Bottini R, Pearce D, Pharis RP, Dann I, Chambers DJ (1987) unpublished Research Results
5. Campell BR, Bonner BA (1986) Plant Physiol 82:909
6. Durley RC, Railton ID, Pharis RP (1973) Phytochemistry 12:1609
7. Evans R, Hanson JR, White AE (1970) J Chem Soc C p 2601
8. Gaskin P, Gilmour SJ, MacMillan J, Sponsel VM (1985) Planta 163:283
9. Graebe JE (1987) Annu Rev Plant Physiol 38:419
10. Hooley R, Beale MH, Smith S, MacMillan J (1989) In: Pharis RP (ed) Plant growth substances 1989. Springer, Berlin Heidelberg New York Tokyo, this vol.
11. Ingram TJ, Reid JB, MacMillan J (1986) Planta (Berl) 168:414
12. Ingram TJ, Reid JB, Murfet IC, Gaskin P, Willis CL, MacMillan J (1984) Planta (Berl) 160:455
13. Ingram TJ, Reid JB, Potts WC, Murfet IC (1983) Physiol Plant 59:607
14. Knox JP, Beale MH, Butcher GW, MacMillan J (1987) Planta (Berl) 170:86
15. Knox JP, Beale MH, Butcher GW, MacMillan J (1988) Plant Physiol 88:959
16. Koshioka M, Jones A, Pharis RP (1988) Agric Biol Chem 52:55
17. Koshioka M, Pharis RP, King RW (1985) Phytochemistry 24:663
18. MacMillan J, Willis CL (1984) J Chem Soc Perkin Trans I p 351
19. Phinney BO (1984) In: Crozier A, Hillman JR (eds) Soc Exp Biol Semin Ser, Cambridge University Press, Cambridge, 23:17
20. Potts WC, Reid JB, Murphet IC (1983) Physiol Plant 55:323
21. Smith VA, MacMillan J (1986) Planta (Berl) 167:9
22. Fujioka S, Yamane H, Gaskin P, MacMillan J, Phinney BO, Takahashi N (1988) Plant Physiol 86:1367
23. Gaskin P, Reid JB, Ross JJ, MacMillan J (1989) Physiol Plant 76:173
24. Albone KS, Gaskin P, MacMillan J, Semenenko FM, Smith ZA (1989) Planta 177:108
25. Albone KS, Gaskin P, MacMillan J, Semenenko FM, Smith ZA (1990) Plant Physiol (in press)
26. Fujioka S, Yamane H, Spray CR, Katsumi M, Phinney BO, Gaskin P, MacMillan J, Takahashi N (1988) Proc Natl Acad Sci USA 85:9031
27. Fujioka S, Yamane H, Spray CR, Katsumi M, Phinney BO, Gaskin P, MacMillan J, Takahashi N (1990) Plant Physiol (in press)
28. Albone KS, Gaskin P, Willis CL (1990) Plant Physiol (in press)

The Dioxygenases in Gibberellin Biosynthesis after Gibberellin A_{12}-Aldehyde

J.E. GRAEBE and T. LANGE[1]

1 Introduction

The pathways and cofactor requirements for the conversion of GA_{12}-aldehyde to C_{20}- and C_{19}-GAs in cell-free systems from different objects are now well known and the purification of the enzymes contained in these systems has begun. In this part of the pathway, the C-20 carbon atom is oxidized step by step and finally lost as CO_2 to yield the γ-lactone typical for C_{19}-GAs. The GA molecule may further become modified by hydroxylations, which profoundly influence its physiological activity. In general, 3β-hydroxylation increases the activity strongly – it may even be essential for activity – whereas 2β-hydroxylation inactivates the GA. The relative rates of C_{19}-GA biosynthesis, 3β-hydroxylation and 2β-hydroxylation determine the amounts of physiologically active GA available to the plant, which explains the interest in the corresponding enzymes and their regulation. The GA biosynthesis pathways have recently been reviewed [4, 5, 9].

2 General Properties of the Enzymes

The enzymes catalyzing the steps after GA_{12}-aldehyde are 2-oxoglutarate-dependent dioxygenases, a type of enzymes known particularly well from the work with prolyl 4-hydroxylase in animal systems. In the case of GA biosynthesis, the participation of this class of enzymes was first shown for the entire pathways in cell-free systems from *Cucurbita maxima* [10] and *Pisum sativum* [12] and for 2β-hydroxylation in cell-free systems from peas and beans [11]. Their properties and distribution in different organisms were surveyed by MacMillan [15].

The 2-oxoglutarate enzymes specifically require 2-oxoglutarate, molecular oxygen, ferrous iron and ascorbic acid for activity. They are stimulated or stabilized by BSA and catalase. DTT or its isomer DTE must be present during purification. In the overall reaction, one oxygen atom of dioxygen oxidizes 2-oxoglutarate to succinate and CO_2 while the other hydroxylates the substrate (Fig. 1). A detailed reaction mechanism and stereochemical concept has been proposed [8] for mammalian prolyl 4-hydroxylase, according to which the reactive center is located in a pocket of the catalytic structure in which the ferrous ion is firmly bound. The formation of an iron-oxygen complex within this pocket and subsequent decarboxylation of 2-oxoglutarate leads to the formation of an even more reactive ferryl

[1]Pflanzenphysiologisches Institut der Universität, Untere Karspüle 2, 3400 Göttingen, FRG

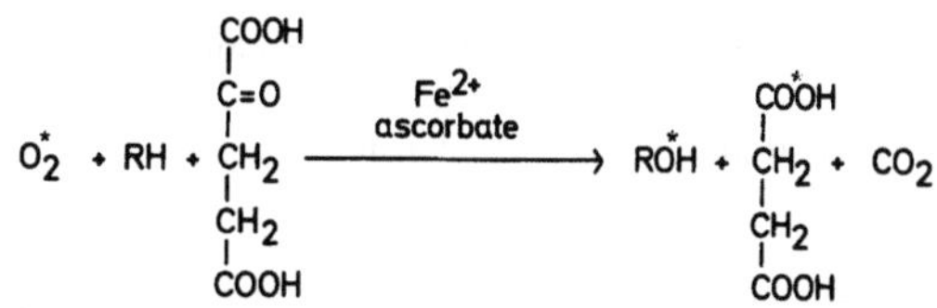

Fig. 1. General reaction of 2-oxoglutarate-dependent enzymes

ion, which in turn hydroxylates the substrate. The iron remains divalent and bound to the active site after the normal reaction cycle, whereas succinic acid, CO_2 and the hydroxylated product are released.

Along with the normal reaction cycle, a certain amount of uncoupled decarboxylation occurs. In this variant, 2-oxoglutarate becomes decarboxylated and a ferryl ion is generated as described before, but the substrate does not become hydroxylated. Instead, the ferryl ion decomposes to Fe^{3+} and an hydroxyl radical, whereby Fe^{3+} remains bound to the active site making the enzyme unavailable for new catalytic cycles. In this situation, ascorbic acid specifically reduces the enzyme-bound Fe^{3+} to Fe^{2+} again. Ascorbic acid is consumed stoichiometrically to the amount of uncoupled decarboxylation, which fully explains this cofactor requirement [1, 17]. The ascorbic acid binding site in prolyl hydroxylase is partially identical with the binding site of 2-oxoglutarate [16] and the oxidation state of the catalytically active iron ion is controlled directly at the active centre. The bound ferric ion does not equilibrate with ferrous iron present in the reaction mixture [7]. The rate of the uncoupled reaction for prolyl hydroxylase is only about 1–4% of that of the complete reaction in the presence of saturating substrate concentrations [17], but at low substrate concentrations it becomes significant and rapidly inactivates the enzyme, unless sufficient ascorbic acid is present in the incubation mixture [19]. If ascorbate is omitted from the incubation mixture, prolyl hydroxylase catalyzes the full hydroxylation at a high initial rate but becomes inactivated after 30–45 catalytic cycles [see Ref. 1]. The hydroxy radicals generated in uncoupled decarboxylation probably inactivate the enzyme by oxidation, which might be prevented to some degree by DTT [19].

As for the purification of the GA biosynthesis enzymes, Table 1 shows the results known to us at the time of writing. Although comparison is difficult because

Table 1. Properties of purified GA biosynthesis enzymes

Plant material	Reaction	Molecular weight	Spec. act.[a] (nmol h^{-1} mg^{-1})	Recovery (%)	Purification factor (fold)	Reference
Spinach leaves	$GA_{53} \rightarrow GA_{44}$	42 500	0.0235			[3]
Spinach leaves	$GA_{44} \rightarrow GA_{19}$	39 500	0.152	0.3	99	[3]
Spinach leaves	$GA_{19} \rightarrow GA_{20}$	38 100	0.0335			[3]
Bean cotyledons	$GA_{1} \rightarrow GA_{8}$	36 000	6.81	4.2	532	[21]
Pea cotyledons	$GA_{1} \rightarrow GA_{8}$	44 000	6.95[b]	37	8.2	[22]
Pea cotyledons	$GA_{20} \rightarrow GA_{29}$		24.80[b]			[22]
Pea cotyledons	$GA_{9} \rightarrow GA_{51}$		22.90[b]			[22]
Pea embryos	$GA_{53} \rightarrow GA_{44}$	44 000	120	15	270	[14]
Pumpkin endosperm	$GA_{12} \rightarrow GA_{15}$	48 000	20 000	15	13	unpublished

[a] At saturating substrate concentration ("V_{max}").
[b] Corrected values (MacMillan, pers. comm. 1988).

Table 2. Changes in the Michaelis constants during purification

Plant material	Purification stage	K_m (μM)	Reference
Pea embryos	Crude extract	2.10	[14]
	Highly purified	0.70	
Pumpkin endosperm	Crude extract	0.55	unpublished
	Highly purified	0.20	

the data have been gained by very different methods, it is clear that the specific activities, recovery rates and purification factors generally are very low. The molecular weights of the different enzymes lie within a narrow range. The available K_m-values vary widely and are not shown here, except for the values in Table 2, which illustrate that the K_m-values may change during purification. This possibility may not have been considered when some of the specific activities shown in Table 1 were calculated. In this case, the values for recovery and degree of purification will be incorrect.

3 Factors Affecting Activity

One very important factor for the activity of 2-oxoglutarate enzymes is the availability of Fe^{2+}. The addition of Fe^{2+} to the reaction mixture is necessary for full activity, but the enzymic activity in the absence of added Fe^{2+} varies considerably for the preparations reported in the literature (Table 3). This was found for prolyl hydroxylase also and attributed to the release of the metal from the active site to different degrees during purification [18]. Another reason for the Fe^{2+} requirement may be the presence of chelating agents, such as phosphate buffer, in the assay mixture. The effect of chelating agents will be discussed first.

Chelating agents do not remove the Fe-atom from purified prolyl hydroxylase, but they can inhibit the enzyme activity by chelating the metal in situ. Thus, considerable activity was obtained with a highly purified preparation in the absence of externally added Fe^{2+} even after dialysis against EDTA and more dialysis to remove EDTA. However, if EDTA was present during the activity measurements, 98–100% of the activity was lost [18]. Chelators capable of inducing a low-spin state at the Fe-centre, such as 2,2′-dipyridyl, are efficient inhibitors of prolyl hydroxylase activity, but only if they are in excess of the molar ratio of complex formation with Fe^{2+} in the external solution [7]. Since each mol of Fe^{2+} can bind 3 moles of 2,2′-dipyridyl the enzyme activity only becomes inhibited if the chelator is more than three times as concentrated as Fe^{2+} in the assay mixture. No titration with 2,2′-bipyridyl has been published for the GA-biosynthesis enzymes, but the reverse case is documented. Thus, Smith and MacMillan [21] inhibited GA 2β-hydroxylase from beans with 50 μM 2,2′-bipyridyl and titrated back the activity with Fe^{2+}. In the presence of catalase, which appears to preserve reduced iron in an unknown way, the activity began to appear, when the concentration of added Fe^{2+} reached 20 mM,

Table 3. Dependency of purified enzymes on added Fe^{2+}

Plant material	Reaction	Residual activity without added Fe (%)	Optimal concentration of Fe^{2+} (mM)	Reference
Bean cotyledons	$GA_1 \rightarrow GA_8$	0	0.1–1.0	[21]
Spinach leaves	$GA_{53} \rightarrow GA_{44}$	51		[3]
Spinach leaves	$GA_{44} \rightarrow GA_{19}$	36		[3]
Spinach leaves	$GA_{19} \rightarrow GA_{20}$	73		[3]
Bean embryos	$GA_{20} \rightarrow GA_1$	12	0.2	[13]
Pea embryos	$GA_{53} \rightarrow GA_{44}$	63	0.008	[14]
Pumpkin endosperm	$GA_{12} \rightarrow GA_{15}$	28	0.008	unpublished

i.e. about one third of the chelator concentration. Ligands not capable of inducing a low-spin state at the Fe-centre, such as EDTA and phosphate, must be present in larger excess over Fe^{2+} in the reaction mixture to inhibit the enzyme activity. Of the preparations shown in Table 3, the preparation in reference [13] was measured in the presence of phosphate buffer; it also has low residual activity without added Fe^{2+} and a relatively high requirement of Fe^{2+} for optimal activity.

How can the Fe-atom be lost to different degrees during purification if it is not even lost by dialysis against EDTA? Perhaps by ion exchange. Metal ions are known to inhibit prolyl hydroxylase [8] as well as the GA biosynthesis dioxygenases [10, 12, 13, 14, 22]. This inhibition is competitive [8, 10], which suggests that other metals can replace Fe at the active centre. For example, it has long been known that Mn^{2+} inhibits the GA biosynthesis enzymes after GA_{12}-aldehyde [6], and that this inhibition can be reversed by Fe^{2+} [10]. Interesting is also the absolute requirement for Fe^{2+} found in the 2β-hydroxylase from bean cotyledons (Table 3). This preparation had been made with 10 mM $MgCl_2$ in the homogenization mixture [21], which might have replaced Fe at the active centre thus making the enzyme dependent on added iron. Seemingly in contrast to this view, neither Mg^{2+} nor Mn^{2+} (both 2.5 mM) was inhibitory to pea seed 2β-hydroxylase in other experiments [22], but in that case the assay was done in the presence of 1 mM Fe^{2+}. Zn^{2+} is perhaps the most potent inhibitor of the GA enzymes, 0.5 mM abolishing all activity, even in the presence of 1mM Fe^{2+} [22]. Zn^{2+} is also a potent inhibitor of prolyl 4-hydroxylase, but only if it is added before 2-oxoglutarate [20], which latter according to the model of Hanauske-Abel and Günzler [8] chelates the Fe-atom at the reactive centre of the enzyme. In the light of these examples it seems reasonable to assume that Fe^{2+} can be exchanged for other metal ions during the purification procedure and that enough Fe^{2+} has to be supplied in the external solution to ensure that it gets exchanged back. In any case it is important to determine the amount of Fe^{2+} needed for optimal activity, since it is often very low and over-optimal amounts may be very inhibitory.

DTT (or DTE) is not required during incubation even of the most purified enzyme preparations [14, 22], but it is absolutely essential for activity that it is present during all purification steps (all references in Table 1, except [21]). Figure 2 shows that the concentration of DTT during purification (here simulated by preincubation) is critical, over-optimal concentrations being as detrimental as sub-optimal concentrations to activity. Ascorbate cannot replace DTT in pre-

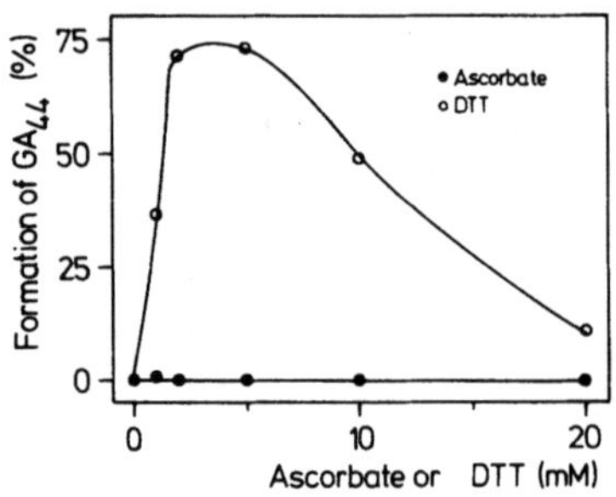

Fig. 2. Protective action of DTT and lack of effect of ascorbic acid during 20 h preincubation at 4°C of GA_{53} C-20 hydroxylase from immature peas. Both cofactors were also present during the subsequent assay [14]

Table 4. Inhibition of GA_{53} C-20 oxidase activity by the simultaneous presence of Fe^{2+} (0.5 mM) and DTT (4 mM) in the incubation mixture [14]

Addition	Activity ($nmol\ h^{-1}\ mg^{-1}$)
None	48.0
$Fe(SO_4)_2$	39.2
DTT	41.6
DTT + $Fe(SO_4)_2$	6.5

serving the activity (Fig. 2). As essential as DTT may be during purification, there are at least two good reasons for leaving it out of the incubation mixture. Firstly, the combination of DTT and Fe^{2+} in the incubation mixture may inhibit the dioxygenases considerably (Table 4). To judge from values in the literature, this inhibition does not occur when DTT has been added during enzyme preparation and Fe^{2+} is added to the assay mixture. In these cases, most of the DTT may have been oxidized before the Fe^{2+} is added. Another reason to be cautious with DTT is that it may cause the formation of artifacts in the form of non-enzymatic products, which chromatograph like true GA products. This occurs when products are extracted from incubation mixtures containing DTT and very low protein concentrations as is often the case when highly purified enzyme preparations are assayed. The artifacts, which are also formed in the complete absence of enzyme (Fig. 3), may be formed in amounts of up to 40–80% of the added substrate at 1–2 mM DTT and may be mistaken for genuine GA products, if the identification is by chromatography only. The addition of small amounts (1 mg ml^{-1}) of ampholine or, presumably, BSA prevents their formation even at high (at least 16 mM) DTT concentrations.

A major loss of activity is due to the instability of many enzymes in very dilute solution. Figure 4 shows the activity of a purified enzyme preparation which is either diluted directly with a solution of BSA in buffer or first diluted with pure buffer. Glycerol can replace BSA to a certain degree, but high concentrations are needed. Full activity is only reached with BSA (Fig. 4).

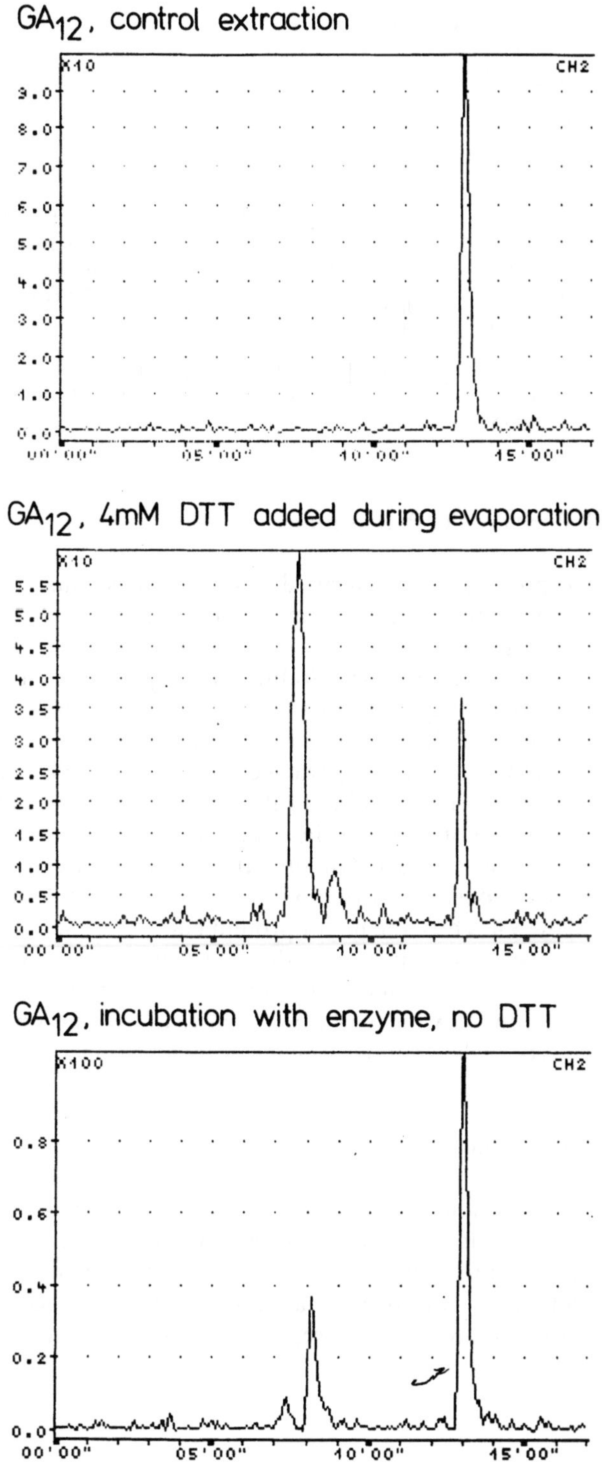

Fig. 3. Formation of artifacts in the presence of DTT in the absence of protein. *Top* (control): $[^{14}C]GA_{12}$ was subjected to the extraction procedure and radiochromatographed (HPLC); *Middle* (artifact formation): Same as before, but 4 mM DTT was added during the evaporation of the solvent; *Bottom* (enzymatic reaction without artifact formation): $[^{14}C]GA_{12}$ was incubated for 5 min at 30°C with pumpkin GA_{12} C-20 hydroxylase (450 ng ml^{-1} protein), ascorbate (4 mM), FeSO4 (8 μM), BSA (2 mg ml^{-1}) and catalase (0.1 mg ml^{-1}) but not DTT

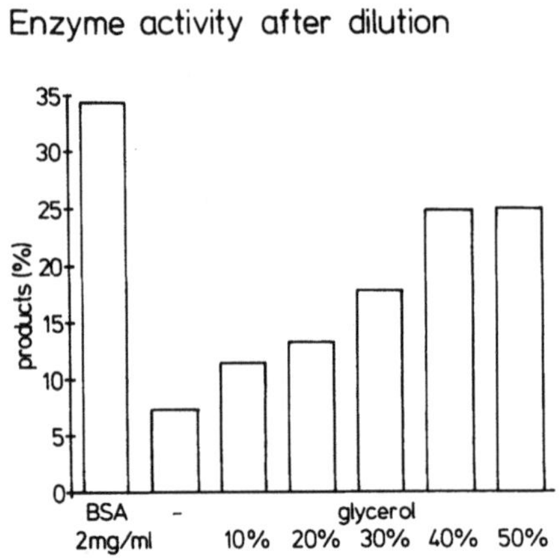

Fig. 4. Effect of dilution on enzyme activity. Samples of purified GA_{12} C-20 hydroxylase (0.45 mg protein ml^{-1}) were 100-fold diluted with Tris-HCl (0.1 M, pH 7) containing BSA, no addition, or glycerol as indicated. After 15 min on ice, the samples were diluted another 10-fold with an incubation mixture as in Fig. 3, adjusting all samples to 2 mg ml^{-1} BSA and 5% glycerol. Incubation for 5 min at 30°C, extraction and HPLC followed

4 Purity of the Preparations

The preparations of purified GA biosynthesis enzymes to date have been obtained by ammonium sulfate or methanol precipitation followed by gel filtration, anion exchange and, in two cases [3, 13], hydrophobic interaction chromatography in different combinations. Sodium dodecyl sulfate electrophoresis has been used to check the purity of the final preparations and sometimes it has been suggested that one or two main protein bands in the proper molecular weight range correspond to the enzyme. In our experience, this is not so. The purified preparation from pea cotyledons yield a sodium sulfate PAGE pattern containing one major band (and several minor ones) in the range of 40 000–50 000 Daltons, in which one would expect the enzyme. On preparative isoelectric focusing, however, the enzyme activity appeared at pI 5.6–5.9, whereas the bulk of the protein appeared at a pI below 5.4. Thus, the major protein band does not represent the enzyme. This is not surprising, since pea cotyledons contain a multitude of proteins, of which the GA biosynthesis enzymes certainly are a very minor part. One therefore would not expect the GA biosynthesis enzymes to become singled out by a few purification steps of such general character as gel filtration and ion exchange chromatography.

5 Conclusions

In conclusion, the enzymes of GA biosynthesis after GA_{12}-aldehyde are unstable and difficult to purify. The aspects discussed in this short essay may help to preserve activity during purification by classical methods. Since these methods lead to a limited purity only, we may have to await the success of affinity chromatography before highly purified preparations can be expected.

Abbreviations

BSA	bovine serum albumin
DTE	dithioerythritol
DTT	dithiothreitol
EDTA	ethylenediaminetetraacetic acid
GA(s)	gibberellin(s)
GA_n	gibberellin A_n
HPLC	high performance liquid chromatography

Acknowledgements. The new data presented in this paper was obtained with support of the Deutsche Forschungsgemeinschaft under grant No. Gr 331/18-3.

References

1. De Jong L, Kemp A (1984) Biochim Biophys Acta 787:105
2. Gilmour SJ, Zeevaart JAD, Schwenen L, Graebe JE (1986) Plant Physiol 82:190
3. Gilmour SJ, Bleecker AB, Zeevaart JAD (1987) Plant Physiol 85:87
4. Graebe JE (1985) In: Bopp M (ed) Plant growth substances 1985. Springer, Berlin Heidelberg New York Tokyo, p 74
5. Graebe JE (1987) Annu Rev Plant Physiol 38:419
6. Graebe JE, Hedden P, Gaskin P, MacMillan J (1974) Phytochemistry 13:1433
7. Günzler V, Majamaa K, Hanauske-Abel HM, Kivirikko KI (1986) Biochim Biophys Acta 873:38
8. Hanauske-Abel HM, Günzler V (1982) J Theor Biol 94:421
9. Hedden P (1983) In: Crozier A (ed) The biochemistry and physiology of Gibberellins, vol 1. Praeger, New York, p 99
10. Hedden P, Graebe JE (1982) J Plant Growth Regul 1:105
11. Hoad GV, MacMillan J, Smith VA, Sponsel VM, Taylor DA (1982) In: Wareing PF (ed) Plant growth substances 1982. Academic Press, London, p 91
12. Kamiya Y, Graebe JE (1983) Phytochemistry 22:681
13. Kwak SS, Kamiya Y, Sakurai A, Takahashi N, Graebe JE (1988) Plant Cell Physiol 29:235
14. Lange T (1986) Diplom-Thesis, Göttingen University, FRG
15. MacMillan J (1984) In: Menhenett R, Lawrence DK (eds) Biochemical aspects of synthetic and naturally occurring plant growth regulators. British Plant Growth Regulator Group, Wantage, p 13 (Monograph No 11)
16. Majamaa K, Günzler V, Hanauske-Abel HM, Myllylä R, Kivirikko KI (1986) J Biol Chem 261:7819
17. Myllylä R, Majamaa K, Günzler V, Hanauske-Abel HM, Kivirikko KI (1984) J Biol Chem 259:5403
18. Nietfeld JJ, Kemp A (1980) Biochim Biophys Acta 613:349
19. Nietfeld JJ, Kemp A (1981) Biochim Biophys Acta 657:159
20. Nietfeld JJ, De Jong L, Kemp A (1982) Biochim Biophys Acta 704:321
21. Smith VA, MacMillan J (1984) J Plant Growth Regul 2:251
22. Smith VA, MacMillan J (1986) Planta (Berl) 167:9

The Action of Plant Growth Retardants at the Biochemical Level

P. HEDDEN[1]

1 Introduction

Xenobiotics which reduce plant growth without affecting the pattern of development have been available for almost 40 years. A group of nicotinium growth retardants was described by Mitchell et al. in 1949 [36], followed shortly by the quaternary ammonium carbamates, including AMO-1618 [47]. Partial reversal of the dwarfing effect of these compounds by GA_3 [32] and the demonstration that the retardants reduced GA production by the fungus *Fusarium moniliforme* [30] implicated GA biosynthesis as a site of action. This was confirmed by Dennis et al. [10], who showed that AMO-1618 and other retardants inhibited the cyclization of geranylgeranyl pyrophosphate (GGPP) to *ent*-kaurene in cell-free homogenates from *Marah macrocarpus* endosperm. A large number of quaternary ammonium, phosphonium and sulphonium retardants have been described [11], although only chlormequat chloride (CCC) and mepiquat chloride have found large-scale application in agriculture, due predominantly to their use on wheat and cotton, respectively. Ironically, the precise mode of action of CCC is in considerable doubt [11].

Within the last 20 years, the discovery of a second group of retardants with much higher potency than the *ent*-kaurene synthetase inhibitors has stimulated renewed interest in the agronomic potential of growth retardants. The first of this class was ancymidol, a pyrimidine, and this has been followed by a large number of similar compounds, many of which are structurally related to fungicides. In 1976, Coolbaugh and Hamilton [8] demonstrated that ancymidol inhibited *ent*-kaurene oxidase and this step has been shown subsequently to be the common site of action for all retardants of this type. The extreme potency of many *ent*-kaurene oxidase inhibitors (KOIs) would appear to enhance their value as agrochemicals; they need be used only very sparingly and this has economic and environmental advantages. However, the specificity of many of these compounds is not well understood and side effects (both desirable and undesirable) may be expected. A complete understanding of their biochemical effects is of prime importance and should aid in the design of retardants with specificity, potency and persistency tailored to particular needs.

[1] Department of Agricultural Sciences, University of Bristol, AFRC Institute of Arable Crops Research, Long Ashton Research Station, Long Ashton, Bristol, BS18 9AF, UK

2 Inhibitors of *ent*-Kaurene Synthetase

Some examples of this group of inhibitors are shown in Fig. 1. They are characterized by the presence of a permanent positive charge or a tertiary N-atom which would be positively charged at physiological pH. AMO-1618, Phosphon D and several others were shown to inhibit activity A, but not activity B, of *ent*-kaurene synthetase from *F. moniliforme* [15] and *M. macrocarpus* [17].

Activities A and B of *ent*-kaurene synthetase from *M. macrocarpus* endosperm are separable enzymes, which appear to form an enzyme complex in vivo [13]. GGPP is converted by a proton-initiated cyclization to a bicyclic intermediate, copalyl pyrophosphate (activity A), which loses pyrophosphate and undergoes a Wagner-Meerwein rearrangement to *ent*-kaurene (activity B). The cyclization of GGPP (Fig. 2) is analogous to the conversion of 2,3-oxidosqualene to lanosterol or cycloartenol in the sterol biosynthetic pathway and indeed the cell-free system from *M. macrocarpus* converts (R,S)-14,15-oxidogeranylgeranyl pyrophosphate to a mixture of 3α- and 3β-hydroxykaurene [6]. Tertiary amine analogs of squalene,

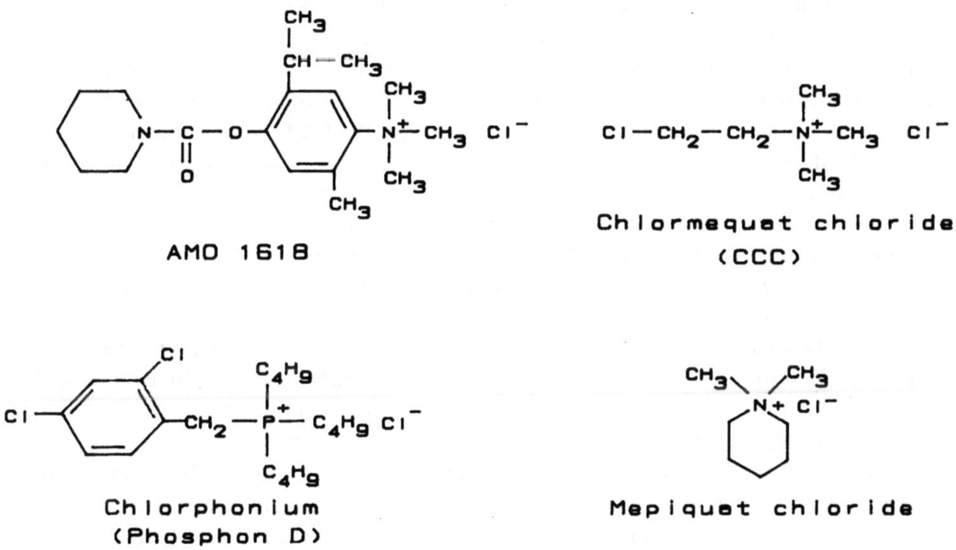

Fig. 1. Structures of some inhibitors of activity A of *ent*-kaurene synthetase

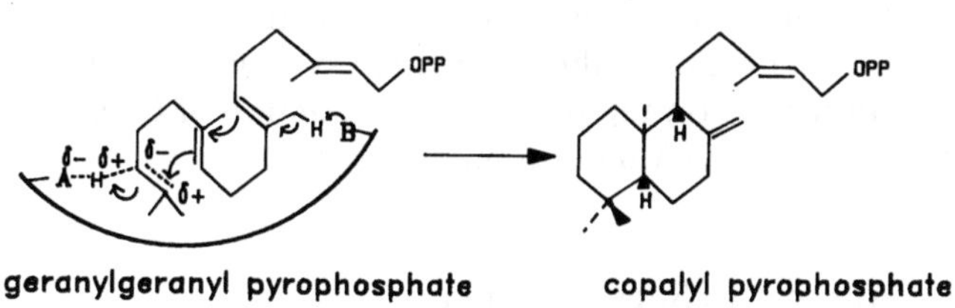

Fig. 2. The proton-initiated cyclization of GGPP to copalyl pyrophosphate involving transient carbocationic intermediates

such as 2-aza-2,3-dihydrosqualene, are efficient inhibitors of oxidosqualene cyclase [14]. It is suggested that these compounds, which would be positively charged at physiological pH, may mimic carbocationic high-energy intermediates in the cyclization. Such intermediates are expected to bind more tightly to the enzyme than the substrate and consequently some of the inhibitors have Ki values several orders of magnitude lower than the Km of oxidosqualene [14]. By analogy, the *ent*-kaurene synthetase inhibitors mimic cationic intermediates in the cyclization of GGPP to copalyl pyrophosphate. Although the retardants are less potent than the oxidosqualene cyclase inhibitors, it may be possible to design much more effective compounds based on the structure of the charged intermediate.

2.1 The Case for CCC

Although CCC has by far the largest usage of the growth retardants there is considerable doubt as to its precise mode of action. There are conflicting reports on the effect of CCC on GA concentrations in tissues and it did not inhibit *ent*-kaurene synthetase activity significantly in vitro at concentrations of 10^{-4} M and below [15, 17] although this enzyme is the expected site of action. However, there are few reports in which GA concentrations in CCC-treated and control plants were compared using methods other than bioassays. In a recent study the dose-response curve for growth inhibition in wheat seedlings by CCC was determined [33]. Growth inhibition occurred at a CCC concentration of 2.5 mM and this inhibition was completely reversed by simultaneous application of 25 μM GA_3 (Temple-Smith and Lenton, unpublished data). However, at higher concentrations of CCC growth inhibition was not completely reversible by GA_3 and additional sites of action must be inferred. The levels of GA_1 in shoots and grains of 6-day-old seedlings of a GA-insensitive dwarf (*Rht3*) wheat were reduced to 3% and 10% respectively, of those in untreated seedlings by 3 mM CCC treatment [33]. Since 3mM CCC had no effect on leaf length on this mutant, the reduction in GA_1 concentration was a direct result of the CCC treatment. These results indicate that in wheat seedlings CCC does act as an inhibitor of GA biosynthesis.

We have recently re-examined the effect of CCC and AMO-1618 on the incorporation of (R)-[2-^{14}C]mevalonic acid into *ent*-kaurene in a cell-free system from *Cucurbita maxima* endosperm. The dose-inhibition curves for the two retardants (Fig. 3) confirm that AMO-1618 with an I_{50} of about 5×10^{-8} M (inhibitor concentration at which the reaction rate is reduced by 50%) is a much more effective inhibitor than CCC (I_{50} about 5×10^{-4} M). However, the concentration of CCC required to retard growth in wheat seedlings would be sufficient to inhibit *ent*-kaurene synthesis in the *C. maxima* system. Thus, the evidence suggests that for growth inhibition the primary site of action of CCC, except at very high concentrations, is the GA biosynthetic pathway.

2.2 Alternative Sites of Action

The simple concept that AMO-1618 and related retardants were growth inhibitors solely by virtue of their effects on GA biosynthesis has been questioned by Douglas

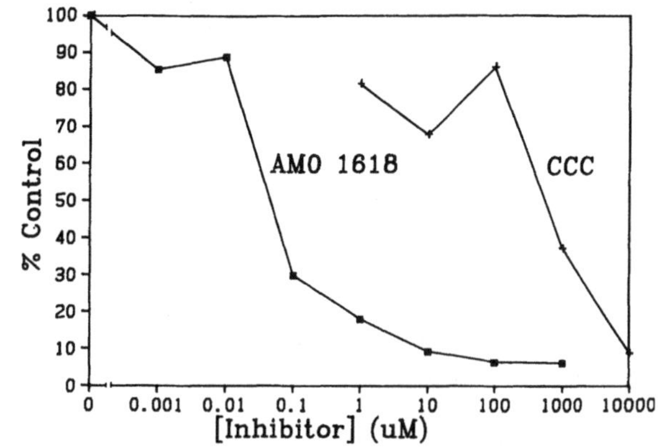

Fig. 3. Effect of concentration of AMO-1618 and CCC on *ent*-kaurene formation from mevalonic acid. [2-^{14}C] Mevalonic acid was incubated with a 40 000 × g supernatant fraction of a cell-free system from *C. maxima* for 1 h at 30°C in the presence of the inhibitors at several concentrations. Control is radioactivity incorporated into *ent*-kaurene in the absence of inhibitor

and Paleg [12]. They showed that AMO-1618, CCC and Phosphon D inhibited the incorporation of radioactivity from [2-^{14}C]MVA into sterols in tobacco seedlings and caused an accumulation of radiolabelled 2,3-oxidosqualene. As already discussed, the cyclizations of oxidosqualene and GGPP are mechanistically similar and hence, they are both potential sites of inhibition by cationic compounds. In addition, later steps in sterol biosynthesis, including those catalyzed by isomerases and methylases, involve carbocationic high energy intermediates [3] and might also be sensitive to this class of retardants. Despite the considerable indirect evidence that AMO-1618 inhibits oxidosqualene cyclase, Duriatti et al. [14] found no effect of 10^{-4} M AMO-1618 on this enzyme isolated from rat liver or pea seedlings. These authors suggested that AMO-1618 might be oxidised in vivo to a more active species or that the substrate must be produced in situ for inhibition to occur. Since Paleg and co-workers often used concentrations of AMO-1618 well in excess of 10^{-4} M, another explanation is that the retardant may inhibit sterol biosynthesis only at very high concentrations.

Douglas and Paleg [12] provided further evidence that inhibition of sterol biosynthesis by AMO-1618 and CCC could contribute to growth retardation by demonstrating that the effects of these compounds on tobacco seedlings could be reversed by spraying with an emulsion of β-sitosterol. However, since restoration of height was also achieved with GA_3, it was not clear which biosynthetic pathway, sterol or GA, was limiting in the treated plants. A cytological examination of the treated seedlings might help resolve this question. The role of sterols in plant growth will be further considered in Section 3.2.1.

3 *ent*-Kaurene Oxidation Inhibitors (KOIs)

The first of a new class of retardants was described in 1970 by Tschabold et al. [45]. The effect of this substituted pyrimidine was shown to be reversed by GA_3 [34], and furthermore, application of 4×10^{-5} M ancymidol to *Phaseolus vulgaris* seedlings caused a dramatic decrease in the levels of extractable GA-like activity [42]. Confirmation that ancymidol inhibited GA biosynthesis was provided by Coolbaugh and Hamilton [8], who showed it to inhibit *ent*-kaurene oxidation in a cell-free homogenate of *Marah oreganus* endosperm. Subsequently, Coolbaugh et al. [9] showed that ancymidol inhibits all three steps in the oxidation of *ent*-kaurene to *ent*-kaurenoic acid, but not later steps in the biosynthetic pathway. These authors found it to be an uncompetitive inhibitor of *ent*-kaurene oxidase with Ki of 2×10^{-9} M in microsomal suspensions from *Marah macrocarpus* endosperm. *ent*-Kaurene oxidase in higher plants is a cytochrome P-450 dependent microsomal mixed-function oxygenase [21, 22]. It is unclear whether a single enzyme catalyzes all three steps in the oxidation of *ent*-kaurene to *ent*-kaurenoic acid or whether this is a multi-enzyme process. On the basis of a type II difference spectrum when ancymidol was added to oxidized *M. macrocarpus* microsomes, Coolbaugh et al. [9] suggested that ancymidol associated directly with the heme moiety of the cytochrome P-450.

A large number of retardants with the same site of action as ancymidol have now been introduced. In addition to the pyrimidines, these compounds include 1,2,4-triazoles, imidazoles, pyridines and a norbornanodiazetine (see Rademacher, this volume, for structures). The common feature of these compounds is an N-containing heterocyclic ring in which a N atom is sp^2 hybridized [39]. As with the sterol biosynthesis inhibiting (SBI) fungicides [28], a lone electron pair on the N atom is thought to interact with the Fe of the heme giving the type II difference spectrum, as discussed above, and preventing O_2 from binding [9].

Due to their general high potency and high stability, and their activity in a wide range of plants, the triazoles are prominent amongst the commercially important KOI growth retardants. Agrochemical companies currently produce triazole retardants that are chemically closely related to each other and also similar to certain SBI fungicides. An important feature of the triazoles, and also of other KOIs, is that the commercial products are usually racemic mixtures. For example, paclobutrazol has two asymmetric carbon atoms and therefore can exist as four isomers, i.e. two diastereoisomeric pairs of enantiomers. The commercial material contains about 98% of the 2*RS*,3*RS* diastereoisomer; and of this the 2*S*,3*S*, enantiomer is the most potent growth retardant, whereas the 2*R*,3*R* compound has considerable fungicidal activity [43]. The relative potencies of all four isomers of paclobutrazol as KOIs in vitro is illustrated in Fig. 4 [4]. Thus, as well as being the most potent retardant, the 2*S*,3*S* compound is likely to be the most effective inhibitor of GA biosynthesis.

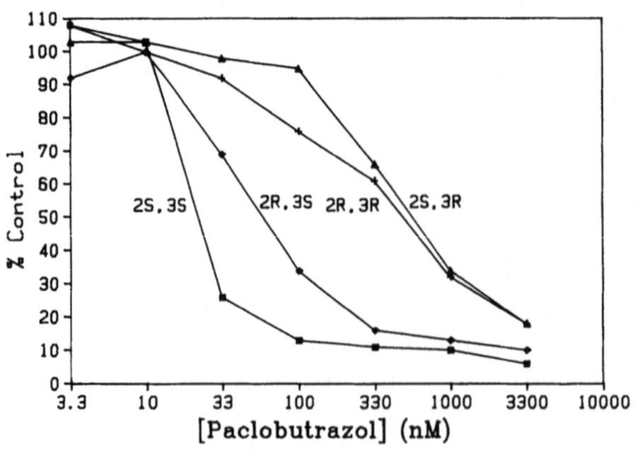

Fig. 4. Effect of concentration of all four stereoisomers of paclobutrazol on the oxidation of *ent*-kaurene by a cell-free system from *C. maxima*. Control is radioactivity incorporated into oxidation products in the absence of inhibitor. Reproduced with permission from [4]

The similarity in structure of KOI retardants and SBI fungicides arises because they inhibit analogous biosynthetic reactions in the GA and sterol pathways. In the fungal pathway to ergosterol, the site of action is the loss of the methyl group from C-14 of lanosterol or 24-methylenedihydrolanosterol, which is initiated by a cytochrome P-450 dependent hydroxylation at C-32. In the case of paclobutrazol, Sugavanam [43], rationalized the relative activities of the 2*S*,3*S* and 2*R*,3*R* enantiomers on the basis of their stereochemical similarities to *ent*-kaurene and lanosterol, respectively, as determined by computer graphics. Katagi et al. [27] showed also that the most effective configuration of the KOI uniconazole superimposed well on *ent*-kaurene. The conclusion from these modelling experiments is that the effectiveness of a triazole will depend on how closely it mimics the substrate which it replaces at the active site.

3.1 Effects on Endogenous GA Levels

Although activity at the enzyme level is an important criterion in determining the mode of action of growth retardants, their effect on the concentrations of endogenous GAs indicates more directly their function in vivo. There are numerous reports of reduced GA-like activity in treated plants. Recently, several authors have used GC-SIM to demonstrate reduced concentrations of several GAs, including GA_1, in plants treated with uniconazole, paclobutrazol or tetcyclacis [26, 33, 48]. In each case normal plant height was restored after treatment with GA_3, and in the case of uniconazole it was shown that the relationship between applied GA_3 and growth-response was independent of inhibitor concentration. This information is consistent with the growth reduction in treated plants being solely a consequence of lowered GA levels.

Lenton et al. [33] demonstrated a dose-related reduction in GA_1 concentration in the lower leaf sheath of wheat seedlings treated with 2*S*,3*S*-paclobutrazol. The final length of the leaf was related linearly to the log of the concentration of GA_1, the purported physiologically-active GA [38], in the sheath. Lenton et al. [33] also showed that the GA_1 content of shoots of the *Rht3*, GA-insensitive, dwarf wheat genotype were reduced after root treatment with 10^{-6} M 2*S*,3*S*-paclobutrazol to 30% of the very high levels found in untreated plants. Since the height of the *Rht3* genotype is unaffected by this concentration of paclobutrazol, it can be assumed that the reduction in GA biosynthesis is a direct result of paclobutrazol treatment rather than a consequence of an altered growth rate.

We have recently examined the effect of the triazole retardant BAS 111..W on the GA levels in oilseed rape seedlings (P. Hedden, S.J. Croker, W. Rademacher and J. Jung, in preparation)[2]. The results of two experiments in which the compound was applied to the foliage or as a root drench are compared in Table 1. It is clear that root application is more effective and this is consistent with BAS 111..W being primarily mobile in the xylem, as noted for related compounds (e.g. [14]). In the case of foliar application, inhibition of GA biosynthesis is only partial and the

[2] Now published (1989) Physiol Plant 75:445

Table 1. Height of and GA concentrations in 3-week-old oilseed rape seedlings treated with foliar- or soil-applied BAS 111..W expressed as percentage of untreated seedlings

	Foliar application[a]	Soil application[b]
Height	49	22
GA_{44}	9	<5
GA_{19}	4	6
GA_{20}	13	4
GA_1	81	21
GA_{29}	64	8
3-epi GA_1	<5	7
GA_8	142	13

[a] 3 mg sprayed over ca. 100 plants grown on a surface area of 500 cm^2.
[b] 3 mg in 25 ml water applied to 500 ml of a peat-based substrate in which ca. 100 plants were grown.

effect on the pool sizes of individual GAs differs substantially. The concentrations of GAs which are turned over rapidly will be more sensitive to inhibition than the relatively long-lived GAs at, or near, the end of the pathway. Thus, in foliar-treated plants, the intermediates GA_{44}, GA_{19} and GA_{20} are reduced to about 10% of the controls, whereas the levels of GA_1, GA_8 and GA_{29} are little changed. Similar observations have been reported by other authors [26, 48].

3.2 Alternative Sites of Action

The triazoles and related growth retardants are not general inhibitors of plant cytochrome P-450 monooxygenases [39]. Their action appears to be confined to methyl hydroxylases, with which they interact to differing extents depending on structure. Apart from *ent*-kaurene oxidase, there are several other such enzymes that are involved in important metabolic pathways (Fig. 5), the products of which may influence growth. These are potential sites of action and will be considered below.

3.2.1 Sterol Biosynthesis

In common with the *ent*-kaurene synthetase inhibitors, KOIs may also affect phytosterol biosynthesis. Indeed, most of these compounds arose from fungicide screening programmes where the principal requirement is for inhibition of lanosterol or 24-methylenedihydrolanosterol 14α-demethylase [28]. The equivalent enzyme in phytosterol biosynthesis, obtusifoliol 14α-demethylase, has been extracted from maize embryos and shown to be a cytochrome P-450 monooxygenase [40], which is inhibited more strongly by certain triazole fungicides than by related growth retardants [44].

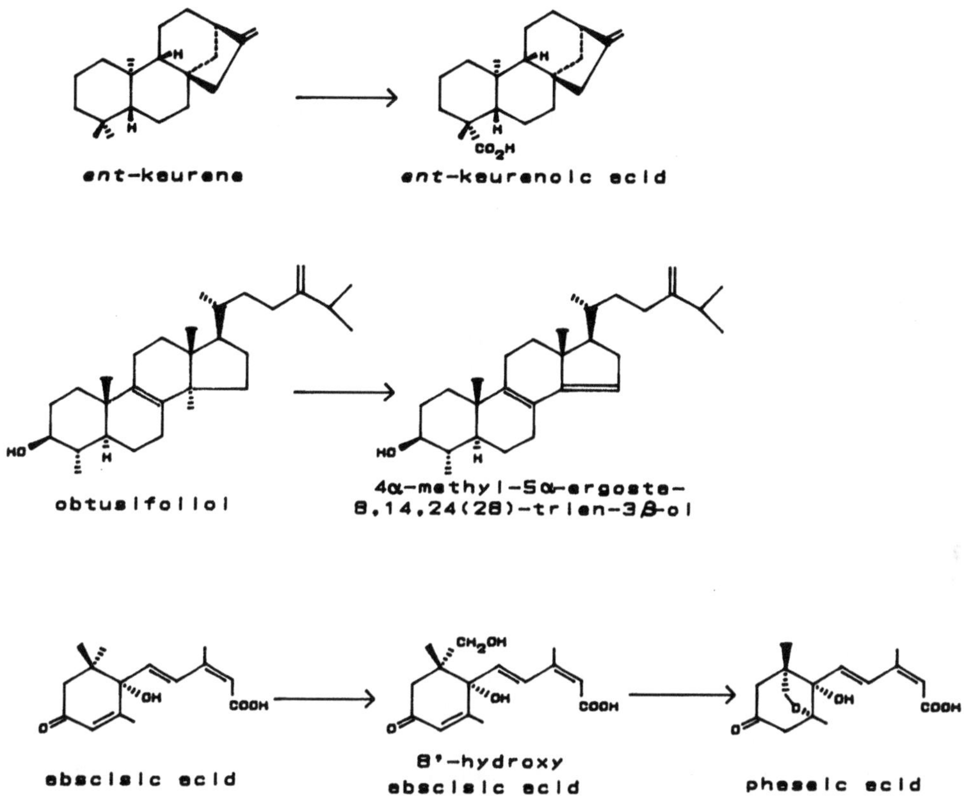

Fig. 5. Methyl hydroxylations catalyzed by monooxygenases that are potential sites of inhibition by the N-heterocyclic retardants. Reproduced with permission from Hedden P, Lenton JR (1988) In: Biomechanisms regulating growth and development. Steffens GL, Rumsey TS (eds) Kluwer Academic Publishers, Dordrecht

Köller [31] reported that growth inhibition of wheat seedlings by all four isomers of the triazole fungicide triadimenol was not abolished by GA_3, but that the effects of fungicide and GA treatments were independent. He found that the most effective isomer as a retardant was one of the most active inhibitors of sterol metabolism and concluded that this was the primary mode of action of triadimenol. In the case of the paclobutrazol enantiomers, however, there is a good correlation between KOI activity and growth retardation [4].

The most fungicidal isomer of paclobutrazol (2*R*,3*R*) is not the most active inhibitor of plant 4-desmethyl sterol synthesis [4] and it would appear that the structural requirements for inhibition of the 14α-demethylase in fungi and plants are different. Burden et al. [5] reported a γ-ketotriazole with high herbicidal activity which caused considerable growth retardation in barley and peas. This compound is a potent inhibitor of higher plant 14α-demethylation, but has little fungicidal or KOI activity. It is structurally related to a group of triazole fungicides [1], which presumably have the same mode of action.

The availability of compounds that are specific inhibitors of *ent*-kaurene oxidase or obtusifoliol 14α-demethylase allows the relative contributions of GAs and phytosterols to plant growth to be examined. Whilst an important part of the physiological effect of KOIs is on cell elongation, SBIs inhibit cell division and for this reason their effect on cell cultures has been studied extensively [19, 37]. Applied sterols, but not GA_3, overcame the growth inhibition by tetcyclacis of several cell suspension cultures [20]. On the basis of experiments with celery cell cultures and paclobutrazol, Haughan et al. [23] suggested two functions for sterols during cell division. A bulk requirement, presumably to maintain the physical properties of the membrane, is satisfied by a range of 4-desmethyl sterols such as cholesterol. In addition, there is a requirement for 24-ethyl sterols, such as stigmasterol, at low concentrations and this may be of a hormonal nature.

3.2.2 Abscisic Acid Metabolism

The metabolism of ABA to phaseic acid is initiated by hydroxylation at C-8′ (Fig. 5). The enzyme catalyzing this reaction in *Echinocystis lobata* was found to be microsomal and to require O_2 and NADPH [18]. Although it was not fully characterized, the enzyme has properties consistent with a monooxygenase which, as a methyl hydroxylase, is a potential site of action of the N-heterocyclics. There is evidence that this reaction may indeed be inhibited in some cases, leading to increased ABA levels [49] and therefore reduced transpiration [2]. Although there is little information on the structural requirements for inhibitors of ABA metabolism, it is of interest to note that inhibition of both this reaction and *ent*-kaurene oxidation would result in growth retardation.

3.2.3 Metabolism of Xenobiotics

Resistance to herbicides has been associated with a high capacity to metabolise these compounds. N-Demethylation, a first step in the deactivation of the phenylureas, is thought to involve a cytochrome P-450 dependent oxygenase [16]. Some inhibitors of this reaction, including triazole derivatives, act synergistically with phenylurea herbicides by reducing their metabolism [7, 29]. The normal function of these monooxygenases is unknown, but could be one or more of the processes discussed above. There is also evidence that cytochrome P-450 activity can be induced in plants treated with some xenobiotics [25] and this may represent a natural defense mechanism.

4 Inhibition at Other Steps in the GA Biosynthetic Pathway

Although growth retardants acting on GA biosynthesis beyond *ent*-kaurenoic acid are not available, there is interest in developing such compounds since they may offer more specificity than the current inhibitors. The enzymes catalyzing the steps between *ent*-kaurenoic acid and GA_{12}-aldehyde are microsomal oxygenases of unknown character, and beyond GA_{12}-aldehyde all steps are catalyzed by soluble α-ketoglutarate-dependent dioxygenases [24]. Dioxygenase inhibitors that bind competitively to the α-ketoglutarate binding site are known and, in particular, the dicarboxypyridines are extremely potent [35] and have been found to inhibit plant growth [46]. Such inhibitors may not, however, have the required specificity and compounds that bind to the GA binding site may prove more valuable, albeit more difficult to prepare.

Acknowledgements. I should like to thank Dr J.R. Lenton for making available unpublished data. Long Ashton Research Council is financed through the Agricultural and Food Research Council.

References

1. Anderson NH, Heritage KJ, Branch SK (1983) In: Miyamoto J (ed) IUPAC Pesticide chemistry, human welfare and the environment. Pergamon, Oxford, p 345
2. Asare-Boamah NK, Hofsta G, Fletcher RA, Dumbroff EB (1986) Plant Cell Physiol 27:383
3. Benveniste P (1986) Annu Rev Plant Physiol 37:275
4. Burden RS, Carter GA, Clark T, Cooke DT, Croker SJ, Deas AHB, Hedden P, James CS, Lenton JR (1987) Pestic Sci 21:253
5. Burden RS, James CS, Cooke DT, Anderson NH (1987) In: 1987 British Crop Protection Conference – Weeds. BCPC Publications, Thornton Heath, p 171
6. Coates RM, Conradi RA, Ley DA, Akeson A, Harada J, Lee S-C, West CA (1976) J Am Chem Soc 98:4659
7. Cole DJ, Owen WJ (1987) Plant Sci 50:13
8. Coolbaugh RC, Hamilton R (1976) Plant Physiol 57:245
9. Coolbaugh RC, Hirano SS, West CA (1978) Plant Physiol 62:571
10. Dennis DT, Upper CD, West CA (1965) Plant Physiol 40:948
11. Dicks JW (1980) In: Clifford DR, Lenton JR (eds) Recent developments in the use of plant growth retardants. British Plant Growth Regulator Group, Wantage, (BPGRG Monograph 4), p 1
12. Douglas TJ, Paleg LG (1974) Plant Physiol 54:23
13. Duncan JD, West CA (1981) Plant Physiol 68:1128
14. Duriatti A, Pierrette B-N, Benveniste P, Schuber F, Delprino L, Balliano G, Cattel L (1985) Clin Pharm 34:2765
15. Fall RR, West CA (1971) J Biol Chem 246:6913
16. Fonne-Pfister R, Simon A, Salaun J-P, Durst F (1988) Plant Sci 55:9
17. Frost RG, West CA (1977) Plant Physiol 59:22
18. Gillard DF, Walton DC (1976) Plant Physiol 58:790
19. Grossmann K, Rademacher W, Sauter H, Jung J (1984) J Plant Growth Regul 3:197
20. Grossman K, Weiler EW, Jung J (1987) Planta 164:370
21. Hasson PJ, West CA (1976) Plant Physiol 58:473
22. Hasson PJ, West CA (1976) Plant Physiol 58:479
23. Haughan PA, Lenton JR, Goad LJ (1988) Phytochemistry 27:2491
24. Hedden P (1982) In: Crozier A (ed) The biochemistry and physiology of the gibberellins, vol 1. Praeger, New York, p 99
25. Hendry GAF, Jones OTG (1984) New Phytol 96:153
26. Izumi K, Yamaguchi I, Wada A, Oshio H, Takahashi N (1984) Plant Cell Physiol 25:611
27. Katagi T, Mikami N, Atsuda T, Miyamoto J (1987) J Pestic Sci 12:627
28. Kato T (1986) In: Hung G, Hoffman H (eds) Chemistry of plant protection 1. Springer, Berlin Heidelberg New York Tokyo, p 1
29. Kemp MS, Newton LV, Caseley JC (1988) In: Proceedings of the European Weed Research Society – Factors Affecting Herbicidal Activity and Selectivity, 1988, (in press)
30. Kende H, Ninnemann H, Lang A (1963) Naturwissenschaften 50:599
31. Köller W (1987) Physiol Plant 71:309

32. Lang A (1970) Annu Rev Plant Physiol 21:537
33. Lenton JR, Hedden P, Gale MD (1987) In: Hoad GV, Lenton JR, Jackson MB, Atkin RK (eds) Hormone action in plant development – a critical appraisal. Butterworth, London, p 145
34. Leopold AC (1971) Plant Physiol 48:537
35. Majamaa K, Hanauske-Abel HM, Guenzler V, Kivirikko KI (1984) Eur J Biochem 138:239
36. Mitchell JW, Wirwille JW, Weil L (1949) Science 110:252
37. Nitsche K, Grossmann K, Sauerbrey E, Jung J (1985) J Plant Physiol 118:209
38. Phinney BO, Spray C (1982) In: Wareing PF (ed) Plant growth substances 1982. Academic Press, London, p 101
39. Rademacher W, Fritsch H, Graebe JE, Sauter H, Jung J (1987) Pestic Sci 21:241
40. Rahier A, Taton M (1986) Biochem Biophys Res Commun 140:1064
41. Richardson PJ, Quinlan JD (1986) Plant Growth Regul 4:347
42. Shive JB, Sisler HD (1976) Plant Physiol 57:640
43. Sugavanam B (1984) Pestic Sci 15:296
44. Taton M, Ullmann P, Benveniste P, Rahier A (1988) Pestic Biochem Physiol 30:178
45. Tschabold EE, Taylor HM, Davenport JD, Hackler RE, Krumkalns EV, Meredith WC (1970) Plant Physiol 46 (Suppl):19
46. Weir AJ (1988) Ph D Thesis, University of Bristol
47. Wirwille JW, Mitchell JW (1950) Bot Gaz 111:491
48. Zeevaart JAD (1985) Plant Physiol 166:276
49. Zeevaart JAD (1990) In: Pharis RP, Rood SB (eds) Plant growth substances 1988. Springer, Berlin Heidelberg New York Tokyo, this vol

Measurement of the in Vivo Rate of Indole-3-Acetic Acid Turnover

H.M. Nonhebel and T.P. Cooney[1]

1 Introduction

There has recently been an increased awareness of the importance of quantitative work in the study of hormone metabolism. This is probably a result of considerable improvements in the sensitivity of accurate, definitive, physico-chemical methods of plant hormone analysis. The purpose of this paper is to describe methods of measuring hormone turnover rate, to review the results obtained with particular reference to IAA, and to examine whether the rates obtained are likely to reflect the normal turnover rate of the endogenous hormone.

The turnover rate of a plant hormone may be useful in several ways:

1. Where there are several possible sources of hormone, e.g. de novo synthesis and conjugate hydrolysis, the rates of individual pathways may be compared with the overall turnover rate, giving a measure of the relative importance of the different routes.
2. It has been suggested that hormone turnover rate may be correlated with biological activity [e.g. 16].
3. Turnover studies may also yield information on compartmentation of hormones.

2 Measurement of Turnover Rates Using Dilution of Specific Activity

Most studies use the method of Zilversmit et al. [21] in which the turnover rate is calculated from the decrease in specific activity of a radiolabelled compound which has been introduced into the organism. The rate constant and half-time for turnover are calculated using the first order rate Eqs. (1) and (2)

$$k = \ln \frac{Co}{C_t} \times \frac{1}{t} \tag{1}$$

$$t_{1/2} = \frac{\ln 2}{k} \tag{2}$$

where k is the first order rate constant, Co the specific activity at zero time, C_t the specific activity at time t, and $t_{1/2}$ the half-time for hormone turnover.

[1]Department of Biochemistry, University of Auckland, Private Bag, Auckland, New Zealand.

As stated by Zilversmit et al. [21] this method makes two major assumptions;

1. That the decrease in specific activity obeys first-order kinetics.
2. That the radiolabelled compound is metabolized at the same rate as endogenous material.

2.1 Assumption of First-Order Kinetics

This assumption has been discussed [2, 19, 21] and will not be dealt with in detail here. The assumption requires that the pool size of the compound assayed remains constant and that the rates of synthesis and degradation remain constant. As well as measuring the amount of hormone present at the beginning and end of the experiment, this requires that the amount of exogenous hormone does not alter the pool size of the hormone significantly. It should also be pointed out that in most biological reactions the decrease in specific activity actually follows pseudo first-order kinetics, although the reaction itself is not usually a true first-order reaction. The pseudo first-order rate constant can be used, however, to obtain true values of the half-time and the reaction rate, v [Eq. (3)]

$$v = k\,[A], \tag{3}$$

where [A] is the concentration of substance in the tissue and k the apparent rate constant.

2.2 Assumption that the Labelled Compound is Metabolized at the same Rate as the Endogenous

This encompasses two assumptions:

1. That there is no significant isotope effect or exchange of isotope with the medium. These potential problems can usually be avoided by ensuring that the isotope is not in a labile position and that bond(s) to the labelled atom are not broken during initial stages of metabolism.
2. That total and rapid equilibration of the labelled compound with the endogenous takes place.

Zilversmit et al. [21] used the original equation to look at the disappearance of a compound from the blood stream where the assumption of quick and efficient equilibration is likely to be valid. The situation is likely to be considerably more complex, however, where a plant hormone is supplied to an intact root or shoot. This paper sets out to examine the factors involved in equilibration of exogenous hormone with endogenous pools and to what extent lack of equilibration affects measurement of turnover rates.

3 Evidence for Different Pools of IAA with Different Rates of Metabolism

There is evidence that plants contain several pools of hormone with different turnover rates. These pools may exist in different cells or tissues, or within a cell in different organelles.

3.1 IAA Metabolism in the Cortex and Stele of *Zea mays* Roots

Segments of *Zea mays* roots can be separated into cortex and stele allowing measurement of the IAA content of each tissue. The stele contains most of the endogenous IAA [1]. However, the rate of metabolism of [2-^{14}C]IAA is much greater in cortical segments than in the stele [12]. *Zea mays* roots thus appear to contain at least two pools of IAA which are turning over at different rates and which are located in different cells. Most of the endogenous IAA is in the pool which is turning over slowly.

3.2 IAA Metabolism in the Chloroplasts and Cytoplasm

IAA has been identified in extracts of a crude chloroplast fraction from pea shoots [3]. Sandberg et al. [17, 18] have shown that whereas both a crude chloroplast fraction and a cytoplasmic fraction were capable of converting [5-^{3}H] tryptophan to [5-^{3}H]IAA, the two fractions differed in the effect of light on IAA metabolism. Different compartments within a cell may therefore metabolize IAA independently with differing turnover rates. The ability of exogenous IAA to equilibrate across plastid membranes must be considered when measuring IAA turnover.

4 Rate of Equilibration: Evidence from Changes in Measured Rates of IAA Turnover with Time

We can obtain some idea of how long equilibration takes by measuring the change in turnover rate with time. If the level of endogenous IAA remains constant but the apparent rate of turnover changes, this suggests a slow rate of equilibration.

4.1 IAA Turnover in *Zea mays* Seedlings

Epstein et al. [5] measured the rate of IAA turnover in kernels of 4-day-old *Zea mays* seedlings. [^{14}C]IAA was applied to the cut surface of the endosperm and the experiment started immediately. The half-time for IAA turnover over an 8-h incubation, calculated from the change in specific activity, was 3.2 h, giving a rate of IAA synthesis of 65 pmol/kernel/h.

In a later experiment using the same variety of *Zea mays* [11], the rate of IAA metabolism was measured indirectly by measuring the turnover rate of oxindole-3-acetic acid (OxIAA), the principal metabolite of IAA in *Zea mays* seedlings [13, 14]. In this experiment a 24 h period for uptake and equilibration was allowed after application of [^{3}H]OxIAA to the cut surface of the endosperm of 3-day-old seedlings, before the measurement of turnover was started. The half-time for OxIAA turnover in the kernels was 35 h. The rate of production of OxIAA (from IAA) was then calculated using equation 3 and the concentration of endogenous OxIAA, 357 pmol/kernel [15]. A value of 7.1 pmol/plant/h was obtained and was taken as an estimate of the rate of IAA metabolism. In a single experiment, using IAA instead of OxIAA but allowing the same 24 h uptake and equilibration period, a half-time of 40 h was obtained.

These results give a rate of IAA metabolism an order of magnitude lower than those of Epstein et al. [5] where the rate was measured immediately after application, suggesting that IAA is moving from a pool with a fast turnover rate to one with a much slower rate.

4.2 IAA Turnover in Tomato Shoots

More recently we have measured the turnover rate of IAA in tomato shoots after different incubation times, as part of a study of IAA biosynthesis in that tissue. IAA was supplied to the surface of the youngest leaf, 1 cm or longer, of excised shoots from twenty, 4-week-old tomato plants, as a 10 μl drop (2:1:1, ethanol: propan-2-ol: H_2O) containing 3711 Bq [^{3}H]IAA (1.11 GBq/μmol). Uptake was allowed for 2 h, surface radioactivity was then removed by several washes with 50% aqueous propan-2-ol followed with water, and the plants equilibrated for 1 h. After 0, 4, 10 or 20 h incubation in continuous light, IAA was extracted, purified, derivatised to form the pentafluorobenzyl ester [10] and the specific activity measured by liquid scintillation counting and GC-ecd. The purity and identity of selected samples was confirmed by combined GC-MS.

No consistent change in turnover rate with increasing incubation time was found (Table 1), indicating that equilibration took place rapidly. In a parallel experiment the amount of endogenous IAA was measured and shown to remain constant at between 9.2 and 9.4 ng/g fw.

Table 1. Change in specific activity of [^{3}H]IAA and $t_{1/2}$ for IAA turnover with increasing incubation time after application of [^{3}H]IAA to the youngest leaf of 4-week-old tomato shoots

Time (h)	Specific activity (Bq/ng)	$t_{1/2}$ (h)
0	9.32	
		7.9
4	6.54	
		10.8
10	4.45	
		9.8
20	2.21	

5 Effect of Site of Application on Distribution of Exogenous IAA and on Measurement of Turnover

It is possible for radiolabelled hormone to equilibrate rapidly with IAA pools in some cells or tissues, but for the overall distribution of exogenous hormone in the plant or organ to be quite different from that of the endogenous hormone. The measured turnover rate will then reflect the rate of metabolism in the compartment where the labelled hormone is found but will not necessarily be an accurate measure of the overall rate of IAA metabolism in the shoot. A second experiment with tomato shoots showed that the distribution of radioactivity and the calculated turnover rate differs with the site of application of [^{3}H]IAA.

The half-life of [^{3}H]IAA supplied to shoots via the transpiration stream was compared with that obtained in the previous experiment. The IAA solution containing 2192 Bq [5-^{3}H]IAA per plant (ca. 11 pg IAA, i.e. approx 0.1% of endogenous amount) was completely taken up in 1 to 1.5 h. The plants were equilibrated for a further 1 to 1.5 h (total time, uptake plus equilibration for all plants was 2.5 h), and incubated for 0, 6, 18 and 30 h before extraction and purification of the IAA.

In three experiments with different incubation times the half-times for IAA turnover varied between 3.1 h to 5.6 h but there was no consistent increase or decrease with incubation time. However, the $t_{1/2}$ values from shoots supplied with [^{3}H]IAA via the transpiration stream were significantly smaller (an average of 4.2 h) than those from plants supplied via the youngest leaf (10.3 h), Table 2. The

Table 2. Comparison of $t_{1/2}$ values for turnover of [^{3}H]IAA supplied to tomato shoots via the transpiration stream or to the surface of the youngest leaf

Expt.	$t_{1/2}$ apex application (h)	$t_{1/2}$ transpiration stream (h)
1	10.4	3.6
2	11.0	4.2
3	9.6	4.8

difference in calculated half-lives reflected a difference in the overall distribution of radioactivity in the shoots. In shoots supplied with [^{3}H]IAA via the transpiration stream, the majority of radioactivity after a 4-h incubation was in the oldest section of internode and the oldest leaf, while in shoots supplied with [^{3}H]IAA via the youngest leaf, nearly half of the radioactivity remained in the apical region, most of the rest being found in the oldest internode. As the major site of IAA synthesis is in the young leaves, the final distribution of [^{3}H]IAA in shoots in which it was applied at this point is more likely to reflect the natural distribution, as it should be transported by the same routes.

6 Alternative Method of Measuring Turnover Rates

The results presented above indicate that exogenous hormone does not necessarily equilibrate effectively with all endogenous pools so that a true turnover rate is obtained. In view of these problems, the turnover rates obtained by measuring the rate of dilution of specific activity of radiolabelled hormone should ideally be compared with those obtained by an independent method.

An alternative method of measuring hormone turnover uses the rate of incorporation of either heavy or radiolabelled isotope into the hormone from small molecules which equilibrate rapidly with all parts of the plant such as $^{2}H_2O$, $^{18}O_2$, $^{14}CO_2$ or $^{13}CO_2$. This method has not been used for IAA but has been used successfully for measuring ABA turnover. The method also has potential sources of error however; isotope effects are more likely to be a problem and a precise rate of metabolism cannot be obtained unless the label is incorporated during the final biosynthetic step.

6.1 Use of $^{18}O_2$ to Measure ABA Turnover

Creelman et al. [4] used the rate of ^{18}O incorporation to calculate the turnover rate of ABA in stressed *Xanthium* leaves, obtaining a half-time of 15.5 h. This value is very close to that calculated by Zeevaart [20] using an independent method.

The rates of ABA metabolism obtained by the two different methods do not always agree however. ^{18}O incorporation was also used to estimate the turnover rate of ABA in turgid *Xanthium* leaves, a half-time of "several days" being obtained. In contrast, [^{14}C]ABA supplied to turgid leaves *via* the transpiration stream was completely metabolized in 48 h [4]. In turgid leaves most of the ABA is located in the chloroplasts while in stressed leaves it is redistributed to the cytoplasm [8]. Creelman et al. [4] suggest that exogenous ABA does not cross the chloroplast membrane and therefore does not equilibrate with endogenous ABA in turgid leaves. These results are somewhat surprising in view of the observation [7] that the membranes of isolated chloroplasts are permeable to ABA. They are also pertinent to a study of IAA turnover, as both molecules are at least partially localized in the chloroplasts [3, 8] and have similar pKa values; 4.2 [9] and 4.7 [6] for ABA and IAA respectively.

If the chloroplast membranes are relatively impermeable to ABA it might be expected to equilibrate only at longer incubation times and the rate of [^{14}C]ABA metabolism might be expected to decrease. This does not occur, at least in tomato shoots supplied with +[^{14}C]ABA via the transpiration stream (Nonhebel, unpublished).

The half-time over the first 4 h incubation was 14.5 h compared with a $t_{1/2}$ of 12 h for the following 16 h. Interestingly, shoots incubated in darkness metabolized ABA more slowly; $t_{1/2}$ was approximately 34 h.

6.2 Incorporation of ^2H from ^2H_2O into IAA

We have measured ^2H incorporation into IAA and tryptophan in tomato shoots equilibrated with 30% ^2H_2O. The amount of ^2H in the m/z 130 ions from IAA and tryptophan was compared by mass spectrometry. In a typical experiment 30% of tryptophan and 18% of IAA molecules were labelled after an 11 h incubation, both compounds becoming labelled in 4 positions. Although a precise turnover rate for IAA cannot be obtained from these results, as the label is incorporated at several points in the biosynthetic pathway, prior to tryptophan, they are consistent with the turnover rates calculated above.

7 Conclusions

Two different methods can be used to measure hormone turnover. Both have sources of error, but as these are different a comparison of values obtained by each method should give a reliable estimate of turnover, as was accomplished successfully for ABA in stressed *Xanthium* leaves [4, 20]. ^{18}O incorporation might also be used to measure IAA turnover, as oxygen is incorporated during the final oxidation of indole-3-acetaldehyde.

If dilution of specific activity alone is used to measure turnover, care should be taken to use small amounts of labelled compound, if possible less than 1% of endogenous. The compound should be applied so that it will be transported by the same routes as endogenous. Finally, sufficient time should be allowed for equilibration before starting the experiment. Using these guidelines we have measured the turnover of IAA in tomato shoots. Confirmation of the order of magnitude of the turnover rate was obtained from ^2H incorporation results. Using either the half-life of IAA turnover (10.3 h) or the rate constant (0.0673 h^{-1}) the rate of IAA metabolism was calculated from Eq. (3) and the IAA concentration of the shoots (9.3 ng/g fresh wt). A value of 3.6 pmol/g fresh wt/h was obtained. This can be compared in future experiments with rates of IAA synthesis from different potential precursors.

References

1. Bridges IG, Hillman JR, Wilkins MB (1973) Planta 115:189–192
2. Broda E (1960) Radioactive isotopes in biochemistry. Elsevier, London
3. Brown BH, Crozier A, Sandberg G, Jensen E (1986) Phytochemistry 25:299–302
4. Creelman RA, Gage DA, Stults JT, Zeevaart JAD (1987) Plant Physiol 85:726–732
5. Epstein E, Cohen JD, Bandurski RS (1980) Plant Physiol 65:415–421
6. Goldsmith MHM (1977) Annu Rev Plant Physiol 28:439–478
7. Heilmann B, Hartung W, Gimmler H (1980) Z Pflanzenphysiol 97:67–78
8. Loveys BR (1977) Physiol Plant 40:6–10
9. Milborrow BV, Rubery PH (1985) J Exp Bot 36:807–822
10. Netting AG, Milborrow BV, Duffield AM (1982) Phytochemistry 21:385–389
11. Nonhebel HM (1986) J Exp Bot 37:1691–1697
12. Nonhebel HM, Hillman JR, Crozier A, Wilkins MB (1985a) Planta 164:105–108

13. Nonhebel HM, Hillman JR, Crozier A, Wilkins MB (1985b) J Exp Bot 36:99–109
14. Reinecke DM, Bandurski RS (1981) Biochem Biophys Res Commun 103:429–433
15. Reinecke DM, Bandurski RS (1983) Plant Physiol 71:211–213
16. Reinecke DM, Bandurski RS (1987) In: Davies PJ (ed) Plant hormones and their role in plant growth and development. Nijhoff, Dordrecht, p 24–42
17. Sandberg G, Jensen E, Crozier A (1982) Planta 156:541–545
18. Sandberg G, Jensen E, Crozier A (1983) Plant Cell Environ 6:111–115
19. Sprinson DB, Rittenberg D (1949) J Clin Invest 28:715–726
20. Zeevaart JAD (1983) Plant Physiol 71:477–481
21. Zilversmit DB, Enterman C, Fishler MC (1943) J Gen Physiol 26:325–331

Relationship Between Stimuli, IAA and Growth

R.S. Bandurski, A. Schulze, M. Desrosiers, P. Jensen, B. Epel, and D. Reinecke[1]

1 Introduction

1.1 The Problem

"Ohne Wuchsstoff kein Wachstum": without auxin no growth [e.g. 32]. A particularly valuable test of this dictum was made by Dolk [e.g. 32] who showed that double decapitation of the *Avena* coleoptile stopped growth. Growth could then be partially restored by means of applied auxin. As beautiful as this experiment is, it is important to test the conclusion by determining if there is a quantitative relationship between endogenous IAA and growth. The relationship between numbers of coleoptile tips placed on a receiver block and growth induced by applying the receiver block to shoot stumps does not establish this relationship, since the tip supplies both free IAA and IAA derived from the seed auxin precursor [2, 23] presumably by hydrolysis of bound IAA [28]. Thus, we wish to make a quantitative test of the dictum by determining the relationship between growth rate and amount of free IAA. We wish further to determine the effect of various environmental stimuli on the amount of IAA in the tissue. In that manner we can determine the extent to which the transduction of environmental stimuli is accomplished by varying the amount of free IAA in the experimental system. Lastly, we wish to understand how the plant can so precisely regulate its IAA levels so as to have more IAA on one side than on the other of a tropically stimulated stem. Lastly, we wish to understand tropic curvatures in terms of IAA metabolism and transport.

1.2 The Experimental System

The best test of the dictum would be a determination of the endogenous IAA content of load bearing tissue, (tissue most resistant to cell expansion), and the growth rate of that same tissue. The epidermal cells have been found to be the cells which most restrict tissue growth [18]. Our *Zea mays* seedling system, shown in Fig. 1, consists of a 1 cm section of mesocotyl cortex plus epidermis cut from 0.4 to 1.4 cm below the node between coleoptile and mesocotyl. We have chosen this section because it is still growing rapidly [25] and because it is possible to remove the vascular stele tissue. IAA and its conjugates are being transported from endosperm

[1] Department of Botany and Plant Pathology, East Lansing, MI 48824-1312, USA

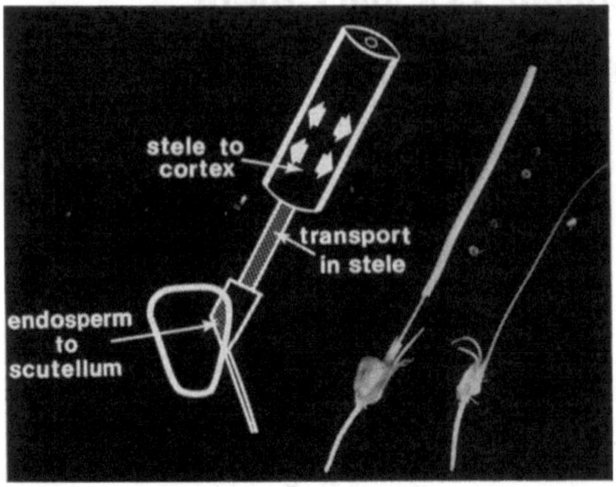

Fig. 1. The maize seedling system. *Left* The putative regulatory sites for movement of IAA from endosperm to shoot. *Right* A photograph showing partially dissected seedlings

to shoot through the stele [4, 8, 9, B. Epel pers. comm.] thus making it desirable to eliminate the vascular stele from the tissue assayed. We will return to a discussion of this experiment following a consideration of use and supply of IAA.

2 Rate of Use of Indole-3-Acetic Acid (IAA)

The rate of use of IAA by 5-day-old (24 h imbibition, 96 h germination) seedlings is discussed elsewhere in this volume. We estimate the rate of use of IAA as 5 pmol shoot^{-1} h^{-1} and 50 pmol kernel^{-1} h^{-1} [9, Nonhebel, this volume]. This estimate is based upon many experiments involving both the rate of IAA production by the tip, the rates of upward transport and hydrolysis of IAA conjugates, and the rate of IAA oxidation. The agreement between these various experiments is better than order of magnitude and we believe are correct to better than fivefold.

3 Rate of Supply of IAA from Endosperm to Shoot

3.1 Sources of IAA

3.1.1 de novo Synthesis

Our method for estimating de novo synthesis of IAA, involving biosynthesis of the indole ring system, is to imbibe and grow plants on 30% D_2O [24]. If the indole ring is being synthesized, deuterium will be incorporated into non-exchangeable positions in the ring system: that is, positions 2, 4, 5, 6 and 7 [20]. We choose to use so general a precursor as deuterium because we need make no assumptions as to the biosynthetic route of IAA synthesis, a route which remains uncertain [1, 11].

The experimental protocol is as follows: The kernels are imbibed with 30% D_2O, then rolled in paper towels and grown in 30% D_2O and in darkness at 25 °C. Following harvest the free IAA is extracted from the shoot tissue, purified by DEAE and HPLC chromatography, methylated, purified again by HPLC, and analyzed by GC-MS.

The data of Table 1 shows ion abundancies for M^+ and the quinolinium ion derived from the methyl ester of IAA [17]. As can be seen from the theoretical abundancies, one expects 88% of the mass of M^+ to occur at 189 with 11% at 190 and 1% at 191. Similarly for the quinolinium ion the expected abundancies are 88% at 130, 11% at 131, and 1% at 132. The ions at one and two mass units above the expected mass are owing to the natural abundancy of the heavy ions, ^{15}N, ^{13}C, ^{18}O and ^2H. As can be seen, no deuterium incorporation has occurred demonstrating that there is no de novo, synthesis of the indole ring in these 5-day-old seedlings of *Zea mays*, var. Silver Queen. This knowledge makes the Silver Queen system attractive since it presents a closed system with no inputs of newly synthesized indole ring system.

Table 1. Mass spectral analysis of the methyl ester derivative of IAA[a]

Mass No.	Relative Abundance[b]			
	Theoretical	Experimental		
		IAA Standard	4-day Shoot	7-day Shoot
130	100	100	100	100
131	10.5	10.19 ± 0.26	10.45 ± 0.14	10.00 ± 0.02
132	0.4	0.59 ± 0.07	0.64 ± 0.04	0.60 ± 0.02
133	0.0	0.04 ± 0.03	0.18 ± 0.21	0.03 ± 0.03
189	100	100	100	100
190	12.9	12.2 ± 0.10	12.16 ± 0.01	12.24 ± 0.08
191	1.1	1.15 ± 0.10	1.03 ± 0.02	1.11 ± 0.05
192	0.2	0.46 ± 0.69	0.41 ± 0.33	0.07 ± 0.25

[a] The IAA was isolated from shoots grown 4 or 7 days in 30% D_2O in the dark.
[b] The base peak (m/z = 130) and M^+ (m/z = 189) of methyl IAA were each normalized to 100%, and the m/z+1, m/z+2, and m/z+3 ions for each are expressed as percent abundance relative to m. Data from P. Jensen [17].

We had earlier reported on the incorporation of deuterium into IAA isolated from 5-day-old seedlings of *Zea mays*, var. Stowells Evergreen [24]. Stowells Evergreen does not hydrolyze its endosperm to the extent observed with Silver Queen. In addition, in these earlier studies our safe light emitted some radiation above 580 nm. These earlier results are shown in Table 2, and as can be seen there is incorporation of deuterium into non-exchangeable positions of the indole ring of IAA. Both experiments are accurate and reliable. We conclude that some factor, such as varietal differences, endosperm reserves, or traces of R can trigger the transition from reliance upon IAA reserves to de novo synthesis. We are attempting to determine what triggers the change from shoot hormone heterotrophy involving reliance of the shoot upon endosperm conjugate reserves, to shoot hormone autotrophy, with the shoot synthesizing its own IAA.

Table 2. Mass spectral analysis of the methyl ester derivative of IAA[a]

Mass No.	Relative Abundance[b]			
	Theoretical	Experimental		
		IAA Standard	Root	Shoot
130	100	100	100	100
131	10.5	10.8	17.3	33.1
132	0.4	0.7	4.6	11.9
133	0.0	0.0	0.8	2.8
189	100	100	100	100
190	12.9	12.7	20.3	34.3
191	1.1	1.1	5.8	10.9
192	0.2	0.1	1.1	2.9

[a] The IAA was isolated from roots and shoots grown 4 d in 30% D_2O in the dark.
[b] The base peak (m/z = 130) and M^+ (m/z = 189) of methyl IAA were each normalized to 100%, and the m/z+1, m/z+2, and m/z+3 ions for each are expressed as percent abundance relative to m. Data from W. Pengelly [24].

3.1.2 Tryptophan and Tryptamine

Both tryptophan and tryptamine have been considered to be precursors of IAA [11]. This may be true for many tissues, but there are reasons for doubting this biosynthetic route for maize seedling tissue. For example, applying either labeled tryptophan or tryptamine to the endosperm of maize seedlings does not result in appreciable amounts of labeled IAA in the shoot [16, 21, 22]. There is also danger that some of the reported synthesis of labeled IAA from labeled tryptophan may represent radiological decomposition of tryptophan to IAA. Simply drying labeled tryptophan in vacuo in Pyrex flasks can form IAA in 30% yields [14]. Thus, owing to the lack of deuterium incorporation into IAA, and the failure to convert tryptophan or tryptamine into IAA, we conclude that Silver Queen seedlings are dependent upon IAA conjugate hydrolysis for an IAA supply.

3.1.3 IAA Esters from the Endosperm

In studies by Chisnell [8, 9], either free 5-[^{3}H]-IAA (IAA) or 5-[^{3}H]-IAA-*myo*-inositol (IAInos) was applied to the endosperm of maize seedlings and the amount and distribution of labeled IAA and IAInos determined in the shoot. IAInos supplied both free and ester IAA to the mesocotyl and coleoptile. Labeled IAA applied to the endosperm supplied some of the IAA in the mesocotyl, but supplied essentially *no* free IAA to the coleoptile. Therefore, free IAA from the endosperm is not a source of IAA for the coleoptile. Neither IAA nor IAInos accumulated particularly in the coleoptile tip. This observation, plus the previous observation of Hall [15], which indicated that the enzyme that hydrolyzes IAInos was not confined to the coleoptile tip, leaves unanswered how the coleoptile tip controls the amount of IAA in the shoot.

It is uncertain where the hydrolysis of IAInos occurs, though it may possibly be in the scutellum. It is certain that the upward transport of IAA is in the stele since following injection of labeled IAA into the endosperm the stele has more radioactivity than the cortex [8, 9]. The amount of radioactivity in the stele then declines as the radioactivity in the cortex increases [4].

A transport study was also made by Komoszynski [19] who observed that intact IAA-*myo*-inositol-galactoside (IAInosgal) was not transported from endosperm to shoot. Rather, the galactose was first hydrolyzed off, presumably in the scutellum, and then the free galactose and IAInos was transported into the shoot.

In both expriments, the ester was observed to yield about 2 pmol-plant^{-1}-h^{-1} of IAA in the shoot. Thus, each ester yielded 40% of the estimated requirement of IAA for the shoot. Since IAInos, IAInosgal, and IAInosarabinose [10] all occur in the endosperm, the esters could account for all of the IAA of the shoot.

3.1.4 Maize Seedlings Represent a Closed System

Whether the maize seedling is an open or closed system is an important consideration. If, as we believe, there is no de novo synthesis of IAA, and if tryptophan, tryptamine, and other preformed indoles are not converted to IAA, then the system is closed, and only transport, conjugate synthesis and hydrolysis and IAA oxidation need be considered as regulators of the amount of IAA in the tissue. On the basis of present evidence we conclude that the maize seedling system is indeed a closed system.

4 Relationship Between IAA Supply and Growth

4.1 General Comments

It has not previously been possible to determine the relationship between the rate of growth and the amount of the growth hormone, IAA. Early experiments involved exodiffusion of coleoptile tips, followed by a bioassay of the amount of IAA in the agar blocks that were used to collect the diffusing growth substance. Such experiments measured the capacity of the tips to supply IAA over a sustained time period and undoubtedly involved conversion of a seed auxin precursor [2, 23]. We now believe that the seed auxin precursor is indole-3-acetyl-*myo*-inositol (IAInos), which upon hydrolysis yields free IAA [8, 9, 15, 22]. Also, it is now possible to extract the tissue rapidly with aqueous acetone and to measure the free IAA by GC-MS.

4.2 Growth Modified by Coleoptile Decapitation

As discussed above, we chose to measure the growth rate of a portion of the mesocotyl cortex and to measure the free IAA in that same section of tissue. To vary the endogenous IAA we chose the time-honored system of decapitating the coleoptile of the growing maize shoot and then waiting varying lengths of time for the free IAA content to decline. The plants were then harvested, the appropriate part of the mesocotyl cortex collected, and the tissue assayed for IAA as described above. Growth was measured over that same time interval by simply measuring the change in length of marked tissue with a ruler. We plotted cumulative growth of that 1-cm section of mesocotyl against the amount of IAA in the equivalent section. At some later date it may be possible to measure growth over a short time period so as to better relate growth rate to IAA amount.

The results of this experiment are shown in Fig. 2. We used a Michaelis-Menten plot, with the reciprocal of the free IAA content of the mesocotyl cortex plus epidermis, plotted against the reciprocal of the growth of the equivalent section. A straight line results, yielding an estimate of K_m as 7×10^{-8} M. The concentration of free IAA in the mesocotyl cortex is 6×10^{-8} M, so the tissue is growing at about one-half saturation with IAA. V_{max} may also be estimated as slightly over 0.1 mm h^{-1}, which is quite reasonable for the 1 cm section of mesocotyl.

4.3 Growth Modified by Red Light

Only three points are available for the decapitation experiment, although each point represents the average of 3 IAA determinations and growth measurements on 60 plants. The fourth point (Fig. 2) is of interest in that, for that experiment, the IAA level was lowered by a 5-min exposure to R, and growth and IAA were then measured after 2 h. The point relating growth to amount of IAA, as regulated by the R exposure, falls exactly on the line relating growth to amount of IAA, following decapitation.

4.4 Conclusions

We conclude that the amount of IAA in the mesocotyl cortex can be reduced either by coleoptile decapitation or by exposure of the seedling to R. Both give exactly the same reduction of growth. This experiment strengthens and confirms the statement, "Ohne Wuchsstoff kein Wachstum," and indicates a primary control of growth by the *amount* of IAA.

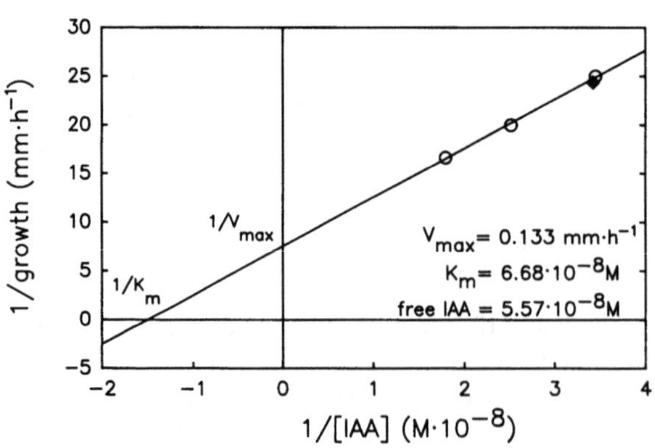

Fig. 2. A double reciprocal plot showing the relationship between the rate of growth of a maize seedling as a function of the amount of free IAA in the mesocotyl cortex plus the epidermal cells

There has been much speculation concerning the question of whether growth is controlled by IAA or by some, as yet, unidentified IAA receptor. Earlier, such speculations were specious in that they were based on unacceptable assays for IAA. Hopefully, the above experiment will stimulate studies of the controls of plant extension growth.

5 Effects of Other External Stimuli on IAA and Growth

5.1 Gravity Effects on Endogenous IAA

Portions of these results have been published [3, 4]. They are reproduced here (Table 3) to demonstrate that placing the plants in a horizontal position, relative to the gravity vector, induces a rapid asymmetric distribution of both free and ester IAA in the mesocotyl cortex. Since both free *and* ester IAA increase on the lower side of the mesocotyl cortex, the gravity effect can not be solely due to increased hydrolysis of IAA ester on the lower side [3, 4]. An acceptable explanation is that either free or ester IAA selectively leaks from the lower side of the stele into the mesocotyl cortex. Since ester and free IAA are interconvertible, movement of either would be equivalent [8, 9, 22], and we conclude that the plant can regulate leakage from one or the other side of a horizontally placed stele.

Table 3. Percent of free and total (free plus ester) IAA in the lower half of the mesocotyl cortex of a *Zea mays* seedling as a function of time after moving the seedling from a vertical to a horizontal position

Time (min)	Free IAA	Total IAA
0	50 ± 0	50 ± 0
1	50 ± 0.1	–
3	56 ± 3	60 ± 1
5	55 ± 3	–
15	56 ± 1	62 ± 5
30	56 ± 2	56 ± 1
90	57 ± 3	54 ± 2

A second example of selected leakage, only this time for labeled IAA injected into the endosperm, is shown in Table 4. For this experiment 5-[^{3}H]-IAA was injected into the endosperm, with the plants held in a vertical position for varying periods of time to permit loading of the stele. The kernel was then severed to remove the source of labeled IAA, and the shoots then placed in a horizontal position. As can be seen, the radioactivity leaks selectively into the lower portion of the cortex, with movement into the upper portion of the cortex occurring to a much lesser extent. We interpret this experiment to again indicate that the plant can regulate the rate of leakage of IAA from the stele, and that more IAA is leaked to the lower side of the cortex. The resultant greater amount of IAA on the lower side could then result in greater growth on that side, and hence the expected gravitropic curvature.

Table 4. Movement of 5-[^{3}H]-IAA from stele to cortex during gravity stimulation[a]

30-min load	Total radioactivity (%)
Cortex, left half	23.1 ± 1.8
Cortex, right half	23.7 ± 2.6
Stele	53.2 ± 3.8
30-min load, followed by a 90-min gravity stimulus	
Cortex, upper half	28.5 ± 1.2
Cortex, lower half	35.9 ± 1.8
Stele	35.6 ± 1.7

[a] Radioactivity was administered by endosperm injection. The plants were kept in a vertical position for a 30-min loading period following injection. The plants that had not been sacrificed were then given a 90-min gravity stimulus by placing them in a horizontal position.

5.2 White Light

A further test of the ability of the plant to regulate movement of IAA from the endosperm into the stele, and finally into the surrounding cortex is shown in Table 5. For this experiment, labeled 5-[^{3}H]-IAA was injected into the endosperm and the plants were then incubated in darkness for 30 min, or given a 27-s light flash (at 2 m with a 250 W quartz iodine lamp), and then incubated in darkness. As can be seen, the light flash significantly reduces the amount of radioactivity that moved into the shoot. Further, the ratio of radioactivity in cortex to stele is less in continuous dark-treated plants than in light-treated plants. Thus, not only is the total radioactivity transported reduced by light, but also light restricts the amount of radioactivity moving from the stele into the cortex.

Table 5. Photoinhibition of the transport of 5-[^{3}H]-IAA from endosperm to mesocotyl

Cortex (dark)/Cortex (light)	= 1.28 ± 0.24, n = 16, t = 3.50
Stele (dark)/stele (light)	= 1.16 ± 0.22, n = 16, t = 2.18

5.3 Applied Voltage

As will be discussed below, the above results suggested a possibly voltage-controlled leakage of IAA from the stele into the surrounding cortical cells. If such is the case, an applied voltage should cause changes in the growth rate of the shoot, depending upon whether the applied voltage opened and/or closed IAA transport channels between stele and cortex. That such is indeed the case is shown Fig. 3 [13]. As can be seen, the application of a 5 V potential along 8 cm of shoot (0.6 mV/10 micrometer cell) causes an almost complete cessation of growth if the tip of the plant is made positive relative to the roots. Reversal of the polarity, with the shoot tip negative, relative to the roots, has almost no effect on growth although the current flow is the same. It is clear that an applied electric potential that is small relative to the cells' normal biopotential, can cause profound changes in growth.

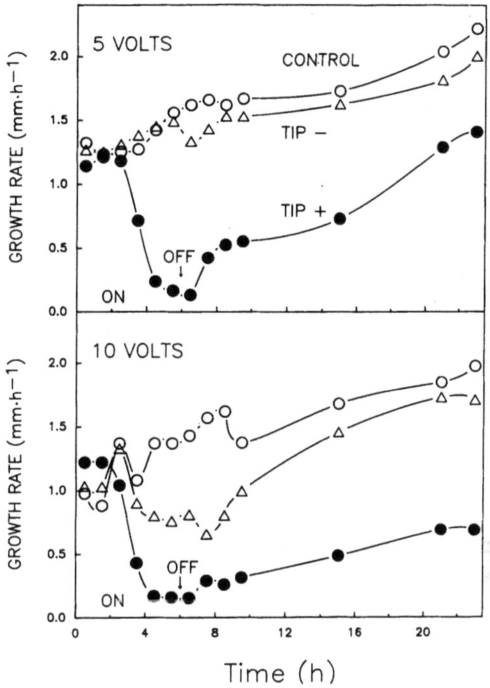

Fig. 3. The effects of an applied DC potential on the rate of growth of 5-day-old seedlings of *Zea mays*. It should be noted that 5 V applied along 8 cm of seedling amounts to a potential drop of about 0.6 mV per 10-μm-long cell

We are currently attempting to discover what changes in chemical composition result from the applied voltage. In results not shown here we find that the applied positive potential causes: (1) an increase in ester IAA in the stele: (2) decreased glucose transport to the cortical cells of the shoot; (3) decreased Ca transport to the cortical cells of the shoot. We do not yet know which, if any, of these changes, might be responsible for the profound reduction in growth caused by making the tip positive, relative to the roots.

6 A Working Theory

6.1 General Comments

Tropisms have been used as an exquisite test of the ability of a plant to selectively control its growth, since one side of a stem grows more rapidly than the other side of that same stem. The traditional explanation is the theory of the Lateral Transport of Auxin proposed by Went and Cholodny [32]. This theory states that the tropism induces lateral transport of IAA, thus causing more IAA to accumulate on one side of the stem than on the other. The theory is adequate, except there has been no explanation of how lateral transport of IAA is attained. We have developed a working theory to explain tropisms and lateral transport based upon our experiments, as described above, and upon the earlier studies of changes in the bioelectric potential of plant tissues induced by environmental stimuli.

6.2 Bioelectric Potential Changes

It has been known since the time of Bose and Brunner [e.g. 32] that external stimuli can induce large changes in the bioelectric potential of plants [27, 32]. Later studies confirmed and extended these observations [26, 27, 30, 31], although some workers felt the changes to be too slow to account for the tropic response, and were rather the result of enhanced growth instead of its cause. Quite recently it has been established by Behrens et al., [6] that a gravity stimulus induces membrane depolarization in *Lepidium* roots within 8 s after the stimulus. Similarly, Tanada [30] has observed changes following a R stimulus within 15 s of the stimulus. It is, as yet, uncertain how the gravity stimulus causes a change in bioelectric potential. Possibly it is the movement of a statholith [5, 7, 29], or the movement of any charged particle in a fixed field [5], or pressure-induced changes in the cells' microtubuler structure [5]. There are also uncertainties regarding the light receptor for phototropism [12]. Even in the absence of such knowledge it is possible to postulate that the stimulus causes a bioelectric potential change, and that this change, in turn, opens and/or closes transport channels for IAA and possibly other messengers such as Ca. The attractiveness of such a postulate is that it would provide a mechanism for the bioelectric change to cause an asymmetric distribution of a hormone, or Ca, thus eventually causing the asymmetric growth.

6.3 The Potential-Gating Theory of Tropic Response

The theory is simple and testable. Its postulates are: (1) an environmental stimulus induces a change in the bioelectric potential of the plant; (2) The bioelectric change represents a membrane depolarization; (3) The depolarization opens and/or closes voltage-gated IAA and Ca transport channels between stele and cortex, and possibly between cortical cells; (4) the resultant asymmetric distribution of IAA and Ca causes the asymmetric growth of the tropism; (5) IAA and Ca metabolism ultimately result in symmetric distribution and symmetric growth. A diagrammatic representation of how potential-gating could result in asymmetric IAA distribution is shown in Fig. 4.

7 Summary and Conclusions

1. The cortex plus epidermis of the mesocotyl of *Zea mays* can be separated from the vascular stele, making it an attractive experimental system.
2. The *Zea mays* seedling is a closed system apparently deriving its free IAA by hydrolysis of IAA conjugates.
3. The apparent K_m for IAA for growth is 7×10^{-8} M and the concentration of IAA in the mesocotyl cortex is 6×10^{-8} M. Thus, the plant grows at about one-half maximal rate and it is IAA, and not binding sites which appear to limit growth rate.
4. Both transport of IAA and shoot growth can be modulated by environmental stimuli such as gravity, R, and white light.

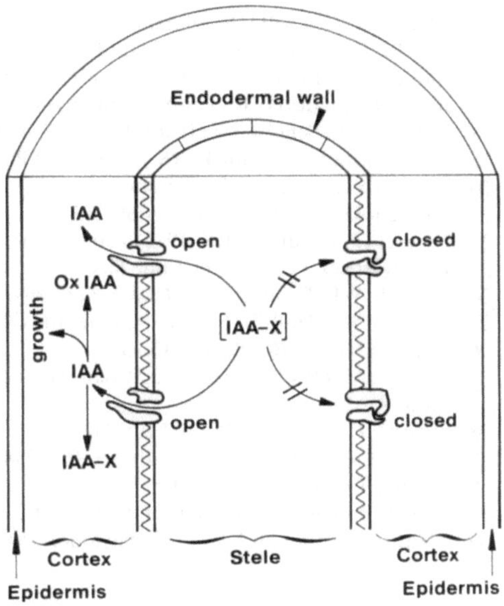

Fig. 4. The potential-gating theory is an attempt to explain how a change in bioelectric potential might result in a change in the distribution of IAA, and thus a resultant change in growth such that the plant responds to the tropic stimulus. It is postulated that the plasmodesmatal channels between the stele and the surrounding cortical cells are voltage regulated such that the channels can open and/or close following any stimulus which elicits a change in the plant's bioelectric potential

5. An applied electrical potential can reduce growth if the tip of the plant is made positive, relative to the roots.
6. A potential-gating theory is proposed to account for the asymmetric distribution of IAA induced by a tropic stimulis.

Acknowledgements. Supported by the National Science Foundation, Metabolic Biology DMB-8504231, by Space Biology, Life Sciences, National Aeronautics and Space Administration, NAGW-97; and by the NASA Flight Program, NAG 2-362. We acknowledge the aid of Ms. Wendy Whitford in manuscript preparation.

References

1. Bandurski RS (1984) In: Crozier A, Hillman JR (eds) The biosynthesis and metabolism of plant hormones. Soc Exptl Biol, Cambridge Press, p 183
2. Bandurski RS, Schulze A (1974) Plant Physiol 54:257
3. Bandurski RS, Schulze A, Dayanandan P, Kaufman PB (1984) Plant Physiol 74:284
4. Bandurski RS, Schulze A, Desrosiers M, Jensen P, Reinecke D, Epel B (1988) In: Boss WF, Morre DJ (eds) Second messengers in plant growth and development. Liss (in press)
5. Bandurski RS, Schulze A, Domagalski W (1986) Adv Space Res 6:47
6. Behrens HM, Gradmann D, Sievers A (1985) Planta 163:463
7. Casper T, Somerville CR, Pickard BG (1988) Planta (in press)
8. Chisnell JR (1984) The presence and translocation of indole-3-acetyl-*myo* inositol in shoots of *Zea mays*. Ph.D. thesis, Michigan State University, East Lansing
9. Chisnell JR, Bandurski RS (1988) Plant Physiol 86:79
10. Cohen JD, Bandurski RS (1982) Annu Rev Plant Physiol 33:403
11. Cohen JD, Bialek K (1984) In: Crozier A, Hillman JR (eds) The biosynthesis and metabolism of plant hormones. Soc Exptl Biol, Cambridge Press, p 165
12. Dennison DS (1984) In: Wilkins MB (ed) Adv plant physiol. Pitman, London, p 149
13. Desrosiers MF, Bandurski RS (1988) Plant Physiol 87:874

14. Epstein E, Cohen JD, Bandurski RS (1980) Plant Physiol 65:415
15. Hall PJ, Bandurski RS (1986) Plant Physiol 80:374
16. Hall PL, Bandurski RS (1978) Plant Physiol 61:425
17. Jensen P, Bandurski RS (1988) Plant Physiol 86 S:69
18. Kutschera V, Schopfer P (1986) Planta 169:437
19. Komosznski M, Bandurski RS (1986) Plant Physiol 80:961
20. Magnus V, Schulze A, Bandurski RS (1980) Plant Physiol 66:775
21. Momonoki Y (1985) Plant Physiol 77(S):2
22. Nowacki J, Bandurski RS (1980) Plant Physiol 65:422
23. van Overbeck J (1941) Am J Bot 28:1
24. Pengelly WL, Bandurski RS (1983) Plant Physiol 73:445
25. Pengelly WL, Hall PJ, Schulze A, Bandurski RS (1982) Plant Physiol 69:1304
26. Racusen RH (1976) Planta 132:25
27. Schrank AR (1947) In: Land EJ, Rosane HF (eds) Bioelectric fields and growth. University of Texas Press, Austin, p 75
28. Skoog F (1937) J Gen Physiol 20:311
29. Song I, Lu CR, Brock TG, Kaufman PB (1988) Plant Physiol 86:1155
30. Tanada T (1983) Plant Cell Environ 6:69
31. Tanada T, Vinten-Johansen C (1980) Plant Cell Environ 3:127
32. Went FW, Thimann KV (1937) Phytohormones. MacMillan, New York

Oxidation of Indole-3-Acetylaspartic Acid in *Vicia*

S. TSURUMI and S. WADA[1]

1 Introduction

Oxidation of IAA catalyzed by peroxidase and IAA oxidase with accompanying decarboxylation has been extensively studied [17, 18, 20, 21] and enzymatic kinetics of horseradish peroxidase and reaction products of IAA have been elucidated [8, 9, 12, 14]. In some plant species, however, peroxidative decarboxylation of IAA was reported to be a minor component in *Zea* seedlings [6, 15] and in *Pinus* seeds [7], where IAA was oxidized to OxIAA [19]. In the present report, we show another IAA oxidation system without decarboxylation in *Vicia* seedlings, where IAA-Asp was oxidized to dioxindole-3-acetic acid (DIA) conjugates, and a possibility that peroxidase could contribute to the IAA-Asp oxidation was also examined.

2 Metabolic Pathway of IAA in *Vicia* Seedlings

Application of [2-^{14}C]IAA to the cotyledon of etiolated 5-day-old broad been (*Vicia faba* L. cv Chukyo) seedlings resulted in accumulation of radioactive substances in the root primordia and in the stele of the basal part of the roots [24, 25]. Two metabolites being more polar than IAA-Asp accounted for 70–80% of total radioactivity in the root after 24-h treatment, and they were not extracted with ether in acid pH. After hydrolysis with 2 M HCl or 7 M NaOH, their radioactive moieties were extracted with ether, but they did not coincide with IAA. We purified the two substances from *Vicia* roots and identified them as 3-(O-β-glucosyl)-2-indolone-3-acetylaspartic acid (Glc-DIA-Asp) and 3-hydroxy-2-indolone-3-acetylaspartic acid (DIA-Asp) [26, 27]. The DIA moiety is converted into 2-quinolone-4-carboxylic acid (QCA) by acid hydrolysis and the UV spectrum of QCA is quite different from that of DIA, which is in contrast with the conversion of OxIAA into 1,2,3,4-tetrahydro-2-quinolone-4-carboxylic acid without accompanying large spectral change.

Figure 1 shows the metabolic pathway of IAA in *Vicia* seedlings. The precursor of the DIA conjugates was investigated using [2-^{14}C]DIA, indole-3-[2-^{14}C]acetylaspartic acid and indole-3-[2-^{14}C]acetyl[2,3-^{3}H]aspartic acid [28]. The radioactive substances were applied to the cotyledon or to the basal part of the roots of *Vicia* seedlings, and their metabolites were analyzed by TLC and reversed phase C_{18} HPLC. [^{14}C]DIA was not metabolized, but [^{14}C]IAA-Asp was converted into

[1] Department of Biological Fundamentals, Division of Science of Biological Resources, The Graduate School of Science and Technology, Kobe University, Rokkodai, Nada, Kobe 657, Japan

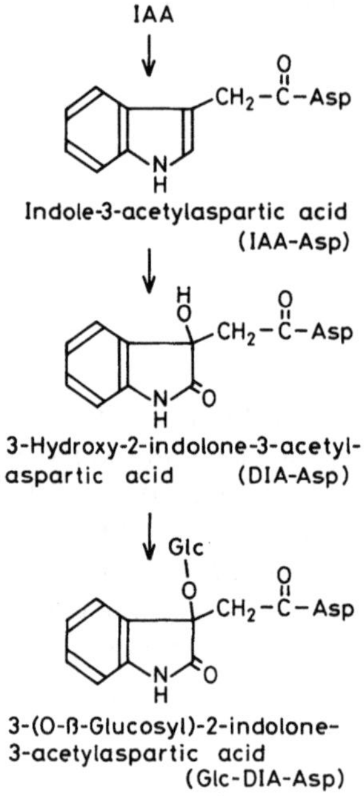

Fig. 1. Metabolic pathway of IAA in *Vicia* seedlings

[^{14}C]DIA-Asp by 39.1% of total activity in the root within 8 h and the formation of Glc-[^{14}C]DIA-Asp was 20.6% in the cotyledon after 24 h. The double labels from [^{14}C]IAA-[^{3}H]Asp were also incorporated into the DIA conjugates. The radioactivity ratio of [^{3}H]/[^{14}C] of donor [^{14}C]IAA-[^{3}H]Asp was 2.22, and the ratios of its metabolites associated with DIA-Asp and Glc-DIA-Asp were 2.30 and 2.36, respectively. Acid hydrolysis of the formed [^{14}C]DIA-[^{3}H]Asp and Glc-[^{14}C]DIA-[^{3}H]Asp gave [^{3}H]Asp and [^{14}C]QCA. These results indicated that the precursor of DIA conjugates was IAA-Asp. Metabolism of DIA-Asp was investigated using [^{14}C]DIA-Asp purified from the roots of *Vicia* seedlings after application of [^{14}C]IAA to the cotyledon for 24 h. [^{14}C]DIA-Asp was efficiently converted into its glucoside by 78.1% of total activity in the cotyledon within 8 h. Application of [^{14}C]IAA to the epicotyl tips produced both labeled DIA conjugates, but detailed studies on their identities have not been done.

IAA-Asp was first found in pea seedlings as an amide conjugate of exogenously supplied IAA [1]. Existence of endogenous IAA-Asp has now been shown in many plant species [4, 22, 23], but metabolism of IAA-Asp to DIA conjugates in plants has not been reported. Reinecke and Bandurski [19] identified OxIAA as a catabolic product of IAA in *Zea* seedlings, and OxIAA was found to be further metabolized to 7-OH-OxIAA and its glucoside [13, 16]. Although OxIAA was not found in *Vicia* roots, the two processes of DIA conjugates formation and OxIAA formation involved oxidation of IAA without decarboxylation. The presence of such a pathway in rice has been suggested by Kinashi et al. [11]. In *Zea* seedlings, IAA was oxidized at the 2-position of the indole ring and subsequently oxidized at the

7-position of the formed oxindole ring. However, in *Vicia* seedlings, the IAA moiety was oxidized at the 2- and 3-positions of the indole ring.

The two DIA conjugates were not found in pea seedlings, suggesting that formation of DIA conjugates might be limited to *Vicieae*.

3 Role of the Cotyledon in Glc-DIA-Asp Accumulation in the Root and Epicotyl of *Vicia* Seedlings

The labeled DIA conjugates, DIA-Asp and Glc-DIA-Asp, accounted for 39.8% and 42.8% of total radioactivity in the roots after 24 h, respectively [28]. However, excising the cotyledon treated with [^{14}C]IAA after 4 h suppressed the accumulation of labeled Glc-DIA-Asp to 10.5% of total activity in the root and increased that of DIA-Asp to 68.2% after 24 h. In the epicotyls, Glc-[^{14}C]DIA-Asp was a major metabolite (79.2%) after the application of [^{14}C]IAA to the cotyledon and accumulation of the glucoside was also suppressed by excising the cotyledon treated with [^{14}C]IAA. [^{14}C]IAA-Asp application to the root produced labeled DIA-Asp as a major metabolite (44.2%) in the root, but Glc-[^{14}C]DIA-Asp was a sole metabolite (80.0%) in the root following [^{14}C]IAA-Asp application to the cotyledon. These results indicated that Glc-DIA-Asp formed in the cotyledon was translocated to the root and epicotyl.

Figure 2 shows the scheme of transport and metabolism of IAA applied exogenously to the cotyledon of *Vicia* seedlings. A part of IAA is transported to the

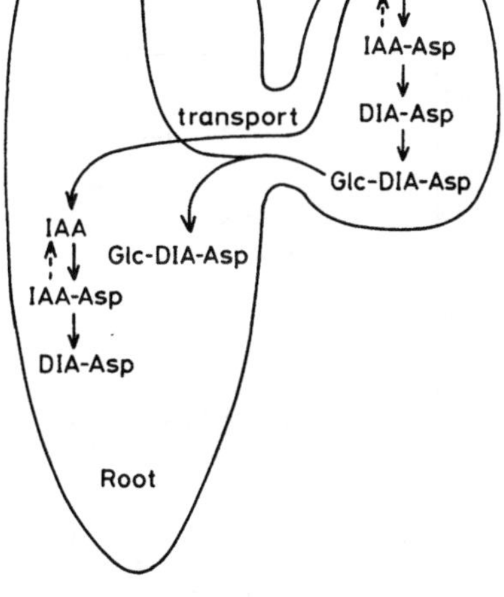

Fig. 2. Scheme showing transport and metabolism of IAA applied to the cotyledon of etiolated 5-day-old *Vicia* seedlings

root through the phloem and metabolized to DIA-Asp via IAA-Asp in the root. Other IAA is metabolized to Glc-DIA-Asp in the cotyledon, and the formed glucoside is transported to the root and shoot. Broken lines indicate slow release of free IAA from IAA-Asp, which also might occur [3].

4 Content of Glc-DIA-Asp in *Vicia* Seedlings

The amount of Glc-DIA-Asp was estimated by UV absorbance (254 nm) on HPLC [26]. Unimbibed seeds were ground with a mortar before homogenization. Seedlings were homogenized with 80% methanol and extracted stirring at 4°C for 12 h, which improved extraction efficiency of native Glc-DIA-Asp from the cotyledons. Dry cotyledons contained about 30 nmol Glc-DIA-Asp per plant and their content decreased gradually in the course of development, while the amount in the epicotyls and roots increased gradually (Fig. 3). These time courses of Glc-DIA-Asp content were consistent with the translocation of Glc-[^{14}C]DIA-Asp from the cotyledon to the root and epicotyl (Fig. 2). Total amount of Glc-DIA-Asp increased to the 4th day and decreased thereafter, suggesting the synthesis of native Glc-DIA-Asp and its further metabolism. On a per organ basis, the greatest amount of Glc-DIA-Asp in etiolated 4-day-old seedlings was found in the cotyledons (29.3 nmol/cotyledon), while lesser amounts were found in the epicotyls and roots (5.1 and 3.9 nmol/organ, respectively) [26]. However, on a per g fw basis, the concentrations of Glc-DIA-Asp in the epicotyl tips and the basal part of the roots (15.6 and 12.2 nmol/g, respectively) were higher than in other parts of their organs (4–8 nmol/g), which was consistent with accumulation of radioactive metabolites in the epicotyl tips and the basal part of the roots of *Vicia* seedlings treated with [^{14}C]IAA to the cotyledons [24].

Andreae and Ysselstein [2] reported that a lag of about 2 h was required for IAA-Asp formation in pea roots. However, in *Vicia* roots, DIA-Asp formation was dected within 2 h of IAA treatment. DIA-Asp was also a naturally occurring compound and its content in *Vicia* roots was estimated to be an order of magnitude

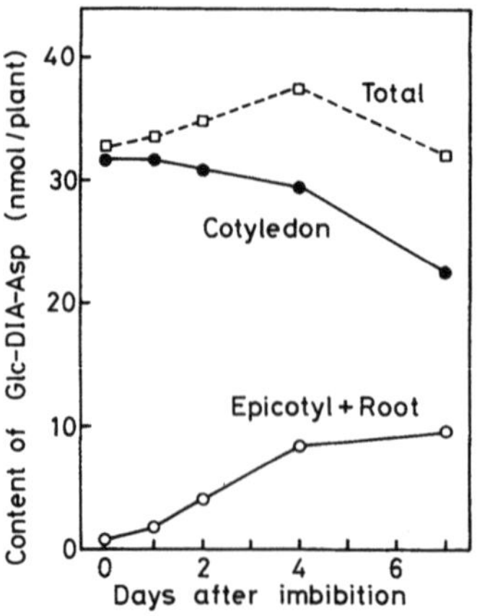

Fig. 3. Time course of endogenous Glc-DIA-Asp content in the cotyledon (●), epicotyl + root (○) and their total (□) during the development of *Vicia* seedlings

lower than that of Glc-DIA-Asp. Although for accurate determination of DIA conjugates content further studies are required, from the natural occurrence of these DIA conjugates and the correlation between the distributions of native Glc-DIA-Asp and of labeled Glc-DIA-Asp, the DIA conjugates formation is likely to be a natural metabolic pathway of endogenous IAA.

5 Physiological Activities of DIA-Asp

DIA-Asp was synthesized by aerobic oxidation of IAA-L-Asp in the presence of bisulfite and Mn^{2+} [10]. The authentic DIA-Asp was separated into two isomers by reversed phase HPLC [27]. Only one isomer coincided with the natural DIA-Asp and it showed a positive Cotton effect. DIA-Asp was tested on stem growth, adventitious root formation, lateral root formation and apical dominance using *Vicia* seedlings and segments, but it was not active in any of these assays.

6 Oxidation of IAA-Asp by Peroxidase

The activity of peroxidase on IAA-L-Asp oxidation was examined by using horseradish peroxidase (Wako, RZ 1.0–2.0). The UV-spectrum of IAA-Asp was not changed by peroxidase without added H_2O_2 [e.g. 5, 17]. Addition of H_2O_2, however, induced oxidation of IAA-Asp in acid pH. Figure 4 shows the comparison of the spectral changes in IAA-Asp and IAA oxidations, which were very similar for the first 20 min, but the spectra of their end products were different. A hydroperoxide derived from IAA was shown [14] to accumulate in the IAA oxidase reaction catalyzed by horseradish peroxidase and this hydroperoxide reacted with perox-

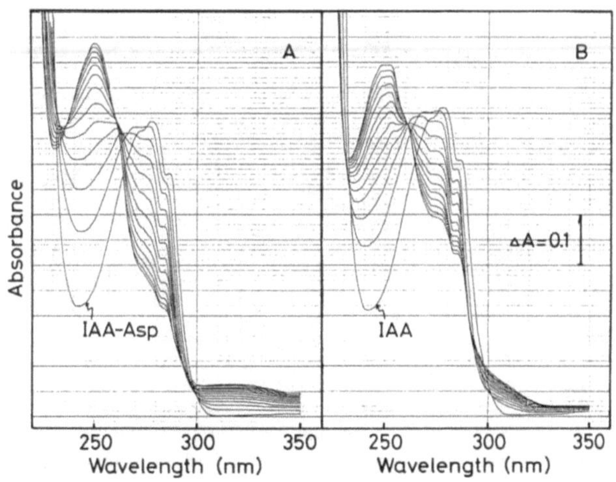

Fig. 4. Repeated scan UV-spectra in IAA-Asp (*A*) and IAA (*B*) oxidase reactions by horseradish peroxidase. A spectrum was scanned every 5 min. Room temperature was 23 °C. 0.4 μM peroxidase, 50 mM sodium acetate (pH 4). A, 0.1 mM IAA-Asp + 0.1 mM H_2O_2. B, 0.1 mM IAA

idase to form compound I. The time course of the hydroperoxide concentration in the IAA oxidase reaction coincided with that of UV absorbance at 274 nm [14]. The temporary increase of absorbance at 274 nm was also found in the oxidation of IAA-Asp (Fig. 4A), suggesting the formation of a hydroperoxide derived from IAA-Asp. However, the hydroperoxide does not seem to react with peroxidase, which may be the reason of stability of IAA-Asp against peroxidase in the absence of H_2O_2.

To collect the oxidation products, IAA-Asp was oxidized overnight in the mixture (30 ml) of 0.2 mM IAA-Asp, 0.4 μM peroxidase, 0.1 mM H_2O_2 and 50 mM sodium acetate (pH 4.0) at room temperature. The oxidation products were applied to a Toyopearl HW 40 column (1.8×35 cm, Toyo Soda) and developed with a MeOH-0.1% HOAc (3:7, v/v) solution. Major peaks (60–95 ml) were collected and analyzed by HPLC (TSK GEL, ODS 120T, 7.8×300 mm, Toyo Soda) using a solvent of MeOH-0.05% HOAc (2:8, v/v), which produced four peaks (A, B, C and D) at Rvs of 40.8, 43.8, 48.7 and 57.2 ml, respectively. The products showed tautomer-like behaviors as follows A ⇆ B and C ⇆ D. Their UV-spectra were similar to that of OxIAA, but their hydrolysis with 2 M HCl at 100 °C for 14 h yielded QCA and aspartic acid. Reasonable structures for the oxidation products would be 2-indolone-3-(2-hydroxy)acetyl-Asp (2-OH-OxIAA-Asp) which has two asymmetric carbon atoms at the 3-position of the oxindole ring and also at the 2-position of the side chain giving four isomers. However, the absolute configuration at the 3-position of the oxindole ring changes through its tautomer of enol-form as shown in Fig. 5. The sites of incorporation of oxygen atoms in 2-OH-OxIAA-Asp were comparable with those in oxindole-3-carbinol which was one of the oxidation products of IAA oxidase reaction by peroxidase [9]. Peroxidase oxidized IAA-Asp without decarboxylation, but the oxidation products were not DIA-Asp.

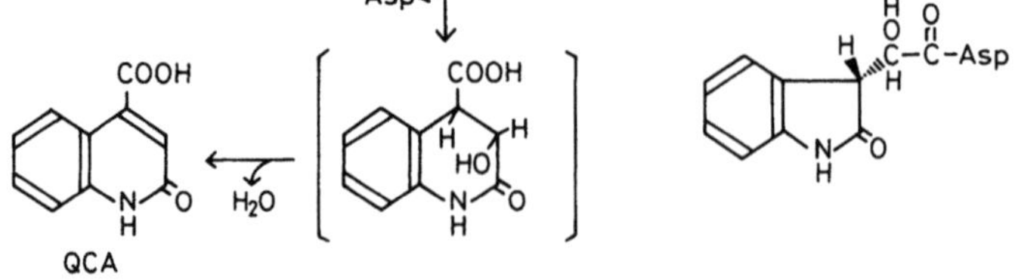

Fig. 5. Scheme showing a reasonable structure for oxidation products of IAA-Asp by peroxidase and its interconversion through the enol-form. The 2-OH-OxIAA moiety is converted into QCA by acid hydrolysis

7 Conclusions

1. The major catabolic pathway of exogenous IAA in *Vicia* seedlings is IAA → IAA-Asp → DIA-Asp → Glc-DIA-Asp.
2. The DIA conjugates formation is also likely to be a catabolic pathway of endogenous IAA.
3. DIA-Asp was not biologically active.
4. IAA-Asp was oxidized by horseradish peroxidase with added H_2O_2 to form four oxindole derivatives, but the reaction products were not DIA-Asp.

Acknowledgements. We wish to thank Prof. I. Yamazaki of Hokkaido University for his invaluable suggestion and Dr. Y Takeuchi of Shionogi Seiyaku Co. Ltd. for his measurements of CD spectra.

References

1. Andreae WA, Good NE (1955) Plant Physiol 30:380
2. Andreae WA, van Ysselstein MW (1960) Plant Physiol 35:225
3. Bialek K, Meudt WJ, Cohen JD (1983) Plant Physiol 73:130
4. Cohen JD (1982) Plant Physiol 70:749
5. Cohen JD, Bandurski RS (1978) Planta 139:203
6. Epstein E, Cohen JD, Bandurski RS (1980) Plant Physiol 65:415
7. Ernstsen A, Sandberg G, Lundstrom K (1987) Planta 172:47
8. Grambow HJ, Langenbeck-Schwich B (1983) Planta 157:131
9. Hinman RL, Lang J (1965) Biochemistry 4:144
10. Horng AJ, Yang SF (1975) Phytochemistry 14:1425
11. Kinashi H, Suzuki Y, Takeuchi S, Kawarada A (1976) Agric Biol Chem 40:2465
12. Kobayashi S, Sugioka K, Nakano H, Nakano M, Tero-Kubota S (1984) Biochemistry 23:4589
13. Lewer P, Bandurski RS (1987) Phytochemistry 26:1247
14. Nakajima R, Yamazaki I (1979) J Biol Chem 254:872
15. Nonhebel HM, Crozier A, Hillman JR (1983) Physiol Plant 57:129
16. Nonhebel HM, Kruse LI, Bandurski RS (1985) J Biol Chem 260:12685
17. Park RD, Park CK (1987) Plant Physiol 84:826
18. Ray PM (1958) Annu Rev Plant Physiol 9:81
19. Reinecke DM, Bandurski RS (1981) Biochem Biophys Res Commun 103:429
20. Schneider EA, Wightman F (1974) Annu Rev Plant Physiol 25:487
21. Sembdner G, Gross D, Liebisch HW, Schneider G (1980) In: MacMillan J (ed) Hormonal regulation of development I. Encyclopedia of plant physiology, new series, vol 9. Springer Berlin Heidelberg New York, p 281
22. Sonner JM, Purves WK (1985) Plant Physiol 77:784
23. Tillberg E (1974) Physiol Plant 31:271
24. Tsurumi S, Wada S (1980) Plant Cell Physiol 21:803
25. Tsurumi S, Wada S (1980) Plant Cell Physiol 21:1515
26. Tsurumi S, Wada S (1985) Plant Physiol 79:667
27. Tsurumi S, Wada S (1986) Plant Cell Physiol 27:559
28. Tsurumi S, Wada S (1986) Plant Cell Physiol 27:1513

Indole-3-Ethanol Metabolism and Its Possible Role in the Regulation of Indole-3-Acetic Acid Biosynthesis

V. MAGNUS and G. LAĆAN[1]

1 Introduction

Indole-3-ethanol was discovered by Ehrlich [5] in 1912, as a metabolite of tryptophan added to yeast cultures, and given the trivial name *tryptophol*. The compound is also present in products of yeast fermentation, such as beer and wine [6] and, as one of the constituents determining flavor, has not ceased to attract the attention of brewers and enologists [e.g. 10]. As tryptophol contains the same C-N backbone as IAA, some common pathways of biogenesis and metabolism have long been anticipated. Details have, however, been slow to emerge. The older literature was reviewed earlier [e.g. 33, 38], and we will try to incorporate recent data into the classical concepts.

2 Tryptophol as a Natural Plant Constituent

While chromatographic evidence points to the presence of tryptophol in a variety of plant tissues [38], the compound was first identified by MS in cucumber seedlings [28]. Other species in which tryptophol has been found by MS, or GC-MS include:

Bacteria (grown on tryptophan-free medium): *Rhizobium phaseoli* [7]

Fungi (media not supplemented with tryptophan): *Phycomyces blakesleeanus* [35]; *Saccharomyces sp.* [6, 10]; *Candida albicans* [17]; *Drechslera* (*Helminthosporium*) *nodulosum* [40]

Ferns: *Matteucia struthiopteris* [34]

Seed Plants: *Pinus silvestris* [30, 32]; *Picea abies* [31]; *Pisum sativum* [3]; *Dalbergia dolichopetala* [22].

3 Tryptophol as a Plant Metabolite under in Vitro Conditions

When the in vitro conversion of indolic compounds in plants was investigated, tryptophol was one of the most commonly detected metabolites. Tryptophan, tryptamine, ILA, and indole-3-acetaldoxime were among its precursors [9, 38]. All these compounds are metabolized via indole-3-acetaldehyde (Fig. 1), which would

[1] Institut Rudjer Bošković, Department of Organic Chemistry and Biochemistry, 41001 Zagreb, PO Box 1016, Yugoslavia

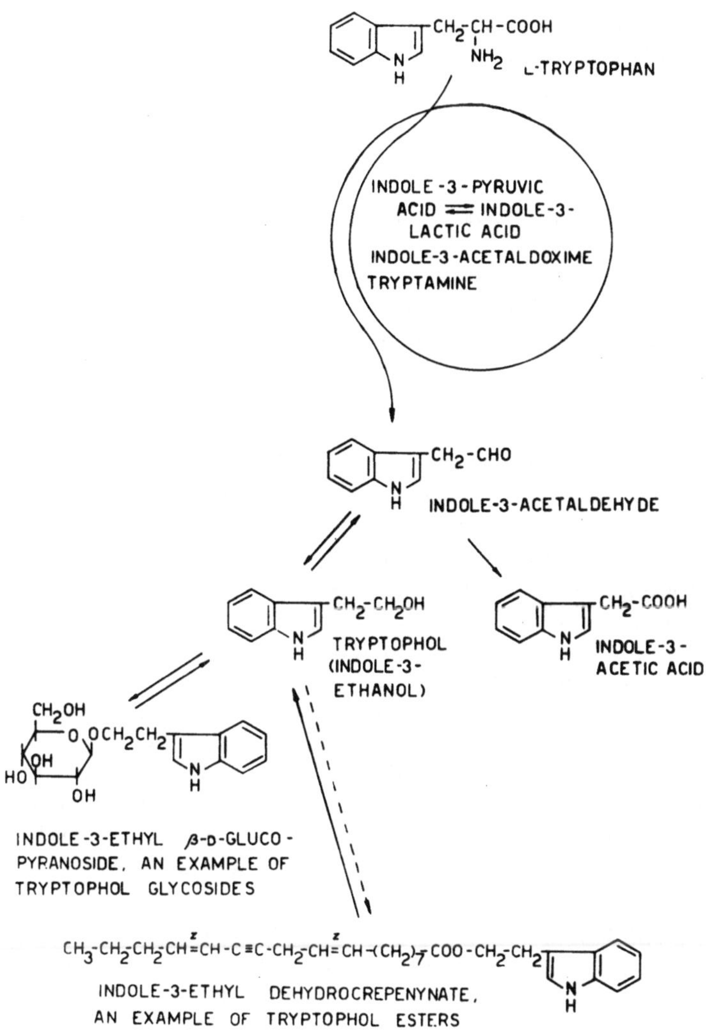

Fig. 1. Pathways of tryptophol (indole-3-ethanol) biogenesis and conjugation

then be the immediate precursor not only of IAA, but of tryptophol as well. Indeed, that aldehyde was reduced to tryptophol by both plants and bacteria [38]. Moreover, when cultures of *Acetobacter xylinum* were supplied with tryptophan, the formation of both tryptophol and IAA was drastically reduced when an aldehyde-trapping agent like $NaHSO_3$ was added, while the bisulfite adduct of indoleacetaldehyde accumulated [16].

Only the plant pathogen *Corynebacterium fascians* has so far been reported to reduce IAA to tryptophol [12]. The inverse oxidation, however, appears to be common in bacteria and eucaryotic organisms [14, 21, 29, 30]. So the reduction of indoleacetaldehyde must be reversible. Tryptophol could, however, accumulate in much higher concentrations than IAA when, for example, tryptamine was supplied in excess to plants [9, 20, 39]. This prompted detailed investigations upon the enzymes interconverting tryptophol, indoleacetaldehyde, and the plant growth hormone, IAA.

4 Enzymatic Interconversion of Indoleacetaldehyde, Tryptophol, and IAA

The simple hypothesis that indoleacetaldehyde dismutates to yield the homologous alcohol and acid [15] was soon abandoned. Not only were the amounts of tryptophol and IAA widely different in most plant systems [26], the respective enzymatic activities were also sensitive to different inhibitors, and they could be separated by fractional ammonium sulfate precipitation [27]. The most complete study on the enzymology of indoleacetaldehyde metabolism was done by Purves' group working with cucumber seedlings [1, 2, 4, 25, 42].

The IAA-forming activity present in crude buffer extracts could unfortunately not be stabilized to permit isolation of the respective enzyme(s) [2]. It could, however, be attributed to a true oxidase (or a set of oxidases) which was at least group specific for aromatic aldehydes. The assay system did not permit a check for product inhibition by IAA. PAA and the auxin analogue 2,4-D did, however, inhibit indoleacetaldehyde oxidation.

Three indoleacetaldehyde reductases were purified from cucumber seedlings [1, 4]. The enzyme requiring NADH as a cofactor occurred in the cytosol; one of the two NADPH-specific reductases was associated with a microsomal fraction. The latter reduced phenylacetaldehyde at about half the rate observed for indoleacetaldehyde and exhibited minor activity on some of the aliphatic aldehydes tested. The NADH-requiring enzyme acted only on indoleacetaldehyde and phenylacetaldehyde. None of the three enzymes would catalyze the reverse oxidation of tryptophol.

There are, however, specific tryptophol oxidases (at least two) in cucumber seedlings [1, 25, 42]. These oxidases are inhibited by indoleacetaldehyde, IAA, and synthetic auxin analogues. So IAA could regulate its own biosynthesis, at least from tryptophol.

A tryptophol oxidase similar to the cucumber enzymes was recently isolated from the mold *Phycomyces blakesleeanus* [36]. The demonstrated occurrence of IAA-sensitive tryptophol oxidases in a fungus and a higher plant strongly suggests that enzymes of this kind may be widely distributed. In fact, even the microorganism *Arthrobacter* responded to tryptophol feeding by formation of a distinct oxidase. This oxidase could also be induced by indoleacetaldehyde, but by no other indoles and, notably, not by aliphatic alcohols such as ethanol [21].

5 IAA-Formation from Excess Exogenous Tryptophol

There is a drawback to studying purified enzymes. The proteins may be modified during isolation, cofactors may be lost, and the concentrations of substrates and inhibitors at the natural sites of action are difficult to determine. We therefore decided to supply excess tyrptophol to plant systems containing undisrupted cell structures [14]. The compound is readily taken up [41], and the concentrations used did not cause obvious damage to the tissues examined. If tryptophol oxidation is regulated in accord with intracellular auxin levels, then the excess tryptophol should *not* be converted to a comparable excess of IAA. This was indeed observed with 50 out of 52 seed plants and many of the thallophytes studied.

6 Tryptophol Conjugation

Even plants which refused to convert exogenous tryptophol to IAA to any significant extent, formed substantial amounts of tryptophol conjugates: glycosides and esters [14]. The β-D-glucopyranoside prevailed in seed plants; but the β-D-galactopyranoside was so far only detected in three species of unicellular algae. Tryptophol acetate was formed by about half of the species examined. A set of highly lipophilic esters was among the most prominent tryptophol metabolites in bacteria, thallophytes, and seed plants. Three such esters were also isolated after tryptophol feeding from fruiting bodies of the basidiomycete *Craterellus cornucopioides*. NMR and mass spectra, as well as comparison with authentic standards obtained by chemical synthesis, permitted identification of tryptophol dehydrocrepenynate (Fig. 1) as the major component, accompanied by smaller amounts of the oleate and linoleate [13, 20a]. The respective fatty acids occur in the triglycerides of *Craterellus*; their acyl residues may then be transferred to tryptophol by transesterification. This must be enzymatic, as boiled fungal material did not esterify tryptophol.

The formation of tryptophol glucoside in etiolated pea seedlings was formally reversible [19]. This was also true for tryptophol oleate in *Craterellus* [20a], and may be expected to apply to other tryptophol conjugates and plant species as well.

As no plant could be found which did not conjugate tryptophol in vitro [14], a search for such conjugates as naturally occurring compounds could be rewarding. In fact, the presence of tryptophol esters in products of yeast fermentation was inferred by Ehrlich [6], and the acetate was recently identified in wine using GC-MS [10]. An alkali-labile tryptophol conjugate, presumably an ester, was also detected in seeds of *Pinus silvestris* [32]. The formation of tryptophol esters by the phylogenetically unrelated yeasts and pines may well indicate that such compounds commonly occur in plants.

7 Tryptophol and Its Conjugates as Reserve Auxins

It would violate the general principles of "economy", if tryptophol was simultaneously formed, conjugated, and reoxidized in the same plant tissue. Etiolated pea seedlings, for example, efficiently transform potential IAA precursors into tryptophol and its glucoside [20]. Both compounds can also be converted to IAA, but in very low yields, as suggested by chromatographic evidence and bioassay [19, 20]. In the pea stem elongation test at about half-optimal response, the activity of

IAA could only be matched by the ~1000-fold concentration of tryptophol and the ~50 000-fold concentration of its glucoside [18]. So etiolated pea seedlings appear to be programmed for tryptophol accumulation, but not for its metabolism to IAA.

On the other hand, pumpkin hypocotyl explants cultured in vitro utilized tryptophol and its glycosides as sources of auxin, presumably by converting them to IAA [11]. IAA, however, was not nearly as good in supporting callus growth and embryogenesis as was tryptophol α-L-arabinopyranoside. A plausible explanation would be that IAA supplied in the medium is metabolized by the explants before it can exhibit its full effect, while the auxin formed continuously from tryptophol and its conjugates is available for a much longer period of time. So it appears that tryptophol and its conjugate(s) *can* be storage forms which are subsequently usable for IAA biosynthesis. However, only few data have so far been obtained which demonstrate that they actually have such a function in the real life of a higher plant.

However, in the fungus *Phycomyces blakesleeanus*, tryptophol accumulated in the mycelium, and was subsequently metabolized during the phase of sporangiophore elongation, a period when extractable tryptophol oxidase activity attains its highest levels [35]. Extractable IAA concentrations in the whole fungus did not rise.

Mature pine seeds contain about 0.5 pmol/seed of each IAA and tryptophol, 2 pmol/seed of ester IAA, and 3 pmol/seed of an alkali-labile tryptophol conjugate [32]. These levels build up gradually during seed development. During germination, the high auxin levels required for rapid hypocotyl elongation are apparently formed simultaneously from the stored conjugates of IAA and tryptophol. The concentration of tryptophol peaked about one day before that of IAA. No tryptophol was detectable in rapidly growing pine seedlings, but it appeared again in the maturing needles [30].

8 Other Pathways of Tryptophol Metabolism

The oxidation of tryptophol to indoleacetaldehyde is not necessarily its only pathway of metabolism. The microorganism *Pseudomonas fluorescens*, for example, contains a "tryptophan side-chain oxidase". This oxidase, at least in vitro, attacks a variety of substrates including tryptophol, to form indole-3-glycol and indole-3-ketol [Fig. 2; 23]:

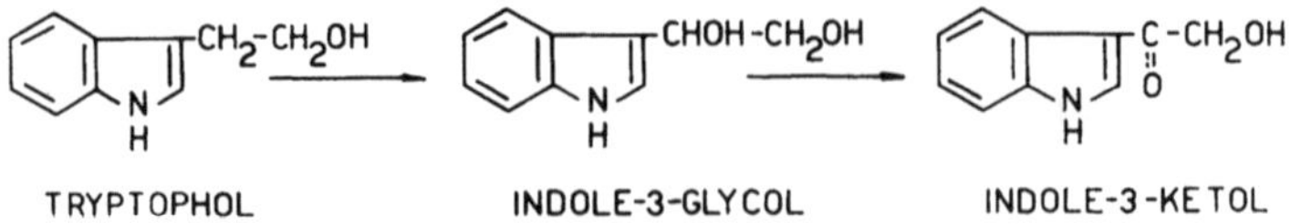

Fig. 2. Tryptophol oxidation catalyzed by "tryptophan side-chain oxidase" from *Pseudomonas fluorescens*

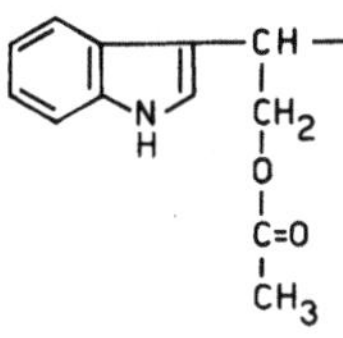

Fig. 3. A genotoxic indole derivative produced by the intestinal bacterium *Streptococcus faecium*. The compound is essentially *O*-acetyl tryptophol substituted at the side chain with an additional indole nucleus

A genotoxic compound termed streptindole has been isolated from cultures of the intestinal bacterium *Streptococcus faecium* [Fig. 3; 24]. The compound contains the carbon backbone of tryptophol acetate, so a close metabolic relationship appears plausible. These kinds of tryptophol metabolites have so far not been looked for in higher plants.

9 Concluding Remarks

There are several potential IAA precursors other than tryptophol which may accumulate in some plants. Fairly abundant in the Brassicaceae and in some related families, although controversial with respect to its relationship to auxin metabolism, is glucobrassicin. Healthy plants that accumulate the compound are apparently unable to convert it to IAA in the same tissue and at the same time [37]. However, might it be reutilized during a later phase in the life cycle of the plant? A less controversial but rarely appreciated reserve pool of IAA precursors is comprised of D-tryptophan and its N-malonyl conjugate. The latter occurs naturally in plants, supports growth in plant tissue culture, and the [^{14}C]-labelled compound has been shown to be converted to [^{14}C]-IAA [8].

Why do plants accumulate reserve pools of IAA precursors, instead of just regulating the first step in IAA biogenesis? More data will be required before this question can be definitely answered. In particular, though, there is much to be learned about possible compartmentalization of tryptophan and of the enzymes which transform that amino acid into indoleacetaldehyde. If these enzymes are uniformly distributed in cells, and are as non-specific as claimed by most authors, then feed-back regulation by IAA would seem to be pointless. The high bulk concentration of L-tryptophan in rapidly growing plant tissues may, however, be misleading. Most of the L-tryptophan may be concentrated at centres of protein synthesis; there could even be a shortage at sites of IAA formation. If so, this would then be the time to mobilize reserve pools of auxin precursors, such as tryptophol and its conjugates, to provide the high amounts of IAA necessary for rapid, coordinated growth.

References

1. Bower PJ, Brown HM, Purves WK (1976) Plant Physiol 57:855
2. Bower PJ, Brown HM, Purves WK (1978) Plant Physiol 61:107
3. Brown BH, Crozier A, Sandberg G, Jensen E (1986) Phytochemistry 25:299

4. Brown HM, Purves WK (1976) J Biol Chem 251:907
5. Ehrlich F (1912) Ber Dtsch Chem Ges 45:883
6. Ehrlich F (1917) Biochem Z 79:232
7. Ernstsen A, Sandberg G, Crozier A, Wheeler CT (1987) Planta 171:422
8. Gamburg KZ, Rekoslavskaya NI (1985) Usp Sovrem Biol 100:44
9. Gibson RA, Schneider EA, Wightman F (1972) J Exp Bot 23:381
10. Güntert M, Rapp A, Takeoka GR, Jennings W (1986) Z Lebensm Unters Forsch 182:200
11. Jelaska S, Magnus V, Seretin M, Laćan G (1985) Physiol Plant 64:237
12. Kemp DR (1978) In: Loutit MW, Miles JAR (eds) Microbial ecology. Springer, Berlin Heidelberg New York, p 341
13. Laćan G, Lewer P, Magnus V, Iskrić S (1987) In: Schreiber K, Schütte HR, Sembdner G (eds) Proc Intern Symp Conjugated Plant Hormones. Structure, Metabolism and Function. 3–5 Nov 1986. VEB Deutscher Verlag der Wissenschaften, Berlin, East Germany, p 69
14. Laćan G, Magnus V, Šimaga Š, Iskrić S, Hall PJ (1985) Plant Physiol 78:447
15. Larsen P (1950) Am J Bot 37:680
16. Larsen P, Harbo A, Klungsöyr S, Aasheim T (1962) Physiol Plant 15:552
17. Lingappa BT, Prasad M, Lingappa Y, Hunt DF, Biemann K (1969) Science 163:192
18. Magnus V (1975) Study of the metabolism of indolic compounds in higher plants (in Croatian). Thesis, University of Zagreb
19. Magnus V (1979) Carbohydr Res 76:261
20. Magnus V, Iskrić S, Kveder S (1973) Planta 110:57
20a. Magnus V, Laćan G, Iskrić S, Lewer P, Aplin RT, Thaller V (1989) Phytochemistry 28:2949
21. Mino Y (1974) Agric Biol Chem 38:211
22. Monteiro AM, Sandberg G, Crozier A (1987) Phytochemistry 26:327
23. Narumiya S, Takai K, Tokuyama T, Noda Y, Ushiro H, Hayaishi O (1979) J Biol Chem 254:7007
24. Osawa T, Namiki M (1983) Tetrahedron Lett 24:4719
25. Percival FW, Purves WK, Vickery LE (1973) Plant Physiol 51:739
26. Rajagopal R (1967) Physiol Plant 20:982
27. Rajagopal R (1968) Physiol Plant 21:1076
28. Rayle DL, Purves WK (1967) Plant Physiol 42:520
29. Rayle DL, Purves WK (1967) Plant Physiol 42:1091
30. Sandberg G (1984) Planta 161:398
31. Sandberg G, Ernstsen A (1987) Tree Physiol 3:185
32. Sandberg G, Ernstsen A, Hamnede M (1987) Physiol Plant 71:411
33. Schneider EA, Wightman F (1978) In: Letham DS, Goodwin PB, Higgins TJV (eds) Phytohormones and related compounds – a comprehensive treatise, vol I. Elsevier, Amsterdam, p 29
34. Schneider EA, Wightman F (1986) Physiol Plant 68:396
35. Schramm P, Rausch T, Hilgenberg W (1987) Physiol Plant 69:99
36. Schramm P, Rausch T, Hilgenberg W (1987) Plant Physiol 84:541
37. Schraudolf H, Weber H (1969) Planta 88:136
38. Sembner G, Gross D, Liebisch H-W, Schneider G (1980) In: MacMillan J (ed) Encyclopedia of plant physiology, new series, vol 9. Springer, Berlin Heidelberg New York, p 281
39. Sherwin JE, Purves WK (1969) Plant Physiol 44:1303
40. Sugawara F, Strobel GA (1987) Phytochemistry 26:1349
41. Sussman MR, Goldsmith MHM (1981) Planta 150:15
42. Vickery LE, Purves WK (1972) Plant Physiol 49:716

The Oxindole-3-Acetic Acid Pathway in *Zea mays*

D. Reinecke[1,2]

1 Introduction

Research interest concerning the catabolism of the plant hormone IAA has continued since auxin's discovery in the 1930s, because catabolism is a way for the plant to regulate hormone level and thus hormone activity. Until recently, it was accepted that IAA was decarboxylated during oxidation. The "indole oxidase" was presumed to be peroxidase due to the results from in vitro studies [9, 25] even though chemical evidence for carboxyl-free IAA metabolites in plants was lacking. However, we know now that several plants oxidize the IAA ring with carboxyl-retention as shown unambiguously with physicochemical methods in corn, broad bean, and Scots pine [18, 29, 5]. There is also chemical evidence that a similar pathway may occur in rice [11].

In corn, Bandurski et al. [1] have identified several inputs to and outputs from the IAA pool including 1) de novo IAA synthesis, 2) conjugate synthesis, 3) conjugate hydrolysis, 4) transport of IAA to and from the actively growing tissue, and 5) catabolism. The enzymes involved in conjugate synthesis of IAA to IAA-glucose, and the catabolism of IAA to OxIAA, are likely to be regulated enzymes, since they are initial reactions in branch points of IAA metabolism. IAA oxidation is unique in that it is the only irreversible output from the IAA pool. It is also interesting to speculate that IAA catabolism may be temporally or spatially close to the initiation of the hormone cascade of reactions, thus keeping the hormone's level growth limiting. The following discussion will examine the route of IAA oxidation by *Zea mays* using the results from feeding studies, quantitative assays, and in vitro assays.

2 IAA Catabolism

2.1 In Situ vs in Vitro Studies

The IAA metabolites from in vitro studies can be very different from in situ studies: with an in vitro corn peroxidase assay 11 decarboxylated IAA products were identified [2]. However, by feeding shoot and root sections radiolabeled IAA, 6 to 11 IAA carboxyl-retaining metabolites were isolated by HPLC [16, 17]. The major

[1] Department of Botany and Plant Pathology, Michigan State University, East Lansing, MI 48824, USA
[2] Present address: Department of Plant Biology, 220 Bio. Sci. Center., 1445 Gortner Avenue, St. Paul, MN 55108, USA

IAA catabolite had chromatographic properties similar to OxIAA; the other metabolites were not analyzed, but presumably consisted of IAA conjugates and other catabolites. In Scots pine, an in vitro system rapidly catabolized IAA to indole-3-methanol plus 3 other decarboxylated catabolites [26]. However, Scots pine protoplasts oxidized IAA more slowly to a carboxyl-retaining catabolite (presumably OxIAA from later work, [5]), and indole-3-methanol as a minor metabolite. The route(s) of IAA catabolism as well as IAA turnover rates need to be unambiguously measured in other plants so that control points for IAA turnover can be identified.

2.2 IAA Turnover and Catabolism Studies

IAA turnover experiments were the first indication that there might be a catabolic pathway other than peroxidase-catalyzed decarboxylation of IAA in corn. Epstein et al. [4] observed that IAA was turning over at about 60 pmol•hour^{-1}•endosperm^{-1}, but only 4 pmol•h•$^{-1}$•endosperm^{-1} of the turnover could be accounted for by decarboxylation following feeding 1-^{14}C-IAA and trapping the $^{14}CO_2$ released [recalculated [23]]. The following experiments were designed to identify the unaccounted for 90% of IAA catabolism in corn.

Feeding studies with 1-^{14}C-IAA [18] showed that IAA was metabolized to a polar carboxyl-retaining metabolite. Identification of the metabolite was hindered by the fact that it could not be easily derivatized for GC-MS. Our initial hypothesis was that the pyrrole ring of IAA would be opened during oxidation in a manner similar to tryptophan metabolism. Also, mixed in with the plethora of peroxidase papers were a few papers that suggested that the indole ring of IAA could be oxidized with the carboxyl carbon remaining intact. These papers included colorimetric evidence by Klambt in *Brassica rapa*, *Ribes rubrum* and *Zea mays* [12], IR, NMR, and MS data for OxIAA and DiOxIAA derivatives in *Oryza sativa* (rice) bran [11], and UV evidence for DiOxIAA derivatives in *Vicia faba* [29]. Synthetic OxIAA [synthesized according to [8]] was shown to have the same GC and HPLC chromatographic properties as the corn IAA catabolite. Finally, GC-MS spectral data gave physicochemical evidence for OxIAA as a major product of oxidation of IAA in corn [18].

2.2.1. OxIAA Quantitation in Corn

To confirm that OxIAA occurred naturally in corn, an isotope dilution assay was developed to measure OxIAA levels using synthetic 1-^{14}C-OxIAA [19]. Any OxIAA naturally present in corn tissue (extract) would thus dilute the specific radioactivity of the re-isolated 1-^{14}C-OxIAA internal standard added to, and re-isolated from the tissue extracts [24]. This method demonstrated that OxIAA was a natural constituent of both corn endosperm and shoot tissues with 357 pmol OxIAA•kernel^{-1} and 49 pmol OxIAA•shoot^{-1} [19], an amount similar to the free IAA present in these tissues (308 and 27 pmol•tissue^{-1} respectively, [4]).

2.3 Biological Activity of OxIAA

The oxidation of IAA at carbon-2 results in the loss of biological activity, as demonstrated by the corn mesocotyl bioassay [21]. This supports other laboratories' data that OxIAA does not stimulate growth in *Avena* coleoptile curvature, *Avena* internode growth, and tobacco tissue culture, or ethylene production in mung beans (cf 19). However, OxIAA was reported to stimulate growth in pea [7], and the 5-hydroxy-analogs of OxIAA and diOxIAA were synergistic with IAA in mung bean ethylene production [27]. A systematic examination of IAA metabolites in several bioassay systems is therefore warranted.

2.4 In Vitro Assay for OxIAA Synthesis

There is approximately seven times more OxIAA in the corn endosperm than in the shoot. The initial IAA turnover study [4] and the IAA to OxIAA precursor-product study [18] employed corn endosperm tissue, so that the origin of OxIAA isolated from the shoot could have been OxIAA transported from the endosperm. To determine whether IAA oxidation to OxIAA occurred in vegetative as well as endosperm tissues, an in vitro IAA oxidizing system was developed [20]. Radiolabeled IAA was incubated with corn enzyme and the resulting products were separated by HPLC from the unreacted IAA. Enzyme activity was absent or low when measured in buffer preparations from fresh tissue. But when detergent was added during enzyme preparation 1 to 10 $pmol \cdot h^{-1} \cdot mg\ protein^{-1}$ of enzyme activity was observed from shoot, root and endosperm. The increased enzyme activity with detergent was not due to increased protein recovery or the solubilization of a microsomal enzyme [the enzyme was soluble after ultracentrifugation]. Rather, the detergent extracted a heat-stable factor from the tissue which stimulated enzymatic activity up to 10-fold [21]. The metabolism of IAA to OxIAA in vegetative corn tissues was also supported by the work of Nonhebel et al. [17], who reported chromatographic data for an "OxIAA-like" metabolite as the major metabolite of radiolabeled ^{14}C-IAA fed to coleoptile or root sections.

2.4.1 Cofactor Requirement

Subsequent experiments showed that the heat-stable factor had chromatographic properties similar to fatty acids on HPLC, and ion exchange chromatography [23]. The unsaturated fatty acids linoleic and linolenic acids substituted for the corn heat-stable lipid-soluble factor in stimulating enzymatic activity, while oleic, stearic and palmitic acids were inactive. The triacylglycerol of linoleic acid is a major component of corn oil [3], and linolenic acid [free and ester] occurs in light- and dark-grown corn seedlings [30]. Whether the natural cofactor [co-substrate] of IAA oxidation to OxIAA is linoleic or linolenic acid, and whether the lipid factor is co-oxidized during IAA oxidation awaits purification of the enzyme and identification of the enzyme mechanism.

The cofactor requirement for the oxidation of IAA was examined by adding cofactors/cosubstrates of peroxidase (H_2O_2, 2,4 dichlorophenol), mixed function oxygenase (NADPH, FAD, pterin), or intermolecular dioxygenase (Fe^+, NADPH, alpha ketoglutarate) to the assay medium (Table 1). None of the cofactors/cosubstrates stimulated IAA oxidation except the unsaturated fatty acids. This indicated that a novel enzyme was involved in IAA's oxidation. Cations, including Fe^{++}, Ca^{++}, Cu^+, and Mn^+, also were not stimulatory [23]. The enzyme reaction requires O_2 for optimal oxidation of IAA since argon sparging of the assay tube reduced enzymatic activity up to 9 fold [21], suggesting that molecular O_2 may be incorporated into the ring. $^{18}O_2$ experiments with the purified enzyme will help clarify the role of O_2 in the reaction.

Table 1. Effect of cofactors and cosubstrates of oxygenases and peroxidase reactions on the oxidation of IAA to OxIAA by endosperm enzyme preparations[a]

Treatment	Tissue
	Endosperm (Triton x100) percent of control[b]
2,4 dichlorophenol 5 μM	87
2,4 dichlorophenol 5 μM + Mn^+ 5 μM	97
Fe^{++} 50 mM + 500 μM Pterin	93
	Endosperm (without Triton x100) percent of control[c]
Pterin 500 μM + 1.2 mM NADPH	42
FAD 10 μM + 1.2 mM NADPH	38
Fe^{++} 1 mM + 1.2 mM NADPH + ketoglutarate 1 mM	21
Fe^{++} 1 mM + 5.7 mM ascorbate + ketoglutarate 1 mM	0
Mn^{++} 0.1 mM + 17 mM H_2O_2	90
NADPH 0.7 mM	90

[a] The enzyme was prepared with or without Triton x100 as indicated, and described in the text [21].
[b] Control value, 8.3 ± 1.8 pmol h^{-1} mg^{-1} protein.
[c] Control value, 3.8 ± 1.1 pmol h^{-1} mg^{-1} protein.

2.4.2 Enzyme Mechanism

The IAA oxygenase was further purified to free it of peroxidase and lipoxygenase since there are reports in which both enzymes can oxidize organic molecules using fatty acid as the co-oxidant [6, 14]. Corn scutellum was used as the source of enzyme since this tissue gave the highest initial specific activity [up to 80 pmol·h^{-1}·mg protein^{-1}].

In initial experiments, the corn enzyme was fractionated by ammonium sulfate. Lipoxygenase [85% with the 42–60% fraction], and lipid peroxide isomerase [90% with the 30 to 42% fraction] did not co-fractionate with the IAA oxygenase which fractioned nearly equally in the 30 to 42% and 42 to 60% fractions [22]. The corn enzyme was then chromatographed on DEAE Sephacryl. Peroxidase, lipoxygenase, and IAA oxygenase had increasing elution volumes, respectively, on the anion exchange column. Further fractionation on gel filtration Sephadex G200 separated lipoxygenase and peroxidase into molecular weights of 27 000, and 77 000, while the IAA oxygenase separated into three peaks, the largest eluting in the exclusion volume [greater than 250 000 Daltons]. The separation of the IAA oxygenase into three peaks of activity may indicate that the IAA oxygenase has isozymes, is partially absorbed to another macromolecule, or has several aggregation states. Nevertheless, the cofactor studies along with the protein fractionation studies indicate that IAA is being oxidized by a novel reaction.

2.5 Further Metabolism of OxIAA

Subsequent experiments [15, 13] demonstrated that OxIAA is further metabolized by oxidation to 7-OH-OxIAA and then glycosylated to 7-hydroxy-OxIAA-glc. Isotope dilution assays showed that 7-OH-OxIAA and 7-OH-OxIAA-glc were naturally occurring compounds in corn endosperm and shoot tissues with 3100 pmol•endosperm^{-1} 7-OH-OxIAA, and 4800 pmol•endosperm^{-1} and 62 pmol•shoot^{-1} 7-OH-OxIAA-glc. IAA, OxIAA or 7-OH-OxIAA will feed into the OxIAA oxidation pathway, unlike IAA oxidation in *Vicia faba* where IAA is initially conjugated to IAA-Asp and subsequently oxidized to DiOxIAA-Asp (free DiOxIAA is not an intermediate). Recently Tateishi et al. [28] have reported isolating 7-hydroxy-DiOxIAA-glc from immature sweet corn kernels. This data along with the enzyme data for corn seedlings indicate that oxidation of IAA with carboxyl-retention occurs during both seed maturation and seed germination. Lewer et al. [13] using NMR observed that the 3-position of 7-OH-OxIAA is labile and will exchange with D_2O. He also observed that 7-OH-OxIAA and its glucoside were non-enzymatically oxidized to putative 3-OH compounds. In light of the Lewer experiments, the isolation of DiOxIAA derivatives from corn can not be ruled out as the products of non-enzymatic conversion of OxIAA derivatives. The addition of radiolabeled OxIAA-glc during isolation of the DiOxIAA derivative would, however, determine whether conversion occurs. Lewer also [13] observed that 5-^{3}H-7-OH-OxIAA-glc fed to corn seedlings released ^3H_2O, indicating that the 5 position on the ring is labile and may be also oxidized.

2.6 Function of IAA Oxidation

Oxidation of IAA to OxIAA in corn appears to be a way of reducing the pool of active auxin since OxIAA is inactive in corn mesocotyl, and several other bioassays. The metabolites 7-OH-OxIAA and its glucoside are expected to be inactive as auxins, but must be shown to be so. The current working hypothesis is that IAA is inactivated by oxidation to OxIAA. OxIAA is futher metabolized to 7-OH-OxIAA and 7-OH-OxIAA-glc for sequestering in vacuoles for further metabolism and possibly carbon recycling. IAA oxygenase activity increases in corn endosperm over 6 days of germination [23] suggesting a role for IAA turnover during germination. A role for fatty acid metabolism in the regulation of IAA levels [turnover] is suggested since certain fatty acids stimulate the catabolism of IAA.

3 Conclusions

IAA is oxidized with the carboxyl group being retained in corn as well as in several other plant species. Corn can oxidize radiolabeled IAA to OxIAA to 7-OH-OxIAA and to 7-OH-OxIAA-7^1-O-B-D-glc. Isotope dilution assays have shown that Ox-IAA, and 7-OH-OxIAA and its glucoside are natural constituents of corn seedlings. IAA can also be oxidized to OxIAA in vitro by shoot, root and endosperm preparations. This oxidation reaction is catalyzed by a soluble enzyme, and is stimulated by O_2, and fatty-acids. Current evidence indicates that the reaction does not involve peroxidase as shown by the lack of stimulation by peroxidase cofactors, and by protein chromatography separations. The putative IAA oxygenase enzyme is being further purified in order to better understand how its activity may be regulated, and to identify how IAA oxidation may regulate IAA levels during plant growth.

Abbreviations

OxIAA, oxindole-3-acetic acid (2-indolinone-3-acetic acid)
DiOxIAA dioxindole-3-acetic acid (3-hydroxy-2-indolone-3-acetic acid
7-OH-OxIAA-glc, 7-hydroxy-oxindole-3-acetic acid-7^1-O-B-D-glucopyranoside

Acknowledgements. Supported by grants from the NSF Metabolic Biology section DMB 8504231, NASA Space Biology Life Sciences NAGW-97, and the NASA Flight Program NAG 2-362.

References

1. Bandurski RS, Schulze A, Reinecke DM (1986) In: Bopp M (ed) Plant growth substances 1985. Springer Berlin Heidelberg New York Tokyo, p 83
2. BeMiller JN, Colilla W (1972) Phytochemistry 11:3393
3. Eckey EW (1954) Vegetable fats and oils. Reinhold, New York
4. Epstein E, Cohen JD, Bandurski RS (1980) Plant Physiol 65:415
5. Ernstsen A, Sanberg G, Lundstrom K (1987) Planta 172:47
6. Gailliard T (1978) In: G Kahl (ed) Biochemistry of wounded plant tissues. de Gruyter, New York, p 155
7. Galston AW, Chen HR (1965) Plant Physiol 40:699
8. Hinman RL, Bauman CP (1964) J Org Chem 29:1206
9. Hinman RL, Lang J (1965) Biochemistry 4:144
10. Ishimaru A, Yamzaki I (1977) J Bio Chem 252:6118
11. Kinashi H, Suzuki Y, Takeuchi S, Kawarada A (1976) Agric Biol Chem 40:2465

12. Klambt HD (1959) Naturwissenschaften 46:649
13. Lewer P, Bandurski RS (1984) Phytochemistry 26:1247
14. Niinomi A, Morimoto M, Shimizu S (1987) Plant Cell Physiol 28:731
15. Nonhebel HM, Bandurski RS (1984) Plant Physiol 76:979
16. Nonhebel HM, Crozier A, Hillman JR (1983) Physiol Plant 57:129
17. Nonhebel HM, Hillman JR, Crozier A, Wilkins MB (1985) J Exp Bot 36:99
18. Reinecke DM, Bandurski RS (1981) Biochem Biophys Res Commun 103:429
19. Reinecke DM, Bandurski RS (1983) Plant Physiol 71:211
20. Reinecke DM, Bandurski RS (1984) Plant Physiol 75(S):108
21. Reinecke DM, Bandurski RS (1988) Plant Physiol 86:868
22. Reinecke DM (1988) Plant Physiol 86(S):114
23. Reinecke DM (1987) PhD Thesis, Michigan State University. The Catabolism of indole-3-acetic acid to oxindole-3-acetic acid
24. Rittenberg D, Foster GL (1940) J Biol Chem 133:737
25. Sembdner G, Gross D, Liebisch H-W, Schneider G (1981) In: MacMillan J (ed) Hormonal regulation of development. I. Molecular aspects of plant hormones. Encyclopedia of plant physiology 9, p 281, Springer Berlin Heidelberg New York
26. Sundberg B, Sandberg R, Jensen E (1985) Plant Physiol 77:952
27. Suzuki Y, Kinashi H, Takuchi S, Kawarada A (1977) Phytochemistry 16:635
28. Tateishi K, Shibata H, Matsushima Y, Iijima T (1987) Agric Biol Chem 51:3445
29. Tsrumi S, Wada S (1980) Plant Cell Physiol 21:1515
30. Vick BA, Zimmerman DC (1982) Plant Physiol 69:1103

Metabolism of Jasmonic Acid

G. SEMBDNER, A. MEYER, O. MIERSCH, and C. BRÜCKNER[1]

1 Introduction

(–)-Jasmonic acid [(–)-JA (Ia)] and its methyl ester (Ib) were found to be widespread in plants [15] and are considered to represent a new type of native PGRs with hormone-like properties [22, 23]. Further members of the group are (+)-7-iso-JA (II), known as the major product of JA biosynthesis in fruits of *Vicia faba* [16] and in the fungus *Botryodiplodia theobromae* [17], as well as cucurbic acid (IV) [6, 7] and a number of structurally related cyclopentanoidal C_{12}-acids isolated more recently [18, 19]. Some studies dealing with the biosynthesis of JA in plants clearly demonstrate a route starting from linolenic acid [25]. However, knowledge of the JA metabolism in plants is very limited [9, 10, 21]. Reports on the physiological activity of 9,10-dihydrojasmonic acid (DJA, III) and recent results demonstrating its natural occurrence in broad bean fruits [19] prompted us to start tracer studies on the metabolism of radiolabelled rac-III, which then were followed by experiments using [2-^{14}C]-(±)-JA (rac-Ia) confirming the pattern of major metabolites. The significance of metabolic routes of exogenously applied DJA and JA, respectively, will be discussed by comparing the known spectrum of naturally occurring JA analogues and derivatives including glucosyl and amino acid conjugates.

2 Metabolic Studies

Both DJA (rac-III) and JA (rac-Ia) were synthesized with [2-^{14}C]-labelling [11, 24] in order to study their biotransformation in plant tissues. Results to be presented here have been obtained using excised shoots of 6-day-old barley seedlings. The plant material was extracted 72 h after application of the compounds and the extracts separated and purified as described elsewhere [14]. The ethyl acetate extract (60% radioactivity) and the remaining aqueous phase (40% radioactivity) gave after purification on DEAE-Sephadex A-25 columns [8], three groups of metabolites: (1) a neutral fraction containing sugar esters, (2) an acidic fraction containing compounds such as free acids and glucosides, and (3) a more polar acidic fraction containing amino acid conjugates. Each group was separated by means of prep. TLC, SEP-PAK C_{18}-cartridges and prep-HPLC. Structural

[1] Institute of Plant Biochemistry, Halle/Saale, Academy of Sciences of the German Democratic Republic, Weinberg, 4010 Halle/Saale, GDR

identification of the metabolites was performed by GC, GC-MS, MS, NMR and ORD [cf. 14].

The results obtained with rac-III are summarized in Fig. 1. As a primary metabolic step hydroxylation took place preferentially at C-11 and to a lesser extent at C-12, resulting in (XXIII) and (XXIV), respectively. The major metabolite (XXIII) was conjugated with glucose in a high degree to give the O (11)-β-D-glucopyranoside (XXII). Metabolic reduction at position 6 of the 11-OH and 12-OH metabolites (XXIII) and (XXIV) led to the cylopentanol compounds (XXV) and (XXVI). As minor metabolites a number of amino acid conjugates have been structurally elucidated. These derived from either

1. The DJA acid (III) itself, giving N-(dihydrojasmonoyl)-isoleucine (XVII), N-(dihydrojasmonoyl)-leucine (XVIIa) and N-(dihydrojasmonoyl)-valine (XVIIb); or
2. The 11-OH and 12-OH metabolites (XXIII and XXIV), yielding N-(11-hydroxydihydrojasmonoyl)-isoleucine (XXIX), N-(11-hydroxydihydrojasmonoyl)-valine (XXIXb) and N-(12-hydroxydihydrojasmonoyl)-isoleucine (XXX); or
3. The cyclopentanol metabolites (XXV) and (XXVI), resulting in N-(tetrahydro-11-hydroxyjasmonoyl)-isoleucine (XXVII) and N-(tetrahydro-12-hydroxyjasmonoyl)-isoleucine (XXVIII).

Metabolic studies using [2-^{14}C]-($\pm$)-JA gave the same pattern of metabolites. Even the favoured hydroxylation at position 11 instead of position 12 could be confirmed.

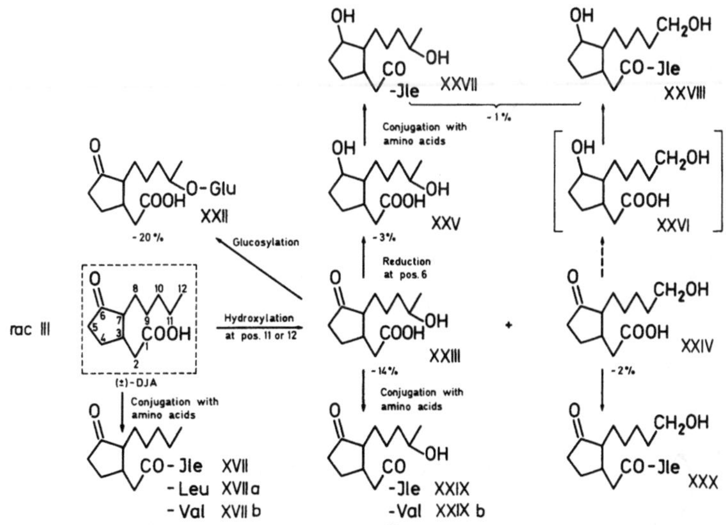

Fig. 1. Metabolic pathway of ($\pm$)-dihydrojasmonic acid in barley shoots. *Full arrows*: established by GC-MS; *broken arrow*: assumed route

3 Occurrence of Jasmonic Acid Analogues in Plants

After the (+)-7-iso-JA (II) had been assumed to be the original biosynthetic product in *Vicia faba* [25], whereas (–)-JA (Ia) is formed by isomerization, the natural occurrence of II was shown in the fungus *B. theobromae* [17] and of Ia and II in the broad bean pericarp [16]. From immature fruits of *Vicia faba* some further JA-like compounds have been isolated and identified as DJA (III), 3,7-didehydro-JA (IX), and (+)-6-epi-7-iso-cucurbic acid (V) [19]. Compound V was found to occur in *Equisetum* species, too [5]; it is considered to be a native JA metabolite because reduction at position 6 took place during biotransformation of both DJA and JA (Fig. 1; XXV and XXVI). The isomeric cucurbic acid (IV) has long been known to occur naturally [6, 7]. In *Equisetum* species, besides the mentioned compound (V) and traces of JA, also the 4,5-didehydro derivative of either JA or 7-iso-JA (VII) was found [5]. During studies on the biosynthetic formation of JA and 7-iso-JA by the fungus *B. theobromae* a number of minor products have been isolated and identified as cucurbic acid (IV), 9,10-dihydro-7-iso-JA (VI), 4,5-didehydro-7-iso-JA (VII), 11,12-didehydro-7-iso-JA (VIII), 3-oxo-2-(2Z-pentenyl) cyclopent-1-yl-propionic acid (X), 3-oxo-2-(2Z-pentenyl)cyclopent-1-yl-butyric acid (XI), and the ethyl ester of 7-iso-JA (XII) [18]. With respect to the position in biosynthetic or metabolic routes of the various JA analogues mentioned, some

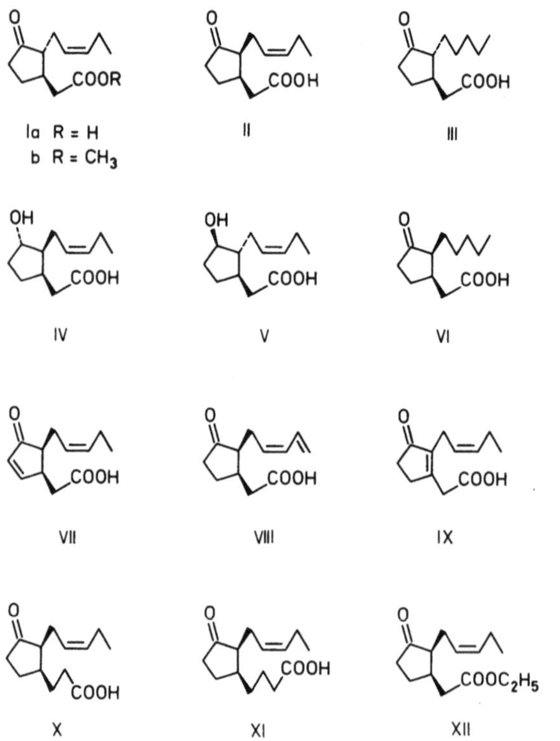

Fig. 2. Jasmonic acid and native analogues

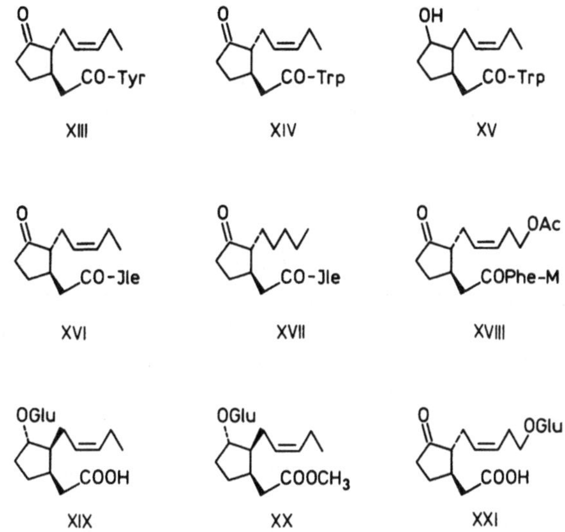

Fig. 3. Naturally occurring JA conjugates

of them might be considered as either precursors (XI) or by-products (III, VI, VII, IX, X) of biosynthesis. It is known that dihydro and didehydro precursors of C_{18}-cyclopentanoidal fatty acids can be converted to the corresponding C_{12}-acids (JA analogues) [25]. Some others seem to be metabolites of 7-iso-JA (IV, VIII, XII) and JA (V), respectively (Fig. 2).

4 Naturally Occurring Conjugates

The O-glucosides of cucurbic acid (XIX) and cucurbic acid Me (XX) had been found in pumpkin seeds [12]. Recently, the O(12)-β-D-glucopyranoside of 12-hydroxy JA and/or the aglycone (XXI) has been described as a native tuber-inducing substance in potato plants [26]. This, again confirms the probable significance of hydroxylation steps during JA metabolism (see Fig. 1). During our studies on the distribution of JA within the broad bean plant by means of RIA, amino acid conjugates have been detected. The structures of these were elucidated by GC-MS, MS, NMR, ORD and CD analyses. The authentic substances have been synthesized chemically [13]. The conjugates isolated from flowers of the broad bean were identified to be N-[(−)-jasmonoyl]-(S)-tyrosine (XIII), N-[(−)-jasmonoyl]-(S)-tryptophan (XIV) and a related tryptophan conjugate, putatively appointed as N-[(+)-cucurbinoyl]-(S)-tryptophan (XV) [2, 3]. Apical leaves of *Vicia faba* contain N-[(−)-jasmonoyl]-(S)-isoleucine (XVI) as the major polar immunoreactive compound [20]. As mentioned above, conjugation with aliphatic amino acids, especially isoleucine, frequently took place at different steps during biotransformation of DJA and JA in barley leaves (see Fig. 1). Furthermore, the isoleucine conjugates of both JA (XVI) and DJA (XVII) had been isolated as products of the fungus *Gibberella fujikuroi* when grown under unusual conditions [4]. The occurrence of a phenylalanine conjugate of 12-acetoxy JA (XVIII) in *Praxelis clematidea* [1] gives further hints that both hydroxylation and conjugation with either amino acids or glucose are important steps within metabolism of JA.

5 Conclusions

Jasmonic acid and related compounds are widely distributed in plants. They possess plant hormone-like properties; nevertheless, their endogenous regulatory roles have to be confirmed further. Biotransformation in barley shoots of radiolabelled DJA and JA, respectively, gave the same pattern of metabolites which could be related to some naturally occurring JA analogues and conjugates. The major metabolic steps are:

1. Hydroxylation, preferentially at C-11, and to a lower extent at C-12;
2. Formation of the 11-O-glucoside;
3. Reduction at position 6 to give the cyclopentanol derivatives
4. Conjugation with amino acids, preferentially isoleucine, of the non-metabolized dihydrojasmonic acid and jasmonic acid, their 11-OH and 12-OH metabolites and the cyclopentanol metabolites.

Thus, hydroxylation, hydrogenation and conjugation with either glucose or amino acids are the most important steps in both the biotransformation of exogenously applied PGRs of the jasmonic acid type and their endogenous metabolism in plants.

References

1. Bohlmann F, Wegner P, Jakupovic J, King RM (1984) Tetrahedron 40:2537
2. Brückner C, Kramell R, Schneider G, Knöfel HD, Sembdner G, Schreiber K (1986) Phytochemistry 25:2236
3. Brückner C, Kramell R, Schneider G, Schmidt J, Preiss A, Sembdner G, Schreiber K (1988) Phytochemistry 27:275
4. Cross BE, Webster GRB (1970) J Chem Soc (C):1839
5. Dathe W, Miersch O, Schmidt J (1989) Biochem Physiol Pflanz 185:83
6. Fukui H, Koshimizu K, Usuda S, Yamazaki Y (1977) Agric Biol Chem 41:175
7. Fukui H, Koshimizu K, Yamazaki Y, Usuda S (1977) Agric Biol Chem 41:189
8. Graebner R, Schenider G, Sembdner G (1976) J Chromatogr 121:110
9. Jacobus A (1986) Dipl.-Thesis, University of Halle, East Germany
10. Kehlen A (1987) Dipl.-Thesis, University of Halle, East Germany
11. Knöfel HD, Gross D (1988) Z Naturforsch 43c:29
12. Koshimzu K, Fukui H, Usuda S, Mitsui T (1974) In: Plant growth substances 1973, Hirokawa, Tokyo, p 86
13. Kramell R, Schmidt J, Schneider G, Sembdner G, Schreiber K (1988) Tetrahedron 44:5791
14. Meyer A, Gross D, Vorkefeld S, Kummer M, Schmidt J, Sembdner G, Schreiber K (1989) Phytochemistry 28:1007
15. Meyer A, Miersch O, Büttner C, Dathe W, Sembdner G (1984) J Plant Growth Regul 3:1
16. Miersch O, Meyer A, Vorkefeld S, Sembdner G (1986) J Plant Growth Regul 5:91
17. Miersch O, Preiss A, Sembdner G, Schreiber K (1987) Phytochemistry 26:1037

18. Miersch O, Schmidt J, Sembdner G, Schreiber K (1989) Phytochemistry 28:1303
19. Miersch O, Sembdner G, Schreiber K (1989) Phytochemistry 28:339
20. Schneider G, Kramell R, Brückner C (1989) J Chromatogr 483:459
21. Schröder B (1982) Dipl.-Thesis, University of Halle, East Germany
22. Sembdner G, Gross D (1986) In: Bopp M (ed) Plant growth substances 1985. Springer, Berlin Heidelberg New York Tokyo, p 139
23. Sembdner G, Klose C (1985) Biol Rundsch 23:29
24. Unverricht A, Gross D (1986) J Labelled Compd Radiopharm 23:515
25. Vick BA, Zimmerman DC (1984) Plant Physiol 75:458
26. Yoshihara T, Omar EA, Shibata H, Koshino H, Sakamura S, Yokota Y, Kikuta Y, Okazawa Y (1988) 13th Intern Conf Plant Growth Subst, Calgary, Abstr No 427

V Hormones: Physiology and Effects

How Abscisic Acid Causes Depressions of the Photosynthetic Capacity of Leaves

K. RASCHKE[1]

1 Introduction

When a solution of ABA is fed into an illuminated leaf, stomata begin to close about 10 min later and the rate of CO_2 assimilation, A, declines. Simultaneous measurements of water-vapor loss and CO_2 uptake allow the computation of the intercellular partial pressure of CO_2, c_i. Such computations showed that often c_i remained constant when the assimilation rate decreased in response to an application of ABA. Apparently, ABA had affected the photosynthetic machinery directly to an extent that the CO_2 available in the leaf was used with reduced efficiency [1, 9]. This phenomenon was observed to occur in leaves of numerous species, monocots as well as dicots, and employing the C3 as well as the C4 mechanisms of photosynthesis [10]. In this contribution will be explained how application of ABA can cause depressions of photosynthesis at apparently constant computed c_i.

2 Carboxylation of RuBP

The first reaction leading to the reduction of CO_2 in photosynthesis is the carboxylation of ribulose 1,5-bisphosphate (RuBP) and the ensuing formation of 3-phosphoglycerate (PGA). It is a slowing of this reaction which we measure after application of ABA:

$$CO_2 + RuBP \xrightarrow{\text{Rubisco}} 2\ PGA$$

A reduction of the velocity of this reaction could have three causes, [1] a reduced supply of CO_2, [2] a reduced supply of RuBP (which in turn could be the result of a diminished rate of reduction of PGA, or of an inhibition on the pentose-phosphate-path), or [3] a reduced catalytic activity of the enzyme, ribulose bisphosphate carboxylase-oxygenase (Rubisco).

If depressions of photosynthesis occurred at constant c_i after applications of ABA, reduced CO_2 supply apparently was not the cause 1. Analysis of quickly killed leaf samples contained more RuBP if they had received ABA than without, in absolute terms as well as relative to the level of PGA [4]. Therefore, rates of photosynthesis cannot have been depressed by cause 2 either. The observed

[1] Pflanzenphysiologisches Institut und Botanischer Garten der Universität Göttingen, Untere Karspüle 2, 3400 Göttingen, FRG

increase of the ratio RuBP/PGA indicated an inhibition of the carboxylation reaction itself. Rises in the RuBP content were confirmed by an indirect approach. Using a gas-analyzer system, designed for capturing transients in gas exchange [7], CO_2 uptake of leaves was recorded after the light was turned off and the gas was switched to an oxygen-free atmosphere (to suppress photorespiration). During this period, the remaining pools of pentose-phosphate-path metabolites were being emptied and used to fix CO_2 while Rubisco was still active. The amount of fixed CO_2 is a measure of the pool sizes of RuBP, as well as of its precursors; it represents the "assimilatory charge" [6]. The assimilatory charge of sunflower leaves increased after application of 10^{-4} M ABA. In the example shown in Fig. 1, ABA had caused a depression of the rate of photosynthesis in the preceding light period from 20.6 to 3.6 μmol m^{-2} s^{-1} and a reduction of stomatal conductance from 0.353 to 0.039 mol m^{-2} s^{-1}. This decrease in conductance reduced the maximum rate of CO_2 uptake in the dark, but the total postillumination CO_2 fixation, as represented by the integration of the transients with respect to time, had increased from 275 to 353 μmol m^{-2}. In this experiment, the assimilatory charge was 1.3 times larger after treatment of the leaf with ABA than before. The recorded depression of photosynthesis cannot have been caused by a reduced supply of RuBP.

There remains an impaired catalytic activity of Rubisco as an explanation. Activities were determined 2 min after extraction of leaf halves, one before, the other after feeding ABA. No significant reduction of these "initial activities" was seen after application of 10^{-5} M ABA but, after adding 10^{-4} M ABA, a reduction to 0.8 of the control value had occurred.

This decline was too small to explain the observed depressions in photosynthesis (in some cases to as low as 0.12 of the control rates; [10]). The reduction of

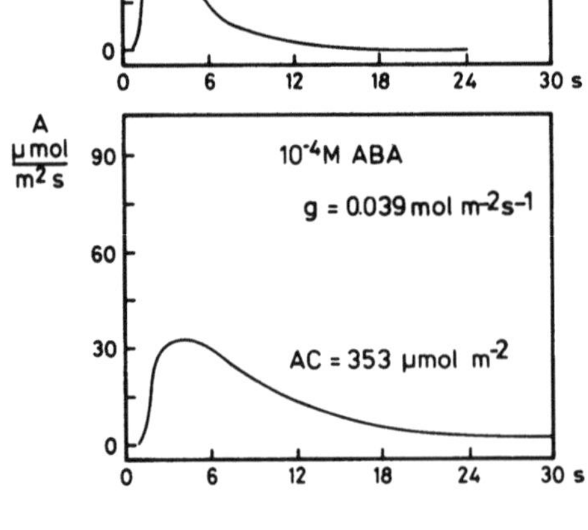

Fig. 1. Postillumination uptake of CO_2 from 1925 μl l^{-1} CO_2 in N_2 by a leaf of *Helianthus annuus* before and after addition of 10^{-4} M (±)-ABA to the transpiration stream. *AC* Assimilatory charge, in RuBP equivalents (6); *g* stomatal conductance. Rates of CO_2 assimilation (in normal air) in the preceding light periods were 20.6 and 3.6 μmol m^{-2} s^{-1}, before and after ABA applications, respectively. Leaf temperature was 23.5 °C, absorbed quantum flux 1.7 mmol m^{-2} s^{-1}

the initial activity of Rubisco could be repaired by a 7-min activation of the enzyme with 20 mM $NaHCO_3$ and 12 mM Mg. It appeared unlikely that cause number 3, a reduced catalytic activity of Rubisco, was responsible for the ABA effect on photosynthesis. Rather, the lessening of the initial activity of Rubisco indicated that the effective partial pressure of CO_2 in the leaf was lower than the computed value. Reduction of c_i below 100 μbar caused in untreated leaves a decline of Rubisco activity. At the CO_2 compensation point (about 50 μbar), activity was reduced to 0.6 of its full value (as a result of a loss of CO_2 required for the activation of the enzyme). The possibility, that a reduced supply of CO_2 had caused the observed depressions of photosynthesis needed to be re-examined.

3 Effects of Changes in the Distribution of Stomatal Apertures on Intercellular Partial Pressure of CO_2 and on Photosynthesis

Laisk demonstrated that changes in the distribution function of stomatal aperture could lead to erroneous interpretations of the relationship between an overall stomatal conductance of a leaf and its rate of photosynthesis [5]. Farquhar et al. reported on stomata responding to ABA in groups (patches) in leaves [3]. In the following, effects of changing stomatal response patterns on c_i, and on rates of CO_2 assimilation, will be examined.

3.1 Proportional and Binary Responses of Stomata

One can think of two ways by which stomata can reduce the gas exchange of a leaf from, for example, a full value to one half: Either all stomata close proportionally to one half of their original aperture (Fig. 2, left), or one half of all stomata maintain their apertures, whereas the other half closes completely (Fig. 2, right). In the simplest form of this second situation, stomatal aperture exists only in two states, closed and open; they represent a binary system. If no lateral diffusion of CO_2 occurred within the leaf, CO_2 assimilation was shut off completely beneath closed stomata. The fraction of closed stomata among all stomata would be identical with the fraction by which photosynthesis had been reduced. In tissue beneath open stomata, photosynthesis would proceed unhindered because CO_2 supply suffices, and water was being lost at a rate, E, depending on the water-vapor pressure difference, Δw, between leaf and air. If rates of gas exchange are being used to compute c_i, the ratio of A/E enters the computation:

$$E = g \quad \Delta w \tag{1}$$

$$A = \frac{g}{1.6} \Delta c = \frac{g}{1.6} (c_a - c_i), \tag{2}$$

where g is the conductance for water vapor; 1.6 is the ratio of the conductances for H_2O and CO_2; and c_a is the partial pressure of CO_2 in the atmosphere.

Combining [1] and [2] yields:

$$c_i = c_a - \frac{A}{E} \cdot 1.6 \, \Delta w. \tag{3}$$

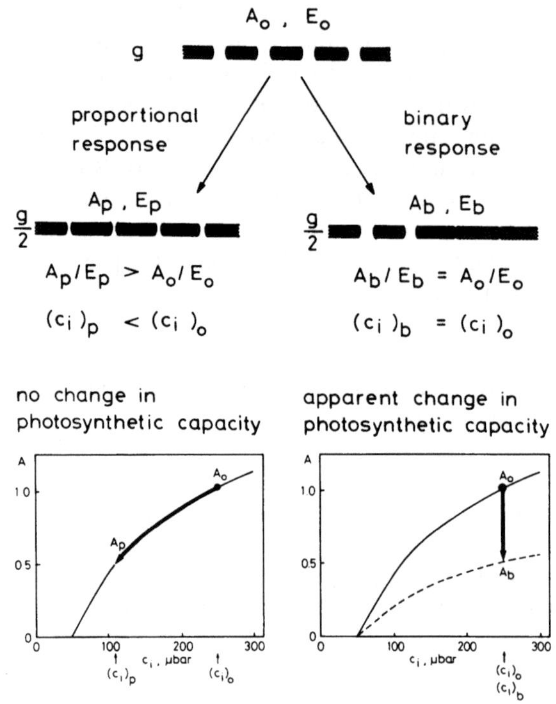

Fig. 2. Reduction of stomatal conductance for gases by a factor of 0.5, and its effect on photosynthesis, accomplished either through proportional responses of all stomata (*left*), or by closure of one-half of all stomata (*right*). *A* CO_2-assimilation rate, *E* transpiration rate; *g* conductance for CO_2 or H_2O; c_i intercellular partial pressure of CO_2; *subscripts: o* initial; *p* after proportional response; *b* after binary response

We recognize that the area participating in gas exchange does not appear in the equation. The computed c_i is correct for that part of the leaf in which stomata are open, but not for the other part in which they are closed. In the latter, c_i should be at the CO_2 compensation point. If ABA caused a binary stomatal response, it would produce a reduction of the photosynthetically active area which would appear as a depression at constant c_i and a reduced photosynthetic capacity of the leaf over the whole range of partial pressures of CO_2 (dashed line in Fig. 2, right).

In real leaves, there will be lateral diffusion of CO_2, and correlated closure may not occur among opposite stomata in the upper and lower epidermis. An exact equality of the fraction of closed stomata with the relative reduction of the rate of photosynthesis is therefore not to be expected; nevertheless, discrepancies between measured rates of photosynthesis and computed values of c_i would appear. In the case of stomata responding proportionally, computed values of c_i will be applicable, and measured rates of photosynthesis represent working points on the saturation curve of CO_2 assimilation.

3.2 Demonstrations of Changes in the Distribution of Stomatal Apertures in Response to ABA

The distribution of stomatal apertures in epidermal strips of *Commelina communis* floating on solutions shifted towards smaller stomatal widths when ABA was added

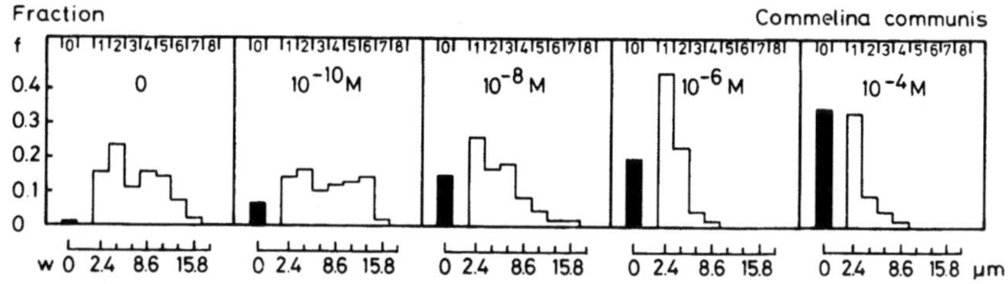

Fig. 3. Frequency distribution of stomatal apertures in illuminated strips from the lower epidermis of leaves of *Commelina communis*, floating on 80 mM KCl plus the indicated concentrations of (±)-ABA. *f* Fraction of stomata in a particular aperture class; *w* stomatal aperture; *upper scales* aperture-class numbers. *Black* fraction of closed stomata

(Fig. 3). This displacement caused an increase in the fraction of closed stomata. At 10^{-4} M ABA the fraction could be as high as 0.6 (not shown in Fig. 3). We recognize that responses of populations of stomata to ABA can be looked upon as having a quasi-binary and a proportional component. In leaves, lateral diffusion of CO_2 would reduce the effect of complete closure of individual stomata on local rates of photosynthesis. However, inhibition would increase if stomata responded in groups, particularly if they shut off the gas exchange of entire alveoli, separated from each other by minor veins. A further requirement for a severe reduction in photosynthesis would be a correlation of stomatal responses in the upper and lower epidermis, or even a complete closure on one side of the leaf. Distinct differences in CO_2 assimilation between individual alveoli were indeed seen in leaves that had received ABA. Terashima et al. [12] used the absence of starch formation to demonstrate this. Downton et al. [2] made grouped stomatal closure visible through the exclusion of $^{14}CO_2$ incorporation it produced. In both, starch prints and ^{14}C autoradiograms, areas in which products of photosynthesis had accumulated were sharply separated by veins from areas with little or no sign of CO_2 assimilation. Fluorescence image analysis of leaves provided a procedure to follow time courses and the spatial distribution of stomatal closure and localized CO_2 depletion. If photosynthetic electron flow produces more reduction equivalents than can be consumed, mechanisms of energy dissipation begin to operate that convert excitation into heat. An energy-dependent quenching of chlorophyll fluorescence from photosystem II (PS II) develops [13]. The degree of this quenching can be determined by saturating PS II with light for about one s, recording the fluorescence emitted in the far red, and expressing the reduction of this emission as a fraction, q_E, of the emission measured after the leaf had adapted to darkness (with minimal energy-related quenching) [11]. At a low partial pressure of O_2, q_E is a measure of the CO_2 deficiency in a leaf, and $1-q_E$ is a measure of photosynthetic electron transport. Fluorescence images of leaves were recorded with a TV camera and the images used to compute topographies of photosynthetic electron transport. The top frame of Fig. 4 shows such a distribution in a *Xanthium* leaf at the time of application of ABA. Ten min later (middle frame), electron transport began to decline in some alveoli, unexpectedly at some distance from the veins. Areas with decreasing electron transport coalesced during the following 10 min until photo-

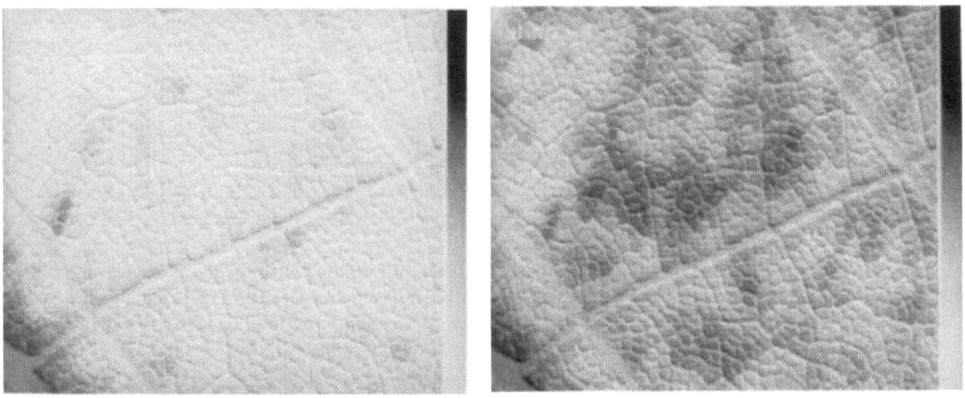

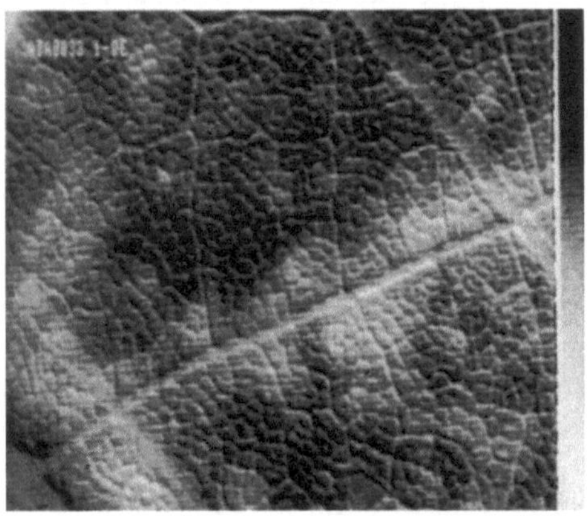

Fig. 4. Changes in the topography of photosynthetic electron transport following an application of 10^{-4} M ($\pm$)-ABA to a leaf of *Xanthium strumarium*. *Top left* At the time of ABA application; *top right* 13 min after ABA; *left* 23 min after ABA. Brightness scales at the *right* margins cover the range from no electron transport (black) to full electron transport rate (white). Computed from images of chlorophyll fluorescence during saturating light pulses of 1-s duration

synthesis was restricted to areas alongside veins. The fluorescence images show (a) that stomata responded in groups, usually restricted to one or several adjoining alveoli, and (b) that a gradual (proportional) stomatal closure was superimposed upon the pattern of closure in groups. The patterns became less expressed or disappeared at increased partial pressures of O_2 or CO_2. The effect of ABA on photosynthesis was mediated by reduced CO_2 supply.

4 Changes in the Distribution of Stomatal Aperture Directly Determined and Depressions of Photosynthesis Computed

The gas exchange of leaves was recorded, then the leaves were frozen within < 0.1 s, freeze-dried, and stomatal apertures were directly measured under the scanning EM. In ABA-treated *Xanthium* leaves, groups of closed stomata appeared which covered areas of up to 0.1 mm^2, which would correspond to an alveolus. The frequency distributions of stomatal apertures resembled those seen in epidermal strips of *C. communis* (Fig. 3). In the example shown in Fig. 5, application of 10^{-4}

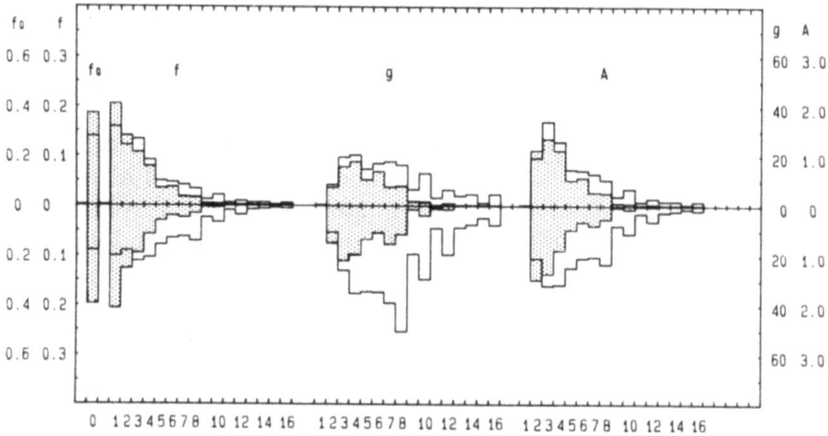

Fig. 5. Frequency distribution (*left histogram*) of stomatal apertures in samples from 4 leaves of *Xanthium strumarium* treated with 10^{-4} M (±)-ABA (*stippled*) and control leaves (*open histograms*). Above zero line: upper epidermis; below: lower epidermis; *j* aperture class (class unit = 0.24 μm). *f* fraction of total number of stomata in one epidermis; f_0 fraction of closed stomata (separate scale!). *Center histogram* (*g*): stomatal conductances computed from stomatal dimensions and the frequency distribution in the left histogram. *Right histogram* (*A*): assimilation rates computed from frequencies of conductances (*center histogram) and responses of a standard Xanthium* leaf for an ambient partial pressure of CO_2 of 340 μbar, a quantum flux of 600 μmol m^{-2} s^{-1} and a leaf temperature of 25 °C. Assimilation rates (control/ABA treatment) by computation: 18.2/10.7 μmol m^{-2} s^{-1}; by gas analysis: 20.6/7.7 μmol m^{-2} s^{-1}. Fraction of stomata directly sampled: 0.26% in the upper, 0.19% in the lower epidermis

M ABA to four leaves had caused an increase of the fraction of totally closed stomata from on the average 0.2 in the control leaf to 0.4, and a shift of the population of open stomata toward classes with small apertures. From these data, stomatal conductances for water vapor (and CO_2) were computed (Fig. 5; center histogram). Based on a saturation curve of photosynthesis of an untreated *Xanthium* leaf, rates of photosynthesis were estimated for each aperture class. Integration yielded total assimilation rates of the control and the ABA-treated leaves (right histogram). The observed increase in the number of closed stomata, in combination with the narrowing of the open stomata, caused in the ABA-treated leaf a computed decrease of the assimilation rate from 18.2 to 10.7 μmol m^{-2} s^{-1}. The directly measured rates had been 20.6 and 7.7 μmol m^{-2} s^{-1}. The agreement is satisfactory in view of the fact that only about 0.2% of the stomatal population had been sampled under the SEM.

5 Conclusion

Abscisic acid supplied to leaves causes a shift in the distribution of stomatal apertures toward narrower pores and an increase in the fraction of closed stomata. The increase of the fractions of completely closed and very narrow stomata leads to a virtual reduction of the photosynthetically active area of a leaf. Stomatal closure occurs in groups, reducing the gas exchange of one or several adjoining alveoli. Because computed intercellular partial pressures of CO_2 are customarily derived from ratios between CO_2 assimilation and evaporation, they are valid only for leaf areas with open stomata and therefore remain high after ABA-treatment. However, they do not apply to areas with closed stomata, where the partial pressure of CO_2 can approach the compensation point. Application of ABA can thus cause apparent depressions of the photosynthetic capacity of leaves, solely through changes in the distribution function of stomatal apertures and the spatial heterogeneity of the stomatal responses.

Acknowledgments. I gratefully acknowledge support by Deutsche Forschungsgemeinschaft, Eesti NSV Teaduste Akadeemia (Estonian Academy of Science), and Stiftung Volkswagenwerk. I acknowledge the use of unpublished data from collaborations with M. Arretz, J.T. Ball, J.A. Berry, P.F. Daley, A. Laisk, V. Oja, and J. Patzke.

References

1. Cornic G, Miginiac E (1983) Plant Physiol 73:529
2. Downton WJS, Loveys BR, Grant WJR (1988) New Phytol 108:263
3. Farquhar GD, Hubick KT, Terashima I, Condon AG, Richards RA (1987) In: Biggins J (ed) Progress in photosynthesis research. Nijhoff, Dordrecht, p 209
4. Fischer E, Raschke K, Stitt M (1986) Planta 169:536
5. Laisk A (1983) J Exp Bot 34:1627
6. Laisk A, Kiirats O, Eichelmann H, Oja V (1987) In: Biggins J (ed) Progress in photosynthesis research. Nijhoff, Dordrecht, p 245
7. Oja VM (1983) Fiziol Rast 30:1045
8. Parlange JY, Waggoner PE (1970) Plant Physiol 46:337
9. Raschke K (1982) In: Wareing PF (ed) Plant growth substances. Academic Press, London, p 581
10. Raschke K, Hedrich R (1985) Planta 163:105
11. Schreiber U, Schliwa U, Bilger W (1986) Photosynth Res 10:51
12. Terashima J, Wong SC, Osmond CB, Farquhar GD (1988) Plant Cell Physiol 29:385
13. Weis E, Berry JA (1988) Biochim Biophys Acta 894:198

The Role of Abscisic Acid in Chilling Resistance

RUI-CHI PAN[1]

1 Introduction

There is growing evidence which indicates that the naturally occurring plant growth regulator, ABA, is involved in stress adaptation. This is principally based on the positive correlation between the accumulation of ABA and the development of stress resistance. It is well documented that exogenous ABA increases chilling resistance in plants. This provides a basis for identifying the factors involved in chilling resistance and selecting chilling resistant plants.

2 Changes in ABA Induced by Low Temperature

A number of researchers have demonstrated increased ABA content following chilling exposure. Levels of free and bound ABA of tomato seedlings subjected to day/night temperatures of 10/5°C for 12 h were significantly higher than in plants grown at optimal diurnal temperatures (21/15°C) [4]. An increase in ABA in warm-season crops such as bean and corn exposed to 10°C was similarly attributed to the low temperature [5]. Guo and Pan [11] have shown that growing rice seedlings at 8–10°C for 9 days increased both ABA content and chilling resistance (Table 1). We have observed that the ABA levels of cotyledons of cucumber seedlings increased 16-fold when these seedlings were exposed to 1–5°C for 3 days. If those seedlings were transferred from a low temperature (1–5°C) to an optimal temperature (28–30°C), the ABA content decreased during the recovery period (Fig. 1).

Table 1. Changes in endogenous ABA and electrolyte leakage of rice seedlings exposed to low temperature 8–10°C [see 10][a]

Duration of low temperature treatment (days)	Content of endogenous ABA in leaf (ng g^{-1} FW)	(%)	Leakage of electrolytes in root ($\mu\Omega cm^{-1} g^{-1}$ FW)	(%)	Leakage of electrolytes in leaf ($\mu\Omega cm^{-1} g^{-1}$ FW)	(%)
0	0.7	100	220	100	360	100
3	13.5*	2100	196	89	396	110
6	20.5*	3200	688*	313	1242*	345
9	58.5*	8600	1177*	535	2016*	560

[a] *$P < 0.01$.

[1] South China Normal University, Biology Department, Guangzhou, 510631, China

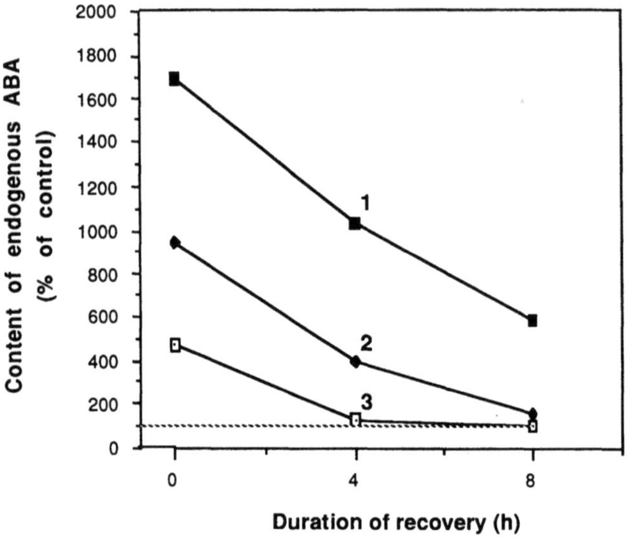

Fig. 1. Change in endogenous ABA content in chilled cucumber cotyledons during recovery from low temperature. *1* chilled 1 day; *2* chilled 2 days; *3* chilled 3 days [see 38]

There are two principal explanations for the increase of endogenous ABA levels during low temperature. Firstly, low temperature per se could cause the increase in ABA levels. Daie et al. [5] showed there were no significant water potential differences between tomato plants subjected to high or low temperature. They concluded that the increase of ABA was the result of the low temperature, and not the result of induced water stress. Secondly, the accumulation of ABA could result from an induced moisture deficit and not as a result of low temperature per se. Eze et al. [8] grew bean seedlings at 5, 25, or 45 °C for 2 h under various moisture conditions. Only under conditions favoring water loss was there a measurable increase in ABA. In *Phaseolus vulgaris, Rhoes discolor* and cucumber, a major cause of the chilling injury was concluded to result from altered stomatal opening accompanying reduced water uptake by the roots due to the low temperature [6]. Similarly, low temperature also altered stomatal functioning in *Zea mays* [30]. At low temperature exogenous ABA resulted in stomatal closure. The chilling resistance of *Hibiscus esculentus* was increased by a foliar application of ABA leading Eamus [7] to postulate that stomatal closure would be of major importance in the prevention of chilling injury. Eamus [7] also attributed the increase in ABA at low temperature to a water stress.

3 The Effect of ABA on Chilling Resistance

The optimal concentration of ABA to increase chilling resistance in plants is about 10^{-6} to 10^{-4} M. Application of ABA to cucumber seedlings reduced chilling injury [29]. Pretreatment of discs of cotton cotyledons with [2-^{14}C] ABA for various periods

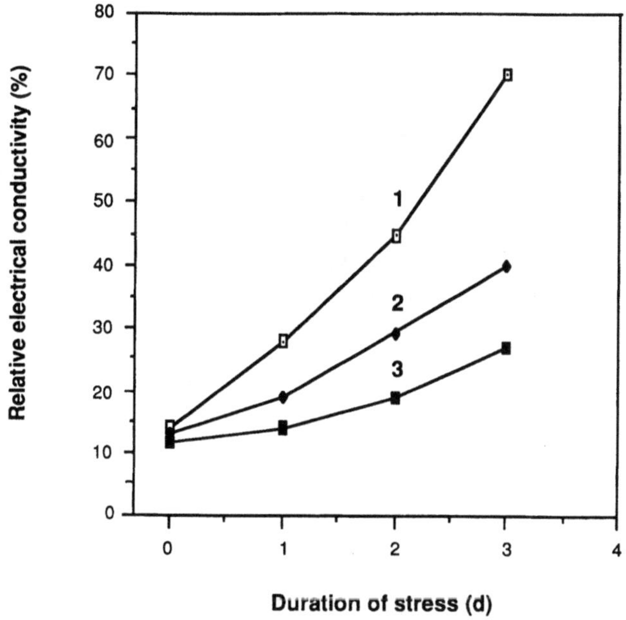

Fig. 2. Effect of ABA solution on the leakage of electrolytes of cotyledons in chilled cucumber seedlings Chilled condition: temp 3°C, RH 80–83%, darkness, *1* Control; *2* 10^{-6} mol/l, *3* 10^{-4} mol/l ABA solution [see 38]

of time at room temperature showed that ABA applied at the inception of the chilling was ineffective [27]. Pretreatment for at least 3 h was required for a significant effect. Rice seedlings required 24 h of ABA treatment prior to the chilling stress in order to increase stress resistance [9]. We have recently observed that ABA increased chilling resistance in cucumber cotyledons [38]. Plants treated with 10^{-4} M ABA and stressed for 3 days had significantly less electrolyte leakage as than controls or plants treated with 10^{-6} M of ABA (Fig. 2).

4 Possible Mechanism of the Reduction of Chilling Injury by ABA

Three theories exist concerning the reduction of chilling in plants by ABA.

4.1 ABA May Reduce Membrane Damage Caused by Chilling

Membranes or membrane function may be injured by a variety of stress conditions including chilling [23]. When cucumber plants were chilled, the membrane phospholipids of their leaves declined [37]. Under water saturated conditions leakage of electrolytes of chilled cucumber seedlings increased, but application of ABA reduced this damage. Thus, ABA apparently protects the membrane from chilling injury [27]. Kasamo [14] reported that the addition of ABA to the leaves resulted in depolarization across the plasmalemma and the exclusion of K^+ within 5 min of application. Shaner et al. [32] also showed that the plasmalemma of root cells became depolarized following exogenous application, and this may result from the reduction ion pumping activity in roots.

ABA-induced chilling resistance may be due to an increase in the mobility of the membrane hydrocarbon acyl chains [18]. In intact seedlings and isolated cotyledonary discs of cotton, ABA reduced chilling injury by preventing a decrease in the content of reduced glutathione and this was closely correlated with membrane stabilization [28]. At an injurious chilling temperature desaturation of leaf polar lipids fatty acids in cucumber leaves was reduced. These observations support the suggestion that ABA stabilizes membranes [35].

4.2 ABA Inhibits Water Loss Induced by Low Temperature

Application of ABA to bean seedlings grown at 25 °C, 80% RH, resulted in stomatal closure after 1 h [36]. When these seedlings were transferred to 5 °C, 85% RH for 24 h, no decrease of fresh weight or wilting occurred. It was concluded that ABA prevented chilling injury by closing the stomata, thus preventing dehydration [36]. Chilling injury in cotton and bean seedlings was reported to be the result of a reduction in root membranes hydraulic conductivity and a loss of stomatal control [19]. Markhart et al. [17] have shown that the application of ABA increased the hydraulic conductivity of soybean root, promoted the water absorption of the roots, did not induce stomatal closure, but eventually increased the resistance to chilling. Collectively, these results support the interaction of water stress and chilling injury and a role for ABA in reducing both.

4.3 The Effect of ABA on Metabolism

ABA may also increase chilling resistance through a number of metabolic process. Bornman et al. [2] observed that the resistance of tobacco tissues to chilling depends upon their proline content. It is probable that the chilling resistance of rice seedlings is not related to changes in proline content, but may be related to an increase of soluble sugars [11]. From our observations the content of soluble sugar in rice seedlings is increased by both a chilling treatment and exogenous ABA treatment, especially on the 4th day after chilling treatment (Fig. 3).

Light also plays a role in chilling resistance. Pretreatment of cotyledon discs of cotton with ABA for 24 h in darkness before exposure to chilling did not alter chilling resistance. However a similar pretreatment in light greatly decreased chilling injury. Addition of 0.5% sucrose during the dark period prevented the chilling injury usually induced in the dark, thereby replacing the light requirement for ABA to improve chilling tolerance. Sucrose may also maintain a high level of NADPH, since sucrose, via the hexose monophosphate shunt, generates NADPH and the latter is thus available for the reduction of protein -S-S-bonds. These results suggest that the photosynthetic product could provide energy for decreasing the chilling sensitivity of the tissue. Further, ABA might increase chilling resistance

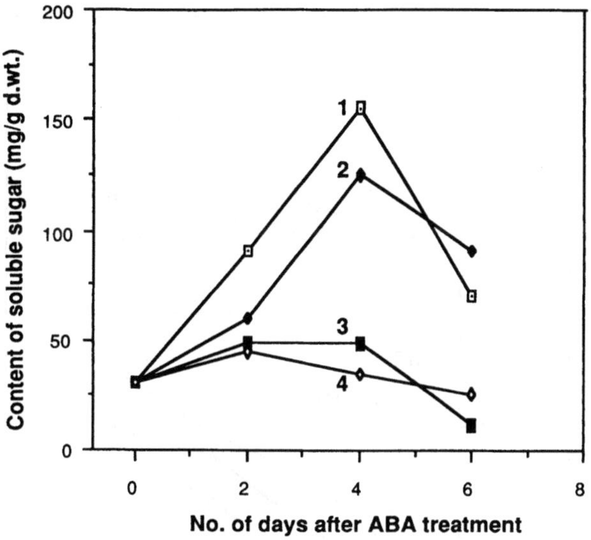

Fig. 3. Effect of ABA (10^{-6} mol/l) on the content of soluble sugar in the shoots of rice seedlings (Shanyou No. 2). *1* Low temperature 8–10°C with ABA; *2* low temperature 8–10°C without ABA; *3* normal temperature 25–27°C with ABA; *4* normal temperature 25–27°C without ABA [see 11]

through a metabolic process which is dependent on photosynthetic activity [30]. It is possible that ABA may increase soluble sugars, and that these then provide energy for the ABA action.

5 The Role of ABA in Cross Adaptation

Exposure of plants to one environmental stress, such as chilling, heat, drought, or salinity may confer resistance to other apparently unrelated stress [3, 15]. This provides a basis for the concept of cross-adaptation, and plants possessing this characteristic have all showed an accumulation of ABA. We observed that chilling injury to rice seedlings was reduced by a drought treatment for 8 h prior to exposure to chilling. Chilling resistance of rice seedlings also increased after the seedlings were grown in the presence of 0.1 M NaCl [Fig. 4; 12]. Both salinity and chilling caused an increase in the ABA content of cucumber cotyledons, and when these seedlings were transferred to a chilling temperature, they were more resistant to chilling (Fig. 5). Hence, ABA may be an essential regulating factor for cross adaptation.

6 Plant Growth Retardants and ABA

Recent research indicates that certain plant growth retardants improve the chilling resistance of plants [24]. Application of paclobutrazol to cucumber, squash [33] and rice [25, 34] will increase chilling resistance. Triadimefon has the same effect on bean [1], cabbage, barley [9] and cucumber [35]. There are three views concerning the relationship between plant growth retardants and endogenous ABA:

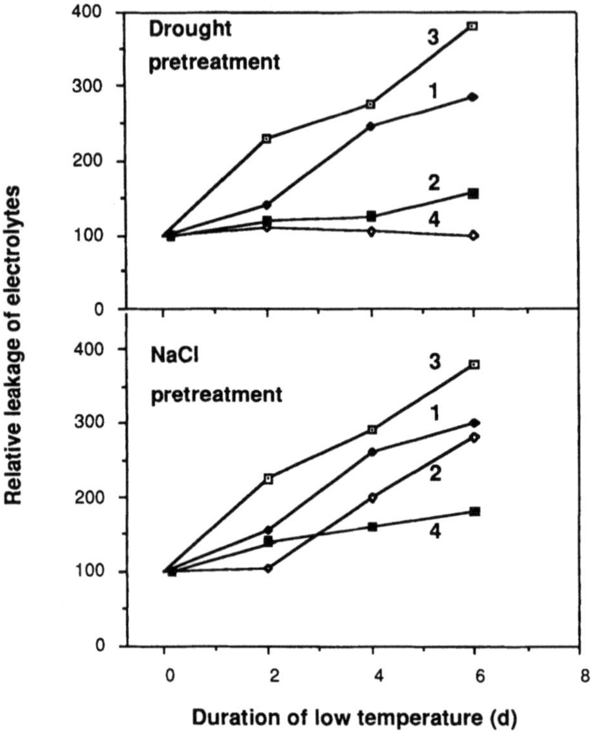

Fig. 4. Effects of drought and NaCl pretreatment on the leakage of electrolytes from leaves of chilled rice seedlings. Drought pretreatment: seedlings were air-dried, maintained in light for 8 h until leaves were rather wilted and the water content of leaves had decreased from 82% to 75%. Plants were then rewatered for 1 h to recovery. NaCl pretreatment: seedlings were removed to 0.1 mol/l NaCl solution for 24 h, then changed to distilled water. Plants from both treatments were then exposed to 8–10°C. *1*, *2* Dansheng No. 1; *3*, *4* Guichao No. 2; *1*, *3* control; *2*, *4* treatment [see 12]

6.1 The Inhibition of ABA Biosynthesis by Growth Retardants

Some plant growth retardants can inhibit ABA biosynthesis in the fungus *Cercospora rosicola* [22]. The most effective were paclobutrazol, ancymidol and decylimidazole [21]. CCC, Alar and paclobutrazol also act as inhibitors of GA biosynthesis and both ABA and GA are synthesized via the isoprenoid pathway. However, different growth inhibitors act at different points on the pathway and thus may inhibit either or both ABA and GA biosynthesis.

6.2 The Promotion of ABA Biosynthesis by Growth Retardants

Spraying apple rootstocks with Alar increased ABA levels and decreased the GA levels in stem tips [13]. We found that peanut seedlings treated with paclobutrazol resulted in a rapid increase in ABA and a reduction in GA in their leaves [Fig. 6; 16]. On this basis, one might propose a competition between ABA and GA

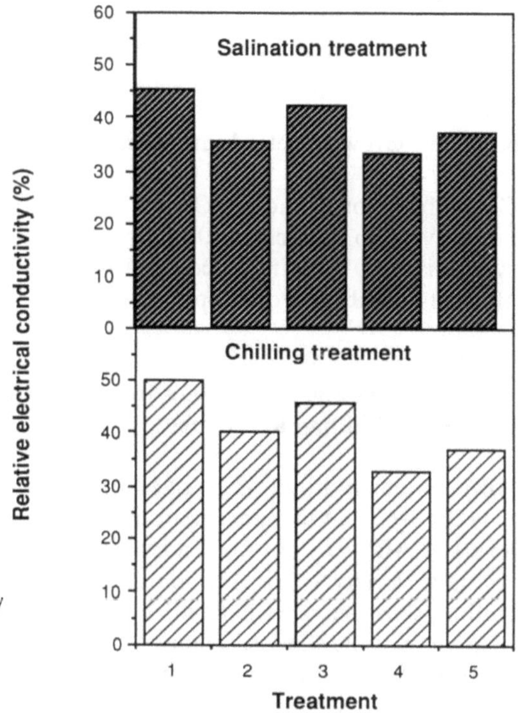

Fig. 5. Changes of relative electrical conductivities of cotyledons in chilled cucumber seedlings after salination treatment (0.25 mol/l NaCl) and chilling treatment (3 °C), respectively. *1* control; *2* 12 h recovery after 24 h treatment; *3* 24 h recovery after 24 h treatment; *4* 12 h recovery after 30 h treatment; *5* 24 h recovery after 30 h treatment; [see 38]

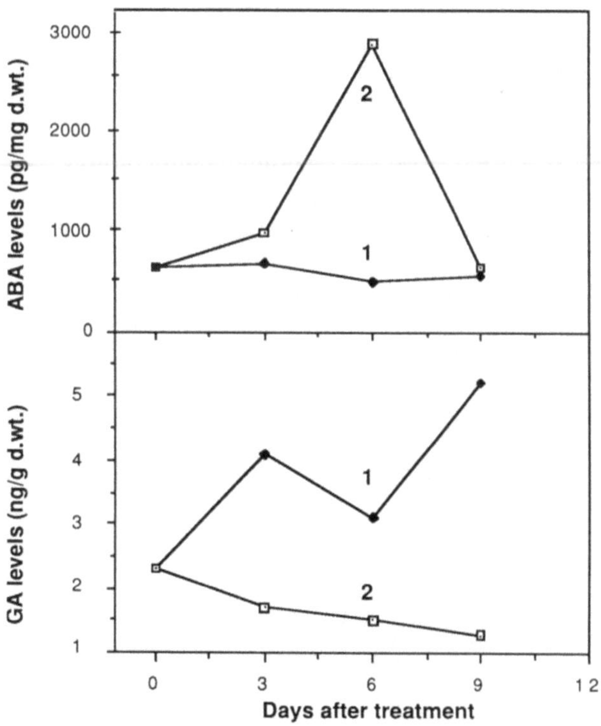

Fig. 6. Effect of paclobutrazol on ABA and GA levels in peanut leaves under normal condition (temperature 28 °C RH 75%, light intensity 20 000–30 000 lux). ABA and GA levels were determined by RIA and biossay methods respectively. *1* 0 ppm; *2* 5 ppm [see 16]

synthesis; biosynthesis of one phytohormone being inhibited while production of the other was apparently promoted.

6.3 Plant Growth Retardants Which do not Affect ABA

CCC, Phosphon D and AMO-1618 did not affect the biosynthesis of mevalonic acid in the avocado mesocarp [20]. Alar inhibited GA synthesis but did not affect ABA synthesis [22]. The difference in results obtained may be due to the concentration of plant growth retardant applied.

7 Conclusions and Suggestions for the Use of Chemicals to Increase Chilling Resistance

ABA content in chilling-sensitive crops exposed to low temperature was higher than in crops grown at warmer temperatures. Chilling-resistant crops exhibited similar levels of ABA at both low and warm temperatures [4]. The leaves of a chilling-resistant variety of rice accumulated more endogenous ABA than those of a chilling-sensitive rice when both are grown at low temperatures [11]. Thus, the level of ABA may be used to identify chilling resistance crops and/or varieties after exposure to chilling temperatures.

While ABA may make plants more resistant to chilling, it is too expensive for broad field application. We suggest some practical and effective alternative measures for application in agriculture:

1. Jasmonic acid, an inexpensive synthetic compound, has some physiological effects similar to those of ABA [26]. Recently we observed that jasmonic acid increased the chilling resistance of rice seedlings. In addition, Flores et al. [10] also showed that LAB 144143 and LAB 173711 increased resistance of cucumber seedlings to chilling temperatures.
2. CCC and plant growth retardants such as Alar and paclobutrazol decrease GA and increase ABA content, in at least some plants. The resistance of plants to chilling may thus be increased consequently by certain plant growth retardant treatments.
3. Plants can be hardened under adverse environments such as exposure to chilling, heat, salinity and drought stress treatments, all of which can result in an increase in ABA in plants. In accordance with the principle of cross adaptation, these hardened plants may have increased resistance to chilling.

References

1. Asare-Boamah NK, Fletcher RA (1986) Physiol Plant 67:353
2. Bornman HC, Janson EVA (1980) Physiol Plant 48:491
3. Boussiba S, Rikin A, Richmond AE (1975) Plant Physiol 56:337
4. Daie J, Campbell WF (1981) Plant Physiol 67:26
5. Daie J, Campbell WF, Seeley SD (1981) J Am Soc Hortic Sci 106:11

6. Eamus D, Wilson JM (1983) J Exp Bot 34:1000
7. Eamus D (1986) J Exp Bot 37:1854
8. Eze JMO, Dumbroff EB, Thompson JE (1983) Physiol Plant 58:179
9. Fletcher RA, Hofstra G (1985) Plant Cell Physiol 26:775
10. Flores A, Grau A, Laurich F, Dorffling K (1988) J Plant Physiol 132:362
11. Guo Q, Pan RC (1984) Acta Phytophysiol Sin (in Chinese with English abst) 10:295
12. Guo Q, Pan RC (1986) Acta Phytophysiol Sin (in Chinese with English abst) 12:396
13. Hoad GV, Monselise SP (1976) Sci Hortic 4:41
14. Kasamo K (1981) Plant Cell Physiol 22:1257
15. Levitt J (1980) In: Levitt J (ed) Responses of plants to environmental stress. 2nd edn, vol I, Academic Press, New York, p 23
16. Li L, Pan RC (1988) Oil crops of China (in Chinese with English abst) 2:36
17. Markhart AH (1984) Plant Physiol 74:81
18. Markhart AH, Fiscus EL, Naylor AW, Kramer PJ (1979) Plant Physiol 64:611
19. McWilliam JR, Kramer PJ, Musser LR (1982) Aust J Plant Physiol 9:343
20. Milborrow BV (1976) In: Sunderland N (ed) Perspectives in experimental biology, vol 2. Pergamon, New York, p 111
21. Norman SM, Bennett RD, Poling ST, Maier VP, Nelson MD (1986) Plant Physiol 80:122
22. Norman SM, Poling SM, Maier VP, Orme ED (1983) Plant Physiol 71:15
23. Pan RC (1979) Physiology of rice plant. Science Press, Beijing, p 260
24. Pan RC (1984) J South China Normal Uni (Nat Sci Edn) 2:121
25. Pan RC (1987) Int Rice Res Newsl 12:36
26. Pan RC, Li HH (1989) Plant Physiol Comm 1 2:78
27. Rikin A, Richmond AE (1976) Physiol Plant 38:95
28. Rikin A, Atsmon D, Gitter C (1979) Plant Cell Physiol 20:1537
29. Rikin A, Blumenfeld A, Richmond AE (1976) Bot Gaz 173:307
30. Rikin A, Gitter C, Atsmon D (1981) Plant Cell Physiol 22:453
31. Rodriquez JL, Davies WJ (1982) J Exp Bot 33:977
32. Shaner DL, Mertz SM, Arntzed CJ (1975) Planta 122:79
33. Wang CY (1985) Sci Hortic 26:293
34. Wang YL, Luo YX, Pan RC, Zhang MG (1987) J South China Normal Uni (Nat Sci Edn) 1:29
35. Whitaker BD, Wang CY (1987) Physiol Plant 70:404
36. Wilson JM (1976) New Phytol 76:257
37. Wright M, Simon EW (1973) J Exp Bot 24:400
38. Wu Y, Pan RC (1988) J South China Normal Uni (Nat Sci Edn) 2:25

Dormancy in Cereals – Levels of and Response to Abscisic Acid

M. WALKER-SIMMONS[1]

1 Introduction

During grain development the level of plant hormone ABA increases until maximum grain fw is reached and then decreases rapidly as the grain begins to desiccate. New advances using ABA-deficient mutants, ABA immunoassays, and measurement of ABA effects on germination and gene expression are advancing our understanding of the role of ABA in induction and maintenance of grain dormancy.

1.1 Dormancy Induction and ABA

ABA-deficient mutants of *Arabidopsis* have provided convincing evidence that the presence of ABA is required for dormancy induction in developing grain. Single gene mutants with severely reduced ABA levels during grain development produced mature grain with reduced dormancy levels [10, 11]. Dormancy induction in *Arabidopsis* is also prevented in mutants with reduced responsiveness to ABA. These ABA-insensitive mutants germinated precociously, even though the mutants accumulated normal levels of ABA during grain development [15].

Viviparous mutants of maize provide a second example of an ABA deficiency during grain development affecting germinability. Precocious germination is observed in the viviparous mutants which are deficient in ABA [7, 25]. ABA levels in developing maize have also been reduced by treatment of the embryonic tissue with fluridone, which blocks the synthesis of the carotenoid precursors of ABA. The lack of ABA synthesis may explain why fluridone treatment of embryonic tissue between 9 and 13 days post-anthesis inhibited induction of dormancy [7, 20].

1.2 ABA Effects on Embryonic Germination and Protein Synthesis

Application of ABA to immature wheat embryos prevents embryonic germination, while embryo culture without ABA results in precocious germination [23]. Incubation of immature embryos in ABA also causes the early accumulation of proteins that are characteristic of later stages of embryo maturation [22]. Some of the maturation proteins have been identified and these include wheat germ

---

[1] USDA-ARS, 209 Johnson Hall, Washington State University, Pullman, WA 99164–6420, USA

agglutin, E_m protein, and a globulin storage protein [22]. The same ABA effect on immature embryos has been observed in other plants including cotton [8], oilseed rape [6], and soybean [5].

2 ABA Levels and Embryo Responsiveness in Developing Grain

2.1 ABA Levels During Grain Development

King [12–14] found that the ABA content of wheat grain increases until maximum fw is reached. Then ABA levels drop sharply upon grain desiccation. A monoclonal ELISA for ABA has recently been developed so that embryonic ABA levels can be measured [27]. ABA changes in the embryo during grain development are essentially the same as for the whole grain, though the ABA levels are 2–3 times higher in the embryo than in the remaining part of the grain. Embryonic ABA levels of a sprouting-resistant and a susceptible cultivar were compared throughout development. ABA levels were found to be about 25% lower in the sprouting-susceptible cultivar but differences between the cultivars were not statistically significant [27].

2.2 Embryonic Responsiveness to ABA

It is important to consider the response of the embryo to ABA, as the embryo is considered to be the site of initial germination and of ABA action. In most seeds there is a decrease in embryonic responsiveness to ABA upon seed maturation or desiccation such as observed in oilseed rape [6] and soybean [1]. However, in wheat prolonged embryonic responsiveness to ABA has been found in dormant grain. Wheat embryonic responsiveness to ABA, as measured by the capability of ABA to block embryonic germination, has been found to correspond with whole seed germination capability during development and in mature grain [28, 33]. ABA (0.5–5.0 μM) is ineffective in blocking embryonic germination once the grain has lost dormancy [27, 28, 29]. McCrate et al. [18] reported that variation in dormancy among wheat cultivars is caused by differential responses of the embryo to endogenous inhibitors in the grain. Recently tryptophan has been identified as an endogenous inhibitor which causes similar differential responses to ABA in dormant and non-dormant grain [19].

2.3 Environmental Effects on Induction of Dormancy, ABA Levels and Embryonic Responsiveness to ABA

Differences in degree of grain dormancy are affected by genotype, the environment during grain development, and the developmental stage of the grain [2]. The temperature during grain development has a strong effect on the degree of dormancy induced in mature grain, particularly in some wheat cultivars such as Brevor [9, 24]. Brevor grain grown under cool conditions will exhibit maximum levels of dormancy at maturation. When the same cultivar is grown at a warm temperature the mature grains are far less dormant.

In order to compare the growth temperature effects on ABA levels and responsiveness with induction of grain dormancy, Brevor wheat plants were grown in our laboratory at a warm (25 °C) or a cool (15 °C) temperature from anthesis until maturity. Embryonic ABA levels were measured throughout grain development (Fig. 1). Grains matured considerably faster at 25 °C reaching final dw by 30 days post-anthesis, while that process took 85 days at 15 °C. At 25 °C embryonic ABA levels increased rapidly to very high levels during early embryonic development and then dropped sharply with grain maturation (Fig. 1, upper). At 15 °C embryonic ABA levels increased and then plateaued for 20 days at about half the maximum ABA level observed for the grains grown at the warmer temperature. Upon maturation ABA levels of grains which had developed at 15 °C were decreased. No apparent differences in the final embryonic ABA levels were observed at the two developmental temperatures.

Germinabilities of the grains which developed at cool or warm temperatures are compared in Fig. 2 at immaturity and upon maturation. At mid-development germinability was low, but was slightly better for grains developing at 25 °C than at 15 °C. By maturation large differences were observed between the two developmental temperatures in the degree of grain dormancy. Grains developed at 25 °C had no dormancy at maturation and were 100 percent germinable. Grains developed at 15 °C were very dormant. The ABA responsiveness of embryos from these grains was determined, as measured by the capability of ABA to block embryonic germination. After development at either temperature embryos dis-

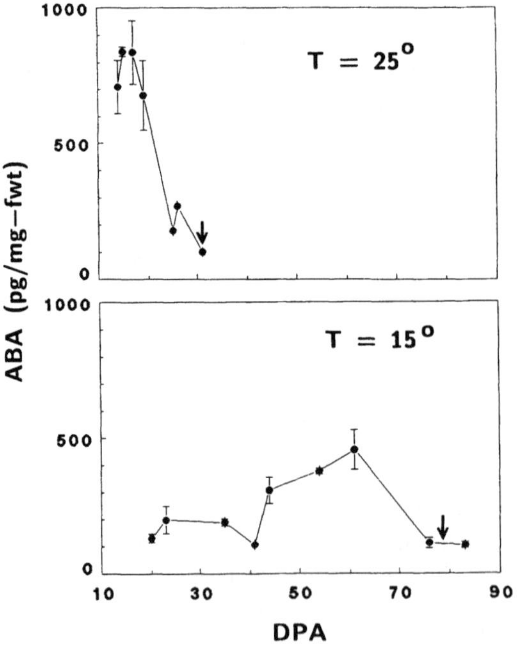

Fig. 1. (+)ABA levels in Brevor embryos during grain development at 25 °C (*upper*) or at 15 °C (*lower*). The *arrow* indicates the time that final dw was reached. *DPA* "Days post-anthesis". Data are the means (±SD) of three replicate extractions

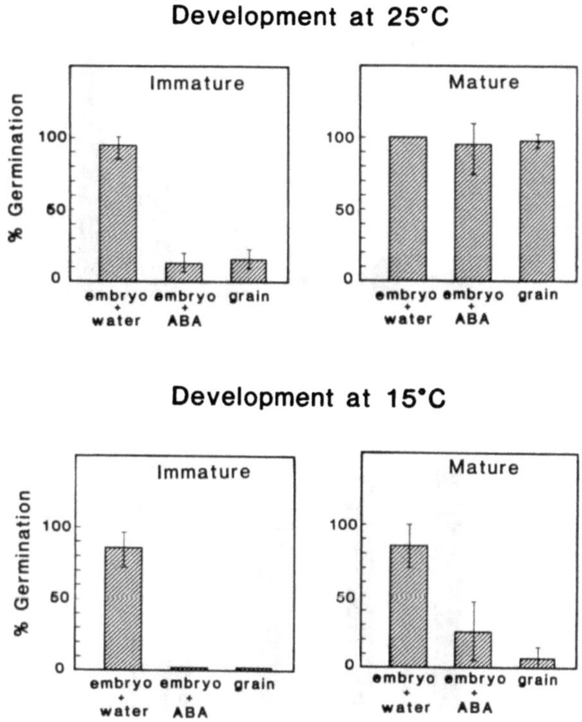

Fig. 2. Percentage germination of embryos or grains from plants grown from anthesis on at 25 °C (*upper*) or 15 °C (*lower*). Embryonic germination was measured upon incubation in water ±5 μM (±)ABA at 15 °C. Immature grains were ca. 40% dw. Data are the means of three replicate germination tests (±SD) of 10 embryos or grains each

sected from immature or mature grain and then imbibed in water had a high ability to germinate. ABA was effective in blocking germination of immature embryos from grains grown at either temperature (Fig. 2). Mature embryos from grain grown at 15 °C were considerably more responsive to ABA and the germination of these embryos in ABA solution was low, similar to that exhibited by the whole grain. Embryos developed at 25 °C lost responsiveness to ABA upon maturation and germination was only slightly inhibited by ABA. It can be concluded that the same environmental conditions that produce grains with a high degree of dormancy also produce embryos with enhanced responsiveness to ABA during development and at maturity.

3 ABA Levels and Responsiveness in Mature Grain

The relationship between ABA and dormancy of embryos and whole grains has been further characterized in mature grains by comparing the responsiveness of the dormant cultivar Brevor with the non-dormant cultivar Greer. The concentration effects of applied ABA on resulting endogenous ABA levels and embryonic germinability are shown for Brevor and Greer embryos in Fig. 3. Endogenous ABA levels in the Brevor and Greer embryos 8 h after incubation are similar (Fig. 3, upper). The small variation in endogenous ABA levels does not appear sufficient to explain the differences in ABA effects on germination capability between the two

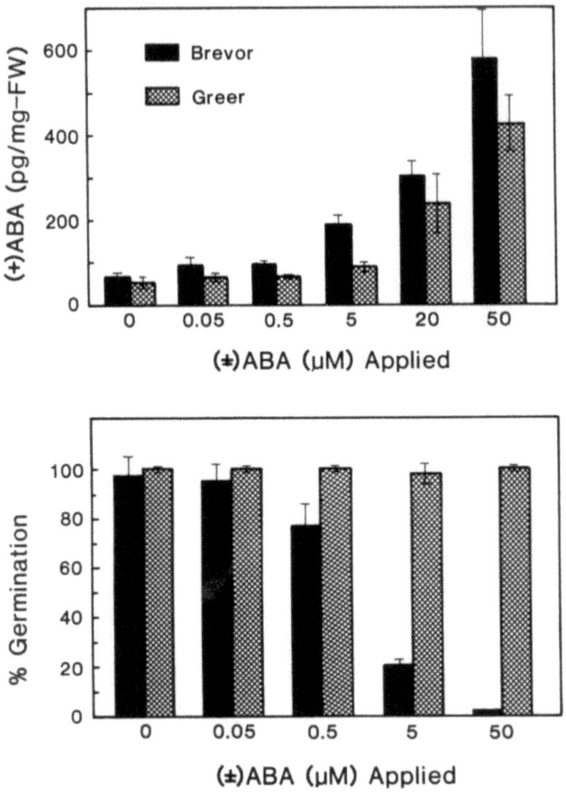

Fig. 3. Measurement of endogenous (+)ABA levels (*upper*) in embryos after incubation in varying concentrations of (±)ABA. Embryos were dissected from Brevor or Greer grains, incubated in ABA for 8 h at 30°C, and assayed for ABA [27]. Data are the means (±SD) of three replicate extractions. Comparison of the percentage germination (*lower*) of Brevor and Greer embryos after 9 days incubation in the indicated ABA concentrations

cultivars (Fig. 3, lower). Exogenous ABA levels at 0.5 μM and above reduced embryonic germinability in Brevor embryos, but even 50 μM ABA did not reduce germination in Greer embryos. It has previously been reported that ABA was more effective in blocking embryonic germination in mature grain from a sprouting-resistant wheat cultivar than from a susceptible cultivar [26].

A higher level of dormancy is expressed in cereal grains at warmer germination temperatures [2, 9, 17]. Warmer germination temperatures also significantly enhance the ability of ABA to block germination of isolated embryos from dormant grains. ABA is 100 times more effective in reducing the embryonic germination of Brevor at 30°C compared to 15°C [30].

Because we have found that more embryonic responsiveness to ABA is maintained in dormant whole grain (even into grain maturity), we have been examining ABA effects on protein synthesis in embryos isolated from mature dormant grain compared to non-dormant grain. Significant differences in the pattern of polypeptides synthesized in response to ABA in wheat embryos from dormant or non-dormant grain have been found [31].

Since embryos from dormant grain are more responsive to ABA it can be suggested that ABA-inducible proteins might be used as biochemical selection markers for dormancy. Possible candidates for such markers include those ABA-inducible proteins observed in immature embryos [22], in mature embryos from dormant grain [31], in ABA-treated barley aleurone [16] and in dehydration-stressed barley [3, 4]. Another potential protein is the ABA-inducible alpha-amylase inhibitor found in barley and wheat [21, 32].

4 Conclusions

Overall, it is not just the final ABA levels in the embryo or whole grain that determines dormancy. Changes in ABA levels and embryo responsiveness to ABA that occur during development influence the final degree of dormancy in the mature grain. ABA content and embryonic responsiveness to ABA during grain development can be altered by both the environment and the genotype.

Acknowledgments. I thank Joan Sesing for excellent technical assistance.

References

1. Ackerson RC (1984) J Exp Bot 35:414
2. Black M, Butler J, Hughes M (1987) In: Mares DJ (ed) Fourth international symposium on pre-harvest sprouting. Westview Press, Boulder, CO, p 379
3. Chandler PM, Ariffin Z, Huiet L, Jacobsen JV, Zwar J (1987) In: Mares D (ed) Fourth international symposium on pre-harvest sprouting. Westview Press, Boulder, CO, p 295
4. Chandler PM, Walker-Simmons M, King RW, Crouch M, Close TJ (1988) J Cell Biol(S) 12C:143
5. Eisenberg AJ, Mascarenhas JP (1985) Planta 166:505
6. Finkelstein RR, Tenbarge KM, Shumway JE, Crouch ML (1985) Plant Physiol 78:630
7. Fong F, Koehler DE, Smith JD (1983) In: Kruger JE, LaBerge DE (eds) Third international symposium on pre-harvest sprouting in cereals. Westview Press, Bouldler, CO, p 188
8. Galau GA, Hughes DW, Dure III L (1986) Plant Mol Biol 7:155
9. George DW (1967) Crop Sci 7:249
10. Karssen CM, Brinkhorst-van der Swan DLC, Breekland AE, Koorneef M (1983) Planta 157:158
11. Karssen CM, Groot SPC, Koorneef M (1987) In: Thomas H, Grierson D (eds) Developmental mutants in higher plants. Society for Experimental Botany, Seminar Series 32, Cambridge University Press, Cambridge, UK, p 119
12. King RW (1976) Planta 132:43
13. King RW (1982) In: Khan AA (ed) The physiology and biochemistry of seed development, dormancy and germination. Elsevier Biomedical, Amsterdam, p 157
14. King RW (1989) In: Derera H (ed) Pre-harvest field sprouting in cereals. CRC Uniscience Books, Boca Raton, FL, p 27
15. Koornneef MG, Reuling CM, Karssen CM (1984) Physiol Plant 61:377
16. Lin L-S, Ho T-H D (1986) Plant Physiol 82:289
17. Mares DJ (1984) Aust J Agric Res 35:115
18. McCrate AJ, Nielsen MT, Paulsen GM, Heyne EG (1982) Euphytica 31:193
19. Morris GF, Mueller DD, Faubion JM, Paulsen G (1988) Plant Physiol 88:435
20. Moore R, Smith JD, Fong F (1985) Am J Bot 72:1311
21. Mundy J (1984) Carlsberg Res Commun 49:439
22. Quatrano RS (1987) In: Miflin BJ (ed) Oxford surveys in plant molecular and cell biology. Vol 3. Oxford University Press, Oxford, UK, p 467
23. Quatrano RS, Ballo BL, Williamson JD, Hamblin MT, Mansfield M (1983) In: Goldberg R (ed) Plant molecular biology. Liss, New York, p 343
24. Reddy LV, Metzger RJ, Ching TM (1985) Crop Sci 25:455
25. Robichaud CS, Wong J, Sussex I (1980) Dev Genet 1:325

26. Stoy V, Sundin K (1976) Cereal Res Commun 4:157
27. Walker-Simmons M (1987a) Plant Physiol 84:61
28. Walker-Simmons M (1987b) J Cell Biol(S) 11B:30
29. Walker-Simmons M, Sesing J (1987) In: Mares D (ed) Fourth international symposium on pre-harvest sprouting in cereals. Westview Press, Boulder, CO, p 591
30. Walker-Simmons M (1988) Plant Cell Environ 11:769
31. Walker-Simmons M, Crane K (1988) J Cell Biol(S) 12C:196
32. Weselake RJ, MacGregor AW, Hill RD, Duckworth HW (1983) Plant Physiol 73:1008
33. Wiedenhoeft MH, Chevalier P, Walker-Simmons M, Ciha AJ (1988) Field Crops Res 18:271

Auxin Regulation of Cell Differentiation in Moss Protonema

M.M. JOHRI and J.S. D'SOUZA[1]

1 Introduction

The transition from chloronema to caulonema marks a major developmental pathway during the formation of protonema in mosses like *Funaria hygrometrica*, *Physcomitrella patens* and *Physcomitrium pyriforme*. Since the first demonstration of the enhancement of caulonema differentiation specifically by exogenously applied auxins [14], strong evidence for the endogenous presence and function of auxins as natural hormones in mosses is gradually accumulating. The current understanding of auxin regulation of caulonema differentiation is reviewed briefly in this paper.

2 Caulonema Differentiation in Response to Auxin

In stationary liquid cultures of *F. hygrometrica*, differentiation and the subsequent maintenance of the primary caulonema (1° caulonema) are independent of external auxin. The initiation depends solely on the inoculum size [12], and 1–10% filaments of the caulonema type (exclusively 1° ones) are formed below a cell density of 0.1–0.2 mg ml^{-1}. Applied IAA evokes two responses; at low levels (apparent *Km* 0.1 μM) it inhibits the formation of secondary (2°) chloronema and at a slightly higher level (apparent *Km* 0.4 μM), it increases the production of 2° or higher order caulonema. By manipulating the level of exogenous auxin and the inoculum size, cultures containing > 65–70% caulonema are readily obtained. The differentiation of 1° and 2° caulonema are independent steps. As the 2° caulonema are formed at IAA concentrations higher than those inhibiting chloronema, both responses occur together in liquid cultures.

2.1 Responses of Other Auxins

NAA also evokes responses comparable to IAA. In *F. hygrometrica* 2,4-D is either less active [Heidelberg strain; 17] or inactive (Himalayan strain J-2). Indoleacrylic acid strongly inhibits chloronema proliferation; and there was no effect on caulonema production at any of the concentrations tested (Table 1). Cells also respond to the ethyl ester of IAA; it is only slightly less active than free IAA in

[1]Molecular Biology Group, Tata Institute of Fundamental Research, Homi Bhabha Road, Bombay 400 005, India

Table 1. Caulonema differentiation with various auxins

Additions	μM	Fresh weight	Caulonema filaments (%)	Additions	μM	Fresh weight	Caulonema filaments (%)
None[a]		100	4	None[b]		100	8
IAA	1.0	62	70	2,4-D	0.5	105	< 1
IAA-gly[c]	10.0	89	< 1	2,4-D	1.0	92	< 1
IAA-gly	50.0	82	< 1	2,4-D	5.0	88	< 1
IAA-gly	100.0	85	< 1	2,4-D	10.0	92	< 1
IAA-L-ala[d]	10.0	75	3	IAcA[e]	0.1	28	7
IAA-L-ala	50.0	62	7	IAcA	0.5	22	2
IAA-L-ala	100.0	68	21	IAcA	1.0	5	2
				IAcA	5.0	2	0

[a] Cells cultured in minimal medium (MM) for 8 days. Fresh weight in control is 1.5 gm L^{-1} and is shown above as % of control.
[b] Cells cultured in MM containing 1% glucose (MMG) for 8 days. Fresh weight in control is 14.2 gm L^{-1} and is shown above as % of control.
[c] IAA-gly, indoleacetylglycine.
[d] IAA-L-ala, indoleacetyl-L-alanine.
[e] IAcA, indoleacrylic acid.

enhancing caulonema. Of the various indoleacetylamino acid conjugates tested, only IAA-L-alanine showed some auxin activity by inhibiting chloronema and increasing caulonema. IAA-glycine and other conjugates (–D-alanine, –L-Asp, –L-phenylalanine, –D-valine and –L-valine) were completely inactive. IAA-L-alanine is the most active in supporting the growth of callus and in stimulating the production of ethylene [9]. In this context it may be mentioned that unlike higher plants, IAA-Asp was not detected in the liverworts *Pellia* and *Plagiochila* [16, 24]. Upon feeding labeled IAA to 11 species of bryophytes, IAA-Asp was not formed, but some species produced an IAA-G conjugate [29].

2.2 Bulk Medium pH and Caulonema Differentiation

During the 6-day period when caulonema are formed in the presence of IAA (2 μM), the pH of the medium changes (Fig. 1) from an initial value of 5.1 ± 0.1 to 6.1 ± 0.2. In these experiments, 8.4 mM nitrate was the sole nitrogen source. The pH range during caulonema differentiation is similar to that reported for other auxin-induced responses. Following the disappearance of most of the exogenously supplied IAA (due to photo-oxidation and degradation), a profuse formation of 2° chloronema occurs. The decline in the proportion of caulonema after 8 days is the reason for this. Nitrate is rapidly utilized and the medium pH rises to 8.5 ± 0.5. To determine the effect of buffering the medium in the pH range where auxin-enhanced caulonema differentiate, cells were cultured in media buffered at pH 5–7. In the liquid medium buffered at pH 5, the response to IAA was delayed by one day, but most unexpectedly caulonema differentiated even in the control cultures after a lag of 6 ± 1 days (Figs. 1 and 2A). A microscopic examination showed that the

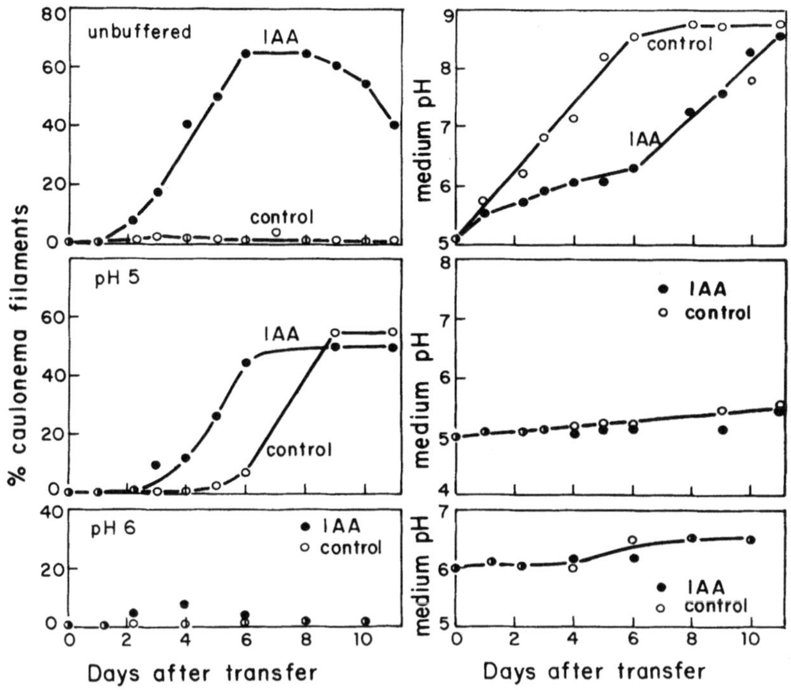

Fig. 1. Caulonema differentiation and medium pH changes in MMG and buffered MMG. Chloronema cells were cultured at an inoculum density of 0.1 mg ml^{-1} in liquid medium. IAA was 2 μM; buffers were 15 mM, pH 5 sodium succinate and 25 mM, pH 6 sodium β-glycerophosphate

conversion of 2° chloronema apical cells into caulonema apical cells started 3 or 4 days after setting up the culture (Fig. 2B). A similar conversion also occurs in the protonema growing on the agar surface [15, 27]. Caulonema differentiation in the buffered medium (pH 5), is independent of inoculum size up to 2.5 mg ml^{-1}. The lag period of 6 days observed at a cell density of 0.05 mg ml^{-1} is reduced to 3 or 4 days at a cell density of 1 mg ml^{-1}. As the cultures become old, differentiation of rhizoids is observed (Fig. 2C).

There was no delay in the differentiation of 1° caulonema in the buffered medium. The formation of 2° caulonema in auxin-free, buffered medium depends only on the pH and is independent of buffer type (e.g. it also occurs with malate or citrate buffers). In the media buffered at pH 6 or 7, the 1° caulonema differentiate normally, but the production of 2° caulonema is greatly reduced (Fig. 1). The observed pattern of caulonema differentiation at different pH values suggests that the transport of IAA occurs as passive diffusion of undissociated molecules. These data do not exclude the possibility of a carrier-mediated auxin influx which has an optimum of about pH 6 and is half-saturated at 1–5 μM auxin [22]. In protonema grown in low light there is strong evidence for the presence of influx and efflux carriers [21]. As IAA accumulation in the protonema of *F. hygrometrica* is strongly pH dependent, being greatly enhanced at pH 4 as compared to that at pH 7.6 [21], the low proportion of caulonema formed at pH 6 or 7 in the IAA-medium could be

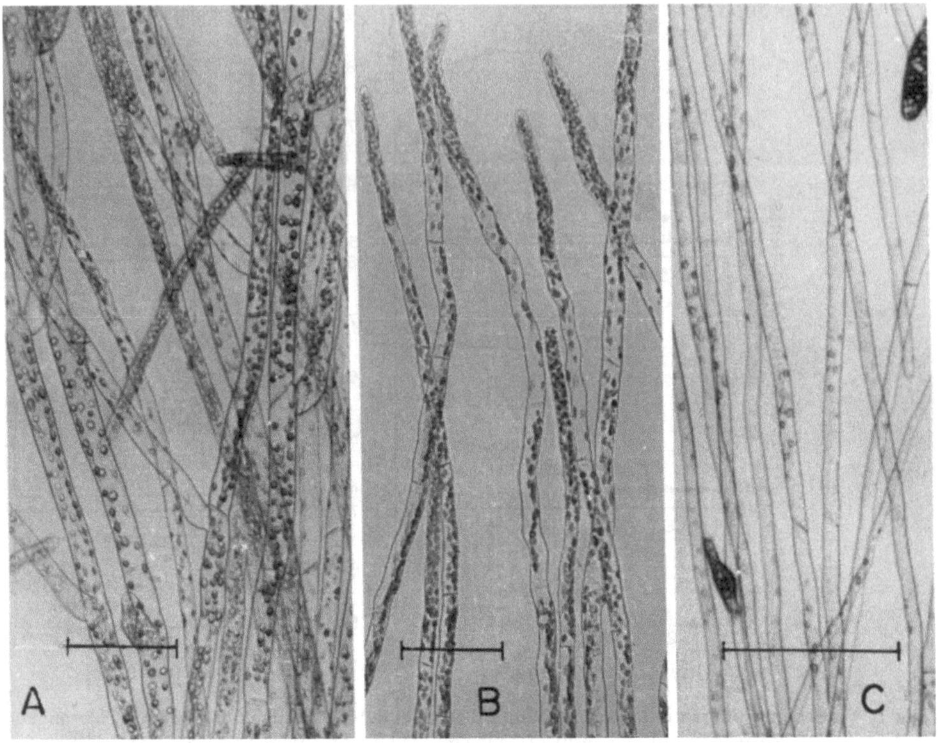

Fig. 2A-C. Cell differentiation in auxin-free, MMG buffered at pH 5. Cells were cultured at an inoculum density of 0.05 mg ml^{-1} in liquid medium. **A** Caulonema filaments 8 days after inoculation; **B** conversion of 2° chloronemal apical cells into caulonemal apical cells in some of the filaments 3 days after inoculation; **C** formation of rhizoids 12 days after the transfer of cells. *Horizontal bars* are 100 μm

due to reduced uptake and accumulation of auxin. The differentiation of caulonema in auxin-free medium buffered at pH 5 could conceivably be either due to a change in the sensitivity of cells to endogenous auxin or due to a build-up of auxin (intracellular or in the medium) above a threshold during the lag period. Even if the auxin synthesized by proliferating chloronema and 1° caulonema diffused into the medium, at pH 5, a passive diffusion back into the cells would be favoured.

2.3 Nature of Lag Period

Upon addition of auxin antagonist PCIB to the buffered medium (pH 5), the lag is prolonged beyond 6 days. The total duration depends on the concentration of PCIB and is more at 15 μM than at 5 μM (Fig. 3). At these levels, PCIB does *not* inhibit the growth of cells, in fact there is a stimulation of growth associated with the excessive proliferation of chloronema. The primary caulonema were not distinct in cultures containing 15 μM PCIB. Their initiation also depends on endogenous auxin. The lack of 1° caulonema initiation at high inoculum densities in unbuffered

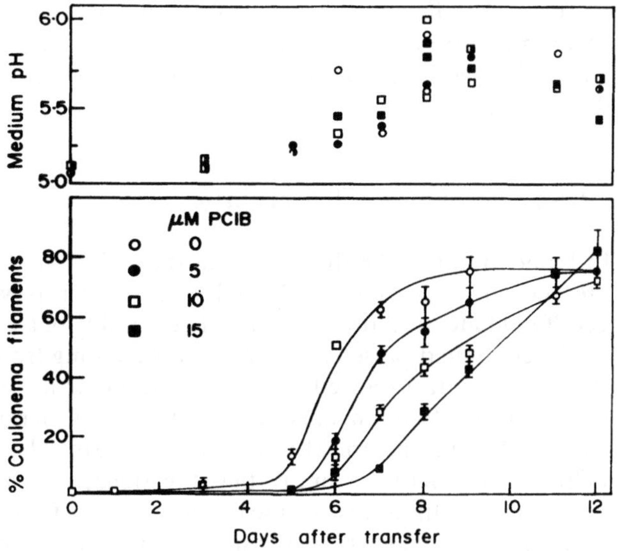

Fig. 3. Effect of PCIB on the duration of lag prior to caulonema differentiation in liquid MMG buffered at pH 5. Changes in the pH of the medium are shown at the *top*. Cells were cultured at an initial density of 0.05 mg ml^{-1}

medium could be related to IAA degradation [11] or to a basification of the medium. A relationship between enhanced IAA degradation and lack of caulonema differentiation is also observed in the NAR2 mutant of *F. hygrometrica* [3]. The prolongation of lag by PCIB is nullified by applying IAA (Table 2). The effect of 5 µM PCIB is completely nullified by 1 µM IAA, but that of 10 µM PCIB is nullified partially even by 2.5 µM IAA.

The precise mode of PCIB action in *Funaria* is unknown as yet. In the rhizoids, 10 µM PCIB reduces the polar, basipetal transport to about 40% of the control [20]. In maize coleoptiles and *Cucurbita* hypocotyls, PCIB specifically competes with IAA for IAA-binding sites in vitro [8, 10]. The delay in caulonema production by

Table 2. Effect of applying IAA on PCIB-prolonged lag in caulonema differentiation[a]

Additions	Caulonema (%)		
	Day 3	Day 5	Day 7
None	3	5	37
1 µM IAA	18	47	67
5 µM PCIB	0	1	13
1 µM IAA + 5 µM PCIB	2	47	67
2.5 µM IAA	17	58	70
10 µM PCIB	0	1	9
2.5 µM IAA + 10 µM PCIB	2	18	48

[a] Cells of *Funaria hygrometrica* protonema were cultured at an initial inoculum density of 0.05 mg ml^{-1} in MMG buffered with 15 mM, pH 5.0 sodium succinate buffer.

PCIB can therefore be either due to, a) a disruption of endogenous auxin transport, b) an inhibition of auxin synthesis and or accumulation, and c) competition for auxin-binding sites. Based on data available at present, these alternatives cannot be distinguished.

2.4 Stability of the Caulonema State

In liquid cultures, the differentiated state of 2° caulonema is stable only in the presence of exogenous IAA and if the latter is withdrawn, the caulonema dedifferentiate to chloronema [14]. The stability of a caulonema filament depends on the continued presence of signal(s) emanating from the caulonema apical cell [15]. Auxin is transported in a basipetal, polar manner. The inhibition of IAA transport and of caulonema differentiation by PCIB raise the possibility that one of the signals from the apical cells could be auxin. The results of experiments with PCIB would therefore imply that the signal (i.e. auxin) must be transported basipetally continuously for the differentiation of subapical cells into caulonema. A relationship between IAA transport and stability of caulonema state has also been proposed earlier [20]. By reasoning along these lines, the problem of caulonema differentiation can be treated or reduced to a problem of creation of new apical cells *de novo*, either from chloronema apical cells or from side branch initials. A few primary caulonema (1–2%) are formed without exogenous IAA under a variety of conditions of medium pH and nutritional status even at high inoculum densities (up to 2.5 mg ml^{-1}). Does this mean that the generation of 1° caulonema apical cells is autonomous? An inhibition of their differentiation by PCIB implies that endogenous auxin is involved. Since the 1° caulonema formation is not accompanied by an inhibition of chloronema, the formation of 1° caulonema apical cells must be under the regulation of extremely low levels of endogenous auxin. The generation of 2° caulonema apical cells can be manipulated by a variety of conditions, and the role of endogenous auxin in their differentiation needs to be defined more clearly.

3 Caulonema and Rhizoid Differentiation in Nitrogen-Limited Medium

As already stated, as liquid cell cultures age, the subapical cells produced by apical cells differentiate into rhizoid cells. The protonemal rhizoids are unbranched, show a further reduction of chloroplasts as compared to caulonema, and do not form buds in response to cytokinin application. In *Funaria*, following spore germination on phosphate or nitrate deficient medium, there is an excessive formation of rhizoids and a retarded development of chloronema [23]. The morphology of the protonemal rhizoids is similar to those arising from the base of buds or gametophores. Auxins also markedly stimulate the formation of rhizoids in liverworts and mosses [7]. In *F. hygrometrica*, mutants producing rhizoids with IAA have been isolated (Table 3). In these, IAA-enhanced rhizoid production occurs at a concentration lower than that stimulating caulonema. Rhizoids and caulonema are similar structures and products of same differentiation process [15]. This gene-

Table 3. Rhizoid formation with IAA in the mutant 72 of *Funaria hygrometrica*[a]

IAA (μM)	Fresh weight % of control	Medium pH	Chloronema (%)	Caulonema (%)	Rhizoids (%)
0	100	6.66	93	7	0
0.01	81	6.39	93	7	0
0.10	62	4.93	55	5	40
0.30	56	4.62	23	22	55
0.60	62	4.74	18	$< 10^b$	Predominant
1.0	61	4.58	8	$< 10^b$	Predominant
2.0	49	4.56	$< 10^b$	$< 15^b$	Predominant
5.0	53	4.61	$< 10^b$	$< 15^b$	Predominant
10.0	43	4.62	$< 10^b$	$< 10^b$	Predominant

[a] Cells of *Funaria hygrometrica* were cultured in MMG at an inoculum density of 0.05 mg ml^{-1}. Responses were scored 6 days after culture. The pH of the medium in IAA-treated cultures of the mutant remains below 5. Predominant implies > 80% filaments. The mutant was isolated following mutagenesis with nitrosomethyl urea and was identified directly on the basis of altered morphology on IAA-containing medium (MMJ unpublished). Fresh weight in control was 0.68 g l^{-1}.
[b] Most of the 2° chloronema and caulonema differentiated into rhizoids.

ralization suggests that there must also be a relationship between starvation and caulonema differentiation.

There is a gradient of nutrients around a 10- to 12-day-old protonema growing on the agar surface. At various distances from the protonema, analysis showed that the levels of nitrate, phosphate and glucose are respectively 63%, 42% and 51% of that present farthest away from the protonema (D'Souza and Johri, unpublished). Thus, the protonema growing on the agar surface form caulonema spontaneously and also experience a nutrient-limited environment. Daily transfer of protonema on to the fresh medium (so as to prevent it from nutrient limitation), reduces drastically the production of caulonema (Fig. 4). This reduction is not due to the act of transfer per se, because on nitrate-limited medium (0.5 mM nitrate as compared to normal level of 8.4 mM) the level of caulonema differentiation is same in the transferred and untransferred protonema. Results with phosphate- or sulfate-limited medium are similar to those obtained with nitrate-limited medium.

The above experiment indicates that a limitation of nutrients (e.g. partial starvation) enhances caulonema differentiation in protonema growing on the agar surface. Similar enhancement is also obtained in auxin-free, nitrate-limited liquid medium. Under the latter conditions, in the presence of 0.5 to 2.5 mM nitrate, both caulonema and rhizoids were produced. Their formation is independent of inoculum size between 0.05 and 2.5 mg ml^{-1}. The cells cultured at a high cell density (0.5 mg ml^{-1}) in the low nitrate medium (LNM, nitrate 1.5 mM) responded to 1 μM IAA (Fig. 5), while cells in the medium with adequate nitrate (MMG, nitrate 8.4 mM) responded very slightly to 3 μM IAA. Upon buffering these media at pH 5, cells responded to lower auxin concentrations and caulonema differentiated at 10–30 nM IAA. The cells grown in LNM or MMG respond to more or less the same

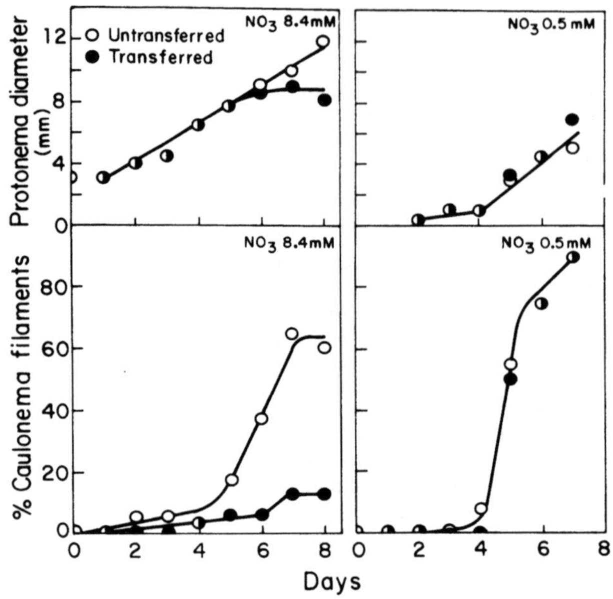

Fig. 4. Differentiation of caulonema on solid MMG (nitrate at 8.4 mM) and low-nitrate medium (LNSM, nitrate at 0.5 mM). Protonema were grown on cellophane discs placed on the agar surface. Protonema were allowed to remain on the media (untransferred), or were transferred daily to fresh MMG or LNSM (transferred). Protonema stopped growing on MMG in the transferred set after 6 days (8% caulonema), while the control (untransferred) showed 40% caulonema. On LNSM protonema grow slowly, but caulonema differentiate more rapidly than on MMG

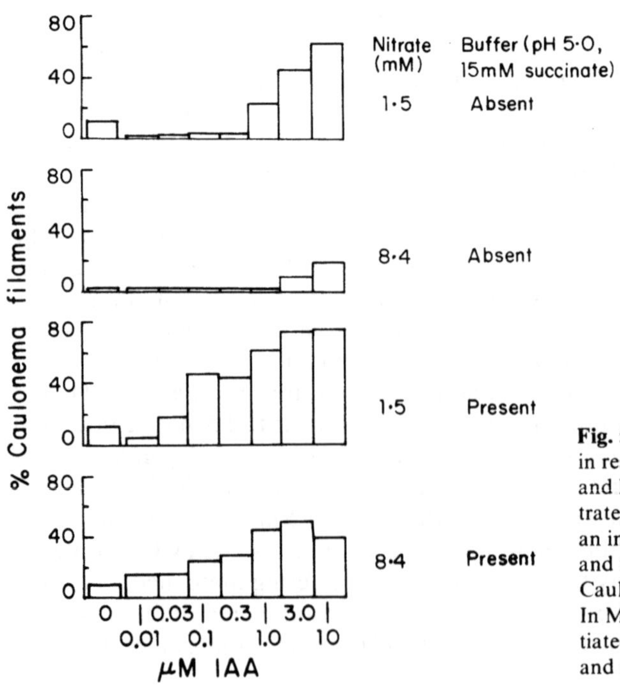

Fig. 5. Differentiation of caulonema in response to IAA in liquid MMG and low-nitrate medium (LNM, nitrate 1.5 mM). Cells were cultured at an inoculum density of 0.5 mg ml^{-1} and the media were buffered at pH 5. Caulonema were scored after 3 days. In MMG only caulonema differentiated while in LNM both caulonema and rhizoids were formed

range of IAA concentration, but the proportion of caulonema and rhizoids was always more in LNM. In these experiments caulonema differentiation with IAA was scored 3 days after transfer of cells because later on, the caulonema differentiate even in the control (buffered) flasks. Unlike the low level of differentiation with IAA in MMG buffered at pH 6 or 7, about 35–40% caulonema and rhizoids differentiated even at high densities in the LNM buffered at the same pH values. Therefore, under sub-optimal growth conditions, cells seem to acquire enhanced sensitivity to respond to exogenous or endogenous auxin. An enhanced caulonema differentiation by 1–10 nM IAA, occurs in the protonema grown in low light [17].

4 Role of Endogenous Auxin

IAA is present in mosses [2, 11]. In the protonema of *F. hygrometrica,* the fluorometric assay indicates an IAA level of 1.9 to 5.0 ng gm^{-1} fw [11]. In gametophytes of *P. patens*, using GC-MS-SIM, an IAA concentration of 2.1 ng gm^{-1} fw was noted [2]. Both determinations are in good agreement, but the concentrations are much lower than those found in the higher plants. It is not known if IAA is the predominant auxin in mosses. Determination of auxin concentration in the medium of liquid cultures is also needed. IAA is synthesized via the transamination pathway with IPA as an intermediate [11]. Caulonema formation is also enhanced by precursors of IAA (e.g. tryptophan and IPA [13, 17]. Based on the analysis of auxin-repairable mutants and a dedifferentiation of caulonema by PCIB, there is unequivocal evidence for the involvement of endogenous auxin in differentiation. However, due to lack of experimental data, it is still not possible to correlate endogenous auxin levels with the status of differentiation. Possibly, the use of GC-MS-SIM with $[^{13}C]$IAA as a quantitative internal standard, could resolve this problem. An attempt using immunological assays [26] was made, and putative IAA levels between 0.4 and 2 nmol gm^{-1} fw (70–350 ng) were observed [5 and Venkatanarayanan and Johri, unpublished]. As these levels are two orders of magnitude higher than the values reported earlier, an independent validation (e.g. GC-MS-SIM) of the immunoassays is needed.

5 Properties of Caulonema Formed in the Buffered Medium

The caulonema formed in auxin-free, buffered medium are highly responsive to various phytohormones. Depending on the particular phytohormone applied, the terminal destiny of the side branch initial can be specified. The multiple differentiation potentialities once again illustrate the undetermined nature of newly-formed subapical or side branch initials. To induce and/or maintain the differentiated state, either an external factor or some endogenously-generated signal must be provided. Upon transfer to fresh buffered medium, the side branch initial differentiates into chloronema. However, if auxin or cytokinin or ABA is added individually, caulonema or bud initial or club-shaped gemmae-like structures differentiate and even the apical cells were transformed into club-shaped structures.

5.1 Bud Differentiation on Chloronema and Caulonema

In mosses like *F. hygrometrica* and *Physcomitrium sphaericum*, bud initials arise as side branches, either from the caulonema cells directly or from the basal cell of the 2° chloronema or caulonema filaments [6, 27, 28]. Prior to caulonema formation, buds are also induced with low concentrations of cytokinin on chloronema cells in *Physcomitrella patens* [18]. Mutants of this moss which form buds exclusively on chloronema have also been reported [19].

In liquid cultures of *F. hygrometrica*, 4 or 5 days after culture in buffered medium, but before the differentiation of caulonema, buds form profusely after a 24-h treatment with 1 μM kinetin. However, these buds were restricted to the 1° caulonema and the basal one to three cells of 2° chloronema. In the PCIB-treated cultures, chloronema form profusely (Fig. 6A). The cytokinin response was so pronounced (Fig. 6B) that besides the basal cells even the chloronema apical cell also differentiated into a bud initial in rare instances. Buds forming on chloronema showed far more chloroplasts than those on caulonema. Under our experimental conditions, the magnitude of bud-forming response was far higher than that

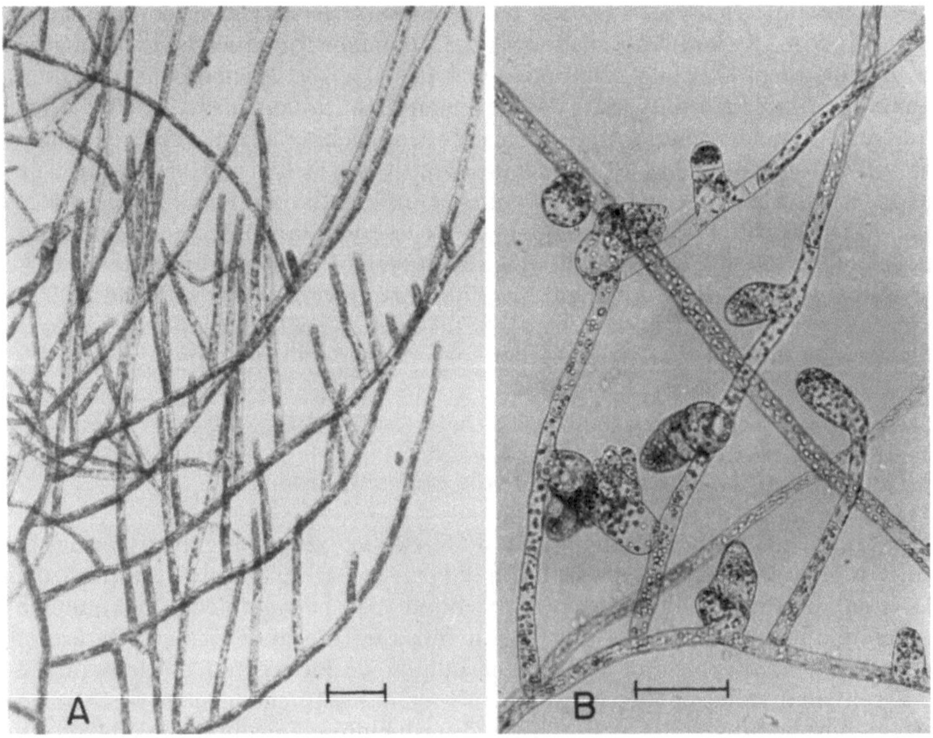

Fig. 6A,B. Bud formation on chloronema in PCIB-treated cultures. Cells were cultured in liquid MMG buffered at pH 5 in the presence of 5 μM PCIB. **A** Profuse chloronema formation occurred 5 days after transfer; **B** bud initials formed 24 h after 1 μM kinetin treatment of 5-day-old protonema. Note the bud initial formation from the apical cell in one of the 2° chloronema branch. *Bars* are 100 μm described in *P. patens* or *F. hygrometrica* [18]. Thus, as well as the basal cell, the second and third cells of a 2° chloronema filament have a potential to form bud initials in response to cytokinin and the buffered medium seems to provide the right conditions for the expression of this potential. The cytokinin-sensitivity phase of 2° chloronema lasts only for a short duration of 2 or 3 days. After differentiation of 2° caulonema, all buds are restricted to caulonema. The reason for the enhancement of cytokinin-induced buds on chloronema by PCIB is not obvious.

6 Conclusions

Compared to liverworts, the moss protonema is far more responsive to auxin. The responses are rapid, well-defined and the protonema is amenable to a variety of manipulations. Some of the more general features of auxin regulation of cell differentiation and development are beginning to emerge. The auxin uptake, transport, biosynthetic and perhaps even action mechanism(s) are turning out to be as complex and evolved as in higher plants. The multiple potential paths for differentiation of the subapical initial cell show the lack of a firm determined state. A relationship between auxin transport (between cells) and the induction of differentiation and its stability seems likely. The auxin level required for bud formation is higher than that required for chloronema growth, but less than that necessary for caulonema differentiation. As the auxin requirement for chloronema growth is the lowest, during protonema regeneration, the differentiation of all cell types to the chloronema stage (the ground state) can be explained.

In addition to determining the endogenous auxin levels in relation to differentiation, an understanding of the conditions evoking the optimal sensitivity to applied auxin is also needed. Specific resource starvation might be one such condition [25]. The enhanced sensitivity of cells to respond to exogenous auxin observed under sub-optimal growth conditions (nutrient limitation or low light intensity) could affect differentiation through a variety of mechanisms, including uptake or auxin-perception ones.

Acknowledgements. We are grateful to Bosco M.A. Henriques for many helpful suggestions. MMJ would also like to thank Prof. M. Bopp, Botanical Institute, University of Heidelberg, for help in various ways.

References

1. Ashton NW, Grimsley NH, Cove DJ (1979) Planta (Berl) 144:427
2. Ashton NW, Schulze A, Hall P, Bandurski RS (1985) Planta (Berl) 164:142
3. Bhatla SC, Bopp M (1985) J Plant Physiol 120:233
4. Bopp M (1980) In: Skoog F (ed) Plant growth substances 1979. Springer, Berlin Heidelberg New York, p 351
5. Bopp M, Bhatla SC (1985) In: Purohit SS (ed) Hormonal regulation of plant growth and development, vol 2. Agro Botanical Publishers, Bikaner (India), p 65
6. Bopp M, Gerhäuser D (1985) Biol Plant (Prague) 27:265

7. Cove DJ, Ashton NW (1984) In: Dyer AF, Duckett JG (eds) Experimental biology of bryophytes. Academic Press, London, p 177
8. Dohrmann U, Hertel R, Kowalik H (1978) Planta (Berl) 140:97
9. Hangarter RP, Peterson MD, Good NE (1980) Plant Physiol (Bethesda) 65:761
10. Jacobs M, Hertel R (1978) Planta (Berl) 142:1
11. Jayaswal RK, Johri MM (1985) Phytochemistry (Oxf) 24:1211
12. Johri MM (1974) In: Plant growth substances 1973. Hirokawa, Tokyo, Japan, p 925
13. Johri MM (1975) In: Mohan Ram HY, Shah JJ, Shah CK (eds) Form, structure and function in plants. Sarita Prakashan, Meerut (India), p 116
14. Johri MM, Desai S (1973) Nature New Biol 245:223
15. Knoop B (1984) In: Dyer AF, Duckett JG (eds) Experimental biology of bryophytes. Academic Press, London, p 143
16. Law DM, Basile DV, Basile MR (1985) Plant Physiol (Bethesda) 77:926
17. Lehnert B, Bopp M (1983) Z Pflanzenphysiol 110:379
18. Reski R, Abel WO (1985) Planta (Berl) 165:354
19. Reski R, Abel WO, Svensson D (1986) Mitt Inst Allg Bot Hamb 21:127
20. Rose S, Bopp M (1983) Physiol Plant 58:57
21. Rose S, Rubery PH, Bopp M (1983) Physiol Plant 58:52
22. Rubery PH, Sheldrake AR (1974) Planta (Berl) 118:101
23. Schoene K (1906) Flora (Jena) 96:276
24. Thomas RJ, Harrison MA, Taylor J, Kaufman PB (1983) Plant Physiol (Bethesda) 73:395
25. Trewavas AJ (1987) Trends Biochem Sci 12:258
26. Weiler EW, Eberle J, Mertens R, Atzorn R, Feyerabend M, Jourdan PS, Arnscheidt A, Wieczorek (1986) In: Wang TL (ed) Immunology in plant science. Cambridge University Press, p 27
27. Yoshida K, Yamamoto K (1982) Plant Cell Physiol 23:737
28. Yoshida K, Yamamoto K (1985) Plant Cell Physiol 26:1549
29. Zenk MH (1963) In: Nitsch JP (ed) Régulateurs Naturels de la Croissance Végétale. CNRS Paris, p 241

The Second Messenger in Apical Dominance Controlled by Auxin

W. RUSSELL[1] and K.V. THIMANN[2]

1 Introduction

In the 1930s the inhibition of the growth of lateral buds by the terminal bud in young plants was shown to be carried out by auxin, using the following sequence of experiments [9]:

1. Bioassay of the auxin diffusing out of buds and leaves of *Vicia faba* showed that the apical bud is by far the largest source of diffusible auxin;
2. Removal of this apical bud allowed rapid elongation of the lateral buds previously inhibited; and
3. Application of an auxin concentrate, or later of IAA, at a concentration somewhat higher than that identified as coming directly from the bud, duplicated the inhibition caused by the apex. Auxin at the same level as that determined by bioassay of diffusate from the apex caused a significant growth inhibition but it was incomplete.

In the immediately subsequent years, Müller, Laibach, Uhrová, Goodwin, Delisle and Wetmore and others extended the observations with different plant materials [for the early literature see 10, 11]. However, it was pointed out in 1939 that "the paradox that auxin, which typically promotes growth. . . should inhibit growth of buds. . . has not been satisfactorily explained. The principal point at issue is whether the inhibition is due to the auxin itself or to some effect of auxin *on the production or movement of other substances*" [10].

It was not until 1968 that Burg and Burg [3] offered the basis of such an explanation. They noted firstly that applied auxin was known to promote the liberation of ethylene in seedlings. Secondly, in their experiments the ethylene was mainly localized at the nodes, and thirdly, applied ethylene inhibited lateral bud development in decapitated pea seedlings. They therefore suggested that the bud inhibition that resulted when IAA was applied to decapitated plants could be actually due to the ethylene production that was induced.

It was later shown that apically applied auxin was also concentrated at the nodes [2]. However, Burg and Burg felt that this general idea might apply better to other inhibitions produced by applied auxin, such as the inhibition of root growth. This reserve was due *inter alia* to their observation that high CO_2 concentrations, which inhibit ethylene action, did not release the buds from inhibition. However,

[1] Thimann Labs., University of California, Santa Cruz, CA 95064, USA
[2] Present address: 3300 Darby Road, Apt. 3314, Haverford, PA 19041, USA

Abeles and Rubinstein (1964) did note that decapitation of *Phaseolus* plants, which caused out-growth of the lateral buds, clearly decreased the evolution of ethylene.

The inconclusiveness of the data on apical inhibition led to the proposal of several other possible agents. Applied GA_3 for instance, did somewhat increase the inhibition by IAA, although it did not exert an appreciable effect by itself. Several workers tried to implicate ABA, but the evidence was unconvincing. In particular, ABA does not appear to be secreted by the apical bud. This later literature was reviewed by Phillips [6, 7] and, with 141 references, by Rubinstein and Nagao [8]. In 1982–84 Yeang and Hillman [12, 13], from ethylene measurements and from applying ethylene to the whole plant, including the apical bud, concluded that ethylene could not be the bud-inhibiting agent in *Phaseolus* or *Pisum*. They reported that AVG (10 μm) applied to the lateral buds of *Phaseolus* "had no effect on apical dominance". Indeed, it actually inhibited outgrowth of the axillary bud. Similarly, Harrison and Kaufman [5] using the basal tiller buds of mature *Avena sativa*, concluded that ethylene actually promoted the swelling of these buds (the first stage of their outgrowth) that was induced by kinetin. On the other hand, AVG and also even CO_2 somewhat inhibited the swelling.

As against these negative findings, Blake et al., who took up the matter again in 1983, favored involvement of ethylene over the other candidates for two reasons: firstly, in the IAA experiments the concentration of IAA needed for complete inhibition was always somewhat higher than that attributable directly to the apex, and secondly because Robin Snow had shown in 1937 [see 10] that in 2-shoot plants the inhibition could travel upwards through an internode, whereas IAA was known to move in a polar manner, downward. The Canadian workers [1] showed that high levels of ethylene did, in fact, inhibit bud growth, and that, as had been shown by several workers earlier, applied IAA did increase ethylene production by a factor of about 5, but only at the peak on the second day. Decapitation lowered the natural ethylene production to about one-third the initial level. Inhibitors of ethylene action clearly increased the bud growth on decapitated plants, the increase being up to 40–50%. These results certainly supported the participation of ethylene, although some of the effects were partial rather than total.

Because the data and the conclusions were thus still in an unsettled state, we undertook new experiments directed towards finally clearing up the matter. It seemed critical to demonstrate three things:

1. that ethylene at low, physiological, levels could duplicate the inhibition caused by auxin;
2. that the inhibition caused by auxin could be nullified or at least decreased by placing the plants under reduced pressure to draw off the ethylene;
3. that inhibitors of ethylene action or of its formation would nullify the inhibition caused by auxin.

2 Materials and Methods

The plants used were *Vicia faba*, as in the original experiments of Thimann and Skoog [9]. A few experiments were also conducted on *Phaseolus vulgaris*. Seeds were germinated on wet filter-paper, planted in soil and grown in the growth chamber under continuous light of 35 μmol m^{-2} s^{-1}. The apex was removed at 1 cm above the third node of *Vicia faba*, or 1 cm above the primary leaves of *Phaseolus*, and in each case bud length was measured daily. For auxin application a narrow plastic tube was slipped over the cut internode, a second slightly larger one placed over that, and a third, with a capacity of about 1 ml, on the second. This arrangement was found necessary to avoid leakage. The auxin (IAA at a range of concentrations) was renewed daily. For experiments under reduced pressure the plants were placed in gas-tight glass chambers (under the same light) with pressure reduced to about one-third atmosphere. The vessels were opened daily for solution renewal and bud measurement. In the early experiments the pressure was reduced only to two-thirds atmosphere, but the results were essentially the same. For experiments with ethylene (at atmospheric pressure) ethylene from a cylinder of purified gas (Matheson Co.) was injected to a level of 15 ppm. The experiments were repeated many times. Preliminary trials established the effectiveness of the concentrations used.

3 Results

The results were, unexpectedly, rather clear. Ethylene at 15 ppm inhibited the outgrowth of the buds on decapitated stems (Fig. 1). The data are the average of 25 trials. Ethylene at 15 ppm is not quite as effective as IAA at 200 μM (=35 mg l^{-1}) applied to the cut surface, but it is about equal to IAA at 140 μM. In any case the trials do show that ethylene can be a potent bud growth inhibiting agent. Figure 2 shows a decapitated plant (L) and one treated with IAA (R). The completeness of the inhibition is a necessary basis for the experiments that follow.

Figure 3 shows that with 200 μM IAA applied to the cut surface, reduced external pressure can relieve most of the resulting inhibition. By the 10th day the

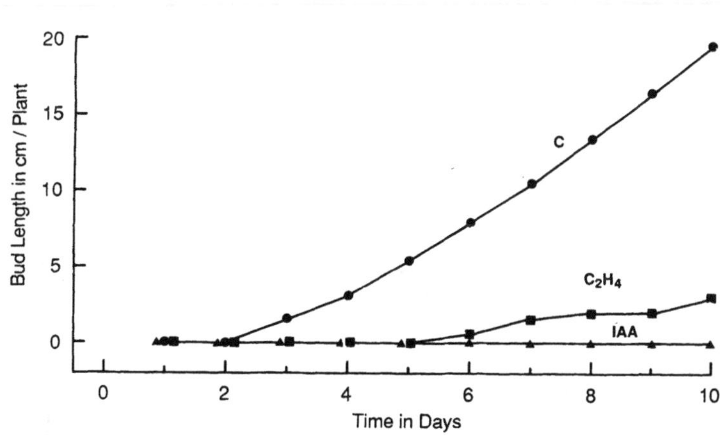

Fig. 1. Average length of the lateral buds per plant of decapitated *Vicia faba* in continuous light at 25±1°C. *C* Controls untreated; C_2H_4 plants in 15 ppm ethylene; *IAA* plants supplied with ca. 1 ml of 200 μM IAA, renewed daily. Average measurements of 25 plants in each group

Fig. 2. *Left* Typical decapitated plant of *Vicia faba*, showing elongated basal bud and second bud beginning elongation. Decapitation at *arrow*. *Right* Similar decapitated plant supplied with 200 μM IAA at the cut surface. Both photographed 10 days after decapitation

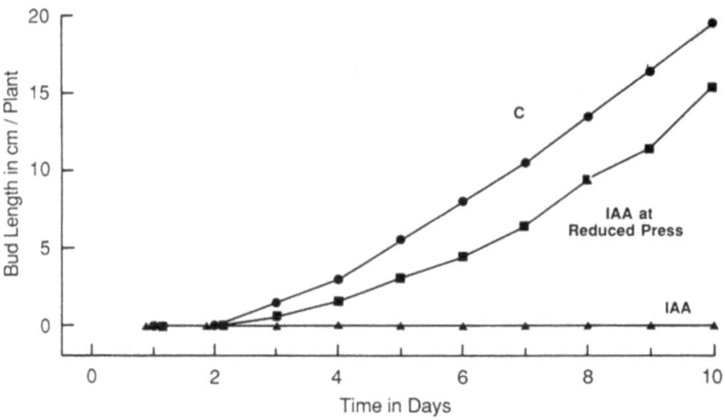

Fig. 3. Average length of the lateral buds per plant of decapitated *V. faba*. *C* Controls; *IAA* IAA-treated plants similar to those of Fig. 1; *IAA at Reduced Press* plants given 200 μM IAA but growing at 1/3 atmospheres. Average of 12 plants in each group

buds under reduced pressure averaged 80% of the length of those on the decapitated untreated controls. Since we could not be sure that two-thirds atmosphere was optimal (for the plant had to receive enough CO_2 for normal photosynthesis and enough water vapor for normal stomatal opening), the result is all the more satisfactory. It leaves no doubt that the auxin-controlled inhibitor has at least *a*

Fig. 4. Typical decapitated *Vicia faba* plant treated with IAA and grown under reduced pressure (as in Fig. 3)

volatile component. IAA is itself totally non-volatile at 25°C (m.p. about 160°C). Figure 4 shows a plant from the reduced pressure vessel with IAA applied; both lateral buds have grown out strongly. Figure 5 shows a similarly treated plant of *Phaseolus vulgaris*.

Decapitated plants (at atmospheric pressure) with IAA supplied to the cut surface were next injected, in the upper node, with 5 nM of AVG, the most highly specific antagonist of ethylene production. Figure 7 shows one of the AVG-treated plants, and Fig. 6 the average measurements of the buds in 13 similar runs.

Figures 8 and 9 show that another specific ethylene antagonist, norbornadiene, similarly allows good lateral bud outgrowth in plants treated with IAA. Norbornadiene tends to produce blackening of *Vicia* cortical tissues, but at 100 μM this toxic effect, though still detectable, is minimized; the good bud growth is thus all the more significant.

While the results with auxin as inhibitor are clear, the release of inhibition by reduced pressure is much less clear when the inhibition is due to the intact apex. The elongation of the upper internodes of the intact *Vicia* plants under reduced pressure was at least as rapid as that of controls, but the liberation of basal buds was relatively weak and began only after a lapse of three days. It may be surmised that withdrawal of nutrients and water by the continued rapid development of the upper parts of the plant makes less xylen sap available to the lateral buds. Under reduced pressure the additional transpiration would further intensify this effect. To some extent this concept comes close to the early interpretation. of apical dominance by Goebel [4].

Fig. 5. Plant of *Phaseolus vulgaris* which was decapitated, supplied with 200 μM IAA and grown at 1/3 atmosphere. The lateral buds in the axils of the primary leaves have grown out asymmetrically. One primary leaf was removed for the photograph, taken after 12 days.

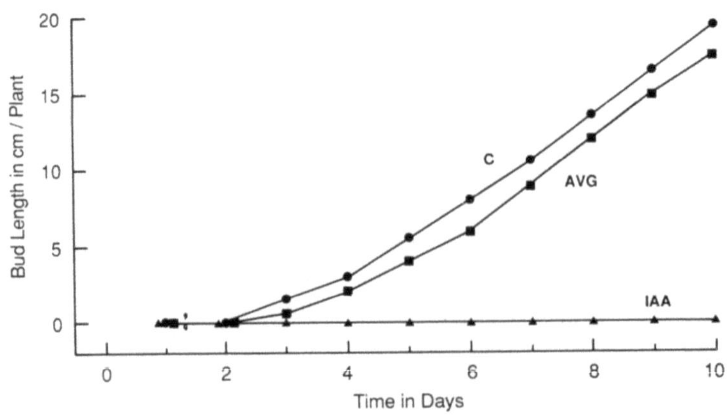

Fig. 6. Average length of the lateral buds (per plant) for *Vicia faba* plants decapitated, treated with IAA at 200 μM, and injected daily in all 3 nodes with 10 nM of AVG for each node. Average of 13 plants

Fig. 7. Typical decapitated *Vicia faba* plant treated as in Fig. 6. with AVG for all 3 nodes. Photographed after 10 days

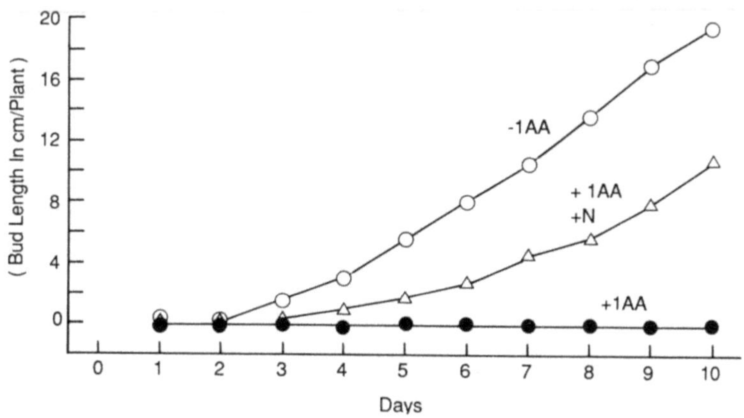

Fig. 8. Average length of the lateral buds of *Vicia faba* plants treated as in Fig. 7 but injected with 2 μM norbornadiene (–1AA = Controls)

Fig. 9. Typical decapitated *Vicia faba* plant given 200 μM of IAA and injected daily with 2 μM norbornadiene in each of the 3 nodes. Photographed after 10 days

Another explanation, based on anatomy, is also possible. When the auxin is applied to a cut surface, most of it may be transported in the cortex. If so, the ethylene that it engenders would be near the stem surface, and thus be readily removed by the reduced pressure. But auxin produced by the living apex may be transported partly within the stele, so that the resulting ethylene would not be so freely exposed to the reduced pressure; however, it would still inhibit the bud development since these primordia originate in the stele.

In any event this behaviour, whichever explanation is favored, can only be a modifying factor, for the three types of results: inhibition by applied ethylene, release of auxin-induced inhibition by reduced pressure, and similar release by ethylene antagonists – leave little doubt as to the major influence. The totality of inhibition by the intact growing apex evidently includes some secondary component, and this may explain some of the inconclusiveness of the earlier work.

References

1. Blake TJ, Reid DM, Rood SB (1983) Physiol Plant 59:481
2. Bourbouloux A (1976) C R Nat Soc Savantes, Sect Sci 101:223, from Chem Abstr (1977) 86:186002
3. Burg SB, Burg EA (1968) In: Wightman F, Setterfield G (eds) Biochemistry and physiology of plant growth substances. Runge Press, Ottawa, p 1275
4. Goebel K (1900) Balfour IB (trans) Organography of plants. Oxford University Press
5. Harrison MA, Kaufman PB (1982) Plant Physiol 70:811
6. Phillips IDJ (1969) In: Wilkins MB (ed) Physiology of plant growth and development. McGraw Hill, London, p 163
7. Phillips IDJ (1975) Annu Rev Plant Physiol 26:341
8. Rubinstein B, Nagao MA (1976) Bot Rev 42:83
9. Thimann KV, Skoog FK (1934) Proc R Soc B 114:317
10. Thimann KV (1939) Biol Rev 14:314
11. Went FW, Thimann KV (1937) Phytohormones. Macmillan, New York, Chapt 12
12. Yeang HY, Hillman JR (1982) J Exp Bot 33:111
13. Yeang HY, Hillman JR (1984) Physiol Plant 60:275

Auxin Transport and Its Regulation by Flavonoids

P.H. RUBERY[1] and M. JACOBS[2]

1 Introduction

There is much debate about the contribution of changes in auxin concentration to the control of auxin responsive processes. Transport of auxin (IAA) could participate in plant signalling networks by influencing its availability to binding sites, whose occupancy is envisaged to modulate coupling and effector processes. However, the evidence for this is still fragmentary. Auxin movement across membranes has been extensively studied at the cellular level as well as in sealed vesicles [22, 38]. Such works lead [34] to the "chemiosmotic" interpretation of polar auxin transport (PAT) which is likely to be an important determinant of IAA's role as a pattern-forming morphogen, particularly in vascular differentiation [15, 39]. PAT is preferentially basipetal in shoot tissue, energy-dependent, and has a velocity of 5–20 mm h^{-1} [18]. The fundamental transport processes can be summarized as follows: IAA can cross membranes by both diffusive and carrier-mediated routes, responding to transmembrane pH (ΔpH) and electrical potential gradients ($\Delta\Psi$). The relatively alkaline cytoplasm (pH 7–7.4) can accumulate IAA from more acidic compartments such as the cell wall because of (i) the high diffusive membrane permeability of undissociated IAAH molecules (pK = 4.7) relative to IAA anions and (ii) a high-affinity, saturable uptake carrier which may operate by electroimpelled $IAA^-/2H^+$ co-transport [22, 38]. There is also a carrier, probably for IAA anions, catalyzing efflux down the electrochemical gradient set up by the accumulative uptake process [33, 38]. The driving forces of polar transport are envisaged as metabolically-maintained pH and electrical potential gradients, while the polarity is most simply obtained by preferential localization of the IAA anion carrier at the basal ends of cells in the transport pathway [12, 36].

Clearly, modulation of transport driving forces and of specific carriers could contribute to the control not only of PAT but also of IAA distribution within and between neighboring cells, for instance in tropic responses [8]. IAA itself can alter ΔpH and $\Delta\Psi$, thus allowing cross-talk between its action and transport. At the physiological level we know that environmental factors such as light [28, 47], gravity [24, 25], wounding [21, 30], ethylene [see 35] and Ca status [8, 45] can affect polar transport. The underlying controls are not understood, but regulation specific to IAA transport would involve carrier catalysis rather than carrier energization.

Powerful synthetic Polar Auxin Transport Inhibitors (PATIS) have been available for many years and have proved valuable investigative tools [35]. They

[1] Department of Biochemistry, University of Cambridge, Cambridge CB2 1QW, UK
[2] Department of Biology, Swarthmore College, Swarthmore, PA 19081, USA

include TIBA, NPA and several fluorescein derivatives. They specifically, reversibly and noncompetitively inhibit PAT [4, 46], probably by blocking the efflux carrier for auxin anions [10, 33, 44]. This inhibition of efflux results in a stimulation of net IAA uptake by tissue segments and membrane vesicles [10, 33, 38]. The most intensively studied transport inhibitor is NPA: a specific "NPA receptor" has been located in the plasma membrane, characterized, and partially purified [14, 20, 43]. Its tissue level is 2 to 20 pmol/g fw, and it reversibly binds NPA with reported K_ds ranging from 2 to 500 nM [43, 48]. Many of the known PATIS have been shown to compete for the NPA receptor, with a specificity pattern generally paralleling their effects on PAT. The conformational features of the receptor have been inferred in a series of structure-activity studies from Katekar's group [e.g. 19]. The mechanism by which NPA inhibits the IAA efflux carrier is unknown, but probably does not involve direct competition for the same binding site as IAA [4, 46; see also 43]. Whether separate binding sites are located on a single polypeptide or on different subunits of a functional complex is not yet known.

The physiological significance of the NPA receptor could be better evaluated if endogenous ligands were identified [35]. Marigo and Boudet [23] suggested that phenolic compounds might affect PAT in tomato plants. They fed tomatoes through their roots with sodium quinate, a carbohydrate feeding into the shikimic acid pathway leading to L-phenylalanine and thence, via the regulatory branch-point enzyme PAL (product: transcinnamic acid), to phenylpropanoids and flavonoids. Plants fed sodium quinate were dwarfed, had high phenolic and IAA levels, and polar transport of IAA and NAA through intact plants was inhibited [23]. Our work developed from this observation and has shown that certain flavonoids have NPA-like effects on auxin transport, and appear to compete for binding to the NPA receptor in zucchini hypocotyl and several other shoot tissues [13]. We would like to discuss this research in more detail here and also discuss the implications it may have for regulation of auxin transport by these naturally-occurring substances.

2 Materials and Methods

2.1 Plant Material

Zucchini hypocotyls (*Cucurbita pepo*, var. All Green Bush) were grown for 5 days in the dark at 25 C or 7 days in white light (14 h/day); maize coleoptiles (*Zea mays* var. Caldera 535) were grown for 5 days in the dark; pea plants (*Pisum sativum* var. Alderman, a kind gift of Dr D.A. Morris) were grown in the light for 21 days and second internodes harvested for use.

2.2 Chemicals and Radiochemicals

Chemicals and phenolic compounds were obtained from Sigma (St. Louis, MO, USA) or Roth (Karlsruhe, FRG). [^{3}H]NPA (2.04 TBq mmol^{-1}) was purchased from Research Products International Corp. (Elk Grove, IL, USA) and [1-^{14}C]IAA (2.18

GBq mmol^{-1}) and [2-^{14}C]DMO (2.04 GBq mmol^{-1}) from Amersham International (UK). [^{3}H]NAA (37 GBq mmol^{-1}) was kindly given by Dr. R.D. Firn. NPA was a gift from Dr. R. Hertel.

2.3 Auxin Net Uptake and NPA Binding Measurements

Experiments measuring both [1-^{14}C]IAA or [^{3}H]NAA net uptake by 2-mm-long tissue segments, and [^{3}H]NPA binding to microsomal membranes were performed as described previously [13]. All uptake data are means of three replicates; binding experiments were conducted in duplicate or triplicate. Tissue segments and membrane preparations from different species were prepared in exactly the same way. For the reversibility experiments in Table 2, zucchini hypocotyl segments were preincubated in an orbital shaker (120 rpm) for 40 min at 25 C in 1.5% (w/v) sucrose alone, or sucrose supplemented with 10 μM quercetin or 0.3 μM NPA. The segments were then washed with 1.5% sucrose, either by ice-cold rinsing alone or by 25 C rinsing (3 × 50 ml), followed by resuspension in 120 ml sucrose for 30 min at 25 C in the orbital shaker. In both cases IAA uptake was then measured at pH 6 after 25 min.

2.4 Auxin Efflux

Two batches of 2 mm zucchini hypocotyl segments (3 g fw/90 ml 1.5% (w/v) sucrose at 25 C) were loaded for 50 min to equivalent levels of [^{14}C] using 0.3 μM [1-^{14}C]IAA in the absence of quercetin and 0.18 μM [1-^{14}C]IAA in the presence of 10 μM quercetin. The segments were collected, rinsed briefly with 50 ml ice-cold sucrose (1.5%), and then resuspended in 150 ml 1.5% sucrose (25 C; conical flask) and placed in an orbital shaker (120 rpm) at 25 C. Samples of medium (1 ml) were withdrawn over a 60 min time course and transferred to a scintillation fluid designed to count aqueous samples [5]. Conventional compartmental analysis by "curve peeling" [3, 29, 37] allowed estimation of first-order rate constants for efflux from the tissue considered to be comprised of a series of three compartments.

2.5 Auxin Transport

Transport assays for Fig. 4 were performed as described previously [14], except that donor blocks contained 6 μM [1-^{14}C]IAA and receiver blocks contained 20 μM quercetin, or ethanol in controls. Basipetal transport was run vertically (against gravity), with receiver blocks on top so that they could be exchanged for fresh receivers every 30 min up to 2 h. For Table 4 zucchini hypocotyl segments (20 mm long) were cut just below the apical hook, and then almost bisected with a transverse cut so as to leave a hinge of tissue. A 1 μl droplet of water, 10 μM NPA or 10 μM quercetin aqueous solution was immediately applied to the exposed surfaces. These were then appressed to restore linear geometry. The treated segments, and unhinged controls (15 in all cases) were then mounted vertically with the apical end of each segment sitting in an aqueous [1-^{14}C]IAA solution (6 μM; 40 μl) in the wells of a microtiter plate. The plate was first covered with stretched Nescofilm pierced to accomodate the segments. After 6 h in darkness at high humidity (ca. 100% RH) the upper half of the segments, distal to the source of labelled ^{14}C-IAA, were transferred individually to scintillant solution [5], extracted overnight, and counted. Samples containing less than 100 cpm were discarded because of probable poor contact at the hinge.

3 Results and Discussion

We have previously reported a striking correlation between the effects of different phenolic compounds on [^{14}C]IAA accumulation in zucchini hypocotyl segments and on [^{3}H]NPA binding to its receptor [13]. Certain widely-distributed flavonoids – for example, quercetin, apigenin and kaempferol – successfully interfere with NPA binding to its receptor and, like NPA, they stimulate net uptake of [^{14}C]IAA by tissue segments. Hydroxycinnamic and benzoic acids were ineffective in both respects. Figure 1A shows a more extensive survey of flavonoids and other phenolic compounds tested at 10 μM for these two effects. Some dose-response curves are shown in [13]. With 7 additional compounds the correlation remains high, although morin (2′,3,4′,5,7-pentahydroxyflavone) [but less so genistein, [13]] appears to be a substantially more effective binding inhibitor than uptake stimulant. The i_{50} value (the concentration that gives a 50% inhibition of [^{3}H]NPA specific binding) for morin is approx. 2 μM (data not shown). A discussion of the general conclusions that can be drawn by relating flavonoid structure (see Fig. 1B) to NPA-like activity may be found in [13].

We have used the widely-distributed and strongly NPA-mimicking flavonoids in further studies of their mode of action (Fig. 2). 1) Quercetin can increase net uptake of [1-^{14}C]IAA throughout a time course (0–60 min; Fig. 2a) and at each of pH 4, 5 and 6 (Fig. 2b). 2) The stimulation by quercetin (Fig. 2c) or apigenin (Fig. 2d) reaches a plateau at 10 μM (pH 6), at a value of about half the maximum stimulation by NPA which is more tissue permeant than flavonoids. IAA uptake is less sensitive to naringenin, the flavanone that corresponds to apigenin (Fig. 2c). The apigenin glycoside, rhoifolin, has no effect (Fig. 2d). 3) Quercetin neither stimulates nor inhibits net uptake in the presence of a maximally stimulatory NPA concentration (10 μM). However, quercetin and NPA effects are additive at suboptimal concentrations (Fig. 2c). This pattern is consistent with a common target for quercetin and NPA. 4) Uptake of [2-^{14}C]DMO (5,5-dimethyloxazolidine-2,4-dione; Fig. 2b,c,d) an indicator of internal pH change, is not increased by up to 30 μM quercetin. This indicates that quercetin does not stimulate net IAA uptake because of a raised intracellular pH increasing the "anion trap" capacity of the tissue, and that general permeabilization has not occurred.

It appears that quercetin does not act simply to create an improved *metabolic* sink, since it is ineffective when 10 μM NPA is present (Fig. 2c). This conclusion is supported by the ability of quercetin to stimulate net uptake of [^{3}H]NAA, a synthetic auxin that is not a substrate for IAA oxidase (Table 1). In any event, the specificity pattern for phenolic modulation of modulation of IAA oxidase [9] is

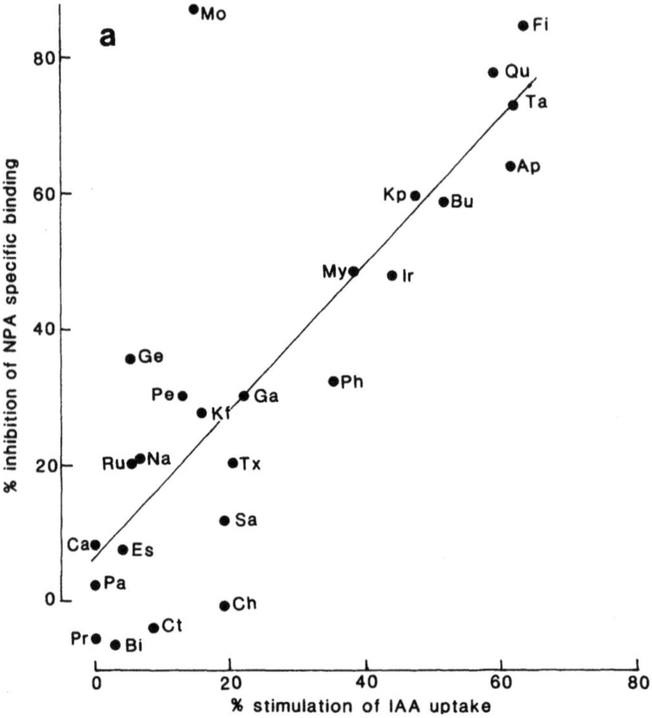

Fig. 1a,b

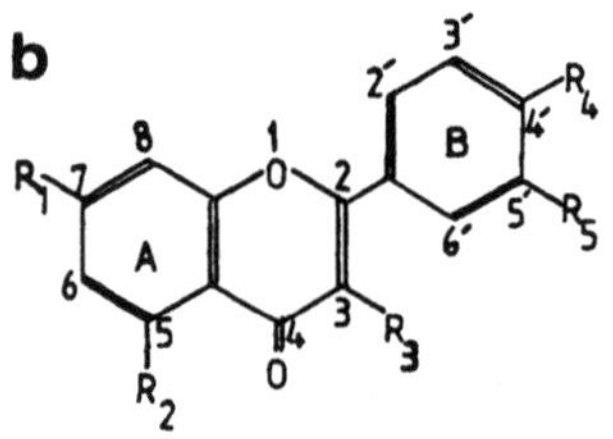

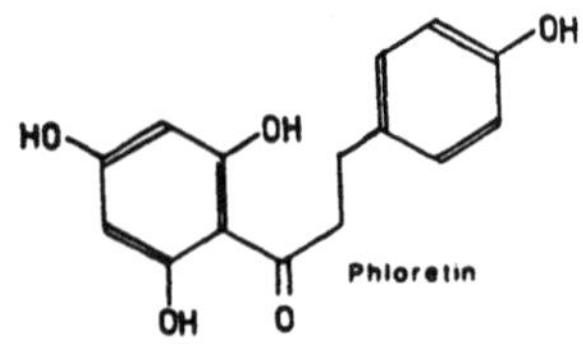

	C_2-C_3	R_1	R_2	R_3	R_4	R_5
naringenin	>-<	OH	OH	H	OH	H
apigenin	>=<	OH	OH	H	OH	H
galangin	>=<	OH	OH	OH	H	H
kaempferol	>=<	OH	OH	OH	OH	H
kaempferide	>=<	OH	OH	OH	OCH_3	H
quercetin	>=<	OH	OH	OH	OH	OH
taxifolin	>-<	OH	OH	OH	OH	OH
isorhamnetin	>=<	OH	OH	OH	OH	OCH_3
tamarixetin	>=<	OH	OH	OH	OCH_3	OH
fisetin	>=<	OH	H	OH	OH	OH

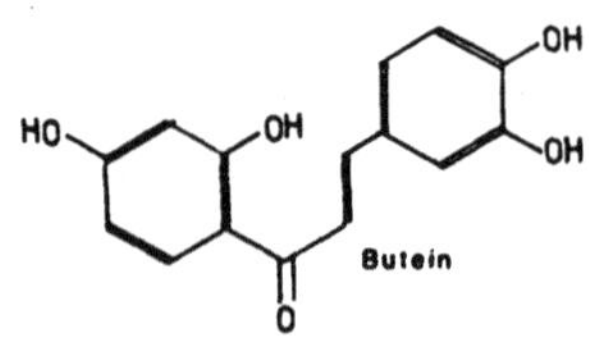

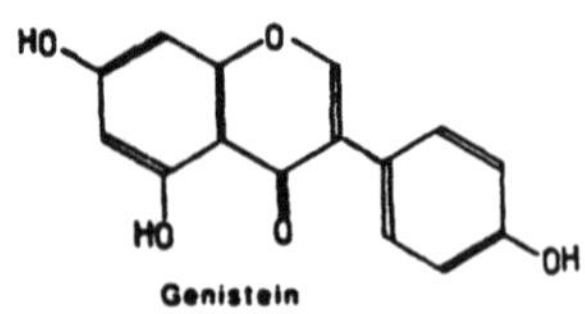

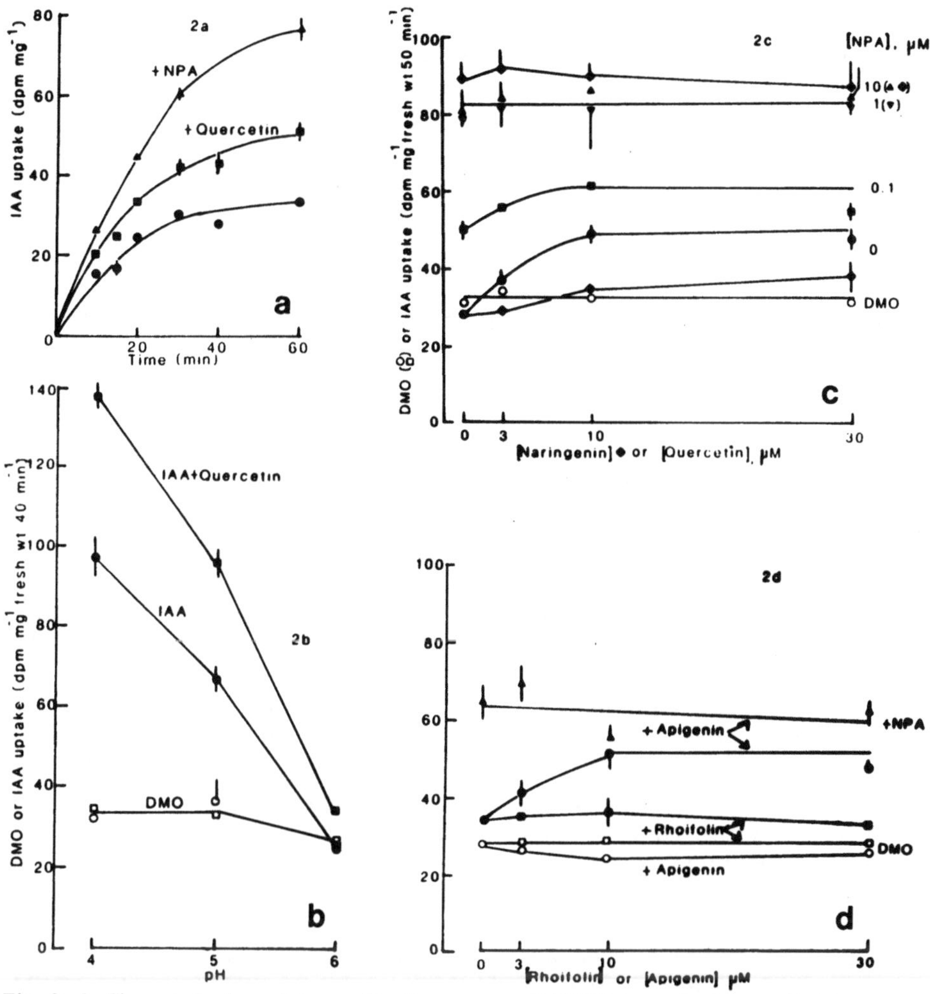

Fig. 2a-d. Characterization of flavonoid effects on net [1-^{14}C]IAA uptake by zucchini hypocotyl segments **(a)** [1-^{14}C]IAA uptake (0.3 μM, 0–60 min) was followed at pH 6.0 in controls (●) and in the presence of 10 μM quercetin (■) or 10 μM NPA (▲). **(b)** pH Dependence (Na phosphate-citric acid buffer) of [12-^{14}C]DMO uptake in the presence (□) and absence (○) of 10 μM quercetin and of [1-^{14}C]IAA plus (■) and minus (●) 10 μM quercetin. **(c)** Effect of varying quercetin concentration on DMO uptake (○) or on IAA uptake in the absence (○) and presence of NPA (0.1 μM, ■; 1 μM, ▼; 10 μM, ▲). Also, effect of 0–30 μM naringenin (♦) plus and minus 10 μM NPA. **(d)** Effect of 0–30 μM apigenin and its 7-O-glycoside rhoifolin on DMO uptake (○ □) and IAA uptake (● ■). Effect of apigenin on IAA uptake plus 10 μM NPA (▲)

◄

Fig. 1a. Correlation of effects of phenolic compounds at 10 μM on [1-^{14}C]IAA net uptake and on specific [^{3}H]NPA binding, assayed as described in Materials and Methods. Uptake (%-stimulation) is compared to a control (no test substance added) and normalized to quercetin stimulation (59 ± 1%; mean of 19 triplicates ± SEM). Inhibition of [^{3}H]NPA binding is relative to the complete inhibition of specific binding by 10 μM NPA. Straight line drawn after unweighted least squares linear regression. Abbreviations: *Ap* Apigenin; *Bi* biochanin A (an isoflavonoid); *Bu* butein; *Ca* caffeic acid; *Ct* (+)catechin; *Ch* chyrsin; *Es* esculetin; *Fi* fisetin; *Ga* galangin; *Ge* genistein; *Ir* isorhamnetin; *Kf* kaempferide; *Kp* kaempferol; *Mo* morin; *My* myricetin; *Pa* phenylacetic acid; *Pe* pelargonidin chloride; *Ph* phloretin; *Pr* protocatechuic acid; *Qu* quercetin; *Ru* rutin; *Sa* salicylic acid; *Ta* tamarixetin; *Tx* taxifolin. Representative flavonoid structures are shown. **b** Structures of some of the flavonoids tested

Table 1. Effects of quercetin on [^{3}H]NAA uptake and on [1-^{14}C]IAA uptake in the presence of nonradioactive IAA to saturate the auxin uptake carrier [5][a]

	Uptake of labelled auxin (dpm mg^{-1} fresh wt 40 min^{-1})[b]				
	[1-^{14}C]IAA	[1-^{14}C]IAA + 10 μM NPA	[1-^{14}C]IAA + 10 μM NPA + 50 μM IAA	[^{3}H]NAA	[^{3}H]NAA + 30 μM NPA
Control	30.4 ± 2.2	77.5 ± 6.1	35.2 ± 1.8	102.0 ± 1.6	136.2 ± 2.5
Quercetin[c]	42.4 ± 2.5	65.8 ± 5.4	37.6 ± 1.0	116.9 ± 4.8	–

[a] Uptake protocol as in Materials and Methods ([^{3}H]NAA was used at 37 GBq $mmol^{-1}$. 1 dpm = 0.45 fmol).
[b] Mean of 3 replicates ± SEM.
[c] 10 μM with [1-^{14}C]IAA, 30 μM with [^{3}H]NAA.

quite distinct from that for phenolic effects on IAA net uptake, and no IAA metabolism could be detected after 1 h in the presence or absence of quercetin [13]. Table 1 also indicates that quercetin does not stimulate the saturable IAA *uptake* carrier: when the efflux carrier is blocked by NPA, quercetin neither increases [1-^{14}C]IAA net uptake (Table 1, Fig. 2c) nor alters the level to which nonradioactive IAA reduces the net uptake of radioactive IAA because of competition for the uptake carrier (Table 1).

A direct investigation confirmed that efflux of [1-^{14}C]IAA from preloaded segments could be inhibited when quercetin was present in either the loading solution alone (Fig. 3), or when added only to the IAA-free efflux solution (data not shown). As is common, three apparent rate constants can be extracted by compartmental analysis. These roughly correspond to cell wall + free space ("fast"), cytoplasm ("medium"), and vacuole ("slow"), respectively. Quercetin appears to affect efflux from the "cytoplasmic" compartment, reducing the estimated rate constant by about 60%.

Since synthetic inhibitors of auxin efflux may act within the cytoplasm [34, 36, 44], we studied the effect of preincubating hypocotyl segments in quercetin or NPA on subsequent [1-^{14}C]IAA uptake after the segments had been either (1) rinsed with ice-cold sucrose, or (2) washed at 25°C before being transferred to additive-free media (Table 2). After the rinse of 0°C, uptake was stimulated compared to controls, with a further increment when quercetin or NPA was present also in the assay medium. In contrast, a thorough washing at 25°C to facilitate efflux reversed the stimulations, although the tissue again retained its sensitivity to quercetin and NPA. These data do not exclude an external site of action. However, they are consistent with quercetin and NPA acting reversibly at the inner face of the plasma membrane to inhibit carrier-mediated IAA efflux. The indirect evidence available indicates that externally applied flavonoids can enter cells. For example, radiolabelled flavonoids are biosynthetic precursors [7] and taxifolin (dihydroquercetin) fed to cut white flowers of a chalcone synthase mutant of *Antirrhinum majus* is converted to anthocyanin [41].

We have previously partially characterized the inhibition of [^{3}H]NPA binding to zucchini hypocotyl membranes by some of the active flavonoids in Fig. 1A [13]. The pattern of quercetin/NPA interaction is suggestive of competitive inhibition

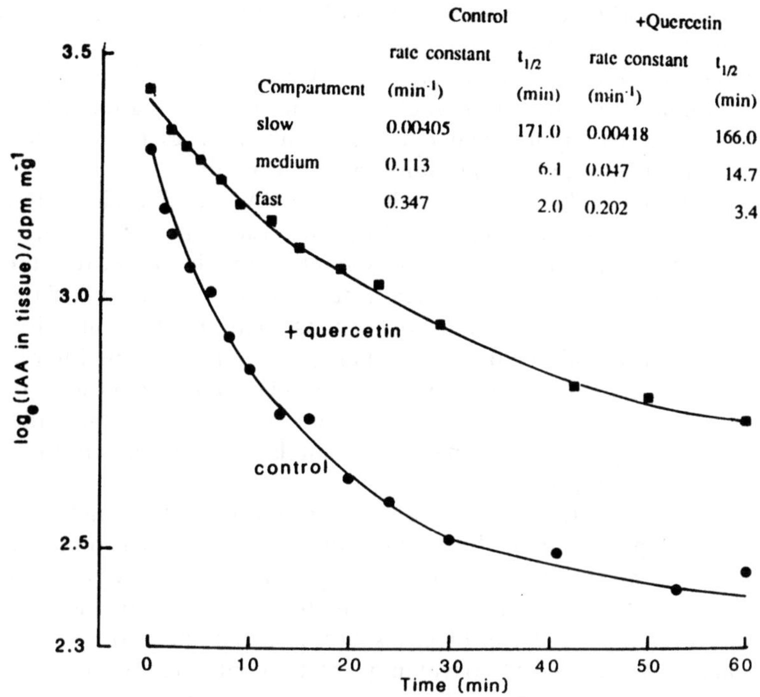

Fig. 3. Inhibition by quercetin of IAA efflux from zucchini hypocotyl segments. Procedures as in Materials and Methods

Table 2. Reversibility of stimulation of [1-^{14}C]IAA net uptake by quercetin and NPA[a]

	[1-^{14}C]IAA net uptake (dpm mg^{-1} fresh wt. 25 min^{-1})[b] Ice-cold rinse			Wash at 25°C		
	Control	Quercetin (10 μM)	NPA (10 μM)	Control	Quercetin (10 μM)	NPA (0.3 μM)
Preincubation (40 min) :						
Sucrose alone	22.9 ± 0.9	34.5 ± 0.1	52.6 ± 2.4	27.2 ± 0.4	44.5 ± 2.4	48.0 ± 2.7
Quercetin (10 μM)	30.5 ± 1.1	38.2 ± 1.8	48.5 ± 1.5	29.3 ± 0.4	43.2 ± 0.6	–
NPA (0.3 μM)	51.3 ± 0.2	60.6 ± 1.4	57.1 ± 4.6	30.0 ± 1.7	–	44.2 ± 3.9

[a] Procedures as in Materials and Methods
[b] Means of 3 replicates ± SEM.

(K_i = 1.9 μM; K_s for NPA = 5.25 nM; [19]) but further analysis is needed for a complete description of the binding kinetics. The inhibition is exerted over a pH range of 5 to 7, and is at least partially reversible [13]. In dose-response experiments, flavonoids less active at enhancing [1-^{14}C]IAA accumulation in the segment uptake assay (e.g. galangin, naringenin) are, with the present exception of morin, less effective than quercetin at inhibiting NPA binding.

The active flavonoids can also have NPA-like effects in [^{3}H]NPA binding and segment uptake tests using light-grown zucchini hypocotyls, light-grown pea internodes, and etiolated maize coleoptiles (Table 3). Because the range of compounds active in these tissues again includes widely distributed flavone and flavonol aglycones, there is potential for response patterns reflecting different degrees of effectiveness both within a particular species and between different species. Thus, for example, apigenin is significantly more effective than quercetin in stimulating IAA net uptake by pea internodes and maize coleoptiles, while in dark-grown zucchini hypocotyls and maize mesocotyls (data not shown) their effects are of a similar magnitude.

NPA and other synthetic auxin transport inhibitors can block PAT when they are incorporated into the receiver blocks in a classical donor-to-receiver polar transport assay. Figure 4 shows that quercetin in receiver blocks can inhibit IAA arrival, but only for the first 2 h of the transport test. Thereafter, the radioactivity delivered to the receiver blocks is increased compared to controls. In some experiments IAA (or NAA) arrival in quercetin-containing receiver blocks is increased at earlier times. A crossover is never observed when NPA is used. Quercetin and other flavonoids bind strongly to cellulose [31] and, unlike NPA, do not diffuse away

Table 3. Effects of flavonoids in different species

	[1-^{14}C]IAA net uptake (dpm mg^{-1} fresh wt. 40 min^{-1})^a Zucchini hypocotyl (light-grown)		Pea internode (light-grown)		Maize coleoptile (dark-grown)	
Substance added (10 μM):	*Uptake*	(%)	*Uptake*	(%)	*Uptake*	(-%)
Control	23.4 ± 0.5	100	31.5 ± 0.5	100	12.3 ± 0.6	100
NPA	67.5 ± 3.5	289	90.9 ± 3.6	289	79.2 ± 6.2	642
Quercetin	43.8 ± 0.3	187	37.0 ± 0.3	118	20.1 ± 0.3	162
Apigenin	–	–	46.9 ± 1.4	149	31.3 ± 1.7	254
Fisetin	–	–	40.8 ± 0.9	130	–	–
	specific [^{3}H]NPA binding (dpm mg^{-1} fresh wt.)^b					
Substance added:	*Binding*	(%)	*Binding*	(%)	*Binding*	(-%)
Control	29.6 ± 1.09	100	13.6 ± 0.55	100	110.5 ± 6.0	100
Quercetin (0.1 μM)	–	–	12.2 ± 0.88	90	–	–
(1.0 μM)	–	–	9.5 ± 0.23	70	–	–
(10 μM)	6.6 ± 0.02	22.3	5.7 ± 0.14	42	34.4 ± 0.9	31
Apigenin (10 μM)	9.7 ± 0.30	32.7	–	–	40.2 ± 2.5	36
Fisetin (10 μM)	–	–	4.5 ± 0.28	33	–	–

^{a} Mean of 3 replicates ± SEM.
^{b} Mean of 2 replicates ± range.

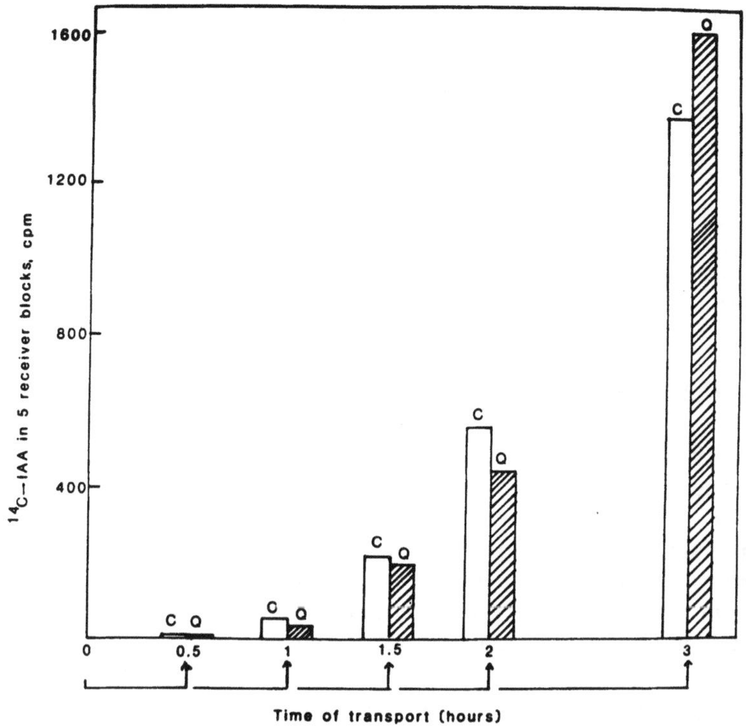

Fig. 4. Effects of quercetin on polar auxin transport by zucchini hypocotyl segments. Data are counts per min collected in the pooled receivers of quintuplicate samples over the half-hour periods shown and for the period from 2 to 3 h of transport. *Open panels*: control treatment, *shaded panels*: quercetin

from a local application site. The contrasting behavior of quercetin and NPA could perhaps be explained if quercetin inhibited carrier-mediated IAA efflux only from cells close to the receiver block. These cells could start to act as local sinks, accumulating IAA delivered from the polar transport system. This system is probably localized in cells adjacent to the vascular bundles [18]. Since IAA is freely membrane permeant as the protonated species, cells neighboring the polar transport route would also increase in IAA content. The augmented movement of IAA from such cells to the receiver block could be responsible for the period of enhanced delivery of IAA in the presence of quercetin, following the initial inhibition. In contrast, NPA is more mobile and it intercepts the polar transport stream well away from its site of application. An alternative explanation could entail quercetin relieving or antagonizing a more potent endogenous inhibitor of IAA efflux.

It is possible to circumvent problems arising from quercetin's apparent immobility when added to the basal end of a transport segment, by applying it or NPA as small local doses half-way up 20 mm long hypocotyl segments (Table 4). With this method both substances can be seen to interrupt the IAA transport stream and to inhibit basipetal auxin transport. Attempts to demonstrate perturbation of seedling gravitropism by quercetin have so far been unsuccessful.

Table 4. Interruption of basipetal IAA transport by quercetin or NPA[a]

IAA transport (dpm 10-mm section^{-1}; means ± SEM) Treatment (1 μl section^{-1})			
Unhinged	Water	Quercetin (10 pmol)	NPA (10 pmol)
1344 ± 311; n = 15	596 ± 51; n = 13	429 ± 52; n = 11 ($p < 5\%$)[b]	414 ± 57; n = 15 ($p < 5\%$)[b]

[a] Procedures as in Materials and Methods.
[b] Probabilities for difference from water-treated segments assessed by Dunnett's Multiple Comparison Test.

4 Conclusions

We have suggested [13] that the flavonoids found to be active in our studies may act as natural regulators of polar IAA transport and of IAA efflux from cells in plants. We have discussed some characteristics of the flavonoids that support such an hypothesis: their ubiquity among plants [9], their likely availability at concentrations that are physiologically effective [42], and their tightly regulated and environmentally-responsive biosynthetic pathways [2, 6, 17]. We also note that free flavonoid aglycones secreted by roots of some leguminous plants appear to regulate rhizobial nodule initiation at the genetic level [32].

There are of course some problems which must be addressed in further testing of the hypothesis. Most estimates of flavonoid levels in the literature do not distinguish between glycosides and aglycones, and probably represent mostly glycosides compartmentalized in the vacuole. Flavonoid glycosides are not active in our assays. It is important to establish the levels and intracellular location of individual free flavonoid aglycones in plant tissues that transport IAA, and to see if control linkages can be demonstrated.

Flavonoids have been reported at high levels in epidermal tissues, particularly in leaves [11]. A recent immunohistochemical study of parsley seedlings provided evidence for chalcone synthase (the entry point to flavonoid biosynthesis) in epidermal tissue, and also in vascular strand tissue of parsley leaves [40]. Polar IAA transport is known to be associated with non-conducting cells of vascular strands [1, 12, 27, 49].

Comparative structure-activity relations for flavonoids and synthetic PATIS (including the compounds designated as phytotropins by Katekar [19, and see 35] will be discussed in a later publication. The apparent affinity of quercetin for the NPA receptor is substantially lower than that of NPA, which itself is one of the less active synthetic phytotropins [19]. This does not, however, argue a priori against lower-affinity natural ligands bringing about physiologically effective occupancy, perhaps by binding to a receptor conformation different from that which permits higher affinity NPA association. Whereas quercetin and other flavonoids which strongly affect IAA transport may be NPA agonists, any compounds which bind to the same receptor but exert relatively little inhibition of IAA efflux could act as antagonists. Further investigation is needed to see if morin is an example of such a compound.

Even if natural ligands to the NPA receptor are used by plants in some circumstances as a means of PAT regulation, other mechanisms are possible. Johnson and Morris have recently reported [16, 26] that the weak natural auxin PAA, although not itself a carrier substrate, appears to inhibit both IAA influx and efflux carriers, and also block PAT in intact pea plants and stem segments. It may do this by competing for the IAA binding sites: we find no stimulation of net uptake of IAA in zucchini by 10 μM PAA (Fig. 1A) or concentrations up to 1 mM. Nor did PAA inhibit NPA binding in pea (up to 50 μM) or zucchini (Fig. 1A).

Abbreviations

IAA indole-3-acetic acid
DMO 5,5-dimethyloxazolidine-2,4-dione
NAA 1-naphthylacetic acid
NPA 1-naphthylphthalamic acid
PAA phenylacetic acid
PAL phenylalanine ammonia lyase
TIBA 2,3,5-triiodobenzoic acid

Acknowledgements. We are grateful to the NSF for a grant (PCM 8314844) that helped support this research and to Sara Ranck and Magdalen Lindeberg for assistance with some of the experiments.

References

1. Bonnemain J-L (1971) CR Acad Sci Paris Ser D 273:1699
2. Bruns B, Hahlbrock K, Schafer E (1986) Planta 169:393
3. Cram WJ (1968) Biochim Biophys Acta 163:339
4. Depta H, Eisele K-H, Hertel R (1983) Plant Sci Lett 31:181
5. Depta H, Rubery PH (1984) J Plant Physiol 115:371
6. Dixon RA (1986) Biol Rev 61:239
7. Grambow H, Grisebach H (1971) Phytochemistry 10:789
8. Gross J, Sauter M (1987) Plant Sci 49:189
9. Harbourne JB (1980) In: Bell EA, Charlwood BV (eds) Encyclopedia of plant physiology, new series, vol 8. Springer, Berlin Heidelberg New York, p 329
10. Hertel R, Leopold AC (1963) Planta 59:535
11. Hrazdina G, Marx GA, Hoch HC (1982) Plant Physiol 70:745
12. Jacobs M, Gilbert SF (1983) Science 220:1297
13. Jacobs M, Rubery PH (1988) Science 241:346
14. Jacobs M, Short TW (1986) In: Bopp M (ed) Plant growth substances 1985. Springer, Berlin Heidelberg New York, p 218
15. Jacobs WP (1952) Am J Bot 39:301
16. Johnson CF, Morris DA (1987) Planta 172:400
17. Jones DH (1984) Phytochemistry 23:1249
18. Kaldewey H (1984) In: Scott TK (ed) Encyclopedia of plant physiology, new series, vol 10. Springer, Berlin Heidelberg New York Tokyo, p 80
19. Katekar GF, Geissler AE, Kennard CHL, Smith G (1987) Phytochemistry 26:1257
20. Lembi CA, Morre DJ, Thomson K-St, Hertel R (1971) Planta 99:37
21. Leopold AC, Lam SL (1962) Physiol Plant 15:631
22. Lomax TL, Mehlhorn RJ, Briggs WR (1985) Proc Natl Acad Sci USA 82:6541
23. Marigo G, Boudet AM (1977) Physiol Plant 41:197

24. Mertens R, Weiler EW (1983) Planta 158:339
25. Migliaccio F, Rayle DL (1984) Plant Physiol 75:78
26. Morris DA, Johnson CF (1987) Planta 172:408
27. Morris DA, Thomas AG (1978) J Exp Bot 29:147
28. Naqvi SM, Gordon AA (1967) Plant Physiol 42:138
29. Poole RJ (1971) Plant Physiol 47:731
30. Rayle DL, Ouitrakul R, Hertel R (1969) Planta 87:49
31. Roberts EAH (1960) Nature 185:536
32. Rolfe BG, Gresshoff PM (1988) Annu Rev Plant Physiol Plant Mol Biol 39:297
33. Rubery PH (1979) Planta 144:173
34. Rubery PH (1986) In: Bopp M (ed) Plant growth substances 1985. Springer, Berlin Heidelberg New York Tokyo, p 197
35. Rubery PH (1987) In: Hoad GV, Lenton JR, Jackson MB, Atkin RK (eds) Hormone action in plant development. Butterworth, London, p 161
36. Rubery PH, Sheldrake AR (1974) Planta 118:101
37. Rugiewicz PT, Bledsoe CS, Glass ADM (1984) Plant Physiol 76:913
38. Sabater M, Rubery PH (1987) Planta 171:501, 507, 514
39. Sachs T (1986) Symposium xxxx Soc Exp Biol, Cambridge, p 181
40. Schmelzer E, Jahnen W, Hahlbrock K (1988) Proc Natl Acad Sci USA 85:2989
41. Stickland RG, Harrison BJ (1974) Heredity 33:112
42. Strack D, Reznik H (1976) Z Pflanzenphysiol 79:95
43. Sussman MR, Gardner G (1980) Plant Physiol 66:1074
44. Sussman MR, Goldsmith MHG (1981) Planta 151:15
45. Tang PM, Dela Fuente RK (1986) Plant Physiol 81:651
46. Thomson K-St, Hertel R, Muller S, Tavares JE (1973) Planta 109:337
47. Thornton RM, Thimann KV (1967) Plant Physiol 42:247
48. Venis M (1985) Hormone binding sites in plants. Longman, London, p 60
49. Wangermann E (1974) New Phytol 73:623

GC-MS Quantifications of Free and Ester Indol-3yl-Acetic Acid in Relation to Root Growth and Gravitropism

M. SAUGY and L. RIVIER[1]

1 Introduction

Studies on phototropism and gravitropism at the beginning of this century led to the discovery of the diffusible growth promoting factor from coleoptile tips [41], later called "auxin" [14]. Indole-3-acetic acid (IAA) has been shown to be one of the main auxins [11], and is known to be present in most of higher plants. This natural hormone appears to be an important endogenous regulator controlling cell enlargement and division, and could be implicated in gravireaction processes [5]. Extensive coverage of the occurrence, biochemistry and physiology of IAA and related auxins were recently published [6, 16, 18, 35].

The use of GC-MS offers the opportunity for unequivocal identification of auxins. The very specific and sensitive detection of trace amounts obtained with this technique allows accurate and precise measurements of IAA levels in plants. This great potential can be exploited only by taking specific precaution for handling and analyzing the samples. By the use of modern method of preparative separation and a well-chosen internal standard (IS), meaningful physiological correlation can be established.

2 General Method for Identification and Quantification

2.1 Extraction and Purification

Extraction of auxins from plant tissues is carried out under reduced temperature. After addition of the suitable IS, the tissue is homogenized in organic solvent (acetone, MeOH, EtOH or diethyl ether) or buffer. The first step of purification traditionally involves partitioning between aqueous phase and organic solvent. A classical purification scheme consists in partitioning buffer at pH 7–8 with diethyl ether. Then the aqueous phase is acidified to pH 3 with 1 N HCl and the indole acids like IAA are extracted with ethyl acetate, whereas the aqueous phase still contains the indole conjugates. IAA conjugates are hydrolyzed in 1 N NaOH for 1 h at 22 °C [27] and the resulting free acid is extracted as above. For further purification, PVP (polyvinyl polypyrollidone) column chromatography, sephadex, ion-exchange, amberlite XAD-7, immunoaffinity, gel permeation, TLC and HPLC were generally used (for extensive review, see [35]).

[1] Institut de Biologie et de Physiologie Végétales, Université de Lausanne, 1015 Lausanne, Switzerland and Institut de Médecine Légale, Laboratoire de Toxicologie Analytique, Université de Lausanne, Bugnon 21, 1005 Lausanne, Switzerland

2.2 Derivatization and Quantification

GC is a separation technique which is based on partition principles between the stationary liquid phase and the mobile gas phase. The derivatization of substances such as auxins and auxin conjugates is needed to increase their volatility and to improve the chromatographic characteristics of the analyte. It is also used to enhance the detector response, then the sensitivity of the assay. The selection of the derivative for MS depends on the method of ionization. A derivative with favorable EI characteristics could be useless for measurements by negative chemical ionization (NCI). Semi-quantitative data were obtained from IAA derivatives mass spectra and their respective merits to GC behaviour and MS ionization are compared in Table 1.

Table 1. Chromatographic behaviour and yield of ionization and fragmentation of fluorinated derivatives of IAA: factors of choice for SIM determinations[a]

Compound	Number of F atoms	Rt[b] (min)	MW[c]	Base Peak (area response) Ionization mode EI	Total ionization (%) PCI	NCI
IAA-Me	0	7.00	189	130 (2144) 39	190 (22 612) 47	188 (2161) 44
5-F-IAA-Me	1	6.40	207	148 (6525) 43	226 (48 570) 46	206 (9228) 73
IAA-Me-TFA	3	5.57	285	226 (2209) 39	304 (10 339) 58	285 (7655) 44
IAA-PFB	5	9.67	355	130 (528) 32	303 (1420) 36	335 (221) 20
IAA-Me-HFB	7	5.62	385	326 (7371) 44	403 (25 820) 65	384 (118) 33
IAA-PFP	10	5.12	453	276 (7655) 46	437 (2036) 22	305 (10 768) 62

[a] Ammonia was used for Cl.
[b] Retention time obtained on a crossed-linked fused silica capillary column of the SE-54 type with temperature program: 100°C for 1 min, then 30°C/min to 220°C. Injector:250°C.
[c] Molecular weight.
The area response is the area of the peak calculated on the SIM chromatogram. TFA:trifluoroacetyl; PFB:pentafluorobenzylester; HFB:heptafluorobutanoyl-1; PFP:pentafluoropropionylester pentafluoropropionyl-1.

In MS of auxins, the widest used method of ionizing was by electron impact (EI). The development and improvement of chemical ionization (PCI and NCI) have accumulated new evidence for identification of endogenous auxins and also has extended IAA detection to a lower limit [33, 34].

Quantitative determinations will be precise only if full corrections for losses occurring during extractions and possible detector instabilities are made. The mass spectrometer measuring the mass of molecules, the use of auxins labeled with stable isotopes provides almost ideal IS for accurate quantifications. The stable isotope dilution method [22] consists in measuring by Multiple Ion Monitoring (MIM) the ratio of normal to heavy isotopes added at the beginning of the analysis. The absolute amount of auxin in plant extract is then obtained [see 19].

To date, most IS available are IAA analogs labeled with deuterium because of the commercial availability of enriched synthetic precursors and chemical reagents

[33]. Number of assays have been described using side chain ^{2}H-IAA as IS [2]. $^2H_2$-IAA (on the methylene carbon of the side chain) is no longer recommended because of exchange of hydrogens at extreme pH and during storage and sample manipulations [1]. Ring labeled IAA ($^2H_4$ and $^2H_5$-IAA) has the advantage of the stability of the label during alkaline hydrolysis of auxin conjugates. Cautions should be taken when using $^2H_5$-IAA under extreme pH conditions whereas $^2H_4$-IAA (not commercially available) has been shown to be the most suitable deuterated IS for IAA quantification, free and conjugated [19].

Recently, J. Cohen and coworkers [9] have synthetized high purity $^{13}C_6$-[benzene ring]-IAA. Compared to the deuterated standard, it has the advantage of no isotope exchange under any conditions of sample preparation and purification. Another important point is that under certain HPLC conditions (Reversed Phase C_{18}, 10% MeOH/water + 1% acetic acid, R.P. Pharis, personal communication), contrary to $^2H_5$-IAA, $^{13}C_6$-IAA did not separate from unlabeled IAA.

3 IAA Pool in Plants

Several authors recently reviewed the reactions and the processes which seem to be determinant for the regulation of IAA content in plant tissues [6, 7, 35]. Bandurski's group [6] has described well the output and input reactions which are regulating the IAA pool size in maize seedlings.

The decrease of this pool size seems to be generated mainly by oxidative catabolism. The oxidation of IAA to OxIAA, then to 7-hydroxy-OxIAA appears to be the major oxidative pathway of IAA in maize even though these products are not accumulating as end products [6].

Both ester hydrolysis and de novo synthesis should increase the amount of IAA in the tissues. Incorporation experiments, of deuterium from ^2H_2O into the indole ring, as measured by MIM [26 and our lab, unpublished data] were exhibiting some labeling into IAA; this could be a promising approach for IAA biosynthesis in seedlings.

Such determinants of IAA turnover (degradation and biosynthesis) should be under metabolic control, and they will establish the measurable IAA content in a tissue. This control is certainly different for each individual plant specimen and each tissue from an organ. Thus, in maize roots, it was demonstrated that curvature could be generated by local application of IAA and that the inhibitory effect of auxin was dependent on the zone treated [30], and on the initial elongation rate of the root [31]. GC-MS quantifications are providing meaningful data on establishing the relation between these two physiological events and the auxin status in the seedling.

4 IAA in Maize Roots: Gravitropism and Growth

4.1 IAA and Gravitropism

Experiments on the implication of IAA in the control of root gravireaction have been accomplished by several authors [3, 29]. A downward lateral transport of applied IAA was observed in the bending zone of horizontal roots [12, 15, 20, 28]. However, results on whether asymmetrical distribution of endogenous IAA occurs during the root gravitropic response seem contradictory. If Suzuki et al. [39] found a clear asymmetry in favour of the lower part of the maize root (75:25) by using bio-assays, Mertens and Weiler [20], using RIA, did not detect any significant difference between the two longitudinal parts.

Using the Me-HFB derivative and $^2H_5$-IAA as IS for the quantification of IAA in root parts, we could not find any asymmetry between the two halves from the bending zone (1.0–5.0 mm from the tip) after 2 h of gravistimulation. But, the IAA level in the stele (from the differentiation zone: 5.0–15.0 mm from the root tip) is about 5-fold that of the cortex [37]. As we postulated that the zone involved in the gravireaction is the cortex, IAA determinations were done in three longitudinal parts of the roots: the upper, the middle and the lower parts. The upper and the lower zones were made up entirely of cortical tissues, whereas the middle zone was composed of both stelar and cortical cells. IAA quantifications (Fig. 1) showed that after 2 h endogenous IAA is distributed asymmetrically in the cortical parts of the bending zone. In the middle part, the level of IAA was not as high as expected, but in this part of the root the stele is not well differentiated, thereby explaining the difference obtained with measurements done directly on the stele which came from the fully differentiated zone.

This IAA asymmetry seems very small, relative to what was observed in lateral transport experiments. But, it is similar to that found between upper and lower halves of gravistimulated *Zea* mesocotyl cortex [5]. These authors postulated that the asymmetry must also occur in the stele itself. However, we could not detect it in our material.

How this asymmetry occurs, and what its main effectors are, are still unanswered questions, but we think that, at least in maize roots, both lateral diffusion and ester hydrolyzis could be implicated.

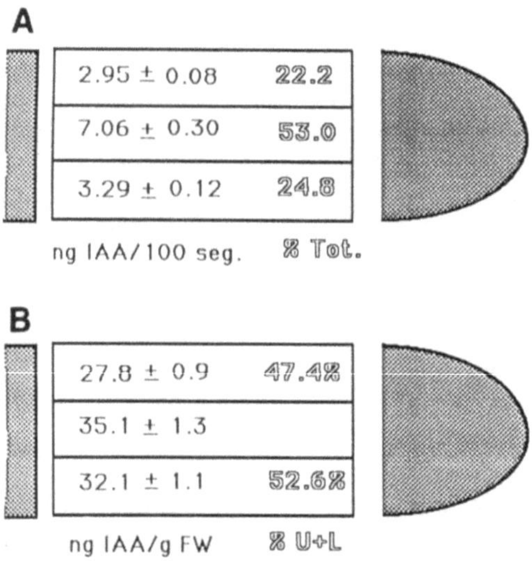

Fig. 1A,B. IAA content in ng ± SE per 100 segments (**A**), and in ng ± SE per g fw (**B**) of three dissected parts of 5 ± 0.1 mm-long apical maize root segments from which the first mm had been removed. Intact 2-day-old-seedlings were maintained horizontally for 2 h, uniformly illuminated (white light). Each value is a mean of 12 samples of 100 segments each. Data obtained by quantification on the β-cleavage fragment (m/z = 326 and 331 for IAA and internal standard derivatives, resp.) [adapted from 37]

All these data support the Cholodny-Went theory which ascribes asymmetry in growth rate to the formation of an auxin gradient across the organ. But this postulate has been disputed, and the direct action of PGRs in tropisms has been questioned. "Sensitivity" of the implicated tissues could be the important factor [17, 40]. This concept, however, is difficult to quantify. But the role of IAA in modulating asymmetic growth of roots, resulting in gravicurvature, has been supported recently by Mulkey and Vaughan [25] and Moore and Evans [24]. The latter proposed an interesting model of action implicating Ca^{++}, which is also distributed asymmetrically in the gravistimulated root. Calcium, as a messenger in the bending zone, could either act as a sink for auxin or sensitize the tissue to auxin. The second proposal was consistent with the early claim that an auxin gradient does not exist in graviresponding roots [20]. However, given the more accurate measurements of endogenous IAA by GC-MS, our data tend to favour the first proposal.

4.2 IAA and Growth Classes

These results led us to wonder if such differences in the auxin content could be related to the growth asymmetry occurring between the two opposite sides of the root.

We used as a research tool the large variability of growth rates inside a root population [31], rather than exogenous application of auxin. A population of 1650 roots (2 days old), with a mean growth rate of 0.64 mm/h, was divided into six main classes between 0.26 mm/h and 1.13 mm/h (for extreme classes).

The endogenous IAA content was measured in the so-called elongation zone (2.5–5.0 mm from the root tip), as it was postulated that only the IAA level in the tissues during elongation could be related with that function. There is a good correlation (Fig. 2) between the growth rate and the IAA content, which is higher in slowly-growing roots. These results are consistent with the generally inhibitory

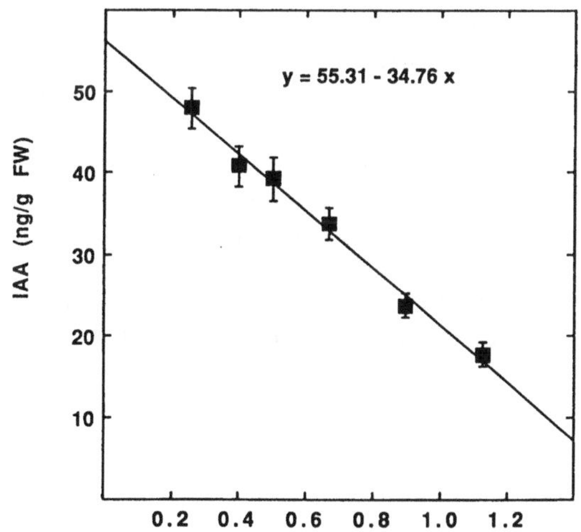

Fig. 2. IAA content (in ng ± SE per g fw) of the elongation zone as a function of the growth rate (in mm/h) for different classes of maize roots. The correlation constant was found to be 0.928 (calculated from 33 samples of 50 segments each) [adapted from 31]

effect of auxin in roots [30]. We have thus demonstrated that endogenous IAA content and a measurable physiological function can be correlated.

Furthermore, a good concordance is obtained with results on IAA content in the upper and lower cortex of gravistimulated roots. The growth rate of these two parts was calculated [36]. With a mean of 0.64 mm/h for the root, the upper and lower cortex grow with a rate of 0.68 and 0.59 mm/h respectively. By reporting these rates on the x-axis of Fig. 2, the ratio between IAA content upper and lower cortex will be 47.6/52.4, which is quite similar to that found by direct measurements (Fig. 1). Although gravireaction cannot be strictly compared to growth [6], we know that these two phenomena are connected [21, 36] and, apparently, the same type of correlation can be established between the hormone content of an elongating tissue and its growth rate.

5 Free and Ester IAA in Roots

There is considerable information on the amount of IAA liberated from esters by alkaline hydrolysis in maize seedlings [4, 13, 23, 27], but little is known about IAA esters in roots. Several roles for esters have been proposed. They apparently do not have auxin activity per se, but could be a source of IAA for growth during germination [10]. Or, IAA esterification could occur as a prevention against peroxidative attack [8]. Two types of experiments were conceived to established the relative importance of IAA and its esters during germination processes [38]. At first, changes in level during germination were measured in the whole root, and in the elongating tissues. Secondly, both were quantified in roots from different growth classes (2-day-old seedlings) to determine whether the IAA pool is influencing the growth rate.

5.1 Free and Ester IAA During Germination

The level of free IAA (expressed per g fr-wt.) is stable throughout the examined period of culture (Fig. 3), whereas ester-linked IAA decreases sharply from day 1 to reach (day 5) the same concentration as free IAA. This could indicate that esters of IAA are a major source of IAA in the root at the beginning of germination, and that, after a certain period, a transition from heterotrophic to autotrophic IAA metabolism will determine the level of free IAA. The same situation occurs for IAA-esters in the elongation zone (Fig. 4). On the contrary, free IAA level which was increasing only slightly between d 1 and d 6, increased dramatically after that time. This higher level after 6 days is corresponds with a lower growth rate that occurs effectively since that time [38]. This increase in the level of IAA cannot be explained by the hydrolysis of IAA ester in this tissue, since the level of the latter is relatively stable after that period; rather, it may be controlled by other factors such as de novo synthesis and transport.

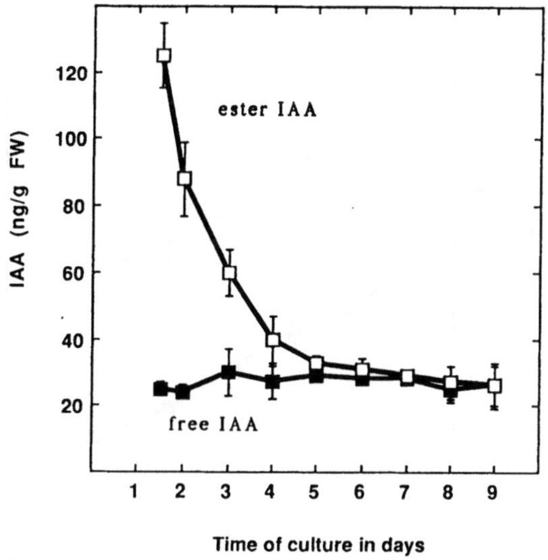

Fig. 3. Free and ester IAA contents (in ng ± SE per g fw) of whole maize roots as a function of time of culture (in days). Each value is the mean of 4 determinations of 50 roots each [adapted from 38]

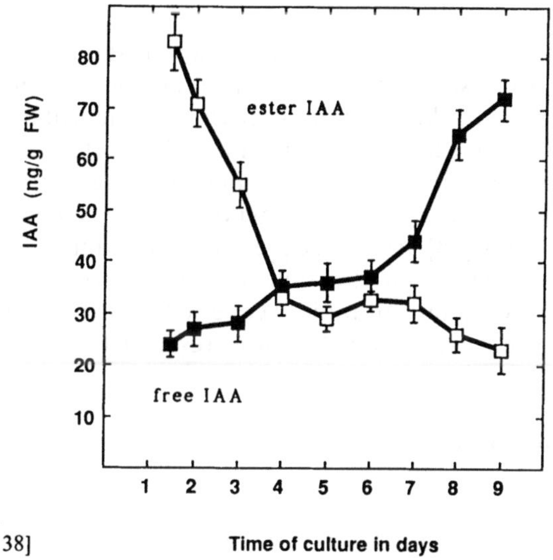

Fig. 4. Free and ester IAA contents (in ng + SE per g fw) at different time of culture (in days) of the elogation zone (2.5–5.0 mm from the tip) of maize roots. [adapted from 38]

5.2 Free and Ester IAA in Growth Classes

Figure 5 demonstrates that 2-day-old maize roots contain larger amounts of esters than free IAA for all six classes. There is an inverse correlation between the level of ester-linked IAA and growth rate as was the case for free IAA. Consequently, differences in growth rates for a population of roots at a certain time are not necessarily controlled by a differential hydrolysis of the ester pool: a root containing large amount of IAA also contains a high level of ester. Thus, this relation is more

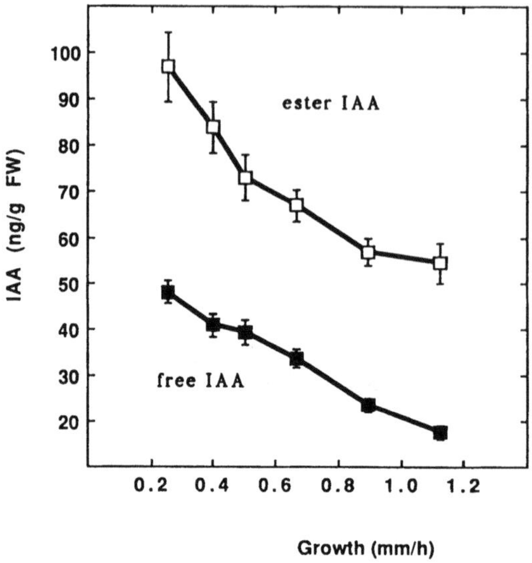

Fig. 5. Free and ester IAA contents of elongation zone (in ng + SE per g fw) as a function of the growth rate for different growth classes of maize roots. [adapted from 38]

characteristic of a general metabolic status (which is defined as free plus ester IAA) of the seedling, rather than their relative amounts. Placed in the gravireaction context, these results suggest that the difference observed in free IAA content between the upper and the lower cortex of the root is not controlled by ester hydrolysis. This is consistent with the fact that the radial distribution of IAA esters in vertical roots is homogenous, whereas the stele contains more free IAA than the cortex [38]. As was suggested by Bandurski et al. [5] in shoots, the IAA could be secreted from the stele into the lower side of the cortex during downward movement, and rapidly coverted into an equilibrium of free plus ester IAA. This would explain why both are present in larger amounts in the lower half of the cortex.

6 Conclusion

GC-MS has proved to be one of the more reliable analytical method for accurate and precise measurements of auxins in plants. GC-MS has allowed us to show that an asymmetry in the endogenous IAA level occurs during the root gravireaction processes. Furthermore, a relation between the growth rate of different groups of roots and the IAA content of their elongation tissues was established. These data showed that the asymmetry in growth rate generated during gravireaction between upper and lower cortex is clearly related to the difference in IAA level.

The content of free and ester IAA in maize seedling changes with time, and with the zone of tissue considered. We must therefore pay more attention to the metabolic turnover of IAA, and take into account all the aspects (transport, conjugates, other hormones, ...) in determining both the auxin level in a specific tissue and its effect on observable physiological events characteristic of that tissue.

Acknowledgements. We gratefully acknowledge Prof. P.E. Pilet at the Institute of Plant Biology of the Lausanne University for his technical support in this work and many helpful discussions.

References

1. Allen JRF, Baker DA (1980) Planta 148:69
2. Allen JRF, Rivier L, Pilet PE (1982) Phytochemistry 21:523
3. Audus LJ (1975) In: Torrey JG, Clarkson DT (eds) The development and function of roots. Academic Press, London, p 327
4. Bandurski RS, Schulze A (1977) Plant Physiol 60:211
5. Bandurski RS, Schulze A, Momonoki Y (1984) Physiologist 27:(S)123
6. Bandurski RS, Schulze A, Reineke DM (1986) In: Bopp M (ed) Plant growth substances 1985. Springer, Berlin Heidelberg New York Tokyo, p 83
7. Cohen J (1983) What's New Plant Physiol 14:41
8. Cohen J, Bandurski RS (1978) Planta 139:203
9. Cohen J, Baldi B, Slovin JP (1986) Plant Physiol 80:14
10. Epstein E, Cohen J, Bandurski RS (1980) Plant Physiol 65:415
11. Haagen-Smit AJ, Dandliker WB, Wittwer S, Murneek AE (1946) Am J Bot 33:118
12. Hestnes A (1979) Ann Bot 14:567
13. Iino M, Carr DJ (1982) Plant Physiol 62:950
14. Kögl F, Haagen-Smit AJ (1931) Proc K Ned Akad Wet 34:1411
15. Konings H (1967) Acta Bot Neerl 13:556
16. Letham DS, Goodwin PB, Higgins TJV (ed) (1978) Phytohormones and related compounds, a treatise, vol 1: The biochemistry of phytohormones and related compounds, and vol 2: Phytohormones and the development of higher plants, Elsevier, Amsterdam
17. MacDonald IR, Hart JW (1987) Plant Physiol 84:568
18. MacMillan J (ed) (1980) Encyclopedia of plant physiology (new series) vol 9, Springer, Berlin Heidelberg New York, p 681
19. Magnus V, Bandurski RS, Schulze A (1980) Plant Physiol 66:775
20. Mertens R, Weiler EW (1983) Planta 158:339
21. Meuwly P, Pilet PE (1987) Plant Physiol 84:1265
22. Millard BJ (1978) In: Quantitative mass spectrometry. Heyden, London, p 171
23. Momonoki YS, Schulze A, Bandurski RS (1983) Plant Physiol 72:526
24. Moore R, Evans ML (1986) Am J Bot 73(4): 574
25. Mulkey TJ, Vaughan MA (1986) In: Bopp M (ed) Plant growth substances 1985, Springer, Berlin Heidelberg New York Tokyo, p 241
26. Pengelly WL, Bandurski RS (1983) Plant Physiol 73:445
27. Pengelly WL, Hall PJ, Schulze A, Bandurski RS (1982) Plant Physiol 69:1304
28. Pilet PE (1971) Bull Soc Bot Suisse 81:52
29. Pilet PE (1977) In: Pilet PE (ed) Plant growth regulation. Springer, Berlin Heidelberg New York, p 115
30. Pilet PE, Meuwly P (1986) Planta 169:16
31. Pilet PE, Saugy M (1985) Planta 164:254
32. Pilet PE, Saugy M (1987) Plant Physiol 83:33
33. Rivier L (1986) In: Linskins HF, Jackson GF (eds) Gas chromatography-mass spectrometry. Modern Methods in Plant Analysis, vol 3, Springer, Berlin Heidelberg New York Tokyo, p 146
34. Rivier L, Saugy M (1986) J Plant Growth Regul 5:1
35. Sandberg G, Crozier A, Ernsten A (1987) In: Rivier L, Crozier A (eds) The principles and practice of plant hormones, vol 2. Academic Press, London, p 1
36. Saugy M (1986) Thesis, Lausanne
37. Saugy M, Pilet PE (1984) Plant Sci Lett 37:93
38. Saugy M, Pilet PE (1987) Plant Physiol 85:42
39. Suzuki T, Kondo N, Fujii T (1979) Planta 145:323
40. Trewavas A (1981) Plant Cell Environ 4:203
41. Went FW (1928) Recl Trav Bot Neerl 25:1

Hemmstoff und Wachstum: Growth Inhibitors, Not Auxin, Regulate Phototropism

J. Bruinsma,[1] M. Sakoda,[2] and K. Hasegawa[2]

1 Introduction

Sixty years ago, a famous botanical dissertation appeared in The Netherlands: 'Wuchsstoff und Wachstum', by Frits W. Went [18]. One of the classical experiments, described in this publication, was the unilateral illumination of an *Avena sativa* L. coleoptile tip, collecting the diffusates from the lighted and shaded sides in agar on either side of a flake of mica, and determining the auxin activities using the bioassay described in the paper (Fig. 1). From the auxin activity larger in the diffusate from the shaded side than from the lighted side, it was deduced that phototropic curvature results from differential flank growth caused by a lateral gradient in the cell-elongating substance. This auxin was later recognized as the ubiquitous IAA. The view was extended to gravitropism as the general Cholodny-Went theory of tropic curvature [19].

For almost 50 years, the evidence for this theory has been indirect. It was derived, either from bioassays that cannot distinguish whether a difference in auxin activity results from differences in the amounts of growth-promoting or of growth-inhibiting substances; or from the distribution of radioactivity from exogenous substances with auxin activity, which was presumed to reflect the distribution of endogenous auxin [13]. The first attempt, in 1975, to obtain direct evidence by spectrofluorometrically determining the amounts of IAA in extracts from the lighted and shaded halves of phototropically curving *Helianthus annuus* L. hypocotyls, failed to support the Cholodny-Went theory [2]. Both sides contained equal amounts of free IAA. Later determinations using GC-ECD [4, 9] and immunoassay [16, 17] rendered similar results.

The present survey will summarize these negative data and demonstrate that, instead, phototropic curvature results from a lateral gradient of growth-inhibiting substances that cause differential flank growth by differential inhibition of cell elongation with unchanged auxin distribution. This also applies to the auxin diffusion from unilaterally illuminated coleoptile tips.

2 The Distribution of IAA in Phototropically Stimulated Organs

Studies on photo- and gravitropic movements have mainly been performed with three objects: the etiolated coleoptile of *Avena sativa* L., the de-etiolated hypocotyl

[1] Department Plant Physiology, Agricultural University, Wageningen 6703 BD, The Netherlands
[2] Biological Institute, Kagoshima University, Kagoshima 890, Japan

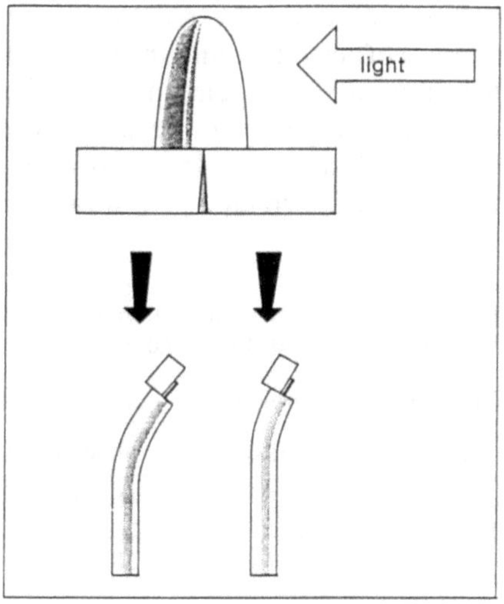

Fig. 1. Scheme of the classical phototropic experiment by Went 1928

Table 1. Distribution of IAA in phototropically responding organs of *Helianthus annuus* L., *Raphanus sativus* L. and *Avena sativa* L.[a]

Organ	No. of Expt.	IAA-distribution (%) Curving			Method of detection	[Ref.]
		Lighted half	Shaded half	Straight half		
Green sunflower	3	51	49	48	SPfl.	[2, 14]
hypocotyl	5	50.5	49.5	50	Imm.	[17]
Green radish hypocotyl	3	51	49	45	ECD	new
Etiolated oat coleoptile	3	49.5	50.5	50	ECD	[9]

[a] Free IAA was extracted from longitudinal halves and determined by: Spfl. = spectrofluorometry, Imm. = immunoassay, or ECD = electron capture detection.

of *Raphanus sativus* L., and the green hypocotyl of *Helianthus annuus* L. Determinations in all three organs show a symmetrical distribution of free IAA during phototropic curvature (Table 1). Weiler was also unable to find a translocation of IAA from the central to the more peripheral tissues [16, 17]. It can be concluded that phototropic movements are not accompanied by detectable changes in the distribution of extractable free IAA.

3 The Distribution of Growth Inhibitors in Phototropically Stimulated Organs

Helianthus annuus L. The phototropic response of the green sunflower hypocotyl is a local phenomenon: apex and cotyledons can be covered or even removed without affecting the reaction to unilateral white or blue light; moreover, the amount of curvature and the rate of elongation growth of the hypocotyl were shown to be mutually independent [3, 5]. This points to a local change in growth conditions, rather than to an overall change in cell elongation as brought about by a changed basipetal flow in IAA. In accordance with the hypothesis of A.H. Blaauw (1915) [1], this light-induced change is a local growth inhibition [2].

Table 2. Inhibitor distribution in straight, curving, and curved sunflower hypocotyls after 0, 30–45, and 60–80 min unilateral white light[a]

Duration of stimulus	Inhibitor (%)		Curvature (°)	n
	Lighted half	Shaded half		
0′	52, 51	48, 49	0, 0	208, 235
30′–45′	72, 61	28, 39	14.4, 16.5	395, 475
60′–80′	62, 57	38, 43	34.4, 34.4	422, 433

[a] Duplicate experiments with n seedlings; average content, in *cis*-xanthoxin-equivalents 60 ng g^{-1}. After [5]

Light-grown sunflower hypocotyls contained considerably more growth-inhibiting substances than etiolated ones. The neutral diethyl ether fraction, particularly, contained large amounts of inhibiting activity in, e.g. the cress-seed germination test. This activity co-chromatographed in several systems with xanthoxin [15]. When the growth-inhibiting activities in purified extracts of hypocotyl halves before, during and after phototropic stimulation were determined with the germination assay, it could be shown that the originally even distribution changed into accumulation of inhibitor(s) on the illuminated side: Table 2. Further purification and GC-flame ionization detection revealed, however, that the contribution of the active isomer, *cis*-xanthoxin, to the total inhibition was only about 10% [14]. Recently, Weiler was unable to detect a shift in immunologically measured xanthoxin [17]. Thus, the nature of the inhibitor(s) involved in the phototropism of green sunflower hypocotyls has still to be elucidated.

Raphanus sativus L. On the contrary, the inhibitors that cause the etiolated radish hypocotyl to curve phototropically have been identified as *cis*- and *trans*-raphanusanin, and raphanusamide [7]. The raphanusanins, particularly, inhibited hypocotyl elongation in vitro and, if applied unilaterally, caused curvature [11].

The rate of flank growth was determined from photographs of seedlings that were provided with small lanolin-coated beads on either side of their hypocotyl. Upon unilateral illumination the growth rate of the irradiated side was reduced, whereas that of the shadow side was barely inhibited [11]. Bilateral illumination with different light intensities caused curvature towards the higher irradiation because of stronger growth inhibition at that side [12].

Etiolated hypocotyls already contained the three inhibitors prior to phototropic stimulation, but illumination caused their contents to increase considerably.

This was determined by UV-detection upon HPLC-purification of the extracts. Unilateral and bilateral illuminations induced such increases in the contents of inhibitors at either side that these contents correlated well, both with the light intensities and with the flank growth rate (Fig. 2) [8, 12].

Together with the even distribution of IAA (Table 1), these data lead to the conclusion that the phototropic movement of the radish hypocotyl is caused by a light-induced lateral distribution of three substances that differentially inhibit the auxin-regulated cell elongation at the two flanks.

Avena sativa L. Flank growth rates of etiolated oat coleoptiles have been frequently measured. Using the bead application and photography techniques, it was shown that the second positive curvature of intact, black-capped, decapitated, or partially illuminated coleoptiles was brought about by cessation of growth of the lighted flank [6, 10]. The opposite flank may show a slight growth acceleration or inhibition, depending on the fluence rate and the position. Also, the fact that the curvature occurred simultaneously over the whole irradiated length does not point to the participation of a longitudinal factor.

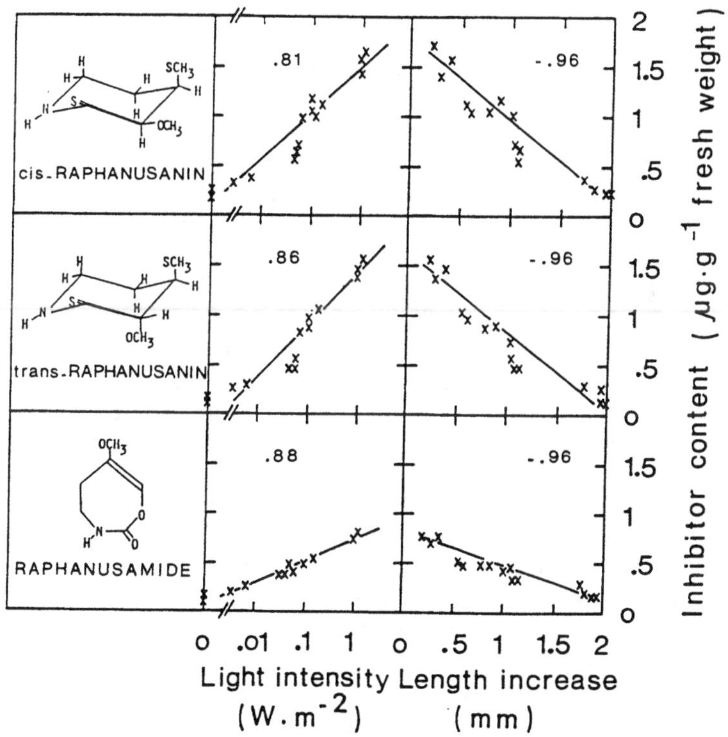

Fig. 2. Contents of the three growth inhibitors in curving radish hypocotyls at either side of the hypocotyl, and the light intensities and growth at these sides, with correlation coefficients. Determinations on samples of 5 seedlings. After [12]

These results have recently been confirmed in Kagoshima, where the distribution of free, extractable IAA was also found not to change during phototropic stimulation (Table 1). Under the experimental conditions used the shaded coleoptile side responded with a growth promotion. Accordingly, the activity of inhibitors in the *Avena* coleoptile straight-growth test increased on the illuminated side and decreased on the shadow side. These inhibitors in the monocotyledonous coleoptile must be different from those in the dicotyledonous hypocotyl, since they occurred in the acidic fraction [9]. Their chemical nature is still unknown.

4 Repetition of the F.W. Went Experiment on Phototropism

In view of these data, all pointing to the regulation of phototropism by a lateral gradient of growth inhibitors rather than of auxin, it became urgent to repeat the classical experiment by Went [18]. In order to obtain a first positive phototropic response of 17° in two h, samples of 2-day-old, etiolated *Avena* seedlings were unilaterally exposed to 1.13 $nmol.m^{-2}$ blue light for 10 s. Two samples of thirty 4-mm tips were cut off under .3 $\mu mol.m^{-2}.s^{-1}$ R directly after the blue illumination. Each sample was placed on two blocks of agar ($2 \times 2 \times 40$ mm^3) separated by a razor-blade in such a way that the irradiated and the shadow sides diffused into different blocks, and incubated for a further 2 h in darkness at 25°C and a high RH. The blocks were then divided into two lots, one lot being immediately used for the *Avena* curvature test; the curvature was determined and calculated as IAA-equivalents from a standard curve.

The other lot was stored for about 30 min at –20°C, the frozen agar then extracted with 80% cold acetone for 30 min at 0°C. Eight ng (2-^{14}C)IAA (6.10^3dpm, 59mCi/mmol) was added as an internal standard. The extract was filtered, the residue rinsed with cold acetone and the combined extracts were evaporated in vacuo at 35°C. The aqueous solution was brought to pH 8.5 with 1 M K_2HPO_4 and partitioned three times with equal volumes of ethyl acetate. The aqueous solution was then brought to pH 2.8 with 2.8 M H_3PO_4 and again partitioned three times with equal volumes of ethyl acetate. The acidic fraction was dried over anhydrous Na_2SO_4 and then evaporated to dryness in vacuo at 35°C. The fraction was treated with diazomethane for 10 min at 0°C and purified by TLC (*n*-hexane:ethyl acetate = 2:3, v:v). The Rf zone where authentic Me-IAA will be located was eluted with 50 ml ethyl acetate, dried in vacuo at 35°C, and further dried in vacuo overnight. It was then treated with 50 µl heptafluorobutyryl imidazole (HFBI) for 2 h at 90°C. The tubes were chilled, 1 ml 0.5 N H_2SO_4 was added. After partitioning against 1 ml *n*-hexane, the *n*-hexane layer was washed with 1 ml H_2O and dried under a stream of N_2. The sample was purified by TLC (*n*-hexane:ethyl acetate = 3:1, v:v). The part equivalent to authentic HFB-Me-IAA was eluted with 50 ml *n*-hexane, evaporated to dryness in vacuo at 35°, dissolved in *n*-hexane, and analyzed by GC-ECD.

Table 3 demonstrates the reproducibility of the Went experiment, the agar blocks from the shaded side showing, on an average, 2.6 times as much auxin activity as those from the illuminated side. However, the interpretation that this would reflect differences in auxin content is apparently wrong, the IAA-diffusion

Table 3. Distribution of auxin activity (10-fold determinations) and of IAA (duplicate determinations) in agar blocks after 2 h diffusion from the lighted and shaded sides of tips of unilaterally illuminated Avena coleoptiles[a]

Expt.	Lighted half	Shadow half	Control half
	Distribution of auxin activity (%)		
1	28.3	71.7	67.9
2	26.1	73.9	69.0
3	28.6	71.4	63.0
Average	27.7	72.3	66.6
	Distribution of IAA (%)		
1	53.4	46.6	49.1
2	53.6	46.4	41.1
3	48.8	51.2	43.8
Average	51.9	48.1	44.7
	Distribution of IAA (pg/half-tip)		
1	275	240	253
2	276	239	212
3	257	270	231
Average	269	250	232

[a] A first positive response of 17°C given in 2 h, half tips of dark control coleoptiles. Half-tip fresh weight 3 mg.

from the two sides into the agar blocks had been the same. In the block underneath the shadow half, both the auxin activity and the IAA-content were slightly higher than in that from the dark control, the strongly decreased auxin activity underneath the illuminated half has therefore to be ascribed to the accumulation of light-induced inhibitor.

The experiment therefore confirms the conclusion obtained from the other experiments on oat coleoptiles and radish and sunflower hypocotyls, viz that phototropic curvature is caused by the light-induced, local accumulation of substances that inhibit cell elongation at an unchanged level of auxin. Sixty years after 'Wuchsstoff und Wachstum' it has become 'Hemmstoff und Wachstum', at least for phototropism.

References

1. Blaauw AH (1915) Z Bot 7:465
2. Bruinsma J, Karssen CM, Benschop M, van Dort JB (1975) J Exp Bot 26:411
3. Bruinsma J, Franssen JM, Knegt E (1980) In: Skoog F (ed) Plant growth substances 1979. Springer, Berlin Heidelberg New York, p 444
4. Bruinsma J, Hasegawa K (1989) Environ Exp Bot 29:25
5. Franssen JM, Bruinsma J (1981) Planta 151:365
6. Franssen JM, Firn RD, Digby J (1982) Planta 155:281
7. Hasegawa K, Noguchi H, Iwagawa T, Hase T (1986) Plant Physiol 81:976

8. Hasegawa K, Noguchi H, Tanone C, Sando S, Takada M, Sakoda M, Hashimoto T (1987) Plant Physiol 85:379
9. Hasegawa K, Sakoda M (1988) Plant Cell Physiol 29:1159
10. Macleod K, Brewer F, Digby J, Firn RD (1984) J Exp Bot 35:1380
11. Noguchi H, Nishitani K, Bruinsma J, Hasegawa K (1986) Plant Physiol 81:980
12. Noguchi H, Hasegawa K (1987) Plant Physiol 83:672
13. Pickard BG (1985) In: Pharis RP, Reid DS (eds) Encyclopedia of plant physiology, new series, vol 11. Springer, Berlin Heidelberg New York Tokyo, p 365
14. Shen-Miller J, Knegt E, Vermeer E, Bruinsma J (1982) Z Pflanzenphysiol 108:289
15. Thompson AG, Bruinsma J (1977) J Exp Bot 28:804
16. Weiler EW (1984) Annu Rev Plant Physiol 35:85
17. Weiler EW (1988) Physiol Plant (in press)
18. Went FW (1928) Recl Trav Bot Neerl 25:1
19. Went FW, Thimann KV (1937) Phytohormones. MacMillan, New York

Cytokinins as Metabolic Stimulants Which Induce Pod Set

D. DYER[1], J.C. COTTERMAN[2], C.D. COTTERMAN[1], P.S. KERR[1], and D.R. CARLSON[3]

1 Whole-Plant Fruit Set Regulation

Bearing numerous fruit, the initiation of which is spread over a period of days or weeks, represents an adaptive advantage by allowing seed to develop and mature during a wider span of time, which reduces the likelihood of pest or environmental stress severely reducing the number of viable propagules that an individual plant produces. To achieve this spread in fruit initiation, many plants have a prolonged flowering period. Normally, those flowers which open first will produce a fruit which will attain maturity if stress is absent. Those flowers opening late in the flowering period are likely to abort, given that the earlier flowers have produced viable fruit. If, however, the early flowers are induced to abort by stress conditions or physical removal, then the likelihood that later flowers will produce viable fruit is greatly enhanced.

This type of fruit set regulation has been well documented in the case of soybean. This plant, as with numerous other species, has a complex floral progression of two types. First, the plant bears numerous floral racemes which flower in sequence from lower to upper plant nodes. Secondly, within an individual raceme, numerous flowers are produced which also open in sequence from the bottom to the top of the raceme. In one report, on intact racemes, flowers at the bottom of the raceme, which open earliest, had 89% pod set, while those at the top of the raceme had 0% set. However, if the bottom flowers were removed the central floral positions were induced to have 88% set while the same position of intact racemes had only 34% set [8].

Aside from being an interesting developmental phenomenon, the regulation of fruit set has significant practical implications. For many multi-fruited plant species, and certainly for soybean, variation of seed yield for a single genotype across environments is predominately related to the number of fruit per plant, with minor contributions from variations in number of seed per fruit and seed size. Furthermore, the plant appears to be able to develop and mature an incrementally larger seed mass, approximately 10 to 20%, than it normally retains. The conservative nature of the plant in restricting its own fruit set has an adaptive advantage. If the plant were to retain the maximum number of fruit which its assimilatory capacity could handle, then any subsequent period of stress would jeopardize the viability

[1] Du Pont, Stine-Haskell Site, P.O. Box 30, Newark, DE 19714, USA
[2] Du Pont, Experimental Station, Route 141, Wilmington, DE 19898, USA
[3] BASF, 26 Davis Drive, Research Triangle Park, NC 27709, USA
Sponsoring Institution: E.I. Du Pont de Nemours and Co., Inc. Wilmington, Delaware, USA

of all seed. Maintaining a reserve capacity allows the plant to continue seed growth and development under sub-optimal growth conditions. For these reasons many researchers have concluded that, if it were possible to increase fruit set, a concomitant increase in yield may be obtained.

2 Attributes of Viable and Aborting Fertilized Ovaries

For these reasons, various groups have investigated the underlying cause of fruit abortion by comparing those fruits destined to mature to those destined to abort. In soybean this is readily done because of the expected pattern of fruit set from the bottom to the top of a raceme, and the ability of the experimenter to influence fruit set by removing the bottom buds.

Anatomically, both types of ovaries are similar for some time after fertilization [1]. At some point early in development of the fertilized egg, cell division of the zygote ceases in the ovaries destined for abortion. The zygote will remain in that quiescent phase for several days before degeneration of the tissue occurs. If, during that quiescent phase, the developing zygotes at other floral positions are killed or removed, then the quiescent zygote will resume development.

From a biochemical perspective, with the exception of a lower level of starch in the aborting ovaries, no substantial differences between the setting and aborting ovaries were found for protein, soluble carbohydrate, RNA, or DNA, while only minor differences in the pools of free amino acids were noted [8, 12]. Abscisic acid levels in seeds and pod walls of setting and aborting pods also did not vary significantly on a tissue weight basis [17]. Contrary to that, however, ABA levels were found to be lower in bean fruits induced to set by removal of fruits lower on the raceme relative to similar fruits on intact racemes [19].

One difference between setting and aborting tissues is that aborting tissues have a reduced capacity to accumulate photosynthate being transported from the leaves. Referred to as "sink strength", the aborting ovaries were shown to accumulate less radiolabeled sucrose than setting ovaries [2, 17]. This does not seem to be the result of inadequate supply, but rather the inability of the fertilized ovary to accumulate and utilize the carbohydrate [18].

3 Influence of Cytokinins on Fruit Set and Yield

As early as 1965, a treatment of cytokinin applied directly to flowers or young fertilized ovaries was demonstrated to promote fruit set of seedless grapes [21]. Since that observation, a similar response has been documented for many other plant species, including pea [11], mung bean [4], and soybean (Table 1) [5, 16]. Analyzing the effect of a cytokinin treatment on yield parameters, it was reported that a 2 mM solution of BA applied to main stem nodes of soybeans had no effect on flower production, but greatly reduced floral abortion (Table 2) [3]. Despite a higher subsequent rate of pod abortion on treated plants, the net effect was more pods per raceme at maturity.

Table 1. The influence of 6-benzylaminopurine on fruiting and seed weight of field-grown Essex and Shore soybeans. [5] Values are means of six replicates (five plants replicate^{-1})

Growing Season	Treatment	Essex Fruit	Essex Seed	Essex Seed wt.	Shore Fruit	Shore Seed	Shore Seed wt.
				mg			mg
1978	Control	7.4	17.2	1950	7.6	16.3	2130
	BA[a]	10.1	22.4	2540	10.7[b]	21.6[b]	2770[b]
1979	Control	6.7	14.0	1320	6.0	10.0	1080
	BA	7.1	14.7	1570	8.0[b]	13.3[b]	1530[b]

[a] Five-tenths milliliter of 2 mM BA applied inflorescence^{-1}.
[b] Denotes significant difference from control at 5% level of probability; fruit and seed number and weight are expressed on a per inflorescence basis.

Table 2. Cumulative effects of BA application on five main stem IX93-100 racemes. Effects were measured as the total accumulated 6 weeks after BA treatment. [3]

Cultivar (Treatment)	Total Production		Total Abscission		Final Pod Number
	Flowers	Pods	Flowers	Pods	
	No.		No.		
IX93-100					
Control	76.0 NS[a]	48.4**[b]	27.6**	29.2*	19.2**
+BA	80.0	76.6	2.8	38.4	38.6

[a] Not significantly different.
[b] *, **Significantly different at the P = 0.05, and 0.001 levels of probability, respectively.

A detailed analysis revealed that the initiation of young pods was promoted at distal raceme locations where pod set was unlikely on untreated racemes [Fig. 1a; 3]. With this single pulse treatment of BA, many of these pods initiated at the distal end of the raceme subsequently aborted (Fig. 1c). The major effect on mature pod load was a large increase in pods in the center of the raceme (Fig. 1d). The increase in likelihood of pod retention with BA treatment is best expressed by reproductive structures ranging from post-anthesis flowers to pods less than 1 cm in length at the time of treatment. Pods which attain a larger size naturally prior to treatment are likely to achieve full development, while buds prior to flowering usually do not respond to the BA treatment (Table 3).

This effect on pod set translates into an increase in yield per plant despite a commensurate decline in seed size, which was observed at all positions on BA treated racemes (Fig. 2). When various adenine cytokinin analogs were used to treat three main stem racemes of soybean plants, a range of pod set effects was obtained [9]. In all cases, stimulation of pod set on treated nodes did not cause an associated decline in pod number on untreated portions of the plant (Fig. 3a). Therefore, variation in pod number per plant was directly associated with variation in the

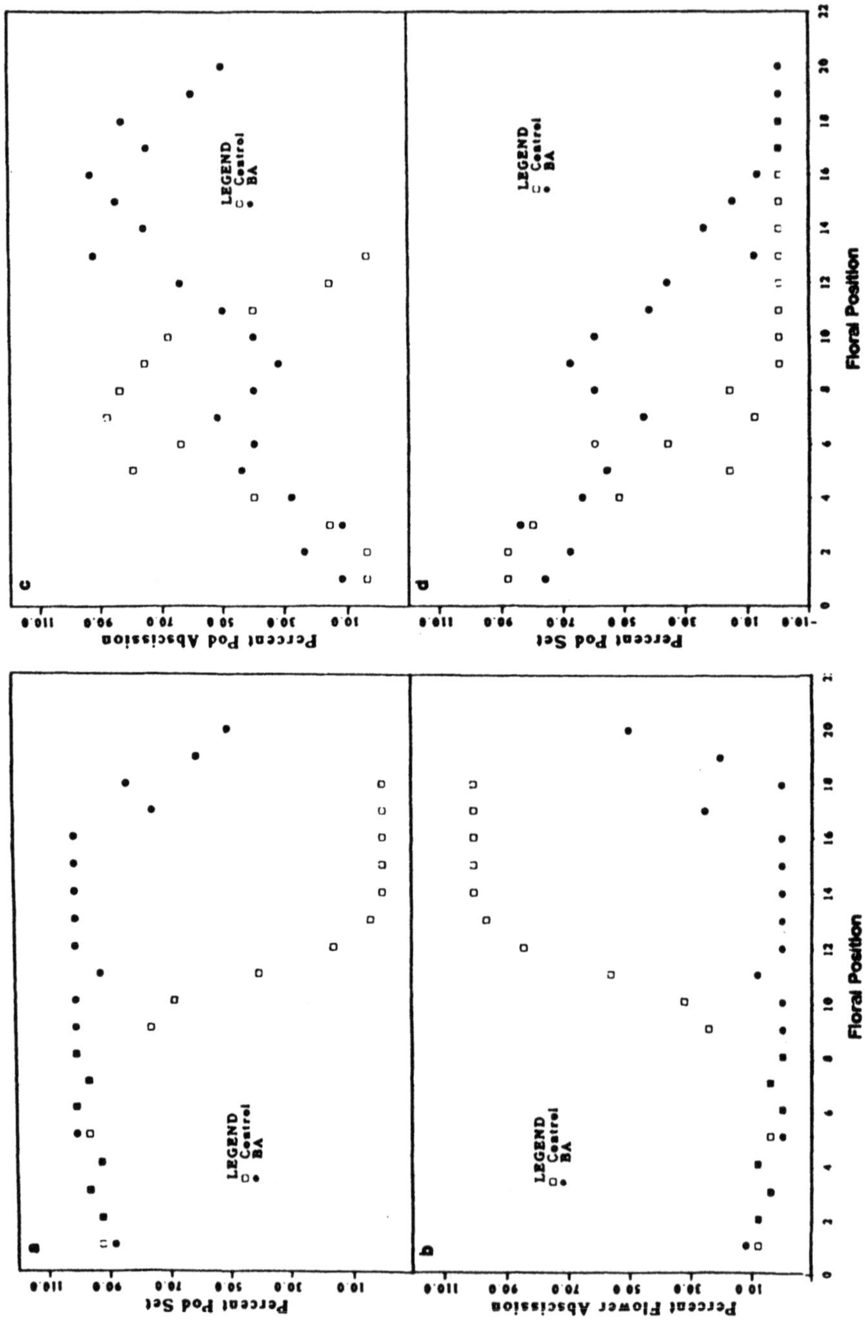

Fig. 1. Effect of benzyladenine treatment on pod set patterns of soybean plants [3]

Table 3. Influence of benzyladenine treatment on retention of reproductive structures at various stages of development when treated

	Percent mature pod set	
	Untreated racemes	BA-treated racemes
Buds	0	15
Open flowers	2	73
Post-anthesis flowers	3	75
0–0.5 cm Pods	44	100
0.5–1.0 cm Pods	75	83
> 1 cm Pods	97	100

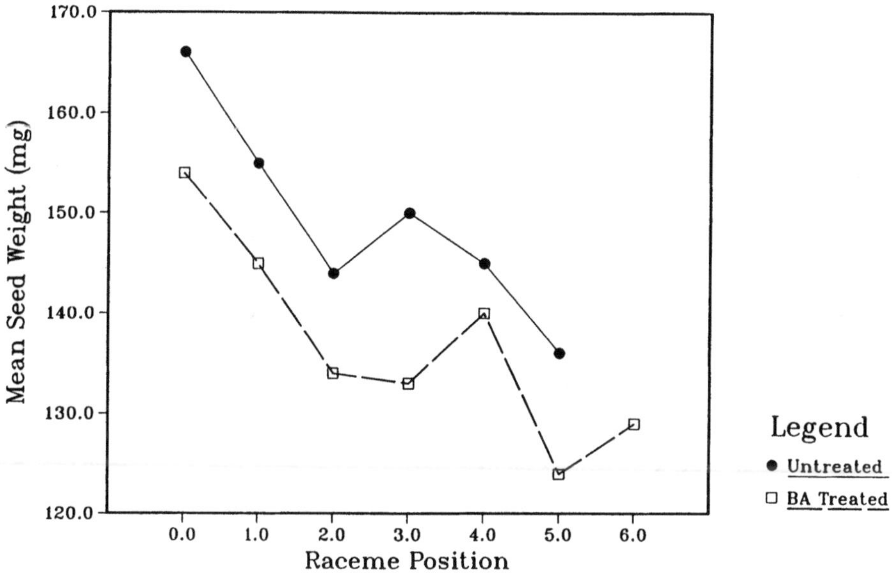

Fig. 2. Effect of benzyladenine treatment on seed size of soybean

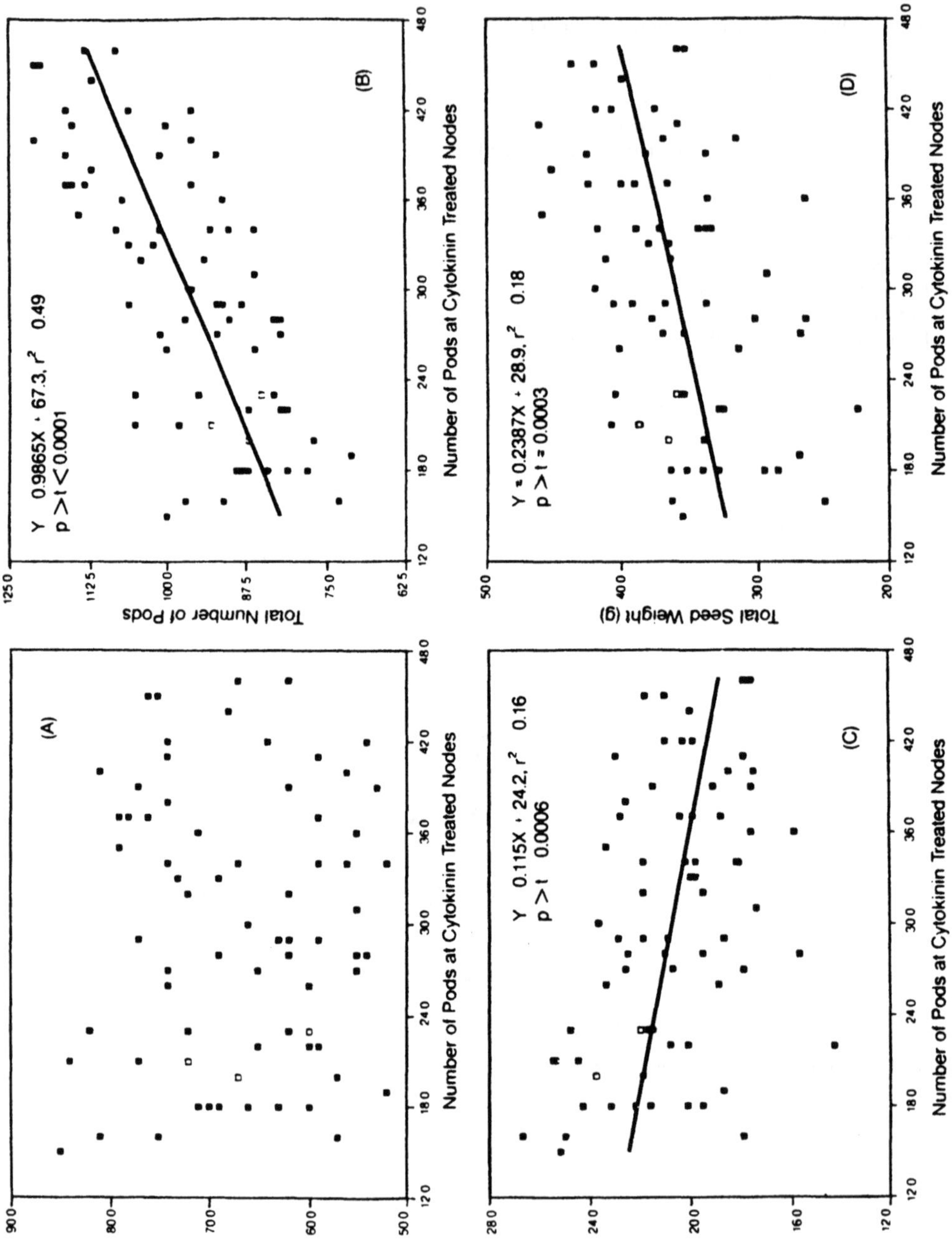

Fig.3. Effect of cytokinin treatment on yield parameters of soybean plants [9]

number of pods at the treated nodes (Fig. 3b). Seed size, however, did decline significantly as pod number increased (Fig. 3c). Despite the decline in seed size there was an overall significant increase in yield per plant as pod number increased with increasingly more effective cytokinin analogs (Fig. 3d). Overall, the percent increase in per plant yield was approximately one-half the percent increase in pods per plant.

3.1 Structure-Activity Relations of Adenine Cytokinins for Pod Set Increase

Several structure-activity conclusions could be drawn from the pod set enhancement which was obtained with various adenine cytokinin analogs. Activity was very dependent on the N-6 and N-9 substituent, with highest activity associated with N-6 alkyl, N-9 hydrogen analogs. N-6 alkyl compounds (i.e. pentyl and hexyl analogs) gave 20 to 25% greater pod set than homologous benzyl compounds (Figs. 4, 5). N-6 hydroxyalkyl compounds [i.e. (diH)Z analogs] were 30 to 40% less active than benzyl compounds, while N-6 alkenyl compounds (i.e. zeatin and iso-pentenyl analogs) and N-6 furfuryl compounds (i.e. kinetin analogs) had no significant activity. In the N-9 position, either ribose or tetrahydropyran reduced activity approximately 35 to 40% relative to the homologous free base [9].

These structure-activity patterns are consistent with the activity of cytokinin analogs in other assays as well as known mechanisms of cytokinin deactivation. Kinetin analogs generally have weak activity on soybean tissues, while N-9 substituted compounds normally have reduced activity associated with the need to metabolically remove the substituent to release the active free base [10]. The higher activity of alkyl analogs relative to hydroxyalkyl and alkenyl compounds may be a result of the resistance of alkyl compounds to deactivation either through cytokinin oxidase or O-glycosylation [14].

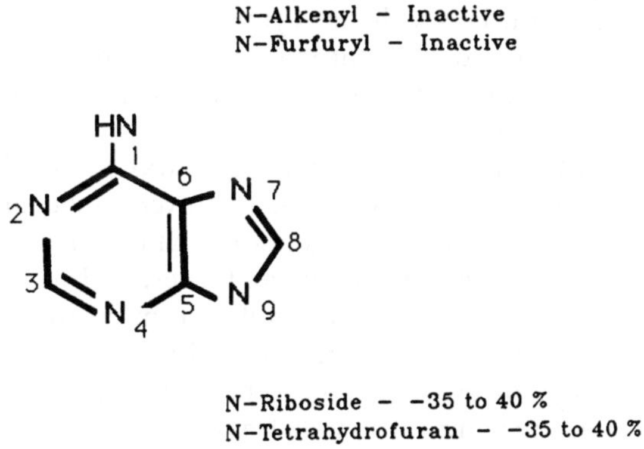

Fig. 4. Structure-activity relations of adenine cytokinins for pod set stimulation of soybeans

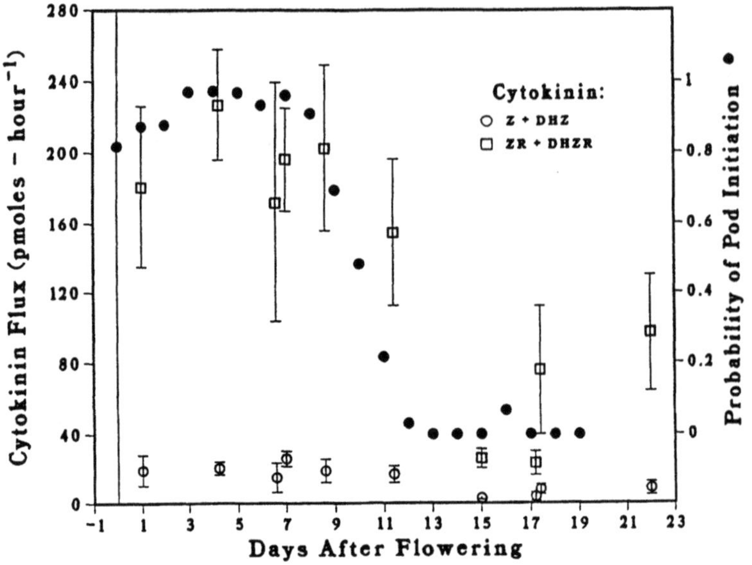

Fig. 5. Relationship between the probability of pod initiation and the cytokinin flux within the root pressure exudate of IX93-100 soybean on the day of anthesis of each raceme position [3]

4 Endogenous Cytokinins in the Natural Regulation of Pod Set

The relationship between the pod set enhancement which can be obtained with exogenous cytokinins and the natural mechanism by which the plant regulates its pod load seems to be based on the temporal pattern of cytokinin flux within the plant. In 1982, Heindl et al. [13] reported that cytokinin flux from the roots of soybean peaked during flowering and pod set. This pattern has also been reported for other species such as white lupine [6, 7]. A subsequent detailed analysis of the contents of Z, (diH)Z, and their ribosides in xylem sap of soybeans during reproductive development identified that cytokinin flux was closely associated with the likelihood of a flower initiating a mature pod (Fig. 5) [3]. Those flowers which opened on days with high cytokinin production had a high probability of set. As pod set progressed, both the cytokinin flux from the roots and the likelihood of pod set declined concomitantly. Pod set was only likely when cytokinin flux exceeded 100 pmol/plant/h.

In a separate experiment, it was found that removing the flowers as they opened, so as to prevent pod set, reversed the normal decline in cytokinin flux (Fig. 6). With depodded plants cytokinin flux continued to increase to even greater levels than the normal peak during flowering. It is an interesting corollary observation that the decline in cytokinin flux from the roots is associated with cessation of root growth, which has been repeatedly documented to occur shortly after flowering. Torrey [20] reviewed many other reports indicating that treatments which increased floral and pod abortion, such as drought, soil salinity, or high root temperatures also cause a decline in cytokinin activity in the xylem sap. Another supporting obser-

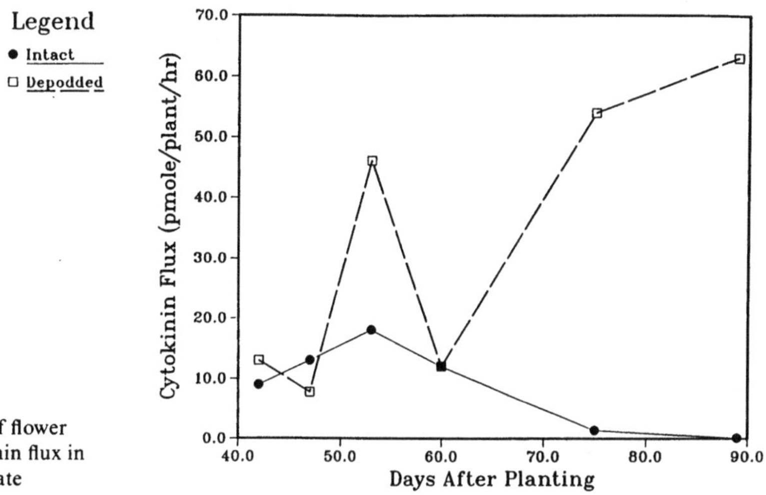

Fig. 6. Influence of flower removal on cytokinin flux in soybean root exudate

vation of the involvement of cytokinins in pod set is that the abortion of interspecific embryos of *Phaseolus vulgaris* × *P. acutifolius* was related to the low level of cytokinins in the young ovules [15].

5 Influence of Cytokinins on Carbohydrate Metabolism

While there is considerable evidence demonstrating a role of cytokinins in regulation of fruit set, there remains little known of the mechanism of this regulation. Earlier, fertilized ovaries which are developing during periods of inadequate cytokinin flux were described as entering a quiescent phase, characterized by a relatively low ability to attract and assimilate radiolabeled sucrose. Removal of subtending fruits or treatment of the ovary with a cytokinin causes the ovary to resume development and, simultaneously restores the carbohydrate sink strength to that tissue (Table 4) [17]. Recent work indicated that treatment of fertilized ovaries destined for abortion with a solution of thidiazuron, a very potent cytokinin-active compound, causes a very rapid and pronounced increase in the

Table 4. Influence of proximal pod removal on flux of [^{14}C] photosynthate to distal pods [17]

	Node	Proximal pods	Specific activity (MBq/kg) Distal pods Intact raceme	Distal pods Proximal pods removed
1982	8	28	24	38
1982	10	15	16	53
1983	8	72	6	101
1983	10	59	27	58
1983	12	37	16	42

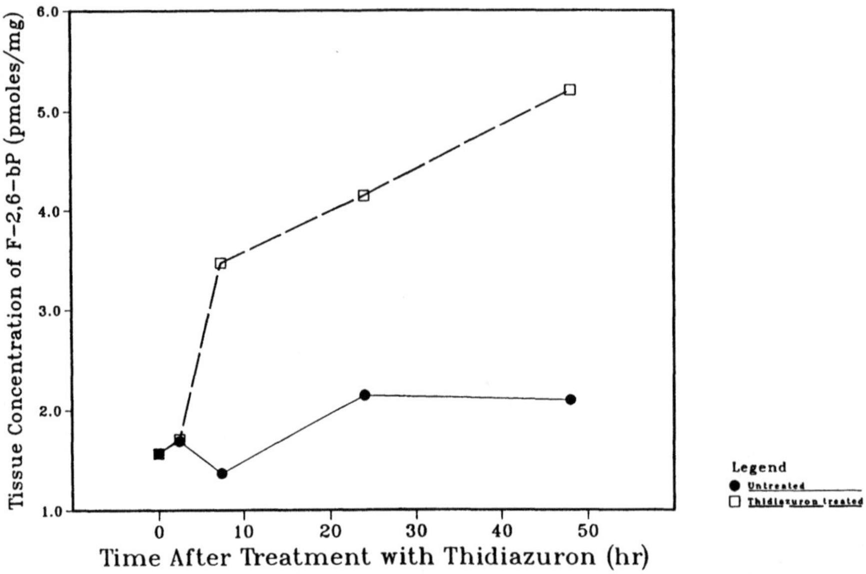

Fig. 7. Response of fructose-2,6-bisphosphate level in soybean buds to treatment with thidiazuron

level of F-2,6-bP in the ovary tissue (Fig. 7). An increase of F-2,6-bP of this magnitude would be expected to dramatically stimulate the sucrose metabolism of that tissue, enhancing its sink strength. The mechanism by which the cytokinin treatment causes a tripling of F-2,6-bP levels within hours is unknown. It is also unknown if this stimulation of F-2,6-bP is a primary response directly regulated by cytokinin. However, both the rapid nature of the effect and the importance of carbohydrate sink strength relative to the development of the fertilized ovary make this response worthy of further investigation.

6 Summary

Cytokinins play a central role in the regulation of fruit set in many multi-fruited species. Because of their primary site of synthesis in the roots, this provides a mechanism for the plant to maintain fruit load consistent with the plant's ability to produce viable seed based on such important soil conditions as fertility and water availability. The complete mechanism by which cytokinins induce development of fertilized ovaries is unknown, but carbohydrate metabolism in the ovary is one process involved. The cytokinins have the ability to promote carbohydrate metabolism and associated sink strength of the ovary by enhancing the level of the regulator F-2,6-bP. How cytokinins cause modulation of F-2,6-bP levels, and what other metabolic or molecular processes are involved are further questions to be answered. In a general sense, cytokinins act as metabolic stimulants of fertilized ovaries which have become quiescent, and, as such, induce their full development into mature seed-bearing fruit.

Acknowledgements. This work was sponsored by E.I. Du Pont de Nemours and Co., Inc., Wilmington, Delaware, USA.

References

1. Abernathy RH, Palmer RG, Shibles R, Andersen IC (1977) Can J Plant Sci 57:713
2. Brun WA, Betts KJ (1984) Plant Physiol 75:187
3. Carlson DR, Dyer DJ, Cotterman CD, Durley RC (1987) Plant Physiol 84:233
4. Clifford PE (1981) Z Pflanzenphysiol 102:173
5. Crosby KE, Aung LH, Buss GR (1981) Plant Physiol 68:985
6. Davey JE, van Staden J (1978) Physiol Plant 43:77
7. Davey JE, van Staden J (1978) Physiol Plant 43:82
8. Dybing CD, Ghiasi H, Paech C (1986) Plant Physiol 81:1069
9. Dyer DJ, Carlson DR, Cotterman CD, Sikorski JA, Ditson SL (1987) Plant Physiol 84:240
10. Fox JE, Sood CK, Buckwalter B, McChesney JD (1971) Plant Physiol 47:275
11. Garcia-Martinez JL, Carbonell J (1980) Planta 147:451
12. Ghiasi H, Paech C, Dybing CD (1987) Plant Physiol 84:91
13. Heindl JC, Carlson DR, Brun WA, Brenner ML (1982) Plant Physiol 70:1619
14. Letham DS, Palni LMS (1983) Annu Rev Plant Physiol 34:163
15. Nesling FAV, Morris DA (1979) Z Pflanzenphysiol Bd 91:5.345
16. Petersen CM, Folsom MW, Dute RR, Dalrymple LM (1986) Research Report, Soybeans, No 4, Alabama Ag Exp Stn, Auburn University, Alabama
17. Spollen WG, Weibold WJ, Glenn S (1986) Crop Sci 26:1216
18. Stockman YM, Shibles R (1986) Iowa State J Res 61(1):35
19. Tamas IA, Wallace DH, Ludford PM, Ozbun JL (1979) Plant Physiol 64:620
20. Torrey JG (1976) Annu Rev Plant Physiol 27:435
21. Weaver RJ, van Overbeck J, Pool RM (1965) Nature 206:952

Gibberellins in Embryo Development

A. Alpi[1]

1 Introduction

After the fusion of egg and sperm, growth of the embryo and of the whole seed takes place.

Shortly after fertilization there is a gradient in the density and distribution of the organelles of the zygote. A large vacuole is often formed and the nucleus migrates to the chalazal end of the cell, where it becomes surrounded by mitochondria and other components [26].

The first division of the zygote results in a cytoplasmatically dense axial cell and a highly vacuolated basal cell. In the dicots, the basal cell immediately divides to produce the suspensor, while the axial cell remains quiescent for a time. When the latter cell begins to divide, it produces a globular embryo.

The initially globular embryo then develops into a heart-shaped embryo, with a great increase in cell number, and finally into a torpedo embryo. The final cell number is reached well before full growth [19]. Differentiation, therefore, starts very early with transition from globular to heart- and torpedo-shaped embryo changing from radial to bilateral symmetry.

2 Hormonal (Gibberellin) Relations in the Developing Seed

In spite of the considerable literature already existing on seed development, the hormonal regulation of the early process mentioned above is still lacking a definitive explanation [5]. Further insight has been provided into the later stage of seed development; but even here, although our knowledge regarding hormones is extensive as far as the biochemistry is concerned, it cannot be generalized and only a hypothesis can be put forward as to the physiological basis.

Unless pollination occurs, growth of the ovary usually stops; hormones, including GAs, have been detected in the pollen of various species. Moreover parthenocarpic development of the fruit can be induced by treating many species with hormones.

Whether brought by the pollen or synthesized in the ovary or elsewhere in the plant and transferred to the growing embryo, hormones (obviously including GAs) seem to be fundamental for seed development.

---

[1] Dipartimento di Biologia delle Piante Agrarie, Università di Pisa, Viale delle Piagge, 23, 56124 PISA, Italy

The information available in the literature comes from research carried out mostly with the aim of answering the following general questions concerning the role of hormones in seed development:

1. The kinds of hormones present in seeds.
2. Whether the levels of these hormones change during seed development.
3. Whether any of the events in seed development are influenced or controlled by endogenous hormones.

The hormonal level is generally much lower in the embryos of seeds with a massive endosperm than in the endosperm itself (avocado, peach, lupin, apple, rye, corn, pea). It is commonly accepted that the developing endosperm can export auxins and GAs to the developing embryo. While it is rather difficult to cultivate pre- or early globular embryos in vitro, there is evidence indicating that many embryos beyond the early globular stage do not require hormones from the rest of the plant to grow [19].

One of the earliest important observations on the time of production of a GA-like substance was made by Mitchell et al. [24] on the Black Valentine bean during fruit development. While pod development began shortly after pollination, the GA-like growth factor was barely detectable until 4 days after pollination. The growth factor was at thc pcak of its activity when the seed began to grow in size.

Seeds are usually the richest source of GAs in plants containing up to several micrograms of GAs per g fw [29].

Seeds from a wide range of species contain GAs detectable by a variety of methods ranging from bioassay to MS and about half of the known GAs and many GA conjugates have been found in developing seeds [29].

Seeds can synthesize GAs by themselves [20]. Most research on GA biosynthesis has been carried out with cell-free extracts from the endosperm of *Marah macrocarpa* and *Cucurbita maxima* and from the embryo of *Pisum sativum*.

A general conclusion can be drawn that GAs synthesized in developing seeds are involved in seed and fruit growth; in some instances growth of the seed tissues correlates to the level of hormones (GAs in *Pisum*) [29]; as maturation proceeds, there is a build-up of conjugated GAs.

Developing seeds or cell-free extracts from seeds actively convert GAs, ultimately to an inactive form [36]. At the early stages of seed development, however, the major GAs are generally ones which are highly bioactive on most bioassays; inactive ones are formed toward the end of seed maturation.

Previous work, which relied on bioassay to detect GA activity, revealed only one broad peak of bioactivity during pea seed formation; now it is clearly established that single GAs peak at different times of seed growth, with non-polar GAs, such as GA_9, being formed early on, while the more polar GAs are mostly formed later, with inactive GAs or GA catabolites (GA_{29}, and GA_{51} catabolites) accumulating toward the end of seed growth [29, 36, 37]. There are two main phases in GA accumulation during the development of the pea fruit. During the first 10 days after anthesis, GAs accumulate in the pod and in the liquid endosperm; later high levels of GAs build up in the cotyledons and testa. Recently it has been observed [18] that the GA in five-day-old seeds are qualitatively different from those present later; evidence was also obtained that GAs, not detectable during the phase of seed enlargement, regulated pod growth [18].

Most of the evidence linking growth regulators with seed development is based on correlations between regulator content and embryo growth. Changes in GAs have been shown to be correlated with the development of plant embryos [16, 21, 28, 29]. The highest concentrations of active GAs in pea seeds (GA_9; GA_{20}) occur during the maximum growth rate stage of the developing embryo. Such data are analogous with what has been shown in *Phaseolus coccineus* [17].

On the other hand, only small qualitative changes in the GAs in seed parts at different developmental stages have been reported, although the amounts of these compounds change markedly during development [22]. In the developing seeds of *Sechium edule* the endosperm contained the greatest amount of free GAs, at the advanced stage of maturation, mostly accounted for by GA_4 and GA_7 [22]. Gibberellins were highest in the *Sechium* cotyledons at the corresponding stages again in the form of GA_4 and GA_7. In the final stage of growth the cotyledons showed a dramatic decrease in the total amount of GAs, while the endosperm showed little change. Therefore, it becomes more difficult to correlate this unchanging hormonal level in the latter tissue with the growth and development of the cotyledons or of the whole seed.

Using cell-free extracts, both endosperm and the cotyledons of *S. edule* have been shown to be the sites of biosynthesis for GA_4, GA_7, GA_9 and GA_{15}. The fact that the same GAs occur and can be biosynthesized simultaneously in the endosperm and in the cotyledons indicates that there is no functional dependence (at least in terms of GAs) of one tissue on the other, but rather suggests a separate functioning of GAs in these two seed components [8].

An unusual GA glucoside with biological activity was also isolated from both endosperm and cotyledons [23] and was present throughout the whole period of seed development, with no appreciable change in quantity; the hypothesis put forward by the authors [23] that such a GA glucoside might serve as a transport or storage form of GA precursors in *S. edule* is a challenging one.

3 The Embryo-Suspensor System

In angiosperms, endosperm growth starts before the zygote begins to divide. The endosperm may be absent in mature seeds, since it has been utilized by the growing embryo and in this case reserves accumulate in the cotyledons (many leguminous seeds); on the other hand, the endosperm may persist in other species and constitute a reserve, as in cereal seeds or castor bean seeds.

With few exceptions, the first division of the zygote produces two cells in both mono and dicotyledons, and the basal one (micropylar) generates the suspensor. The suspensor is almost ubiquitous in angiosperms. Its structure and morphology varies from rudimentary to massive.

According to an old interpretation, the suspensor has the function of pushing the embryo into the endosperm to facilitate its nutrition. On the other hand, a massive suspensor develops in some plants characterized by scarcity of endosperm (*Sherardia*, *Tropeolum*, *Medicago*, *Phaseolus*). These massive suspensors show conspicuous cell growth through chromosome endoreduplication or, much less frequently, multinucleation [14]. Very early on a hypothesis was put forward that the suspensor might take over the functions of the endosperm [35]. To support this idea, Corsi (1972) showed that excised *Eruca sativa* embryos, could grow in vitro only if intact (with suspensor) at the early heart stage, while at the late heart stage the embryo could be easily grown in vitro, even though it was deprived of the suspensor. More detailed studies have subsequently been carried out on *Phaseolus coccineus*.

After the pioneering work of Guignard, the suspensor of *Phaseolus coccineus* has been thoroughly studied by Nagl [25] and by D'Amato [13], who were mostly concerned with the anatomy and cytology of this tissue; this is the most extensively investigated angiosperm suspensor. When completely formed, the suspensor consists of 200±25 cells that undergo chromosome endoreduplication, which increases progressively from the embryonal to the basal pole. At maximal growth the suspensor is a club-shaped structure with a "handle" portion, made of 4C to 128 C cells and a "Knob" portion, whose terminal cells attain the highest level of polytenization (12 endoreduplications that result in 8192 C DNA amount). The nuclei of the giant suspensor cells are in a permanent polytenic condition and this highly polytene structure has favoured the application of such techniques as RNA-DNA hybridization in situ for the localization of the chromosome regions bearing the genes for all types of rRNA: 25S, 18S, 5.8S and 5S. [15]. These results can be utilized to correlate eventual DNA amplification with functional activities of the suspensor. Suspensor cells show localized accumulation of extra DNA (amplification) [4] and, more precisely, such cells contain a satellite DNA which is not detectable in either shoot or root. It is worth considering that the rate of RNA synthesis (transcription) of the polytene suspensor cells of *P. coccineus* is hundreds of times higher than that of embryonic cells.

The above observations, together with the occurrence of wall ingrowths (similar to transfer cells), suggests that the *P. coccineus* suspensor plays a major role in embryo development [40].

4 *P. coccineus* Embryo-Suspensor System and Gibberellins

Early analyses of GA-like activity in embryo and suspensors of *P. coccineus* were conducted at two stages of embryogenesis. Two categories of seeds [3] were therefore used: a) 5-mm seeds containing a heart-shaped embryo (Stage A); b) 12-mm seeds containing a cotyledonary embryo with suspensor in the initial stage of degeneration (Stage B). The results obtained by extracting 500 suspensor and embryos confirmed that the tissues analyzed were among those containing the highest levels of GA-like substances, and particularly the suspensor was a very rich source of such hormones [3].

The next logical step was to identify the GAs in the suspensor. The extraction of 2000 suspensors at stage A led to the characterization of GA_1 as the main GA. Based on this result and earlier ones [17] that examined quantitative variations of the main GAs in seeds of *P. coccineus* at various ages, we concluded [2] that the amount of GA_1 found in the suspensors of heart shaped embryos (9 $\mu g\ g^{-1}$ fw) accounted for a high percentage of the GA (15–40 μg g^{-1} fw) found in the entire seeds of the corresponding age. Given these results we suspected that the suspensor was a site of GA biosynthesis. This was confirmed when a cell-free system of the suspensor tissue incorporated MVA into kaurene, 7-β-hydroxy-kaurenoic acid, and GA_1, GA_5 and GA_8 [9, 10].

The hypothesis that the suspensor might play a role in early embryogenesis by providing the embryo with GAs was supported by results obtained from growing intact embryos (with suspensor) and embryos deprived of suspensor in vitro. Removal of the suspensor had no effect on the further development of embryos which had reached a 5-mm length. However, for younger embryos removal of the suspensor diminished subsequent development, the negative effect being more pronounced in the younger embryos. GA_3, at concentrations of 10^{-8} to 10^{-6} M, can replace the suspensor for heart-shaped and early cotyledonary embryos (0.5 to 1.5 mm in length) [11]. This confirms preliminary results [25] and ascribes a major role to GAs in characterizing the physiological function of the suspensor in early embryogenesis of *P. coccineus*.

However, more evidence is needed before final conclusions can be drawn on the function of the endogenous GAs in the suspensor and embryo per se. The GAs have been thoroughly analyzed both at stages A and B, and this work found that GA_1 was the only endogenous GA present in detectable amount in the suspensor, when 2,000 were extracted. However, when the same number of suspensors was used for the cell-free study [10] three GAs (GA_1, GA_5 and GA_8) were biosynthesized.

Therefore, a higher number of suspensors (6000) from 10 g of cotyledonary tissue was extracted, purified and analyzed for GAs [31, 32]. From this extraction, the cotyledonary embryo (stage B collected from 12-mm long seeds) was found to contain one C-20 GA, namely GA_{44}, and four C-19 GAs, GA_1, GA_4, GA_5 and GA_6. These results are similar to those reported by Durley et al. [17] for the endogenous GAs in whole immature seeds, except for the absence of GA. When the suspensors at this same stage of development were analyzed GA predominated, but GA_4, GA_5, GA_6, GA_8 and GA_{44} were also found.

Thus both embryo and suspensor contain a very high concentration of GAs, some 239.5 μg g^{-1} fw in the suspensor at stage B (Table 1). This is among the highest concentrations reported in literature. However, high percentage of the total GAs is due to a large accumulation of the biologically inactive GA_8.

In describing the localization of GAs in embryos and testas of developing *P. coccineus* seeds, Albone et al. [1] noted that GA_8 accumulates in the testa. Since our data show massive accumulation of GA_8 in the suspensor, and since the degenerated suspensor is probably a component of the testa, it is likely that the suspensor is the major site of GA_8 production at this stage of *P. coccineus* seed development. This supposition was confirmed by Turnbull et al. [38] who showed that the embryo tissue (from 26-mm-long seeds) was not able to biosynthesize GA_8.

In terms of biologically active GAs, the suspensor shows the same qualitative spectrum and about the same amount (about 100 μg g^{-1} fw) at stages A and B, with the exception of GA_8. The suspensor at this stage is metabolically very active; only when it starts to degenerate are some of the biologically active GAs (probably GA_1 and/or GA_5) converted to GA_8 which is 2-hydroxylated and not biologically active.

Table 1. The concentration of native gibberellins in the suspensor and embryo of *P. coccineus*[a]

Gibberellins	Gibberellin content			
	A[b]		B[c]	
	Suspensor	Embryo	Suspensor	Embryo
	µg/g fw		µg/g fw	
C-19 -GAs				
GA_1	51.0	4.1	80.3	18.4
GA_4	12.8	1.6	27.7	2.3
GA_5	29.8	3.8	16.6	2.6
GA_6	+	+	10	1.9
GA_8			100	
GA_{60}		+		
C-20 -GAs				
GA_{44}	13.2	0.6	5	1.3
GA_{19}		+		

[a] + Indicates presence of the GA the amount of which could not be calculated.
[b] Stage A; heart-shaped embryo from 5-mm seeds.
[c] Stage B; cotyledonary embryo from 12-mm seeds.

More recently [30] two other GAs were identified in young embryos i.e. C-20 GA_{19} and the C-19 GA_{60}. GA_{60} is the first C-1 hydroxylated GA reported in *P. coccineus* seeds; all but GA_8 (C-2, 3, 13 hydroxylated) are C-3, or C-3, 13 hydroxylated. The *P. coccineus* embryo thus has a complex hydroxylating system.

At stage A the embryo contains all the GAs found at stage B, and their amount is very similar. As far as dependence on the suspensor is concerned, we suspect, but cannot prove, that these GAs are provided by the suspensor. It is clear that developing cotyledons from 16 mm seeds [38] have acquired the ability to biosynthesize their own GAs. However whether this ability exists in the embryo since the first division of the zygote, or is acquired later, still needs to be clarified.

The partitioning of the biologically active GAs (Table 2) between embryo and suspensor depends on the stage. From stage A to stage B the embryo fr. wt. increases 30-fold while the amount of GAs in the embryo shows an 80-fold increase. This could imply that the ability for GA production progressively increases with embryo growth and a dependence on the suspensor for GA transfer no longer seems

Table 2. Fresh weight and amounts of biologically active gibberellin (GA 8 excluded) per individual structure in *P. coccineus* embryo-suspensor system at the two stages (A and B) of development

Stage A	*fw (mg)*	*ng GAs*
Suspensor	0.21	22
Embryo	1.81	18
Stage B		
Suspensor	0.19	27
Embryo	55.56	1472

reasonable. However, although the embryo already has a fr. wt. 10-fold higher at stage A, its GA amount is comparable to that of suspensor and therefore the hypothesis that the suspensor may provide the embryo with GAs could still be considered as valid.

In vitro culture study of the embryo-suspensor system showed that the embryo begins to be autonomous rather early, e.g. when it was longer than 2 mm (seeds of 6 mm or more) [11]. In other words, cotyledon explants were able to grow without suspensors only when they were longer than 2 mm. Whether this development corresponds to "precocious germination" [7] or to true embryogenesis has yet to be assessed. Our data on endogenous GAs refer to heart-shaped embryos longer than 2 mm. Recent results on the effect of the suspensor and GA_3 on protein synthesis in *P. vulgaris* embryos [39] showed that the suspensor strongly affects protein synthesis in the embryo at the early heart stage, and that this effect can be mimicked by GA_3.

Several lines of evidence obtained with somatic embryos of carrot [27, 28] suggest that endogenous GAs, although required in high amount by proliferating carrot cells, should decline to allow embryogenesis; it should be taken into consideration that the work done with embryos excised from *P. coccineus* [11, 40] showed that the supply of GAs for embryo growth is critical in its amount and timing.

Besides the *P. coccineus* suspensor, other suspensors have been analyzed with the object of testing the hypothesis that all massive suspensors have high GA content. Work carried out on *Tropeolum majus* and *Cytisus laburnum* confirmed that this was indeed the case [34]: GA-like activity even higher than in *P. coccineus* suspensor was found, and GA_{63} (15β hydroxy GA_4) was identified as the main GA in the *T. majus* suspensor [33].

All of these results lead us to conclude that a very thorough future examination of the hormone-supplying role of the angiosperm suspensor could provide important insights into the understanding of both seed and embryo development.

References

1. Albone KS, Gaskin P, MacMillan J, Sponsel VM (1984) Planta 162:560
2. Alpi A, Lorenzi R, Cionini PG, Bennici A, D'Amato F (1979) Planta 147:225
3. Alpi A, Tognoni F, D'Amato F (1975) Planta 127:153
4. Avanzi S, Cionini PG, D'Amato F (1970) Caryologia 23:605
5. Bewley JD, Black M (1985) Plenum, New York
6. Blechschmidt S, Castel U, Gaskin P, Hedden P, Graebe J, MacMillan J (1984) Phytochemistry 23:553
7. Bulard C, Le Page-Degivry Mth (1985) In: Bopp M (ed) Plant growth substances (1985). Springer, Berlin Heidelberg New York Tokyo, p 308
8. Ceccarelli N, Lorenzi R (1987) Phytochemistry 22:2203
9. Ceccarelli N, Lorenzi R, Alpi A (1981) Plant Sci Lett 21:325
10. Ceccarelli N, Lorenzi R, Alpi A (1981) Z Pflanzenphysiol 102:37
11. Cionini PG, Bennici A, Alpi A, D'Amato F (1976) Planta 131:115
12. Corsi G (1972) Giornale Bot Ital 106:41
13. D'Amato F (1978) Giornale Bot Ital 112:407
14. D'Amato F (1984) In: Johri BM (ed) Embryology of angiosperms. Springer, Berlin Heidelberg New York Tokyo, p 519
15. Durante M, Cionini PG, Avanzi S, Cremonini R, D'Amato F (1977) Chromosoma 60:269

16. Dure LS, Jensen WA (1957) Bot Gaz 118:254
17. Durley RC, MacMillan J, Pryce RJ (1971) Phytochemistry 10:1891
18. Garcia-Martinez JL, Sponsel VM, Gaskin P (1987) Planta 170:130
19. Goodwin PB (1978) In: Letham DS, Goodwin PB, Higgins TJV (eds) Phytohormones and related compounds, vol II. Elsevier/North Holland Biomedical, Amsterdam
20. Graebe JE (1987) Annu Rev Plant Physiol 38:419
21. Kefford NP, Rijven AHGC (1985) Gibberellin and growth in isolated wheat embryos. Science 151:104
22. Lorenzi R, Ceccarelli N (1983) Phytochemistry 22:2189
23. Lorenzi R, Ceccarelli N (1986) Phytochemistry 25:817
24. Mitchell JW, Skaggs DP, Anderson WP (1951) Science 114:159
25. Nagl W (1974) Z Pflanzenphysiol 73:1
26. Naylor AW (1984) In: Scott TC (ed) Hormonal Regulation of Development. Encyclopedia of Plant Physiology new series, vol 10. Springer, Berlin Heidelberg New York Tokyo
27. Noma M, Huber J, Phareis RP (1979) Agric Biol Chem 43:1793
28. Noma M, Huber J, Ernst D, Pharis RP (1982) Planta 155:369
29. Pharis RP, King R (1985) Annu Rev Plant Physiol 36:517
30. Piaggesi A, Picciarelli P, Lorenzi R, Alpi A (1989) Plant Physiol 91:362
31. Picciarelli P, Alpi A (1985) Plant Cell Physiol 26:1233
32. Picciarelli P, Alpi A (1986) Plant Physiol 82:298
33. Picciarelli P, Alpi A (1987) Phytochemistry 26:329
34. Picciarelli P, Alpi A, Pistelli L, Scalet M (1984) Planta 162:566
35. Schnarf K (1929) Handbuch der Pflanzenanatomic 10/2. Borntraeger, Berlin
36. Sponsel VM (1985) Physiol Plant 65:533
37. Sponsel VM (1983) In: Crozier A (ed) The biochemistry and physiology of gibberellins, vol 1; Praeger, New York, p 151
38. Turnbull CGN, Crozier A, Schwenen L, Graebe JE (1985) Planta 165:108
39. Walthall ED, Brady T (1986) Cell Differ 18:37
40. Yeung EC, Sussex IM (1979) Z Pflanzephysiol 91:423

Gibberellins and Flower Initiation in Herbaceous Angiosperms[1]

J.D. METZGER[2]

1 Introduction

The developmental changes that a vegetative apex undergoes during flower initiation rank among the most spectacular in biology. And yet despite its obvious importance to agriculture, the fundamental processes underlying flower initiation remain enigmatic. Although we have been able to precisely define the environmental conditions (photoperiod, temperature, etc.) under which flowering for a given species will or will not occur, we do not know much about the mechanism(s) by which inductive stimuli cause flower initiation.

More than 50 years ago, Chailakhyan [2] first proposed the existence of a specific flower-inducing stimulus in photoperiodically sensitive plants that is produced in the leaves following photoinduction and subsequently translocated to the apex where it causes the transition to reproductive development. The fact that production of the flowering stimulus, or florigen, often resides in an organ remote from the site of action suggests a hormonal character. Furthermore, the results of numerous grafting experiments by various workers indicate that a similar or perhaps identical hormonal agent exists in all response types (summaries of successful transmission of the floral stimulus across graft unions can be found in [1, 12, 24]. But while the physiological evidence for the existence of the floral stimulus is compelling, its nature and identity remains unknown.

One approach in obtaining information about the chemical properties of the floral stimulus has been to apply substances to plants in an effort to find a chemical substitute for the inductive stimulus. Of the hundreds of compounds examined, only GAs have consistently promoted flowering in a variety of species under strictly non-inductive conditions, suggesting a regulatory role for GAs in the transition of the vegetative apex to reproductive development. Nevertheless, exogenous GAs often fail to substitute wholly or partially for inductive stimuli. Since the floral stimulus appears to be essentially the same in different response types, it has been argued that GAs are not the floral stimulus [12, 25, 27]. For reasons to be discussed later, this conclusion has not gained universal acceptance [2, 16]. Thus, the main focus of this report is to summarize the evidence – both pro and con – for the floral stimulus being composed, at least in part, of GAs. Special emphasis will be placed

---

[1] Mention of a trademark or proprietary product does not constitute a guarantee or warrantly of the product by the US Department of Agriculture and does not imply its approval to the exclusion of other products that may also be suitable.

[2] USDA, ARS, Biosciences Research Laboratory, State University Station, Fargo, ND 58105, USA

on physiological attributes of the floral stimulus and the experimental approaches that flower physiologists have at their disposal to determine if GAs have qualities identical to the floral stimulus. Only herbacious angiosperms will be considered here; other plant groups such as woody angiosperms and conifers have been discussed in detail in [16, 18].

2 GAs and the Floral Stimulus

If GAs function as the floral stimulus, then several criteria must be fulfilled:

1. Exogenous GAs must substitute for the inductive conditions whether they are photoperiodic, temperature, or other stimuli.
2. Under non-inductive conditions, the GAs in question should be limiting in the apex. Conversely, experimental reduction of endogenous GA levels should inhibit or prevent floral initiation under inductive conditions.
3. The sites of production and action of GAs and of the floral stimulus must be identical.

2.1 Are Exogenous GAs Effective Substitutes for Inductive Stimuli?

It is well known that application of a variety of GAs promote or induce flower initiation in many LDP and CRP, particularly those that maintain a rosette growth habit during the vegetative phase of their life cycle. Nevertheless, application of GAs (usually GA_3) to caulescent LDP and SDP in non-inductive conditions more often than not fails to elicit a flowering response. Moreover, not all rosetted LDP and CRP flower in response to GA treatments. Most notable of these are the CRP *Geum urbanum, Lunaria annua*, and *Oenothera biennis*, and the LDP *Blitum virgatum, Beta vulgaris* cv Maritima (an annual variety), and certain inbred lines ("GA^-") of *Silene armeria*. In all these examples, GA_3 (at least) is unable to elicit a flowering response when applied to the plants in non-inductive conditions, although bolting (stem elongation) occurs, indicating that it is the control of this aspect of reproductive development rather than flower initiation that is mediated by GAs [25, 27].

Although the above evidence, based on the use of GA_3, argues against GAs serving as the floral stimulus, an alternative explanation may be found in the possibility that GAs may differ in their specificities for eliciting responses. It may be that of the GAs native to a given species, one regulates flower initiation, while another controls a different process such as stem growth. Since GA_3 was used in most experiments, the failure of the treatments to cause flowering may due to the fact that the "wrong" GA for initiation was applied. Evidence for this idea has surfaced recently with work on the LDP *Lolium temulentum* [17, 18]. The effects of various GAs on flower formation (as measured by shoot apex length) and stem growth were compared [17, 18]. There were striking differences in the effects of the various GAs. GA_{32} (a compound with hydroxyl groups at carbons 3β, 12α, 13 and 15β and a C 1,2 double bond) strongly promoted flowering with a relatively small effect on stem growth. This GA most closely mimicked the effect of 1 LD. In contrast, GA_1 (a C-3β, 13-dihydroxylated GA) had the opposite effects on the two processes. Other GAs, which can be considered as modified GA_1 molecules containing certain features of GA_{32}, gave intermediate responses. Thus, GA_3 (1,2 di-dehydro GA_1) promoted both flowering and stem elongation more than GA_1. Addition of another hydroxyl group to the C-15 carbon of GA_3 (15β-OH GA_3) resulted in a compound with a greater ability to induce flowering in SD. However, 15β-OH GA_3 was also very active in promoting stem elongation. These results indicate a relationship between GA structure and specific functions. In other words, there may be specific GAs that regulate stem elongation while others function in the control of floral initiation [17, 18].

While such an interpretation is certainly plausible, considerable caution is warranted in generalizing the results of experiments on *Lolium* to other species. First of all, work on the identification of the endogenous GAs in *Lolium* is at a very preliminary stage, and attempts to characterize those GAs (in leaves, phloem exudate of leaves, and apices) which change shortly after LD induction are ongoing. Thus far no native *Lolium* GAs which specifically promote flowering without also promoting significant stem elongation have been identified (R. Pharis, pers. comm.). Additionally, the biological activity of an applied compound is a complex function of (1) the ability to bind to a receptor; (2) translocation from the point of application to the site of action; and (3) the rate of metabolism (deactivation). The interpretation of results from dose-response experiments for two different developmental processes in terms of specific structure-function relationships assumes the latter two factors to be relatively unimportant. Furthermore, such an interpretation also assumes the existence of at least two GA receptors, each having distinct affinities for GAs with different structural features. Unfortunately, this aspect of the hypothesis is impossible to experimentally validate at present.

Although exogenous GAs apparently elicit stem growth by acting on the shoot sub-apical meristem [19], the site(s) of GA action for floral initiation is not entirely clear (see discussion below). Nevertheless, GAs probably act on different tissues to affect the two processes. It is therefore entirely possible that the relative abilities of various classes of GAs to cause either stem growth or flowering may be more of a reflection of differences in translocation to the sites of action and/or tissue specificities for GA metabolism (deactivation). The situation becomes even more complex when GAs that are not native to the species in question are used, since they may contain the structural features necessary to elicit biological responses, but are resistant to deactivation by the plant's normal GA metabolic pathways. Thus, while the occurrence of specific florigenic GAs cannot be discounted, evidence based solely on comparative dose-response data is not sufficient proof for their existence.

Another possible reason why GA treatments do not induce flower initiation may be due to improper timing and/or supraoptimal doses. For example, exogenous GAs either promoted or inhibited flower formation (as measured by number of flower buds produced) in plants of the dwarf strain "Kidachi" of the SDP *Pharbitis nil* that were marginally induced by 1 inductive long night. Promotion of flowering was observed when the GA treatment occurred 11 to 17 h before the inductive dark period, while GA application immediately after the inductive dark period caused a reduction in the flowering response [9, 18].

When examining the effects of exogenous GAs, experimenters are faced with the problem of assessing floral initiation. The most accurate and precise way is to examine the apex microscopically for the presence of flower primordia. However, this method is both time consuming and destructive. In many cases, some aspect of flower development such as the appearance of flower buds or an inflorescence is used as a measure of floral initiation. While this may save time and effort, there is a risk that what is only an effect on flower and/or inflorescence development will be mistaken for flower initiation, especially in plants with quantitative (facultative) requirements for inductive stimuli. In these types of plants, flower primordia will form under non-inductive conditions if given enough time, although macroscopic signs of flowering such as inflorescence development, flower buds, etc. may not be observed. For example, the Savoy Hybrid 612 cultivar of the LDP *Spinacia oleracea* grows as rosettes under continuous SD. This cultivar requires approximately 7 LD before macroscopically visible inflorescenses appear. Application of a variety of GAs in SD also promotes inflorescence development. Thus, if the measure of flower initiation in this case were the appearance of visible inflorescences, one would logically conclude that exogenous GAs can substitute for LD in the promotion of flower initiation. However, Savoy Hybrid 612 flower primordia eventually form under continuous SD, but further development is severely limited until the plants are transferred to LD. In addition, application of GAs in SD promotes inflorescence development, but does not induce flower initiation [23].

A related problem in interpretation arises when GAs are applied to plants maintained under conditions in which flower initiation will occur. Flower formation is promoted in many quantitative LDP when treated with GAs in SD. This promotion is usually the result of a reduction in time necessary to observe the appearance of some structural feature unique to flowering or to an increased number of flowers. It is difficult from such data alone to distinguish between direct GA effects on floral initiation (i.e. behaving like the floral stimulus), or modifying processes that are induced by the floral stimulus. Similarly, flowering is not induced in the SDP *Pharbitis nil* (dwarf strain "Kidachi") by GA applications to plants under non-inductive conditions, although such treatments do modify floral initiation in plants marginally induced by one SD. Depending on the timing and the dose the effects can range from promotion to inhibition [9, 18]. Promotion of flowering in this case is the result of increased number of flower buds observed 10–14 days after treatment. Since control plants also produce some flower buds as well [9], it may be that exogenous GAs and the floral stimulus are acting differently.

2.2 Are Endogenous GA Levels Correlated with Flower Initiation?

Any attempt to rigorously define the role of a hormone in the regulation of a developmental process requires determination of the quantitative relationships between hormone levels and the process. To date, there are few studies in which changes in GA levels in relation to flower initiation have been carefully monitored. Part of the problem has been that most investigations rely on bioassays for quantitative analyses; the reliability of this method of quantitation has been discussed in numerous review articles [e.g. 3, 7]. However, a more serious problem lies not with bioassays per se, but instead with the fact that, more often than not, the identities of the GA-like substances are not known. In any given species there are a number of endogenous GAs that are related metabolically. Within a given family of GAs, there is probably one "effector" GA, while the others are either precursors to, or deactivation products of that one GA. Thus, it is the quantitative relationship of the effector GA to flower initiation that is really important. An additional concern is the localization of changes in GA levels. In many studies, the entire shoot was analyzed for GA content. However, it is imperative to establish that changes in GA levels occurred in the apex where flower initiation takes place.

With these caveats in mind, few generalizations about the relationship between GA levels and floral initiation can be made. The following examples show why. The LDP *Lolium temulentum* can be induced by a single LD. Measurement of GA levels in the apex by bioassay showed a transient 3-to-5-fold increase in the concentration of GA-like substances with chromatographic properties similar to GA_{32} [17]. Since application of GA_{32} and 15β-OH GAs (neither of which are known to be native GA to this species) in SD most closely mimicked the response to one LD, the authors suggested that highly hydroxylated GAs may play a central role in the regulation of flower initiation [17, 18].

However, similar correlations have not been observed in other species, although inductive conditions often lead to both qualitative and quantitative changes in GAs and GA-like substances. As measured by GC-SIM, the level of GA_{20} (the probable effector GA, at least for stem elongation) in *Spinacia oleracea* shoots increased substantially when plants were subjected to LD. But these changes were more closely associated with stem elongation than with flower initiation [14, 27]. More complex changes in endogenous GA levels were observed in shoots of the LDP *Agrostemma githago* using GC-SIM [8]. Transient increases in various endogenous GAs were observed after 8–12 LD; these changes were correlated with the onset of rapid stem elongation, not flower initiation [8, 27].

Another method that is used to establish relationships between GAs and developmental phenomena is experimental reduction of GA levels. This can be accomplished through the use of inhibitors of GA biosynthesis (growth retardants), or with dwarf mutants that have reduced endogenous GA levels. Use of this type of an approach has provided contradictory results. On the one hand, treatment with growth retardants prevents flower formation under inductive conditions in the LDP *Samolus parviflorus* and *Rudbeckia bicolor*, the SDP *Pharbitis nil*, and LSDP *Bryophyllum daigremontianum*; treatment with GA_3 effectively reverses the inhibition [16, 25, 27]. In contrast, growth retardants have little effect on flower initiation while completely suppressing stem elongation in a number of species including *Spinacia oleracea*, *Agrostema githago*, *Lolium temulentum*, and *Silene armeria* [16, 25, 27]. In addition, dwarf mutants of maize and rice with highly reduced endogenous GA levels are also capable of initiating flower primordia. In these species, then, it would appear that GAs are not the controlling factors for flower initiation. It can be argued, though, that these systems are "leaky"; that is, very low amounts of bioactive GAs are produced, perhaps below the threshold required for stem growth, but sufficient for flower initiation. Consistent with this notion is the observation that approximately 100 times more GA_3 is required to reverse the inhibition of stem growth than flower initiation in induced plants of *Pharbitis nil* treated with growth retardants [25, 27].

There are other difficulties in the interpretation of data from experiments using growth retardants. First of all, the site of action of the compounds should be determined, since this will have important implications for the role of GAs in flower initiation. For example, CCC inhibits flower initiation following photoinduction in both the LSDP *Bryophyllum daigremontianum* and the SDP *Pharbitis nil*, but in the former case the growth retardant acts in the leaf whereas it is the apex that is the affected tissue in the latter [25, 27].

A second problem encountered in studies using inhibitors of GA biosynthesis or GA-deficient mutants – and one that is often not fully appreciated – is that flower primordia may have formed although further development to macroscopically visible flower buds is blocked. We have observed this in our studies on the CRP *Thlaspi arvense*. Application of CCC to thermoinduced plants blocks both stem elongation and the appearance of flower buds. However, microscopic examination of apices showed that flower primordia were formed about the same time after the end of the cold treatment as the non-CCC treated plants (K. Dusbabek and J. Metzger, unpublished data). We have also observed a similar phenomenon in a GA-deficient dwarf mutant of this species.

2.3 What are the Sites of GA Production and Action in Relation to Flower Initiation?

The floral stimulus acts at the apex and therefore determination of the site of GA action is important in formulating logical hypotheses about the role of GAs in flower initiation. Although not examined in many species, the evidence indicates that GAs can affect flower initiation at different sites. Flower induction brought about by GA_3 treatments in *Bryophyllum daigremontianum* and *Hyoscyamus niger* is the result of GA action in the leaves, not the apex where the floral stimulus acts [21, 22]. The situation is quite different in the SDP *Pharbitis nil* and *Impatiens balsamina*, where it was determined that the apex is the site of GA action [16, 25, 27].

If GAs are part of the floral stimulus, then the site of production and transport patterns of the two must be identical. In photoperiodically sensitive plants this means production in the leaves and transport via the phloem to the apex. To date there are very few studies that definitively pinpoint the site of GA production and phloem transport in relation to photoperiod. In *Spinacia* leaves, LD treatment causes an increase in the level of GA_{20} (the presumed effector GA for stem elongation in this species) and a concomitant decline in its precursor, GA_{19}, suggesting that the conversion of GA_{19} to GA_{20} is under photoperiodic control [14]. That this is indeed the case has been shown directly in metabolism studies [4, 5, 6]. Moreover, the change in the GA_{20} content of phloem sap under different photoperiodic conditions closely parallels those observed in the leaves [13]. This indicates that in *Spinacia*, GA_{20} could be a photoperiodic signal emanating from the leaves and transported via the phloem to the shoot apex. Whether this compound is the floral stimulus is, however, questionable. As discussed before, GA_{20} probably controls LD-induced stem growth in *Spinacia* rather than flower initiation.

2.4 Conclusions

When considering the mass of experimental data that has accumulated over the past 30 years, the bulk of the evidence does not strongly support the notion that GAs are the floral stimulus. The duality of the nature between the two has been clearly demonstrated in *Bryophyllum* employing grafting experiments. Exogenous GAs, (GA_1, GA_3 and GA_{20}) substitute for the LD portion of the sequence LD → SD; application during LD is ineffective. Flowering of receptor plants maintained in LD or SD is observed when grafted onto donor stocks induced with GA_3. Inasmuch as exogenous GA_3 is unable to induce flowering on *Byrophyllum* in LD, the most logical interpretation is that the GA_3 treatment of the donor induced the production of the floral stimulus, and that it is different from GA_3 [22]. Consistent with this is the observation that receptor scions of the SLDP *Echeveria harmsii*, a plant that cannot be induced to flower with GA_3 under any circumstance, will flower when grafted onto *Bryophyllum* donor stocks induced with GA_3 [26].

Grafting experiments with two inbred lines of *Silene armeria* that differ in their sensitivities to GA_3 demonstrate that the above results are not unique. In one line, application of GA_3 in SD induces both flower initiation and stem elongation, while in the other ("GA^-" line), only bolting is observed. However, flowering of the GA^- line under SD can be obtained when grafted onto donor stocks of the GA sensitive line that had been induced with GA_3 [20]. Again, it is difficult to see how GA_3 and the floral stimulus could be identical.

Nevertheless, one cannot totally dismiss the possibility that, in the above cases, GA_3-treatment induced the production of a heretofore unidentified florigenic GA. The direct approach to addressing this question has been to attempt identification of all the endogenous GAs in a given species and then assess the biological activity of each native GA. However, there are two problems with this. First, a particular important GA may be overlooked because of low levels, lack of biological activity in standard GA bioassays, or unusual chromatographic properties. Second, even if all of the endogenous GAs were identified, some may not be available in quantities to allow testing for biological activity. One possible solution to these difficulties is to generate a series of mutants in which aspects of reproductive development are altered. If different GAs control flower initiation and some other process such as stem growth, it should be possible to select for a mutant that is deficient in one GA and not the other. The phenotypic expression of such a mutation will result in the absence of one of the two processes. For example, in a rosette plant one could look for a mutant that responds to inductive conditions with normal stem growth but no flower initiation. If the biochemical basis for the lack of flower initiation is due to an alteration in GA metabolism, one would have very good evidence for a specific GA being florigenic.

3 GA Regulation of Other Aspects of Floral Initiation

As pointed out earlier there are a number of examples in which inhibitors of GA biosynthesis block flower initiation in inductive conditions, indicating that under certain circumstances GAs are limiting for flower initiation. There are at least three possible points of control. First, as has already been discussed in some detail, a specific GA may act as the floral stimulus; second, GAs may regulate the production of the floral stimulus; and third, a GA may mediate the action of the floral stimulus of the apex; i.e., act as a secondary messenger.

3.1 GA Regulation of Floral Stimulus Production

In the previous section, several examples were described in which physiological evidence indicated that GAs induce flowering in unfavorable conditions by causing the production of the floral stimulus. In the case of *Byrophyllum*, it is also known that growth retardants effectively block flower initiation when applied to the SD portion of the LD → SD inductive sequence. When plants are transferred from LD to SD, the level of GA_{20} in the leaves increases dramatically. Plants maintained permanently under either LD or SD have little or no detectable GA_{20}. Since GA_{20} can substitute for the LD portion of the LD → SD inductive sequence (albeit only 1/20th as effective as GA_3, a non-native GA), it appears that a high level of GA_{20} is required for the production of the floral stimulus [25, 27]. The biochemical basis for the regulation of GA_{20} levels is unknown. It would be interesting to apply several of the 13-C hydroxylated precursors of GA_{20} (e.g., GA_{53}, GA_{44}, or GA_{19}) to *Bryophyllum* plants in SD in an effort to pinpoint possible metabolic block(s) in the biosynthesis of GA_{20} that are removed by the LD portion of the inductive sequence.

It remains to be seen if similar regulatory controls on the production of the floral stimulus are unique to *Bryophyllum* or are of more widespread distribution.

3.2 GAs and Floral Stimulus Action

In other species, the site of GA action is the apex and not the leaves. In the SDP *Pharbitis nil*, for example, growth retardants effectively block flowering and this inhibition is reversible with GA_3. Exogenous GAs also promote flowering in a GA-deficient, dwarf cultivar of this species [9]. In both instances, the site of GA action is the apex, not the leaves where photoinduction occurs [16, 25, 27]. Two interpretations of these results are possible. First, GAs may be acting as the floral stimulus [16]; and, second, GAs may enable the apex to respond to the floral stimulus. In other words, GAs may mediate the action of the floral stimulus upon its arrival at the apex [25, 27]. At present, it is impossible to distinguish between these two possibilities. Perhaps one approach to this question would be through the use of a graft between a non-induced shoot tip (receptor) to an induced donor stock that is depleted of its endogenous GAs. This could be accomplished by either pre-treating the plants used for donor stocks with a growth retardant that is not readily translocated in the phloem or by using a GA-deficient dwarf mutant. Since the site of floral stimulus production is the leaf, one would predict no flowering will be observed in the receptor if GAs are the floral stimulus.

4 Future Prospects

It is quite clear from the foregoing discussion that GAs have a variety of roles in the regulation of flower initiation, depending on species and response type. Although we can distinguish, on a gross physiological level, that differences exist, we have not been able to precisely define them on biochemical or molecular grounds. Part of the reason for this is that in most cases the identity of the GAs native to the plant under study are unknown or only partially known. Thus, it is difficult to design and perform meaningful experiments relating GA levels, metabolism, bioactivity, etc., to flower initiation. However, this is not quite the formidable task that it was only a few years ago. Increasingly, plant physiologists have access to the instrumentation necessary to do much of the analytical work necessary for these types of studies. The advent of relatively inexpensive bench-top mass spectrometers, along with recently developed chemical procedures for the synthesis of labeled and unlabeled GAs and GA precursors has allowed a number of laboratories to perform experiments on a routine basis that heretofore were restricted to a few centers with expertise in organic chemistry.

A credible argument can be made that the greatest impediment in delineating the role(s) of GAs in flowering does not lie in the difficulties associated with GAs themselves, but, instead, with our almost total ignorance of the molecular basis for flower initiation. The processes that occur during flower induction and culminate in the initiation of flower primordia are almost certainly the result of selective changes in gene expression – both switching on and turning off specific genes. Thus, in order to rigorously define the role of GAs in this process, we will have to first identify and characterize the genes that are a part of flower initiation and then determine in which ones GAs regulate expression. Such an approach has furthered our understanding of developmental phenomena in insects and vertebrates. Fortunately, we are now beginning to see more work in which a molecular approach to understanding flowering is being used [e.g. 10, 11, 15].

References

1. Bernier G, Kinet JM, Sachs RM (1981) The physiology of flowering, vol 1. CRC, Boca Raton
2. Chailakhyan Mkh (1936) Dokl Acad Sci USSR 12:443
3. Crozier A, Durley RC (1983) In: Crozier (ed) The biochemistry and physiology of gibberellins, vol 1. Praeger, New York, p 8
4. Gianfagna TJ, Zeevaart JAD, Lusk WJ (1983) Plant Physiol 72:86
5. Gilmour SJ, Zeevaart JAD, Schwenan L, Graebe JE (1984) Plant Physiology S75:186
6. Gilmour SJ, Zeevaart JAD, Schwenen L, Graebe JE (1986) Plant Physiol 82:190
7. Graebe JE, Ropers HJ (1978) In: Letham DS, Goodwin PB, Higgins TJV (eds) Phytohormones and related compounds – a comprehensive treatise, vol 1. Elsevier, Amsterdam, p 107
8. Jones MG, Zeevaart JAD (1980) Planta 149:174
9. King RW, Pharis RP, Mander LN (1987) Plant Physiol 84:1126
10. Lay-Yee M, Sachs RM, Reid MS (1987) Plant Physiol 84:545
11. Lay-Yee M, Sachs RM, Reid MS (1987) Planta 171:104
12. Metzger JD (1987) In: Davies PJ (ed) Plant hormones and their role in plant growth and development. Nijhoff, Boston, p 411
13. Metzger JD, Zeevaart JAD (1980) Plant Physiol 66:679
14. Metzger JD, Zeevaart JAD (1980) Plant Physiol 66:844

15. O'Neil SD, Bicknell RA, Reid MS, Sachs RM (1988) Plant Physiol 86:16
16. Pharis RP, King RW (1985) Annu Rev Plant Physiol 36:517
17. Pharis RP, Evans LT, King RW, Mander LN (1987) Plant Physiol 84:1132
18. Pharis RP, Evans LT, King RW, Mander LN (1989) In: Lord E, Bernier G (eds) Plant reproduction: from floral induction to pollination. Am Soc Plant Physiol Symp Ser I:29
19. Sachs RM (1965) Annu Rev Plant Physiol 16:73
20. Van dePol PA (1972) Meded Landbouwhogesch Wageningen 72:1
21. Warm E (1980) Z Pflanzenphysiol 99:325
22. Zeevaart JAD (1969) In: Evans LT (ed) The induction of flowering: some case histories. Cornell University Press, Ithaca, NY, p 435
23. Zeevaart JAD (1971) Plant Physiol 47:821
24. Zeevaart JAD (1976) Annu Rev Plant Physiol 27:321
25. Zeevaart JAD (1978) In: Letham DS, Goodwin PB, Higgins TJV (eds) Plant hormones and related compounds, vol II. Elsevier, Amsterdam, p 291
26. Zeevaart JAD (1982) Ann Bot 49:549
27. Zeevaart JAD (1983) In: Crozier A (ed) The biochemistry and physiology of gibberellins, vol 2, Praeger, New York, p 333

Cytokinins in Flower Initiation

G. BERNIER, P. LEJEUNE, A. JACQMARD, and J.-M. KINET[1]

1 Mitotic Activation During the Floral Transition

Quite generally, mitotic and DNA synthetic indices are far higher in shoot meristems during the floral transition than in both vegetative and reproductive meristems [5]. In several species this activation at floral evocation was shown to be due to the shortening of the cell cycle [1, 8]. Another remarkable feature of the floral transition, caused by a transfer of plants from non-inductive to inductive daylength conditions, is the transient synchronization of the division cycle of part, at least, of the meristem cell population [1, 5, 8].

Work with a variety of tissue and cell cultures has indicated that cytokinins are generally promoters of the cell division cycle in plants [19]. In some cultured materials, cytokinin withdrawal and refeeding cause an arrest, followed by a synchronous recovery of cell divisions [19]. Moreover, during the course of partially synchronized cell divisions in cultured tobacco cells, there is a dramatic increase in the amount of endogenous cytokinins [18].

Put together, these observations suggest that cytokinins may be involved in the control of the mitotic stimulation during the floral transition. This possibility was investigated in 2-month-old *Sinapis alba* plants grown in 8-h SD and induced to flower by a single 22-h LD.

2 The Sinapis Story

2.1 The Early Mitotic Activation

An early wave of mitoses was observed in the shoot meristem 26–30 h after start of the LD. This wave resulted from both a shortening of the G_2 phase of rapid cycling cells and a return to fast cycling of slow cycling G_2 cells [9]. As a result there was a near-synchronization of the cell population in G_1 at 30 h. This synchronization was kept during the next cell cycle which was the last before flower initiation. All attempts to dissociate this early mitotic activation from flowering have so far been unsuccessful [3], suggesting that this activation is an essential component of evocation.

[1] Laboratoire de Physiologie Végétale et Centre de Physiologie Végétale Appliquée (I.R.S.I.A), Département de Botanique, Université de Liège, Sart Tilman, 4000 Liège, Belgium

2.2 The Mitotic Stimulus

Defoliation experiments showed that the stimulus responsible for the early mitotic wave in the meristem came from mature leaves, and that the stimulus began to move out of these leaves at about 16 h after the start of the LD [2].

2.3 Effects of Exogenous Cytokinins

Cytokinins were applied directly to the apical bud to avoid translocation problems. A single application of BA to plants exposed to one suboptimal LD promoted flower initiation at the concentrations of 5×10^{-6}M and 2.5×10^{-5}M, but inhibited it at 5×10^{-5}M and greater (Fig. 1).

However, when applied to plants kept strictly vegetative by continuous growth in SD, BA was unable to cause flower initiation, even when the treatment was repeated [4]. However, a single low-dose application in SD, timed to correspond with the time of movement of the mitotic stimulus in LD-induced plants, caused an early mitotic activation in the meristem [4]. Interestingly, the timing, magnitude and characteristics of this activation were the same as in meristems of LD-induced

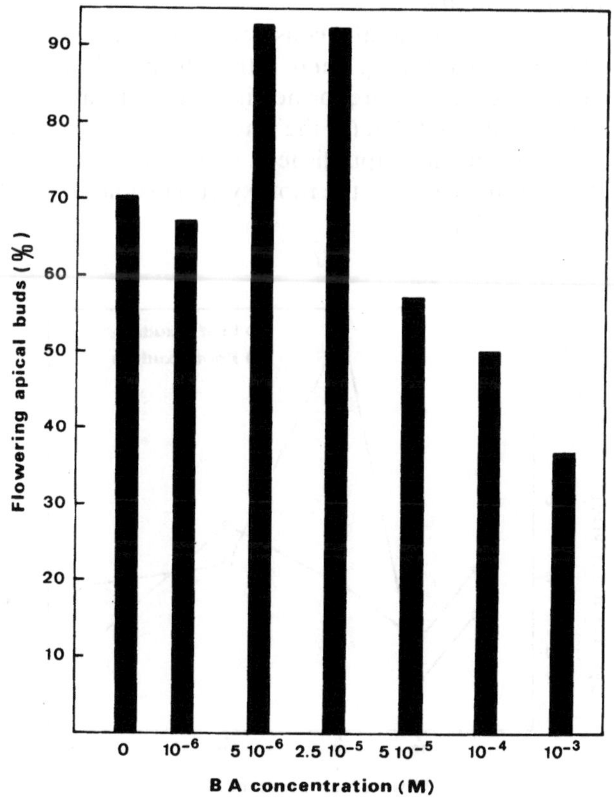

Fig. 1. Effect on flower initiation of various concentrations of BA applied in 0.3 ml to the apical bud of *Sinapis* plants induced by one 13-h LD. Applications were at the start of the LD

plants [4,5, and Houssa, unpublished]. Other chemicals, including sucrose, GA_3, etc. are unable to substitute for BA [4]. Exogenous BA did not cause, however, all of the changes that normally occur in response to LD [12]. Hence cytokinins appear to fulfil only some of the factors required for floral evocation [5]. Even so, cytokinins are the best candidates for the mitotic stimulus produced in LD-exposed leaves of *Sinapis*. If indeed they stimulate the increase in mitotic activity, then their endogenous status should change in both leaves and in phloem sap at the appropriate time, i.e. about 16 h after start of the LD.

2.4 Cytokinin Levels and Fluxes

Tissues were extracted in Bieleski's medium [6]. Techniques to collect exudates and to analyze cytokinins using HPLC methodology and bioassays are described elsewhere [15]. These analyses clearly indicated that the cytokinin content in leaves of induced plants increased above the control level, starting at 16 h. This increase was most marked at 16 h, and was observed mainly in the fraction having the same Rt as Z (Lejeune, unpublished).

However, a marked increase in all cytokinin fractions (Rts of iP, iPA, Z, ZR) was also recorded in the leaf (phloem) exudate of induced plants at 16 h (Fig. 2) [15]. All these data are consistent with the view that cytokinins are the transmissible mitotic stimulus.

Roots are generally considered as the major site of cytokinin biosynthesis in vegetative plants [23]. There is thus the possibility that the extra-cytokinins found in induced leaves were formed in roots and translocated in the xylem sap. Indeed, analyses showed that (a) the cytokinin levels extracted from root tissue decreased at 16 h (Lejeune, unpublished), and (b) the cytokinin activity (mostly at the Rt of ZR) was increased in the root (xylem) exudate at 9, 16 and 22 h, but not at 12 and 30 h (Fig. 2) [15].

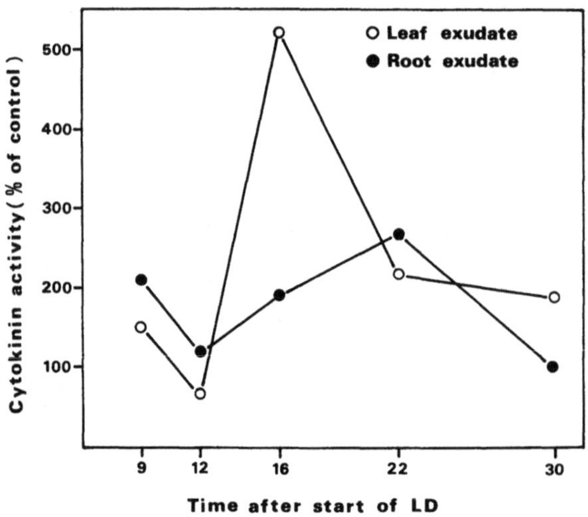

Fig. 2. Total cytokinin activity of butanol-soluble fraction of leaf (○) and root (●) exudates collected at various times after the start of the 22-h inductive LD. Exudation started at the indicated times and lasted 16 h. Exudates showed no activity in the aqueous fraction

So far the *Sinapis* story can be interpreted as follows: The shoot apical meristem of 2-month-old plants is presumably cytokinin-limited. Exposure to LD causes the production, in the leaves, of a signal which is then transported to the root system. There it alters the course of cytokinin metabolism and/or release. The increase in cytokinin levels in the transpiration stream (xylem) causes an increase in leaf cytokinin levels by 16 h. Some of the leaf cytokinins are then re-exported in the phloem sap to the apical meristem, where they cause a mitotic activation at 26–30 h. Since the cytokinin level in root exudate is altered as early as h 9 (Fig. 2), the initial leaf-to-root signal is apparently produced and transported extremely rapidly, i.e. within the first hour of the photo-extension period of the LD. The nature of this signal is unknown, but bark-ringing experiments indicate that it moves in living tissues (Lejeune, unpublished).

3 The Situation in Other Species

3.1 Effects of Exogenous Cytokinins

Applied cytokinins may promote or inhibit flower initiation in a variety of plants [reviewed in 5]. In cases where an inhibition was observed, relatively high and presumably supraoptimal concentrations were applied, or the plant materials used were young seedlings (*Brassica*, *Chenopodium*) or cuttings (*Anagallis*, *Scrophularia*) in which the endogenous cytokinin levels in shoot meristems might have already been optimal, or even supraoptimal due to proximity of the root system. Promotive effects were most often observed with plants exposed to marginal or suboptimal inductive conditions. Also, promotion could be dramatically reinforced by combination of the cytokinin with a GA, as in *Chrysanthemum* [20], conifers [21] and several other species [see 5]. These results support the above idea that cytokinins act in conjunction with other factors in the control of floral evocation.

Cytokinin application caused an increase of the mitotic index in the meristem of several vegetative plants, e.g. in the SDP *Perilla* (Fig. 3) and other species listed in [5]. This effect was not observed in *Xanthium* (Jacqmard and Bernier, unpublished).

3.2 Cytokinin Levels and Fluxes

From the available data it can be tentatively concluded that the level and/or metabolism of endogenous cytokinins changes markedly, often transiently, at the time of floral transition in many plant species [1]. The trend of changes was, however, opposite in species with different photoperiodic requirements: an increase in levels was recorded, for example, in leaves of the LDP *Hyoscyamus* and *Nicotiana sylvestris* and the SDP *Begonia*, whereas a decrease was observed in the SDP *Xanthium* and *Chenopodium* [see 5, 11, 16].

In the SDP red *Perilla*, preliminary results (Grayling and Hanke, unpublished) showed that cytokinin levels were higher in leaf and root exudates of induced plants than in non-induced plants. The increase in ZR found in leaf exudate at induction

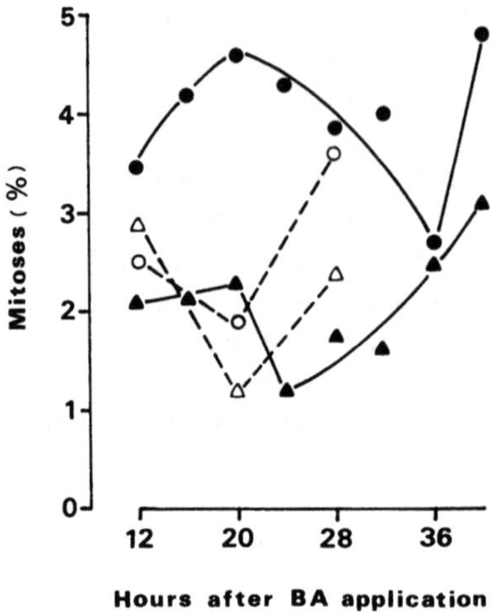

Fig. 3. Mitotic index in the peripheral (○, ●) and central (△, ▲) zones of the apical meristem of red *Perilla* plants kept in non-inductive LD, at various times after application of BA (0.3 ml; 4.3 × 10^{-5}M) to the apical bud. *Open symbols* untreated plants; *solid symbols* treated plants. The activation was essentially in the peripheral zone

was sustained after a long period in LD, thus correlating well with the stable induced state [24].

In cold-requiring plants an increased level of cytokinins is usually found in response to a chilling treatment [see 5, 10, 14]. A similar situation was reported for buds and shoots of *Pinus* and Douglas fir during transition to reproductive growth [13, 17, 22]. In Douglas fir the increase was mostly in an iP-like fraction and was correlated with enhanced female flowering [13].

One could use the opposite variations in cytokinin levels from various photoperiodic species to dismiss these compounds as regulatory agents of the floral transition. However, we interpret the evidence at hand as showing that cytokinins are involved in the control of evocation, but that there is a permissive range of concentrations: the endogenous levels at the vegetative stage seem to be subthreshold in many plants, whereas in others, particularly in seedlings as *Chenopodium*, endogenous levels are apparently supraoptimal. Altered cytokinin metabolism prior to or during flower formation, as suggested by results on *Pinus*, chicory and Douglas fir [7, 14, 22] is also a possibility.

Acknowledgements. Our research work was supported by grants from the FRFC of Belgium (2.9009.87 and 2.4507.87), the Belgian Government and the University of Liège (Action de Recherche Concertée 88/93-129). One of us (P.L.) is grateful to IRSIA for the award of research fellowship.

References

1. Bernier G (1988) Annu Rev Plant Physiol Plant Mol Biol 39:175
2. Bernier G, Bodson M, Kinet JM, Jacqmard A, Havelange A (1974) In: Plant growth substances 1973. Hirokawa, Tokyo, p 980
3. Bernier G, Kinet JM, Bodson M, Rouma Y, Jacqmard A (1974) Bot Gaz 135:345
4. Bernier G, Kinet JM, Jacqmard A, Havelange A, Bodson M (1977) Plant Physiol 60:282
5. Bernier G, Kinet JM, Sachs RM (1981) The physiology of flowering, vol II. CRC, Boca Raton, FL

6. Bieleski RL (1964) Anal Biochem 9:431
7. Doumas P, Morris JW, Chien C, Bonnet-Masimbert M, Zaerr JB (1986) In: Proc 9th North Am Forest Biol Workshop, Oklahoma State Univ, Stillwater, p 285
8. Francis D, Lyndon RF (1985) In: Bryant JA, Francis D (ed) The cell division cycle in plants. Cambridge University Press, Cambridge, p 199
9. Gonthier R, Jacqmard A, Bernier G (1987) Planta 170:55
10. Gregorini G (1983) Sci Hortic 21:155
11. Hansen CE, Kopperud C, Heide OM (1988) Physiol Plant 73:387
12. Havelange A, Bodson M, Bernier G (1986) Physiol Plant 67:695
13. Imbault N, Tardieu I, Joseph C, Zaerr JB, Bonnet-Masimbert M (1988) Plant Physiol Biochem 26:289
14. Joseph C (1986) J Plant Physiol 124:235
15. Lejeune P, Kinet JM, Bernier G (1988) Plant Physiol 86:1095
16. Lozhnikova VN, Krekule J, Vorob'eva LV, Chailakhyan MKh (1985) Dokl Akad Nauk SSSR 282:1021 (in Russian)
17. Miginiac E, Pilate G, Bonnet-Masimbert M (1987) 14th Int Bot Congr, West Berlin, Abstracts, p 105
18. Nishinari N, Syōno K (1980) Plant Physiol 65:437
19. Péaud-Lenoël C (1977) In: Pilet PE (ed) Plant growth regulation. Springer, Berlin Heidelberg New York, p 240
20. Pharis RP (1972) Planta 105:205
21. Ross SD, Pharis RP, Binder WD (1983) In: Nickell LG (ed) Plant growth regulating chemicals, vol II. CRC, Boca Raton, FL, p 35
22. Taylor JS, Koshioka M, Pharis RP, Sweet GB (1984) Plant Physiol 74:626
23. Van Staden J, Davey JE (1979) Plant Cell Environ 2:93
24. Zeevaart JAD (1962) Science 137:723

Hormones and *Cuscuta* Development: Influence of Hormones on Secondary Xylem Differentiation, Phenylalanine Ammonia Lyase (PAL) Activity and Lignification

I. RAJAGOPAL,[1,2] S. RAMACHANDIRAN,[1] and S. MAHADEVAN[1]

1 Introduction

As with other plant growth and development processes, all five classes of plant hormones appear to influence xylem differentiation in whole, wounded, amputated or decapitated plants, isolated plant organs, explants or callus and cell cultures in a species and stage-specific manner, [1, 9, 16, 17, 22, 24, 25, 26]. Of these, IAA is generally regarded to be the main regulatory factor [1]. During regenerative xylem differentiation the pattern of tracheary element formation is apparently determined by the path and intensity of auxin flux [24]. A cytokinin is often required with the auxin to stimulate xylogenesis in explant, callus or cell cultures [22], and the need for auxin may precede that for cytokinin in the process [18]. The role of GAs in xylem differentiation thus far appears to be ambiguous, and both stimulation and inhibition of the process by exogenous GA have been reported [16, 26]. IAA and GA synergistically promoted secondary xylem differentiation in the cambium of woody species [30]. Pearce et al. [16] demonstrated correlative changes for endogenous GA's during auxin:cytokinin-mediated xylogenesis in lettuce pith cultures, and noted significant promotion of xylogenesis by GA_1 at a very low dose (0.003 μM). Ethylene promoted xylogenesis and reaction wood formation in several plant systems [1, 23], in the presence of auxin and cytokinin [12]. Limited information indicates that ABA may inhibit xylogenesis, as in artichoke explants [13] or in bean callus where PAL activity is also inhibited [7].

In the morphologically simple rootless and leafless parasitic *Cuscuta*, successive regions of the vine from the apex downwards are sequentially responsive to exogenous GA_3, cytokinin and auxin, singly or in combination, in eliciting in vitro all the growth phenomena observed in vivo, namely apical growth, subapical 'straight' or 'coiled' growth, and the induction (and suppression) of haustoria formation [10, 11, 19, 21]. When growth ceases, and auxin transport capacity is reduced to a minimum [15], an increase in PAL activity occurs and the formation of a ring of secondary xylem can be seen in the post-150-mm region [14]. This suggests a causal connection between the events. An increase in PAL during lignification has been shown in other systems [6]. We report here the nature of secondary xylem development in *Cuscuta* and the influence of hormones on PAL activity and lignification in segments from the 'prelignification' growth region

[1] Department of Biochemistry, Indian Institute of Science, Bangalore-560 012, India
[2] Present Address: Department of Biology, University of California at San Diego, La Jolla, CA 92093, USA

(60–120 mm). This region is growth responsive to IAA and has low PAL activity. In the intact vine which has an average growth rate of 70–100 mm/day, this region becomes the lignification region during the next 24 to 48 h.

2 Materials and Methods

Free-hanging vines of *Cuscuta reflexa* were obtained and segments from the desired regions were treated with hormones or inhibitors, and incubated as described earlier [19, 21]. Procedures for the determination of PAL or α-amylase activity and lignin content are given in Tables 1, 6 and 5. Lignification was microscopically visualized as described in Fig. 2.

3 Results

3.1 Nature of Secondary Xylem in Free-Hanging Vines of *Cuscuta*

The vascular system in the subapical growth region (20–150 mm) consisted of a poorly developed primary xylem of 8 to 11 protoxylem clusters arranged in a ring, each with about 2–5 xylem elements, and a better developed primary phloem outside the xylem, comprising a ring of 16 to 20 discrete bundles, half of which were collateral with the xylem. A ring of thin-walled cells, usually 3 to 4 cells deep, separated the xylem clusters from the phloem bundles. These cells showed no tangential division in any region, growing or non-growing, to identify them as cambium, either fascicular or interfascicular. Increasing in diameter as elongation growth began to decrease, their walls thickened and lignified in the region where growth ceased and where PAL activity increased, i.e. beyond 150 mm, to form a complete ring of secondary xylem. The term 'secondary' for the xylem is used not because of its cambial origin, but because it is formed after elongation growth ceased. Lignification is thus a deferred terminal differentiation event in these cells which appear to have been 'determined' early in development to become xylem. However, lignification does not occur until elongation growth has ceased. In the absence of any cambial activity, no secondary phloem is formed.

3.2 PAL Activity in the 'Ends' and 'Centres' of Incubated Segments

In light of an earlier observation that PAL activity fell rapidly in host-detached vines [14], its activity was compared in 20-mm segments from the apical (0–60 mm), prelignification (60–120 mm) and lignification (180–240 mm) regions either immediately after excision or after 24-h incubation. PAL activity decreased in the initially high apical and lignification regions, while its activity in the 60–120 mm region increased (Table 1). The suspicion that this anomalous increase may be due to an injury effect at the cut ends, as reported in certain other plant tissues [28], was confirmed when PAL activity in the 'ends' and 'centre' of segments were compared with whole segments (Table 1). PAL activity increased to 540% of the 0 h value in

Table 1. PAL activity[a] in whole, "center" or "ends" of excised segments from the tip (0–60 mm), prelignification (60–120 mm) and lignification (180–240 mm) regions[b]

Segment (incubation, h)	PAL activity; nmol/gfw/h		
	Region (mm)		
	0–60	60–120	180–240
Whole (0)	950	176	595
Whole (24)	282	276	293
Centre (24)	48	56	180
Ends (24)	814	949	860

[a] PAL activity was determined in crude supernatant of segments prepared by homogenizing (Polytron, Kinematica Gmbh, fitted with 10-mm head) about 1 gfw. of tissue ('ends' about 200 mg) in 6 ml of cold 0.1 M Tris-HCl buffer, pH 8.2, containing 5 mM potassium metabisulphite for 3 × 30 s, followed by centrifugation at 10 000 × g for 20 min.

Reaction mixture (1 ml) contained 0.1 mM Tris-HCl buffer, pH 8.2, 0.5 mM B-mercaptoethanol, 1.2 mM cold L-phenylalanine, 200,000 cpm of 1-[^{14}C]-DL phenylalanine (Sp. Act. 23.55 mCi/mmol) and 0.5 ml of crude enzyme. Following incubation for 3 h at 30°C, the reaction was terminated with 0.1 ml of 5 N HCl and cinnamic acid (product) was extracted with 5 ml toulene. One ml of the toulene layer was counted by liquid scintillation spectrometry. Reactions were run in duplicate. Activity expressed as nmol cinnamic acid /gfw/h

[b] 20 mm segments were excised and incubated in buffer (cf. Table 2) for 24 h as described [19]. "Ends" were a 2-mm piece cut from the end of the segment; "centers" were the middle portions.

the 'ends', while it dropped to 32% of the zero time value in the 'centres' in the 60–120 mm region. The time course of PAL activity decay, maximal during the first 8 h, was similar in segments from both the apical and prelignification regions (Fig. 1).

3.3 Influence of Hormones on PAL Activity

3.3.1 Effect on Hormones on "Centers" of Segments

Table 2 gives the effect of 24 h of hormone treatment on PAL activity in centres following incubation. In untreated segments PAL activity fell to 31% of initial value. However, PAL activity significantly increased in BA-treated (2.9-fold) and in GA_3-treated (3.9-fold) segments of 24-h controls. The 50% increase observed from NAA treatment was variable and therefore not significant. However, of the three hormones only GA_3 brought about a net stimulation of PAL activity over the initial (0 h) value. ABA, Ethrel and FC were without effect. However, BA + GA_3 showed a 7.3-fold increase over the 24-h control which was equal to the sum of the stimulation by BA alone and GA_3 alone. The net stimulation of PAL activity by

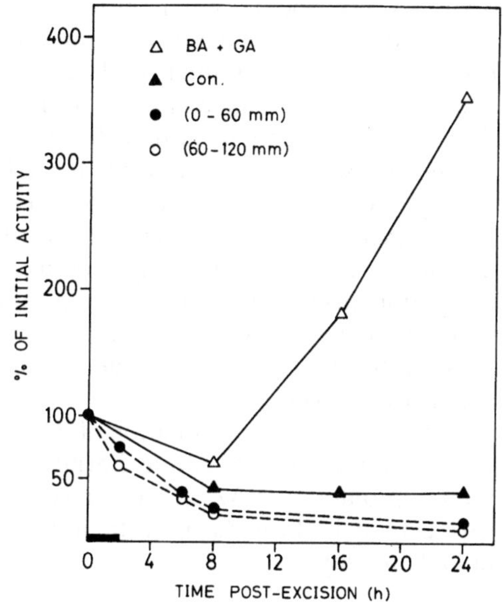

Fig. 1. Time course of excision drop and BA + GA_3-induction of PAL activity. "Excision drop" (*circles*): Segments incubated without shaking in solution. BA + GA_3-induction (*triangles*): Segments from (60–120 mm) region shaken in buffer or hormone solution for 2 h (*black bar* on abscissa) before rinsing and incubation

Table 2. Influence of hormones on PAL activity in segments from the growth responsive, prelignification (60–120 mm) region

Hormone[a]	PAL activity	Segment Length
	nmol/gfw/h + SE (No. of Expts.)	Percent control (24 h)
None (0 h)	267 ± 19 (30)	-
None (24 h)	84 ± 8 (27)	100
BA	240 ± 22 (9)*	99
GA_3	328 ± 37 (9)*	101
NAA	131 ± 35 (5)	122***
FC	88 ± 16 (2)	-
ABA	80	-
Ethrel	71	-
BA + GA_3	610 ± 28 (16)	101
BA + GA_3 + NAA	214 ± 21 (13)**	120***
BA + GA_3 FC	119 ± 9 (2)**	-
BA + GA_3 + ABA	645	-
BA + GA_3 + Ethrel	627	-

*Sig. different at P = 1% of None (24 h); **Sig. different at P = 1% of BA + GA_3 (t-test); ***Sig. 1% of None (0 h)

[a] Hormone concentrations were: BA 50 μM; GA_3 100 μM; NAA 500 μM; FC 10 μM; ABA 100 μM; Ethrel 100 ppm (active ingredient). Segments (20 mm) were shaken in buffer solution (0.1 M potassium phosphate-citrate, pH 5, 1 mM KCl and 100 μg chloramphenicol, with or without hormones) for 2 h before being rinsed and set out for incubation. Total period of treatment and incubation was 24 h. Segments were trimmed before extraction for PAL assay.

$BA + GA_3$ was about three times the initial (0 h) value. Addition of NAA (or FC) along with $BA + GA_3$ significantly inhibited the increase in PAL activity caused by $BA + GA_3$. Again ABA and Ethrel were without effect on the stimulation caused by $BA + GA_3$. A time course analysis of $BA + GA_3$-induced increase in PAL activity showed that PAL activity initially fell during the first 8 h (as in untreated segments), but then subsequently increased (Fig. 2). PAL activity at 48 h in all treatments was usually lower than at 24 h (data not given).

Segment length measurements (Table 2) after 24 h incubation showed that treatment with NAA either singly, or in combination with other hormones, caused elongation. In other experiments FC promoted elongation, often at levels 2–3 times that caused by NAA. It appears therefore that inhibition of the PAL activity increase by NAA may partly be associated with continued elongation in this auxin-responsive region.

In contrast to the stimulation of PAL activity by $BA + GA_3$ in the centres there was no stimulation by these hormones in the "ends" (Table 3). Use of the growth-promoting NAA or FC, however, partially inhibited the expected rise in PAL activity in the ends just as their use had inhibited the $BA + GA_3$-induced rise in PAL activity.

Table 3. Influence of hormones on 'cut-end' induction of PAL activity in segments from 60–120 mm region during 24 h incubation

Treatment	PAL activity (nmol/gfw/h)
Control	1214
$BA+GA_3$	1105
Control	1747
NAA	1187
FC	587

Treatments and concentrations of hormones as in Tables 1 and 2.

3.4 Influence of Translation and Transcription Inhibitors on PAL Activity

Cycloheximide (CHI) infiltration of *Cuscuta* tissue was earlier shown to partially suppress the 'excision drop' of PAL activity, suggesting a need for protein synthesis in the inactivation process [14]. The effects of CHI and actinomycin-D (Act-D) were therefore tested for their effects on both 'excision drop' and $BA + GA_3$-induced PAL activity increase. Table 4 shows that both inhibitors do suppress excision drop by about 85%, suggesting that de novo transcription and protein synthesis are required in the process. However, CHI or Act-D suppression of the PAL induction occasioned by use of $BA + GA_3$ was almost complete. Thus, the increase of PAL activity on treatment with the hormones apparently involves de novo PAL synthesis.

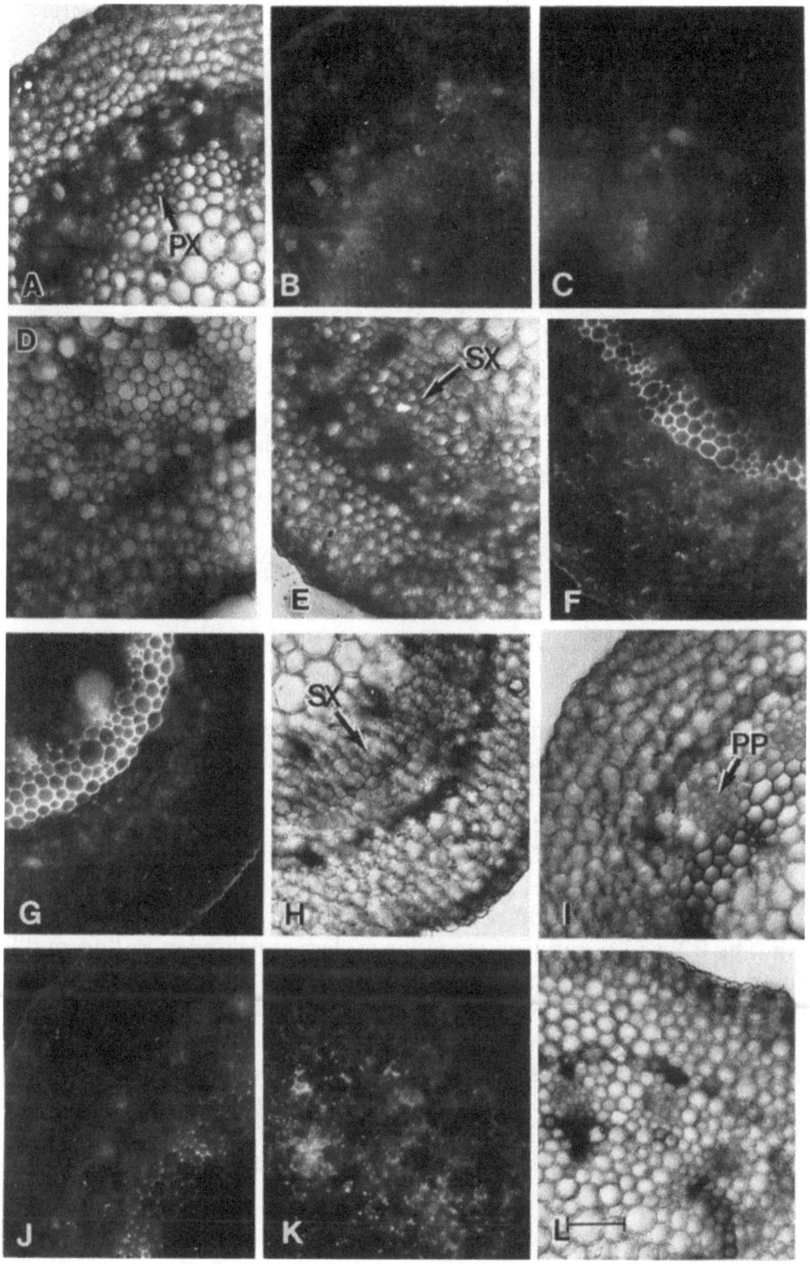

Fig. 2A-L. Influence of hormones on lignification in segments from (60–120 mm) region. 20-mm segments were treated with hormones and incubated for 40 h as described in Table 2, except that NAA was 100 μM. Photographs are of free-hand sections of the stem mounted in about 50% glycerol and taken in a Carl Zeiss photomicroscope with Neofluor optics and epifluorescence attachment and 436 BP, 460 FT and 470 LP filters to provide violet excitation. Under fluorescence lignified cells appear white against dark background. Photographs **B**, **C**, **F**, **G**, **J**, **K** are fluorescence pictures of segments **A**, **D**, **E**, **H**, **I**, **L**, respectively, the latter being viewed under transmitted light. *Bar* 100 μM; *PX* primary xylem; *PP* primary phloem; *SX* secondary xylem. **A**, **B** No hormone (control); **C**, **D** BA; **E**, **F** GA_3; **G**, **H** (BA + GA_3); **I**, **J** NAA; **K**, **L** (BA + GA_3 + NAA)

Table 4. Effect of cycloheximide (CHI) and actinomycin-D (Act-D) on "excision drop" or BA + GA_3-induction of PAL activity in segments from the 60–120 mm region

	Treatment (incubation, h)	PAL activity (nmol/gfw/h) CHI	Act-D
"Excision drop"	None (0)	251	197
	None (24)	111	54
	CHI (24)	214	–
	Act-D (24)	–	176
BA + GA_3 induction	BA + GA (24)	606	615
	CHI + BA + GA_3 (24)	192	–
	Act-D + BA + GA_3 (24)	–	254

Segments (20 mm) were treated with or without inhibitors or with BA + GA_3 for 2 h before rinsing and subsequent incubation for 24 h. Concentrations of inhibitors were: CHI 20 μg/ml and Act-D 250 μg/ml. BA and GA_3 concentrations as in Table 2.

3.5 Influence of Hormones on Lignification in Segments from the Prelignification Region

3.5.1 Visualization of Lignification by Fluorescence Microscopy

The effect of hormones on lignification during secondary xylem differentiation was followed by observing lignin's native fluorescence in hand-cut sections of segments incubated for varying periods of time after hormone treatment. Representative sections from the centers of 40-h incubated segments from the (80–100 mm) region are given in Fig. 2. Sections from GA_3 or BA + GA_3-treated segments invariably exhibited a complete ring of fluorescent cells constituting the lignified secondary xylem (Fig. 2F and G). Lignification could be seen as early as 18–24 h, and quite clearly by 40 h. The BA + GA_3-induced secondary xylem appeared more "normal" than that caused by GA_3 alone, and often matched the in vivo lignified secondary xylem in cell wall thickness and number of rows of fully lignified cells. Sections from controls, or from segments treated with BA or NAA (Fig. 2B, C and J) showed little or no lignification, up to 40 h. However, weak fluorescence could be seen by 72 h in most segments including controls. Addition of NAA along with BA + GA_3 completely inhibited the lignification (Fig. 2K) induced by BA + GA_3, just as it inhibited the BA + GA_3-induced PAL activity increase. GA_3-induced lignification was also similarly inhibited by NAA.

3.5.2 Lignin Content in Segments Following Hormone Treatment

Lignin was extracted with acetyl bromide and spectrophotometrically estimated [4, 5] in control or hormone-treated segments following 40 h of incubation. Use of GA_3, and better still BA + GA_3, increased tissue lignin content, whereas the addition of NAA along with BA + GA_3 decreased it to control or NAA-treatment levels. These results thus confirm conclusions reached by fluorescence microscopy estimates of lignification.

4 Discussion

4.1 Adaptive Significance of Deferred Secondary Xylem Lignification

Evolutionarily and developmentally, the absence of leaves and roots in *Cuscuta* may be the reason for scanty primary xylem formation and the absence of a cambium. The primary phloem can transport host-derived assimilate all along the region of growth. Hence, in the absence of bidirectional development the cambium and its derivatives have become redundant in this dicot.

The option to either grow 'straight' as a free-hanging vine, or to coil around a host and form haustorial contacts remains open for the subapical growth region of the vine. Deferred lignification of the ring of cells that are destined to form secondary xylem until after this option is closed makes adaptive sense, since the exact siting of haustoria (upon coiling) is unknown until actual contact with the host is made. Flexibility in deferring lignification is thus the key to survival. Once haustorial contact with the host is made, a xylem bridge spanning the haustorium eventually connects the host's xylem with that of the parasite's, and secondary xylem development at this stage helps make the appropriate union.

4.2 GA_3 and BA + GA_3-Induced PAL Activity and Lignification

Though either GA_3 or BA was able to promote PAL activity, and thus reverse its decrease that is caused by excision drop, only GA_3 was effective in bringing about near-normal lignification in the putative xylem cells. BA-induced PAL increase could thus be non-specific, and may not be directed to any particular cell type. However, these xylem cells appear to be specifically the target of GA_3.

In combination, BA + GA_3 additively stimulated PAL increase and BA augmented GA_3's effect in bringing about normal level of lignification in the xylem cells (Table 5). BA has been shown to stimulate the formation of xylem elements in the base of preformed haustorial 'mounds' in *Cuscuta campestris* [29]. Taken together,

Table 5. Relative content of lignin in hormone-treated segments from the 60–120 mm region after 40-h incubation

Treatment	% of control
Control	100
GA_3	198
BA + GA_3	262
BA + GA_3 + NAA	105
NAA	126

NAA was 100 μM; all other hormone concentrations and treatment as in Table 2. Lignin content in the lignified (240–300 mm) region of the vine in vivo was about 390% of that in the prelignification growth zone (60–120 mm).

the increase of PAL activity associated specifically with lignification induced by GA_3, or better still by GA_3 + BA, provides evidence for a role for GAs in the terminal stages of xylem differentiation, where prior events such as cell division are not involved. That de novo synthesis of PAL occurs in these hormone-stimulated events is also evidenced by the inhibitory effect of the transcription and translation inhibitors on PAL induction.

In a sense, xylem and lignin are the hallmark of Tracheophyta. The use of GAs as a regulatory molecule (hormone), may be unique to vascular plants [2, 3]. The control of the first enzyme (PAL) involved in the biosynthesis of an unique product (lignin) during the final differentiation of an unique cell type (xylem) by a hormone (GA) unique to vascular plants may not be just fortuitous, but perhaps has significance in vascular plant evolution.

4.3 Auxin-Promoted Elongation Growth and Inhibition of Lignification

Auxin (NAA) inhibited the BA + GA_3-promoted induction of PAL and lignification. The continued elongation growth in those segments whose growth is responsive to auxin was apparently the reason for the inhibition, since growth-promoting FC was similarly able to inhibit PAL induction. In vivo, PAL induction and lignification occurred only after the cessation of elongation, and at least one reason for the deferral of lignification in the putative secondary xylem cells appears to be due to the continued elongation of these cells. Auxin-promotion of xylem differentiation, so commonly observed in other plant systems, apparently occurs at the early cell division or the "determination" stage. In final stages of the process auxin is inhibitory to xylem differentiation, apparently by promoting elongation.

4.4 Absence of Influence of Ethrel and ABA on PAL-Induction by BA + GA_3

Ethrel and ABA had no influence on PAL activity by themselves, nor on PAL activity induced by BA + GA_3. *Cuscuta* is reported to be ABA rich [8] and possibly ABA insensitive, since growth processes such as elongation or cytokinin-induced haustoria formation are relatively insensitive to exogenous ABA [20, 21].

4.5 Absence of Enhanced Amylase Activity During (BA + GA)-Induced Lignification

Since new secondary cell wall formation accompanies lignin deposition during secondary xylem differentiation, the possibility exists that GA_3 or BA + GA_3 enhanced lignification by enhancing amylase activity for starch mobilization. Preliminary experiments (Table 6) showed, contrary to expectation, that all the hormones GA_3, BA and NAA, singly or in combination, actually suppressed the enormous (56-fold increase) α-amylase induction that occurs in untreated control segments. The increased amylase activity seen in control segments after 48 h was

Table 6. Influence of hormones and cycloheximide on α-amylase activity in segments from 60–120 mm region

Treatment	α-Amylase activity mg starch hydrolyzed/gfw/30 min		
	0 h	24 h	48 h
Control	2.9	61.6	168.9
BA	–	6.8	16.9
GA_3	–	11.6	19.9
NAA	–	5.2	31.0
ABA	–	69.0	86.6
BA + GA_3	–	8.0	6.6
BA + GA_3 + NAA	–	9.9	2.6
CHI	–	2.3	–

Values average of 3 (0 h) or 2 (24 h, 48 h) experiments.
Concentrations: BA 50 μM; GA_3 100 μM; NAA 100 μM; ABA 100 μM; CHI 5 μg/ml.
α-Amylase activity was determined essentially by the procedure of Smith et al. [27]. Crude enzyme was a centrifuged tissue homogenate in 20 mM $CaCl_2$.

apparently due to de novo synthesis (e.g. it was completely inhibited by cycloheximide). Also, ABA was ineffective in suppressing amylase synthesis, at least during the initial 24 h. Nor was amylase activity stimulated over the initial level by BA, GA_3 or BA + GA_3, although it was somewhat enhanced by NAA. The super-induction of amylase in the control segments may be related to osmotic control, or to reserve mobilization of sugars to apical regions in detached vines – a control mechanism with obvious survival value in this literally 'infectious' parasite.

5 Conclusions

Secondary xylem in *Cuscuta* is not cambium derived, but arises from a ring of cells 'determined' early in development. Final differentiation (including lignification) of this putative xylem occurs after cessation of elongation growth, and is accompanied by an increase of PAL activity. PAL activity declines in the centre but increases in the cut ends in segments from the 60–120-mm 'prelignification' region. This region is still auxin-responsive with regard to cell elongation growth. PAL in the centres of such segments could be reinduced by BA, by GA_3 or by BA + GA_3 in an additive manner. However only GA_3 (or better still, BA + GA_3) induced near-normal lignification and secondary xylem formation. NAA inhibited both the PAL induction and lignification that is induced by GA_3 or BA + GA_3, apparently by sustaining elongation growth, since FC (which promoted growth) was similarly inhibitory to PAL induction. The GA_3 or BA + GA_3 effect did not have a component of increased starch mobilization by the induction of amylase. Rather, GA_3, BA, NAA (or their combination) actually suppressed the rapid induction of amylase normally observe in excised control segments. Thus, during xylem differentiation the role of native GAs, probably in association with cytokinins, may be in the final stages when both the secondary cell wall and lignin are deposited.

Acknowledgements. This research was supported by the Department of Science and Technology, Government of India, Grants HSC/848/80 and (11)4/84-STPII. Fusicoccin was a gift of Dr. Michielli, Montedison, S.p.A., Milan. We thank Ms. Veena for expert technical assistance.

References

1. Aloni R (1987) Annu Rev Plant Physiol 38:179
2. Bopp M (this volume)
3. Bopp M, Gerhauser D, Kessler U (1986) In: Bopp M (ed) Plant growth substances 1985. Springer, Berlin Heidelberg New York Tokyo, p 263
4. Cowles JR, Scheld HW, Lemay R, Peterson C (1984) Ann Bot 54:33
5. Fukuda H, Komamine A (1982) Planta 155:423
6. Gross GG (1980) Adv Bot Res 8:26
7. Haddon L, Northcote DH (1976) J Cell Sci 20:47
8. Ihl B, Jacob F (1980) Biochem Physiol Pflanz 175:611
9. Jacobs WP (1984) In: Scott TK (ed) Encyclopedia of plant physiol: hormonal regulation of development II (NS), vol 10. Springer, Berlin Heidelberg New York Tokyo, p 149
10. Mahadevan S (1983) New Sci 98:164
11. Maheshwari R, Shailini C, Veluthambi K, Mahadevan S (1980) Plant Physiol 65:186
12. Miller AR, Crawford DL, Roberts LW (1985) J Exp Bot 36:110
13. Minocha SC, Halperin W (1974) Planta 116:319
14. Nagaiah K, Kumar SA, Mahadevan S (1977) Phytochemistry 16:667
15. Paliyath G, Rajagopal I, Unnikrishnan PO, Mahadevan S (1989) J Plant Growth Regul 8:19
16. Pearce D, Raymond Miller A, Roberts LW, Pharis RP (1987) Plant Physiol 84:1121
17. Phillips R (1980) Int Rev Cytol Suppl 11A:55
18. Phillips R (1987) Ann Bot 59:245
19. Rajagopal I, Ramasubramanian TS, Paliyath G, Mahadevan S (1988) J Plant Growth Regul 7:121
20. Rajput BS (1987) Ph D Dissertation, University of Indore, India
21. Ramasubramanian TS, Paliyath G, Rajagopal I, Maheshwari R, Mahadevan S (1988) J Plant Growth Regul 7:133
22. Roberts LW (1976) Cytodifferentiation in plants. Cambridge University Press, Cambridge
23. Roberts LW, Miller AR (1982) What's New Plant Physiol 13:13
24. Sachs T (1981) Adv Bot Res 9:151
25. Savidge RA, Wareing PF (1981) In: Barnett JR (ed) Xylem cell development. Kent-Castle House, p 192
26. Shinniger TL (1979) Annu Rev Plant Physiol 30:313
27. Smith MA, Jacobsen JV, Kende H (1987) Planta 172:114
28. Tanaka Y, Uritani I (1974) Plant Cell Physiol 15:843
29. Tsivion Y (1979) In: Musselman LJ, Worsham AD, Eplee RE (eds) Proc Second Symp Parasitic weeds, North Carolina St Univ, Raleigh, p 296
30. Wareing PF, Haney CEA, Digby J (1964) In: Zimmerman MH (ed) The formation of wood in forest trees. Academic Press, New York, p 323

The Regulation of Tumor Morphology in Crown Gall

W.L. PENGELLY[1], L.-Y. SU[1], and B.R. CAMPELL[1,2]

1 Introduction

Crown-gall tumors in higher plants are caused by the soil bacterium *Agrobacterium tumefaciens* [25]. The response of plants to *Agrobacterium* infection can vary from small swellings to large tumor masses, and in some cases tumor development includes the formation of supernumerary roots or shoots [for a review see 12]. Variation in tumor response has been shown to depend on both the strain of *Agrobacterium* and the inherent reactivity of the host [10].

The role of *Agrobacterium* in tumorigenesis is known to involve the transfer of oncogenes from tumor-inducing (Ti) plasmids to the host genome (for reviews see [6, 18]). Two of these oncogenes, *tms1* (tryptophan monooxygenase) and *tms2* (amidohydrolase), code for IAA biosynthesis [15, 24, 27, 28], and a third gene, *tmr* (isopentenyl transferase), codes for cytokinin biosynthesis [2, 4, 13]. Tumor growth and morphology in many plants are affected by mutations in *tms* or *tmr* [14, 19, 22], and correlate well with the auxin/cytokinin balance in the tissue [1, 3, 23, 29].

On the other hand, many plant species do not require a complete set of hormone oncogenes for rapid tumor growth and will show a fully virulent tumor response when infected by *Agrobacterium* strains containing mutant oncogenes [6, 7, 19]. How plants compensate for defective hormone oncogenes is not known, but the phenomenon is of great interest because of its relevance to host functions controlling growth and development.

In this chapter we describe studies of crown-gall tumors of *Nicotiana tabacum* and *N. glutinosa*, two closely related species which differ in their capacity to compensate for mutant *tms* genes. We provide evidence that compensation for defective *tms* genes is related neither to auxin production nor to the auxin receptivity of the cells.

2 Compensation for *tms* Mutations in *N. glutinosa*

Transformation of *N. glutinosa* by the *tms*-mutant octopine strain A66 gives rise to rapidly growing, unorganized tumors which are indistinguishable in the plant and in culture from tumors induced by the wild-type strain A6 [7]. Binns et al. [8] further

[1] Department of Chemical and Biological Sciences, Oregon Graduate Center, 19600 NW Von Neumann Drive, Beaverton, OR 97006-1999, USA

[2] Present Address: Department of Biology, Case Western Reserve University, Cleveland, OH 44106, USA

showed that transformation of this species with a DNA fragment containing only the *tmr* and nopaline synthase (*nos*) genes from the nopaline plasmid pTiT37 induced hormone-independent tumors which produced auxin in culture. Since opine synthesis is not related to tumorigenesis, these results indicate that transformation by *tmr* is sufficient to induce auxin synthesis in the host. Interestingly, this effect could not be mimicked by feeding cytokinin, and non-transformed *N. glutinosa* cells required both auxin and cytokinin to proliferate in culture [8].

In earlier studies of ethylene biosynthesis [16], we obtained evidence inconsistent with the view that *N. glutinosa* compensates for mutant *tms* genes by increased auxin production. We showed that levels of the ethylene precursor, ACC, were about 50-fold higher in A6-transformed cells than in cells transformed by the *tms*-mutant A66, a result in accord with the well-known induction of ACC synthesis by auxin. Results of ACC analysis were quantitatively similar for *N. glutinosa* and the non-compensating *N. tabacum* and *Lycopersicon esculentum*, indicating that A66-transformed *N. glutinosa* did not accumulate auxin.

Direct measurements of IAA by RIA further support this view. Table 1 shows that levels of IAA and ACC in *N. glutinosa* and *N. tabacum* are affected similarly by *tms* genes. Thus differences in the capacity of these species to compensate for mutant *tms* genes do not result for differences in the accumulation of IAA, and some mechanism other than auxin biosynthesis appears to be involved.

Table 1. IAA and ACC concentrations in A6- and A66-transformed cell lines of *N. tabacum* and *N. glutinosa* after 14 days in culture.

Cell line	Origin	Transforming strain	IAA [pmol (g fw)$^{-1}$]	ACC [nmol (g fw)$^{-1}$]
NgluA6-35	*N. glutinosa*	A6	457	227
NgluA6B-5	*N. glutinosa*	A6	1142	58
TA6-5	*N. tabacum*	A6	439	190
NgluA66-3	*N. glutinosa*	A66	19	2
NgluA66B-2	*N. glutinosa*	A66	52	7
TA66C3-78	*N. tabacum*	A66	40	2

3 Compensation for *tms* Mutations in *N. tabacum*

Nicotiana tabacum shows an attenuated response to *tms* mutant strains. This weakened reaction is characterized by a slower rate of tumor development and by the formation of shoots (teratomas). Cloned teratoma lines are hormone independent in culture [11], but hormone autonomy of the teratoma is dependent on shoot formation [7, 30]. It appears, therefore, that compensation for *tms* mutations in *N. tabacum* teratomas occurs via growth factors, possibly auxin, produced in developing shoots.

We have isolated, from an A66-transformed teratoma line of *N. tabacum*, variants which no longer required shoots to show hormone-independent growth in culture [21]. These variant lines, TA66-D1 and TA66-D2, grew rapidly as unorganized, friable tissues on hormone-free medium and were highly sensitive to auxin

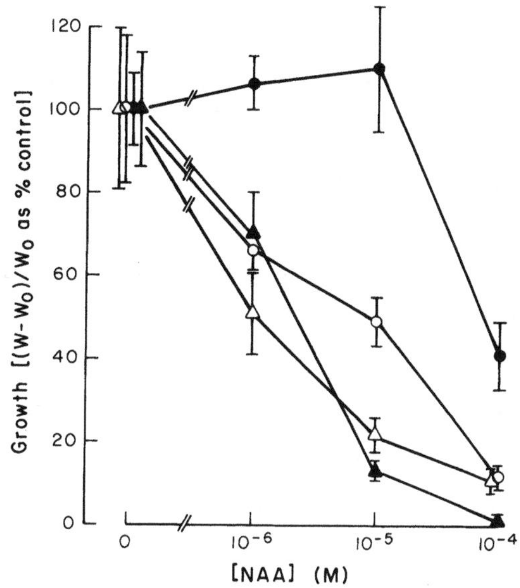

Fig. 1. Dose response of crown-gall tumor lines TA6-5 (○), TA66C3-78, (●), TA66-D1 (△), and TA66-D2 (▲) to NAA. *Bars* 1 SE$\bar{x}$ (n = 5). Data from [21]

feeding (Fig. 1). The sensitivity of the variants to auxin was similar to cells transformed by strain A6 (line TA6-5), whereas the parental teratoma line (TA66C3-78) was relatively resistant to auxin by comparison.

The morphology and auxin sensitivity of the variant lines suggest increased auxin production, but Southern blot analysis of DNA isolated from TA66-D1 showed that the *tms* locus was still mutated [21]. Although the variants mimicked the phenotype of the A6-transformed line TA6-5, RIA analysis showed that IAA levels in the variants were much lower than in TA6-5 and similar to IAA levels in the teratoma line TA66C3-78 (Table 2). The variants and the teratoma line also had similar ACC contents, which were much lower than ACC levels in TA6-5; and ACC levels in the variants increased markedly with auxin treatment [21]. Thus, like *N. glutinosa*, these variant lines of *N. tabacum* compensated for a mutant *tms* locus by some mechanism other than increased auxin accumulation.

Table 2. The IAA concentration of cultured crown-gall tumor tissues of tobacco [from 21]

Days in Culture	IAA [pmol (g fw)$^{-1}$)]			
	TA6-5	TA66C3-78	TA66-D1	TA66-D2
6	1190	42	45	30
14	439	40	54	15
23	151	38	3.3	3.0

4 Auxin Autonomy and Auxin Sensitivity Can Be Separate Phenomena

With further cloning experiments, we isolated several more cell lines from TA66C3-78 which did not form shoots and which grew rapidly as friable tissues on hormone-free medium. Like the aforementioned variants, these new cell lines contained low IAA levels (Campell and Pengelly, in preparation); but unlike the earlier variants, they were resistant to auxin feeding (Table 3). Even cell lines which grew very rapidly (e.g., TA66B-27) showed increased growth rates with auxin treatment, a result in striking contrast to the severe growth inhibition of TA66-D2 with auxin treatment.

Table 3. The effect of 10 μM NAA on the growth of crown-gall teratoma line TA66C3-78 and several unorganized variants isolated from it. Growth was measured after 21 days in culture

Cell line	Growth [W − Wo)/Wo][a]		Ratio (+NAA/-NAA)
	-NAA	+NAA	
TA66C3-78	10.1 ± 3.7[b]	64.0 ± 21.9	6.3
TA66-D2	45.7 ± 6.8	1.5 ± 0.3	0.03
TA66B-27	94.4 ± 13.4	170 ± 22.0	1.8
TA66B-54	38.4 ± 4.3	81.9 ± 13.6	2.1
TA66B-64	12.1 ± 2.9	199 ± 12.0	16.4

[a] W and Wo are the final and initial fresh weights of the explant, respectively.
[b] Mean values expressed ± 1 SE$\bar{x}$ (n = 9).

These results show that auxin autonomy, indicated by rapid and unorganized growth on hormone-free medium, is not necessarily linked to the auxin sensitivity of the tissue and, hence, is a separate phenomenon. This has important implications with regard to the molecular mechanisms underlying phenotypic switching, particularly with respect to auxin receptors. If phenotypic switching resulted from a change in the auxin receptor, such as the appearance of a receptor of high-binding affinity, then one might expect autonomy and sensitivity to be linked, assuming that both phenomena are mediated by receptor occupancy. Our ability to distinguish autonomy from sensitivity implies that auxin receptors are not responsible for phenotypic switching and provide an opportunity to study these phenomena separately.

5 The Physiological Basis for Auxin Autonomy in Tobacco Teratomas

Hormone-independent growth of cultured crown-gall teratomas of tobacco depends on shoot formation [7, 30]. This hormone requirement is demonstrated by the fact that small, unorganized pieces taken from the complex teratoma fail to grow when transferred to hormone-free medium [20]. We found that these unorganized tissue pieces can be rescued and maintained indefinitely as friable tissues by subculture on growth medium containing with either auxin or GA_3. Neither auxin nor GA_3 induced a permanent change in the cells, and tissues would revert progressively to the teratoma phenotype with serial subculture on hormone-free medium.

Tobacco tumor cells of teratoma origin maintained as unorganized tissues on GA_3 medium contained low levels of IAA (10–20 pmol/g fw) and ACC (5–10 nmol/g fw) comparable to the shooty teratoma (cf. Table 1), and GA_3-grown cells also remained resistant to auxin feeding [20]. Studies with non-transformed tobacco pith cells showed that GA_3 could not replace auxin as a growth factor, but rather acted synergistically with auxin. Thus, pith tissues, which grew poorly with suboptimal auxin supplements, would proliferate rapidly if GA_3 was also provided [20]. Therefore, endogenous GAs might account for hormone autonomy in tumor cells containing low auxin and may be the shoot-derived factor important in the growth of teratomas. This hypothesis is supported by the recent studies of Nakagawa et al. [17] who found higher endogenous GA levels in tobacco teratomas than in unorganized tumors. We are currently investigating the possibility that GA production supports hormone autonomy in our variant lines.

6 The Physiological Basis for Auxin Sensitivity in Crown Gall

Plant growth generally shows a biphasic dose response to applied auxin with a phase of increasing growth promotion followed by a phase of progressive growth inhibition. Dramatic differences among tissues in their dose response might be explained by the fact that some tissues, such as certain tumors, already contain optimal or supraoptimal auxin concentrations. Any further addition of auxin to these cells will only inhibit growth, whereas the growth of cells containing lower auxin levels may be promoted by the same applied dose. However, our results show that marked changes in the auxin dose response can occur without a detectable change in the auxin content of the cells. This suggests that the responsiveness of cells to auxin must play a role as well.

In an attempt to explain the biphasic response in terms of auxin receptors, early workers proposed a two-site attachment which resulted in self-inhibition of binding at higher auxin concentrations [9]. This proposal was immediately criticized, since histological examination of tissues showed that high auxin doses were obviously toxic [5]. Thus, cells with similar growth rates but on opposite sides of the dose-response curve are in quite different physiological states. It is not known why applied auxins become toxic at higher concentrations.

One well-known response of plants to auxin is increased ethylene biosynthesis (for a review see [31]). Studies of ethylene biosynthesis in A6- and A66-transformed cell lines of several species showed that many tumor lines, particularly unorganized tumor lines, did not respond to growth-inhibiting concentrations of auxin with increased rates of ethylene production [16]. We also found that supplementing the growth medium with the ethylene-releasing agent, ethephon, was not effective at inhibiting the growth of our tumor lines (Table 4). In these experiments, the rate of ethylene evolution by cultures provided the highest ethephon dose (1 mM) ex-

Table 4. The effect of ethephon on the growth of crown-gall cell lines after 21 days of culture

Tumor line	Growth [(W – Wo)/Wo][a]			
	Ethephon (mM)			
	0	0.01	0.1	1
TA6–5	78.7 ± 3.4[b]	101 ± 10.2	80.0 ± 9.6	69.8 ± 4.2
TA66C3–78	18.5 ± 2.4	26.4 ± 4.6	24.6 ± 3.6	13.0 ± 1.3
TA66-D1	36.4 ± 5.9	50.0 ± 11.2	56.6 ± 10.3	38.6 ± 3.5
TA66-D2	44.7 ± 10.5	39.8 ± 8.7	68.9 ± 14.7	49.4 ± 5.3

[a] W and Wo represent the final and initial fresh weights of the explant, respectively.
[b] Mean values expressed ± 1 S.E.$\bar{x}$ (n = 9).

ceeded that which could be obtained by feeding tissues auxin or ACC. Therefore, inhibition of growth by auxin was not mediated by ethylene.

Although ethylene production is not increased by auxin in many tumor lines, we found that auxin markedly increased ACC levels in all tumor lines examined so far [16, 20, 21]. In contrast to ethephon feeding, treatment with ACC proved to be inhibitory to growth (Fig. 2A). Growth of the auxin-sensitive line TA6–5 and the two variant lines TA66-D1 and TA66-D2 was progressively inhibited by ACC at concentrations ranging from 0.01 to 1.0 mM, whereas growth of the auxin-resistant teratoma TA66C3–78 proved to be more resistant to ACC and was inhibited only at the highest dose. Similar but somewhat weaker growth responses were obtained with the ACC analogue α-aminoisobutyric acid (AIB), which does not serve as an ethylene precursor, providing further evidence that growth inhibition is not mediated by ethylene (Fig. 2B). Since growth-inhibiting concentrations of applied ACC were in the same range as endogenous ACC levels in auxin-treated tissues, these results suggest that ACC mediates growth inhibition by auxin.

We also found that auxin-sensitive variants accumulated higher levels of ACC in response to auxin than did the auxin-resistant teratoma [21]. Analysis of the accumulation of MACC showed this difference to result from variation in the

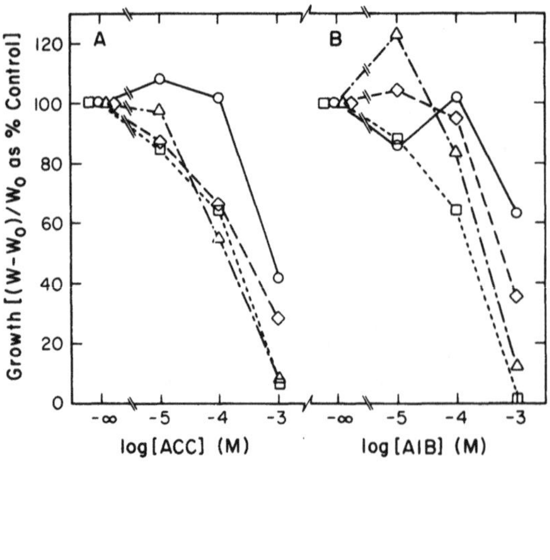

Fig. 2A,B. The effects of ACC (**A**) and AIB (**B**) on the growth of crown-gall tumor lines TA6–5 (□), TA66C3–78 (○), TA66-D1 (△), and TA66-D2 (◊)

Table 5. The effect of 1 μM NAA on ACC and MACC contents of tobacco crown-gall tumor lines after 6 days in culture

Tumor line	ACC [nmol (g fw)$^{-1}$]		MACC [nmol (g fw)$^{-1}$]		Total (ACC + MACC)		MACC as % of total	
	−NAA	+NAA	−NAA	+NAA	−NAA	+NAA	−NAA	+NAA
TA6-5	191 ± 31	330 ± 73[a]	104 ± 12	81 ± 10	295	411	35	20
TA66C3-78	5.5 ± 0.9	17 ± 0.3	19.2 ± 8.2	155 ± 52	24.7	172	78	90
TA66-D1	12.5 ± 3.0	184 ± 38	11.5 ± 2.8	99 ± 46	24.0	283	46	51
TA66-D2	2.4 ± 1.2	86 ± 32	12.9 ± 5.2	58 ± 10	15.3	144	84	40

[a] Mean values expressed ± S.E.$\bar{x}$ (n = 3).

capacity of the cell lines to produce MACC (Table 5). Although the teratoma line accumulated less ACC in response to auxin treatment than did the auxin-sensitive variants, it accumulated more MACC; and levels of total ACC (ACC + MACC) were similar in all A66-transformed cell lines. Since the effect of auxin is to induce ACC synthesis, the similar accumulation of total ACC in response to auxin indicates that auxin receptivity is similar in these cell lines as well.

How ACC acts to inhibit growth is not known. One site which might mediate ACC toxicity is ACC-malonyltransferase itself. Su et al. [26] provided evidence indicating that ACC and D-amino acids were malonylated by the same enzyme activity in mung bean extracts. Hence, saturation of malonyltransferase activity by ACC might lead to the accumulation of toxic amounts of D-amino acids. We are currently examining our cell lines for D-amino acid accumulation and sensitivity.

7 Conclusions

Both *N. tabacum* and *N. glutinosa* can compensate for mutant *tms* genes of *Agrobacterium tumefaciens* and grow rapidly as friable and unorganized tissues on hormone-free growth medium. In neither species does compensation result from increased IAA accumulation, and low levels of ACC in A66-transformed cells provide further evidence that auxin activity remains low. Hence, auxin autonomy of cultured crown-gall cells is not necessarily related to the auxin content of the tissue.

Analysis of several clones of A66-transformed *N. tabacum* showed that cells exhibiting a high degree of auxin autonomy could be either auxin sensitive or auxin resistant. Thus, auxin autonomy and auxin sensitivity can be separate phenomena and are likely controlled by different mechanisms. Gibberellin A_3 can substitute for auxin and support the growth of auxin-dependent teratoma cells, and therefore GAs may be the shoot-derived growth factors accounting for hormone independence of complex teratomas in culture.

Auxin sensitivity, on the other hand, appears to be mediated by ACC, but not by ethylene. Thus, ACC may play a role in plant development independent from its role in ethylene biosynthesis. The sensitivity of cell lines to ACC was related to their capacity for ACC malonylation, and competition for malonyltransferase by ACC may lead to the accumulation of D-amino acids. Inhibition of growth by ACC may provide a general mechanism for auxin toxicity in plants.

Acknowledgements. Support from the National Science Foundation, DMB-8417087, and the U.S. Department of Agriculture, USDA-86-CRCR-1-2150, is gratefully acknowledged.

References

1. Akiyoshi DE, Morris RO, Hinz R, Mischke BS, Kosuge T, Garfinkel DJ, Nester EW (1983) Proc Natl Acad Sci USA 80:407
2. Akiyoshi DE, Klee H, Amasino RM, Nester EW, Gordon MP (1984) Proc Natl Acad Sci USA 81:5994
3. Amasino RM, Miller CO (1982) Plant Physiol 69:389
4. Barry GF, Rogers SG, Fraley RT, Brand L (1984) Proc Natl Acad Sci USA 81:4776
5. Bennet-Clark TA (1956) In: Wain RL, Wightman F (eds) The chemistry and mode of action of plant growth substances. Academic Press, New York, p 310
6. Binns AN (1984) In: Miflin BJ (ed) Oxford surveys of plant molecular and cell biology. Clarendon, Oxford, p 133
7. Binns AN, Sciaky D, Wood HN (1982) Cell 31:605
8. Binns AN, Labriola J, Black RC (1987) Planta 171:539
9. Bonner J, Foster RJ (1956) In: Wain RL, Wightman F (eds) The chemistry and mode of action of plant growth substances. Academic Press, New York, p 295
10. Braun AC (1953) Bot Gaz 114:363
11. Braun AC (1959) Proc Natl Acad Sci USA 45:932
12. Braun AC, Stonier T (1958) Protoplasmatologia 10(5a):1
13. Buchmann I, Marner FJ, Schröder G, Waffenschmidt S, Schröder J (1985) EMBO J 4:853
14. Garfinkel DJ, Simpson RB, Ream LW, White FF, Gordon MP, Nester EW (1981) Cell 27:143
15. Follin A, Inzé D, Budar F, Genetello C, Van Montagu M, Schell J (1985) Mol Gen Genet 201:178
16. Miller AR, Pengelly WL (1984) Planta 161:418
17. Nakagawa S, Tjokrokusumo DS, Sakurai A, Yamaguchi I, Takahashi N, Syono K (1987) Plant Cell Physiol 28:485
18. Nester EW, Gordon MP, Amasino RM, Yanofsky MF (1984) Annu Rev Plant Physiol 35:387
19. Ooms G, Hooykaas PJJ, Moolenaar G, Schilperoort RA (1981) Gene 14:33
20. Pengelly WL, Su LY (1988) In: Kutáček M, Bandurski RS, Krekule J (eds) Physiology and biochemistry of auxins in plants. Academia, Prague, p 401
21. Pengelly WL, Vijayaraghavan SJ, Sciaky D (1986) Planta 169:454
22. Ream LW, Gordon MP, Nester EW (1983) Proc Natl Acad Sci USA 80:1660
23. Rüdelsheim P, Prinsen E, van Lijsebettens M, Inzé D, van Montagu M, De Greef J, Van Onckelen H (1987) Plant Cell Physiol 28:475
24. Schröder G, Waffenschmidt S, Weiler EW, Schröder J (1984) Eur J Biochem 138:387
25. Smith EF, Townsend CO (1907) Science 25:671
26. Su LY, Liu Y, Yang SF (1985) Phytochemistry 24:1141
27. Thomashow LS, Reeves S, Thomashow MF (1984) Proc Natl Acad Sci USA 81:5071
28. Thomashow MF, Hugly S, Buchholz WG, Thomashow LS (1986) Science 231:616
29. Van Onckelen H, Rüdelsheim P, Hermans R, Horemans S, Messens E, Hernalsteens JP, Van Montagu M, De Greef J (1984) Plant Cell Physiol 25:1017
30. Van Slogteren GMS, Hoge JHC, Hooykaas PJJ, Schilperoort RA (1983) Plant Mol Biol 2:321
31. Yang SF, Hoffman NE (1984) Annu Rev Plant Physiol 35:155

Cytokinin: Evidence for Spatial Control of Signal Transduction

M.J. Saunders[1]

1 Introduction

1.1 Cytokinin Signal Transduction

Although it is well established that in animal cells Ca^{2+} plays a role in signal transduction leading to the triggering and or mediation of cell division [see 14], we have limited understanding of how plant cells translate hormonal signals into developmental change [see 10–12, 15, 21]. Several models have been proposed to explain stimulus-response coupling in both plants and animals [15, 17, 20]. These models all consist of a sequential four-component system that include (1) perception of the stimulus by a receptor, (2) activation of an intracellular second messenger, (3) cascade and amplification of the message, and (4) physiological response.

If we look at cytokinin stimulation of cell division as fitting this generalized model [as proposed in 9, 10, 20, 21], we can see that there is supporting evidence for all components of the model. Firstly, several cytokinin binding proteins have been isolated although their relationship to physiological response is not well characterized [see 12, 18]. Secondly, there is increasing evidence that Ca^{2+} plays a role as a second messenger in plant signal transduction in general [see 10, 15], and in cytokinin-mediated events in particular [9, 20]. Thirdly, there is also evidence for a cellular response system which may include a calmodulin cascade system in response to a cytokinin-stimulated Ca^{2+} message [8, 24], phosphatidylinositol (PI) breakdown [4, 21], and protein phosphorylation [16, 21].

However, just the identification of the elements of a signalling system in plants is not sufficient to lead us to an understanding of plant growth and development. The temporal and spatial control of these elements may be the important regulatory control points in plant morphogenesis as it relates to the establishment of cell polarity, asymmetrical cell division, directed cell expansion or cytoplasmic microdomains. The ability to perceive positional information is a central tenet for modeling plant growth [28]. We must understand how hormonal signals in plants can give rise to a morphologically asymmetrical response in the absence of an asymmetrically applied or oriented signal. One way to explain a spatially controlled response to a general signal is that the cells are pre-polarized with respect to that signal (i.e., that receptors or binding proteins have an asymmetrical distribution on or in target cells), thereby limiting the response to either specific cells or to regions

[1] Biology Department, University of South Florida, Tampa, FL 33620, USA

of a particular cell type. To explore the relationship between cytokinin receptor distribution and second messenger activation, I have used a model system that exhibits an asymmetrical hormonal response (i.e., cytokinin-induced bud formation in the moss *Funaria hygrometrica*).

1.2 Model for Spatial Control of Signal Transduction

Funaria grows as a filamentous mat of cells composed of three cell types: (1) actively dividing tip cells, (2) chloronema cells with large chloroplasts and transverse cross walls, and (3) caulonema cells with small chloroplasts and oblique cross walls. The latter are the target cells for bud induction by exogenous cytokinin. Previous research has established that cytokinins induce a change in the morphological symmetry and polarity in caulonema cells of *Funaria* that is first evident approximately 12 h after treatment [2, 6, 22]. The mechanism leading to these cytological rearrangements is not well understood, although a gradient in free Ca^{2+} as a result of differential activation of plasma membrane ion channels has been proposed to effect these changes [19]. The zone of greatest inward current, detectable with a vibrating microelectrode, shifts from the nuclear region to the distal end of target cells within a few minutes after cytokinin treatment [19]. There are three additional lines of evidence that suggest that it is Ca^{2+} uptake from the extracellular milieu that mediates the cellular response to cytokinins. (1). Bud formation can be stimulated by the Ca^{2+} ionophore A23187 [23], and the Ca^{2+} channel agonists CG P28392 and (+)202-791 [5]. (2). Bud formation is inhibited by the competitive Ca^{2+} uptake inhibitors, lanthanum and gadolinium, and the Ca^{2+} channel blockers D 600, verapamil [24], nifedipine and (–)202-791 [5]. (3). Bud formation is correlated with an increase in chlortetracycline fluorescence [22] and an inward Ca^{2+} current [19]. However, TMB-8, an intracellular Ca^{2+} antagonist, has also been shown to inhibit bud formation [24], implying that intracellular Ca^{2+} stores may be released (or alternatively, that TMB-8 somehow affects extracellular Ca^{2+} uptake or intracellular Ca^{2+} utilization).

Since chloronema cells do not divide in the presence of cytokinin it may be that the two cell types may differ in cytokinin binding and/or Ca^{2+} uptake properties. To explore these possibilities, I have immunocytochemically localized cytokinin binding on chloronema and caulonema cells using an antibody to benzyladenine (BA) riboside. In addition, I have physically separated the two cell types and monitored Ca^{2+} uptake in the presence and absence of hormone using the extracellularly localized metallochromic indicator arsenazo III. This dye changes its spectral characteristics upon binding Ca^{2+} [25] and has been used to measure Ca^{2+} transport in isolated mitochondria [1], skeletal muscle [29], sarcoplasmic reticulum [3] and intact chloroplasts [13]. One of the advantages of the use of arsenazo III is its high affinity toward Ca^{2+}, which results in large absorbance changes following relatively small [Ca^{2+}] changes, making possible measurements of Ca^{2+} in the nM range.

The results presented here indicate that chloronema and caulonema cells differ in both cytokinin binding properties and Ca^{2+} uptake after hormone treatment. These cells are preprogrammed to respond asymmetrically to a cytokinin signal.

2 Material and Methods

2.1 Culture of Protonemata

Protonemata of *F. hygrometrica* Hedw. were grown as previously described [19–24] except that 1 μM NAA is added to the growth medium to promote the formation of caulonema cells. Ten-to 14-day-old protonemata (cultured on cellophane on agar-solidified growth medium) were used.

2.2 Indirect Immunofluorescence

Polyclonal antibodies to BA riboside linked to bovine serum albumin were raised in female white New Zealand rabbits, purified [as described in 31], screened for activity by immunodiffusion against BA and further purified by affinity chromatography against BA riboside linked to Sepharose 6B (Pharmacia) [30].

Protonemata were processed for immunocytochemistry using the technique described earlier for *Physocomitrella patens* [7]. Briefly, protonemata are fixed in 2% formaldehyde in 0.1 M phosphate buffer (pH 7.2), attached to glutaraldehyde-derivatized glass coverslips, wall material enzymatically digested in 2% Driselase (Sigma), and permeabilized and extracted in 1% Triton X-100 and 5% DMSO.

Protonemata were incubated in 10 μM BA riboside and then antibenzyladenine riboside antibodies were applied at 5 μg/ml in phosphate buffered saline (PBS) for 60 min followed by FITC-conjugated anti-rabbit IgG, diluted 1:300, for 60 min. The coverslips were rinsed in PBS before mounting in 10% glycerol and observed using Nikon epifluorescence optics; photographs were taken using Tri-X film at ASA 400.

2.3 Controls

There is only a small amount of autofluorescence (primarily from unextracted chloroplasts) from cells incubated in either primary antibody or secondary antibody when added alone, or when the pretreatment with hormone is eliminated.

2.4 Calcium Uptake Studies

The mature protonemata were separated into the two cell types by teasing away the peripheral caulonemata from the central chloronemata using sterile forceps. They were then divided into four groups (two of each type) of equivalent fw (0.02–0.05 g). The four sets were then placed into 15 ml of dye solution [50 μM arsenazo III, 20 mM HEPES (pH 6.8), 100 mM KCl, and 3% sucrose] and equilibrated for 45 min. The resulting [Ca^{2+}] (carried over from the growth medium or released by the cells) was monitored spectrophotometrically. The cells were transferred to 5 ml fresh dye solution (to lower extracellular Ca^{2+} to within dye sensitivity levels), equilibrated for 15 min and remeasured. The cell suspension were transferred to four disposable chromatography columns (Bio-rad) equipped with stopcocks. The volume in the tubes reduced to 1 ml and 1 ml of the initial effluent was used for background measurements. Ten μl of a BA stock solution (100 μM in H_2O) were added to one chloronemata and one caulonemata suspension and 10 μl of H_2O) to the other samples. Equivalent amounts were added to 1 ml of initial effluent to determine the effects of BA or H_2O on the dye solution itself. The dye solution was drained from around the filaments at different time points and measured spectrophotometrically. The solutions were then poured back into the appropriate column and equilibrated before the next time point.

Absolute and differential absorption spectra were obtained on a Hewlett-Packard photodiode array spectrophotometer. Difference spectra were generated by subtracting the absorbance spectra of the test solution from the initial effluent and then subtracting the absorbance spectra of the solution surrounding the cells containing BA from its counterpart which did not contain BA. Difference spectra generated for (1) the dye solution with and without BA and (2) the initial effluent over time, show no absorbance differences. In addition, spectra were run of dye solution with known [Ca^{2+}] to generate a standard curve.

3 Results

3.1 Indirect Immunofluorescence

Chloronema cells exhibit slight surface fluorescence after immunocytochemical processing to localize BA binding (Fig. 1a). In contrast, caulonema cells exhibit a zone of bright fluorescence at the distal end of these cells and a slight surface fluorescence along the rest of the cell (Fig. 1b). The fluorescent region is correlated with the presumptive bud site on these target cells.

3.2 Ca^{2+} Uptake Studies

A standard curve was generated treating the dye solution with increasing concentrations of Ca^{2+} from 10 to 200 μM (Fig. 2). The wavelengths of interest are: (1) 600 nm and 658 nm where the maximum absorbance increase takes place in the presence of Ca^{2+}, (2) 572 nm, an isosbestic point where no spectral changes take place as a result of [Ca^{2+}] changes, and (3) 710–800 nm which is used as a baseline because the dye does not absorb at these wavelengths.

Difference spectra of the medium surrounding chloronemata and caulonemata with and without BA were taken at several time points after addition of BA (Figs. 3, 4). The first detectable change in the absorption spectra is visible 2 h after BA treatment of caulonema cells and indicate that Ca^{2+} is being taken up from the medium. This depletion of extracellular Ca^{2+} continues for up to 5 h. In contrast there are only slight differences in the absorption spectra of chloronema cells after BA addition as compared to controls.

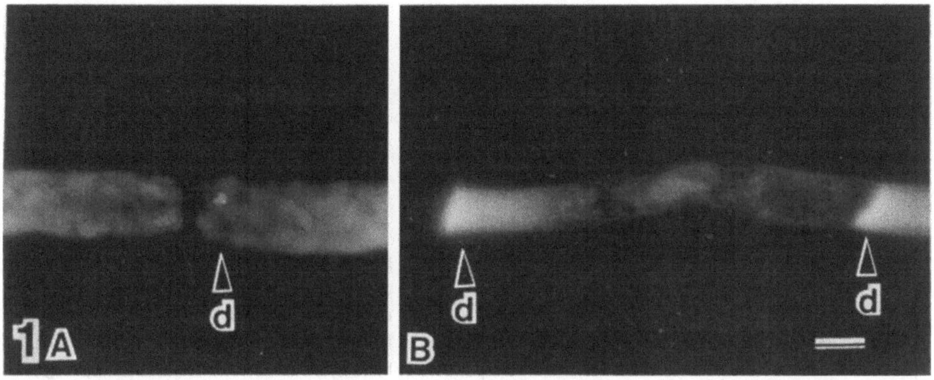

Fig. 1A,B. Fluorescence micrographs of *Funaria* processed for immunocytochemical localization of BA. **A** Chloronema cells exhibit little fluorescence along the length of the cell or at the distal (*d*) end. There is some punctate autofluorescence from chloroplasts. **B** In contrast, target caulonema cells have bright fluorescence at the distal (*d*) end, at the presumptive bud site. *Bar* 50 μm

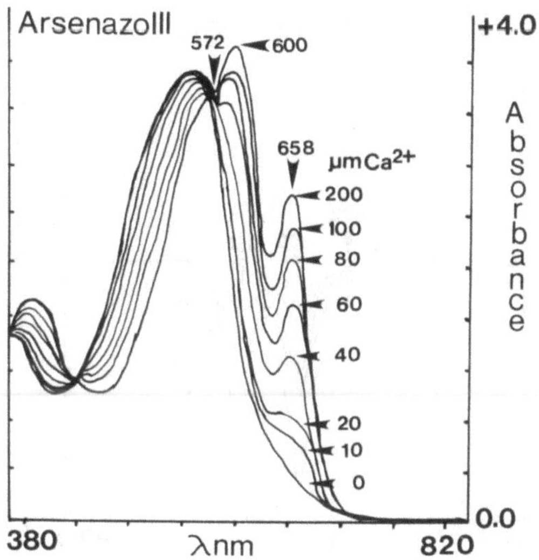

Fig. 2. Absorbance spectra of 50 μM arsenazo III dye solution with and without added Ca^{2+}. Note absorbance increases at 600 and 658 nm with increasing [Ca^{2+}], and isosbestic point at 572 nm (*arrows*)

4 Discussion

The mode and sites of action of cytokinins are not well understood, although cytokinin-binding proteins have been isolated and characterized from several higher plants [see 11, 12, 18], and there is an increasing body of evidence that implicates a rise in intracellular [Ca^{2+}] as one part of the cytokinin signal transduction chain [8, 10, 20]. It has been suggested that cytokinin stimulates asymmetrical cell division of caulonema cells in *Funaria* by activating or concentrating Ca^{2+} channels at the presumptive bud site [19]. This model requires that there be either (1) differential activation of randomized cytokinin receptors, (2) a

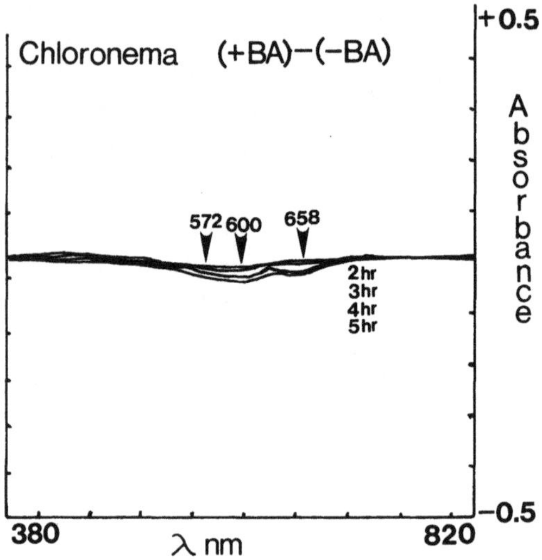

Fig. 3. Differential absorption spectra of arsenazo III solution surrounding chloronema cells incubated in BA minus solution surrounding chloronema cells with no added BA over time. Note only slight differences in the absorption at 600 and 658 nm as compared to isosbestic point at 572 nm (arrows) indicating that the [Ca^{2+}] in both samples is similar and does not change over time

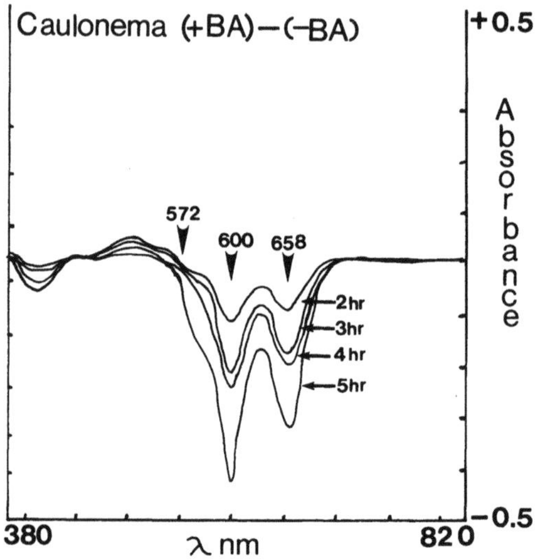

Fig. 4. Differential absorption spectra of arsenazo III solution surrounding caulonema cells incubated in BA minus solution surrounding caulonema cells with no BA added over time. Note decrease in absorption at 600 and 658 nm as compared to the isosbestic point at 572 nm (*arrows*) indicating an uptake of Ca^{2+} by cells incubated in BA that is greatest at 5 h after BA treatment

localization of receptors after binding, or (3) preprogrammed asymmetrical distribution of receptors or ion channels on target cells. Previous research has established that ion channels appear to be initially concentrated at the nuclear region on caulonema cells and migrate to the presumptive division site after cytokinin treatment [10], indicating that these cells are not preprogrammed to respond asymmetrically to a cytokinin signal by the distribution of ion channels in the plasma membrane of quiescent cells. There is also little or no evidence of a cytoplasmic gradient in caulonema cells before cytokinin treatment [2, 6, 26], in contrast to tip cells which exhibit a distinctive polar organization [27, 28].

The results presented here indicate that the asymmetrical response may be a result of preferential binding of cytokinins to the distal end of target caulonema cells and subsequent Ca^{2+} uptake by these cells. Localized uptake of cytokinin by caulonema cells of *Funaria* has previously been demonstrated by Brandes and Kende [2] using [^{14}C] BA. They showed that after a 12-h incubation, very little label was detectable in chloronema and tip cells, whereas caulonema and bud initial cells are heavily labeled. The subcellular localization of the labeled cytokinin appears to be at the distal end of caulonema cells as the initial bud cell develops. The conclusion from both studies is that target caulonema cells have specific cytokinin-binding sites that are lacking in chloronema cells. It also appears as if these binding sites have a specific subcellular distribution that is spatially related to the subsequent morphogenesis of caulonema cells.

The results presented here also indicate that Ca^{2+} uptake by target cells can be detected by spectral changes in arsenazo III 2 hr after BA treatment. If there are earlier increases in intracellular [Ca^{2+}] before that time, they may come from internal stores (which is indicated by the TMB-8 studies [24]). An alternative explanation is that the initial Ca^{2+} uptake is undetected by the dye (which may be saturated until the extracellular Ca^{2+} falls below a detectable level).

5 Conclusion

These data support the hypothesis that cytokinin exerts at least part of its effect by stimulating Ca^{2+} uptake by responsive cells. It can be hypothesized that the polarized response of cytoplasmic rearrangement and asymmetrical division that occurs in *Funaria* caulonema cells after cytokinin treatment is a result of localized cytokinin binding to the distal end of these cells. This may have direct results on specific ion channels leading to a spatially controlled change in intracellular Ca^{2+}. Localization of receptors may be a general mechanism to establish subcellular positional information, which is essential for plant morphogenesis.

Acknowledgements. Supported by National Science Foundation Grants PCM84-8496, DCB88-2011; and United States Department of Agriculture Competitive Grant 8701001.

References

1. Bernardi P, Azzone GF (1982) FEBS Lett 139:13
2. Brandes H, Kende H (1968) Plant Physiol 43:827
3. Chiu WCK, Haynes DH (1977) Biophys J 18:3
4. Conrad PA, Hepler PK (1986) Plant Physiol 80:60
5. Conrad PA, Hepler PK (1988) Plant Physiol 86:684
6. Conrad PA, Steucek GL, Hepler PK (1986) Protoplasma 131:211
7. Doonan JH, Cove DJ, Lloyd CW (1985) J Cell Sci 75:131
8. Elliott DC (1983) Plant Physiol 72:215
9. Elliott DC (1986) In: Trewavas AJ (ed) Molecular and cellular aspects of calcium in plant development. Plenum, New York, p 285
10. Hepler PK, Wayne RO (1985) Annu Rev Plant Physiol 36:397
11. Kende H, Gardner G (1976) Annu Rev Plant Physiol 27:267

12. Klambt D (1981) In: Guern J, Peaud-Lenoel C (eds) Metabolism and molecular activities of cytokinins. Springer, Berlin Heidelberg New York, p 172
13. Kreimer G, Melkonian M, Latzko E (1985) FEBS Lett 180:253
14. Morrill GA, Kostellow AB (1986) In: Cheung WY (ed) Calcium and cell function, vol VI. Academic Press, New York, p 209
15. Poovaiah BW, Reddy ASN (1987) Crit Rev Plant Sci 6:47
16. Ralph RK, Buillivant S, Wojcik SJ (1976) Bichem Biophys Acta 421:319
17. Rasmussen H, Barrett PQ (1984) Physiol Rev 64:938
18. Romanov GA, Taran VY, Chvijka L, Kulevȧ ON (1988) J Plant Growth Regul 7:1
19. Saunders MJ (1986) Planta 167:402
20. Saunders MJ (1986) In: Trewavas AJ (ed) Molecular and cellular aspects of calcium in plant development. Plenum, New York, p 188
21. Saunders MJ (1990) In: Chopra RN, Bhatla SC (eds) Bryophytes: physiology and biochemistry (in press)
22. Saunders MJ, Hepler PK (1981) Planta 152:272
23. Saunders MJ, Hepler PK (1982) Science 217:943
24. Saunders MJ, Hepler PK (1983) Dev Biol 99:41
25. Scarpa A (1979) Methods Enzymol 56:301
26. Schmiedel G, Schnepf E (1979) Protoplasma 100:367
27. Schmiedel G, Schneph E (1980) Planta 147:405
28. Schnepf E (1986) Annu Rev Plant Physiol 37:23
29. Vergara J, Delay M (1985) Cell Calcium 6:119
30. Vretblad P (1976) Biochem Biophys Acta 434:169
31. Zavala ME, Brandon DL (1983) J Cell Biol 97:1235

Plant Wound Signals and Translation

E. DAVIES[1]

1 Introduction: Significance and Difficulties of Wound Research

In Nature, plants are subjected to wound stress by biotic agents such as large herbivores, insects, and fungi and abiotic agents such as hail and wind. Wound stress caused by such agents may be complicated by other factors such as loss of photosynthetic tissue, production of elicitors especially fragments cleaved from host or pathogen cell walls, inoculation with viruses; the role of wounding as a stress in its own right is rarely considered.

In the laboratory, in efforts designed to remove hormone-producing regions or to enhance uptake of labelled or other exogenous substances, plants are punctured, abraded or excised, the latter furnishing stem segments, leaf discs, and storage tissue slices [4]. It does not necessarily follow that results from such wounded tissues need duplicate responses which occur in the intact plant. In addition, protocols requiring intracellular recordings of, for instance, action potentials, must be avoided. Otherwise the act of measurement changes the response being measured – a situation not unknown to quantum physicists.

2 Early Studies with Auxin in Decapitated, yet Otherwise Intact Plants

Our original focus was on auxins and how they elicit growth in intact plants. Accordingly, we employed a semi-intact pea epicotyl system in which the plumule and hook were cut off to remove meristematic cells and the major supply of endogenous auxin, and the cut stump was painted with lanolin containing various additives. Our earlier findings with this system showing that auxin induces cellulase synthesis in vivo and in vitro have been reported [12].

3 Later Studies with Auxin on Aged, Intact Plants

We became aware that this system, even though almost intact, suffered from being wounded immediately prior to receiving auxin treatment. Anticipating by about 10 years the objections raised by Hanson and Trewavas [11] to using wounded tissues to study hormone responses, we modified our system such that after the plumule and hook were excised, the tissue was allowed to age for 3 days, partly to remove all traces of IAA and the growth system it modulated, and partly to permit recovery

[1] School of Biological Sciences, University of Nebraska, 348 Manten Hall, Lincoln, NE 68588-0118, USA

from the wound. We were then able to ask the question "How does auxin stimulate long-term growth in an almost intact, non-growing system in which there is no fast growth response?" This was in contrast to the question being asked by many others, i.e., "How does auxin stimulate a transient change in growth rate in excised, abraded (wounded) stem tissue floating in solution and already growing in response to the hormone?" Using this aged sytem we found that IAA treatment stimulates the formation of free polysomes (FP) [6], membranebound polysomes (MBP) [3], and cellulase activity [7] prior to any measurable stimulation of growth.

4 More Recent Studies on Wound Responses in Aged, Intact Plants

Unfortunately, the very property of the above tissue most necessary for our auxin studies (its integrity) was itself an impediment to measurement of processes such as protein synthesis in vivo, since labelled amino acids were not taken up through the cut apex. In order to facilitate uptake into the tissue, therefore, apical segments from aged epicotyls were excised and placed with their bases in solution.

4.1 The Existence of Wound Signals

Surprisingly, this treatment (excision of the apex) led to polysome formation which was greater 1 h after excision than 10 h after auxin treatment of the intact plant [8]. In efforts to circumvent this wound-evoked polysome formation, we excised the tissue at a point 5 cm below the apex and were again surprised. Polysome formation in the apical 1 cm was almost as massive as in tissue wounded at the 1 cm point. In fact, polysome formation took place in the apical 1 cm within 15 min of inflicting a wound at a point 20 cm distant. This led to our realization that there was a rapidly-generated (and bidirectionally-transmitted) wound signal that could very rapidly elicit polysome formation in distant tissue [8, 18].

Numerous physical, chemical and electrochemical wound signals have been described [4], with the two major candidates here being electrochemical (presumably action potentials) and hormonal (presumably ethylene).

4.2 Action Potentials as Wound Signals

Action potentials (APs) are generated in numerous plant tissues in response to various stresses including wounding, UV, electrical stimulation and cold treatment [5, 14]. Electrodes, gently placed against the pea epicotyl at various points do indeed detect APs in the underlying tissue when it is wounded by cutting with a razor (or any other device). Table 1 gives a summary of their properties (frequency of generation, rate of transmission, magnitude). Table 1 also shows that APs are generated and transmitted sufficiently rapidly for them to qualify as the major intercellular wound signal.

Table 1. Properties of wound-induced action potentials in plants[a]

Location of electrode	Plants responding (%)		Velocity (mm/min)		Magnitude (mV)	
	Direction of transmission (towards)					
	Base	Apex	Base	Apex	Base	Apex
30 mm Distant	87	95	77	81	35	25
60 mm Distant	53	39	45	55	31	19
90 mm Distant	33	24	41	47	22	9

[a] Etiolated 8-day-old pea epicotyls were placed in a Faraday cage and monitored for action potentials by KCl/agar impregnated surface electrodes placed at strategic locations previously moistened with 10 mM KCl for at least 20 min prior to recording. Electrical changes were recorded as differences between events occurring at the measuring electrode and a reference electrode placed at the distal end of the plant. They were monitored on Curken recorders after passing through pre-amps (specially designed by W. Pickard and furnished through B. Pickard, Washington University, St. Louis). Wounding was inflicted with a clean razor cut, similar results were obtained with glass and plastic. Data were collected over a 2-year period from more than 100 plants.
Note: Plants exhibiting no action potential were not included in the calculations for average velocity and magnitude.

We have not yet been able to prove (or disprove) conclusively that APs are the signals which evoke polysome formation. Attempts to mimic wounding electrically show that stimulations of sufficient voltage to pass the cuticle do, in fact, elicit an AP and evoke the formation of polysomes. Unfortunately, such treatments also cause severe damage and even death of the cells in the region stimulated (data not shown). Thus AP generation and polysome formation may be the result of wounding, and not of electrical stimulation directly.

As shown in Fig. 1, exposure of a 5-mm region of the epicotyl to UV 20 cm distant from the apex results in polysome formation in the apical 2 cm which is more massive than that induced by wounding. Others have also shown that UV evokes APs [5]. Furthermore, cold treatment of the middle of a stem (which we had hoped would prevent transmission of APs) did not prevent polysome formation at one end when the tissue was wounded at the other end. In fact cold treatment alone elicited polysome formation (not shown). Again, others have shown that cold treatment can elicit (not inhibit as we had originally presumed) APs [5].

Finally, numerous collaborative experiments with Pickard have failed to show a substantial correlation between the polysome content in individual apices and the magnitude of the AP each apex experienced as a result of wounding. This failure results at least in part from the large variability in the polysome content of apparently identical epicotyls: using 1-cm apical tissue, we found that the proportion of ribosomes existing as polysomes ranged from about 40–70% in unwounded tissue and from about 50–80% in wounded tissue. We now use a minimum of 10 epicotyls per treatment to obtain representative polysome extractions.

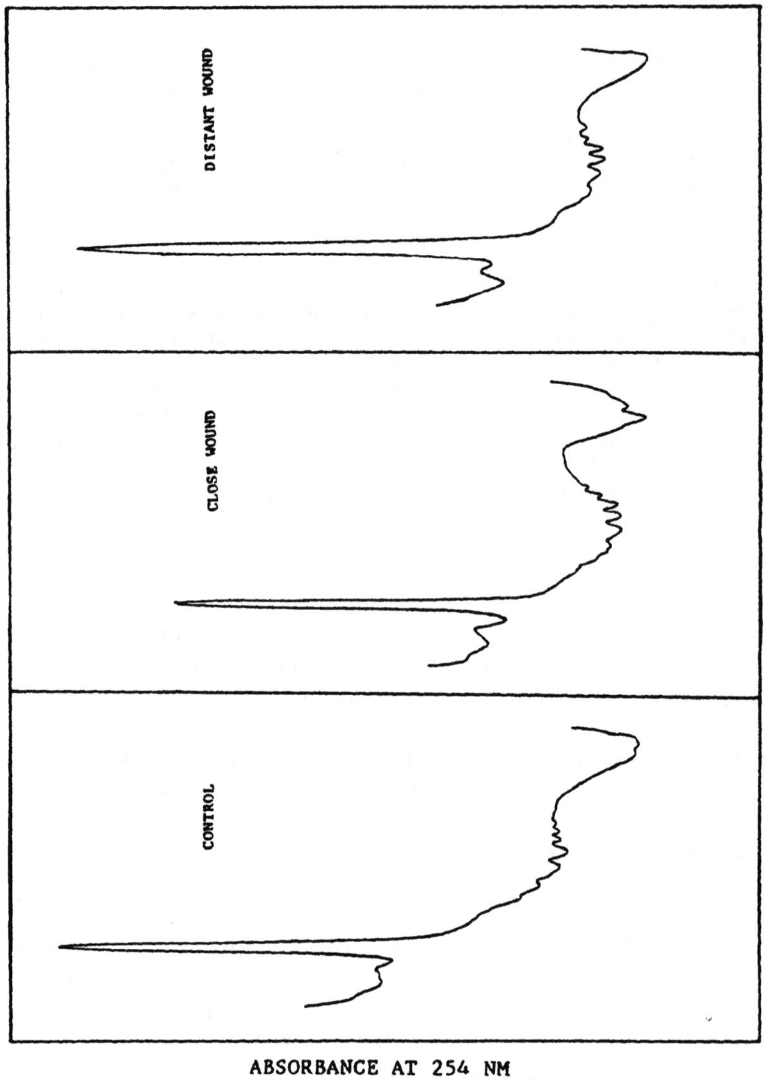
DISTANT WOUND
CLOSE WOUND
CONTROL
ABSORBANCE AT 254 NM

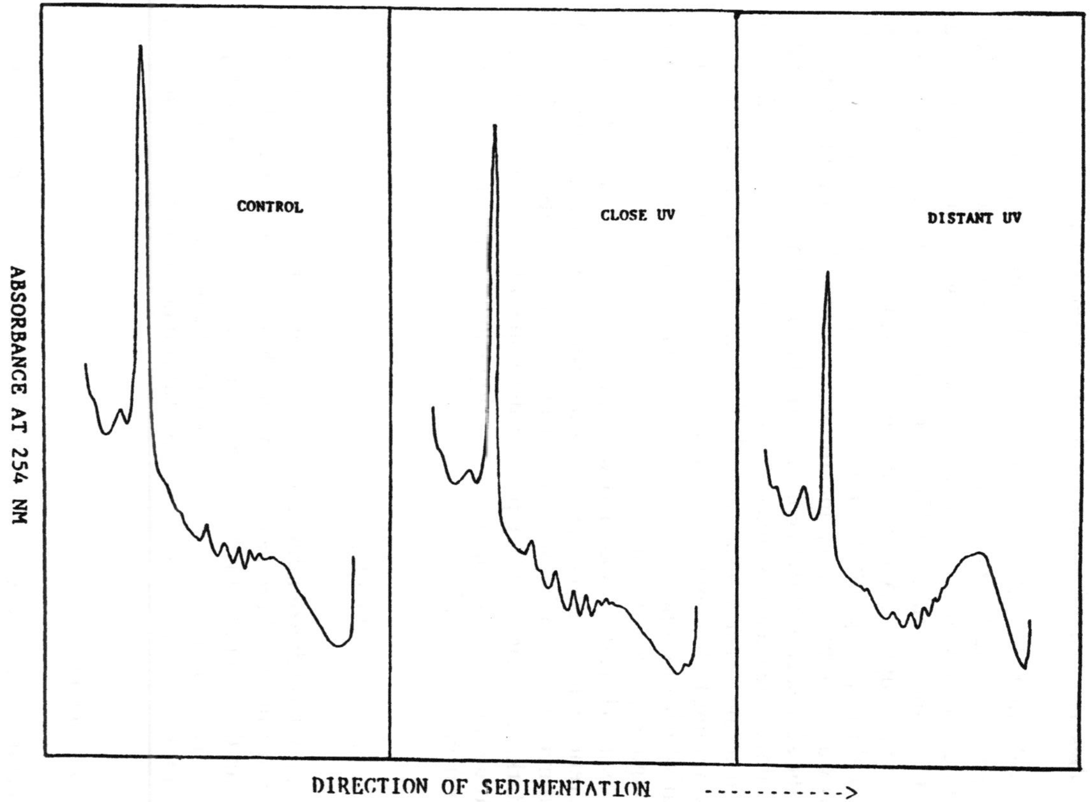

Fig. 1. Polysome formation in aged peas in response to UV irradiation and wounding. Aged pea plants were either left untreated (*control*), wounded at the tip (*close wound*), or at 20 cm (*distant wound*), or irradiated with UV for 15 min at the tip (*close UV*) or at 20 cm from the tip (*distant UV*). After 1 h, polysomes were isolated from the tip region using normal procedures [1]. A sterility lamp provided the 254 nm UV. Note: Close UV treatment caused polysome dissociation (direct effect of UV), whereas the distant UV treatment caused polysome formation (UV signal)

5 Current Studies on Wound Responses in Aged, Excised Tissue

It is highly likely that the subcomponent of an AP which evokes the major intracellular response is the AP-induced transient increase in cytosolic Ca^{2+} [5]. The intact, aged system we have described is not amenable, however, to studies looking at the role of Ca^{2+} and the effects of Ca^{2+} antagonists, since its integrity acts as a major impediment to uptake. Accordingly, we modified our experimental system to allow for uptake by pea epicotyls without the need for a newly cut surface (fresh wound). This was done by excising the apical 5 cm from the 3rd internode (or occasionally the almost entire 20 cm epicotyl) painting lanolin or vaseline over the cut apices, standing the excised segments with their bases in water or buffer, and allowing them to age for a 3–4-day period. This yielded aged tissue which responded to wounding by forming polysomes to an extent very similar to intact aged tissue [9], but which was able to take up labelled amino acids (and other compounds) through its previously cut base.

It must be stated here that, although excised, aged tissue has already undergone severe stress (etiolation, excision, aging) it is far from being incapacitated and it does respond massively to being wounded. The excised segments, after about 10 days in the dark with their bases in tap water, develop roots, and if stems bearing nodes are brought out into the light they develop leaves. Such stressed tissue is, therefore, far from being dead.

5.1 The Polysome/Protein Synthesis Paradox

Surprisingly, even though wounding causes an *increase* in polysomes in this aged, excised tissue, it causes a *decrease* of 75% within 5 min in protein synthesis [9]. Despite the fact that wounding increases ribosome initiation rate in vitro [15], in vivo it seems to act by inhibiting ribosome movement along polysomal mRNA [9] or causing dislocation of the polysomes from a functional to a non-functional site.

5.2 A Role for Calcium

Experiments employing aged, excised tissue treated at the base with Ca^{2+}, a Ca^{2+} ionophore, a Ca^{2+} chelator (ethylene glycol-bis-aminoethyl ether N,N,N′N′-tetraacetic acid – EGTA), a Ca^{++} channel blocker (Lanthanum), an inhibitor of plasma membrane ATPase (vanadate), and inhibitors of calmodulin (chlorpromazine and trifluoperazine) have yielded perplexing results. Ca^{2+}, with or without ionophore has little or no effect on unwounded tissue, whereas virtually all other treatments mimic the wound response in stimulating polysome formation and inhibiting protein synthesis in vivo (Davies, Abe and Ramaiah, unpublished results). While some of these results (with calmodulin inhibitors especially) might be explained by elevated levels of cytosolic Ca^{2+} [10] or (with vanadate) by direct inhibition of translation [16], it is not immediately obvious how these results support an AP-evoked increase in cytosolic Ca^{2+}.

Indeed, at first sight these results might be taken as grounds to disqualify a wound-induced transient increase in cytosolic Ca^{2+} as being the primary intracellular wound signal. Nevertheless, the fact that four agents (wounding, electric shock, UV and cold), all evoke action potentials and also all stimulate polysome formation provides correlative evidence for a role for APs in modulating translation. An explanation does exist, however, for why exogenous chemical treatments do not yield anticipated results. The effects of a substance (Ca^{2+}) whose concentration changes transiently in vivo during an AP may not be duplicated satisfactorily by chemical treatments which need to be of extended duration so as to permit sufficient uptake through a limited surface area.

6 A Role for Ethylene as the Wound Signal

6.1 Kinetics of Synthesis and Response

Numerous authors have shown that wounding stimulates ethylene evolution [21]. Data in Fig. 2, obtained in studies using dark-grown Alaska pea epicotyls, are based partly on ethylene measurements reported by others [17] and partly on polysome analyses shown here. These studies show that in apical tissue, wounding does not evoke an increase in ethylene until at least 25 min have elapsed [17], whereas polysome formation begins in 15 min, even in tissue 15 cm distant from the wound [8]. It is hard to imagine a response being governed by a signal which has not yet been generated (let alone transmitted). In addition, in basal tissue wounding does not elevate ethylene above the very low ambient levels, yet polysome formation is as massive in basal tissue as it is in apical tissue. Again, it is difficult to imagine a response governed by a signal which is not generated.

6.2 Effects of Ethylene on Polysome Formation

Ethylene gas (1000 ppm in helium), supplied to 5-cm-long excised, aged segments by injecting the appropriate volume into an air-tight container to furnish final concentrations ranging from 1–100 μl/l has little or no effect on polysome formation in the upper half (apical) or lower half (basal) region of unwounded tissue, and only a slight effect on the apical half of tissue wounded at the base (Table 2). Ethylene supplements the effects of wounding in the wounded basal region, however, perhaps because it is taken up more readily through the wound.

6.3 Effects of Endogenous Wound-Induced Gaseous Hormones and Ethylene Inhibitors on Polysome Formation

Incubation of unwounded tissue in a closed container in the presence of several pieces of wounded tissue does not cause polysome formation, whereas treatment with the gaseous inhibitor of ethylene action, norbornadiene does (not shown). This

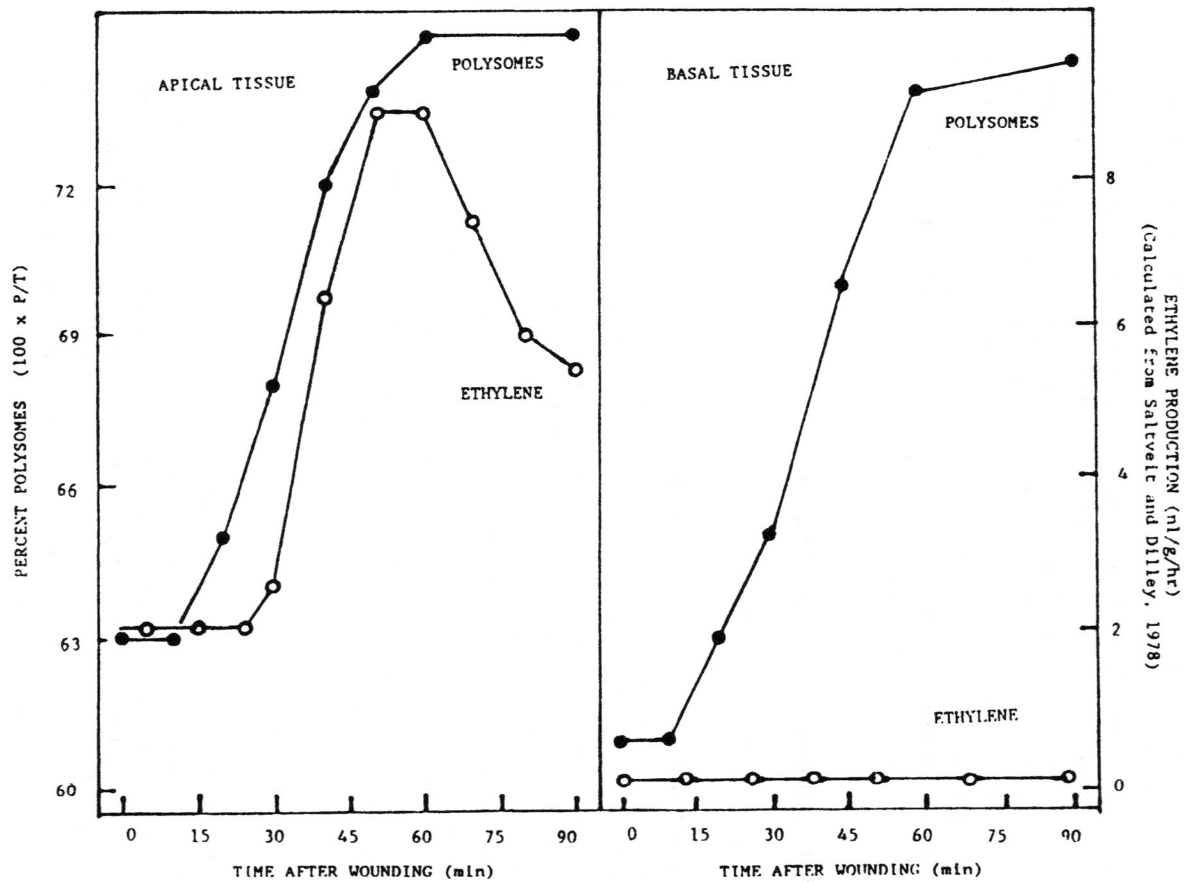

Fig. 2. Time course for wound-induced ethylene evolution and polyribosome formation in apical and basal regions of etiolated Alaska pea epicotyls. Polyribosomes were isolated by published methods [1] from apical and basal regions of aged pea epicotyls wounded for the periods indicated. Data on ethylene evolution was recalculated from [17]

Table 2. Effect of ethylene on polysome formation in aged, excised pea epicotyls[a]

Tissue and region	Polyribosome content (100 × P/T)			
	Ethylene treatment (μl/l)			
	0	1	10	100
Unwounded, upper half	51	48	51	49
Unwounded, lower half	55	58	57	56
Wounded, upper half	63	66	64	66
Wounded, lower half	68	71	70	77

[a] Aged, excised epicotyls 5-cm-long were either left intact (unwounded) or wounded by trimming 1–2 mm from the base and oriented with their bases in water in an enclosed 2.5 l chamber, into which ethylene was injected to the concentrations indicated. After incubation for 1 h the segments were quickly cut into an upper lower half and frozen in liquid N_2 prior to polysome extraction by normal methods [1]. "T" denotes the total area under the profile, "P" denotes the area under the polysome region. Values are from one typical experiment. Duplicate samples differed in polysome content by 3% or less.

substance stimulates polysome formation both in unwounded tissue (with or without wounded pieces added) and also in wounded tissue. These results could be taken to imply that ethylene is not the polysome-forming wound signal, indeed ethylene might even be inhibitory to this signal. Again, as with APs, the results to date are equivocal with regard to identifying ethylene as the major wound signal.

7 Alternative Approaches

7.1 The Existence of Cytoskeleton-Bound Polysomes in Plants

If Ca^{++} is involved in wound-induced changes in polysome function, what processes might it modulate? Among its many functions in animal tissues it causes destabilization of the cytoskeleton. In many animal tissues the cytoskeleton appears to be a major subcellular site for polysome attachment and disturbing this attachment disrupts polysome function, i.e., inhibits protein synthesis [13]. Such an occurrence could explain the rapid, massive inhibition of protein synthesis that occurs upon wounding.

In spite of the vast amount of research performed on cytoskeleton-bound polysomes (CBP) in animals, there are few reports showing CBP in plants [e.g., 2]. We have begun developing methods for the isolation of a fraction biochemically-definable as CBP. Figure 3 shows that pea epicotyls do contain a fraction of polysomes (FP) that are not released by grinding buffer alone, nor by non-ionic detergents at up to 2% concentration (MBP), but are released to a certain extent by amiprophosmethyl (APM), a disrupter of the microtubule system, more so by cytochalasin B (CB), a disrupter of actin filaments, and even more so by protease K. We have also noted that there is a transient decrease in the population of "CBP"

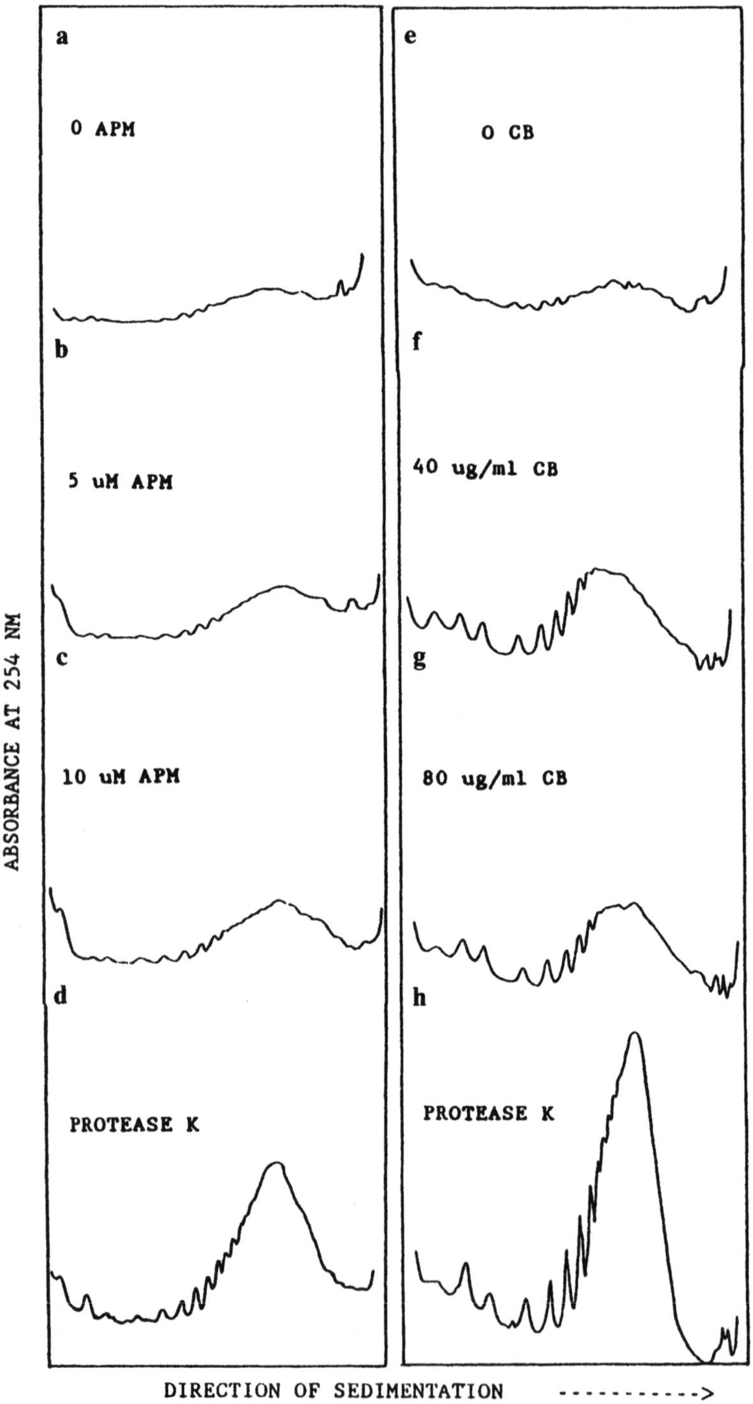
a
0 APM
b
5 uM APM
c
10 uM APM
d
PROTEASE K
e
0 CB
f
40 ug/ml CB
g
80 ug/ml CB
h
PROTEASE K
ABSORBANCE AT 254 NM
DIRECTION OF SEDIMENTATION ---------->

after wounding (data not shown). However, we can not yet claim that this is meaningful in terms of the drastic inhibition of protein synthesis which occurs after wounding, since the CBP seem to comprise only 1–2% of the total polysome population when typical extraction buffers are used.

Recent experiments, however, (data not shown), indicate that when the ionic strength of the extraction buffers is decreased to about 1/10th of the normal level, the proportion of detergent-stable, protease K-releasable polysomes (i.e., CBP) increases to 10–15% of the total polysomes.

7.2 A Role for Cytoskeleton-Bound Polysomes in the Wound Response

Cytoplasmic streaming is a Ca^{++}-modulated process which requires a functional cytoskeleton and which can be visualized in many aquatic plants, such as *Elodea* and *Vallisneria* [19]. Both of these organisms exhibit an increase in polysomes and a decrease in protein synthesis within 15–30 min of wounding in a manner similar to (but less rapid than) that occurring in pea epicotyls. Accordingly, studies on the effects of wounding, CB, and APM on cytoplasmic streaming and protein synthesis have been performed and results of a typical experiment are shown in Table 3. Cytoplasmic streaming occurs in about 90% of the control cells throughout the 60-min period, streaming is reduced to about 50% after wounding and by APM and to less than 15% by CB. These treatments also reduce protein synthesis, although the

Table 3. Effect of wounding and cytoskeleton disrupters on cytoplasmic streaming and protein synthesis in *Elodea*[a]

Incubation conditions	Time after treatment (min) 0–10	10–20	20–30	30–40	40–60
	Number of cells streaming (%)				
Intact	70	75	95	95	90
Wounded	40	30	35	45	45
Amiprophosmethyl	45	25	45	45	50
Cytochalasin B	15	10	5	0	0

[a] Leaves, excised from *Elodea* plants grown in the greenhouse, were incubated under normal laboratory conditions for 2 days in artificial pond water and random regions were then examined for streaming under the light microscope [19]. Similar leaves were incubated in [^{35}S]-methionine (25 μCi/ml) for 1 h and values for protein synthesis (100 × incorporation/uptake) were: intact, 38%; wounded, 16%; amiprophosmethyl-treated, 24%; and cytochalasin B-treated, 17%

Fig. 3a-h. Release of detergent-insoluble polysomes by cytoskeleton disrupting agents. Apical pea epicotyls were ground to release free and membrane-bound polysomes [1]. The insoluble debris was washed in the same buffer and then re-extracted in the same buffer containing: **a** no additions (0 APM); **b** 5 μM amiprophosmethly (APM); **c** 10 μM APM; **d** 50 μg/ml protease K, and, in a separate experiment; **e** no additions (0 CB); **f** 40 μg/ml cytochalasin B (CB), **g** 80 μg/ml CB; **h** 50 μg/ml protease K, and clarified by spinning in a microfuge and the supernatant analyzed for released polysomes inhibition of cytoplasmic streaming does not correspond exactly to the inhibition of protein synthesis. It must be noted that the inhibition of cytoplasmic streaming evoked by wounding was less reproducible and occurred over a much narrower range of external Ca^{++} concentrations than did inhibition induced by APM or CB.

While it is premature to define a role for CBP in plants, especially on wound-induced changes in their function, it is tempting to speculate that the wound-evoked transient increase in cytoplasmic Ca^{++} could cause a temporary disruption of the cytoskeleton such that the associated CBP exhibit transient malfunction.

It is of interest, and perhaps some significance, that the transient effect of auxin on both cytoplasmic streaming [20] and protein synthesis is opposite to the effects of wounding. This perhaps indicates that the cytoskeleton may play a role in auxin responses as well as in wounding. If so, this would be an excellent reason for not using wounded tissue to study auxin responses.

References

1. Abe S, Davies E (1986) Plant Cell Physiol 26:1499
2. Cox G, Vesk M, Juniper B (1986) Nord J Bot 6:641
3. Davies E (1976) Plant Physiol 57:516
4. Davies E (1987a) Biochem Plants 12:243
5. Davies E (1987b) Plant Cell Environ 10:623
6. Davies E, Larkins BA (1973) Plant Physiol 52:339
7. Davies E, Ozbay O (1980) Z Pflanz 99:461
8. Davies E, Schuster A (1981) Proc Natl Acad Sci USA 78:2422
9. Davies E, Ramaiah KVA, Abe S (1986) Plant Cell Physiol 27:1377
10. Gilroy S, Hughes WA, Trewavas AJ (1987) FEBS Lett 212:133
11. Hanson JB, Trewavas AJ (1982) New Phytol 90:1
12. Maclachlan GA, Davies E, Fan DF (1968) In: Wightman F, Setterfield G (eds) Biochemistry and physiology of plant growth substances. Runge, Ottawa, p 443
13. Ornelles DA, Fey EG, Penman S (1986) Mol Cell Biol 6:1650
14. Pickard BG (1973) Bot Rev 39:172
15. Ramaiah KVA, Davies E (1985) Plant Cell Physiol 26:1223
16. Ranu RS (1983) Proc Natl Acad Sci USA 80:3148
17. Saltveit ME Jr, Dilley DR (1978) Plant Physiol 61:447
18. Schuster A, Davies E (1983) Plant Physiol 73:817
19. Takagi S, Nagai R (1986) J Cell Sci 62:385
20. Thimann KV, Sweeney BM (1937) J Gen Physiol 21:123
21. Yang SF, Hoffmann NE (1984) Annu Rev Plant Physiol 35:155

The Effect of Growth Regulators on the Winter Survival of Winter Wheat

L.V. Gusta, B.J. O'Connor, and M.J.T. Reaney[1]

1 Introduction

Insufficient winter hardiness is the main factor limiting the expansion of winter wheat on the northern great plains. The winter wheat cultivars grown today are of similar cold hardiness as the Crimean cultivars introduced from Russia in the late 1880s. The most cold hardy winter wheat cultivar grown today contains the major cold hardiness genes available in the winter wheat gene pool [6]. Thus, little or no progress has been made in enhancing the genetic potential of cold hardiness in winter wheat.

The winter hardiness of winter wheat is not a static entity but varies depending upon the cultivar, cultural practices and environment. In most years, cold hardy winter wheat cultivars do not attain their full hardiness potential or do not maintain a maximum level of freezing tolerance for an extended period of time. The development of cold hardiness and its maintenance is controlled by both inducible genes and regulatory genes. How the expression of these genes is controlled by both hormones and plant growth regulators is the subject of this report.

2 Development of Freezing Tolerance

A period of acclimation is required for winter cereals to attain their maximum winter hardiness potential. Depending upon environmental conditions at the time of emergence in the field, winter wheat crowns and leaves only tolerate -5 to -8 °C. However, cool fall conditions, (soil temperature cooler than 10 °C) provides the environment for winter wheat plants to acclimate. The morphology of the very hardy winter wheats changes from an upright growth habit to a prostrate growth habit. Generally, very hardy wheats, upon completion of acclimation, have very dark green narrow leaves and a water content of 2 to 2.5 g H_2O/g dry wt. in the crowns and leaves, whereas the less hardy cultivars at full hardiness are tall, upright and have a leaf and crown water content greater than 2.5 g H_2O/g dry wt. [7]. These characteristics can be used to select the more cold hardy winter wheats in the field.

The crowns and leaves of the less winter hardy cultivars can tolerate -12 to -15 °C for 1 to 2 hours after a period of 6 to 8 weeks of hardening. The crowns of very hardy winter wheats, e.g. Norstar can tolerate temperatures as low as -29 °C for 1

[1] Department of Crop Science and Plant Ecology, University of Saskatchewan, Saskatoon, Sask. S7N OWO, Canada

to 2 h if hardened under long, bright, sunny and dry conditions in the autumn. However, under wet, cloudy conditions, with a short autumn, Norstar will only acclimate to tolerate -20°C.

3 The Interaction Between ABA and GA on Freezing Tolerance

Several research groups have demonstrated that a hardiness-promoting factor(s) is(are) produced in leaves of plants exposed to a hardening environment while hardiness-inhibiting factor(s) is(are) produced under a non-hardening environment [8, 10, 12, 19, 20]. Irving and Lamphear [12] quantitated endogenous GA-like and ABA-like compounds in box elder (*Acer negundo* L.) exposed to both long days (hardiness inhibition) and short day (hardiness promotion). Leaves exposed to short days were enriched in ABA-like compounds while long-day leaves were enriched in GA-like compounds. Exogenous ABA increased the cold hardiness of box-elder leaves and behaved similarly to the natural hardiness promoter. There is considerable evidence to support the hypothesis that GA interferes with cold acclimation and/or decreases the cold hardiness of plants. This has been demonstrated both in trees [8, 12, 17] and in herbaceous plants [5, 16, 19, 23].

Conflicting results have been obtained in regard to the effect of ABA on cold acclimation. Gusta et al. [9] found ABA inhibited growth but had little or no effect on enhanced freezing resistance of winter wheat crowns if applied as a foliar spray or added to a nutrient hydroponic solution. Irving and Lamphear [12] induced *A. negundo* to harden from -15 to -21.4°C by inserting leaves into vials containing 100 mg/l of ABA. The most dramatic increases in freezing tolerance induced by ABA have been reported for potato leaves and discs [2, 3, 4] and in cell suspension cultures (*Bromus inermis*) [4]. Suspension cells treated with 7.5×10^{-5} M ABA for 4 days at 20°C could tolerate -30°C as compared to -8°C for the control cells [4]. The optimum temperature for ABA induction of cold hardiness is 25° to 30°C. Temperatures cooler than 20°C were not as effective for inducing freezing tolerance. Maximum hardiness is attained after seven days in either the light or dark. Thereafter, freezing tolerance declines either due to a depletion of nutrients, ABA, cell cycle or a combination of factors.

The relationship between ABA inducing cold hardiness and GA inhibiting cold hardiness is complex. Waldman et al. [22] hypothesized that ABA induces cold hardiness by inhibiting the synthesis of GA. These authors demonstrated that both cold acclimating conditions and exogenous ABA dramatically reduced the levels of GA-like activity in a cold hardy cultivar of alfalfa (*Medicago sativa*). Therefore, factors which alter the ratio of ABA to GA may have a profound effect on cold hardiness.

The ratio of ABA to GA can be altered by exogenous applications of GA, by anti-GA compounds such as Cycocel, and by compounds which inhibit the synthesis and degradation of ABA [1].

Reaney [18] investigated the effects of GA_3, GA_4, GA_7, GA_9 and a mixture of $GA_{4,7,9}$ on the ABA-induced cold hardening and dehardening in a suspension culture of bromegrass. Bromegrass cells treated with 7.5×10^{-5} M ABA for 7 days at 25°C could tolerate a slow freeze to -37°C. However cells treated with a mixture of ABA (7.5×10^{-5} M) and $GA_{4,7,9}$, 10^{-4} M did not cold harden and were similar in hardiness to the controls (-10°C). Individually GA_4, GA_7 and GA_9 were all equally effective at inhibiting the ABA induced freezing tolerance, however GA_3 had little effect. A mixture of $GA_{4,7,9}$ at 10°C did not effect the dehardening of bromegrass cells induced to cold harden with ABA.

4 Plant Growth Regulators and Cold Hardening

The synthetic plant growth regulators which have resulted in increased cold hardiness are mainly growth retardants. These compounds are either anti-gibberellins or are ethylene-releasing agents. Cycocel (CCC), an anti-gibberellin and ethephon, an ethylene releasing agent, have been the most effective in increasing frost tolerance. For comprehensive reviews on the effect of plant growth regulators on cold hardiness, the reader is referred to the articles by Carter and Brenner [1] and Howell and Dennis [11].

Cycocel has been shown to increase the frost tolerance of winter wheat [19], winter rape [14], alfalfa [16], cabbage [15], potato [3] and box elder [12]. In several studies, Cycocel was only effective when applied under hardening conditions, whereas a measurable degree of hardiness under non-hardening conditions has been reported for box elder [12] and winter wheat [19].

A fall foliar application of Cycocel, Terpal C, triadimefon and Mefluidide to winter wheat and barley enhanced the winter survival of winter wheat and winter barley grown in Saskatchewan. From controlled freeze tests, it was determined that the plant growth regulators retarded the loss of cold hardiness in the crowns from January to March. In Saskatchewan, winter cereals attain their maximum level of freezing tolerance from November to December, thereafter the crowns steadily lose freezing tolerance. Winter ryes, e.g. Puma, are the most cold hardy of the cereals and maintain a high level of freezing tolerance throughout the winter months. Norstar is one of the most winter hardy winter wheats. The cultivars Winalta, Sundance, and Agassiz, under good hardening conditions, can attain the same level of freezing tolerance in late fall and early winter as Norstar. However, these cultivars are less winter hardy because they lose freezing tolerance at a faster rate in January and February than Norstar. January and February are considered to be critical months because the lowest soil temperatures have been recorded during this period. The loss in freezing tolerance in the winter months is also influenced by cultural practices: late seeding, excessive fertility, sowing depth exceeding 5 cm.

Plant growth regulators were fall applied to winter wheats to determine if they would delay or arrest the loss in cold hardiness. Norstar winter wheat plants, at the five- to six-leaf stage were sprayed with three concentrations of Cycocel-Extra (2.6, 1.3 and 0.66 l/ha), Terpal C (2.6, 1.3 and 0.66 l/ha) and Mefluidide (0.04, 0.002 and 0.001 l/ha) on October 11. Plants were removed from the field at the beginning of November and stored at -4°C for further testing. The freezing tolerance of the crowns was determined in December, January, February, March and April as outlined by Gusta et al. [9] (Table 1).

At the beginning of December, the crowns of Norstar attained the maximum level of freezing tolerance (LT_{50} = -27°C). By the middle of January, the crowns

Table 1. The effect of Cycocel-Extra, Mefluidide and Terpal-C applied in October on the cold hardiness of Norstar crowns

		LT_{50} (°C)				
Treatment		07 Dec	16 Jan	20 Feb	21 Mar	18 Apr
Control		–27 a[d]	–18 b	–15 c	–11 c	–9 d
CCC[a]	H	–27 a	–22 a	–23 a	–20 a	–9 d
	M	–27 a	–20 ab	–22 a	–20 a	–16 abc
	L	–26 a	–22 a	–22 a	–20 a	–14 abc
Mef[b]	H	–27 a	–22 a	–17 bc	–14 b	–17 a
	M	–27 a	–22 a	–18 abc	–22 a	–12 bc
	L	–27 a	–22 a	–20 ab	–20 a	–17 a
Ter[c]	H	–24 a	–22 a	–22 a	–22 a	–15 ab
	M	–24 a	–20 ab	–22 a	–22 a	–10 cd
	L	–27 a	–20 ab	–22 a	–20 a	–17 a

[a] CCC-Cycocel-Extra applied as a foliar spray at 0.66, 1.31 and 2.63 l/ha.
[b] MEF-Mefluidide applied as a foliar spray at 0.004, 0.002 and 0.001 l/ha.
[c] TER-Terpal-C applied as a foliar spray at 0.66, 1.31 and 1.97 l/ha.
[d] Means followed by the same letters in the same column are not significantly different at $P = 0.05$ level according to Duncan's multiple comparison procedures.

had an (LT_{50} of –18 °C, which represents a loss of 9 °C in frost tolerance. The crowns continued to lose hardiness and by April they could only tolerate –9 °C. Plants treated with the growth regulators, Cycocel-Extra, Mefluidide and Terpal-C, had a similar level of freezing tolerance as the controls in December. However, in February, plants treated with Cycocel-Extra and Terpal C had an LT_{50} of –22 to –23 °C, in contrast to an LT_{50} of –15 °C for the controls. These results demonstrate that Norstar winter wheat has the genetic potential, under ideal conditions in the fall, to harden to a similar freezing tolerance as Puma rye. However, in contrast to Puma rye, Norstar lacks the genetic potential to regulate a high level of cold hardiness from January to March. The loss in freezing tolerance from December to March may be under hormonal control since the plant growth regulators delay this loss in hardiness.

Seven cultivars of winter wheat (Norstar, Agassiz, Norwin, Redwin, Houser, Cimmaron and Monopol; listed in descending order of hardiness) were sprayed with Bayleton, Cycocel-Extra, Mefluidide and Terpal C at the beginning of October and at the middle of October. Two rates were used for each chemical: Bayleton at 0.13 and 0.26% (1/ha); Cycocel-Extra at 0.66 and 1.3 (1/ha); Mefluidide at 0.07 and 0.13 (1/ha) and Terpal C at 0.66 and 1.31 (1/ha). The plants were tested for cold hardiness as described above. All plant growth regulators tested significantly retarded the loss of cold hardiness on the February sampling date (Table 2). Cycocel-Extra applied early at the low rate and Terpal C applied early at the high rate increased yields by 18 and 21%, respectively. In five years of testing, fall-applied growth regulators increased yields in plants exposed to a severe stress from floral induction to anthesis. Generally, if yields are high, no beneficial effect of fall

Table 2. The effect of PGR on cold hardiness and yield of seven[a] winter wheat cultivars

			December	February	May	Yield
				LT^{b}_{50} (°C)		(kg/ha)
	Control		-22.3	-20.6	-10.3	949
Bay[c]	0A[d]	L[e]	-22.1	-24.4	-12.7	1014 *
	A	H	-23.7	-21.4	-12.2	987 NS
	B	L	-21.6	-22.6	-12.0	860 NS
	B	H	-23.9	-21.9	-10.5	1071 *
CCC	A	L	-22.4	-22.1	-11.3	1123 **
	A	H	-21.7	-22.7	-11.4	916 NS
	B	L	-21.9	-23.0	-10.7	948 NS
	B	H	-22.3	-23.1	-11.1	913 NS
MEF	A	L	-22.3	-21.3	-11.1	841 NS
	A	H	-23.0	-23.3	-11.8	912 NS
	B	L	-24.0	-22.7	-10.2	1090 **
	B	H	-21.7	-20.1	- 9.6	978 NS
TER	A	L	-22.6	-22.4	-12.3	954 NS
	A	H	-24.4	-21.3	- 9.8	1149 **
	B	L	-23.7	-22.7	-11.7	985 NS
	B	H	-22.6	-22.1	-10.3	949 NS

*,**Significantly different from the control at the 0.10 and 0.05 levels of probability, respectively.
[a] Norstar, Agassiz, Norwin, Redwin, Houser, Cimmaron, Monopol
[b] LT_{50} – temperature at which 50% of the population is killed
[c] Bay-Bayleton at 0.13 (L) and 0.26 (H) l/ha.
CCC-Cycocel Extra at 0.66 (L) and 1.31 (H) l/ha.
MEF-Mefluidide at 0.07 (L) and 0.13 (H) l/ha.
TER-Terpal C at 0.66 (L) and 1.31 (H) l/ha.
[d] Date of application of PGR, Oct. 15 (A) and Oct. 22 (B).
[e] Low rate (L) and high rate (H).

Table 3. The effect of fall-applied PGR on the cold hardiness of tender and hardy winter wheat cultivars

	Hardy[a]	Tender[b]
Bayleton	**	**
Cycocel Extra	**	NS
Mefluidide	**	NS
Terpal C	**	NS

Values combined for all concentrations.
*,**Significantly different from control at the 0.10 and 0.05 levels of probability, respectively; NS, not significant.
[a] Hardy Cultivars: Norstar, Norwin, Agassiz, Redwin.
[b] Tender Cultivars: Monopol, Houser, Cimmaron.

applied growth regulators are measurable. Of the four growth regulators tested, only Bayleton had a significant effect on the hardiness of the tender and hardy cultivars (Table 3). Cycocel Extra, Mefluidide and Terpal C only affected the hardiness of the hardy cultivars.

6 Conclusions

The cold hardiness of winter wheat is strongly influenced by both gibberellins and abscisic acid. Factors which increase the gibberellin to abscisic ratio reduce the freezing tolerance of the crowns. The hardy winter wheats have the capacity to cold harden from -25 to -29°C in December but rapidly lose their hardening capacity in January and February. Synthetic growth regulators which affect gibberellin and abscisic acid levels can enhance the freezing tolerance of winter wheat.

Acknowledgements. The work of the authors was supported by a NSERC Strategic Grant G 1919 to L.V. Gusta.

References

1. Carter JV, Brenner ML (1985) In: Pharis RP, Reid DM (eds) Encyclopedia of Plant Physiology, vol XI. Springer, Berlin Heidelberg New York Tokyo, pp 418–443
2. Chen HH, Li PH, Brenner ML (1983) Plant Physiol 71:362
3. Chen P, Li PH (1976) Bot Gaz 137:105
4. Chen THH, Gusta LV (1983) Plant Physiol 73:71
5. Corns WM (1959) Can J Plant Sci 39:293
6. Fowler DB, Gusta LV (1979) Crop Sci 19:769
7. Fowler DB, Gusta LV, Tyler NJ (1981) Crop Sci 21:896
8. Fuchigami LH, Weiser CJ, Evert DR (1971) Plant Physiol 47:98
9. Gusta LV, Fowler DB, Tyler NJ (1982) Can J Bot 60:301
10. Howell GS, Weiser CJ (1970) Plant Physiol 45:390
11. Howell GS, Dennis FG Jr (1981) In: Olien CR, Smith MN (eds) Analysis and improvement of plant cold hardiness. CRC, Boca Raton, Fl, pp 175–204
12. Irving RM, Lamphear FO (1968) Plant Physiol 43:9
13. Irving RM (1969) J Am Soc Hortic Sci 94:419
14. Kasperska-Palacz A, Egierszdorff S (1972) Bot Gaz 133:355
15. Kacperska-Palacz A, Blaziak M, Wcislinska B (1969) Bot Gaz 130:213
16. Paquin R, Belzile L, Willemot C, St Pierre J-C (1976) Can J Plant Sci 56:79
17. Probesting EL Jr, Mills HH (1974) J Am Soc Hortic Sci 99:464
18. Reaney MJT, Gusta LV, Abrams SR, Robertson AJC (1989) Can J Bot 67:3640
19. Roberts DWA (1971) Can J Bot 49:705
20. Timmis R, Worrall J (1974) Can J For Res 4:229
21. Tumanov II, Trunova TI (1958) Sov Plant Physiol 5:108
22. Waldman M, Rikin A, Dorvat A, Richmond AE (1975) J Exp Bot 26:853
23. Wunsche U (1966) Naturwissenschaften 53:386

Hormonal Control of Senescence

L.D. Noodén[1], J.J. Guiamét[1], S. Singh[2], D.S. Letham[2], J. Tsuji[1], and M.J. Schneider[1]

1 Introduction

1.1 Correlative Controls

Senescence and other developmental processes are often or perhaps always under correlative control [17]. These controls appear to be mediated mainly by hormones as opposed to nutrient fluxes or other influences [6]. Determining the correlative controls is an important first step in understanding the hormonal controls of senescence, and together these will provide "handles" for analyzing the biochemistry of senescence.

Monocarpic (whole plant) senescence is among the most prominent examples of correlative control of senescence [17, 19]. In soybean at least, the primary components are the leaves (targets), the developing fruits (controllers) and the mediating influence (senescence signal).

1.2 Hormonal Controls of Senescence

At the outset, it should be noted that the hormonal controls for a particular senescence process may not be universal; some differences exist among different tissues in the same species and even among the same tissues in different species [18, 19, 21].

In general, the evidence implicating particular hormones in the endogenous regulation of senescence consists of : (a) effects of exogenous hormone treatments on senescence and (b) simple correlation of endogenous hormone levels with senescence [6, 17]. While these, especially the endogenous correlations, are useful indicators, they do not constitute proof of causality. Stronger evidence for causality can be obtained by separating the hormone target from the source, and replacement of the source by defined hormone solutions [6]. This can be done quite readily with soybean explants as described in Section 2.1 below. Other types of probes which induce parallel variation between endogenous hormone levels and senescence processes can also provide evidence for causality (or against, if nonparallel).

Senescence-regulating hormones appear to fall into two general classes, promoters and retardants. Ethylene is the best known senescence-promoting

[1] Biology Department, University of Michigan, Ann Arbor, MI 48109-1048, USA
[2] Research School of Biological Sciences, Australian National University, Canberra, ACT 2601, Australia

hormone [10], but it may not control senescence in all tissues, particularly leaves [21]. By contrast, the role of ABA is less well understood, though considerable circumstantial evidence exists to implicate it in leaf senescence [18]. However, some striking noncorrelations of ABA with senescence exist. Many of these discrepancies may be manifestations of a balance between senescence promoters and retardants. For example, young leaves may have high levels of ABA relative to older leaves, but they may also have even higher levels of senescence retardants. In any case, the role of ABA in senescence is not clear. Although a number of other naturally-occurring compounds can promote senescence under various circumstances [18], little evidence exists to implicate them in the natural control of senescence.

CKs stand out as the most important class of senescence-retarding hormones [21, 27]. While CKs may function in a wide range of tissues, they may not be universal anti-senescence hormones, for other hormones, particularly auxin and the GAs, sometimes may retard senescence instead of, or together with, CKs [18].

Our knowledge of the chemistry of the natural CKs and the methods for their measurement have advanced rapidly over the past two decades [7]. However, this has complicated the once simple picture of CK control by increasing the number of natural CKs requiring consideration. Most of the studies which correlate endogenous CK levels and senescence predate the newer, more definitive analytical methods, and many deal with unresolved mixtures [27]. Although the general principles will probably not be altered, further analysis will produce a better quantitative picture of the ways the numerous forms of CK change and a better understanding of their roles in senescence.

1.3 Hormonal Control Systems which Regulate Senescence

Although considerable data now implicate particular hormones in the control of senescence, data which demonstrate and integrate the specific hormonal control systems that function in intact plants are limited. A model for hormonal control of monocarpic senescence proposed by Woolhouse will not be considered here, because it is mostly incorrect and it is discussed elsewhere [19]. The hormonal regulation of leaf senescence as a component of monocarpic senescence in soybean is a case where we can begin to relate specific hormones to the correlative controls. At this time, the role of cytokinins (Section 2) and the nature of the senescence signal (Section 3) warrant special consideration.

2 Monocarpic Senescence: The Cytokinins

2.1 Implication of Cytokinins as Senescence-Retarding Hormones

Active roots inhibit leaf senescence and that effect is mediated mainly by CKs produced in the roots [27]. Our previous studies showed that the monocarpic senescence signal in soybean does not consist of a diversion of CK flux from the root system to the pods at the expense of the leaves [22, 23]. Nonetheless, CKs do play an important role in soybean monocarpic senescence. Experiments with podded explants (a leaf, one or more pods and a subtending stem segment) allow defined solutions to be substituted for the roots [12, 15]. Since the mode of application influences CK activity [27], it is important to apply the CKs in a manner which mimics the natural supply, as we have done with explants. In this way, the roles of different CKs in leaf senescence and pod development can be studied directly. Such experiments indicate that a decline in CK flux from the roots may not itself cause senescence of leaves in plants undergoing monocarpic senescence, but it must decrease in order for the pods to induce leaf senescence [12, 15, 19]. These observations are supported by the ability of foliar CK applications to delay leaf senescence in whole plants [8, 13, 28]. In order to reconstruct the intact plant, the data obtained from feeding CK to podded explants will be matched with the data on xylem sap CK composition in Sections 2.2–2.4.

2.2 Dose-Response Relations for Cytokinin Supplied Through the Xylem in Explants with Pods

As noted above, the explant system readily serves to analyze the effects of the CK and mineral supply from the roots on leaf senescence. The mineral mixture used approximates the elemental levels in xylem sap during early pod development [3, 24]. Minerals enhance the ability of CK to delay leaf yellowing, especially at lower CK concentrations (Table 1) [11]. Thus, the xylem sap mineral constituents need to be included along with any CKs that are to be tested for activity in the explants.

The threshold for the ZR response with minerals is about 10 nM. The thresholds appear to be slightly above 10 nM for Z and (–)DZ [23]. Above the threshold, the activity hierarchy in this assay is (±)DZR> BA>(±)DZ>ZR>Z [3] and (+)DZ>(±)DZ>Z>(–)DZ [23]. Thus, the natural dihydro isomer (–)DZ is less active than the unnatural isomer (+)DZ, and the same is probably true for (–)DZR and (+)DZR [9]. (–)DZR is probably more active here than ZR [9].

Several potential sources of artifacts can also be ruled out through the use of soybean explants. CK has little delaying effect on pod development in these explants indicating that the CK does not retard leaf senescence indirectly through action on the pods. Because (a) the explants placed on mineral medium tend to produce adventitious roots (while those cultured on water tend not to) and (b) roots

Table 1. Delay of leaf yellowing in explants[a] by cytokinin (ZR) and mineral nutrients supplied through the cut base

	Days of delay[b] in leaf yellowing due to mineral nutrients and/or cytokinin						
(ZR)	0	10^{-9} M	10^{-8} M	10^{-7} M	10^{-6} M	3×10^{-6} M	10^{-5} M
Without minerals	0	–	0.0	0.4	9.4	12	15
With minerals	4.6	4.6	6.0	11.0	>16	>16	>16

[a] The explants were excised at early-mid podfill and placed with their bases in defined media as described by elsewhere [3, 12, 15].

[b] Delay in reaching 50% leaf yellowing relative to explants on water only; these reach 50% leaf yellowing 13 days after excision. Comparable leaves on intact plants reach this point 14 days later than explants on water [3].

are known to retard leaf senescence, the minerals could delay leaf senescence through promotion of root development [27]. CK concentrations above 10 nM inhibit rooting, and 100 nM or greater suppress it completely (data not shown). Thus, the senescence-delaying effects of CK on explants cannot be due to promotion of root development.

2.3 Cytokinins in the Xylem Sap

Table 2 shows the levels of CKs present in soybean xylem sap and their changes during pod development. The main CKs are ZR, DZR, Z and DZ. Smaller amounts of DZMP as well as the O-glucosides, OGZR + OGZ and OGDZR + OGDZ, are also present. ZMP does not occur in detectable amounts. In general the major

Table 2. Changes in the levels of cytokinin bases, ribosides, ribonucleotides and O-glucosides in xylem sap[a] collected from soybean rootstocks taken at various stages during pod development

Developmental stage	Short day	Cytokinin level (nM, with standard errors in parentheses)[b]							
		ZR	DZR	Z	DZ	DZMP	OGZR + OGZ[c]	OGDZR + OGDZ[d]	Total
Pods 1 cm	22	64.2 (10.8)	74.9 (9.3)	39.7 (6.4)	50.7 (5.4)	6.2 (2.0)	17.1 (6.8)	17.0 (5.4)	270
Pods at full extension	28	7.0 (1.4)	13.0 (2.5)	9.6 (2.3)	14.9 (2.7)				
Early-mid podfill	36	11.2 (1.7)	12.1 (2.0)	12.7 (2.7)	21.2 (8.6)	1.7 (0.6)	9.7 (2.9)	6.5 (2.4)	75
Late podfill	47	16.3 (1.1)	12.1 (1.7)	12.3 (4.1)	32.6 (9.9)				
Leaves yellow with a trace of green, pods yellow	57	20.4 (3.7)	18.9 (1.1)	26.3 (5.9)	48.6 (11.8)	4.2 (1.1)	21.4 (7.2)	22.4 (3.5)	162

[a] Sap collected at 100 kPa pressure for 50 min from plants grown in environmental chambers as described by Noodén and Mauk [24].
[b] The cytokinins were fractionated and purified by chromatography on phosphocellulose. The runoffs (nucleotides) were evaporated, dissolved in water, extracted with water-saturated *n*-butanol (extracts discarded), re-evaporated, hydrolyzed with alkaline phosphatase and partitioned into *n*-butanol for radioimmunoassay. The bases, ribosides and O-glucosides were eluted with 0.5 M NH_4OH, partitioned against *n*-butanol, dried down and fractionated by silica gel TLC with *n*-butanol, conc. acetic acid, H_2O (450: 113: 188, v/v) with marker dyes to produce the O-glucoside and base-riboside zones [22] which were eluted. The mixtures of bases and ribosides were then resolved into a base fraction (Z + DZ) and a riboside fraction (ZR + DZR) by chromatography on a column of dihydroxyboryl polymer [5]. The glucoside fraction was hydrolyzed with β-glucosidase and the resulting CKs were partitioned into *n*-butanol and determined as base-riboside mixtures by radioimmunoassay. Hence, the results are expressed as ZR and DZR equivalents. Radioimmunoassay was performed as described by Badenoch-Jones et al. [1,2]. Losses were corrected by estimating the recovery of comparable amounts of Z, DZ, ZR and DZR. ZMP was below the detection threshold.
[c] Expressed as ZR equivalents.
[d] Expressed as DZR equivalents.

CKs decline rapidly during pod extension. However, DZ increases again during podfill.

Both our results and those reported by Heindl et al. [4] for chamber-grown plants indicate that ZR, DZR, Z and DZ are significant CKs in xylem sap and their concentrations decline during pod development. Their plants showed this decline later (during podfill) than ours (pod extension); however, this may simply reflect differences in root activity which is known to vary with conditions [14, 19, 27]. The CK concentrations cited by Heindl et al. [4] were also considerably lower. For example, their peak concentrations, which were observed at full pod extension, were 6.4, 11.4, 1.2 and 0.1 nM for ZR, DZR, Z and DZ, respectively, totaling about 19 nM and a similar total, about 20 nM, earlier at full bloom. At full pod extension, we observed concentrations of 7.0, 13, 9.5 and 14.9 nM, respectively for the same four CKs, totaling 44 nM and a higher concentration (total 229 nM) earlier (1 cm pods). If the O-glucosides and DZMP were included, our totals would be even higher (Table 2). The values of Heindl et al. [4] may be lower due to different conditions or, more likely, differences in procedure. We corrected for losses during purification, but our shorter (50 min compared to 8 h) sap collection period is probably a more important factor (see 24 for discussion). Very likely, a reduction in root metabolism during the longer sampling periods [24] causes a significant decrease in CK production. In any case, the decline in xylem sap CKs during pod extension (Table 2) precedes a decrease in foliar chlorophyll levels in similar plants [26]. It is also significant that depodding just before podfill, a treatment which prevents the rapid leaf yellowing, also causes a substantial rise in the levels of some CKs, particularly DZR (data not shown).

2.4 Correlating Xylem Sap Cytokinin Levels with Their Effects in Explants

When supplied in conjunction with minerals, 100 nM ZR is able to maintain the greenness of the leaves of explants at a level similar to that in leaves on intact plants (Table 1). A sustained supply of ZR at higher concentrations may actually override the effect of the pods and prevent leaf yellowing (Table 1) [3]. Above the threshold in this system, Z is less active than ZR, while (+)DZ is about the same as ZR [3, 23]. (−)DZR should be somewhat more active than ZR [9]. OGZR can delay leaf yellowing; however, at the concentrations of OGZ + OGZR in xylem sap, OGZR has little or no activity (Table 3). (±)DZMP seems inactive. Thus, the ZR + DZR + Z + DZ (229 nM) in xylem sap at the start of pod development (pods 1-cm-long) would substantially inhibit leaf senescence, whereas the CKs (44 nM) in the sap at full pod extension would have much less influence. Even the peak values (19–20 nM or less) reported by Heindl et al. [4] for plants at full bloom and full pod extension appear to be near the threshold for activity, while the concentrations at other stages seem to be below the threshold. Nonetheless, other factors such as GA from the roots [16] could amplify the effects of these low CK levels. The rise in CK concentration (to 114 nM, Table 2) in the xylem sap of plants with yellowing leaves may be too late and not enough. Moreover, this increase in CK concentration may be due in part to decreased transpiration [15, 19], which would result in less dilution of the CKs in the xylem. Thus, CK flux to the yellowing leaves

Table 3. Delay of leaf yellowing in explants by the cytokinins OGZR and (±) DZMP supplied through the cut base[a]

Days of delay in leaf yellowing due to cytokinin		
	10^{-8} M	10^{-7} M
ZR	2	4
OGZR	0	2
(±)DZMP	0	0

[a] The cytokinin were supplied in mineral nutrient solutions, and the explants were handled as in Table 1.

may actually be less than at earlier stages. In any case, our observations indicate that high CK levels in the xylem sap of intact plants must decrease in order to permit the pods to induce leaf senescence (yellowing).

Previously, we [8, 13] noted changes in the pattern of CK activity (based on bioassays of partially purified material) in soybean leaves during pod development and monocarpic senescence. In spite of differences in conditions, the foliar CK activity declined during pod elongation and early podfill, the same period when xylem sap CK declines. Now that the assay methods and our understanding of the identity of the natural CKs have improved, these studies should be repeated to determine how the levels of the different foliar CKs change.

Since the metabolism of xylem-supplied Z and ZR in the leaves does not change as senescence progresses [23], and foliar transpiration rates remain high until late in leaf senescence [15, 19], the decrease in foliar CK activity appears to result from changes in CK production by the roots. Because CKs are rapidly metabolized to adenine and adenosine in the leaves, there is little opportunity for accumulation of the active CKs and a continuous influx of CKs is needed.

3 Monocarpic Senescence: The Senescence Signal

3.1 Behavioral Characteristics of the Senescence Signal

The simplest explanation for the mechanism (senescence signal) by which the developing seeds induce leaf senescence is that the seeds produce a senescence-promoting hormone [13, 14]. Even if this view is correct, it does not exclude other factors from background roles. A number of the behavioral characteristics are known for the senescence signal in intact soybean plants [13, 19]. First, it appears to originate from the seeds. Second, it is exerted relatively late in seed development. Third, it shows limited mobility, acting mainly on the nearest leaf and to a lesser extent on those below. Fourth, it travels into the nearest leaf via the xylem. For example, the pods are able to induce senescence of the target leaf even if its petiole has been steam girdled [25]. Unlike the leaves of many species, soybean leaves do not yellow normally when detached [12]; however, the fourth observation suggests that any putative senescence signal should still act on detached leaves when supplied through the xylem. This provides the basis for a bioassay of the senescence signal.

3.2 Testing Known Compounds for Senescence Signal Activity

Senescence signal activity can be assayed by allowing transpiration to pull the test solution into the xylem through the base of the petiole of an excised leaf or through the base of a depodded explant. In order to simulate the flow of xylem sap into the intact leaf during senescence induction, we dissolved the test substances in mineral solution [3] plus CK. The depodded explants were used rather than excised leaves, because they were less prone to wilting and water stress after prolonged culture. Maintenance of a high humidity (80 + % R.H.) also helped to prevent chlorophyll loss in the control leaves during prolonged culture.

Table 4 shows that a wide range of chemicals do not exert senescence signal-like effects on the leaves. Of particular note here is inactivity of ACC, a precursor of ethylene. In addition, dipping leaves of podded explants in 1 mM $CoCl_2$ or $Ag_2S_2O_3$ (both antiethylene agents [10]) does not delay leaf yellowing. (data not shown). Leaf abscission is delayed by the $Ag_2S_2O_3$ but not by $CoCl_2$. Furthermore, the inability of 2% (v/v) CO_2 to counteract the senescence signal in intact plants [19] and the inactivity of 5% CO_2 on leaves of podded explants (data not shown). These observations plus the restricted movement of the senescence signal within the plant argue against its being either ethylene or ACC. While ABA is able to accelerate monocarpic senescence when applied to the leaves of intact soybean plants, the data shown in Table 4 and other lines of evidence [13] suggest that ABA is not the senescence signal. Methyl jasmonate, jasmonic acid, serine, linolenic acid and even IAA have been implicated as senescence promoters in other tissues [see 18, 19], but none of these substances show senescence signal activity. In fact, IAA even causes the leaves to become a darker green. Linoleic and linolenic acid are only very sparingly soluble in water without help from high levels of detergent, so earlier reports claiming effects from concentrations higher than 10^{-5} M probably reflect the action of lower concentrations.

Table 4. Bioassay of known compounds for senescence signal activity in depodded soybean explants[a]

Compounds and concentrations tested	Response
Serine (10^{-3}, 10^{-4}, 10^{-5} M)[b]	No visible effect on leaf chlorophyll levels within 20 days[c]
Linoleic acid (10^{-5}, 10^{-6}, 10^{-7} M)	No visible effect on leaf chlorophyll levels within 20 days[c]
Linolenic acid (10^{-5}, 10^{-6}, 10^{-7} M)	No visible effect on leaf chlorophyll levels within 20 days[c]
Abscisic acid (10^{-5}, 10^{-6}, 10^{-7} M)	No visible effect on leaf chlorophyll levels within 20 days[c]
Aminocyclopropane (10^{-5}, 10^{-6}, 10^{-7} M) 1-carboxylic acid	No visible effect on leaf chlorophyll levels within 20 days[c]
Methyljasmonate (10^{-5}, 10^{-6}, 10^{-7} M)	Slight promotion at 10^{-5} M
Indoleacetic acid (10^{-5}, 10^{-6}, 10^{-7} M)	The leaves become a darker green

[a] Explants were excised at midpodfill and their bases were placed in test solutions containing half-strength mineral nutrient solution [3] and 10^{-7} M ZR in order to approximate the composition of the xylem sap which carries the senescence signal. The explants were incubated inside clear plastic bags through which humidified air (80 + % R.H.) was blown.

[b] Must be kept axenic.

[c] Within this time, podded explants and comparable parts of intact plants turn yellow.

3.3 Detection of Senescence Signal Activity

Organic solvent extraction of soybean leaves yields high levels of toxic activity in the oat-leaf chlorophyll-retention assay (Lindoo and Noodén, unpublished data 1974). The advent of a better bioassay system (Section 3.1) prompted a renewed search for endogenous senescence signal activity. Since (a) the yellowing of the cotyledons progresses inward from the seed coat and (b) the seeds cause the yellowing of the neighboring pod wall [26] as well as the nearest leaf, we reasoned that the senescence signal might be diffusing out of the seed coat and we tried washing excised seeds with water to leach out senescence signal activity. After testing a variety of systems, we found the best results with the continuous flow siphon system shown in Fig. 1. In many ways, this simulates what happens in the intact plant. In the siphon system as in explants and intact plants, three seeds are sufficient to induce yellowing of the target leaf, and the seeds are most active toward the end of their filling period. The bases of excised leaves or depodded explants can also be immersed in test solutions, but this requires greater amounts of wash solution. These experiments now provide the first direct evidence that the senescence signal is a senescence-promoting hormone. We are also using the depodded explants as a bioassay in the purification of this unknown hormone. During the induction phase, the pods also transmit some unknown ^{3}H-labelled compounds to the leaves [20], and in the future these various findings may converge to further clarify the nature of the senescence signal.

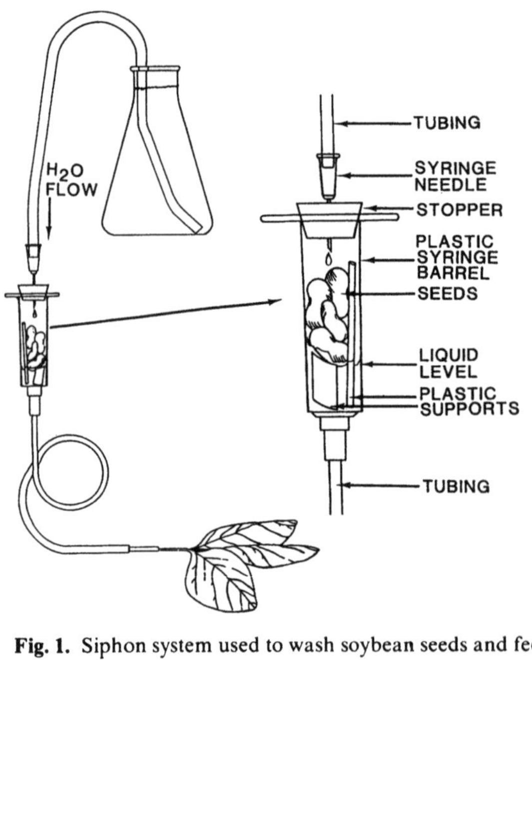

Fig. 1. Siphon system used to wash soybean seeds and feed the wash into a detached leaf

Abbreviations

ABA abscisic acid
ACC 1-aminocyclopropane-1-carboxylic acid
CK cytokinin
GA gibberellin
IAA indoleacetic acid
Z zeatin
ZR zeatin riboside
ZMP zeatin 5'-riboside monophosphate
DZ, DZR and DZMP dihydroderivatives of Z, ZR and ZMP
OGZ, OGZR, etc. O-glucosides of the corresponding cytokinins

Acknowledgements. We gratefully acknowledge the support of NSF grant DCM-830207 to LDN, the assistance of Ms. Patricia Arscott in the collection and preliminary purification of the xylem sap cytokinins and a scholarship from Consejo Nacional de Investigaciones Cientificas y Técnicas de la República Argentina to JJG.

References

1. Badenoch-Jones J, Letham DS, Parker CW, Rolfe BG (1987) Plant Physiol 75:1117
2. Badenoch-Jones J, Parker CW, Letham DS (1987) J Plant Growth Regul 6:159
3. Garrison FR, Brinker AM, Noodén LD (1984) Plant Cell Physiol 25:213
4. Heindl JC, Carlson DR, Brun WA, Brenner ML (1982) Plant Physiol 70:1619
5. Jameson PE, Letham DS, Zhang R, Parker CW, Badenoch-Jones J (1987) Aust J Plant Physiol 14:695
6. Leopold AC, Noodén LD (1984) In: Scott TK (ed) Encyclopedia of plant physiology, new series, vol. 10. Hormonal regulation of development II. Springer, Berlin Heidelberg New York Tokyo, p 4
7. Letham DS, Palni LMS (1983) Annu Rev Plant Physiol 34:163
8. Lindoo SJ, Noodén LD (1978) Plant Cell Physiol 19:997
9. Matsubara S, Shiojiri S, Fujii T, Ogawa N, Imamura K, Yamagishi K, Koshimizu K (1977) Phytochemistry 16:933
10. Mattoo AK, Aharoni N (1988) In: Noodén LD, Leopold AC (eds) Senescence and aging in plants. Academic Press, San Diego, CA, p 241
11. Neumann PM, Noodén LD (1983) J Plant Nutr 6:735
12. Neumann PM, Tucker AT, Noodén LD (1983) Plant Physiol 72:182
13. Noodén LD (1980) In: Corbin FT (ed) World soybean research conference II: Proceedings. Westview Press, Boulder, CO, p 139
14. Noodén LD (1984) Physiol Plant 62:273
15. Noodén LD (1985) In: Shibles R (ed) Proc world soybean res conf III. Westview Press, Boulder, CO, p 891
16. Noodén LD (1986) Plant Cell Physiol 27:577
17. Noodén LD (1988a) In: Noodén LD, Leopold AC (eds) Senescence and aging in plants. Academic Press, San Diego, CA, p 1
18. Noodén (1988b) Ibid, p 329
19. Noodén LD (1988c) Ibid, p 391
20. Noodén LD, Finkelstein D, Wetzel P (1987) Plant Physiol 83:(Suppl):121
21. Noodén LD, Leopold AC (1978) In: Letham DS, Goodwin PB, Higgins TJ (eds) Phytohormones and related compounds – a comprehensive treatise, vol II. Elsevier/North-Holland Biomedical, Amsterdam, p 239

22. Noodén LD, Letham DS (1984) J Plant Growth Regul 2:265
23. Noodén LD, Letham DS (1986) In: Bopp M (ed) Plant growth substances 1985. Springer, Berlin Heidelberg New York Tokyo, p 324
24. Noodén LD, Mauk CS (1987) Physiol Plant 70:735
25. Noodén LD, Murray BJ (1982) Plant Physiol 69:754
26. Okatan Y, Kahanak GM, Noodén LD (1981) Physiol Plant 52:330
27. Van Staden J, Cook EL, Noodén LD In: Noodén LD, Leopold AC (eds) Senescence and aging in plants. Academic Press, San Diego, CA, p 281
28. Zhang R, Letham DS, Wong OC, Noodén LD, Parker CW (1987) Plant Physiol 83:334

The Possible Role of Protein Kinases in the Plant Cell Response to Phytohormones

O.N. KULAEVA[1]

1 Introduction

The reversible phosphorylation of proteins plays an important role in eukaryotic cell regulatory systems. Phosphorylation of seryl, threonyl or tyrosyl residues of proteins triggers conformational changes which modulate their biological properties [2, 3, 20]. Phosphorylation/dephosphorylation of proteins depends on protein kinases and protein phosphatases in the cell. Sufficient evidence has now accumulated to suggest that these two types of enzymes are involved in transduction and amplification of hormonal signals in animal cells. Hormonal regulation of these enzymes is transmitted by second messengers such as cyclic nucleotides, IP_3, DAG, Ca^{++} or the Ca^{++}-calmodulin complex [2, 3].

Protein phosphorylation is a widespread phenomenon in plant cells and phosphorylated proteins and protein kinases have been detected in all compartments of plant cells [16]. These data support the investigation of the role of protein kinases in plant cell response to phytohormones.

2 Phytohormones in Modulation of Nuclear Protein Phosphorylation

It is clear that the regulation of gene expression is involved not only in long-term plant cell response to phytohormones, but also in short-term responses [21]. Trewavas [23] was the first to suggest that phosphorylation of chromatin-associated protein may mediate the hormonal regulation of transcription in plant cells.

Various data indicate that phytohormones modulate protein phosphorylation in plant cell nuclei. For example, 2,4-D-pretreatment of soybean hypocotyls activated in vitro protein phosphorylation in isolated nuclei, and this was consistent with an in vivo increase of RNA synthesis [12]. Other phytohormones also alter nuclear protein phosphorylation in plants, and these include ABA [23], GA_3 [25] and cytokinins [15]. However, in all these experiments phytohormones were applied to intact plants or to isolated plant organs. Therefore, the hormonal modulation of nuclear protein phosphorylation could be an indirect result of earlier cell response to the phytohormone.

Recently an increase in nuclear protein phosphorylation was demonstrated in vitro in isolated nuclei incubated with polyamines, substances which appear to play

[1] K.A. Timiriazev Institute of Plant Physiology, Academy of Sciences of the USSR, Botanitcheskaj 35, Moscow, 127276 USSR

an important role in plant growth regulation [4]. For example, spermine induces increases in phosphorylation of several proteins in nuclei isolated from peas [4].

However, in none of the above cases have hormone-responsive protein kinases been isolated from nuclei of the plant cells.

3 *In Vitro* Effects of Cytokinin on RNA-Polymerase Associated Protein Kinase

It is known that RNA polymerase I is associated in animal cells with protein kinase, which phosphorylates RNA-polymerase and induces an increase of its activity [20]. RNA polymerase I isolated from barley leaf chromatin is also associated with protein kinase activity [19], which is dramatically increased by the addition of BA into the reaction medium (Fig. 1). The dose response curve of this effect was typical of other phytohormone-induced reactions. The effect was specific for biologically active cytokinins (Z, kinetin, BA) and could not be demonstrated with their non-active analogues, adenine or 6-methyladenine. These results were confirmed in experiments with RNA-polymerase I and II isolated from nuclei of lupin cotyledons. Both enzymes possessed low protein kinase activity, but this was increased in vitro 10-fold by BA.

In all of these experiments cytokinins stimulated in vitro RNA polymerase associated protein kinase activity. Thus, this enzyme may be one of the molecular targets for cytokinins. However the direct action of cytokinin on RNA polymerase in vitro activates only protein kinase without an effect on RNA polymerase. For revealing cytokinin effects on RNA polymerase activity in vitro, an additional cytokinin dependent cytoplasmic protein transcriptional factor was necessary [8].

Recently, non-enzymatic radiolabelling of proteins by ^{32}P nucleotides has been observed [18]. Although this process differs from an enzymatic one in time course (up to 20 h) and temperature condition (increased up to 70°C), it must be taken into account in discussions of in vitro protein phosphorylation.

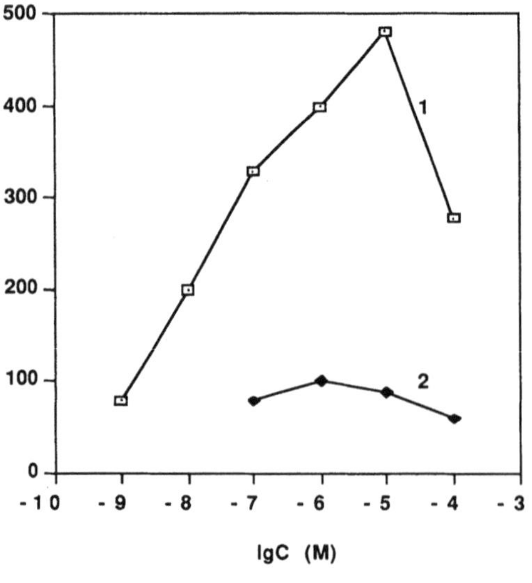

Fig. 1. Effect of BA and adenine on protein kinase activity associated with RNA-polymerase I from barley leaves. The reaction mixture (100 μl) for protein kinase estimation contained: 50 mM Tris-HCl buffer, pH 8.0; 10 mM $MgCl_2$; 0.1% Triton X-100; 10 mM NaF; 20 nM γ^{32}-P-ATP; 30 μg of RNA polymerase protein. BA was added directly to the reaction medium. Incubation for 15 min at 37°C. Details were described earlier [19]. Data obtained are given as percentage of control data (without BA) *1* BA; *2* adenine

4 Roles of Cytokinin and ABA in Modulation of Ribosomal Protein Phosphorylation

The role of reversible phosphorylation of ribosomal proteins in hormonal regulation of protein synthesis in animal cells has been widely discussed [22]. Phosphorylation of the ribosomal protein S_6 located inside the ribosome initiating center can play an important role in the early step of the initiation of translation, although the exact molecular mechanism of such regulation is still not clear [22]. Trewavas has shown an influence of cytokinin and ABA on total ribosomal protein phosphorylation in plants [23]. We have established that BA activated in vivo S_6-ribosomal protein phosphorylation in isolated pumpkin cotyledons (Fig. 2), a system in which the BA also stimulated polysome formation and protein synthesis [5, 6, 26]. ABA inhibited the activation of S_6-protein phosphorylation caused by BA (Fig. 2). Since the action of BA and ABA on the S_6-protein phosphorylation correlates with these phytohormones' effects on polysome formation in cotyledon cells [7, 26], it is possible that S_6 phosphorylation is involved in the mechanism of hormonal regulation of RNA translation in plants. If so, it would give a key to understanding transcription-independent hormonal regulation of translation [6, 7].

Since hormonal regulation of S_6 protein phosphorylation was observed only in vivo and was absent in vitro, it is probable that some cofactors or second messengers are necessary for transduction of the hormonal signal to the protein kinase phosphorylating ribosomal proteins.

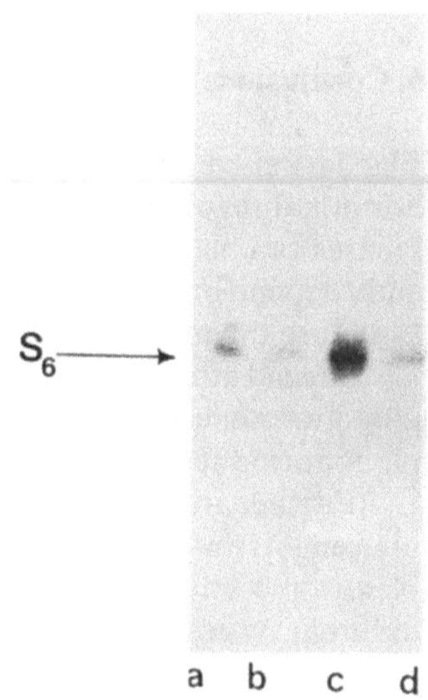

Fig. 2. Effect of phytohormones on phosphorylation of ribosomal protein S_6 in detached pumpkin cotyledons. The autoradiograms of one-dimensional SDS gel electrophoresis of ribosomal proteins isolated from pumpkin cotyledons. Cotyledons were incubated for 6 h in water *a* and in solution of: *b* ABA (10^{-4} M); *c* BA (5×10^{-5} M) or *d* BA + ABA in the same concentrations respectively. Discs from cotyledons were then labeled for 2 h with ^{32}P (50 μCi/ml), and isolated ribosomal proteins were subjected to electrophoresis [26]

5 Signal Transduction System Involving Protein Kinase C

Very important regulatory systems in animals involve the receptor mediated activation of phosphoinositidase which hydrolyzes Ptd Ins4,5 P_2, yielding IP_3 and DAG, two types of secondary messengers controlling separate, but interacting signal transduction in cells. IP_3 controls the intracellular level of Ca^{++}, which regulates a wide range of cellular processes. DAG acts in concert with Ca^{++} to activate protein kinase C and then the kinase amplifies the primary signal by phosphorylation of specific proteins. Thus, there is a feedback control of signal-receptor induced mechanisms. These two signal pathways cooperate with each other in the regulation of many processes in animal cells, and are involved in the molecular mechanisms of insulin, mammalian growth factors and neuromediator action [2].

Recent reports suggest the existence in plant cells of all of the main components of this IP_3/Ca^{++} and DAG/protein kinase C signal transduction system. A protein kinase C-like enzyme was partially purified from zucchini [17] and wheat cells [14]. It was also recently reported that IP_3/Ca^{++} and DAG/protein kinase C pathways are involved in carrot cell response to an elicitor [9].

Preliminary data further suggest the involvement of phosphoinositides in GA_3-induced synthesis and secretion of α-amylase in aleurone cells [13]. GA_3 activation of protein kinase C-like enzyme has been shown in potato tubers [10]. The in vitro effect of 2,4-D on protein phosphorylation in a membrane fraction isolated from etiolated soybean hypocotyls may also be associated with this system [24]. Further investigation in this field may be valuable in the elucidation of molecular mechanism of phytohormone signal transduction and amplification.

6 Conclusions

The data cited give support for the suggestion that protein kinases are involved in hormonal regulation of transcription and translation in plant cells. Some of these kinases can be assumed to be a primary target for phytohormones, others apparently depend on second messengers, among which Ca^{++}, Ca^{++}-calmodulin, IP_3 and DAG are the most attractive for further investigation.

Special attention should be paid to the investigation of the phosphorylation of phytohormone binding proteins, to elucidate changes in plant cell sensitivity to phytohormones [1, 11].

Further investigation into the involvement of protein kinases and second messengers in hormonal signal transduction in plant cell is needed to explain pleiotropic effect of phytohormones, and also the quite distinct responses in different types of plant cells to phytohormones.

References

1. Aducci P, Ballio A, Federico R, Montesano L (1982) In: Wareing PF (eds) Plant growth substances 1982. Academic Press, London, p 395
2. Berridge MJ (1987) Annu Rev Biochem 56:159
3. Cohen PH (1985) BioEssay 2:63
4. Datta H, Roux SY (1986) BioEssay 5:120
5. Kulaeva ON (1980) In: Skoog F (eds) Plant growth substances 1979. Springer, Berlin Heidelberg New York, p 119
6. Kulaeva ON (1981) In: Guern J, Péaud-Lenoël C (eds) Metabolism and molecular activities of cytokinins. Springer, Berlin Heidelberg New York, p 218
7. Kulaeva ON (1982) XLI Timiriazev Lecture Nauka, Moscow
8. Kulaeva ON (1985) In: Proceedings of the 16th FEBS Congress, p 391
9. Kurosaki F, Tsurusawa Y, Nishi A (1987) Plant Physiol 85:601
10. Ladedgenskaya EP, Korableva NP, Morozova TM, Sidorkina OM, Salganik RI (1987) Dokl Akad Nauk SSSR 292:763
11. Libbenga KR, van Telgen HY, Mennes AM, van der Linde PCG, van der Zaad EY (1986) In: Fox YE, Jacobs M (eds) Molecular biology of plant growth control. Liss, New York
12. Murray MG, Key JL (1978) Plant Physiol 61:190
13. Murthy P, Renders J (1987) Plant Physiol (Suppl) 83:132
14. Oláh Z, Kiss Z (1986) FEBS Lett 195:33
15. Ralph RK, Wojcik SY (1981) Plant Sci Lett 22:127
16. Ranjeva R, Boudet AM (1987) Annu Rev Plant Physiol 38:73
17. Schäfer A, Bygrave F, Matzenayer S, Marme D (1985) FEBS Lett 187:25
18. Schmidt MC, Hanna MM (1986) FEBS Lett 194:305
19. Selivankina S Yu, Romanko EG, Novikova GV, Muromtseva DG, Kulaeva ON (1988) Sov Plant Physiol 35:266
20. Severin ES, Kochetkova MN (1985) Role of phosphorylation in regulation of cell activity. Nauka, Moscow
21. Theologis A (1986) Annu Rev Plant Physiol 37:407
22. Traugh J (1981) In: Litwack (ed) Biochemical actions of hormones, vol 8. Academic Press, New York, p 167
23. Trewavas S (1976) Annu Rev Plant Physiol 27:349
24. Varnold RL, Morré DJ (1985) Bot Gaz 146:315
25. Weilgat B, Klechkowski K (1981) Plant Sci Lett 21:381
26. Yakovleva LA, Kulaeva ON (1987) Biochem Physiol Pflanz 182:359

VI Practical Applications and Economic Implications

New Types of Plant Growth Regulators of Microbial Origin: The Likelihood of Practical Use

H.G. CUTLER[1]

1 Introduction

With the exception of the brassinosteriods, all of the 'major' plant growth regulators have been isolated from microorganisms: IAA from *Rhizopus suinis* [38] and *Neurospora crassa* [39]; GA_3 from *Gibberella fujikuori* (*Fusarium moniliforme*) [3] and other GA's from *Sphaceloma* and *Elsinoe* species [29]; ABA from *Cercospora cruenta* [15, 27, 28], *Botrytis cinerea* [19], *Rhizoctonia solani*, *Fusarium oxysporum* f. sp. *lycopersici*, *Ceratocystis coerulescens*, and *C. fimbriata* [8]; ethylene from myriad fungi [16]; trans-zeatin, trans-zeatin riboside, and 6-(4-hydroxy-1,3-dimethyl-but-trans-2-enylamino)-9-β-D-ribofuranosylpurine from *Pseudomonas syringae* pv. *savastanoi* [31]. However, there exist many microbial metabolites that have either plant growth regulatory or phytotoxic activity, e.g., the cytocholasins are a potent group of compounds. Cytochalasin H inhibits flowering in tobacco [43] while its congener, pyrichalasin H, inhibits root and shoot growth in rice [25]. Hexylitaconic acid significantly promotes growth in rice seedlings [17]. Neovasinone stimulates lettuce root growth 70% at 100 ppm while neovasinin inhibits root growth 50% at 300 ppm, yet the structures differ by a carbonyl function [24]. Herbicidins are potent, selective toxins that kill weeds [2, 12, 34, 35]. With respect to synthetic modifications, deoxyabscisic acid has been made with a view to regulating plant growth [32].

More complex molecules, specifically AM toxins I, II, and III are selectively phytotoxic at 0.1 ppb against apple and represent a class of cyclic peptides that are under investigation [41, 42]. Tentoxin, while it induces chlorosis in certain plants, is a cyclic peptide whose synthetic fragments induce auxin-like activity in plants [9]. Moniliformin, a phytotoxic metabolite of *Fusarium moniliforme* has been derivatized and its analogs patented as herbicides [4]. These examples, taken from an ever growing list of biologically active products, suggest that parent molecules and their synthetic, or biotransformed products may be useful in agriculture and plant physiology.

[1] US Department of Agriculture, ARS, Richard B. Russell Center, P.O. Box 3677 Athens, GA 30613, USA

2 Biologically Active Natural Products

2.1 The Cytochalasins

Many of the cytochalasins inhibit the extension of etiolated wheat coleoptiles and these include cytochalasins A, B, E, H [7, 44, 7, 43] and epoxycytochalasin H [6], epoxydeacetylcytochalasin H [6] (Fig. 1) and deacetylcytochalasin H [5]. Of these, cytochalasin H (Fig. 2) has been isolated in fairly large quantities, making it available for testing in higher plants. In greenhouse experiments with bean, corn, and tobacco plants, no effects were noted in corn plants, while petioles of mature bean first true leaves were strongly bowed upward at rates of 83 g/ha. All leaves were longitudinally rolled making the compound a possible anti-drought candidate, but within a week all plants appeared normal and flowering was also normal some weeks later. The most dramatic responses were noted in tobacco plants that had been treated in the seedling stage when they were approximately 4 cm in diameter, at 83, 8.3, and 0.83 g/ha. One month after treatment, control plants were

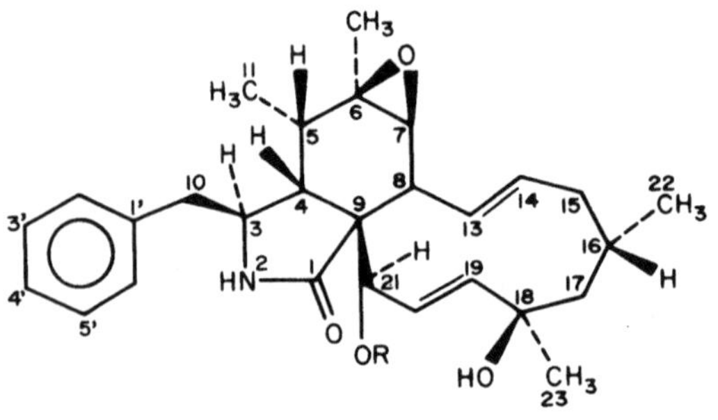

EPOXYCYTOCHALASIN H: R = C(=O)CH$_3$
EPOXYDEACETYLCYTOCHALASIN H: R = H
(Phomopsis sojae)

Fig. 1. Epoxycytochalasin H and Epoxydeacetylcytochalasin H

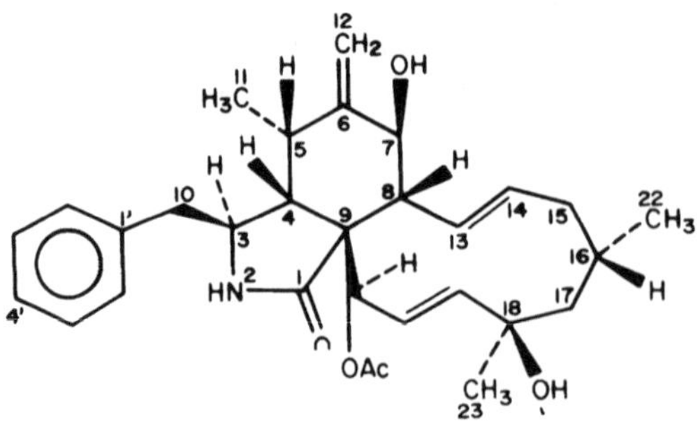

CYTOCHALASIN H
(Phoma sp.)

Fig. 2. Cytochalasin H

in full flower; in the 0.83 g/ha treatments, flowers were starting to open. At 8.3 g/ha, plants were in the compact button stage and at the highest concentration, no flowers were visible [43]. In other experiments with peach trees, flowering was not inhibited (unpublished data), thus, a certain amount of selectivity was seen with the anti-flowering activity.

A limited number of synthetic derivatives have been made of cytochalasin H. These include the 7-acetoxy; 2,7-diacetoxy; and 2,7,18-triacetoxy cytochalasin H (Fig. 3). All were assayed in the etiolated wheat coleoptile assay and in day-old chicks [7]. The C7 acetate significantly inhibited coleoptile growth ($P < 0.01$) at concentrations as low as 10^{-3}, 10^{-4} and 10^{-5}M, while the C7, N2 diacetate inhibited growth only at 10^{-4}M or higher. The C7, N2, C18 triacetate was inactive. The C7 acetate was inactive in chicks. Thus, the C7 hydroxyl is necessary for toxicity in chicks, whereas neither the C7 hydroxyl nor the N2 proton is essential for plant growth inhibition, while the C18 hydroxyl is. Also, these experiments suggest that the toxic properties of cytochalasins may be deleted or reduced for animals without negating the plant growth regulatory properties.

A new cytochalasin, pyrichalasin H (Fig. 4), has recently been isolated from *Pyricularia grisea*, wherein the 4′ position has an OCH_3. Rice seedling shoots and roots were greatly inhibited at 1.9×10^{-6} M and there was characteristic curvature of the shoots at lower concentrations relative to controls [25]. Two points of interest make this molecule desirable for further examination. First, the existence of the methoxy cytochalasin H implies that the 4′ hydroxy species may also exist in nature. Second, either substitution or synthetic manipulation of the phenyl ring suggests that biologically active products may be derived from natural product progenitors.

2.2 Hexylitaconic Acid

Because of its structural simplicity and odd history, one of the more remarkable natural products of microbial origin is hexylitaconic acid from *Aspergillus niger*

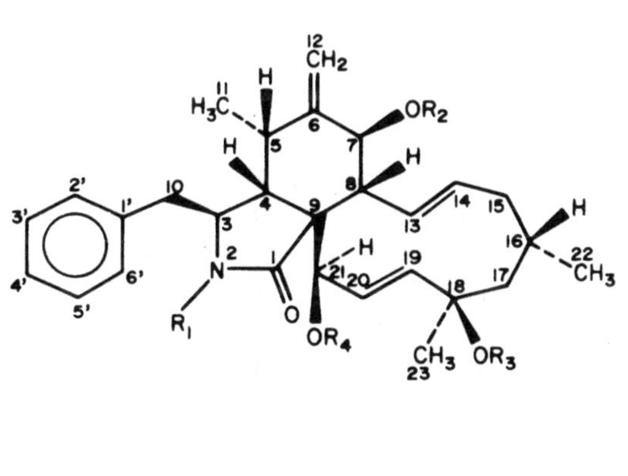

	R_1	R_2	R_3	R_4
7-ACETOXY	H	Ac	H	Ac
2,7-DIACETOXY	Ac	Ac	H	Ac
2,7,18-TRIACETOXY	Ac	Ac	Ac	Ac

Fig. 3. 7-Acetoxy; 2,7-diacetoxy; 2,17,18-triacetoxy cytochalasin H

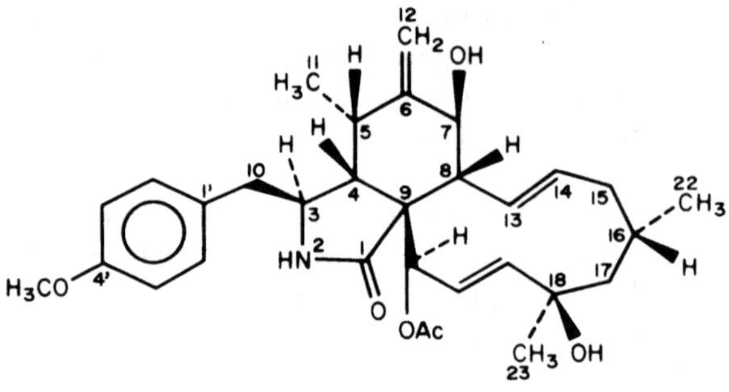

Fig. 4. Pyrichalasin H

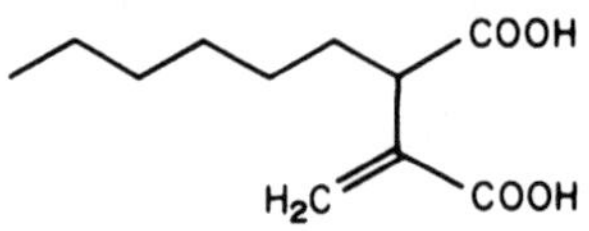

Fig. 5. Hexylitaconic acid

K-88 [17] (Fig. 5). In an unrelated but brilliant study, McCorkindale postulated that hexylitaconic acid must be the precursor of the antibiotic dilactone, canadensolide, from *Penicillium canadense* [22]. Accordingly, he synthesized the radiolabelled species of hexylitaconic acid and, as predicted, *P. canadense* incorporated this precursor into canadensolide. Unfortunately, hexylitaconic was not tested in biological systems until, by chance, the material was isolated from *A. niger*. In lettuce, root growth was promoted to about 250% of controls. In rice, roots were promoted with concentrations of 2–20 ppm, the greatest extension being at 20 ppm. Rice shoots were promoted 20–30%, with 20 ppm being the optimum concentration. Above 20 ppm shoots and roots were inhibited with complete cessation of growth at 500 ppm.

The molecule, because of its simplicity and relative ease of synthesis lends itself to further development. For example, the disodium or dipotassium salt is an obvious candidate, as is saturation of the only double bond. Addition on the acyl chain is also obvious and reduction of the carboxyl groups either together, or individually, is another alternative. The possibility of amino acid derivatives (*vide infra*, oligopeptides) is intriguing. The almost auxin-like activity makes it an appropriate candidate for development.

2.3 Neovasinone and Neovasinin

Both α-pyrone substances were discovered during the search for plant growth regulators from microorganisms (Fig. 6). The producer organism, *Neocosmospora*

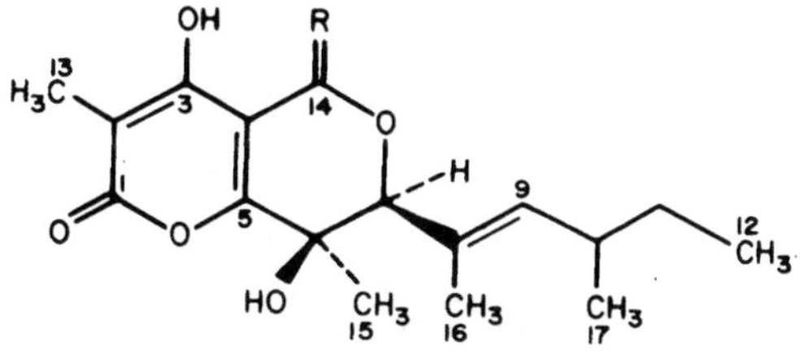

Fig. 6. Neovasinone and neovasinin

vasinfecta, a pathogen of many economic crops, generated a growth accelerator of lettuce roots in the culture filtrate. Subsequent extraction led to the isolation of neovasinone which increased root length of lettuce 20 to 65% at concentrations that ranged from 5 to 300 ppm and neovasinin that inhibited lettuce root growth 10 to 50% at concentrations ranging from 5 to 300 ppm [23, 24]. The authors point out that the carbonyl group at C14 forms a chelate with the hydroxyl group at C3 and that these functions, "played an important role in the stimulation of lettuce growth" [24]. No doubt the C14 carbonyl is important as may well be the C1 carbonyl. It would be significant, in determining the sites necessary for biological activity, if the C14 carbonyl could be selectively reduced followed, subsequently, by selective reduction of C1 carbonyl and, eventually, reduction of both groups. With selective reductions, hydroxyl groups become available for further synthetic propositions.

2.4 The Herbicidins

Bacteria also offer unique sources of potential herbicides and some of the most potent and target specific are the herbicidins from *Streptomyces saganonensis* which are purine analogs (Fig. 7). While only five have been elucidated to date, and these are A, B, E, F, and G, the implications are that at least two more exist (C and D). Their structures have been fully reported [2, 12, 34, 35] but revisions of the structures have been made [37]. Unfortunately, only the effects of herbicidins A and B have been reported but these have been extensively tested against both monocotyledonous and dicotyledonous weeds and economic crops. Barnyard grass, goose grass, manna grass, and green panicum were controlled with treatments of herbicidin A and B at concentrations from 37.5 to 300 ppm. Both compounds controlled common purslane, *Achyranthes*, white goose foot, smart weed, wild amaranth, Asiatic day flower, tomato and radish when they were treated with applications from 30 to 300 ppm. An additional but very important feature of these metabolites is that they were not toxic to rice plants and they controlled a major rice pathogen, *Xanthomonas oryzae*. Hence, these compounds are tailor-made for specific use in a major crop to control both weeds and a plant pathogen. The many functional groups lend themselves to synthetic derivatization.

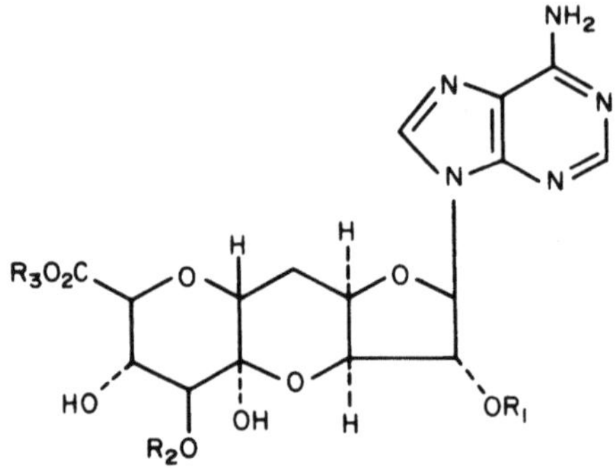

	R_1	R_2	R_3
HERBICIDIN A	CH_3	$CO(CH_2OH)C{=}CHCH_3$	CH_3
HERBICIDIN B	CH_3	H	CH_3
HERBICIDIN E	CH_3	$COCH(CH_3)_2$	CH_3
HERBICIDIN F	CH_3	$CO(CH_3)C{=}CHCH_3$	CH_3
HERBICIDIN G	H	$CO(CH_3)C{=}CHCH_3$	H

(*Streptomyces saganonensis*)

Fig. 7. Herbicidins

2.5 The AM Toxins

The oligopeptides, especially the cyclic members, are especially interesting to synthetic chemists and to those using plant growth regulators for a number of reasons. First, they may be composed of both D and L amino acids as in the cases of the cyl-1 and cyl-2 toxins, from *Cylindrocladium scoparium* [13, 14, 33] and HC toxin, from *Helminthosporium carbonum* [18]. Second, fragments of these peptides may produce unexpected plant growth regulatory responses. Third, the number of synthetic permutations that may generate new compounds is staggering. And fourth, in specific instances the quantity necessary to induce a response is extremely small. The AM toxins are one class of compounds that fall into this category, though their amino acids are all L. They are tetrapeptides obtained from the phytopathogen *Alternaria mali* [41, 42]. All the AM toxins differ only by the substitution on the phenyl ring and that substitution may be either a proton, hydroxyl, or methoxyl function. Thus, AM toxin I is composed of L-α-hydroxyisovaleric acid, L-alanine, α-aminoacrylic acid and L-α-amino-δ-(p-methoxyphenyl)-valeric acid. AM toxins II and III are identical except for the L-α-amino-δ-phenylvaleric acid and L-α-amino-δ-(p-hydroxyphenyl)-valeric acid residues, respectively (Fig. 8). While all the AM toxins produce leaf spot and necrosis in susceptible apple varieties, as does the pathogen, even resistant varieties appear not to be immune to the toxin. For example, "Indo" which is very vulnerable to *A. mali* is also affected by toxins I and III so that within 18 h of application of 0.1 ppb interveinal necrosis develops. In contrast, the 'resistant' apple variety "Jonathan" requires 1 ppm of AM toxin I, and 10 ppm of AM toxin III to induce identical responses. If, in the case of AM toxins I and III, only 0.1 ppb produces a response when applied exogenously to apple leaves it is inferred that less than 0.1 ppb reaches the active site and the amount of material necessary for necrotic induction is far less. In addition, fragments of the peptide have, apparently, not been tested for biological activity.

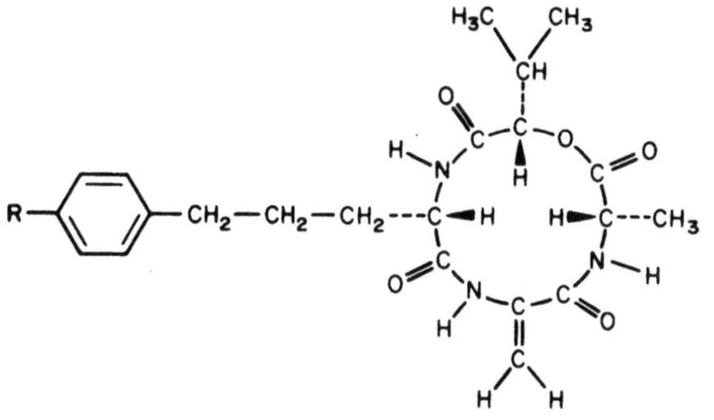

AM TOXIN I, R = CH_3O
II, R = H
III, R = OH
(*Alternaria mali*)

Fig. 8. AM Toxins I, II, III

2.6 Tentoxin

This cyclic tetrapeptide from *Alternaria tenuis* (*A. alternata*) and *A. mali* [21, 26] induces selective chlorosis in certain plants such as lettuce and mung bean but does not, apparently, affect corn and tomato [20, 11, 36] (Fig. 9). The structure, composed of L-amino acids, has been used as a template for the production of potential herbicides with some surprising results and these examinations include the testing of fragments for plant growth regulatory effects. Also, modified cyclic structures have been synthesized and evaluated. The cyclic derivative {Pro^1} tentoxin, wherein proline replaces N-methyl alanine, produces intense chlorosis in barnyard grass and lettuce, but moderate chlorosis in pitted morning glory [9]. Tentoxin was highly active in all three species, eliciting an intense chlorosis. Specific fragmentation, and substitution, of a tripeptide of tentoxin has been accomplished and those of intense interest are *tert*-butyloxycarbonyl (Boc)-Val-N-(C_2H_5)Δ^z Phe-Gly-OMe; Boc-Leu-N(C_2H_5)Δ^z Phe-Gly-OMe; and Boc-Leu-N(CH_3)Δ^z Phe-Gly-OMe. In etiolated wheat coleoptile bioassays, the first two analogs, which possess the C_2H_5 substitution at N-Δ^z-Phe, significantly inhibited extension at 10^{-3} but, surprisingly, significantly ($P < 0.01$) promoted coleoptile extension at 10^{-4}M as did the standard, IAA. At 10^{-5}M, the first analog did not promote growth, but the second did. The third tripeptide analog, which contained the N-CH_3 substitution and, therefore, resembled more closely a tripeptide fragment of tentoxin, inhibited growth at 10^{-3} and 10^{-4}M [9]. In cress and lettuce seedlings assays both C_2H_5 containing fragments promoted cress roots (40 and 19%, respectively for the first and second fragments) while the second fragment only promoted lettuce roots 40–50%. Furthermore, the N-CH_3 fragments inhibited cress roots 50% and lettuce roots 60% [9]. Again, the opportunity for other chemical substitutions, whether amino acids or other molecules, becomes an interesting permutation relative to biological activity. Some of the possibilities have been addressed by Edwards [10] and include such questions as the alignment and stereochemistry of the peptide backbone, that relationship to enzymatic degradation, the flexibility or rigidity of the backbone and relative biological activity.

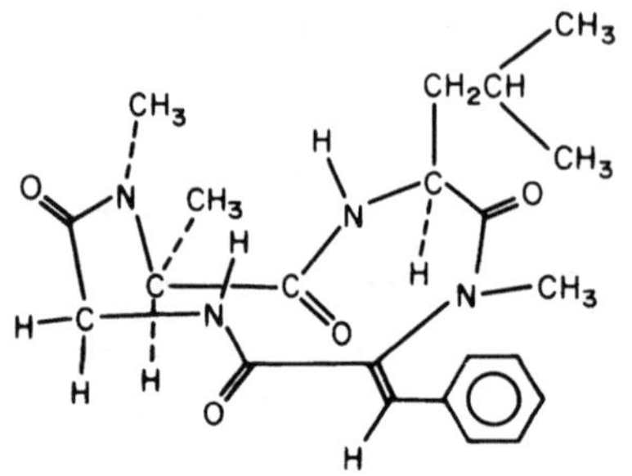

Fig. 9. Tentoxin

2.7 Abscisic Acid Derivatives

ABA has been disappointing as an agrochemical for field application. However, the recent discovery of ABA and its metabolites in fungi has led to synthetic modification of the molecule primarily for establishing the exact conformation of these novel congeners. This, in turn, has led to derivatives possessing biological activity. Of these, the (2Z) deoxy-ABA and (2E) deoxy-ABA have been tested on lettuce and rice seed (Fig. 10). The (2Z)-isomer weakly inhibited rice seedling growth at 1×10^{-3}M while germination of lettuce seed was greatly inhibited at 1×10^{-4}M(~100%). The (2E)-isomer inhibited rice ~30% at 10^{-3}M and inhibited lettuce ~100% at 5×10^{-4}M(32). Because of the increased activity in the synthetic modification of ABA and its analogs the possibility of finding a suitably active product for practical application is possible.

2.8 Moniliformin

With the exception of ethylene, the simplest structure yet discovered that has plant growth regulatory activity is moniliformin, a water-soluble metabolite from *Fusarium moniliforme* [4] (Fig. 11). Moniliformin inhibits spindle fibre formation in mitosis producing a c-metaphase plate [30]. Application of moniliformin to greenhouse-grown tobacco and corn plants at 266 and 26.6 g/ha caused significant growth inhibition for at least 3 weeks. In mature field grown tobacco from which the floral apices had been removed, application of moniliformin inhibited axillary shoot growth for 3 weeks when applied at 266 and 26.6 g/ha. The standard chemical, which was used to control axillary shoot growth, was MH 30 at 3.39 kg/ha and it controlled shoot growth for 5–6 weeks.

The structurally related analog, squaric acid, 3,4 dihydroxy-3-cyclobutene-1,2-dione, which has two hydroxyl groups which replace the proton and salt of moniliformin is biologically inactive. Swiss patent 609,836 was awarded to Fischer and Bellus (Ciba-Geigy) in 1979. An abstract of the patent states, "The cyclobutene-3,4 diones(R-alkyl, substituted Ph,CH_2CH_2SEt,$CH_2CH{:}CH_2$, etc) are her-

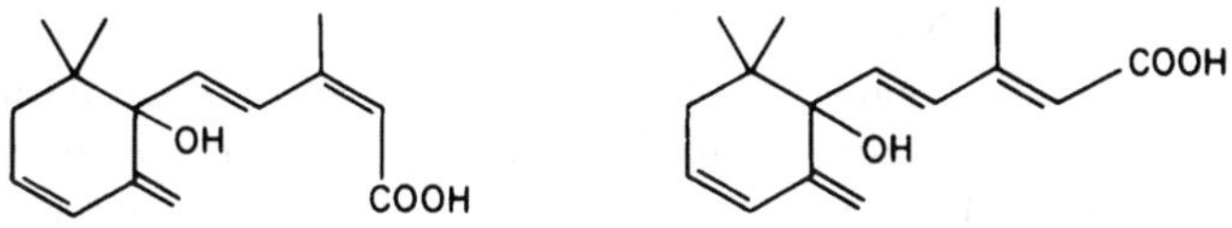

(2Z) DEOXY-ABSCISIC ACID (2E) DEOXY-ABSCISIC ACID

Fig. 10. 2Z and 2E deoxy-abscisic acid

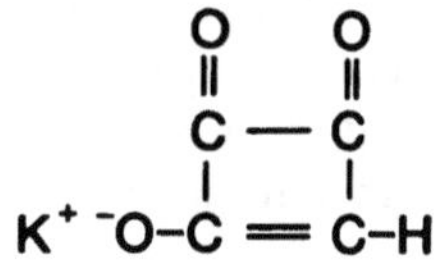

MONILIFORMIN
(*Fusarium moniliforme*)

Fig. 11. Moniliformin

bicides and plant growth regulators. Thus, postemergence application of 4kg 1-octyloxycyclobutene-3,4 dione/ha controlled *Setaria italica*, *Lolium perenne*, *Sinapis alba* and *Stellaria media* more than did the standard moniliformin Na Salt." (Chem. Abstracts Vol 90:1979 Indent 90:1988 82p). There are still a number of derivatives of moniliformin to be made.

3 Discussion: The Likelihood of Practical Use

The six sets of compounds discussed represent only a small portion of natural products from microorganisms described as having biological activity in the literature and have been chosen because they show, among other things, the diversity of structure. This diversity ranges from the simple building blocks of life, the amino acids which are common to all species, to those molecules that seem to be highly specialized. The latter is not intended to imply that the genetic information for the production of certain natural products may be deleted during evolution. Indeed, it may only remain suppressed. For example, ABA, which has been discovered in the brains of pigs and rats [1], is either compartmentalized from dietary intake, which would require passage across the blood brain barrier, or is synthesized de novo and is, therefore, genetically controlled. IAA is also found in lower and higher life forms. However, there are many tantalizing structures that have been isolated from microorganisms [40] that have not been assayed for biological activity and which appear to have been elucidated for sheer chemical sport. When one includes products from higher plants, both those that have been tested as agrochemicals and those that have not, the numbers increase.

Perhaps the question about likelihood of the practical use of fermentation products should be in two parts. First, will fermentation products alone suffice or, second, will those products have to be synthetically elaborated or modified? Most of the examples illustrated contain references to possible synthetic modifications.

Ideally, it is hoped that fermentation products would alone suffice and there is evidence to suggest that they would. However, each natural product represents a template for elaboration. Since it is in the nature of man to tinker and to obtain well-defined patents for specific purposes, the outcome will probably be for modified natural products. But for all these changes to occur there has to be a change in philosophy. Many natural products appear to have high specificity when it comes to target pests and this selectivity may be perceived as being a small market. In fact, the pesticides of the future may contain a mixture of chemically unrelated natural products for protecting an individual crop and, ideally, that mixture will be non-persistent in the environment. Ultimately, the search for new biologically active templates and a major concern for the environment will be the driving forces that will secure the practical use of natural products, and derivatives, as agrochemicals.

4 Conclusions

Microorganisms are sources of biologically active, secondary metabolites that are diverse in structure. These are fermentation products that may be used as pesticides or which may be used as novel templates for the synthetic production of agrochemicals. A limited number of examples are given and included are cytochalasins, hexylitaconic acid, neovasinone and neovasinin, the herbicidins, AM toxins, tentoxin, ABA derivatives, and moniliformin. There are many biologically active natural products to be discovered and the impetus for accessing new compounds will depend upon two factors; the protection of the environment and the necessity for new agrochemical templates.

References

1. Anonymous (1986) Science News, March 29. Reference to experiments conducted at the University of Nice, France, p 202
2. Arai M, Haneishi T, Kitahara N, Enokila R, Kawakubo K, Kondo Y (1976) J Antibiot 29:863
3. Brian PW, Elson GW, Hemming HG, Radley M (1954) J Sci Food Agric 5:602
4. Cole RJ, Kirksey JW, Cutler HG, Doupnik BL, Peckham JC (1973) Science 179:1324
5. Cole RJ, Wells JM, Cox RH, Cutler HG (1981) J Agric Food Chem 29:205
6. Cole RJ, Wilson DM, Harper JL, Cox RH, Cochran TW, Cutler HG, Bell DK (1982) J Agric Food Chem 30:301
7. Cox RH, Cutler HG, Hurd RE, Cole RJ (1983) J Agric Food Chem 31:405
8. Dorffling K, Petersen W (1984) Z Naturforsch 39:683
9. Edwards JV, Cutler HG, Zorner PS, Coffman CB (1987) In: Baker DR, Fenyes JG, Moberg WK, Cross B (eds) Synthesis and chemistry of agrochemicals. ACS Symposium Series 355, Washington DC, p 151
10. Edwards JV, Dailey OD Jr, Bland JM, Cutler HG (1988) In: Cutler HG (ed) Biologically active natural products: potential use in Agriculture. ACS Symposium Series 380, Washington DC, p 35
11. Fulton KND, Bollenbacher K, Templeton GE (1965) Phytopathology 55:49
12. Haneishi T, Terahara A, Kayamori H, Yabe J, Arai M (1976) J Antibiot 29:870
13. Hirota A, Suzuki A, Suzuki H, Tamura S (1973) Agric Biol Chem 37:643
14. Hirota A, Suzuki A, Tamura S (1973) Agric Biol Chem 37:1185
15. Ichimura M, Oritani T, Yamashita K (1983) Agric Biol Chem 47:1895
16. Ilag V-L (1970) PhD Thesis, Purdue University, Lafayete, Indiana 93

17. Isogai A, Washizu M, Kondo K, Murakoshi S, Suzuki A (1984) Agric Biol Chem 48:2607
18. Kawai M, Rich DH, Walton JD (1983) Biochem Biophys Res Commun 11:398
19. Marumo S, Kohno E, Natsume M, Kanoh K (1987) In: Cooke AR (ed) Proc 14th Ann Meeting Plant Growth Reg Soc America, LK Alfred, Florida, USA, 146
20. Meyer WL, Kuyper LF, Lewis RB, Templeton GE, Woodhead SH (1974) Biochem Biophys Res Commun 56:234
21. Meyer WL, Templeton GE, Grable CI, Jones R, Kuyper LF, Lewis RB, Sigel CW, Woodhead SH (1975) J Am Chem Soc 97:3802
22. McCorkindale NJ, Blackstock WP, Johnston GA, Roy TP, Troke JA (1978) 11th Int Sym Chem Nat Prod (IUPAC) vol-1, p 151
23. Nakajima H, Nishimura K, Hamasaki T, Kimura Y, Yokota T, Udagawa S (1987) Agric Biol Chem 51:1221
24. Nakajima H, Nishimura K, Hamasaki T, Kimura Y, Udagawa S (1987) Agric Biol Chem 51:2831
25. Nukina M (1987) Agric Biol Chem 51:2625
26. Okuno T (1978) Chem Abstr 88:73011
27. Oritani T, Ichimura M, Yamashita K (1982) Agric Biol Chem 46:1959
28. Oritani T, Ichimura M, Yamashita K (1984) Agric Biol Chem 48:1677
29. Rademacher W, Jung J (1985) Poster presentation 12th IPGSA Conference, Heidelberg, West Germany (Abstract No P0260)
30. Styer CH, Cutler HG (1984) Plant Cell Physiol 25:1077
31. Surico G, Evidente A, Iacobellis NS, Randazzo G (1985) Photochemistry 24:1499
32. Takahashi S, Oritani T, Yamashita K (1986) Agric Biol Chem 50:3205
33. Takayama S, Isogai A, Nakata M, Suzuki H, Suzuki A (1984) Agric Biol Chem 48:839
34. Takiguchi Y, Yoshikawa H, Terahara A, Torikata A, Terao M (1979) J Antibiot 32:857
35. Takiguchi Y, Yoshikawa H, Terahara A, Torikata A, Terao M (1979) J Antibiot 32:862
36. Templeton GE, Grable CI, Fulton ND, Bollenbacher K (1967) Phytopathology 57:516
37. Terahara A, Haneishi T, Arai M, Hata T, Kuwano H, Tamura C (1982) J Antibiot 35:1711
38. Thimann KV (1935) J Biol Chem 109:279
39. Tomita K, Kitsuwa T, Murayama T, Nakamura T (1987) Agric Biol Chem 51:2633
40. Turner WB, Aldridge DC (1983) Fungal Metabolites vol II. Academic Press, London
41. Ueno T, Nakashima T, Hayashi Y, Fukami H (1975) Agric Biol Chem 39:1115
42. Ueno T, Nakashima T, Hayaski Y, Fukami H (1975) Agric Biol Chem 39:2081
43. Wells JM, Cutler HG, Cole RJ (1976) Can J Microbiol 22:1137
44. Wells JM, Cole RJ, Cutler HG, Spalding DH (1981) Appl Environ Microbiol 41:967

Can PGRs Alleviate the Recently Evolved Cross Resistances to Herbicides?

J. GRESSEL[1]

1 Introduction

1.1 The Evolution of Single Herbicide Resistances

Herbicide resistances have evolved in a large variety of weeds to two types of herbicides having single modes of action:

a) Herbicides inhibiting electron transport in PSII (mainly s-triazines, uracils and some but not all phenylureas),
b) Herbicides inhibiting microtubule formation by complexing with tubulin (dinitroanilines).

One type of cropping system coupled with one particular property of the herbicides seem to be responsible for the "rapid" evolution of resistance. "Rapid" is a relative term; it took 8 to 10 generations of selection with both DDT on flies, and atrazine on weeds to evolve resistant populations; in the first case this happened in a year, in the second 8–10 years. All cases of resistance appeared in either monoculture where a single herbicide was used, or in rotation where only herbicides with the same site of action were used. In all cases the herbicide had high persistence and thus a high season-long selection pressure that did not allow later germination of weeds.

Resistance has yet to evolve where crop-rotations were used; where herbicides of different sites of action were used as mixtures or sequentially, or where herbicides with low persistence were used not more than few times per season. Resistance to atrazine had been predicted to evolve in rotational situations in maize, based on the paucity of information available a few years ago [16]. It is now clear that the genetic unfitness of triazine-resistant weeds causes a decimation of selected plants during the rotational periods. This results in a much lower rate of evolution than had previously been considered possible, and accentuates the value of rotating both crops and herbicides to delay resistance [17].

1.2 The Evolution of Sequencial Resistances

All but one of the 50 weed species that evolved resistance to atrazine or simazine in maize, in orchards or along roadsides, have a chloroplast genome mutation in the

[1]Department of Plant Genetics, Weizmann Institute of Science, Rehovot, 76100, Israel

psbA gene that precludes triazine binding. Such mutant plants are exceedingly rare in untreated populations due to the large number of chloroplast DNA molecules per cell; so rare that only guesses can be made as to the natural frequency of plastome mutations.

In one case, it was found that atrazine resistance evolved in plants having a plastome mutator gene [2]. This nuclear gene increases the frequency of plastome genes by order(s) of magnitude. It is possible that this happened in most cases of triazine-resistance, and that triazine-resistant populations bear this gene in much higher frequencies than wild type populations. This would explain the recent reports of rapid evolution of resistance to photosystem II inhibiting herbicides, to which the triazine resistant plants had previously been sensitive [41]. While these reports still need external verification, it is clear that it is unwise to replace atrazine only with other PSII inhibiting herbicides when resistant populations evolve.

Some genera seem more prone than others to rapidly evolve resistance, as they have evolved resistance to more than one herbicide in many areas. These include *Lolium* spp, *Conyza* = *Erigeron* spp among others. There is even one case of sequentially appearing resistance to herbicides with different sites of action. Paraquat (a photosystem I inhibitor) was used to eliminate atrazine-resistant *Conyza* in vineyards, but resistance to the paraquat used to control the triazine-resistant weed then evolved [32].

1.3 Evolution of General Cross Resistances

Resistance was not initially expected to evolve to paraquat, as this herbicide has no biological persistence; it is only a contact herbicide, and is immediately bound to soil colloids. Farmer persistence made up for this; resistant weed populations evolved where paraquat was applied 6–10 times annually in a variety of weeds. It was clearly shown that PSI was totally suppressed in isolated thylakoids [cf. 11] in paraquat resistant *Conyza*, yet extreme membrane damage appeared only in the paraquat treated, intact chloroplasts of the sensitive biotype [37]. Photosynthesis in intact leaves was immediately inhibited in both biotypes, but the resistant biotype recovered in a few hours [37].

Paraquat resistance in *Conyza* was traced to an elevated level of the "Halliwell-Asada" enzyme pathway responsible for the degradation of active oxygen species [36]. Resistance in *Conyza* is controlled by a single dominant nuclear gene that pleiotropically increases the level of the whole pathway [38].

As resistance seemed to be due to elevated levels of a pathway that generally degrades active oxygen species, it was hypothesized that weeds resistant to paraquat might have co-tolerance to other oxidant stresses [19]. Herbicides that inhibit photosystem I and II, as well as herbicides that bring about protoporphyrin IX accumulation (diphenylethers) all cause the generation of active oxygen species. Many fungal toxins have the same effect, as does drought, when stomates are closed. It was shown that there is a modicum of co-tolerance among species that had or recently evolved resistance to one of these stresses (Table 1). Still, it is expected that the greatest modicum of resistance will be to paraquat, as this herbicide has the shortest half-life in plants. It is rapidly degraded, and thus the resistant-plant has

Table 1. Herbicide cross tolerances to oxidative stresses and relations with enzymes of the Halliwell-Asada active oxygen detoxification pathway [cf. 36–39]

Species	Primary tolerance	Cross tolerance	Enzymatic correlations[a]	ref.
Conyza bonariensis	Paraquat	Atrazine, SO_2	High SOD/GR/AP	[39]
Lolium perenne	Paraquat	SO_2	Elevated SOD/GR	[39]
Lolium perenne	SO_2	Paraquat	Elevated SOD/GR	[39]
Nicotiana tabacum cv. Florida	O_3	Paraquat	Elevated SOD/GR	[39]
Cotton	Drought	Paraquat	Elevated GR	[5]

[a] Data are relative to the sensitive biotype. SOD – the chloroplast isozyme of a Cu-Zn superoxide dismutase, GR-glutathione reductase, AP-ascorbate peroxidase.

only to be capable of tolerating its toxic effects for the shortest of all the active-oxygen generating herbicides.

2 The Evolution of Cross Resistances to Wheat – Selective Herbicides

The first highly selective herbicides for wheat – the phenoxyacetates (2,4-D and MCPA) were released in the 1940s, and have remained the major broad leaf weed herbicides to date. They have low selection pressure on weeds, and no cases of resistance have evolved. They are being heavily replaced by the highly persistent sulfonylurea types, especially chlorsulfuron. The persistence of chlorsulfuron is especially great in the alkaline soils of the northern plains, such that sugar beets cannot be cultivated for more than 3 years after chlorsulfuron application [31], and thus it is not widely used in the spring wheat growing areas. Chlorsulfuron has replaced about 40% of the phenoxyacetate usage in the winter wheat growing areas of the US central plains [1]. Resistance is expected to rapidly evolve in these areas. Some cases will be due to mutations in the gene for the target enzyme and some cases will be resistant to virtually all wheat selective herbicides (see below).

With the advent of the phenoxyacetates, grassweeds became the major weed problem in wheat; the grass weeds evolved during wheat domestication in the Middle East and spread with wheat throughout the world. The grass weeds evolved phenological and morphological 'mimicries' such that they ripened with wheat, and their plants and seeds were hard to distinguish from wheat seeds. The grass weeds are highly competitive with wheat; a few plants per meter can drastically lower yields. Some of the grass weeds can utilize fertilizers better than wheat and are even more competitive under intensive agriculture. The lowering of wheat height to increase the proportion of photosynthate going into the grains (by PGRs or by breeding) requires good weed control, so that the weeds do not tower above the grain [cf 15].

It was a daunting task for industry to find a few compounds that could distinguish between wheat and the related weeds from the Graminae. The first carbamates and thiocarbamate herbicides for this purpose were only marginally selective and often damaged wheat. A few compounds were developed with the desired selectabilities, although they were often efficacious on only one or two grass weeds. Chlorotoluron was found to be especially useful for controlling *Alopecurus myosuroides* (blackgrass or slender foxtail) in Northern Europe. Diclofop-methyl was especially developed to control *Avena fatua* and related *Avena* spp. among the worst worldwide problem in wheat. The control of *Avena* spp. is not perfect; i.e. the selection pressure is low, which is probably the reason why there is only one major case of resistance reported in *Avena* [cf 33].

The high selection pressure exerted by chlorotoluron on *Alopecurus* brought about the evolution of resistance in one area of England (so far). This resistance was of a new kind. Chlorotoluron acts on photosystem II, yet photosynthesis of isolated chloroplasts of the resistant biotype is impaired. The researchers discovered that the plants of the resistant biotype was co-resistant to every wheat selective herbicide that was previously inhibitory to this weed (Table 2). It was even resistant to some herbicides that are injurious to wheat. This is not a sequential evolution of resistance; the resistant *Alopecurus* populations had no history of being treated with any of these other herbicides.

Diclofop-methyl is widely used in Australia to control *Lolium rigidum* (annual ryegrass) both in broadleaf crops and in wheat. This *Lolium* weed is often cultivated there as a pasture grass in rotation with wheat and other crops, guaranteeing a huge reservoir of seeds from which to select resistance. A loss of control of diclofop-methyl was reported a few years ago in one population [22]. Since then, resistance has concurrently evolved in innumerable foci in four Australian states. Resistance is characterized by a simultaneous cross resistance to wheat herbicides that previously controlled it, and to other herbicides that damage wheat (Table 2). There are some differences in the quantitative levels of cross resistances in the Aus-

Table 2. Cross resistances resulting from repeated selection pressure of chlorotoluron on *Alopecurus myosuroides* and diclofop-methyl on *Lolium rigidum*

Herbicide group	Site of Action[a]	Multiple resistance
	Alopecurus myosuroides[b]	
Phenylureas	PSII	Chlorotoluron, isoproturon, metoxuron
Phenoxypropanoates	ACoAC	Diclofop-methyl, *fluazifop-butyl*
Cyclohexene-1-one	ACoAC	Tralkoxydim
Thiocarbamate	Lipids(?)	Triallate
Sulfonylurea, imidazolinone	ALS	Chlorsulfuron, imazamethabenz
Triazines	PSII	*Terbutryn*, *simazine*, SMY 1500
Dinitroaniline	Tubulin	Pendimethalin
	Lolium rigidum[c]	
Phenoxypropanoates	ACoAC	Diclofop-methyl, *fluazifop*, *haloxyfop*, Chlorazifop
Cyclohexene-1-ones	ACoAC	*Sethoxydim*, tralkoxydim, alloxydim
Sulfonylureas	ALS	Chlorsulfuron, metsulfuron, triasulfuron
Acetanilide	PSII	Propanil
Dinitroaniline	Tubulin	Trifluralin

Herbicides in italics are not recommended for use with wheat.
[a] PSII – photosystem II; ACoAC-acetyl CoA carboxylase; ALS – acetolactate synthase.
[b] [29, 30];
[c] [22, 23, 33, 34] and Powles (pers. comm.).

tralian biotypes (Powles, pers. comm.), further suggesting that each biotype evolved separately, i.e. the spread of resistance is not due to seed dissemination. These *Lolium* populations had never seen these other herbicides, many of which were introduced into Australia after diclofop-methyl resistance evolved.

In the two cases described in Table 2, there are basically no major chemical methods of selective weed control for controlling these grass weeds in wheat. The few herbicides that control these grasses also kill wheat. The only chemical methods available would be herbicides with placement selectivity. Such herbicides can be used in only certain soil types and the planting depth of wheat must be carefully controlled.

Mecoprop-resistant *Stellaria media* evolved in England, and the biotype is co-tolerant to all phenoxyacetate and propionate herbicides, but not to other wheat herbicides [28].

It is highly likely that these are not the only cases of cross-resistance to appear in wheat. It is expected that more resistances will evolve to the newer, very effective grass killers as well as to the highly persistent sulfonylureas and imidazolinones that are mainly broad-leaf-killing herbicides. This latter group has some effect on grasses, and the recently evolved resistant grasses are also resistant to sulfonylureas (Table 2). Two types of resistance have evolved to sulfonylureas in laboratory and field selections; modified target site resistance at the level of acetolactate synthase and cross resistance, as shown in Table 2 [cf. 15]. There are wheat selective sulfonylurea herbicides available with far less persistence and sufficient weed control efficacy. It is very worth while considering abandoning the highly persistent ones in favor of those newer ones. The situation is exacerbated by the possibility of use of other acetolactate synthase inhibitors in rotational crops.

3 The Metabolic Limitations of Wheat

It is hypothesized that it is highly improbable that cross resistance to herbicides of many sites of action could occur in most crops, and that wheat is a special case. This is based on the mode of selectivity in wheat in contrast to the modes of selectivities of other herbicides. Most crop selectivities are based on the ability of the crop to degrade the selective herbicide. For example, maize naturally possesses three different glutathione -*S*-transferases with different herbicide specificities, glycosyl transferases and many oxidases. Different herbicides are degraded by different enzyme systems in maize.

In an extensive review of the literature on the degradation of selective herbicides in wheat it was noted that essentially all the herbicides for which information was available, are oxidized [15]. This oxidation consists mainly of aryl or alkyl hydroxylations at different points of the herbicide molecules (Table 3). In all of these cases, the oxidation is by addition of an atom of oxygen; typical of the reaction of a monooxygenase. Monooxygenases are often highly specific, but there are types, especially in liver, which have very broad substrate specificities. At present we cannot say if wheat has one or many such enzymes, or how they evolved. However, wheat does seem to use only a monooxygenase system to degrade herbicides. Many researchers assume, but based on analogies and rather on slim

Table 3. Types of monooxidation of wheat selective herbicides

Herbicide	Ring	Alkyl	Ref.
	Hydroxylation		
Chlorotoluron		+	[13]
Chlorsulfuron	+		[3]
2,4-D	+		[40]
Dicamba	+		[9]
Diclofop-methyl	+		[14]
Isoxaben		+	[6]
Imazamethabenz		+	[4]

evidence, that the monooxygenase system is a NAD(P)H dependent cytochrome P450 mixed function oxidase system.

4 Biochemical "Mimicry" in the Weeds

As the grass weeds in Table 2 evolved resistance to all the wheat selective herbicides, irrespective of their site of action, the most likely hypothesis is that they evolved the same biochemical detoxification mechanism as wheat, i.e. evolved a biochemical "mimicry". This is supported by evidence that compounds that suppress the herbicide degradation in wheat also suppress their degradation. Little is known about the monooxygenases of wheat, and less is known about those in weeds, and presently not too much can be said. Still, it is clear from the data in Table 2 that biochemical mimicries need not be absolute. The weeds evolved spectra of cross resistances that are broader than the resistance in wheat.

5 Overcoming the Problem: Can Monooxygenase Inhibiting PGRs be Selective Synergists?

This author is of the opinion that the wisest and most effective solutions will be by increasing the genetic capacity of wheat to degrade herbicides. This can best be done by genetically engineering genes for herbicide degradation into wheat. The economics of doing this, especially into hybrid wheats, are discussed elsewhere in depth [15].

Many PGRs have been developed that dwarf plants because they inhibit kaurene oxidase or other cytochrome P450 monooxygenases in the GA pathway [cf. 25, 35], as well as other cytochrome P450's [42]. It has recently been shown that these inhibitors are not as specific as had been thought; they also can inhibit the monooxygenases that degrade various herbicides (Fig. 1, Table 4). This can be measured as the actual rate of degradation of the herbicide (Fig. 1A), or by measuring using whole plants when the herbicide has lost its effect (Fig. 1B). Various researchers have indicated that ancymidol and paclobutrazol have similar effects on herbicide degradation (pers. comm.). These compounds, by inhibiting

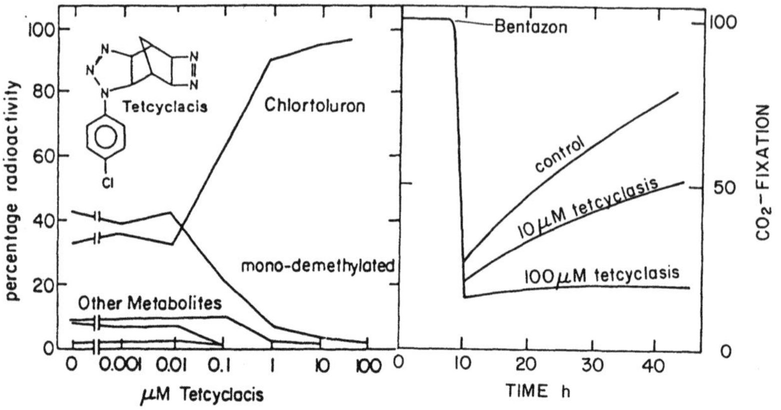

Fig. 1. The suppression of herbicide detoxification by tetcyclasis. The inhibition of chlorotoluron monooxygenation in cotton (*left*). Redrawn from Cole and Owens [8]. The prevention of loss of bentazon inhibition of photosynthesis (*right*). CO_2 fixation is measured as the percent of the control rate prior to treatment with bentazon. Redrawn from Fritsch et al. [10]

Table 4. The suppression of herbicide degradation by monooxygenase inhibitors

Herbicide	Presumed cytochrome-P450 inhibitor	Species	Ref.
EPTC	SKF 525A[a], PBO[b]	Maize	[26]
Tebuthiuron	EPTC/butylate with dichlormid	Maize	[20]
Chlorotoluron	Aminobenzotriazole	Wheat	[12]
	2,4-DP[c]	Cotton, maize	[8]
	Tetcyclasis	Cotton, maize	[8]
Metolachlor	2,4-DP	Cotton	[8]
Chlorotoluron, isoproturon	Aminobenzotriazole	*Alopecurus myosuroides*	[24]
Diclofop-methyl	Aminobenzotriazole	*Lolium rigidum*	[d]
Bentazon	Tetcyclasis	Wheat	[10]
2,4-D	Aminobenzotriazole	Potato	[7]

[a] SKF 525A-diethylaminoethyl-2-2-diphenylvalurate HCl.
[b] PBO-piperonylbutoxide.
[c] 2,4-DP-2,4-dichlorophenoxy-1-propyne.
[d] S. Powles (pers. comm.).

degradation, keep the herbicide active for a sufficient length of time for it to have an irreversibly toxic effect on the plant. This is a positive effect for weed control but would be devastating to a crop. Such compounds are commonly referred to as herbicide synergists [cf 18].

The problem is selectivity; to suppress herbicide degradation in the weeds but not in wheat. As the biochemical mimicries are not absolute, even small differences between monooxygenase systems can be amplified by synergists. By analogy we may consider that this may be possible.

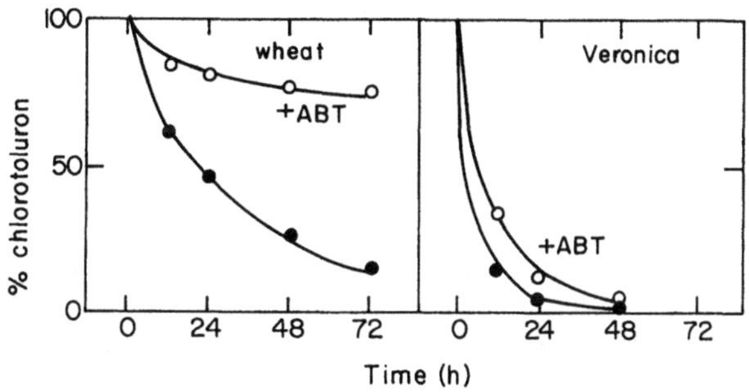

Fig. 2. Differential effect of aminobenzotriazole on chlorotoluron degradation in wheat and *Veronica persica*. Modified and redrawn from Gonneau et al. [13]

For example: atrazine is degraded by glutathione -*S*-transferase action in maize and grass weeds. The synergist tridiphane forms a suicide inhibitor complex with glutathione, inhibiting the enzyme, thereby suppressing atrazine degradation in both maize and weeds. There are differences, however; the inhibitor-complex is less inhibitory to the maize enzyme and the complex is thus degraded more rapidly in maize than weeds [27]. Together, this results in death in weeds, and life in maize.

Different species degrade the same herbicide by the same type of enzyme, but at different points on the molecule. For example, chlorsulfuron and other sulfonylureas are attacked by monooxygenases at different positions on the molecule [cf. 3]. This suggests different enzymatic specificities. Indeed, inhibitors of monooxygenases can also show different specificities. Aminobenzotriazole (ABT) suppresses chlorotoluron degradation in wheat with hardly any effect on the weed *Veronica* (Fig. 2). This is of course the antithesis of the herbicide synergist desired to kill resistant weeds; such a synergist must not affect degradation of the herbicide in wheat while suppressing weed degradation. Still, the data show that there are differences between wheat and weeds that can be amplified. This may be hard in reality. It is expected that there will be fewer differences in the monooxygenases of wheat and related grass weeds, than between wheat and the dicot *Veronica*.

Most of the work on synergizing monooxygenase-degrading herbicides has been done with wheat. Thus, we have little information on such synergisms in weeds. It thus seems advisable for industry to re-screen the azole type compounds (originally synthesized as GA biosynthesis inhibiting PGRs, and sterol biosynthesis inhibiting fungicides [cf. 25, 35, 42]) for differential herbicide synergisms in wheat and its weeds. The systemic fungicides inhibit monooxygenases in fungi with little effect on plants; again indicating different inhibitor specificities. Some compounds that were inactive, as fungicides or PGRs may have desirable properties as selective synergists. This approach may supply the necessary suppression of "evolved multiple herbicide resistance" to wheat herbicides via biochemical mimicries.

6 Concluding Remarks

As wheat is the major crop for feeding people, it would be unwise to overlook the potential of cross-resistance to wheat herbicides evolving at many points over the globe. It is an all too common practice to avoid such issues just because they have only occurred elsewhere (AIDS was hardly studied when only indigenous to Africa). The advent of more effective herbicides, effecting high selection pressures, will increase the frequency of occurrence of such resistances. There are ways of overcoming such resistances. At present we know too little about the mechanisms of resistance and how they evolved. This precludes making more accurate predictions about how resistance will evolve in the future, and where.

The mechanisms are becoming available to broaden the genetic base of wheat by genetic engineering. Still, a five-to ten-year lead time is necessary to put the right genes in the right varieties so as to be prepared for the advent of more resistance. A similar lead time will be required to screen, test, develop and register a synergist for use to suppress herbicide resistance in wheat. The world cannot afford to have more weed competition in wheat; the reduction in food supply would be too drastic. Herbicides have been a major factor in gaining the high wheat yields presently achieved. We cannot allow ourselves to lose them to the effects of cross resistances. Why wait until it is too late?

Acknowledgements. Much of the information collated and ideas developed in this manuscript were elucidated while on sabbatical leave at Biotechnology Affiliates. The author is indebted to them for providing this opportunity and to the University of Reading for their facilities.

References

1. Anonymous (1988) In: Agricultural resources, inputs, situation and outlook report AR-9. USDA Econ Res Serv, Washington, p 16
2. Arntzen CJ, Duesing JH (1983) In: Ahmand F, Downey K, Schultz J, Voellmy RW (eds) Advances in gene technology. Academic Press, New York, p 273
3. Beyer EM, Duffy MJ, Hay JV, Schlueter DD (1988) In: Kearney PC, Kaufmann DD (eds) Herbicides: chemistry, degradation and mode of action. Vol 3. Dekker, New York, p 117
4. Brown MA, Chiu TY, Miller P (1987) Pestic Biochem Physiol 27:24
5. Burke JJ, Gamble PE, Hatfield JL, Quisenberry JJ (1985) Plant Physiol 79:415
6. Cabanne F, Lefebvre A, Scalla R (1987) Weed Res 27:135
7. Cole DJ, Loughman BC (1985) Physiol Veg 23:879
8. Cole DJ, Owens WJ (1987) Plant Sci 50:13
9. Frear DS (1976) In: Kearney PC, Kaufmann DD (eds) Herbicides, chemistry, degradations and mode of action. Vol 2. Dekker, New York, p 542
10. Fritsch H, Rademacher W, Retzlaff G (1987) In: Greuter W, Zimmer B, Behnke HD (eds) Proc XIV Int Bot Cong, Berlin 2-11 3b-5
11. Fuerst EP, Nakatani HY, Dodge AD, Penner D, Arntzen CJ (1985) Plant Physiol 77:984
12. Gaillardon P, Cabanne F, Scalla R, Durst F (1985) Weed Res 59:397
13. Gonneau M, Pasquette B, Cabanne F, Scalla R (1988) Weed Res 28:19
14. Gorecka K, Shimabukuro RH, Walsh WC (1981) Physiol Plant 53:55
15. Gressel J (1988) Wheat herbicides: the challenge of emerging resistance. Biotechnology Affiliates, Checkendon/Reading (UK), p 247
16. Gressel J, Segel LA (1978) J Theor Biol 75:349
17. Gressel J, Segel LA (1989) In: Green MB, Moberg WK, LeBaron HM (eds) Fundamental and practical approaches to combating resistance. Am Chem Soc Symp Ser, Washington DC
18. Gressel J, Shaaltiel Y (1988) In: Hedin PA, Menn JJ, Hollingworth RM (eds) Biotechnology for Crop Protection. Am Chem Soc Symp Ser 379, Washington DC, p 4
19. Gressel J, Ezra G, Jain SM (1982) In: McLaren JS (ed) Chemical manipulation of crop growth and development. Butterworth, London, p 79
20. Hatzios KK (1981) Weed Sci 29:601

21. Heap IM (1987) In: Lemerle D, Leys AG (eds) Proc 8th Aust Weeds Conf, p 114
22. Heap I, Knight R (1986) Aust J Agric Res 37:149
23. Heap J, Knight R (1982) J Aust Inst Agric Sci 48:156
24. Kemp MS, Caseley JC (1987) Br Crop Prot Conf – Weeds, Thornton Heath, UK, p 895
25. Koller W (1987) Pestic Sci 18:129
26. Komives T, Dutka F (1980) Cereal Res 8:627
27. Lamoureux GR, Rusness DG (1986) Pestic Biochem Physiol 26:323
28. Lutman PJW, Snow HS (1987) 1987 Br Crop Prot Conf – Weeds, Thornton Heth, UK, p 901
29. Moss SR (1987) Br Crop Prot Conf – Weeds, Thornton Heath, UK, p 879
30. Moss SR, Cussans GW (1987) In: Ford M, Hollomon D, Khambay B, Sawicki R (eds) Combating resistance to xenobiotics: biological and chemical approaches. Horwood, Chichester, p 200
31. Petersen DE (1987) 1988 Agricultural weed control guide, Circ. W-253 rev, NDSU Extension Serv, Fargo, p 71
32. Pölös E, Mikulas J, Szigeti Z, Matikovics B, DoQuy H, Parducz A, Lehoczki E (1988) Pestic Biochem Physiol 30:142
33. Powles SB (1987) In: Lemerle D, Leys AG (eds) Proc 8th Aust Weeds Conf, p 107
34. Powles SB, Liljegren D (1988) Weed Sci Soc Am Abstr 28:187
35. Rademacher W, Fritsch H, Graebe JE, Sauter H, Jung J (1987) Pestic Sci 21:241
36. Shaaltiel Y, Gressel J (1986) Pestic Biochem Physiol 26:22
37. Shaaltiel Y, Gressel J (1987) Plant Physiol 85:869
38. Shaaltiel Y, Chua NH, Gepstein S, Gressel J (1988) Theor Appl Genet 75:850
39. Shaaltiel Y, Glazer A, Bocion PF, Gressel J (1988) Pestic Biochem Physiol 31:13
40. Shimabukuro RH (1985) In: Duke SO (ed) Herbicide physiology, vol 2. CRC, Boca Raton, p 215
41. Solymosi P, Lehoczki E (1987) Novenyvedelem 23:439
42. Vanden-Bossche H, Marichal P, Gorrens J, Bellens D, Verhoeven H, Coene M-C, Lauwers W, Janssen PAJ (1987) Pestic Sci 21:289

Economic Aspects of Plant Growth Regulators

A.D. BAYLIS[1]

1 Introduction

PGRs have had a chequered history on the world agrochemical scene. Over-optimism in some quarters in the early 1970s led to predictions that PGRs would soon outstrip herbicides in sales [cited by 4]. Clearly, this has not happened and several companies in the agrochemical industry have pulled out of extensive PGR research, disillusioned after making substantial investments in the area.

Presently, although PGR sales are a small portion of the agrochemical market, several new compounds are in late stages of development or early commercialization. In 1987, the global sales of agrochemicals amounted to some $20 billion, this figure having almost doubled over the past 15 years (Table 1). The contribution made by PGRs, including desiccants and defoliants, accounted for only 3.6% of sales. It is predicted, however, that the real annual growth rate of the PGR sector will be 5.9% in the short-term, almost twice that of the agrochemical market as a whole. The success of the current generation of new PGRs, mainly retardants based on triazole chemistry (eg paclobutrazol, uniconazol, triapenthenol), will be crucial both to the realization of these forecasts and the commercial future of the science. Also important will be the continued expansion into non-agricultural areas such as growth control in amenity grass and trees.

Excluding desiccants and defoliants, 1987 sales of PGRs were $430 million. The major products are ethephon, accounting for about 20% of the total (over $70 million), CCC, GA_3 and mepiquat. The latter three have smaller, but significant, sales of $30 million or more. Available data (excluding Eastern Europe) show that the largest sales are of anti-lodging products for use on wheat and barley (30%). Cotton and orchard fruit both have expenditures of $30–40 million, and grapes, tobacco and potatoes $15–20 million. Geographically, western Europe has the largest sales of PGRs, due to their use in intensive cereal production. Regional distribution of sales are shown in Table 2. The prominence of the Far East as a PGR market is likely to increase substantially in the near future as the market for anti-lodging products in Japanese rice develops. Compounds such as paclobutrazol are also stimulating markets in tropical fruit, giving regulation of flowering [2] and increases in yield [5], in addition to growth control.

Having described the current position of PGRs on the world agrochemical scene, the rest of this paper will consider the economics of PGRs from the points of view of the agrochemical industry and the customer, and then discuss the prospects for PGRs in the changing fields of agriculture and plant science.

[1] ICI Agrochemicals, Jealott's Hill Research Station, Bracknell, Berkshire, RGI2 6EY, UK

Table 1. Value of the world agrochemical market (Source: Wood Mackenzie)

	Sales in $ Million of 1987		
	1972	1987	1990
Herbicides	4015	8600	9205
Fungicides	2440	4100	4585
Insecticides	3945	6100	6680
PGRs	360	720	855
Others	390	480	525
Total	11 150	20 000	21 850

Table 2. Regional distribution of PGR sales (excluding desiccants and defoliants) in 1986 (Source: Wood Mackenzie)

Annual expenditure on PGRs	$ Million
West Europe	120.0
East Europe	100.0
North America	100.0
Far East	22.5
Others	17.5

2 Agrochemical Industry Perspective

2.1 Phases of Product Invention

The sequence of events in the progression of an idea through to a product can be divided into phases of targeting, to determine a value; research, to discover an active molecule; and development, to establish safety to the user and the environment, define the scope of efficacy, generate guidelines for use of the product, and institute a production process.

2.2 Targeting

A pre-requisite to commencing a research programme is a targeting exercise to establish that a sufficiently valuable opportunity exists to justify the large investment necessary.

With conventional agrochemicals, the technical profile describing the efficacy of a new product can usually be compared to that of a relevant precedent on the market, and a value per hectare estimated. For a new PGR, with so few product precedents, a bearable price usually has to be estimated de novo. This involves calculating the net financial benefit of the prospective product to the end-user and predicting the price he would be prepared to pay for it. Most often a minimum cost: benefit ratio of 1:3 is used. An estimate of the volume of the sales must then be made in order to arrive at the gross returns. Usually, the potential market for a new product will be segmented. Each sector may be characterized by geographic or agronomic differences, which may have implications for the rate of user uptake of a new product, or the robustness required to attain a sufficiently large market size. In addition, especially in the case of PGRs, various degrees of sophistication of agricultural practices will determine the rate of growth in sales. The early adopters will be those already working near to the optimum efficiency with respect to yield and management. An important strategy in PGR marketing can be to use such pioneers to develop the market.

A commonly used concept in targeting is the *net present value* (NPV). This is a forecast of the overall profitability in real terms allowing for the costs of investments that must be made in order to achieve and support sales. An important part of the calculation of the NPV is the estimate of the time taken from launch to reach maximum sales. With any new PGR, except those in the best established markets, e.g. anti-lodging in cereals, more gradual uptake by users can be expected than with a conventional agrochemical. Therefore, the payback period on the investment in research and development for PGRs can be expected to be longer the more novel and unprecedented the invention.

2.3 Research

PGR research by the agrochemical industry is a high risk business. The risks stem from anticipating the user demand for PGR products, and from the difficulties inherent in the invention process. As Lürssen [6] pointed out, in a situation where continuously more compounds need to be screened per product, PGRs do hold an advantage in that there are few established products to better. However, this is probably more than counterbalanced by screening problems resulting from that same novelty. Geissbuehler et al. [3] have outlined some of the difficulties. Whole plant screens derived from herbicide screening are of limited value and versatility when used in isolation from other testing procedures, e.g. in vitro biochemical tests, and when using a random chemical input. That screening for PGRs other than retardants is inherently problematical is clearly demonstrated by the paucity of such products reaching the market place. A better understanding of specific PGR targets at the biochemical and molecular level is needed to facilitate both the rational design of chemical input and the development of suitable screens. This knowledge is essential in order encourage the invention of marketable molecules showing effects in whole plants beyond those of retardants. In fact, research into PGRs has now entered a more realistic phase where these needs are fully recognized.

2.4 Development

PGRs going into development, where a company is committed to bringing a new product to the market, face ever more stringent environmental and toxicological testing necessary for registration with regulatory authorities. A minimum of 6 to 8 years is now required to obtain sufficient data to establish that a product is safe to the environment, the applicator and the consumer.

Alongside such testing must be a very large programme of trials to demonstrate the degree and consistency of biological effect. In the case of PGRs, where variable performance in the field is often cited as a problem, these trials can be expected to be more extensive than those of other agrochemicals. Inhibitors of GA biosynthesis, and other compounds affecting such fundamental plant processes, can be expected to have utilities on a broad spectrum of crop and amenity species. So, in addition to testing across a wide range of environments, a diverse range of species must also become involved in the efficacy programme. Experience with paclobutrazol has shown that uses never envisaged in the original targeting can become an important part of the project, especially when the compound reaches the, perhaps more adventurous, hands of university scientists and other co-operators. However, tight regulatory controls that restrict uses to those stated on the product label, may make this means of expanding product use more difficult in future. The cost of destroying treated crops for which there is no experimental use clearance can be a significant deterrent to working in high value crops such as fruit. Moreover, in such perennial species, continuing trials over a number of years is necessary to determine the reliability of effect and its persistence from season to season, making them substantially more expensive than trials with annual crops.

Compounds initially tested under conditions requiring mg quantities go through stages demanding g and kg quantities, to the several hundred tonnes of active ingredient eventually needed for commercially viable sales of product. The chemical routes to the active molecule will change with the scale of synthesis, usually necessitating research into process chemistry. If a new process is discovered to produce a molecule more cost-effectively, it will be patented as further protection to an invention.

2.5 Economic Overview

The overall dynamics of cash flow in the research, development, and sales phases have been described for typical agrochemicals by Lever [4]. These show that a break-even point will not be reached for several years into sales, when there may be only a few years of patent life left on a molecule. At least $40 million is required to cover the research and development costs of a new agrochemical. The cost of bringing a PGR to sales may be significantly greater. The strategy for most of the PGRs marketed so far has been to adopt a sales plan that will generate some early cash from markets with smaller registration costs, typically ornamentals and other non-food outlets, before moving into the far more costly areas of agriculture and perennial horticulture. New PGRs must have opportunities in major crops to be viable. Smaller product niches can only be filled as upsides to the main uses.

There is, therefore, a very large degree of risk associated with agrochemical research in general, and PGRs in particular. The implications are clearly that only very large companies (or other institutions) can be expected to cope with the size of investment needed in PGR research and the time-scale of the economics. The attraction of PGRs, however, is that they offer potentially new markets for agrochemicals, while the conventional ones head inexorably towards becoming commodities.

3 Grower Perspective

In agriculture, the most successful PGRs have been those which have become integrated into crop management systems as key inputs. These are well illustrated by the use of CCC and ethephon-based products in winter wheat in western Europe, and by mepiquat and ethephon in US cotton.

3.1 Wheat

Imperial Chemical Industries plc (ICI) conducts an annual survey amongst its cereal farmer customers in the UK. Data on all aspects of crop husbandry, from cultivations through to yields, including a financial breakdown, are held on a searchable database. It has, therefore, been possible to examine the impact of PGRs used in wheat, primarily as anti-lodging agents, using data from several thousand fields over five seasons. Table 3 shows that the use of more or less intensive PGR regimes is justified in terms of yield and profitability on groups of varieties with contrasting degrees of lodging resistance. If these PGRs were acting soley as anti-lodging agents, the benefits of treatments to the stronger strawed varieties are perhaps surprising. In the absence of lodging, small increases in yield attributed to CCC have often been noted, as have occasional decreases. In the ICI surveys, fields treated with PGRs have consistently yielded around 0.5t/ha more than untreated fields. There are no side-by-side comparisons of treated or untreated fields otherwise identical, but the large numbers of crop fields involved would suggest the differences to be real. An examination of the financial details gives a clearer

Table 3. Effect of PGR regime on yield and profitability of crops of strong and weak strawed UK winter wheat varieties (1986) [Source: ICI (original data is converted at 1.60 $/£)]

	No. Crops		Yield (t/ha)		Gross margin ($/ha)	
	Strong	Weak	Strong	Weak	Strong	Weak
No PGR	256	205	7.43	7.16	989	869
Early application only[a]	416	509	7.96	7.75	1040	941
Early + later application[b]	89	105	8.54	8.49	1102	1067

[a] CCC at Zadoks growth stage 30–31.
[b] CCC at Zadoks growth stage 30–31 followed by ethephon + mepiquat at Zadoks growth stage 32–39.

Table 4. Effect of PGR use on the yield, costs and profitability of winter wheat[a]

	No. Crops		Increase in yield (t/ha)	PGR cost ($/ha)	Increase in variable costs ($/ha)	Increase in gross margin ($/ha)
	−PGR	+PGR				
1983	1486	1561	0.51	9.31	34.98	59.01
1984	1053	2025	0.57	7.78	30.08	60.27
1985	770	1874	0.51	7.98	45.81	44.62
1986	456	1461	0.64	8.21	36.21	64.69
1987	233	914	0.42	7.87	34.26	50.46

[a] Data shown are increases above untreated fields in the UK [Source: ICI (original data in £ converted to $ at 1.60 $/£)]

indication of how PGR regimes are supporting substantial increases in profitability. Table 4 illustrates how the total variable costs (attributed to seed, fertilizer and agrochemicals) increase when PGRs are included. More detailed analyses show PGR use to be associated with increased rates of N fertilizer and greater use of fungicides. PGRs give the farmer confidence to optimize the use of N. Greater fungicide use may come from a greater attention to agronomic details by those prepared to adopt a more complex growing system. Or, it may come from necessity due to a lusher more disease susceptible crop! The overall effects of the more intensive husbandry are clearly shown as increased gross margins. In this situation, cost: benefit ratios are in the order of 1:6 across the UK national wheat crop.

3.2 Cotton

In the USA, cotton growing is a highly sophisticated agriculture. Ethephon and mepiquat are each used on about 15% of the total acreage. However, their use can be far greater locally where their effects are particularly beneficial. Ethephon is principally used as an aid to more efficient harvesting, aiming to save the expense of a second harvest by opening up the bolls earlier. The earlier harvest brought about by ethephon use may avoid losses due to insects or frost at the end of the season. It may also allow less expenditure on defoliants by predisposing leaves to drop. Mepiquat is used to control the excessive vegetative growth of rank cotton encouraged by irrigation, tall varieties, high fertility land and wet weather. The measurable advantage is a yield increase through more efficient distribution of growth, reduction in boll rot and allowing better insect control in a crop canopy more accessible to sprays of pesticides.

The success of these more important PGRs can be attributed to two main factors. First, there can be a synergy between the use of PGRs and other inputs. Secondly, PGRs can overcome adverse effects of the environment. These have given farmers clear increases in profitability.

4 The Future

4.1 A Changing Agriculture

What are the future prospects for PGRs in a world where the economics of crop production will change with those of a rapidly changing agriculture? The structure of agriculture in the developed world will continue to move towards larger farming units. Paradoxically, the number of small part-time farmers may also rise as farming on small acreages can no longer fully provide a living and additional full-time work must be sought. This is what has happened in Japan, where now almost 60% of people involved in rice farming are mainly engaged in other employment [7]. In both cases, increasingly efficient production practices will be important. Agrochemicals showing greater versatility and flexibility will command a premium.

In recent years, areas under cultivation to major crops have become static or declined and prices have fallen when inflation is taken into account (Fig. 1). Gone is the general encouragement to farmers in the developed world to produce more food. Farmers had risen to this incentive, aided by the advances in agricultural technologies provided as a result of similar encouragement to research. Even in Japan with its highly protected internal rice market, where prices have made the agrochemical business so attractive, the rice price has recently been cut by 6%. More dramatic measures to reduce surpluses are in the process of being implemented in the USA and EEC. Set-aside schemes have reduced the area of US maize and similar schemes, or quotas and price restraint policies determining the quantity and quality of saleable produce, are coming into effect in Europe. These measures will further increase the pressure on farmers to maintain profitability by reducing the unit cost per tonne of production. They will also mean that quality of produce will assume a greater importance, as achieving a quality premium may make the difference between profit and loss.

There are many implications for PGRs in the changing strategies of land-use. They may be considered as those within agriculture and those in non-agricultural

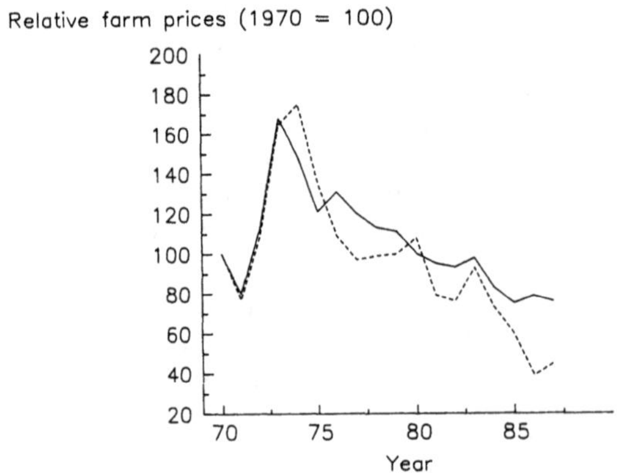

Fig. 1. Average annual farm prices of UK wheat and US maize adjusted to 1970 values: 1970 100. Sources: UK Home Grown Cereals Authority (wheat), United States Department of Agriculture (maize)

or amenity areas. In crop husbandry, PGRs will have roles in maintaining or increasing yield and quality, and as management aids enabling, for instance, easier and earlier harvesting. The potential for retardants in Japanese rice provides a good example where there is a premium for producing high quality 'delicious tasting' varieties, particularly cv Koshihikari. However, these varieties are very susceptible to lodging. More direct means of enhancing yield or quality will be more difficult to achieve, but will be well rewarded. Inevitably, the use of such products in the developing world will follow later, but the needs of the rapidly escalating populations of such countries must not be forgotten.

The present recession in agriculture also serves to highlight opportunities in amenity species. In North America, several hundred million dollars per annum are spent on pruning amenity trees. The newer triazole retardants are far superior in efficacy and symptomology to older anti-pruning agents and have only begun to tap the potential market. Similarly, the potential to vastly reduce mowing costs on golf courses, road-side verges and other amenity grass areas is potentially large for products with the right technical profile.

4.2 Biotechnology

The future of PGRs will also be affected by advances in biotechnology, particularly when applied to plant breeding. Biotechnology, as a means to producing desirable improvements to plants, holds several advantages over the chemical approach: stimulation is more feasible, more precise effects are possible, and registration may be notionally cheaper; PGRs, however, do allow the grower to choose if and when to modify plants [1].

Biotechnology may also, however, offer attractive solutions to PGR screening problems. One example would be the use of recombinant DNA technology to engineer specific genes to overexpress target enzymes in suitable organisms, in order to obtain sufficient material for crystallization and x-ray diffraction studies to determine the nature of the active site. Molecular modelling might then be used as a route to a rational design of inhibitors. Those engaged in PGR research will need to work increasingly closely with those working in biotechnology in order to avoid conflicting approaches and seek ones which will be complementary.

There are several indications of what the future holds in the light of advances in biotechnology, but in the end it is salutory to look back to at the history of CCC as a straw shortener in intensive wheat production. Sales were initially slow after being launched in the late 1960's, particularly in the UK. Here, when the first wheat with an Rht semi-dwarfing gene was introduced in 1975, far shorter and higher yielding than the currently grown varieties, the prospects for CCC seemed grim. However, Fig. 2 shows CCC has between extremely successful, even though in 1987 93% of the winter wheat acreage was planted to semi-dwarf varieties. The factor explaining this apparent paradox was the increasing use of nitrogen fertilizer, which pushed even short-strawed varieties to the point of lodging.

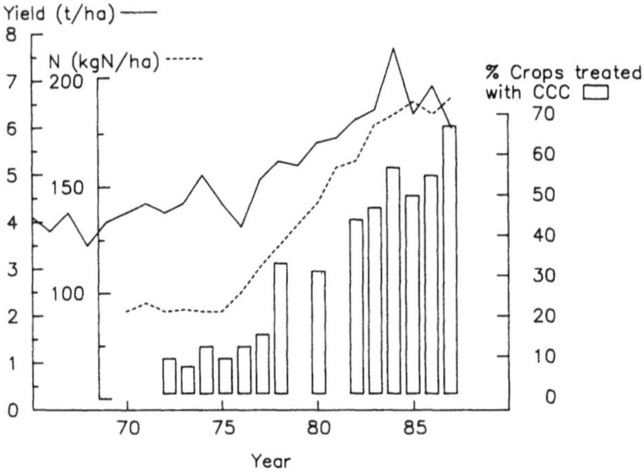

Fig. 2. UK national average winter wheat yield, nitrogen fertilizer usage and proportion of crops treated with CCC (Data for 1979 and 1981 are not available). Sources: Ministry of Agriculture, Fisheries and Food (yields), Fertilizer Manufacturers Association (nitrogen), ICI (CCC)

5 Conclusions

In conclusion, PGR research is a high-risk investment for the agrochemical industry, but several new products seem capable of being significant improvements in profitability over their forerunners. The promise of PGRs is to create new markets as those for conventional agrochemicals dwindle. Experience and technological advances also should make the prospects for the discovery of novel PGRs appear more attainable. To the grower, PGRs that increase or maintain profitability as management aids, or as yield and quality enhancers, remain attractive if they can be shown to be economically viable.

References

1. Bridges IG (1987) In: Hawkins AF, Stead AD, Pinfield NM (eds) Plant growth regulators for agricultural and amenity use, Monograph 36. British Crop Protection Council, Lavenham, p 175
2. Edgerton LJ (1986) Acta Hortic (179) 1:467
3. Geissbuehler H, Kerber E, Mueller U, Pfister K (1987) In: Hawkins AF, Stead AD, Pinfield NM (eds) Plant growth regulators for agricultural and amenity use, Monograph 36. British Crop Protection Council, Lavenham, p 11
4. Lever BG (1985) In: Menhenett R, Jackson MB (eds) Growth regulators in horticulture, Monograph 13. British Plant Growth Regulator Group, Bristol, p 15
5. Lever BG, Shearing SJ (1987) In: (ibid), p 155
6. Lürssen K (1980) In: Jeffcoat B (ed) Aspects and prospects of plant growth regulators, Monograph 6. British Plant Growth Regulator Group, Wantage, p 241
7. Takano M (1986) History, geography and economics of Japanese agriculture. Chemistry and Industry, 20 January 1986. Society of Chemical Industry, London, p 46

Practical Considerations in Using Growth Regulators on Turfgrass

J.E. KAUFMANN[1]

1 Introduction

Several 100 papers have been written on the turfgrass PGR effects and have been reviewed in detail by Elkins [2] and Watschke [6]. A major conclusion that can be drawn from the literature on growth retardants is that turfgrass responses have not been consistent and many of the responses have been aesthetically undesirable.

PGRs are reported to suppress grass growth anywhere from 4 to 10 weeks after application, occasionally with a period of post-retardation enhancement. Seedhead suppression has ranged from none to complete season long control. The various types of foliar injury reported include chemical phytotoxicity, discoloration, droughty appearance, off color and tip dieback. However, injury often does not occur, but has not been predictable. Suppression of new leaves, tillers, stolons, rhizomes and roots have been observed in some cases while enhancements of nearly all of the above have also been observed. Because biological responses have been erratic, commercial acceptance of PGRs has been limited.

This chapter will not repeat or extend previous reviews, but rather develop the importance of turfgrass biology and ontogeny in understanding erratic turfgrass responses to PGRs. The information on turfgrass biology has been accessed from Beard [1], Langer [5] and Etter [3]. This paper also characterizes the PGRs according to types of biological action and discusses how grasses, at selected times in their life-cycle, respond to each PGR type. Other details of the practical implications of these concepts have been previously published [4].

2 Growth and Growth Regulation

2.1 Cool-Season Grass Growth

Cool-season grasses in the cool humid region of the United States exhibit a typical growth pattern shown in Fig. 1. The long daylength during spring orients growth vertically resulting in a greater percentage of vegetation removal. Often more than one-half of the vegetation grown for the year is removed during a six-week period in the spring. Summer heat and dry conditions slow the rate of growth to a minimum. Growth resumes in the fall, but short daylength orients growth more horizontally and the amount of clippings is often only half of that in the spring.

[1] Science Fellow. Turf. Industrial and Residential Products Research. Monsanto Company. St. Louis. MO 63167. USA

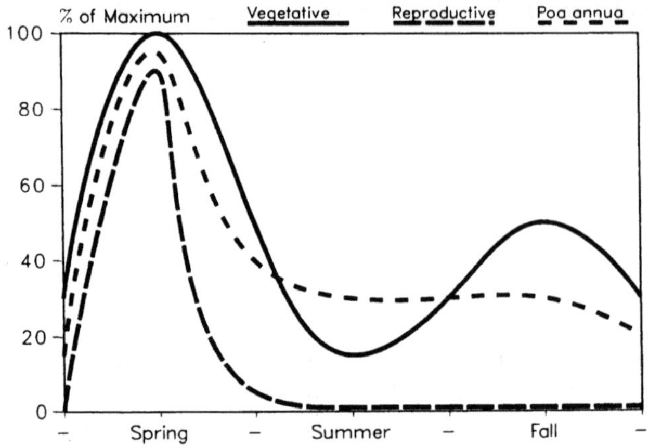

Fig. 1. Cool-season grass vertical growth rate

The reproductive structures of cool-season grasses grow only once in the spring of the year. Elongation of the seedhead usually occurs in May, but may extend into June. In mowed turfs, grasses that grow seedheads are described as being stemmy and are noted for poor turf quality during that time. Within the cool-season grasses one anomoly exists in *Poa annua* which produces abundant seedheads in the spring and then maintains a reduced level of production throughout the year.

The ideal time to apply a PGR to cool-season grasses is early spring to prevent the most vertical growth (Fig. 2). In nature, post-retardation growth enhancement is usually of little consequence, as the hot summer environment is already reducing growth. In fact, improved green color often observed at this time is desirable. Because cool-season grasses only grow seedheads once per year, properly timed application of PGRs that control seedheads can create the effect of season-long control without residual chemical action.

2.2 Warm-Season Grass Growth

Warm-season grasses exhibit peak vegetative growth in the summer (Fig. 3). Growth rate is primarily influenced by the total amount of solar radiation and not by specific daylength responses. Cool temperatures in the fall cause growth cessation and the grasses remain dormant until late spring.

Reproductive structures of warm-season grasses grow abundantly in early summer and then are maintained at a reduced level throughout the growing season. Once again, there is an anomaly in the warm-season grasses in that zoysiagrass exhibits seedhead development similar to cool-season grasses. In areas where zoysiagrass exhibits winter dormancy, seedhead appearance often preceeds vegetative green-up. Certain zoysiagrass varieties also exhibit some seedheads in the fall.

The ideal time to apply a PGR to warm-season grasses is in early summer as growth begins (Fig. 4). For most of the warm-season region, a single application is often not sufficient to regulate growth for the entire period of high vertical shoot growth, and reapplication is necessary for season long control.

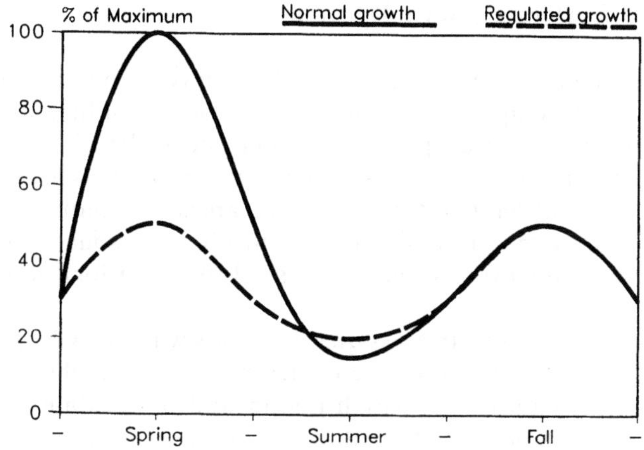

Fig. 2. Chemical regulation of cool-season grass growth

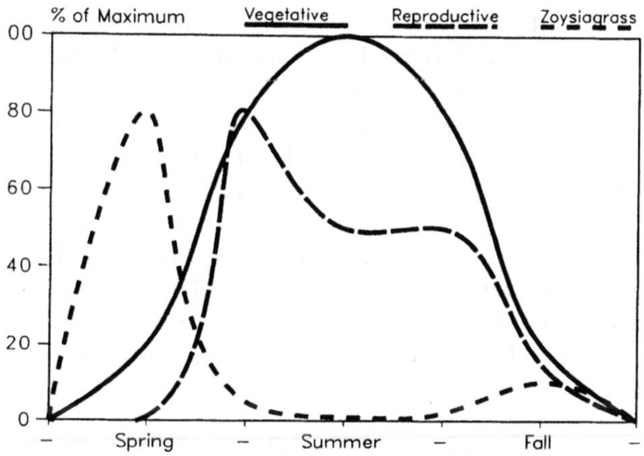

Fig. 3. Warm-season grass vertical growth rate

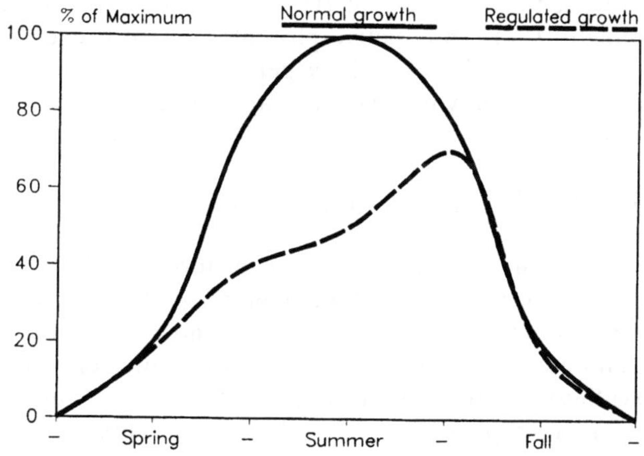

Fig. 4. Chemical regulation of warm-season grass growth

3 Turfgrass Growth Dynamics

Growth curves in Figs. 1 and 3 are drawn from the areas of the world where high or low temperature is the principle factor resulting in growth reduction. Certainly in parts of Europe, the Mediterranean, Pacific Northwestern US and various elevations of the tropics, either warm or cool-season grasses can experience uniform seasonal temperatures. In these areas the frequency of rainfall or irrigation is usually the principle factor controlling seasonal growth rate. Regardless of the factor involved, when turfgrasses do not grow for a period of time the plants become dormant.

Superior turfgrass quality occurs when new leaf appearance rate is high and exceeds leaf senescence or death rate. Under these conditions turfgrasses build density and exhibit a high amount and high visibility of juvenile tissue. The process of dormancy begins with a lack of new leaf initiation followed by highly visible natural aging of existing leaves. Once natural aging occurs, dormant turfgrasses exhibit poor turfgrass quality.

The interval of time between the appearance of each new leaf is the plastrochron [5]. For *Poa pratensis L.* growing in average conditions, the plastochron has been calculated to be about 5 to 7 days. Additionally, the average tiller has about four to five leaves at any one time. Thus, the expected duration of any one leaf is about 20 to 35 days before the onset of senescence. Application of PGRs that shut down growth of the turf sward can only be expected to maintain acceptable quality for a period of 20 to 35 days before "dormancy" becomes evident.

The life span of turfgrass leaves is very short compared to broadleaf plants [5]. In aging grass leaves, cell contents are translocated to other portions of the plant and the old leaves begin to lose weight. The leaf exhibits a dull grayish appearance, eventually discolors and dies back from the tip to the base. It is important to understand that tip die-back is an inevitable event for all turfgrass leaves.

4 Cool-Season Grass Life Cycle

Figure 5 outlines the annual life-cycle of cool-season grasses and identifies proposed growth stages: (I) cold dormancy; (II) greenup; (III) rapid vertical growth; (IV) reproductive physiology; (V) revegetation; (VI) heat and drought dormancy; and (VII) fall revegetation. Since spring is the preferred time for PGR application, only the first five stages are discussed.

4.1 Stage I, Cold Dormancy or Pre-Greenup

Pre-greenup is the appearance of the turf immediately following loss of snow cover. Appearance varies with kind of grass, quality or color of turf the previous fall, and severity of the winter. Soon after the snow melts and under full sun, existing leaves that were not excessively damaged from winter effects greenup through chlorophyll synthesis. Leaves damaged beyond repair remain brown and fully visible until warmer temperatures hasten their degradation.

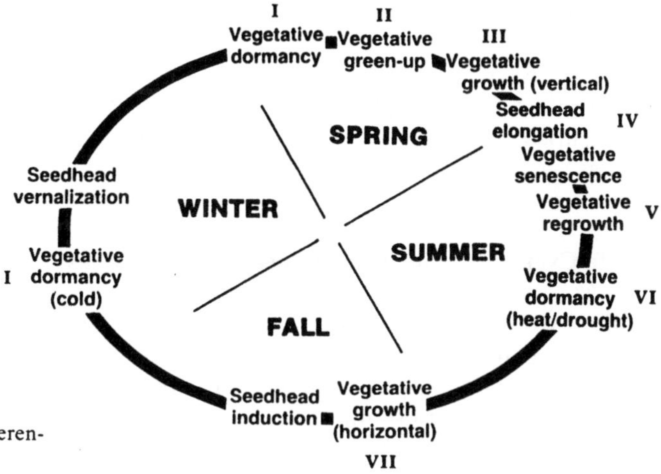

Fig. 5. Annual life-cycle of perennial cool-season grasses

4.2 Stages II, Greenup and Initial Growth

As temperatures increase, new leaves grow from the crown apex within existing leaves while old leaves degrade. Greenup may occur over a period of several weeks depending on rate of soil temperature increase. If this stage is prolonged by continued cool temperatures, the turf may reach 100 percent greenup while achieving only minimal vertical growth.

4.3 Stage III, Rapid Vertical Growth

The beginning of Stage III is most easily characterized by the need to mow. The grass begins to grow so fast that weekly mowings often remove much more than the recommended one-third to one-half of the existing leaf height. If spring temperatures warm rapidly and consistently, this stage can be entered before 100% greenup, and more than one mowing may be required before complete greenup has been achieved. Stage III ends when the first young, short seedheads appear in the turf area. While it is too late to control those seedheads, a high number of later forming seedheads can still be controlled with a PGR. The duration of Stage III varies with climate, but usually lasts 2–4 weeks.

4.4 Stage IV, Reproductive Physiology

In this stage, the seedstalk below the seedhead has begun to elongate. In many cool-season grass species, about the time the seedhead becomes visible in a mowed turf, the leaves on the tiller that bears the seedstalk stop growing and provide nutrients and energy to the developing seedstalk. Aging leaves on this tiller discolor, senesce and die very rapidly. At the same time lateral buds start developing into tillers, the plant forms a new crown apex and a new vegetative plant emerges.

4.5 Stage V, Revegetation

The turfgrass sward eventually replaces all the original plants through rapid growth of new tillers. The dead plants degrade and fall into the thatch. Thus, the green color of the lawn is maintained through development of new crown apices and new leaves.

5 Life Cycle Variation Among Species and Varieties

Normally this transition (life-cycle) occurs in a lawn with minimal disruption of turfgrass quality. However grass varieties or species that have difficulty maintaining quality during the reproductive phase are referred to as the "stemmy" types. In the cool-season region, May and June are known as the stemmy months for the stemmy varieties. Certain cultivars of *Lolium perenne* L. and common *Poa pratensis* L. exhibit poor quality at this time of the year while improved *Poa pratensis* L. cultivars, often poor in seed production, maintain superior quality.

As a cool-season species, *Festuca arundinaceae Schreb.* is best adapted to the transition zone of the United States largely due to summer survival. Yet unmowed tall fescue develops a seedhead, matures, and browns off while mowed tall fescue remains green. Thus, it appears that a major contribution to summer "tolerance" of *Festuca arundinaceae Schreb.* is related to frequent mowing that removes the emerging seedhead before reproductive physiology kills the associated leaves.

Variation among species and varieties, in relative ease or difficulty surviving reproductive physiology, appears to be associated with two factors: 1) the overall tendency of the species or variety to produce seedheads (percentage of the plant apices with potential to flower) and 2) the tendency of those plants to follow through with flowering physiology in spite of frequent seedhead removal through mowing.

6 Characterizing the Growth Regulators

Growth is defined as irreversible enlargement in size, while development is the transformation of apparently identical cells into diversified cells and plant organs. Based on these definitions, turf PGRs can be divided into three types (Table 1).

6.1 Type I Growth Regulators

These PGRs suppress both growth and development including most of the stages outlined in figure 5. The chemicals in this group act as inhibitors of cell division or cell cycle. Examples include amidochlor, EPTC, mefluidide, chlorflurenol, and maleic hydrazide. These regulators suppress leaf growth and initiation of new turfgrass leaves for about six weeks. High doses do not kill the meristems of the plant, but some result in direct and immediate injury to the leaves.

Other chemicals known to inhibit growth and development of cool-season grasses are labelled herbicide Type I regulators which inhibit enzymes of amino acid or organic acid biosynthetic pathways critical to growth. Thus, low rates can act

Table 1. Types of growth regulators

1. TYPE I. Mitotic suppression of growth and development.
   - Amidochlor, EPTC, Mefluidide, Chlorflurenol,
   - Maleic hydrazide ...
2. HERBICIDE TYPE I. Amino acid or organic acid inhibition.
   - Glyphosate, Imidazolinones, Sulfonyl ureas,
   - Sethoxydim, Fluazifop-butyl ...
3. TYPE II. Gibberellin and elongation inhibition of growth.
   - Paclobutrazol, Flurprimidol, Uniconazole,
   - Tetcyclasis ...
4. FUNGICIDE TYPE II. Suppression from host modification.
   - Triadimefon, Fenarimol ...
5. TYPE III. Inhibition of reproduction to maintain vegetation.
   - No examples ...

as PGRs. Examples of this type of chemical include glyphosate, imidazolinones, sulfonyl ureas, sethoxydim, and fluazifop-butyl. These compounds have a very narrow margin of safety on cool season grasses and accidental overdoses easily kill turf.

6.2 Type II Growth Regulators

These PGRs suppress growth only. The developmental sequence of the plant continues, however, new plant organs develop in miniature size. Examples of this type include paclobutrazol, fluprimidol, uniconazole, and tetcyclasis. These compounds have primary action as GA biosynthesis inhibitors and are far more effective in suppressing internode elongation than leaf size.

Fungicide Type II growth regulators are those that are primarily used as a fungicide but have been observed to modify the host. Darker green color and growth reduction has been observed with triadimefon and fenarimol. The latter is recommended for use on putting greens for selective suppression of annual bluegrass.

6.3 Type III Growth Regulators

These PGRs will inhibit development only. There are no examples to date, but the possibilities warrant a continued interest in their discovery. A Type III PGR would be extremely valuable in turf culture as proper timing would result in elimination of the stemmy phase of mowed turfs and the unsightly seedheads of unmowed turfs.

7 Life Cycle Responses to the Regulator Types

If a Type I PGR is applied at Stage I, the most noticeable effect is a delay of spring greenup. Since development is slowed as well as growth, the rate of appearance of new green leaves is slowed and leaf size is diminished. Root active PGRs are effective in reducing growth when applied at this stage while foliarly active PGRs require green leaves to absorb the product. Application at Stage II results in delay of further greenup and subsequent growth suppression. Application at this stage is desirable since the turf often has greened sufficiently and rapid spring growth has not yet begun.

Stage III is considered the optimum time for application of Type I PGRs to provide good turfgrass quality and the normal 5-6-week duration of vegetative suppression or inhibition. Often there is a slight loss of turf quality during the 3rd and 4th week from leaf aging, and enhanced dark green color from the 7th to the 10th week or longer. Seedhead control is usually greater than 90% for applications made during this stage.

Applications of any Type I PGR at Stage IV is often detrimental to the appearance of the turfgrass area especially if the grass is a stemmy type. PGRs do not reverse the negative effects of reproductive physiology but exacerbate the situation to completely inhibit growth of existing leaves. Likewise, they greatly slow tiller development, at least for a time. Eventually one or more lateral buds, deep in the thatch, begin to grow very rapidly and develop into elongated tillers. Thus application of Type I PGRs at Stage IV results in undesirable turfgrass responses: (1) excessive growth inhibition for a short period, (2) severe loss of turfgrass quality as leaves senesce and die, and (3) early termination of activity due to rapid growth of escaped tillers not affected by the product.

The problems encountered in Stage IV reinforce the fact that Stages II or III are the preferred time for application. Since developmental inhibitors applied at these stages prevent seedheads from developing, they also prevent reproductive physiology and associated negative turfgrass quality consequences, thereby providing for improved turfgrass quality compared to a nontreated area undergoing the "stemmy" reproductive physiology phase. Further, the effect of preserving leaves seems to be accompanied by a preservation of existing roots. As a result, improved summer growth, color, rooting, and tolerance to summer stresses (heat, drought and diseases) have been observed when using some Type I growth regulators.

Because Type II PGRs do not suppress plant development, applications at any of the stages from I through IV can result in; (1) diminutive seedhead expression below the mowing height, (2) natural senescence and death of the main tiller, and (3) suppression of the size of new tillers that normally grow large enough to mask the dying leaves. Therefore, no stage of application on stemmy varieties in the spring is acceptable for Type II PGRs.

It is important to state that the Type II PGRs do show acceptable results on non-stemmy, highly vegetative species and varieties. For instance, *Festuca arundinaceae Schreb.* seedheads apparently can quite easily be mowed off prior to the reproductive physiology, even when stunted by a Type II PGR, and good results have been achieved. Type II PGR use on *Poa pratensis* L. cv. Baron, however, has not been as successful. Apparently when the seedhead height is stunted, the mower does not remove the seedhead soon enough to prevent natural reproductive physiology and leaves usually senesce and brown off rapidly during Stage IV. Finally, it should also be noted that Type II PGRs have shown excellent performance in the fall when perennial cool-season species do not exhibit a reproductive growth stage.

8 Classifying Turf Areas Relative to PGR Types

A most important step in understanding where to use growth regulators is to classify the areas according to level of management. Table 2 outlines a classification scheme as one method of describing relative management intensities.

Class A turf is that receiving high levels of input. Mowing is done on a frequent basis to maintain a groomed appearance at all times. Fertilizers are usually applied two to four times per year. Pests are generally controlled on a curative program and the areas are often, but not necessarily, irrigated. Examples of Class A turf include golf greens, tees and fairways, sportsfields, high quality home lawns, and improved sections of industrial grounds, parks, and cemeteries. PGR use on Class A turfs has been limited due to off-color associated with leaf aging. Use in hard-to-mow areas such as on steep slopes and around obstacles has continued to grow.

Class B turfs are those that for reasons of aesthetics need to be mowed on a frequent basis but generally do not have other management inputs. The key objective remains vegetation height control similar to that required for Class A turfs. Perhaps once a year or every two years, these areas are fertilized and broadleaf weeds are controlled. Mowing frequency is equal to that for Class A turf when based on turfgrass growth rate, but may be somewhat less based on calendar days. Examples of these areas include the major portion of industrial grounds, parks, cemeteries, golf course roughs, and home lawns. Class B turfs represent the greatest potential for cost effective PGR use. Both Type I and II PGRs can be used successfully. It is important to select PGRs with minimum overlap effects where large areas require numerous passes with the equipment.

Class C turf is mowed two to three times per year, usually never fertilized but control of certain broadleaf weeds may occasionally occur when infestations become severe. The key objective with this mowing frequency is to cut down seedheads (which result in brown color) and excessive vegetation which may harbor unwanted animals. An example of this type of turf area would be highway roadsides and remote industrial sites. Type I, herbicide Type I and Type II PGRs can be used effectively all be used effectively. One pass of the spray boom along each side of the highway provides sufficient vegetation control without the possibility of overlap injury.

Class D areas can no longer be called turf and the mechanical or chemical brush and weed control cannot truly be called mowing. Vegetation control along these areas is usually done with a "brush-hog" or chemicals known as "total veg" control materials. Examples of this class include railroad and power line right-of-ways as well as the more obscure parts of highway right-of-ways. Herbicide Type I PGRs are widely used on Class D vegetation, including those with residual action. Mobility of residual chemistry to surrounding desired vegetation has resulted in the use of mixtures of residual and non-residual chemistry to reduce the chance of injury.

Table 2. Vegetation management classification

Class A	Frequent mowing, fertilization, and pest control; often irrigated
Class B	Frequent mowing; occasional weed control and fertilization
Class C	Infrequent mowing; occasional weed control
Class D	No mowing, occasional brush and weed control

References

1. Beard JB (1973) Turfgrass: science and culture, Prentice Hall, Englewood Cliffs, NJ
2. Elkins DM (1983) In: Nickell LG (ed) Plant growth regulating chemicals, vol II. CRC, Boca Raton, FL, p 113
3. Etter AG (1951) Ann Mo Bot Gard 38(3):293
4. Kaufmann JE (1987) In: Hawkins AF, Stead AD, Pinfield NJ (ed) Plant growth regulators for agricultural and amenity uses, Symposium Proceedings. British Crop Protection Council and the British Plant Growth Regulator Group, Monograph No. 36, p 99
5. Langer RHM (1979) Second Edition, The Institute of Biology's Studies in Biology No. 34. Arnold London, England
6. Watschke TL (1985) In: Lemaire F (ed) Proceedings, Fifth International Turfgrass Research Conference, Avignon, France, p 63

New Aspects of the Practical Use of Ethylene-Releasing Compounds

M.S. Reid[1]

Compounds that release ethylene when sprayed on plants have become of major economic importance, being used to accelerate diverse ethylene responses such as induction of flowering, stimulation of latex flow, leaf and branchlet abscission, fruit ripening, fruit abscission, and pod dehiscence. Novel aspects of application of these materials in agriculture are reviewed.

1 Introduction

Although horticulturists have used the beneficial effects of ethylene and suffered from its adverse effects for centuries, it was only 80 years ago that ethylene was demonstrated to be an active PGR. A Russian graduate student, Neljubow [42] demonstrated that the active component in the gas leak which rendered his elongate and slender etiolated pea seedlings squat, thickened, and ageotropic was ethylene. Cheapest, easiest to apply, and most readily available of all the PGRs, ethylene has been a favorite tool in 'spray and pray' plant physiology ever since. The exponential increase in ethylene-related research in the past 40 years [1] has demonstrated situations in which ethylene alters almost every phase of plant growth and development, from seed germination to seed dormancy. Curious, then, that the most recent research developments might be construed as indicating that ethylene is not normally involved in plant growth and development. Bleeker et al. [6] selected mutant plants of an annual weed, *Arabidopsis thaliana*, that were resistant to ethylene, using the striking growth responses observed by Neljubow as their selection tool. The selected seedlings, although resistant to ethylene in every aspect that was tested, grew perfectly normally, flowering, setting seed, and ripening and abscising fruit just like wild-type plants. Even economic plants whose response to ethylene has been inhibited by application of silver thiosulfate (STS) grow quite normally [8].

Although these data call into question ethylene's present position as the 'do it in almost all' plant hormone, there is no denying the manifold effects of applied ethylene in plant growth and development. For many years ethylene itself has been used commercially to elicit some of these effects, for example, fruit ripening. Compounds which initiate ethylene responses by inducing the plant to produce its own ethylene (a wound response) have been used in the field. Cycloheximide application, for example, stimulates abscission of citrus by stimulating ethylene production at the button [15]. The possibilities for harnessing ethylene responses in

[1] Department of Environmental Horticulture, University of California, Davis, CA 95616, USA

Table 1. Common and chemical names, and structures of some important ethylene-releasing chemicals

Common name	Chemical name	Structure
Ethephon	(2-chloroethyl)phosphonic acid	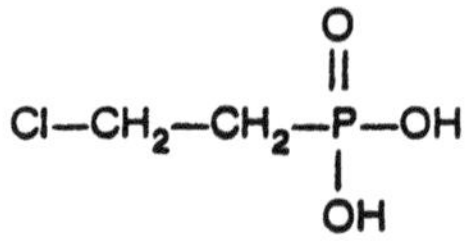
Silaid	(2-chloroethyl)methylbis (phenylmethoxy)silane	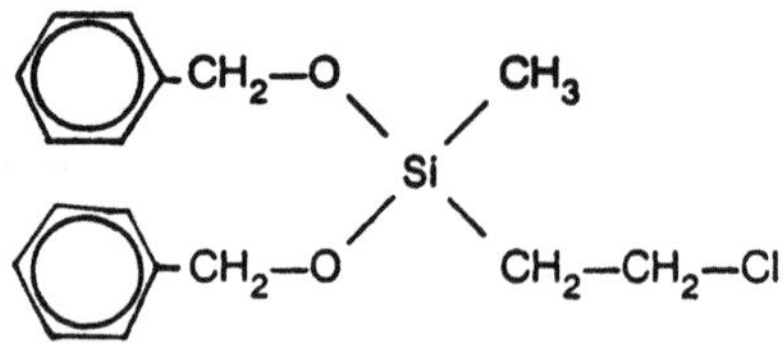
Alsol	(2-chloroethyl)tris (2-methoxyethoxy)silane	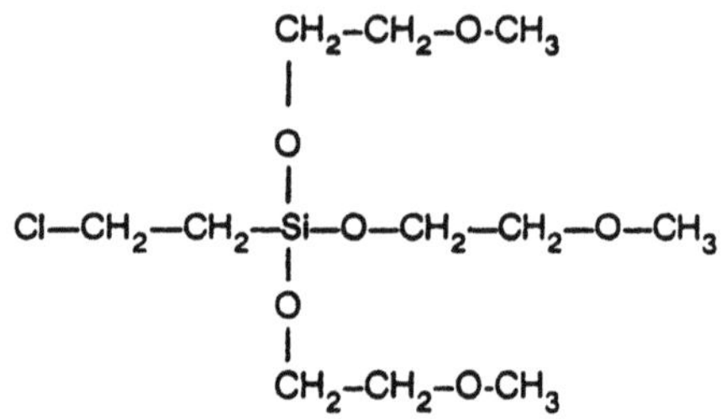
ACC	1-aminocyclopropane-1-carboxylic acid	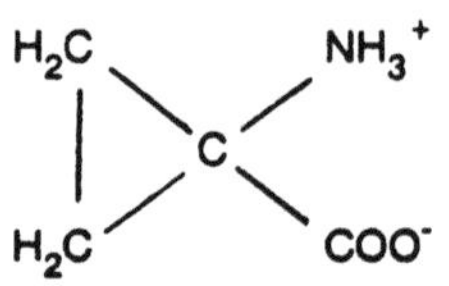

commercial agriculture have been greatly increased by the development, since the early 1960s, of a number of chemicals which, when applied to plant materials, release ethylene as they decompose. The most important of these chemicals are shown in Table 1. The action of these materials, and their uses in agriculture, have been reviewed comprehensively by Kays and Beaudry [30]. An indication of their usefulness is given by the wide range of applications that have been granted registration in USA (Table 2). Recent research that indicates new directions for the possible use of ethephon in agriculture are reviewed briefly here.

2 Growth and Development

The stunting of ethylene-treated plants noted by Neljubow is now the basis of one of the most important uses of ethephon, the prevention of lodging in small grains and other crops. The regulator treatment not only reduces length of the stems, but also stiffens the straw. Knapp et al. [32] examined the effects of ethephon treatment on carbohydrate content of the straw. Although straw from ethephon-treated plants contained more water soluble carbohydrates, they did not find the anticipated increase in structural carbohydrates. They concluded that the arrangement and interaction of the various structural carbohydrates and lignin in the culm cell walls may be more important in lodging resistance than concentrations of these compounds. Researchers have now shown that similar benefits can be obtained from ethephon application to field corn [20] where application greatly reduced lodging and accelerated maturation; a reduction in yield was also noted in some varieties. The floral axis in bulb-type flowers is often too long for their use as potted flowers. Application of ethephon to hyacinth and narcissus plants when the leaves were 10 cm long reduced the length of the floral scape by 20–30% [29]. The reduction in scape length was associated not only with a reduction in cell size, but, surprisingly, also with an *increase* in cell number, indicating that ethephon application had actually stimulated cell division.

Hollow heart and brown center of potatoes are disorders that accompany rapid plant and tuber growth. Although application of ethephon at tuber initiation reduced yield and mean tuber size, it also substantially reduced the incidence of these disorders [36], leading to increased yield of marketable tubers, relative to control fields which had high disorder incidence.

3 Flowering Control

For many years it has been known that ethylene stimulates flowering of Bromeliads, and treatment of pineapple plants with calcium carbide (which releases acetylene on hydration) was widely used commercially to stimulate flowering of pineapples. This response is now elicited in pineapples and flowering bromeliads by application of ethephon. A comparison of the effectiveness of ACC and ethephon in eliciting this response [17] suggested that ACC was the preferred ethylene-releasing compound, but because of differences in application method and concentrations, this interpretation needs confirmation. Other tropical fruit crops are also stimulated to flower by treatment with ethylene. Smoky fires are, for example, commonly lit in mango orchards to stimulate flowering [10]. Application of ethephon has also been shown to stimulate flowering in *Litchi* [11, 50], although the results vary widely, depending on application time and cultivar.

A less familiar phenomenon (reported long ago by Vacha and Harvey [54]) is the effect of ethylene in stimulating or accelerating flowering of many geophytes. In *Iris*, for example, a short treatment of harvested bulbs with ethylene will stimulate flowering in small-sized bulbs that would not normally flower [26]. Japanese bulb growers still treat harvested *Iris* and *Freesia* with ethylene by placing them in a barn and lighting a smoky fire in the enclosed space. Ethephon application has been shown effectively to replace smoke or ethylene treatments [9]. For example, flower formation in *Iris* is greatly enhanced by spraying the fields with ethephon prior to bulb harvesting [23]. The mechanism of action of ethylene in these flowering responses is still uncertain, although our data suggest that it involves stimulation of cell division in the apical meristem, resulting in a larger meristematic dome [24].

Ethephon application has been used to extend the flowering season of plantations of *Leucospermum*, a woody proteaceous ornamental [7]. Application to the developing flowering shoot at an appropriate time caused abortion of the terminal inflorescence, and growth of a new flowering shoot (sometimes two) from axillary buds. This manipulation effectively delayed flowering by many weeks, permitting a longer harvest season. Because treatment of young apple trees (cvs Bramley and Cox's Orange Pippin) with 250 ppm ethephon in summer induced more spur buds (flowering shoots) to form without affecting shoot growth [55], ethephon may be useful in increasing precocity in apples.

4 Leaf and Branch Abscission

The effects of ethylene in stimulating abscission [44] are the basis of important commercial uses of ethylene-releasing compounds. Enhancement of fruit abscission in sour cherry, nut fall in macadamia [53], the removal of leaves from potato plants, cotton, and nursery stock [18], and the dehiscence of the shucks of walnuts and pecans are but a few examples of the potential uses of these materials in agriculture. Defoliation of rutabaga required very high concentrations of ethephon [43]; addition of 1% ammonium peroxydisulfate greatly increased the effectiveness of the ethephon treatment, perhaps due to accelerated oxidation of IAA in the abscising organ.

We studied the effects of ethylene in abscission of leaves and berries of mistletoe during commercial handling of mistletoe sprigs for the holiday season [28]. Moderate concentrations of ethylene caused abscission, not only of leaves and berries, but also of primary and secondary branchlets. We reasoned that ethephon application might therefore be a potential method of controlling broad-leaved mistletoe infestations. Broad-leaved mistletoe (*Phoradendron tomentosum*) is a major pest of shade trees, and is suspected of causing premature limb death, and loss of tree vigor. Application of ethephon in the winter or early spring, when the host leaves are absent, permits the use of the high concentrations of ethephon necessary to ensure abscission of mistletoe branches. Application of ethephon to clusters of mistletoe on black walnut caused rapid abscission, and little of the regrowth which is a common problem with other methods (pruning, application of NAA) used to control mistletoe infestations [27]. The ethephon is apparently absorbed rapidly by the mistletoe leaves, but ethylene is released only slowly if temperatures are low. Using harvested branchlets as a model system, we showed, however, that ethylene release was quite rapid once the treated, cooled, branchlets were returned to room temperature. This indicates that the response to ethephon will probably occur regardless of ambient temperatures at the time of spray application. We are continuing to investigate the implications of using ethephon as a practical tool in the control of broad-leafed mistletoe. Similar studies have shown its usefulness in controlling dwarf mistletoe, a serious pest in coniferous forests [39].

Guava trees (*Psidium guajava*) bear fruit on current season's growth, and under tropical conditions such growth follows the rainy season. In order to spread the harvesting season, producers use cultural techniques (pruning, irrigation, application of fertilizers, or chemical defoliation) to induce flushes at other times of the year. Kobayashi [34] demonstrated that ethephon, particularly in combination with GA_3, helped defoliate the plants, and also stimulated the new vegetative growth which precedes flowering.

6 Fruit Ripening

Ethylene gas is widely used commercially for ripening a variety of climacteric fruits and decoloring non-climacteric citrus fruits. However, the use of ethylene-releasing compounds to effect this response is confined to relatively few crops (Table 2), and normally when they are used as a preharvest application. This is probably due to the ease with which ethylene gas can be applied to harvested fruit, and the dangers of applying aqueous solutions to these disease-susceptible organs.

The major commercial application of ethephon for ripening fruits is in processing tomatoes. Ethephon treatment of tomato plants was shown to have adverse effects on the germinability of older seeds, indicating that there was no benefit of ethephon application to tomatoes destined solely for seed production [37].

Table 2. Uses approved for (2-chloroethyl)-phosphonic acid in one or more States of the USA (adapted from Kays and Beaudry [30])

Use	Approved crops and states (no postscript = all states)
Plant height control	Barley, daffodils, hyacinth, wheat
Stimulation of branching	Azaleas, geraniums
Flower induction	Pineapple and other bromeliads
Sex expression	Cucumber, squash
Flower bud development	Apple
Defoliation	Cotton, roses, buckthorn, apple
Fruit loosening	apples, blackberries (WA, OR), cantaloupes, cherries (CA, AZ, TX) tangerines
Fruit removal	Apples, crabapples, carob, olive
Dehiscence	Walnuts
Postharvest fruit ripening	Bananas, tomatoes (FL)
Preharvest fruit ripening	Tomatoes, peppers
Maturity and/or color development	Apples, cranberries (MA, JH, WI) figs (CA), filberts (OR), grapes, peppers, pineapple, tomatoes (CA, TX, etc.)
Degreening (preharvest)	Tangerines
Degreening (postharvest)	Lemons
Leaf curing	Tobacco

6 Fruit Abscission

Promotion of abscission as an aid in thinning or harvesting fruit crops is an important horticultural use of ethylene-releasing compounds [44]. A recent study by Takeda and Peterson [51] on mechanical harvesting of blackberries provides a good example of the benefits ethylene-releasing compounds provide in stimulating this response. These researchers found that the separation pull force of thornless blackberries (*Rubus* spp.) decreased too slowly to allow differentiation between black ripe and red fruit by mechanical shaking. When they applied sufficient force to remove 80% of the ripe fruit, 31% of the harvested fruit were red and green. When the vines were shaken four days after application of 500 ppm ethephon, only 7% of the harvested fruit were red and green. The ethephon treatment doubled the number of ripe berries on the plants and this increase, as well as a concomitant reduction in attachment force, presumably explains the satisfactory results of the treatment.

For many years, researchers have been investigating ways to promote abscission of olive fruits by using ethylene-releasing compounds. Although ethylene does loosen olive fruits, ethephon application causes undesirable leaf abscission [4]. Several workers have tested the application of ethephon to accelerate shuck loosening in pecan. Trunk injection of ethephon accelerated shuck loosening [49], but caused unacceptable leaflet abscission, and reduced fruit set in the following year. Variable results have also been obtained with pistachio. Toriba [52] found that ethephon application advanced ripening and splitting of the nuts; Crane et al. [16], working with a different variety and in a different production environment, found no increase in shell dehiscence, but did observe excessive exudation of gum from the trunk and major limbs, and severe flower bud abscission.

Workers seeking to dissociate desirable effects of ethephon on fruit ripening or fruit loosening from undesirable leaf abscission have tested a range of adjuvants. The excessive leaf drop in ethephon-treated olive trees was overcome by neutralizing the ethephon solution (with Na_2CO_3). The rapid production of ethylene from the neutralized ethephon solution stimulated fruit abscission, but elevated ethylene concentrations were not present for long enough to cause leaf abscission [5]. Even this expedient did not provide results that were sufficiently consistent to form the basis of a commercial technique. In further study of olive abscission, Ben-Tal and his colleagues have determined that occasional failures of the neutralized ethephon to elicit rapid abscission are due to rapid drying of the applied spray, and consequent failure of ethephon to penetrate the tissues of the olive. They therefore added 1% glycerine to the ethephon formulation. With the drying of the spray thus delayed, the force required to remove the fruits was reduced substantially [4]. In experiments examining the use of ethephon to hasten coloring of Tabasco pepper, Conrad and Sundstrom [14] found that the concentrations they used (0.5% and greater) caused rapid leaf and fruit abscission. When they added 0.1 M $Ca(OH)_2$, leaf abscission was much reduced. These authors did not determine whether the reduction in leaf abscission was due to an effect of calcium on maintaining cell wall integrity in the abscission zone, or in neutralizing the ethephon, thus reducing the period of ethylene release.

Combination with other PGRs has also proved to be useful in modifying plant responses to ethephon. Application to pecans of a spray combining 6 mM NAA with 3 mM ethephon resulted in accelerated shuck dehiscence without major leaflet abscission [56]. Similarly, successful thinning of apples without undesirable leaf drop has been achieved by bloom-time sprays of ethephon combined with an auxin [33, 35, 38].

7 Seed Dormancy

Scarification of seed (mechanically, and/or with H_2SO_4) is a common method of promoting germination of seeds with hard coats. Sometimes this may release a mechanical inhibition of expansion, but in some cases it overcomes a physiological dormancy. Such is the case of *Pelargonium* seed, where germination is improved simply by pricking the seed coat with a needle [45]. Typical germination percentages for unscarified seed of inbred lines of *Pelargonium hortorum* range from 30% to 50%. Soaking imbibed seed for 12 h in 50 ppm ethephon increased germination to 94% [45]. Unimbibed seed responded similarly, but required higher concentrations (>250 ppm).

The stimulation of seed germination by ethephon also suggests the possibility of using this material as part of an herbicidal program, using it to ensure pre-plant germination of dormant weed seeds. Researchers have therefore been studying ethephon effects in the germination of weed seeds [22, 31, 47, 48]

8 Stress Responses

The fact that ethylene-resistant plants grow perfectly normally raises the questions of the role of ethylene as a PGR. It has been suggested that ethylene's gaseous nature makes it an ideal regulator for enabling plants to respond to environmental stresses and attacks from pathogens and insect pests. Several recent studies have addressed the possibility that ethylene-releasing compounds might stimulate defense reactions and thus prevent pathogen attack. The stimulation of phenolic biosynthesis by ethephon applications to grapes [46] and apples [41] may represent such a reaction. Treatment of mung bean seedlings with ethephon decreased their susceptibility to infection by *Rhizoctonia solani*, concomitantly with increased activities of peroxidase and polyphenoloxidase [3]. In contrast, ethylene *stimulates* germination of teliospores of *Urocystis agropyri* [21]. Low-temperature stress is an important limitation in the production of many deciduous fruit trees. Ethephon application to peach trees during the fall delayed bloom in the following spring (thereby reducing the potential risk of frost damage), and also increased the intrinsic winter hardiness of the buds [19].

A wound response also seems implicated in the stimulation by ethephon of latex flow during tapping of rubber trees, a response which has been shown to be associated with a change in the trans-tonoplast membrane potential [13]. The increased production of secondary metabolites such as latex is probably a common response to stress in plants, and ethephon may prove to be useful in increasing their concentration in important economic plants. Cho et al. [12] found that addition of ethephon to cell suspension cultures of *Coffea arabica* and *Thalictrum rugosum* stimulated alkaloid production in both species, even though the biochemical pathways were widely different. They suggest that ethylene-releasing compounds may be valuable in increasing secondary metabolite levels in a variety of cell-suspension culture systems.

A disadvantage of increased secondary product production in response to ethephon was reported by Arita et al. [2], who showed that the Chinese rose bettle (*Adoretus sinicus* Burmeister) fed preferentially on leaves of ginger plants that had been treated with ethephon. The feeding stimulus was not ethylene itself, but presumably a volatile produced by the plant in response to the ethephon treatment. An interesting use of ethephon for insect control is its late-season use during cotton production to remove squares (young flowers), and reduce the opportunity for insects to overwinter [25]. This treatment reduced the number of infested bolls and squares to less than 5% of those in the control plots.

9 Future Prospects

The diversity of present commercial uses of ethephon and other ethylene-releasing compounds indicates the substantial contribution that stimulation of ethylene production is making to agricultural productivity. The many other possible uses shown by the recent research results reviewed here also indicate that there is yet greater commercial potential for the use of these compounds. The variety of time/concentration relationships in the response of plants to ethylene suggest that more specific and efficient use of ethylene-releasing compounds may result from modulating the kinetics of ethylene release. An understanding of the movement of the compounds into leaves [40], and of the factors affecting their decomposition, will permit horticulturists to modify application techniques to elicit specific responses. Such modifications may permit even more widespread use of these benign and useful PGRs.

References

1. Abeles FB (1973) Ethylene in plant biology. Academic Press, New York
2. Arita LH, Furutani SC, Moniz JJ (1988) J Econ Entomol 81:1373
3. Arora YK, Bajaj KL (1985) Phytopathol Z 114:325
4. Ben-Tal Y (1987) HortScience 22(5):869
5. Ben-Tal Y, Lavee S (1976) HortScience 11:489
6. Bleeker AB, Estelle MA, Somerville C, Kende H (1988) Science 241:1086
7. Brits GJ (1977) Agroplantae 9:127
8. Cameron AC, Reid MS (1983) Sci Hortic 19:373
9. Cascante XM, Doss RP (1988) Am Soc Hortic Sci 23(6):1006
10. Chandler WH (1950) Evergreen orchards. Lea & Febiger, Philadelphia
11. Chen WS, Ku ML (1988) HortScience 23:1078
12. Cho GH, Kim DI, Pedersen H, Chih CK (1988) Biotechnol Prog 4:184
13. Chrestin H, Gidrol X, Marin B (1987) NATO Adv Sci Inst Ser Ser A Life Sci 134:467

14. Conrad RS, Sundstrom FJ (1987) J Am Soc Hortic Sci 112(3):424
15. Cooper WC, Henry WH (1967) Proc Fl State Hortic Soc 80:7
16. Crane JC, Iwakira BT, Lin TS (1982) HortScience 17(3):383
17. De Proft MP, Mekers O, Jacobs L, De Greef JA (1986) Acta Hortic Wageningen 181:141
18. Dozier WA Jr, Gilliam CH, Knowles JW (1987) J Environ Hortic 5(3):116
19. Durner EF, Gianfagna TJ (1988) J Am Soc Hortic Sci 113:404
20. Gaska JM, Oplinger ES (1988) Crop Sci 28(6):981
21. Goel RK, Jhooty JS (1987) Ann Appl Biol 111(2):295
22. Goudey JS, Saini HS, Spencer MS (1987) Plant Physiol 85(1):155
23. Halevy AH (1985) In: Bopp M (ed) Plant growth substances. Springer, Berlin Heidelberg New York Tokyo, p 391
24. Han SS, Halevy AH, Sachs RM, Reid MS (1989) J Am Soc Hortic Sci (in press)
25. Henneberry TJ, Meng T, Hutchison WD, Bariola LA, Deeter B (1988) J Econ Entomol 81(2):628
26. Imanishi H, Yue D (1985) Acta Hortic 177:141
27. Joyce DC, Rein K, Berry AM, Reid MS (1987) Acta Hortic 201:141
28. Joyce DC, Evans RY, Reid MS (1989) HortScience (in press)
29. Kamp M, De Hertogh AA (1986) Sci Hortic 29(3):263
30. Kays SJ, Beaudry RM (1987) Acta Hortic 201:77
31. Kepczynski J (1986) Physiol Plant 67(4):588
32. Knapp JS, Harms CL, Volenec JJ (1987) Crop Sci 27(6):1201
33. Knight JN, Spencer JE, Looney NE, Lovell JD (1987) J Hortic Sci 62(2):135
34. Kobayashi KD (1987) Acta Hortic 201:145
35. Koen TB, Jones KM, Longley SB (1988) J Hortic Sci 63:31
36. Koller DC, Hiller LK (1988) Am Potato J 65:529
37. Kwon OS, Bradford KJ (1987) HortScience 22(4):588
38. Lehman LJ, Unrath CR, Young E (1987) HortScience 22(2):214
39. Livingston WH, Brenner ML (1983) Plant Dis 67:909
40. Miller CH, Sheets SM (1986) HortScience 21(2):276
41. Murphey AS, Dilley DR (1988) J Am Soc Hortic Sci 113:718
42. Neljubow D (1901) Beih Bot Centralbl 10:128
43. Poapst PA, Anderson MG, McRae KB (1987) HortScience 22(4):583
44. Reid MS (1985) HortScience 20:45
45. Rogers OM (1987) Acta Hortic 201:165
46. Roubelakis-Angelakis KA, Kliewer WM (1986) Am J Enol Vitic 37(4):275
47. Saini HS, Bassi PK, Spencer MS (1986) 34(1):43–47
48. Saini HS, Bassi PK, Spencer MS (1986) Weed Sci 34(4):502
49. Stein LA, McEachern GR, Storey JB (1986) HortScience 21 (1, section 1):73
50. Subhadrabandhu S, Koo-Duang A (1987) Acta Hortic 201:181
51. Takeda F, Peterson DL (1988) HortScience 23:120
52. Toriba M (1980) HortScience 15(4):521
53. Trochoulias T (1986) Acta Hortic 175:199
54. Vacha GA, Harvey RB (1927) Plant Physiol 2:187
55. Volz RK, Knight JN (1986) J Hortic Sci 61(2):181
56. Wood BW (1986) J Am Soc Hortic Sci 111(4):533

Commercial Uses of Gibberellins and Cytokinins and New Areas of Applied Research

R.D. Carlson and A.J. Crovetti[1]

1 Introduction

Since the discovery of GAs in the 1920s [15] and cytokinins (CKs) in the 1950s [20], a considerable body of information has emerged regarding the roles these hormones play in plant growth and development. As these compounds and analogs became available, either by fermentation or chemical synthesis, a number of potential practical applications were identified. However, despite extensive research with GAs and CKs, the number of commercial uses has fallen short of their predicted impact and growth in agriculture [19]. Unique opportunities may still result from new research and development with GAs and CKs to meet the future demands for increased food quality and production.

2 Historical Uses

2.1 Gibberellins

Although over 70 different GAs have been identified in plants [36], only three (GA_3 and a mixture of GA_{4+7}) are available via fermentation by the fungus *Fusarium monilifome* (= *Gibberella fujikuroi*) in reasonable quantity for extensive applied research. The complexity of GA chemistry has precluded the entirely synthetic routes of production. With the diversity of physiological processes in which GAs are involved, it is not surprising that their practical application spans the entire range of plant growth and development (Table 1). The majority of these uses however, represent specialty niche markets in high cash value horticultural commodities. Despite the extensive research with GA_3 over the past 30 years, no defined GA product has been registered for use on a major agronomic crop. This may be due in part to a lack of specificity of response where multiple effects, often detrimental (i.e. lodging), are manifested from exogenous application of GA.

GA_3 has been used in the seedless table grape industry since 1960 [41, 42] to manipulate three physiological events: rachis cell elongation, flower thinning, and berry enlargement. The first is a prebloom spray to promote elongation of the rachis to provide a larger and looser framework for the grape cluster. GA_3 is also used to thin approximately 60% of the flowers because almost every flower tends to set, resulting in a tightly packed cluster. The physiological basis for grape flower

[1] Abbott Laboratories, Chemical and Agricultural Products Division, 6131 RFD (Oakwood Rd.), Long Grove, IL 60047, USA

Table 1. Physiological response and current commercial uses of gibberellin an cytokinin products registered in the USA and Europe

Physiological response	Crop	PGR Used[a]
Breakage of dormancy	Seed potatoes	GA_3
	Rhubard	GA_3
Acceleration of seed germination/hydrolytic enzyme production	Barley malting	GA_3
Stimulation of bud break and increased lateral branching	Tart cherries	GA_3
	Apples	GA_{4+7}/BA
	Conifers	BA
	Carnations/Roses	PBA
Stimulation of fruit set	Tangerine hybrids	GA_3
	Blueberries	GA_3
Reduction of fruit set	Grapes	GA_3
	Tart & sweet cherries	GA_3
Increased fruit size and quality	Grapes	GA_3
	Apples	GA_{4+7}/BA
	Sweet cherries	GA_3
	Apples	GA_{4+7}
Increased vegetative growth and yield	Sugar cane	GA_3
	Spinach	GA_3
	Hops	GA_3
Delayed senescence/ fruit ripening	Banana	GA_{4-7}
	Sweet Cherries	GA_3
	Citrus	GA_3
	Bermuda grass	GA_3
Stimulation of abscission/ senescence	Cotton	TDZ

[a] Abrreviations: GA_3: gibberellic acid; GA_{4-7}: gibberellin A_4 and A_7; BA: benzyladenine; PBA: tetrahydropyranylbenzyladenine; TDZ: thidiazuron.

thinning by GA is unknown, but could be due to pollenicidal activity. The result of adequate thinning almost appears excessive, but with two subsequent applications of GA, berry size is increased by 60%. The final product is a loose cluster that does not require hand thinning and is less susceptible to *Botrytis* bunch rot.

GA_3 also inhibits flowering on several stone fruits. In young sweet and tart cherries, GA_3 is applied to prevent excessive flowering to minimize the competitive effect of early fruiting on vegetative growth [1]. Since cherry has true flower buds, the node following flowering will not support vegetative growth. If flowering is excessive, fewer spurs develop, thereby resulting in lower yield potential during the production life of the tree. The manipulation of flowering by GA_3 has also been integrated into a management program to suppress a pollen transmitted ring spot virus infection as well as the maintainance of the desired ratio of vegetative to flower bud formation in mature tart cherries. The consequence of continuous GA_3 applications over a five year period has been a 22% increase in yield [1]. An additional application of GA_3 to sweet cherry when fruit is light green to straw colored will delay harvest, increase fruit size and firmness [31]. While this application is not new, the recent development of an export market to Asia for cherries has significantly increased this use.

The effect of delayed fruit senescence by GA_3 has also been exploited in the citrus industry [2]. GA_3 delays maturity of citrus rind without altering the maturation process of the internal fruit, thus preventing several rind disorders associated with fruit ripening. In lemons, this facilitates scheduling of harvesting to synchronize with market demands. As harvest date is delayed in grapefruit, there is added benefit in combining 2,4-D to retain the fruit on the tree. Recent findings [7, 8] suggest that the delay in citrus rind maturity is associated with the quantity and quality of epicuticular wax deposition and its influence on respiratory gas exchange.

The other commercially used GAs produced by *F. moniliforme* cultures are a mixture of GA_4 and GA_7. An application of GA_{4+7} of increasing importance is the prevention of fruit russet, a superficial disorder in which the exocarp is interrupted by raised corky outgrowths over the surface of green and yellow apple cultivars. Although somewhat obscure, a number of factors have been implicated for russet, such as high humidity, presence of free water, light frosts, pesticides and powdery mildew [3, 6]. How GA_{4+7} reduces russeting is unknown, but several studies [37, 43] determined that multiple applications on 'Golden Delicious' effectively increase the fruit finish quality and this corresponded with higher packout grades. Recent work in the US and Canada [16, 27] has suggested that the GA_4 component of GA_{4+7} may be the primary active ingredient in the suppression of russet, while GA_7 appears to inhibit flowering in apple and reduce return bloom [38]. In these studies, GA_4 appeared to have no effect on return bloom, and more recent results [16, 27] suggest that GA_4 may actually promote apple flowering. The differential effect of GA_4 and GA_7 in apple serves to illustrate the specificity associated with a specific GA. However, separation of these compounds on a large scale is presently cost prohibitive. Thus, only the mixture of GA_{4+7} is commercially available. A similar specificity is shown for GAs of varying structure in the flowering of *Lolium temulentum*, a LD-requiring plant and *Pharbitis nil*, a SD-requiring plant [27], as well as for flowering of Cupressaceae and Pinaceae conifers [27]. While it is too early to predict commercial implications of such specificities in flowering response for GAs of differing structure, the above results (discussed in [27]) indicate that one should generalize with extreme care regarding flowering/vegetative elongation responses within the GA class of phytohormones from experiments where only one GA (e.g. usually GA_3) has been applied.

There are also minor practical uses of GAs in seed production of commercially important conifers. GA_3 can be used to promote flowering, and thereby significantly increase female cones and seed per tree in seed production orchards of Cupressaceae and Taxodiaceae species, and is used commercially in Japan and western British Columbia, at least, for this purpose [cited in 26 and 30]. For Pinaceae species, which include most of the commercially important conifers, only the GA_{4+7} mixture can significantly promote flowering [25, 26, 27, 30]. The GA_{4+7} mixture is being used commercially in private and provincial conifer seed production orchards in North America, Europe, and New Zealand. However, much of the present use could best be termed "experimental", and amounts sold for this purpose are modest.

2.2 Cytokinins

The addition of BA to GA_{4+7} in 1977 represented the first registered commercial use of a cytokinin on agricultural commodities. Shortly thereafter, the tetrahydropyranyl derivative of BA (PBA) was registered on ornamentals for induction of lateral branching. The mixture of GA_{4+7} and BA is used to induce the extension of apple calyx lobes for the increased size and length/diameter ratio associated with a high quality 'Red Delicious' apple fruit [39]. Additional research with this mixture demonstrated that applications to non-bearing apple trees increased lateral bud break and improved branch angles to provide a better framework and bearing surface for early cropping. Bud break and branch angles were primarily due to the BA component, but the combination with GA_{4+7} promoted shoot growth to a much greater extent than either of the two compounds alone [18, 44]. Both of these applications are used worldwide and demonstrate the practicality of identifying defined mixtures of PGRs to elicit specific biological response. The synergistic combination of BA with other GAs (especially GA_5) has also been demonstrated in the promotion of flowering of *Chrysanthemum* under non-inductive conditions [24].

BA is registered for use in promoting fasicular bud break in conifers to reduce the required mechanical shearing and altering the rotation period required to grow marketable Christmas trees. While this use has not been commercially significant, the compound's availability has encouraged research in a number of areas. Potential applications such as the extension of vegetable shelf-life [40], prevention of crop senescence [23], and increased soybean pod set [5] have not been pursued due to cultivar differences or lack of field efficacy.

In contrast to GAs, a number of compounds have been made synthetically which exhibit cytokinin-like activity including N^6-substituted adenine analogs, DPUs, pyridylureas and thiadiazolylureas and pyrimidines [14]. Of these the highly active thiadiazolyl urea, thidiazuron (TDZ) was found to have cotton defoliant properties and was subsequently the first cytokinin-like product to be successfully developed for widespread major crop use. TDZ is unique as a harvest aid in cotton because it promotes green leaf abscission and prevents regrowth, thus allowing growers more flexible harvesting schedules. Its mode of action in the cotton leaf abscission process appears to be associated with a sustained increase in ethylene production coupled with the disruption of auxin transport [34, 35].

3 New Opportunities for Gibberellins and Cytokinins

3.1 Expanded Uses in Niche Markets

Increased consumer demands for a greater variety of high quality fruit will continue to provide potential new uses for GA and CK products. An area of considerable promise is in tropical fruit production where there is evidence that GAs and CKs may be effective in the manipulation of flowering and fruit set, promotion of lateral branching and tree structuring, and, the extension of postharvest quality. Uses considered historical, also represent areas of expanding opportunities. With the recognized need for crop diversification, acreage of new and old varieties of seedless grapes have increased in different geographic regions within the U.S. and worldwide. For example, in northern Florida, where the citrus industry was devastated by recent harsh winters, the fruit quality and weight of an emerging grape variety, 'Orlando Seedless', was markedly enhanced by GA_3 application [12]. 'Flame' seedless represents another GA responsive variety which has experienced significantly increased acreage over the past 5 years in Chile, and in western states of California, Arizona, and Texas of the U.S.A.

In wine grapes, there is a need for cluster loosening by either rachis elongation and/or flower thinning to prevent *Botrytis* bunch rot. To date, there is no chemical thinning agent for wine grapes. Results with GA_3 have been variable with some varieties exhibiting a dramatic reduction of return bloom the year following application; others are relatively insensitive to increasing rates of GA_3 [13, 42]. The response of wine grapes to other GAs certainly needs to be examined.

The use of GA to modify growth for protection against plant pathogens was described previously in cherries and grapes and is becoming an increasingly recognized potential benefit of PGRs in a number of crops. The application of GA_3 to citrus for rind preservation, discussed previously, may also have promise as a component of an integrated pest management program for the control of the Caribbean and other fruit fly species. With the greatly curtailed use of ethylene dibromide, there is currently no method available to increase fruit fly egg and larval mortality in the rind. Recent laboratory studies [10] have shown that GA_3 application significantly reduces the number of probes and egg fecundity of the fruit fly due to increased exposure to toxic peel oils contained in the juvenile rind. If this effect is manifested in the field, a high level of resistance may preclude the need for postharvest treatment with an insecticide/larvicide.

Recently, promising results were reported for thinning of apples with BA alone or in combination with carbaryl [11]. Another cytokinin-like compound currently under development which also thins apples is 2-chloropyridylphenylurea (CPPU). A number of potential applications have been examined with CPPU which overlaps with several GA effects including grape berry enlargement, and increased fruit size and weight in apple and kiwi [21, 22]. The future of this compound as a commercial product is yet to be determined and may require an identified use in a major agronomic crop to justify further development.

3.2 Agronomic Crops

The successful introduction of a GA or CK product in an agronomic crop is dependent upon the specificity of a physiological response over a wide range of environmental/cultural conditions, and is complicated by the inherent genetic variability amongst cultivars. For example, the genetic reduction of plant height has led to dramatic increases in cereal grain yield, due principally to increased reduction of lodging. Consequently, these varieties have poor emergence rates and poor seedling vigor. In rice, GA application promotes seedling establishment in the principle semi-dwarf commercial varieties utilized [4]. However, in wheat, the incorporation of the Rht1 and Rht2 dwarfing genes have rendered many commercial wheat varieties to be insensitive to applied GA [9]. The high productivity of hybrid maize and GA-insensitive wheat varieties may be associated with enhanced endogenous levels of GAs [29, 32]. However, the differences in the response of maize to exogenous GA_3 may be a function not only of the endogenous concentrations of GAs [29], but also of varying rates of GA metabolism [28]. In the GA insensitive wheat varieties compartmentalization may influence response to exogenous GA [32]. The current characterization of endogenous GAs and CKs in correlation with developmental events in major crop plants will certainly aid in the identification of application timings for candidate compounds.

A key factor in the successful development of TDZ in cotton was the availability of a large number of synthetic cytokinin-active compounds (over 150 in the patent literature alone). With different classes of cytokinin chemistries, a systematic approach toward structure activity relationships could be employed. On the other hand, synthesis of GAs is complex and considered impractical [36]. A number of GAs, obtained either by extraction or through semi-synthetic routes using the parent GA molecule, have been tested using standard elongation assays. However, without sufficient quantities of these compounds to conduct a series of detailed biological evaluations, it will be difficult to predict useful responses and economic benefits in agronomic crops. The discovery that a substituted phthalimide (AC-94, 377) induced bolting in lettuce [17] represented the first synthetic compound to have GA-like activity. Regardless of its agricultural implications, this compound may be invaluable in the exploration of GA binding sites [33].

The legacy of GA and CK applications to major crops has always been the expectation of increased yield. More realistic opportunities may reside in identifying developmental modifications that are associated with improved production efficiency, thus reducing cost inputs.

4 Conclusions

Gibberellins have been primarily used for manipulating production practices and insuring the quality of high value specialty crops such as grapes, citrus, cherries, and apples. Increasing market demands for a greater variety of quality fruit may provide new opportunities for the expanded use of GAs (i.e. tropical fruits, thinning and sizing wine and new table grape varieties, and as components of integrated pest management programs for postharvest quality of fruits). Despite considerable research, GAs have not found utility in major agronomic crops. However, increased understanding of the interrelationship between GAs and other PGRs, concurrent with the development of new GAs or GA-mimics, may offer the unique activities necessary to penetrate agronomic markets. Unlike GAs, cytokinins have only been used commercially within the past decade. BA and PBA are used principally to promote lateral bud development in ornamentals, and in conjunction with GA_{4+7} for size and improved quality in apples. While adenine cytokinins are not currently used in large acreage crops, the phenyl urea cytokinin TDZ has been successfully marketed as a harvest aid in cotton.

References

1. Bukovac MJ, Hull J, Kesner CD, Larsen RP (1987) Mich State Hortic Soc 116:122
2. Coggins CW, Lewis LN (1965) Proc Am Soc Hortic Sci 86:272
3. Creasy LL, Swartz HJ (1981) J Am Soc Hortic Sci 106:203
4. Dunand RT (1987) Proc Plant Growth Regul Soc Am 14:485
5. Dyer D, Carlson DR, Cotterman CD, Sikorski JA (1986) Proc Plant Growth Regul Soc Am 13:130
6. Eccher T (1978) Acta Hortic 80:381
7. El-Otmani M, Coggins CW (1985) J Am Soc Hortic Sci 110:371
8. El-Otmani M, Coggins CW (1986) J Am Soc Hortic Sci 111:228
9. Flintham JE, Gale MD (1982) Theor Appl Genet 62:121
10. Greaney PD, McDonald RE, Shaw PE, Schroeder WJ, Howard DF, Hutton TT, Davis PL, Rasmussen GK (1987) Trop Sci 27:23
11. Greene D (1989) J Am Soc Hortic Sci (in press)
12. Halbrooks MC, Mortensen JA (1988) Hortscience 23:409
13. Karle RD (1965) Unpublished data
14. Koshimizu K, Iwamura H (1986) In: Takahashi N (ed) Chemistry of plant hormones. CRC, Boca Raton, p 153
15. Kurosaw E (1926) Natl Hist Soc Formosa 16:213
16. Looney N, Pharis RP, Noma M (1985) Planta 165:292
17. Los M, Kust CA, Lamb G, Diehl RH (1980) Hortscience 15:22
18. Luckwill LC, Silva JM (1979) J Hort Sci 54:217
19. Martin GC (1983) Commercial uses of gibberellins. In: Crozier A (ed) The biochemistry and physiology of gibberellins, vol 2. Praeger, New York, p 395
20. Miller CO, Skoog F, Okumura FS, Von Saltza MH, Strong FM (1956) J Am Chem Soc 78:1375
21. Nickell LG (1986) Proc Plant Growth Regul Soc Am 13:236
22. Nickell LG (1987) Proc Plant Growth Regul Soc Am 14:404
23. Nooden LD, Kahanek GM, Okatan Y (1979) Science 206:841
24. Pharis RP (1972) Planta 105:205
25. Pharis RP, King R (1985) Annu Rev Plant Physiol 36:517
26. Pharis RP, Webber JE, Ross SD (1987) Ecol Manage 19:65
27. Pharis RP, Evans LT, King R, Mander LN (1989) Am Soc Plant Physiol Symp Ser I:29
28. Rood SB, Pharis RP (1987) In: Davies PJ (ed) Plant hormones and their role in plant growth and development. Nijhoff, Dordrecht, p 463
29. Rood SB, Buzzell RI, Mander LN, Pearce D, Pharis RP (1988) Science 241:1216
30. Ross SD, Pharis RP, Binder WD (1983) In: Nickell LG (ed) Plant growth regulating chemicals, vol II. CRC, Boca Raton, Florida, p 33
31. Stang E, Weidman R (1986) Hortscience 21:78
32. Stoddart JL (1984) Planta 161:432
33. Suttle JC (1983) New Bull Br Plant Growth Regul Group 6:11
34. Suttle JC (1985) Plant Physiol 78:272
35. Suttle JC (1988) Plant Physiol 86:241
36. Takahashi N (1986) In: Takahashi N (ed) Chemistry of plant hormones. CRC, Boca Raton, p 114
37. Taylor B (1975) J Hortic Sci 50:169
38. Tromp J (1982) J Hortic Sci 57:277
39. Unrath CR (1974) J Am Soc Hortic Sci 99:381
40. Weaver RJ (1972) Plant growth substances in agriculture. Freeman, San Francisco, p 275
41. Weaver RJ, McCune SB (1961) Hilgardia 30:425
42. Weaver RJ, Kasimatis AN, McCune SB (1962) Am J Enol Vitic 13:78
43. Wertheim S (1982) J Hortic Sci 57:283
44. Williams MW, Billingsley HD (1970) J Am Soc Hortic Sci 95:659
45. Wittwer SH, Bukovac MJ (1957) Q Bull Mich Agric Exp Stn 39:3

New Types of Plant Growth Retardants: Additional Perspectives for Practical Application in Agriculture and Horticulture

W. RADEMACHER[1]

1 Introduction

Plant growth retardants reduce the shoot length of plants without being phytotoxic or causing malformations. This is primarily achieved by reduced cell elongation but also by a lowered rate of cell division. Some growth retardants are of economic importance in agriculture and horticulture, since they can be employed to affect the morphology of cultivated plants in a desired way. Reviews on plant growth retardants have been presented by Dicks [14] and Nickell [34].

At present the most relevant practical applications of growth retardants are:

1. Control of lodging in small grains such as wheat, barley and rye with CCC, mepiquat chloride and ethephon,
2. Reduction of vegetative growth and enhanced uniformity of ripening in cotton with mepiquat chloride,
3. Regulation of the ratio between vegetative growth and fruit production and improvement of fruit quality in orchard trees with daminozide,
4. Control of excessive vegetative growth of peanuts with daminozide,
5. Reduction of stalk length or shoot height in ornamentals with compounds such as CCC, daminozide and ancymidol.

Several additional indications are, however, conceivable which cannot be covered by the established compounds. Therefore, continuous efforts are being made to find and develop new and better suited retardants, also as possible substitutes for existing products.

Several new types of plant growth retardants have been detected and developed during recent years. Norbornanodiazetines (e.g. tetcyclacis), triazoles (e.g. paclobutrazol, uniconazole, triapenthenol, BAS 111 . . W), isonicotinanilides (e.g. inabenfide) have to be mentioned here as well as flurprimidol (a pyrimidine derivative) which is closely related to ancymidol (Fig. 1). In general, the growth-retarding activity of these compounds is relatively high in a far broader spectrum of plant species as compared, for instance, with CCC, mepiquat chloride and daminozide [30, 35, 38]. As their basic mode of action, the norbornanodiazetines, triazoles, isonicotinanilides and pyrimidines inhibit the endogenous formation of GAs by blocking cytochrome P-450-type oxygenases that catalyze the reactions between *ent*-kaurene and *ent*-kaurenoic acid. Clear evidence is available that other cytochrome P-450 oxygenases are also affected by these compounds, although most

[1] BASF Agricultural Research Station, P.O. Box 220, 6703 Limburgerhof, FRG

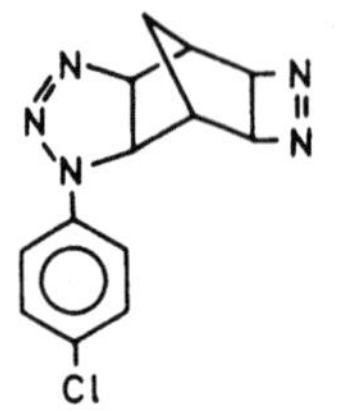

I

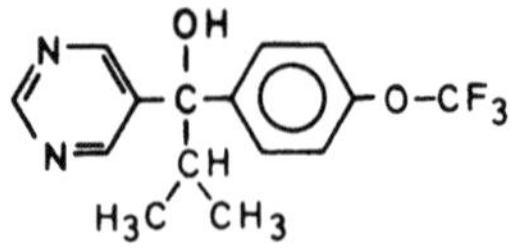

II

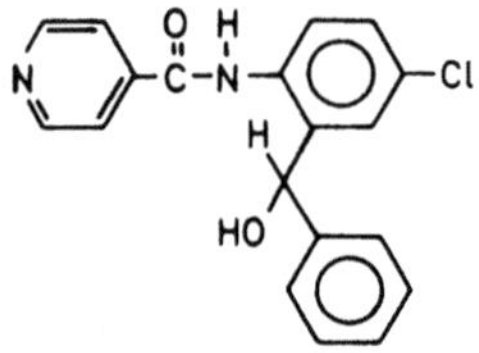

III

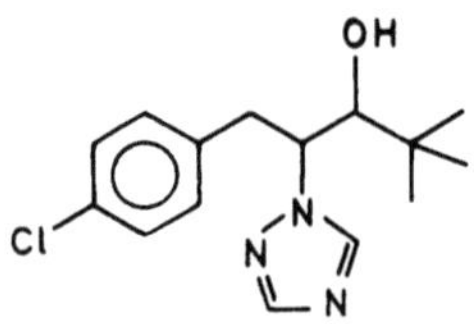

IV

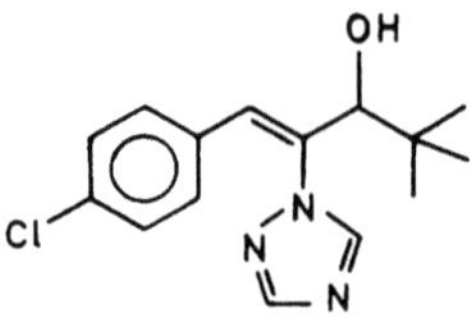

V

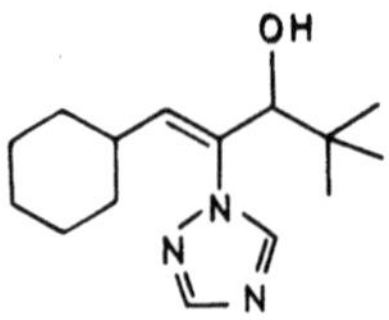

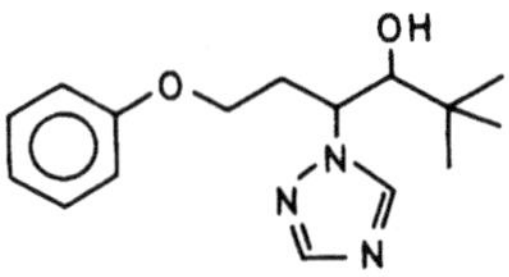

Fig. 1. New types of plant growth retardants. *I* Tetcyclacis (BAS 106 .. W, LAB 102 883), 5-(4-chlorophenyl)-3,4,5,9,10-pentaazatetracyclo-5,4,1$0^{2,6}$,$0^{8,11}$-dodeca-3,9-diene. *II* Flurprimidol (EL 500), α(1-methylethyl)-α-[4-(trifluoromethyloxy)-phenyl]-5-pyrimidine-methanol. *III* Inabenfide (CGR-811), [4-chloro-2-(α-hydroxybenzyl)]-isonicotinanilide. *IV* Paclobutrazol (PP 333), 1-(4-chlorophenyl)-4,4-dimethyl-2-(1H-1,2,4-triazol-1-yl)-pentan-3-ol. *V* Uniconazole (S-3307, XE-1019), (E)-1-(4-chlorophenyl)-4,4-dimethyl-2-(1H-1,2,4-triazol-1-yl)-1-penten-3-ol. *VI* Triapenthenol (RSW 0411), (E)-1-cyclohexyl-4,4-dimethyl-2-(1H-1,2,4-triazol-1-yl)-1-penten-3-ol. *VII* BAS 111 .. W, 1-phenoxy-3-(1H-1,2,4-triazol-1-yl)-4-hydroxy-5,5-dimethyl-hexane

of them at a far lower degree of activity. In particular, the inhibition of sterol formation by the blocking of 14-demethylation and the inhibition of the oxidative inactivation of ABA must be mentioned. It has also been reported that the oxidative metabolism of certain xenobiotics can be inhibited and one may forecast that further such effects will be detected in the future. The situation is even more complex since these side activities can be fairly different from compound to compound and from species to species. For further details and references see [22, 39].

In general the new compounds possess properties distinct from those of most hitherto known growth retardants, which may be the basis for advancement in practical uses. Possibilities for such applications will be described below in selected case studies. Some of these possibilities have already been put into practice, or will most likely be realized in the near future. Other observed effects are probably of academic interest only. They might, however, serve as a starting point for specifically looking for more advanced solutions.

2 Potentials for Practical Uses

In most cases the growth-retarding effect is the only point of interest. This would primarily exploit the potential of the new growth retardants to lower the endogenous levels of vegetative-growth active GAs. In addition to this, other physiological processes also affected by the new types of compounds may cause beneficial effects as well. The usefulness of a given compound in a given crop may therefore be determined by the combination of such side effects with morphoregulation. In certain cases growth retardation may even be of secondary importance.

2.1 Improving the Quality of Seedlings Used for Transplanting

It is a general observation that seedlings raised under the influence, for instance, of tetcyclacis or triazoles exhibit a shoot-root-ratio which has shifted in favor of the root. In many cases, root growth is stimulated and longer and thicker main roots can be observed, while the formation of adventitious roots and root hairs is initially inhibited [38]. The formation of an improved root system may be one of the reasons why most of these plants show a better performance at later stages of development.

Tetcyclacis applied as a seedsoaking considerably improves the quality of rice seedlings used for mechanical transplanting: the shortened seedlings can be kept at a size optimal for transplanting for a longer period of time; they are more resistant to transplantation stress; performance of these plants under field conditions is superior to that of non-treated ones [41]. Similar effects have also been observed in seedlings of maize and other plant species used for mechanical transplantation (Schott and Luib, unpubl, Limburgerhof, FRG).

2.2 Control of Lodging in Small Grains and Rice

The control of lodging is by far the most important indication for which plant growth regulators are being used at present [5]. CCC, mepiquat chloride and ethephon have become an integral part of intensive cereal production in many areas. These compounds are, however, not suited for use in rice, where lodging is also a problem. The triazoles paclobutrazol [19], uniconazole [35], triapenthenol [31], BAS 110 . . W and BAS 111 . . W [25] and the isonicotinanilide inabenfide [33] have shown good results in different species of small grains and in rice.

2.3 Growth Regulation in Oilseed Rape

Oilseed rape has become a major crop in the European Community over the last 10 years and is also of importance in other countries. In order to optimize the productivity of this crop, the use of plant growth regulators can be of relevance. The main objectives for such compounds have been defined by Dawkins [13]. Among these are the reduction of stem length to make the plants less susceptible to lodging and to provide opportunities for easier crop management. Yield increases may be obtained by optimizing the canopy structure of the crop, by improving winter hardiness or by reducing pod abortion or pod shatter. The triazoles triapenthenol and BAS 111 . . W apparently have a positive influence on almost all of these objectives. These compounds reduce the height of the oilseed rape plants and lead to a canopy structure that allows a better penetration of light to lower parts of the plants. Less susceptibility to frost damage, reduced lodging and fewer fungal infections also occur. Independent of the latter effects, yield increases in the range of 10% are commonly observed [7, 25, 32].

2.4 Control of Excessive Shoot Growth in Orchard Trees

The pruning of orchard trees requires skilled labor and is time consuming and costly. Effectively reducing the number and length of vegetative shoots by suitable growth retardants would therefore be of considerable benefit. Reduction of shoot growth would also be of relevance in fruit trees, such as sweet cherries or plums, of which no dwarfing rootstocks or compact scion cultivars are available. This would allow for efficient high-density plantations as, for instance, in apples.

Most of the new triazole-type growth retardants and flurprimidol can be used to control vegetative shoot growth in different pome and stone fruits and in citrus [2, 11, 29, 37, 43]. A general difficulty encountered with the use of these compounds in orchard trees lies in the type of application. Foliar treatments are often less efficient than soil treatments, since most compounds are more readily transported acropetally in the xylem. However, soil applications make precise dosaging relatively difficult, since the availability of a compound to the tree will be strongly influenced by parameters such as moving velocity in the soil and the degree of its persistency. An alternative type of application may be seen in trunk painting [11].

2.5 Growth Reduction in Other Trees

Trees that grow, for instance, under power or telephone lines have to be trimmed regularly. Just as in the case of orchard trees, some of the expenses can be saved by using growth retardants such as paclobutrazol or uniconazole. Application by stem injection has been proposed for administering the active ingredients in order to avoid chemical contact with non-target plants and reduce environmental residues [44, 45].

2.6 Growth Regulation in Amenity Grasses

Suitable PGRs are being sought for amenity grasses, in order to save mowing costs while maintaining an aesthetically pleasing lawn. Such compounds would be of relevance to manage grass growth for instance in home lawns, sportsfields, golf-courses, parks, industrial ground, cemeteries and roadsides. Good reduction of growth has been achieved with compounds such as paclobutrazol and flurprimidol. However, these substances did not control unwanted seedhead formation. Combinations with compounds that have more a herbicidal nature, e.g. mefluidide or amidochlor, may offer more advanced solutions [15, 26, 27].

2.7 Improvement of Quality in Ornamental Plants

Height reduction can be achieved with virtually all of the new compounds in many ornamental plant species such as chrysanthemum, poinsettia, tulip, azalea, and lily [4, 20, 23, 35, 49]. The formation of flower buds was promoted by uniconazole in azalea and rhododendron [35].

2.8 Improved Resistance to Drought and Low Temperatures

CCC is known to enhance the resistance of plants againts drought and frost damage [cf. 24]. The latter effect is used in the USSR, where seed treatment with CCC is carried out over an area of 2 million hectares of winter and spring wheat [36]. Zadoncev et al. [50] emphasize particularly the lowering of the tillering node as the reason for improved wintering of wheat under the influence of this compound.

Improved resistance to low temperatures has also been found, for instance, in cucumber and zucchini squash treated with paclobutrazol [48], in oats and rice under the influence of tetcyclacis [1, 8] and in winter oilseed rape that had been treated with BAS 111 . . W [32]. Similarly, plants performed better under conditions of drought after they had been treated with compounds such as triadimefon, flurprimidol, paclobutrazol, uniconazole and tetcyclacis [3, 16, 42, 47].

There may be several reasons for these observations. First, one should consider a purely physical effect: plants with retarded shoot growth may have their growing points protected in the ground and expose less surface to the attack of unfavorable conditions. Improved root growth may be capable of absorbing increased amounts of water. In addition to this, it must not be overlooked that increased levels of endogenous ABA will also improve the resistance of plants to drought [12] or low temperatures [21, 40]. Significantly lowered transpiration, most likely caused by raised levels of ABA, could, for instance, be induced in detached leaves of barley by tetcyclacis [6].

2.9 Improved Resistance to Fungal Infections

In oilseed rape the rate of infection with different fungi was reduced by treatments with triapenthenol and BAS 111.. W [7, 32, 46]. Paclobutrazol and ancymidol improved the resistance of melon seedlings to fusarium wilt [9]. The effects observed are thought to be mostly due to direct fungicidal effects via inhibition of fungal ergosterol biosynthesis [17]. Evidence is, however, also available that other effects such as delayed senescence have led to increased resistance of the plants [32].

2.10 Enhancement of Herbicidal Activity

Several pesticides and other xenobiotics are metabolized in plants by mixed-function oxidases which are cytochrome P-450-dependent [28]. Inhibitors of such enzymes might therefore serve as synergists of herbicides or other crop protection compounds which are oxidatively degraded. Therefore, it has been logical to test whether the growth retardants believed to be oxygenase inhibitors would interfere with xenobiotic metabolism. Under in vitro conditions it has indeed been observed that tetcyclacis at relatively low concentrations inhibits the metabolism of the herbicide chlortoluron [10]. Similarly, the herbicidal activity of bentazon could be enhanced by tetcyclacis which, most likely, reduced its oxidative inactivation [18]. Further studies will, however, be required to find out whether such effects can be used under practical conditions.

3 Conclusions

The new types of plant growth retardants dealt with in this contribution clearly offer new possibilities for practical application. Considerable interest will have to be paid to beneficial side effects of these compounds which are not related to growth retardation, since these features will also determine the degree of usefulness of a given compound for a specific indication.

References

1. Anderson HM, Huband NDS (1987) In: Hawkins AF, Stead AD, Pinfield NJ (eds) Plant growth regulators for agricultural and amenity use'. British Crop Protection Council, Mono No 36, Thornton Heath, p 45
2. Aron Y, Monselise SP, Goren R, Costo J (1985) HortScience 20:96
3. Atkinson D (1986) Acta Hortic 179:395

4. Barrett JE (1982) HortScience 17:896
5. Baylis A (1990) In: Pharis RP, Rood SB (eds) Plant growth substances 1988. Springer, Berlin Heidelberg New York Tokyo
6. Carlson DR, Rademacher W, Jung J (1988) In: Pharis RP, Rood S (eds) Book of abstracts 13th Intl Conf on Plant Growth Substances. Calgary, Alta, Canada, Abstract No 67
7. Child RD, Butler DR, Sims IM, Johnson W, Thorn M (1987) In: Hawkins AF, Stead AD, Pinfield NJ (eds) Plant growth regulators for agricultural and amenity use. British Crop Protection Council, Mono No 36, Thornton Heath, p 21
8. Chu C, Hwang SJ, Lee TM (1987) In: Proceedings of the 5th Seminar on Science and Technology – Phytohormones, Nara, Japan, Interchange Association, Japan, p 125
9. Cohen R, Yarden O, Katan J, Riov J, Lisker N (1987) Plant Pathol 36:558
10. Cole DJ, Owen WJ (1987) Plant Sci 50:13
11. Curry EA, Williams MW, Reed AN (1987) Bull Plant Growth Regul Soc Am 15:4
12. Davies WJ, Mansfield TA (1983) In: Addicott FT (ed) Abscisic acid. Praeger, New York, p 237
13. Dawkins TCK (1986) News Bull Br Plant Growth Regul Group 8:1
14. Dicks JW (1980) In: Clifford DR, Lenton JR (eds) Recent developments in the use of plant growth retardants, Mono No 4. British Plant Growth Regul Group, Wantage, p 1
15. Elkins DM (1983) In: Nickell LG (ed) Plant growth regulating chemicals, Vol II. CRC, Boca Raton, Fl, p 113
16. Fletcher RA, Nath V (1984) Physiol Plant 62:422
17. Fletcher RA, Hofstra G, Gao J (1986) Plant Cell Physiol 27:367
18. Fritsch H, Rademacher W, Retzlaff G (1987) In: Greuter W, Zimmer B, Behnke HD (eds) Book of abstracts, XIV Intl Bot Congr, Berlin (West), Abstract No 2-113b-5
19. Froggatt PJ, Thomas WD, Batch JJ (1982) In: Hawkins AF, Jeffcoat B (eds) Opportunities for manipulation of cereal productivity, Mono 7. British Plant Growth Regul Group, Wantage, p 71
20. Gianfagna TJ, Wulster GJ (1986) HortScience 21:463
21. Gusta LV, O'Connor BJ, Reaney MJ (1990) In: Pharis RP, Rood SB (eds) Plant growth substances 1988. Springer, Berlin Heidelberg New York Tokyo
22. Hedden P (1990) In: Pharis RP, Rood SB (eds) Plant growth substances 1988. Springer, Berlin Heidelberg New York Tokyo
23. Holcomb EJ, Ream S, Reed J (1983) HortScience 18:364
24. Jung J, Rademacher W (1983) In: Nickell LG (ed) Plant growth regulating chemicals, Vol 1. CRC, Boca Raton, Fl, p 253
25. Jung J, Luib M, Sauter H, Zeeh B, Rademacher W (1987) J Agron Crop Sci 158:324
26. Kaufmann JE (1990) In: Pharis RP, Rood SB (eds) Plant growth substances 1988. Springer, Berlin Heidelberg New York Tokyo
27. Kavanagh T (1987) In: Hawkins AF, Stead AD, Pinfield NJ (eds) Plant growth regulators for agricultural and amenity use. British Crop Protection Council, Mono No 36, Thornton Heath, p 135
28. Lamoureux GL, Frear DS (1979) In: ACS Symposium Series 97. American Chemical Society, Washington DC, p 78
29. Looney NE, McKellar JE (1987) J Am Soc Hortic Sci 112:71
30. Luerssen K (1987) Pestic Sci 21:310
31. Luerssen K, Reiser W (1987) Pestic Sci 19:153
32. Luib M, Koehle H, Hoeppner P, Rademacher W (1987) In: Hawkins AF, Stead AD, Pinfield NJ (eds) Plant growth regulators for agricultural and amenity use. British Crop Protection Council, Mono No 36, Thornton Heath, p 37
33. Nakamura K (1987) Jpn Pestic Inf 51:23
34. Nickell LG (ed) (1983) Plant growth regulating chemicals, Vol 1+2. CRC, Boca Raton, Florida
35. Oshio H, Izumi K (1986) In: Plant growth regulators in agriculture, food and fertilizer technology. Center for the Asian and Pacific region, Taipei, p 200
36. Pikus GR, Grincenko AL (1977) In: Zadoncev AI, Pikus GR, Grincenko AL (eds) CCC in der Pflanzenproduktion. VEB Deutscher Landwirtschaftsverlag, Berlin (East), p 5
37. Quinlan JD (1987) In: Hawkins AF, Stead AD, Pinfield NJ (eds) Plant growth regulators for agricultural and amenity use. British Crop Protection Council, Mono No 36, Thornton Heath, p 149
38. Rademacher W, Jung J, Graebe JE, Schwenen L (1984) In: Menhenett R, Lawrence DK (eds) Biochemical aspects of synthetic and naturally occurring plant growth regulators, Mono No 11. British Plant Growth Regul Group, Wantage, p 1

39. Rademacher W, Fritsch H, Graebe JE, Sauter H, Jung J (1987) Pestic Sci 21:241
40. Rikin A, Richmond AE (1976) Physiol Plant 38:95
41. Schott PE, Knittel H, Klapproth H (1984) In: Ory RL, Rittig FR (eds) Bioregulators – chemistry and uses. ACS Symp Series 257. American Chemical Society, Washington DC, p 45
42. Shanahan JF, Nielsen CD (1987) Agron J 79:103
43. Steffens GL (1988) J Plant Growth Regul 7:27
44. Sterrett JP (1985) J Am Soc Hortic Sci 110:4
45. Sterrett JP (1988) J Plant Growth Regul 7:19
46. Stinchcombe GR, Hutcheon JA, Jordan VWL (1986) In: 1986 British Crop Protection Conference – Pests and Diseases, Vol 3. British Crop Protection Council, Croydon, p 1009
47. Vaigro-Wolff AL, Warmund MR (1987) HortScience 22:884
48. Wang CY (1985) Sci Hortic 26:293
49. Wulster GJ, Gianfagna TJ, Clarke BB (1987) HortScience 22:601
50. Zadoncev AI, Pikus GR, Grincenko AL (1977) CCC in der Pflanzenproduktion. VEB Deutscher Landwirtschaftsverlag, Berlin (East)

Subject Index

ABA
biochemistry 241-253
(-)-R-ABA from diol epimerization (?) 243
ABA diols 241-5
1,4 diols 241-3
glucosyl conjugates of 245-6
4-O-β-D glucopyranosides 245
β-D-glucosyl esters 245
epimerization of hydroxyl groups 242
especially *cis*-diol 242
in *Avocado*, pea, tomato 244
Vicia 243
^{18}O enrichment & exchange 242
GC-MS (negative ion, chemical ionization) 244-5
(-)-7′-hydroxy ABA 243
from (-)-R-ABA (?) 243
inactivation of ABA via 8′methyl hydroxylation 245
-induced proteins in barley aleurone layers 138-143
cotton embryogenesis 142
a similar induction by osmotic or salt stress 142
major metabolites of ABA
4-0-β-D glucopyranoside of DPA (most plants) 245
PA glucose ester (mango) 245
2,4-octadienedoic acid (from violazanthin) 251-2
phaseic acid formation 139-40, 247-8
via cyclization (nucleophilic attack) of 8′-OH ABA 247
use of ^{1}H NMR spectrometry in structural determination 247
xanthoxin as possible ABA precursor 250-52
$^{2}H_{2}0$ incorporation into ABA precursors 250-2
biosynthesis 249-52
in ABA-deficient mutants 24-31
Antherceros 5
rice 14-20
indirect pathway 233-6
in *Avocado* mesocarp 235-7
apple and corn 237
Xanthium 234-6
mevilonin effects 236
use of $^{18}0$ in study of 234-6
cold-hardiness/resistance 234-6, 532-5
cross adaptation 395, 398
derivatives 562
& dormancy in cereals
levels of, and response to ABA 400-405
& embryo responsive in cereal grains 401-2
effects on photosynthetic capacity of leaves 383-90
PAL in *Cuscuta* 495
:GA ratio in cold-hardiness 536
-glucosyl ester 237
in lower plants 4
*Rht*3 wheat mutant 86
ABA inactivation, inhibition thereof by plant growth retardants 237-9, 329-30, 613
-inducible proteins in mature embryos 404-5
barley aleurone 139-43
-induced cDNA clones in barley aleurone 141-3
interaction with GA in cold-hardiness 532-3
levels during grain development 401
metabolism (including catabolism) 233-239
effects of growth retardants 329-30
in barley aleurone layers 139-40
tissue specific in legume seeds, of (s)ABA 224-7
to ABA-GE and PA 139-40, 237-9
- metabolites, effect on GA_3-induced α-amylase 138-43
molecular action of 31
mucilage layer (seed) – effect on 29-30
mutants
aba in *Arabidopsis* 24-31
ABA deficient- & ABA response- 23-31
concept of regulation (synthesis or sensitivity 29-31
uniculm (stunting) mutants in cereals 84-92

regulation of gene expression by, in barley aleurone 139-43
responsiveness to (ABA) in mature grain 403-4
roles (questionable) from use of ABA-mutants 29-30
& root gravitropism 213-4
seed development in cereal grains 402
seed set in legumes 458
senescence 538
turnover 338
7'hydroxy (-)-R ABA
as catabolite of (-) ABA 239
7'hydroxy (s)ABA in *Vicia* (?) 239
abscisic acid (see ABA)
abscission
& calcium, role in auxin-induced abscission 223-4
induced by ethylene-releasing agents
mistletoe 598
fruit 600-1
leaf defoliation & branch, induced by ethephon 598-9
AC-94 609
ACC
degradation to ethylene, CO_2, HCN 295
in crown gall tumours 504, 508-9
oxidase 294-7
& ethylene production (ACC & O_2 dependence) 294-5
N-malonyl transferase 296-8
subcellular compartmentation 296-7
D-amino acid malonyl transferase 297-8
senescence signal contribution 543
synthase
changes in coded mRNA after wounding 115
immunoaffinity purification 114
immunochemical comparison
wound-induced vs auxin-induced 118-9
induction at transcriptional level 120
reactions (catalyzed by), sterochemical courses 292-3
SAM-induced activation 293
wound vs auxin-induced 113-121
O-acetyl-9-beta-D-ribofuranosyl DHZ 5'-MP 277-8
action potentials, as wounding signals 520-1
adenine cytokinins in pod set 463
S-adenosylmethionine (see SAM)
adenylate anticytokinins 180
agonists of hormonal regulation
various phytohormones 204
phytochrome, polyamines 204
Agrobacterium 2-3, 63
auxin biosynthetic genes 40
auxin regulation of crown gall tumours 504-510
HSP-*ipt* chimeric gene – use to induce alterations in CKs in *Arabidopsis* & *Petunia* 186-91
agronomic crops
effects of GAs and CKs 608-9
Alar (see daminozide)
aleurone insensitivity to GA in *Rht* dwarfs 86
α-amylase gene families 130-7
inhibition during seed development 142
behavior in slender mutant of barley 85-90
gene expression, regulation by ABA 138-43
low temperature effects 86
plasma membrane, GA receptors 150
proteins, ABA – & osmotic & salt-induced 142
protoplasts, GA receptor research 145-53
algae, non-green
cytokinins 2-3
rhodomorphin 3
amide-bound IAA 48-9
aminobenzotriazole 573
AM toxins 560
1-aminocyclopropane-1-carboxylic acid (see ACC)
2-aminoethoxyvinyl-glycine (see AVG)
alpha-aminoisobutyric acid 509
AMO-1618 323, 321-31, 398
mode of action 324-5
α-amylase
activity in *Cuscuta* during hormone-induced lignification 500-1
GA_3-induced in barley, effect of ABA on 139-43
gene families in barley 130-7
transcription 138-143, 145
induction by immobilized GAs 150
in grains of *Rht* mutants of wheat 86
inhibitor, during seed development 142
proteins, low & high pI 139-43
ancymidol 32-3, 396, 611
Anemia 5-6
anther GA biosynthesis (rice) 19
antheridic acid 5-6
antheridiogens 5-6
Antherceros, ABA in 5
anticytokinins 179-185
antibodies for IAA-peptides 47
GA receptors 150-1
apical dominance
auxin & ethylene interaction 424-27
effects on by *ipt* chimeric gene in *Arabidopsis* & *Petunia* 188-91

release from, by CKs/*Rhizobium* inoculation 278-9
second messenger, in auxin-controlled 419-27
use of reduced pressure to remove ethylene 420-1
ethylene inhibitors/antagonists 424-7
apigenin 431
apples
ABA biosynthesis in 236
effect of GAs on 604-610
Arabidopsis
& ABA-deficient mutants 24-31
ABA re seed development 26-31
ABA re late abundant proteins 26-31
fusicoccin binding sites 155
in ethylene research 595
increased CKs from HSP-*ipt* chimeric gene 186-91
aresenazo III 512-5
aryazides, photoaffinity labeling 146
Asparagus 184
assimilatory charge, Rubisco pool size 384
ATP 173-7
auxin(s)
action, role of calcium in 220-7
analog (toxic) 5-Br-IAA 35-36
autonomy 506-510
-binding 106-112
assay 124
receptors (membrane-bound) 106-112
high affinity binding sites 126-7
receptor gene 111-2
of G-protein coupling 278
scheme for 128
conjugate mutants 45-6
-controlled apical dominance -second messenger 419-27
effects on growth & development of *Cuscuta, in vitro* 492
efflux 430-1
genetic approaches 32-34, 45-55
Hemmstoff und Wachstum 450-55
-induced ACC synthase 114-21
cell division in cell cultures 102
immunochemical comparisons, wound-induced 118-9
lateral transport and tropisms 349-50
lignification, effects on/in *Cuscuta* 498-501
metabolic mutants 45-55
Lemna 45
metabolism, genetic and analytical approaches 45-55
„Ohne Wuchsstoff Kein Wachstum" 341, 346
perception and transduction 104-5
phototropism and role of inhibitors 450-55
precursor in seed 341-51
- promoted elongation in *Cuscuta*
-regulated cell (caulonema) differentiation 407-18
gene expression 95-99, 100-105
relation between stimulation and growth 341-351
- resistant mutants 32-44
senescence 538
sensitivity
of crown gall tumors 505-7
of the moss, *Physcomitrella* 57-64
signal transduction on membranes 122-9
and PI turnover 122-7
transport
in stele from apical bud 426
root cap 210-12
lateral, and tropisms 349-50
polar, inhibitors (NPA, TIBA) thereof 428-9
& regulation by phenolics 528-39
vascular differentiation 428
bundles 438
wound signals in decapitated plants 519-20
auxin-auxotroph mutants 32-44
enrichment & total isolation techniques 35-44
auxin/cytokinin balance
& tumour growth/morphology 503-510
Avena
aleurone protoplasts-GA receptors 145-53
auxins and growth 341-351
herbicide control of 566-75
phototropism 450
AVG, in apical dominance 423-4
Avocado
ABA biosynthesis in mesocarp 234-5, 244

B-995 (see daminozide)
BA 59,283, 605-7, 539
and flowering 487-90
fruit set and yield 458-62
lignification of *Cuscuta* 498-500
PAL activity in *Cuscuta* 494
signal transduction 514-7
binding with CCKB 173-7
bacterial production of indole-3-ethanol 360-5
BAP (see BA)
barley
ABA regulation of gene expression in aleurone 138-43
α-amylase gene families 130-7
($\pm$) DJA metabolism 377

slender mutant 85-90
BAS-111 327-8, 612
bean
ABA biosynthesis 234
apical dominance in 419-27
metabolism of MVA to $GA_{1,5,8}$ in cell-free extracts of suspensors 472
GA_{20} 3β-hydroxylase from seeds 309-11
IAA-containing peptide conjugates 46
levels in seeds 47
zeatin metabolism in embryos 267-73
bentazon 572
benzyladenine (see BA)
bioelectric potential & IAA distribution 350-1
biotechnology & PGRs & plant breeding 583
Botryodiploidia
JA and analogues & metabolites 374-6
Botrytis, GA_3 effects on 605
brassinosteroid(s) (Br) 18, 204
5-bromodeoxyuridine 35-36
Bryophyllum re GA-induced flowering 480-3
bud formation & calcium uptake/current 512

Cajanus 278-280
calcium 216-229, 511-17
& auxin-action, role in 222-8
gravitropism 212-14
& ionophores 213-4
-binding 512
channel agonists & blockers 223-4, 166-8, 512
current and bud formation 512
gravitropism, role in light-regulated 209-215, 217-222
ion uptake with BA treatment by target cells 517
ionophores and auxin action 223-4
(A23187) & root gravitropism 213-4
phosphorylation, calcium-regulated 205, 220-2
release induced by IP_3 209
blocked by TMB8 209
-requiring protein kinases 205-6
second messengers 547
in CK-mediated cell division (?) 161-171
hormonal regulation 204-7
stimulus-response coupling 216-229
transport, control by CKs (?) 162-4
across root cap 210-212
calmodulin 209, 220, 511
as second messenger 204-7, 550
in CK-mediated cell division (?) 161-171
gravitropism 209-15, 220
inhibitors 204
-requiring protein kinases 205-6
& wound signals 524-5
carbamates (as anticytokinins & florigenic compounds) 175-185
carbohydrate metabolism & cytokinins 465-6
castasterone 18,19
Caulerpa, IAA & GA-like substances 4-5
caulonema differentiation 58-64, 407-18
auxin (endogenous), role of 415
nutrient effects 412-3
PCIB effects on 410-11
pH effect on 408-9
stability of filament 412
CCC 88, 322-31, 323, 396, 398, 481, 533-4, 580-3
mode of action 324-5
CCKBP 173-7, 511
cDNA
auxin-induced sequences from soybean 95-99
barley α-amylase genes 131-7, 141-3
clone for wound-induced ACC synthase 116-7
characterization of mRNA induction by auxin 95-99,101-5
potato calmodulin cDNA clone nucleotide sequences 221
encoding maize auxin receptor 111-2
library for IAA peptide conjugates 48
cell cycle & division & flowering 486-7
CK mode of action in 161-171
cell differentiation in *Physcomitrella* protonema 58-9, 407-18
cell wall linkage turnover, in dwarf mutants 89-90
chalcone synthase 439
cherry 604-606
Chlormequat, (see CCC)
2-chloroethylphosphonic acid (see Ethephon)
2-chloroethyltrimethyl ammonium chloride (see CCC)
4-chloro-2-methylphenoxyacetic acid (MCPA) 568-9
chloronema 58-64, 416-7
(1-(4-chlorophenyl)4,4-dimethyl-2-(1,2,4-triazoll-yl)pentan-3-ol (see Paclobutrazol)
chlorophyll fluorescence from Photosystem II 387
chloroplast(s) 173-7
chloroplast ctyokinin binding protein (see CCKP)
chlorotorulon 569, 572
chlorsulfuron 568
Cholodney-Went theory on phototropism
opposing view based on growth inhibitors 450-55
CKs
alterations in level, by gene transformation 186

effect of heat induction on 190
bases, *O*-acetyl derivatives 261
cyanoethyl derivatives 265
biosynthesis 258-266, 281
in *Datura* crown gall tumor 259-262
Ove-mutants (2-iP) of moss 7-8
germinating lupin seeds 262-4
tobacco leaves 264-5
use of xanthine oxidase in 265
cis- and *trans*-isomers, in rice 17
calcium interactions 161-71
anti-calmodulin drugs 161
calcium channel blockers (verapamil, nifedine) 162
commercial uses 604-610
agronomic crops 608-9
apple thinning 608
flowering in *Chrysanthemum* 607
promotion of lateral branching 607
tropical fruits 607,8
Cuscuta, effects on growth & development *in vitro* 492
effect on carbohydrate metabolism 465-6
H^+ transport (?) 163
effect on phospholipid composition 164-6
root elongation in transgenic *Arab.* 189
nil effect on flowering in *Arabidopsis* *189*
flux from roots, and senescence 489-90, 539, 541
in relation to photoperiodic flowering 489-490
genetic approaches 32-44, 186-91, 267-73
HSP-*ipt* genes, use to alter CK levels 186-91
in algae 2-3
bean embryos 267-73
flower initiation 486-90
lower green plants 4
rice 16-20
seed 276
in senescence 537-545
in xylem sap 275-281, 540-2
inhibition of rooting 540
mediation of action by ethylene (?) 161-171
metabolism 267-73, 275-81, 282-9, 542,
mode of action 161-71
calcium as second messenger (?) 166-71
in soybean callus with requirement for CK 164-71
PI cycle, role of (?) 169
via control of calcium transport 162
protein phosphorylation (?) 163
nucleotides 276-281, 540-2
oxidase 289-9
& pod set 457-67
polyamine increases 163
-resistant mutants 32-44, 59
ribosides, isopropylidene derivatives 263-4
-sensitivity mutants 57-64
signal transduction, control by CKs 511-7
synthase 258
translocation 275-281
CO_2
from ACC degradation 295
intercellular partial pressure 383
& senescence signal bioassay 543
cold & CKs in flowering 490
cold-requiring plants 477-85
cold-hardiness & chilling resistance 391-80
acclimation requirement 531-2
& ABA 532-5
ABA and cross-adaptation in 395,398
GAs 532-6
GA:ABA ratio 536
genetic potential for 534
inhibitory factors 532-3
& PGRs in winter wheat 530-5
& plant growth retardants 615g
promotive factors 532-3
retardation of loss of hardiness, by growth retardants 534-5
coleoptile(s) 86, 110, 341-351
Commelina 154
compactin, effect on MVA levels 259
Con A-Sepharose 287
conjugation of GAs in corn 300-305
copalyl-PP (CPP) 322-3
correlative controls of senescence 537-545
corn
ABA biosynthesis in 236
*an*1, *dwarf* 1, lesions 71
CK oxidase and CK catabolism 282-9
dwarf 5 lesion 68-70
heterosis and GAs 609
GA conjugation 300-305
GA_{20} metabolism and enzymology in 307-313
gravitropism in roots, & calcium 217
IAA & stimuli in relation to growth 341-51
IAA catabolism 367-73
membrane receptors, (purification) for auxin 107-8
metabolism and turnover 333-9
mutants *d*1, *d*2, *d*3, *d*5 307-313
dwarf, various 65-73
„leakiness“ thereof re GA biosynthesis 309
viviparous, ABA-deficient 400
cotton 581

cotyledon(s)
ABA and GA_{12}-ald metabolism 254-7
GAs 470-475
CPPU 608
crown gall tumor morphology, auxins and CKs 503-510
cucurbic acid 374
Cucurbita 114-121, 429-439
Cuscuta 492-501
cycloheximide
stimulation of abscission, via ethylene 595
suppression of PAL activity 496-8
Cycocel (see CCC)
cytochalsins 527, 529, 556-7
cytochrome P-450-dependent mixed function oxygenases 138-9, 237-8, 616
inhibition by tetcyclasis 237-9
cytokinin(s) (see CKs)
2,4-D 568-9
-induced mRNAs 95-9, 102-5

DAG 168, 205, 550
daminozide 396, 398, 611
Daucus 122-9
Datura 261
decapitation & wound signals 519-20
declyimidazole 396
de-etiolation of coleoptiles to modify growth 345-6
defoliation, practical use, with ethephon 598
delayed light emission effects of CCKB and BA on 174-7
deuterium use in IAA biosynthesis studies 342-51
diacylglycerol (see DAG)
2,4-dichlorophenoxyacetic acid (see 2,4-D)
diclofop-methyl 566-75
9,10-dihydrojasmonic acid (DJA) 374-9
glucose & amino conjugates thereof 375-7
dihyrophaseic acid (see DPA)
dihydrozeatin 267-73, 275-281, 539-542
analogues and pod set 463
dihydrozeatin-0-glucoside 275-281, 540-2
dihydrozeatin monophosphate 540
dihydrozeatin riboside 275-281, 539-542
O-acetyl 9 beta-D, 5'monophosphate 277
dihydroxylosylzeatin 271
(diH)Z (see dihydrozeatin)
(diH)[9R]Z (see dihydrozeatin riboside)
dioxygenases in GA biosynthesis 314-21
1,3-diphenylurea (DPU) 289
DJA (see 9,10-dihydrojasmonic acid)
DLE (see delayed light emission)
DNA (complementary) see cDNA
in vitro transcription, auxin-treated soybean 96
polytenization in giant suspensor cells 471
sequence analysis for Aux soybean genes 97
(transferred) (see T-DNA)
tranforming, in the moss, *Physcomitrella* 64
recombinant techniques and role of calcium in signal transduction 225-7
dolichosterone 18, 19
dormancy
in cereal grains, levels of & response to ABA 400-405
induction & ABA 400
seed, breaking by ethylene 601
DPA 239, 244-5,
biological activity of, in barley aleurone 138-143
conjugate 255
droopy mutant of potato 24-31
drought resistance and plant growth retardants 615
dwarf(s)
barley, Cb3014 85-90
cereals (Norin 10, *Rht*, uniculm) 84-92
GA-sensitive, rice and corn 86
pea mutants 74-83
Zea mays, use in GA conjugation studies 303
research tools in growth analysis 65-73

economic aspects of plant growth regulators (PGRs) 576-584
& biotechnology 583
cotton, wheat 580-2
effects of CCC, mepiquat, ethephon on yield & profit 580-1
Paclobutrazol (case study) 579
registration of PGRs 579
research, invention, development & marketing phases 577-80
sales of PGRs 576-7
screening of PGRs 578
targeting the PGR 577-8
use in a changing agriculture 582-3
rice (Japanese), wheat, amenity spp. 580-3
EDTA use in gravitropic experiments 210
EGTA 212-3, 223
embryo
ABA levels 401
effects on
germination 400-1
of embryos from mature grains 404
maturation proteins 400-1
& dormancy induction 402
environmental effects (temperature) on 401-2

ABA-inducible embryo proteins from
mature grains 404-5
responsiveness, in cereal grains 401-2
embryogenesis & embryo development
ABA-inducible proteins in cotton 142
genetic differences in enzyme regulation of
Z metabolism 267-73
role of GAs in 469-75
endoglycosidase H (use in auxin receptor
research) 110
endosperm GAs 469-75
enzymology
of GA glucosylation 303-5
GA_{20} and GA_9 metabolism 307-313
zeatin metabolism in bean em-
bryos 267-73
Equisetum, ABA & JA and various
metabolites therein 375-6
Erigeron (conzya) re herbicide resistance 567
Ethephon 507-8, 580-1 611
Ethrel 500, 595-603
ethylene
action/synthesis inhibitors and senescence
signal bioassay 543
as second messenger of auxin in apical
dominance 419-27
wounding signal 520, 525-7
binding
receptor(s) 192-99
constant Kd 193
exchange between bound & un-
bound 196-7
Km for physiological response 193
long-time binding component 197
number of ligands bound 198
to mung bean sprouts 194
biosynthesis
-auxin increase in 114
dependence on ACC and O_2 concentra-
tion 294-5
in crown gall tumor lines 507-8
methionine cycle 298-9
collection by $Hg(ClO_4)_2$ 193
diffusion from binding sites, and kinetics
of 194-8
half lives 194-5
effect on lateral bud growth in decapitated
Vicia 421-2
from ACC degradation 295
metabolism aspects 291-9
ethylene-releasing compounds, practical
uses 595-603
Ethrel, Silaid, Alsol, ACC 596-7
flowering 597-8
fruit ripening & abscission 599-60
leaf & bud abscission 598-9
lodging prevention 596
secondary metabolite production 602
seed dormancy 601
sex expression 599
stress response, induction of 601-2
wound response (latex flow) mimicking
of 601
ethylenediaminetetraacetic acid (see EDTA)
etiolation, cause of in pea 81
evolution of plant hormone system 1-9

fern antheridiogens 5-6,
spore germination & antheridium forma-
tion 5-6
fisetin 436
flacca mutant of tomato 24-31
flavonoids & regulation of auxin trans-
port 428-39
glycosides 439
& rhizobial nodule initiation 438
floral evocation
& cell cycle/mitotic activation 486-7
stimulus & GAs 477-485
flowering
in herbaceous angiosperms, & GAs 476-85
CK effects on flower initiation 486-90
in transgenic *Arabidopsis* 189
control by ethylene-releasing com-
pounds 597-8
GA effects on 605-6
in cold-requiring plants 490
induction by anticytokinins in
Asparagus 184-5
fluridone 400
flurprimidol 612
fructose-6-biP, influence of CKs 465-6
fruit
abscission (removal), promotion by
ethephon 600
ripening, promotion by ethephon 599
russet, control of by $GA_{4/7}$ 605-6
set and yield, influences of CKs 458-463
Funaria 7-8, 407-18, 512-17
fungal infection resistance improved by plant
growth retardants 616
fusicoccin (FC)
-binding protein, solubilization & purifica-
tion 155-6
photoaffinity labeling 156-8
sites in higher plants 154-60
effects on PAL in *Cuscuta* 495

GA(s) (see gibberellins and specific gibberellin
numbers [e.g. GA_1]
galangin 436
gametophore buds in *Physcomitrella*
58-64
GC-mass spectrometry (see GC-MS)

GC-MS 49-50, 233-239, 244-5, 268, 301-305, 307-313,374-9, 343, 441-9
negative ion, chemical ionization 233-6
GC-MS-selected ion monitoring (GC-SIM) 49, 307-313, 341-351
genetic transformation
insertion of HSP-*ipt* chimeric gene 186-91
of CK levels in *Arabidopsis* & *Petunia* 186-191
of *Physcomitrella* 63
gene expression
& auxins 95-99, 100-105, 225-7
of α-amylase genes 134-6
rapid alterations with auxins 95-99
regulation by ABA in barley aleurone layers 138-43
transcription 145-53
auxin-regulated 103-4
in vitro, re-initiation by auxin 103-4
genes clustered, for SAURS in soybean 97
le & *na* in pea 77
lk, *lm*, *lw*, *lv* in pea 80
genistein 431
germination
of embryos, precocious 400-1, 474
prevention by ABA 400-1
promotion by ethylene 601
GGPP to CPP cyclization step 323-4
Gibberella
DJA & JA as natural products 377
metabolism of GA_4 to GA_3 via GA_7 311
gibberellin(s)
ABA ratio in cold-hardiness 536
as the „floral stimulus" (?) 476-85
binding proteins, photoaffinity labeling 145-53
in aleurone protoplasts 145-53
biosynthesis
effects of plant growth retardants on 322-31
Paclobutrazol on/in *Rht* 3 wheat 301-327
in bean suspensor 472
maize 65-73
pea 74-83
rice 12-20
biosynthetic pathway
enzymes in cell-free extracts of rice 19
system 314-321
dependence on Fe^{2+} 317
2,2 dipridyl (inhibitor) 316
effect of dilution on enzyme activity 320
Km values 315-6
molecular weight 316
enzyme(s) purification 316, 320
BSA
use to retain high activity 318-20
prevent artifacts 319
DTT
-produced artifacts in incubation 319
requirement in enzyme purification 317
EDTA use 316
2-oxyglutarate-dependent oxygenases 314-321
prolyl 4-hydroxylase (mammalian) 314-16
uncoupled decarboxylation 315
gibberellins & chilling resistance 396, 398
cold-hardiness 532-5
commercial uses 604-610
conjugation, in *Zea mays* 300-305
reversibility thereof 300-301
-deficient genotypes, 65-73, 74-83
-dose response curves 78-9
expression of α-amylase genes 130-7
immobilized, α-amylase induction by 150
& flower initiation 476-85
in *Caulerpa* 4-5
cotyledons 470-5
dwarf maize mutants 65-73
*an*1, *dwarf* 1,5,8 lesions 68-72
non-GA responding mutants 72-3
embryo development 469-75
endosperm 470-75
ferns 4
gibberellin-0-glucosides
GC-MS as permethyl derivatives 301-2
metabolism in dwarf mutants of corn 303-4
hypersensitive pea, gene *lv* 81
gibberellin-insensitive mutants, in cereals 84-92
interaction with ABA in cold-hardiness 532-3
gibberellin metabolism and enzymology 307-313
& dwarfism in rice 18-19
corn 65-73
C-13 hydroxylation pathway 12-20, 312
C-13 non-hydroxylation pathway 12-20
differential, in seed coats & cotyledons 256-7
2β-hydroxylation 87
3β-hydroxylation 309-11
phyllogenetic significance of GA pathways 312
gibberellin overgrowth mutants in cereals 84-92
gibberellin(s) & senescence 538

structure:function specificity in flowering 606
receptors
functional assays for 149-50
in *Avena* 145-53
pea 82
on aleurone plasma membranes (?) 150
sensitivity & synthesis mutants in pea 74-83
gibberellin(s) sites of production & action re floral initiation 481-2
tissue specific in legume seeds 224-7
gibberellin(s)
GA_1 74, 303, 328, 473, 482
conjugates 303
& elongation in pea 78
genetic block for 74
2 β-hydroxylation 87
in rice 12-20
Rht mutants of wheat 84-92
slender pea 87-8
suspensor & embryo of bean 472-3
$\log_{10}$ of endogenous level & internode growth in pea 309-10
metabolism from GA_{20} in corn & pea 309-11
enzymology of metabolism from GA_{20} 309-11
3-epi GA_1 328
and the *le* pea mutant 309-11
synthesis from GA_{20}
GA_3
as a native GA of maize (corn) 310-311
biosynthetic origin in higher plants & *Gibberella* 307-313
commercial uses 604-610
on grape clusters 604
horticultural crops (various) 605
diminishment of *Botrytis* in grape 605
virus (ring spot) in cherry 605
conjugates 303-5
effects on crown gall tumors 507-510
growth & development of *Cuscuta, in vitro* 492
lignification in *Cuscuta* 498-91
in flowering induction 477-84
- induced α-amylase synthesis in barley aleurone 139-43
metabolism from GA_{20} via GA_5 in corn 307-313
stimulation of PAL activity in *Cuscuta* 494
15 β-hydroxy GA_3 478
GA_4 473
conjugate 305
in photoaffinity labeling 147
rice 12-20
metabolism to GA_3 via GA_7 in *Gibberella* 311
$GA_{4/7}$
commercial uses 604-610
on agronomic crops & tropical fruits 607-9
to prevent fruit russet 605-6
GA_5 473
conjugate 305
metabolism from GA_{20}, & to GA_3 in corn 309-11
& enzymology thereof 309-11
GA_6 473
GA_7
biosynthesis from GA_9 via 2,3 dehydro GA_9 307-11
conjugate 305
GA_8 19, 87, 309, 328, 473
photoaffinity labeling 148-9
GA_9 5
in rice 19
metabolism to GA_7 via 2,3-dehydro GA_9 in corn 307, 310-11
2,3 dehydro GA_9
metabolism to GA_7 & other GAs in enzyme preps of *Marah* and *Malus* 307, 311
GA_{12} 19
GA_{12} ald 17, 256-7
conjugate 257
dioxygenases in metabolism of 314-21
GA_{17} 19
GA_{19} 257, 328, 473, 481
in rice 12-20
in slenderpea 87-8
GA_{20} 19, 76, 87, 256-7, 328, 473, 481-3
3-β hydroxylation 309-11
in C-13 -OH pathway, central control role for 309-11
slender pea 87-8
metabolism to GA_1, & to GA_3 via GA_5 in corn 307-13
GA_{20}-13-O-glucoside 303-5
GA_{24} 19
GA_{29} 19, 303, 328
GA_{32} 477-8
GA_{34} 19
GA_{44} 19, 257, 328, 473
GA_{53} 19, 257
GA_{53}-ald-conjugate 257
GA_{60} 473
glucobrassins 365
glucosylating enzymes for GAs 303-5
glutathione 394
-s-transferase 570
grape 604-5

grass-clump dwarf mutants 84-92
grass, turf
classifying turf areas re PGR types & mowing & fertilizing frequency 593
cool season growth 585-6, 588-9
annual life cycle 589
greenup 588, 591-2
seedhead production control 586-7, 589, 591
growth dynamics (plastachron & leaf duration) 588
life cycle variation among species/varieties 590
practical uses of PGRs 585-593
warm season growth 586-8
peak vegetative growth 586-7
control by PGRs 590-3
Type I (inhibitors cell division or cell cycle) 590-1
Type II (growth suppressors/GA biosynthesis inhibitors) 591
gravitropism
calcium role, in light-regulated gravitropism 209-215, 217
IAA localization & quantification 347-8, 441-9
in roots 209-15
ABA substitution for R light 213
experiments using plasmolysis of root cap 210-12
interactions of calcium and light 213-4
phytochrome-mediated R response 213-4
movement of calcium & IAA across root cap 210-215
scheme for interactions of light, Ca, ABA, IAA 215
use of EDTA & EGTA to influence auxin movement via calcium 210-4
phosphoinositides in light-regulated gravitropism 217-8
growth classes of roots
re graviresponse and IAA content 445-8
growth inhibitors as regulators in phototropism 450-55
patterns, differential (flank) in photoropism 450-55
retardants (see plant growth retardants)
guanosine triphosphate (GTP)
binding assay & high affinity binding 123-5

[^{2}H] incorporation from water to measure IAA turnover 339
„Halliwell-Asada“ enzyme pathway 567
HCN, ability by higher plants to metabolize 295
from ACC 295
heat shock HSP-*ipt* chimeric gene
use to modify CK levels 188-91
effect on CK levels with and w/o heat induction 190
Helianthus 450-55
herbicide(s)
activity, enhanced by plant growth retardants 330, 616
evolution of cross & sequential resistance 566-73
herbicidins 559
heterosis in maize, & GAs 609
heterotropic plants, phytohormones in 2-3
hexylitaconic acid 557-8
HMG-CoA reductase 259
hormone receptors, up & down regulation 204-5
horticultural crops, GA and CK use thereon 604-610
hs-ipt (see heat shock isopentenyl transferase gene)
hydraulic conductivity 394
3-hydroxy-2-indolone-3-acetylaspartic acid (DIA-Asp) 354-9
4-hydroxyl-5-isopropyl-2-methylphenyl trimethyl-ammonium chloride, 1-piperidine carboxylate (see AMO-1618)
7-hydroxyoxindole-3-acetic acid (7-OH-OxIAA) 371, 443
7-OH-OxIAA-glucopyranoside 371
Hyoscyamus 35-44
hyponasty & CKs 278

IAA, amide-bound 48-9
anion carrier 428
bioelectric potential & IAA distribution 350-1
biosynthesis
D-tryptophan use 54
genes coding for 503-510
-binding proteins 100
caulonema differentiation 408-18
[$^{13}C_6$] IAA as quantitative internal standard for GC-MS 49
catabolism 367-72
conjugates, hydrolysis 51
in germinating corn 342-51, 446
rice 12
-containing peptides in *Phaseolus* 46
polyclonal antibodies for 47
de novo synthesis 342-51
derivitization for GC-MS 442
esters from endosperm 344-5
conjugates, quantification of 441-9
in roots, & during germination 446-7
gravity effects on 347-8
IAA in auxin autotroph mutants 40-1

Caulerpa 4-5
coleoptile tips 344-5
lower green plants 4
Physcomitrella 57
rice 11-12
roots of corn, re growth & gravitropism 441-9
senescence signal bioassay 543
IAA-induced cell elongation, role of calcium 223-4
Km for growth 350
levels in wheat caryopses & bean seed 47
light effects on 346-8
IAA metabolism 45-55
in auxin autotrophs 43-44
Vicia 355-6
isotope method to study 53-4
mutants, use to study IAA metabolism 51-55
Lemna gibba jsR, 53-4
IAA myo-inositol (as IAA precursor) 344-5
in coleoptile tips 344-5
myo-inositol-galactoside 345
[^{15}N] IAA, use with high resolution MS 53-4
oxidation, co-factor requirment 369-70
function of 371-2
perception & auxin-regulated gene expression 100-105
polysome production, promotion of 520
pool size 344-5, 443
precursors, reserve pools of 365
quantification by GC-MS-SIM 441-9
rate of use in germination 342-51
relation between stimuli and growth 341-51
total, methods for obtaining 47-8
transport & metabolism in *Vicia* 355-6
turnover 333-9, 368, 443-4
voltage (applied) effects on IAA leakage from stele 348-51
IAA ethyl ester
effect on callus growth 41-3
caulonema differentiation 407-18
immunoaffinity purification of ACC synthase 114
immunocytochemistry, in signal transduction research 512-3
immunofluorescence (to localize BA binding) 513-7
inabenfide 612
indoleacetylamino conjugates 408
indole-3-acetic acid (see IAA)
indole-3-acetyl-L-alanine (IAA-L-ala) 408
indole-3-acetylaspartic acid (IAA-asp) 43
metabolism 354-9
to dioxindole acetic acid conjugates 354-9
oxidation by peroxidase 357-8
indole-3-carboxylic acid (ICA), in rice 11, 12
indole-3-ethanol (tryptophol)
from various bacteria, yeast & higher plants 360-1
gene toxic indole derivative strepindole from *Streptococcus* 365
metabolism 360-5
in *Pseudomonas* 362
to indole-3-glycol & indole-3-ketol 364
tryptophol oxidases 362
tryptophol conjugation
glycosides & esters & as reserve auxins 363-4
in *Craterellus* and *Pinus* 363
via indole-3-acetadehyde 360-2
possible role in IAA biosynthesis
indole ring *de novo* synthesis (IAA) 342-3
2-indolinone-3-acetic acid (see OxIAA)
2-indolone-3-(2-hydroxy)acetylaspartic acid (see 2-OH-OXIAA-Asp)
inhibitor(s) from microbial sources 555-565
role in phototropic curvature 450-55
raphanusamide, raphanusamins & xanthoxin 452-5
inositol 1,4,5-trisphosphate [(IP_3 or 1,4,5)IP_3] 550
blockage by TMB8; calcium release 208-9
pathway, control of calcium levels 208
insect control by ethephon in cotton 602
internode extension
& 3 β-hydroxylation of GAs in pea & maize 307-13
internode length mutants in pea 74-83
ionophore A23187 512, 223
iPA monophosphate 16
[9R]iP (see iPA)
ipt (isopentenyl transferase) 63
gene 186-91
isopentenyl adenine (iP) 8, 16-17, 272, 282-9
isopentenyl adenosine (iPA) 16-17, 58, 282-9
isoprenoid pathway 68-9

(-)jasmonic acid/methyl jasmonate (JA) 543
analogues, naturally occurring 375-6
& chilling resistance 398
in senescence signal bioassay 543
metabolism 374-9
DJA & JA amino acid conjugates 375-8
JA-11-OH & 12-OH metabolites 375-6
amino acid & glucose conjugates thereof 375-6
JA cyclopentanol metabolites, glucose conjugates 375-6
(+)-7-iso-JA 374

kaempferol 431
kanamycin resistance, use in
  Physcomitrella 63-4
kaurene synthase, inhibitors of, & oxidation inhibitors 322-6

lanthanum (calcium uptake inhibitor) 512
latex flow, induced by ethephon 601-2
LDP 476-85, 487-490
leaf CKs & flowering 488-9
  curl syndrome in pigeon pea 278-280
  senescence and CK translocation in xylem 280-1
legume seeds, GA & ABA metabolism 224-7
Lemna 45, 51-3
light & calcium in root gravitropism 213-4
  & chilling resistance 394
  interaction with auxin & cytokinins in moss 59-60
lignification in *Cuscuta*: influence of phytohormones 492-501
linolenic acid in senescence signal bioassay 543
Lockhart equation re growth/turgor relations 89
lodging, prevention with ethephon 596-7
Lolium re evolution of herbicide resistance 567
  re flowering & GAs 477-8
long-day plant (see LDP)
lower plant pheromones, sexual hormones, phytohormones 1-9
lupin 262-4, 275-281
Lycopersicon 504-510
Lygodium 5-6

MACC 508, 9
  subcellular compartmentation 296-7
maize (see corn)
N-malonyl-ACC (see MACC)
Malus & *Marah* & GA_9 metabolism 307-13
mass spectrometry, high resolution, re IAA metabolism 53-4
  of CKs from xylem exudate 281-1
membrane depolarization, hydrocarbon acyl chain 394
  phospholipids, & their stabilization 393-4
  -bound auxin receptors 106-112
  PI turnover & signal transduction 122-9
mepiquat chloride 580-1, 323, 611
mesocotyl 341-351
messenger ribonucleic acid (see mRNA)
methionine cycle in ethylene production 298-9
methyl hydroxylases 328-9
mevalonic acid (see MVA)
mevinolin (inhibitor of MVA formation (?)) 236
  inhibitor of other steps in ABA biosynthesis 236
microorganism(s)
  & production of plant growth substances 555-565
Michaelis-Menten plot re IAA & growth of mesocotyl 346
mimicry, biochemical re herbicide resistance 571
mistletoe, control by ethephon 598
moniliformin 562-3
monoclonal antibodies (use in auxin receptor purification) 107-9
monooxygenases & herbicide resistance 570-2
  in ABA metabolism 138-9
morin 431
moss protonema, auxin regulation of cell differentiation 407-18
mosses, growth hormones in, including IAA & CKs 6-8
  Ove-mutants 7-8
mRNA
  auxin-regulated 95-99
  cDNA probes in auxin induction 95-99, 102-5
  changes in mRNA coded for ACC synthase re wounding 115
  hybridization in GA-treated aleurone 135-6
  ipt mRNA induction by heat shock 187-8
  levels for α-amylase gene families 130-7
  pI families for α-amylase 134-6
mutants
  ABA-sensitivity & synthesis 24-31
  auxin-resistant 32-44,
  -sensitive 32-44, 57-64
  cell wall extensibility & plasticity 89-90
  CK-resistant 32-44, 59
  CK-sensitivity 57-64
  GA sensitivity & synthesis 74-83
  IAA, for study of IAA metabolism 51-3
  „leaky“, in pea 78
  Lemna, variant lines 52-3
  maize (corn) *d*1, *d*2, *d*3,*d*5 and/or *d*8 65-73, 303-4, 307-313
  moss (*Physcomitrella* re cell differentiation 57-64
  pea 87-8
  rice; Tan-ginbozu, Tong-il, Waito-C 11-20
  wheat, Norin 10, *Rht*1,2,3 84-92
MVA 68-69, 325
  as ABA precursor 236-7
  biosynthesis 259
  level reduced by compactin 259
  metabolism into CKs 258-66

naphthaleneacetic acid (NAA)
auxin-regulated mRNAs 95
effects on growth of auxin mutants 36-44
crown gall tumors 506-7
lignification in *Cuscuta* 498-501
PAL activity in *Cuscuta* 494
zeatin riboside metabolism 284
use in selecting auxin sensitivity mutants in moss 59-60
1-naphthylphthalamic acid (NPA) 429-39
naringenin 436
neovaninin & neovasinone 558-9
Nicotiana (see tobacco)
nifedipine (calcium channel blocker) 512
2,5 norbornadiene 116, 423
norbornanodiazetines 611-618
Northern blot analysis for ACC synthase 116-7
HS-*ipt* chimeric gene 187-8
nutrients and senescence 539

^{18}O, use to measure ABA turnover 338
2,4 octadienedoic acid 251-2
oilseed rape endogenous GAs, effects of BAS 111 on 328
oligopeptides (cyclic) AM toxins 560
oncogenes from Ti plasmids 504-510
ovary abortion & CKs 464-6
OxIAA 443
biological activity of 369
pathway & oxidative metabolism 367-73
quantitation in corn 368-9
synthesis (*in vitro* assay) 369
turnover 335-9
2-OH-OXIAA-Asp 371
oxindole-3-acetic acid (see OxIAA)
(OX)Z (see 0-xylosylzeatin)

PAA 439
Paclobutrazol 79-80, 88, 326-7, 395-6, 398, 579, 612
pathogenic, heterotrophic plants, phytohormones in 2-4
PBA 607
PCIB competition for IAA-binding 411
effects on caulonema differentiation 410-11
pea 244
& ABA-deficient mutants 24-31
auxin action, role of calcium 222-7
GA synthesis & sensitivity mutants 74-83
GA metabolism & enzymology 307-13
GA_{12}-ald metabolism 256-7
permethylation of GA-0-glucosides
fragmentation scheme & selected key ions 301-3
peroxidase, oxidation of IAA-asp 357
Persea 234
Petunia 186-91
CK alteration with HSP-*ipt* chimeric gene 181-91
Pharbitis 478-81
phaseic acid (PA) 237-9, 244-5, 247-8
biological activity of, in barley aleurone 139-43
effects of plant growth retardants on synthesis 329-30
metabolism of 138-43, 255
Phaseolus (see bean)
phenolics & polar auxin transport 428-39
phenylacetic acid (see PAA)
phenylalanine ammonia lyase (PAL) 429, 492-501
n-phenylcarbamates
as anticytokinins & flowering inducers 183-5
phosphatidylinositol (PI) 122-9, 217-220, 511
cycle involvement in CK effects (?) 164-171
PI C assay 123, 127
phosphatidylinositol 4,5 biphosphate (PIP_2) 16
Phosphon D 323, 396, 398
photoaffinity labeling of GA receptors 148-9
fusicoccin-binding proteins 156-8
photoinactivation assays in GA receptor research 147-8
photosynthesis (Ps), effects of ABA on
electron transport topographies & depression of Ps 383-390
phototropism & growth inhibitors, including raphanusamide, raphanusanins & xanthoxin 450-55
Physcomitrella 7, 57-64, 513-4
phytotropins 439
phthalimide (AC-94) 609
phytochrome
as agonist in hormone regulation & phosphorylation 204-206
influence on GA sensitivity 81
-mediated moss differentiation 61
root gravitropism 213-4
pine, flowering in 606
$PI(4,5)P_2$ (see PIP_2)
Pisum (see pea)
plant growth substances in microbial organisms 555-565
plant growth retardants 611-18
action of 322-31
compactin 259
effects on endogenous GA levels 327-8
cold-hardiness, & retardation of loss thereof 533-5
flowering 480-1

lodging in grains and rice 614
PA biosynthesis 328-9
enhancement of herbicide activity 615
growth regulation in oil seed rape 328, 614
orchard trees 614
inhibition of sterol biosynthesis 325
practical uses on turf grass 585-93
resistance to chilling 395-6, 398, 614
drought, fungal infection 614-5
root growth 613
suppression of herbicide detoxification 572
pod set in legumes, & CKs as metabolic stimulants 457-67
polar auxin transport 428-39
polyamine increases by CK application 163
polyclonal antiserum (use in auxin receptor methods) 107-8
polysome formation, cytoskeleton bound 527-30
in response to wounding or UV treatment 520-4
potato 605
& ABA-deficient mutants 24-31
proline 394
protein(s)
ABA- & osmotic/salt-induced in barley aleurone 141-3
in hormonal regulation 205-6
isozymes for α-amylase synthesis 130-7, 140-3
kinase & calcium requirement 205-6
late abundant 2S and 12S in *Arabidopsis* ABA mutants 26-31
phosphorylation 511, 548-50
& auxins 225
S-S bonds in chilling response 394
protein kinase C 550
role in response of cell to plant hormones 547-551
PP333 (see Paclobutrazol)
pSS1, 63
PtdIns4,5P_2 (see PIP_2)
pyridol & pyrrolol [2,3-d] pyrimidines (as anti-CKs) 181-85

2-quinolone-4-carboxylic acid (QCA) 324
quercetin 430
quinolineum ion (IAA) 343

R (see red light)
radish 450-55
raphanusamide & raphanusanins in phototropism 452-5
receptors
auxin (purification) 106-7
concentration in auxin-binding 110-2
CK-binding 517
ethylene-binding 192-99
binding constant Kd 193
Km for physiological responses 193
gene (for auxin) 111-2
red light (see R)
and spore germination 58
monochromatic vs polychromatic
in auxin sensitivity mutants (moss) 58-61
use to modify growth 346
RFLP analysis of α-amylase gene variation 133-4
rhizoid, protonemal differentiation, nutrient effects 412-3
Rhizobium & CK production 278-80
rhodomorphin, in algae 3
Rht mutants 84, 327
ribonucleic acid (see RNA)
ribosomal ribonucleic acid (rRNA) protein 56, 549
rice
brassinosteroids, & CKs, in shoots 16-18
dwarf mutants; Tan-ginbozu, Tong-il, Waito-C 11-20
regulation of life cycle by plant hormones 11-20
RNA
polymerase-associated protein kinase activity 548
target for CKs 548
translation, hormonal regulation 549-50
RNA (messenger RNA) (see mRNA)
RNA (transfer RNA) (see tRNA)
root(s)
gravitropism, role of calcium in 209-215
root cap, role in gravitropism 210-215
root exudate (CKs) & flowering 488-9
root growth
plant growth retardants, effects on 613
rooting, CK inhibition of 540
Rubisco 383-4

SAM-induced activation of ACC synthase 293
SAURS, small auxin-up regulated RNAs 96-99
rapid redistribution & sequences in gravistimulation 98
SDS-PAGE 286-8
second messenger in signal transduction/CKs 511-17
auxin-controlled apical dominance 419-27
secondary xylem formation in *Cuscuta*, & phytohormones 492-501

seed(s)
(s)ABA & GA_{12}-ald metabolism in coat & cotyledons 254-5
testa GAs in bean 472
development & ABA and vivipary 26-31
GAs 469
IAA turnover in maize 333-9
in tomato, ABA-deficient mutants 26-31
dormancy
breaking by ethylene 601
in seeds & in ABA-deficient mutants 26-31
set in legumes & ABA & CKs 457-67
senescence & ABA, auxin, CKs & GAs 537-45
& nutrition & CK flux 539
monocarpic & senescence signal 537-45
short day plant(s) (SDP) 478-485, 487-490
signal transduction, role of calcium in 216-229
Sinapis 486-90
slender pea 79-80, 87
sln allele in barley 87
Southern blot 187
soybean
auxin-related mRNA 95-99
ABA metabolism 244-5
callus & degradative CK metabolism 283-4
podset and CKs 457-67
use to study CK mode of action 164-71
spore germination & R light 58
stele, vascular & IAA & growth stimuli 341-351
Stellaria, herbicide resistant 570
stem elongation & GA level, in pea 78
steroid receptors 204
sterol(s)
biosynthesis inhibition by growth retardants 324-5, 328-9
by compactin 259
via 14 demethylation 329, 613
stimulus-response coupling, role of calcium 216-229
stomatal
apertures (ABA effects on, & photosynthesis) 385-390
closure patterns, influences of ABA & CO_2 38-9, 392, 394
by ABA in *Antherceros* & *Funaria* 5
control in ABA-deficient mutants 24-31
frequency distribution in response to ABA 387
structure-activity relations
anti-CKs (*S*-triazines and carbamates) 179-185
sugars, & chilling resistance 394
sunflower 450-5
suspensor cells
& DNA polytenization 471
endogenous GAs 472-3
role in early embryogenesis as source of GAs 472-3

taxifolin 434
T-DNA & CK levels 284
temperature-sensitive auxin-autotroph mutants 35-44
tentoxin 561
teratomas on crown gall tumors 504
tetcyclasis 572, 612
inhibition of ABA inactivation step 237-9
thidiazuron 289, 607
& carbohydrate metabolism 465-6
Thlaspi 481
Ti plasmids & oncogene transfer 63, 504-510
TIBA 428-9
tissue culture 102-5
tobacco 32-44, 100-105, 182, 234, 264-5, 281, 284, 503-10
tomato 118, 244
& ABA-deficient mutants & seed development 24-31
transgenic *Arabidopsis* & *Petunia*, to alter CK levels 186-91
triadimefon 395
triadmenol 329
triapenthenol 612
s-triazines, as anti-CKs & flowering-inducers 179-185
triazoles 326-7, 573, 611-18
tributyl 2,3-dichlorobenzyl phosphonium chloride (see Phosphon D)
tRNA, metabolism in Ove-mutants of *Funaria* 8
tropical fruit production 607-8
tropism(s), potential-gating theory 350-1
curvature & IAA metabolism/transport 341-351
tryptamine 344-5
tryptophan 344-5
analogs (toxic) for mutant selections re IAA metabolism 53
D-tryptophan, as IAA precursor & reserve pool thereof 53-4
& N-malonyl conjugate 365
tryptophol (see indole-3-ethanol)
turf grass (see also grass, turf), practical uses of PGRs on 585-93
turgor pressure & growth, 89-90
probe 89

uniconazole 612

Vallisneria re wounding 529
vascular differentiation/bundles
& polar auxin transport 428, 438
verpamil (calcium channel blocker) 512
Vicia, ABA biochemistry 243
apical dominance in 420-7
DJA, JA, & various analogues & conjugates 374-9
fusicoccin binding sites 154
glc-DIA-asp accumulation, IAA metabolism 355-6
IAA-asp metabolism 354-9
virus, GA_3 effects on 605
vivipary & ABA & seed development in *Arabidopsis* 26-31
voltage, effect on IAA leakage from stele & growth of corn 348-51

Western blotting, re IAA-containing peptides 46-7
wheat
caryopses – IAA levels 47
cold-hardiness, effects of PGRs 530-5
dwarfs *Rht*1, *Rht*2, 327, 608
economic aspects of PGR use 580-1
selective herbicides use on 568-70
wilty mutant of pea 24-31
wounding
& ACC synthase 114-121
immunochemical comparison with auxin-induced 118-8
at transcriptional level 120
action potential & ethylene 520
changes in mRNA coded for ACC synthase 115
cytoskeleton bound 527-30
polysome formation, promotion of 520
in intact plants, in aged, excised tissue 519-5
response mimicked by ethylene 601-2
role for calcium 524-5
-signals and translation 519-30
studies with auxin, in decapitated plants 519-20

xanthine oxidase in CK synthesis 265
xanthoxin
ABA precursor (?) 234-237, 250-2
as possible regulator of phototropism 452
xenobiotic metabolism
cytochrome P-450-dependent mixed function oxidases
interference by plant growth retardants 330, 616
xylem sap CKs & leaf curl syndrome 275-81
lateral movement into bark 277-8
& senescence 540-2
xylem differentiation, secondary, in *Cuscuta* 492-501
0-xylosylzeatin, biological significance of 269
in bean embryos & enzymology of 268-270

zeatin (Z) 16, 17, 57, 267-73, 275-281, 282-9, 539-42
genetic difference in reductive metabolism 271-2
metabolism in higher plants, in bean embryos 267-73
ZR monophosphate (ZMP) (*cis*- and *trans*-) 16-17, 542-5
Zea mays (see corn)
zeatin-7-glucoside ([7G]Z) 284
zeatin-9-glucoside ([9R]Z) 275-81
zeatin-O-glucoside (*cis*- and *trans*-) 16-17, 267-73, 275-81, 540-2
zeatin-0-glucoside riboside (*cis*- and *trans-)* 16-17, 275-81, 540-2
zeatin riboside (ZR, [9R]Z) 16, 17, 270, 275-81, 282-9, 539-42

MIX
Papier aus verantwortungsvollen Quellen
Paper from responsible sources
FSC® C105338

If you have any concerns about our products,
you can contact us on
ProductSafety@springernature.com

In case Publisher is established outside the EU,
the EU authorized representative is:
Springer Nature Customer Service Center GmbH
Europaplatz 3, 69115 Heidelberg, Germany

Printed by Libri Plureos GmbH
in Hamburg, Germany